Important Information

This book belongs to: _____

_____

_____

_____

phone: _____

PE examination date: _____ time: _____

location: _____

_____

_____

Examination Board Address: _____

_____

_____

phone: _____

Application was requested on this date: _____

Application was received on this date: _____

Application was accepted on this date: _____

Information received during the examination (such as booklet numbers) follows:

_____

_____

"Why do engineers prefer GLP study guides?"

Some comments about the GLP approach...

"In your FE book you cut to the chase and don't embellish it. Yours is the only review that doesn't scare the pants off people. It's exactly what I want for teaching my review course."
—*John Thorington*

"I passed the first time I took the test. Potter's book is a great, great review. It's right to the point and doesn't include extra stuff you don't need. All I did was work through the book. My friends took review courses and didn't use your book—they failed! Do you have a PE review? I don't want to take the exam until you do!"
—*Dana, calling to back-order our new PE Civil Review*

"I used your FE book and thought it was far superior to any other. I took the exam and did great. Now I need your PE Review!"
—*Tom Roach, Indiana Midwest Steel*

"Your FE book kept straight to the point, was very informative and more than adequate to pass the exam. I'm calling to see if you have a review for the PE exam."
—*Burton Mills, Fruehauf Company*

"When you're working 50 hours a week, you don't have a lot of time to prepare. The Potter FE book is condensed and focuses on everything that is on the exam. It has a good strategy: it concentrates on the basics—just what you need to pass. It doesn't overwhelm you."
—*James Alt, graduate of University of Wisconsin*

"We have been using your FE review manual for several years with great success. Congratulations on a job well done!"
—*Ted Huddleston, review course coordinator, University of S. Alabama*

"The FE Reference is very concise. It does a good job with respect to the time that you have to study, getting you through the material for the test. It's right where it should be in terms of length and depth of review. It doesn't overwhelm those studying for the exam. I agree with the approach you took. It's a very useful book."
—*Dr. Packard, University of California*

"Several of the engineers I sold a copy of the FE book to passed the exam after trying and failing repeatedly. They all said "What is it with this book?" They didn't even attend my course and they passed with the book alone."
—*Shahin Mansour, S&R Engineering Foundation*

International Standard Book Number 1-881018-15-6

The authors and publisher of this publication have given their best efforts in the development of this book. They make no warranty of any kind regarding the documentation contained herein. The authors and publisher shall not be liable in any event for damages arising from the use of this publication.

Second Edition Copyright © 1996 by Great Lakes Press, Inc.

All rights are reversed.
No part of this publication may be reproduced in any form or
by any means without prior written permission of the publisher.

All comments and inquiries should be addressed to:
    Great Lakes Press—Customer Service
    PO Box 172
    Grover, MO 63040-0172
    Phone (314) 273-6016

Library of Congress
Cataloging-in-Publication Data

Printed in the USA by Braun-Brumfield, Inc., Ann Arbor, Michigan.

10 9 8 7 6 5 4 3 2 1

...from the Professors who know it best...

PRINCIPLES & PRACTICE OF CIVIL ENGINEERING
—2nd Edition—

The most efficient and authoritative review book
for the PE License Exam

Editor: **MERLE C. POTTER**, PhD, PE
Professor, Michigan State University

Authors:		
	Mackenzie L. Davis, PhD, PE	Water Quality
	Richard W. Furlong, PhD, PE	Structures
	David A. Hamilton, MS, PE	Hydrology
	Ronald Harichandran, PhD, PE	Structures
	Thomas L. Maleck, PhD, PE	Transportation
	George E. Mase, PhD	Mechanics
	Merle C. Potter, PhD, PE	Fluid Mechanics
	David C. Wiggert, PhD, PE	Hydraulics
	Thomas F. Wolff, PhD, PE	Soils

The authors are professors at Michigan State University, with
the exception of R. W. Furlong, who teaches at the University of
Texas at Austin and D. A. Hamilton who is employed by the
Michigan Department of Natural Resources.

published by:
GREAT LAKES PRESS
P.O. Box 483
Okemos, MI 48805-0483

Table of Contents

Preface I-1
Introduction I-3
      State Boards of Registration I-7
      English and SI Units I-11

1. **MATHEMATICS—Merle C. Potter**
    1.1 Algebra 1-1
        1.1.1 Exponents 1-1
        1.1.2 Logarithms 1-1
        1.1.3 The Quadratic Formula and the Binomial Theorem 1-2
        1.1.4 Partial Fractions 1-2
    1.2 Trigonometry 1-4
    1.3 Geometry 1-7
    1.4 Complex Numbers 1-11
    1.5 Linear Algebra 1-13
    1.6 Calculus 1-17
        1.6.1 Differentiation 1-17
        1.6.2 Maxima and Minima 1-18
        1.6.3 L'Hospital's Rule 1-18
        1.6.4 Taylor's Series 1-18
        1.6.5 Integration 1-19
    1.7 Vectors 1-23
    Practice Problems 1-25
    Solutions to Problems 1-30

2. **STATICS—George E. Mase**
    2.1 Forces, Moments and Resultants 2-1
    2.2 Equilibrium 2-4
    2.3 Trusses and Frames 2-6
    2.4 Friction 2-10
    2.5 Properties of Plane Areas 2-12
    2.6 Properties of Masses and Volumes 2-17
    Practice Problems 2-19
    Solutions to Problems 2-26

3. **THE MECHANICS OF MATERIALS—George E. Mase**
    3.1 Stress and Strain, Elastic Behavior 3-1
    3.2 Torsion 3-4
    3.3 Beam Theory 3-5
    3.4 Combined Stress 3-9
    3.5 Composite Bars and Beams 3-13
    3.6 Columns 3-15
    Practice Problems 3-17
    Solutions to Problems 3-22

4. **FLUID MECHANICS—Merle C. Potter**
    4.1 Fluid Properties 4-1
    4.2 Fluid Statics 4-2
    4.3 Dimensionless Parameters and Similitude 4-7
    4.4 Control Volume Equations 4-9
    4.5 Open Channel Flow 4-16
    Practice Problems 4-18
    Solutions to Problems 4-25

Table of Contents

5. OPEN CHANNEL FLOW—David C. Wiggert
- 5.1 Basic Concepts ... 5-1
  - 5.1.1 Channel Geometry ... 5-1
  - 5.1.2 Uniform Flow ... 5-2
  - 5.1.3 Energy Principle ... 5-3
  - 5.1.4 Momentum Principle ... 5-5
  - 5.1.5 Comment ... 5-6
- 5.2 Rapidly-Varied Flow ... 5-7
  - 5.2.1 Hydraulic Jump ... 5-7
  - 5.2.2 Channel Expansions and Contractions ... 5-8
  - 5.2.3 Measurement of Discharge ... 5-10
- 5.3 Nonuniform, Gradually Varied Flow ... 5-11
  - 5.3.1 Water Surface Profiles and Controls ... 5-11
  - 5.3.2 Equation for Gradually Varied Flow ... 5-12
- Practice Problems (Essay) ... 5-15
- Practice Problems (Multiple Choice) ... 5-16
- Solutions to Problems ... 5-17

6. PIPING SYSTEMS—David C. Wiggert
- 6.1 Losses in Piping ... 6-1
  - 6.1.1 Frictional Losses ... 6-1
  - 6.1.2 Minor losses ... 6-3
- 6.2 Energy and Continuity Equations ... 6-4
  - 6.2.1 Single Pipe Systems ... 6-5
  - 6.2.2 Series and Parallel Piping Systems ... 6-9
  - 6.2.3 Branching Pipe Systems ... 6-11
- 6.3 Use of Centrifugal Pumps in Piping Systems ... 6-13
  - 6.3.1 Matching pump to system demand ... 6-14
  - 6.3.2 Cavitation and Net Positive Suction Head ... 6-15
  - 6.3.3 Pump Similarity Rules ... 6-16
- Practice Problems (Essay) ... 6-17
- Practice Problems (Multiple Choice) ... 6-19
- Solutions to Problems ... 6-20

7. HYDROLOGY—David A. Hamilton
- 7.1 Water Balance ... 7-2
- 7.2 Precipitation ... 7-3
- 7.3 Precipitation Losses ... 7-5
- 7.4 Runoff ... 7-5
- 7.5 Runoff Hydrograph ... 7-10
- 7.6 Effects of Storage ... 7-12
- 7.7 Flood Probability ... 7-13
- 7.8 Flood Flow Calculation Techniques ... 7-14
  - 7.8.1 Rational Method ... 7-14
  - 7.8.2 Unit Hydrograph Method ... 7-18
- 7.9 Flood Routing ... 7-24
- 7.10 Groundwater ... 7-26
- Practice Problems (Multiple Choice) ... 7-30
- Practice Problems (Essay) ... 7-31
- Solutions to Problems ... 7-33

8. STRUCTURAL STEEL—Richard W. Furlong
- 8.1 Safety Concept ... 8-1
- 8.2 Tension Members ... 8-3
  - 8.2.1 Effective Net Area ... 8-3
  - 8.2.2 Tension Member Example Problems ... 8-4
- 8.3 Compression Members ... 8-7

Table of Contents

        8.3.1 LRFD Compression Strength Determination 8-9
        8.3.2 ASD Allowable Compression Strength 8-9
        8.3.3 Design of Compression Members 8-10
        8.3.4 Local Buckling and Width-to-Thickness Ratios 8-10
    8.4 Beams 8-14
        8.4.1 Flexure of Beams 8-14
        8.4.2 Lateral Stability of Beams 8-16
        8.4.3 Beam Shear 8-18
    8.5 Combined Axial Force Plus Flexure 8-25
        8.5.1 Cross Section Strength 8-25
        8.5.2 Moment Magnification 8-25
        8.5.3 ASD Allowable Combined Stress 8-27
    8.6 Connections 8-31
        8.6.1 Welded Connections 8-32
        8.6.2 Bolted Connections 8-34
        8.6.3 Block Shear 8-36
        8.6.4 Combined Shear and Tension on Bolts 8-37
        8.6.5 Slip-Critical Connections 8-37
        8.6.6 Eccentric Shear on Connections 8-38
        8.6.7 Other Connectors and Special Considerations 8-38
  Practice Problems 8-41
  Solutions to Problems 8-47

9. **CONCRETE—Richard W. Furlong**
    9.1 Strength Design Concept 9-2
        9.1.1 Load Effects 9-2
        9.1.2 Strength Effects 9-3
    9.2 Flexural Forces—Beams 9-3
    9.3 Compression Forces—Columns 9-7
    9.4 Shear Forces 9-14
    9.5 Bar Development—Concrete/Steel Force Transfer 9-18
    9.6 Components of Concrete Structures 9-22
        9.6.1 Slabs 9-22
        9.6.2 Joists 9-22
        9.6.3 Beams—Reinforcement for Structural Continuity 9-26
        9.6.4 Columns—Slenderness Effects and Reinforcement Details 9-27
        9.6.5 Walls 9-28
        9.6.6 Footings 9-30
        9.6.7 Cantilever Retaining Walls 9-35
  Practice Problems 9-39
  Solutions to Problems 9-41

10. **INDETERMINATE STRUCTURES—Ronald S. Harichandran**
    10.1 Basic Concepts 10-1
        10.1.1 Units 10-1
        10.1.2 Degree of Static Indeterminacy 10-1
        10.1.3 Degrees-of-Freedom 10-3
        10.1.4 Fixed-End Forces 10-5
        10.1.5 Equivalent Joint Loads 10-5
        10.1.6 Displacement Calculations by the Unit Load Method 10-6
    10.2 The Flexibility Method 10-9
    10.3 Moment Distribution 10-15
        10.3.1 No Joint Translation 10-16
        10.3.2 Joint Translation 10-19
    10.4 The Stiffness Method 10-22
    10.5 Support Movements 10-26
        10.5.1 Flexibility Method 10-26

10.5.2 Moment Distribution	10-27

      10.5.2 Moment Distribution..10-27
      10.5.3 Stiffness Method..10-27
  10.6 Selection of Analysis Method 10-28
  10.7 Calculation of Stresses 10-28
  Practice Problems (Essay)..10-29
  Practice Problems (Multiple Choice)..10-33
  Solutions to Problems..10-35

11. WATER TREATMENT—Mackenzie L. Davis
  11.1 Water Source Evaluation..11-1
  11.2 Demand Estimates..11-2
  11.3 Coagulation/Flocculation 11-2
      11.3.1 Coagulation Physics 11-2
      11.3.2 Coagulation Chemistry 11-3
      11.3.3 Selection of Coagulant..11-4
      11.3.4 Rapid Mix..11-5
      11.3.5 Flocculation...11-8
  11.4 Softening 11-11
      11.4.1 Definition of Hardness 11-11
      11.4.2 Softening Chemistry 11-11
      11.4.3 Mixing..11-15
  11.5 Sedimentation 11-16
      11.5.1 Sedimentation Physics...11-16
      11.5.2 Sedimentation Tank Design...11-22
      11.5.3 Sludge Volume Calculations 11-23
      11.5.4 Sludge Dewatering 11-24
  11.6 Filtration 11-25
      11.6.1 Media Selection 11-25
      11.6.2 Grain Size Distribution 11-26
      11.6.3 Head Loss During Filtration 11-26
      11.6.4 Backwash Hydraulics 11-27
      11.6.5 Filter Box Design 11-28
  11.7 Stabilization..11-31
  11.8 Disinfection...11-34
      11.8.1 Disinfectant Selection 11-34
      11.8.2 Free and Available Chlorine..11-34
      11.8.3 Safety..11-36
  Practice Problems 11-37
  Solutions to Problems..11-41

12. WASTEWATER TREATMENT—Mackenzie L. Davis
  12.1 Water Quality Management..12-1
      12.1.1 Biochemical Oxygen Demand 12-1
      12.1.2 DO Sag Curve...12-2
  12.2 Pretreatment 12-5
      12.2.1 Bar Racks...12-5
      12.2.2 Grit Chambers 12-5
      12.2.3 Equalization 12-6
  12.3 Primary Treatment 12-8
      12.3.1 Sedimentation Physics...12-8
      12.3.2 Primary Settling Tank Design 12-8
  12.4 Secondary Treatment 12-10
      12.4.1 Activated Sludge 12-10
      12.4.2 Trickling Filter 12-17
      12.4.3 Lagoons..12-20
      12.4.4 Secondary Settling..12-23
  12.5 Tertiary Treatment...12-24

Table of Contents

    12.5.1 Filtration 12-24
    12.5.2 Phosphorus Removal 12-24
    12.5.3 Nitrogen Control 12-25
    12.5.4 Carbon Adsorption 12-26
  12.6 Sludge Treatment 12-27
    12.6.1 Conventional Aerobic Digestion 12-27
    12.6.2 Completely Mixed High Rate Anaerobic Digestion 12-30
    12.6.3 Sludge Dewatering 12-32
  Practice Problems (Essay & Multiple Choice) 12-36
  Solutions to Problems 12-40

13. HIGHWAY DESIGN—Thomas L. Maleck
  13.1 Capacity 13-2
    13.1.1 Uninterrupted Flow 13-3
    13.1.2 Interrupted Flow 13-3
  13.2 Highway Safety and Accident Analysis 13-8
  13.3 Sight Distance 13-9
    13.3.1 Stopping Sight Distance 13-9
    13.3.2 Passing Sight Distance 13-12
    13.3.3 Decision Sight Distance 13-14
  13.4 Horizontal Curves 13-14
    13.4.1 Superelevation and Friction 13-14
    13.4.2 Design Criteria 13-16
  13.5 Vertical Curves 13-18
    13.5.1 Crest Vertical Curve 13-18
    13.5.2 Sag Vertical Curve 13-20
  13.6 Cross Section Design 13-22
  13.7 Geometric At-Grade Intersection Design 13-23
    13.7.1 Design Of Auxiliary Lanes 13-23
    13.7.2 Design of Intersection Sight Distance 13-25
    13.7.3 Design of Turning Radii 13-29
  13.8 Corridor Design 13-32
    13.8.1 Data Required 13-32
    13.8.2 Space-Time Diagram 13-32
  13.9 Drainage 13-36
    13.9.1 Runoff 13-36
  Practice Problems 13-40
  Solutions to Problems 13-42

14. SOILS—Thomas F. Wolff
  14.1 The Phase Diagram 14-1
  14.2 Weight-Volume and Mass-Volume Relationships 14-2
  14.3 Specific Gravity 14-3
  14.4 The Use of a Phase Diagram 14-4
  14.5 Other Useful Equations for Weight-Volume Problems 14-8
  14.6 Grain-Size Characteristics of Soils 14-9
  14.7 Atterberg Limits and Plasticity 14-10
  14.8 Soil Classification 14-10
  14.9 Compaction and Compaction Quality Control 14-14
  14.10 The Effective Stress Principle 14-17
  14.11 Water Flow Through Soils 14-19
  14.12 Consolidation Behavior of Saturated Clay 14-23
  14.13 Consolidation Settlement Calculations 14-25
  14.14 Shear Strength of Soils 14-27
  14.15 Triaxial Tests and the Unconfined 14-29
  Practice Problems (Multiple Choice, Essay & PE-Format) 14-31
  Solutions to Problems 14-37

Table of Contents

15. FOUNDATIONS—Thomas F. Wolff
- 15.1 Estimating Soil Strength from Standard Penetration Test Data 15-1
- 15.2 Estimating Clay Strength from Cone Penetration Test Data 15-3
- 15.3 Types of Foundations 15-3
- 15.4 Bearing Capacity of Shallow Foundations 15-4
  - 15.4.1 Net bearing capacity in clay 15-6
  - 15.4.2 Square and rectangular footings 15-7
  - 15.4.3 Effect of water table 15-8
  - 15.4.4 General bearing capacity equation 15-9
  - 15.4.5 Effect of eccentricity 15-9
- 15.5 Stress Increase under Shallow Foundations 15-11
  - 15.5.1 Vertical stress increase due to vertical point load 15-11
  - 15.5.2 Vertical stress increase due to uniform surface pressure 15-12
  - 15.5.3 Vertical stress increase outside limits of foundation 15-15
  - 15.5.4 Vertical stress increase due to non-uniform surface pressure 15-15
  - 15.5.5 Average vertical stress increase under a uniformly loaded rectangular area 15-15
- 15.6 Settlement of Shallow Foundations 15-16
  - 15.6.1 Immediate Settlement Calculations for Shallow Foundations 15-17
  - 15.6.2 Consolidation Settlement of Shallow Foundations 15-18
  - 15.6.3 Settlement vs. Time Calculations for Shallow Foundations 15-20
- 15.7 Deep Foundations 15-22
  - 15.7.1 Pile Foundations in Clay 15-23
  - 15.7.2 Deep Foundations in Sand 15-24
- 15.8 Lateral Earth Pressure 15-24
- 15.9 Retaining Walls 15-29
- 15.10 Braced Excavations 15-31
- Practice Problems (Multiple Choice) 15-34
- Practice Problems (Essay) 15-36
- Solutions to Problems 15-41

16. Engineering Economy—Frank Hatfield
- 16.1 Value and Interest 16-1
- 16.2 Cash Flow Diagrams 16-1
- 16.3 Cash Flow Patterns 16-2
- 16.4 Equivalence of Cash Flow Patterns 16-3
- 16.5 Unusual Cash Flows and Interest Periods 16-6
- 16.6 Evaluating Alternatives 16-9
  - 16.6.1 Annual Equivalent Cost Comparisons 16-9
  - 16.6.2 Present Equivalent Cost Comparisons 16-10
  - 16.6.3 Incremental Approach 16-11
  - 16.6.4 Rate of Return Comparisons 16-11
  - 16.6.5 Benefit/Cost Comparisons 16-12
  - 16.6.6 A Note on MARR 16-13
  - 16.6.7 Replacement Problems 16-13
  - 16.6.8 Always Ignore the Past 16-14
  - 16.6.9 Break-Even Analysis 16-14
- 16.7 Income Tax and Depreciation 16-15
- 16.8 Inflation 16-17
- Practice Problems 16-20
- Solutions to Problems 16-26
- TABLES: Compound Interest Factors 16-29

INDEX

Preface

This book was written to provide the most efficient review possible for the P.E./Civil, Principles and Practice of Engineering Examination. Eight currently lecturing professors from major universities have done us the courtesy of delivering this review. We feel that a professor who lectures in a specialized area of engineering is the person most qualified to write a review of the pertinent material from that area. Until now, concise materials from lecturing professors have not been available to those wishing to prepare for the PE. You are now getting ready to use the result of the efforts of our distinctive team; we feel that it is an authoritative, highly efficient review. Credit is also due Jeff Potter, John Gruender, and Michelle Gruender for their diligence and hard work in bringing the strengths of this unique manual to the engineering community.

A passing grade on the FE exam is the first step for all engineers in the registration process. The second exam, The Principles and Practice of Engineering Examination, is presented in each major area of engineering. It may be taken after varying amounts of engineering apprenticeship, depending on state requirements. To practice engineering in some states an engineer must pass an examination in the particular area in which the engineer wishes to practice. Consequently, most states require each engineer to specify the particular area of engineering for testing; this allows for reciprocity in the event that an engineer moves from one state to another.

We published a review book in 1984 for the FE (EIT) exam and by 1994 over 100,000 engineers had used this review in their preparation. Many of the users of our FE Review have requested a review book for the PE exam. This PE/Civil Review is the first of a series we plan to publish.

As with our FE Review, we include detailed reviews, numerous examples, and problems with full solutions in one book. Since the PE exams include both essay design problems (partial credit) and multiple-choice problems, we have included both types of problems at the end of most chapters. This should provide sufficient practice for the actual exam. There seems to be an excellent correlation between an engineer's performance on the machine-graded, multiple-choice problems and that same engineer's performance on the partial-credit essay problems graded by professional engineers. Since a machine avoids the subjectivity of humans, there is a tendency to move in the direction of more multiple-choice problems. It's worth noting that a multiple-choice question sometimes helps lead the test-taker through a tough problem that may otherwise be quite formidable.

We have included in this review the major subjects encountered in most design problems that civil engineers are required to solve. You may find types of problems on the exam that we have not included in this review book. We have not attempted to include all areas of review, only the major ones. Of the numerous problems that can be selected, there will always be several that are thoroughly reviewed in this book. A good understanding of these subjects will ensure passing.

Make sure you're proficient in the basics—math, statics, solid mechanics, and fluid mechanics—these basic subjects are used in most civil engineering design problems; they form the first four chapters in this review book. Then select those areas in which you're most familiar and emphasize those areas in your review. Because you may not find four problems in the morning session and four problems in the afternoon session in one or two areas, you will probably have to work problems in three or perhaps four areas. So, to be safe, you should familiarize yourself with several areas.

Preface

The PE exam may also contain some economics problems, so we have added a chapter on engineering economics as Chapter 16. An economics problem may be one of the essay problems of the morning session. Some of the multiple choice questions of the afternoon session may require economic analysis.

Use your time wisely in your review—start early—perhaps several months early. Sign up for a review course if there's one in your area; that'll force you to review. Get a good night's sleep before the examination, and we here at Great Lakes Press wish you well!

Dr. Merle C. Potter
Okemos, Michigan

P.S. Your comments are a vital part of our user-oriented publishing method and are greatly appreciated. With this in mind, note that a book has its most errors in the first printing of its first edition even though we've reviewed each chapter several times. Please fill out and return the comment and errata cards included in this volume.

Introduction

The NCEES

The National Council of Examiners for Engineering and Surveying (NCEES) develops both the FE/EIT exam and the various Principles and Practice of Engineering examinations (the PE exams). The PE exams are given in a variety of major areas including Civil (Sanitary/Structural), Mechanical, Electrical, and Chemical Engineering. The NCEES develops the examinations using standard guidelines that allow for maximum fairness and the highest quality. Experienced testing specialists and professional engineers from across the nation including those from private consulting, government, industry and education, prepare examinations which are valid measures of competency.

Applications

Each state establishes the education and experience level required for eligibility to take the PE exam. Application procedures and fee requirements vary for each state. Sufficient time must be allotted for the application process. Consult the Board of the state in which you desire registration for the details of the registration process and an application for the PE exam. The address of each state is given in a table that follows.

Major Area

Some states permit an engineer to practice only in the major area of the PE exam that was passed, while other states allow engineers to practice in all areas of engineering by passing a PE exam in only one major area. The Board of a state must be consulted for the requirements in that particular state. To allow for reciprocity, a particular state may require that the PE exam be taken in one major area only.

When and Where

The PE exam is offered twice a year: in mid-April and late October. The exact dates are selected by each state Board. The location may also change year to year. Be sure and consult the state Board for the exact time, date and location of the exam.

Content

The PE exam is an eight-hour examination consisting of a morning session and an afternoon session. The four-hour morning session consists of twelve essay type problems and the four-hour afternoon session consists of twelve multiple-choice type problems. Four problems are to be solved in each session. This provides about one hour for each problem.

The essay problems are similar to the long problems given in an undergraduate hour-long exam in a design course. The situation is described in detail and the expected response is carefully outlined. Partial credit is awarded if mistakes or errors in judgment are made.

The multiple-choice problems each consist of a particular situation that is again described in detail, followed by ten multiple-choice questions. There is no partial credit, however, guessing is permitted.

There is some advantage in this type of problem in that the ten questions may lead a candidate through the steps to a successful solution, even though the problem appears unsolvable without the "hints" contained in the wording of the ten multiple-choice questions.

Scoring

The four problems selected in the morning session are each worth ten points for a maximum score of forty points, with partial credit awarded on each problem. Scores of 0, 2, 4, 6, 8 or 10 points are earned on each problem with 6 being the score of minimum competency.

The four multiple-choice problems selected in the afternoon session are also worth ten points each for a maximum score of forty points. Each of the ten questions in a particular problem is worth one point with 6 again representing minimum competency.

A combined score of 48 is needed to pass the exam. This is the score of minimum competency established by the NCEES. Much effort is made through workshops composed of practicing engineers to assure the quality and fairness of the exam. The engineers are representative of the profession regarding gender, ethnic background, geography and area of practice.

Minimum competency is defined by NCEES as:

> "The lowest level of knowledge at which a person can practice professional engineering in such a manner that will safeguard life, health and property and promote the public welfare."

The NCEES, with the aide of committees of professional engineers, continues to develop examinations that assure minimum competency of all who pass.

The Examination

For the morning session proctors distribute an exam booklet and a solution pamphlet to each candidate. Detailed instructions are given before the booklets are to be opened. Your name must be recorded on the front cover. In addition, your board code and identification number, the number of the problem you are solving and the number of pages in the solution must be recorded in the upper right corner on each page of the solution pamphlet. You must also write the numbers of the problems that you solved on the front of the solutions pamphlet. If you do not enter this information in the appropriate places, errors may result in the scoring process.

There is no partial credit for the machine-scored answers of the multiple choice questions of the afternoon session. As in all machine-scored answers, you must use a #2 pencil to avoid any improper scanning by the machine. Remember to answer all the questions, even if you must guess, since there is no penalty for a wrong answer.

If you complete either session of the examination with more than 30 minutes remaining in the session, you will be free to leave. If there is less than 30 minutes remaining in the session, you must wait until the end of the session to permit an orderly collection of exam materials and to avoid disruptions of candidates. Give full attention to the proctors' instructions at the end of each session as well as at the beginning.

What to Bring

The PE exam is open-book. You may bring review books, text books, handbooks, and other bound material into the exam. You should probably predetermine the areas in which you will most likely select problems to solve. The handbooks that are used for design purposes in those areas (at most 4 areas) should be part of your selection. Select one handbook, if one exists, for each area. Use the handbook you're most familiar with even when a new edition has just been released. A good, battery-operated, silent, non-printing calculator that you can easily operate is a must. Don't make last minute selections of materials. Work with your selected materials all through the review process. Familiarity is very important.

HP-48G—the ideal PE Exam Calculator

You really should use a silent, pre-programmed calculator at this exam. A calculator is essential when solving many problems. In fact, with the exception of a couple states, the premier engineering calculator, the HP-48G, is allowed into the exam (check with your state board). This calculator is a hand computer which has hundreds of basic equations and constants preprogrammed. If you need extra help with exam calculations, the HP-48G will be very useful. We at GLP offer this calculator and its higher-powered downloadable relative, the GX, for sale at substantial discounts. (Our 48G price $105; list $165. the GX is $215; list $350.)

New 'Jump Start the HP-48G/GX' Manual—Quick'n'Easy Primer for Engineers

We offer a unique and concise manual to guide you effectively through the steps to using the powerful HP-48 G/GX calculators The manufacturer's manual is known to be difficult to use for even basic operations. Our engineering-oriented manual will help you make the most of these calculators in the PE exam room. It is the only one like it on the market and is available for $19.95. Call us at 1-800-837-0201 for ordering or info.

Special Requirements

If you require special accommodations due to some condition that would put you at a disadvantage, you should consult your state board. Special arrangement can be made if you communicate such requirements well in advance of the exam.

What's in this Review

This book presents an efficient review of the major subject areas covered on the PE exam and also serves as an excellent desk reference.

Select those chapters that are related to the subject area of the PE exam that you are most familiar with. Select at least four of the major design areas:

- Structures (steel, concrete, indeterminate)
- Water (supply and waste)
- Hydraulics (open channel, piping)
- Hydrology
- Soils and Foundations
- Highways

Note that problems may require you to perform an economic analysis of a proposed design; in some cases this may be the primary focus of the problem. Be sure and be prepared by studying our Chapter 16.

To help you in your overall preparation, this book contains:

- Succinct reviews of both theoretical and practical aspects of the subject areas
- Tables, charts and figures pertinent to the subjects
- Example problems with detailed solutions illustrating important concepts
- Practice problems similar to those that may be encountered on the PE exam
- Solutions to all practice problems

Why Our PE Review Stands Apart from Other Approaches

We at Great Lakes Press developed this reference with the cooperation of a team of professorial colleagues in response to the existing prep situation. The kind of concise material that an engineer naturally desires to use was needed. We were also motivated by the fact that over the years, we've received hundreds of requests for a PE title from users of our FE review. So we recruited popular lecturers from the university campus and prepared this study guide. The design of this review is to deliver an adequate preparation without excess. Our goal as publisher is always to act as an advocate for the test-taker in our interactions with engineering departments, associations, review courses and registration governing bodies. We are dedicated to making the entire licensing and registration process as reasonable as possible.

How To Become A Professional Engineer

To become registered as an engineer, a state may require that you:
1. graduate from an ABET-accredited engineering program
2. pass the *Fundamentals of Engineering* (FE) exam
3. pass the *Principles and Practice of Engineering* (PE) exam after several years of engineering experience

Requirements vary from state to state, so you should obtain local guidelines and follow them carefully.

Why Get an Engineering License?

Registration is necessary if an engineer works as a consultant, and is highly recommended in certain industries—especially when one is, or hopes to be, in a management position.

Handling Multiple Choice Problems

The afternoon portion of the PE exam consists of 4 multiple choice problems, each having 10 questions. For those of us with less immediate experience with this sort of test question, a few points of strategic advice will make for much less confusion and frustration, and help you maximize your score for any given amount of preparation.

Process of Elimination

- Sometimes the best way to find the right answer is to look for the wrong ones and cross them out. On questions which are difficult for you, wrong answers are often much easier to find than right ones!

- Answers are seldom given with more than three significant figures, and may be given with two significant figures. The choice **closest** to your own solution should be selected.

- Use the **process of elimination** when *guessing* at an answer. If only one answer is negative and four answers are positive, eliminate the one odd answer from your guess. Also, when in doubt, work backwards and eliminate those answers that you believe are untrue until you are left with only 2 or 3 and then guess. By using a combination of methods, you greatly improve your odds of answering correctly.

Should I Guess?

- **Do not leave any answers blank** on your answer key. A guess *cannot* hurt you, only help you. Your score is based on the number of questions you answer correctly. An incorrect answer does not harm your score.

Getting the Most from Your 'Uncertains'

- **Place a question mark** beside choices you are uncertain of, but seem correct. If time prevents you from re-working that problem, you will have at least identified your best guess.
- **Be sure to make a best guess** if you have spent some time on a problem. Then circle the problem number, so you can go back to it later if time permits. But your guess is most "educated"—and most likely to be correct—if you make it *immediately* while working on the problem.

State Boards of Registration

All State Boards of Registration administer the National Council of Engineering Examiners and Surveyors (NCEES) uniform examination. The dates of the exams cover a span of three days in mid-April and three days in late October. The specific dates are selected by each State Board. To be accepted to take the PE exam, an applicant must apply well in advance. For information regarding the specific requirements in your state, contact your state board's office. Any comments relating to the exam should be addressed to the Executive Director of NCEES, POB 1686, Clemson, SC 29633-1686, ph: (803) 654-6824.

State Boards of Registration Addresses and Phone Numbers

ALABAMA: State Board of Registration for Professional Engineers and Land Surveyors, P. O. Box 304451, Montgomery 36130-4451. Executive Secretary, Telephone: (205) 242-5568.

ALASKA: State Board of Registration for Architects, Engineers and Land Surveyors, Pouch D, Juneau 99811. Licensing Examiner, Telephone: (907) 465-2540.

ARIZONA: State Board of Technical Registration, 1951 W. Camelback Rd., Suite 250, Phoenix 85015. Executive Director, Telephone: (602) 255-4053.

ARKANSAS: State Board of Registration for Professional Engineers and Land Surveyors, P. O. Box 2541, Little Rock 72203. Secretary-Treasurer, Telephone: (501) 324-9085.

CALIFORNIA: State Board of Registration for Professional Engineers and Land Surveyors, 2535 Capitol Oaks Dr., Sacramento 95853. Executive Secretary, Telephone: (916) 263-2222.

COLORADO: State Board of Registration for Professional Engineers and Professional Land Surveyors, 1860 Broadway, Suite 1370, Denver 80202. Program Administrator, Telephone: (303) 894-7788.

CONNECTICUT: State Board of Registration for Professional Engineers and Land Surveyors, The State Office Building, Room G-3A, 165 Capitol Avenue, Hartford 06106. Administrator, Ph: (203) 566-3290.

DELAWARE: Delaware Association of Professional Engineers, 2005 Concord Pike, Wilmington 19803. Executive Secretary, Telephone: (302) 577-6500.

DISTRICT OF COLUMBIA: Board of Registration for Professional Engineers, 614 H Street, N.W., Room 910, Washington 20001. Executive Secretary, Telephone: (202) 727-7468.

FLORIDA: Department of Business and Professional Regulations, Board of Professional Engineers, Northwood Centre, 1940 N. Monroe Street, Tallahassee 32399-0755. Executive Director, Telephone: (904) 488-9912.

GEORGIA: State Board of Registration for Professional Engineers and Land Surveyors, 166 Pryor Street, SW, Atlanta 30303-3465. Executive Director, Telephone: (404) 656-3926.

GUAM: Territorial Board of Registration for Professional Engineers, Architects and Land Surveyors, Department of Public Works, Government of Guam, P. O. Box 2950, Agana 96911. Chairman, Telephone: (671) 646-3115.

HAWAII: State Board of Registration for Professional Engineers, Architects, Land Surveyors and Landscape Architects, P. O. Box 3469, (1010 Richards Street), Honolulu 96801. Executive Secretary, Telephone: (808) 586-2693.

IDAHO: Board of Professional Engineers and Land Surveyors, 600 S. Orchard, Suite A, Boise 83705. Executive Secretary, Telephone: (208) 334-3860.

ILLINOIS: Department of Registration and Education, Professional Engineers' Examining Committee, 320 West Washington, 3rd Floor, Springfield 62786. Unit Manager, Telephone: (217) 782-0458.

INDIANA: Indiana Professional Licensing Agency, 302 W. Washington St., Room E034, Indianapolis 46204. Executive Director, Telephone: (317) 232-2980.

IOWA: State Board of Engineering Examiners, Capitol Complex, 1918 S.E. Hulsizer, Ankeny 50021. Executive Secretary, Telephone: (515) 281-5602.

KANSAS: State Board of Technical Professions, 900 Jackson, Room 507, Topeka 66612. Executive Secretary, Telephone: (913) 296-3053.

KENTUCKY: State Board of Registration for Professional Engineers and Land Surveyors, 160 Democrat Dr., Frankfort 40601. Executive Director, Telephone: (502) 573-2680.

LOUISIANA: State Board of Registration for Professional Engineers and Land Surveyors, 1055 St. Charles Avenue, Suite 415, New Orleans 70130. Executive Secretary, Telephone: (504) 568-8450.

MAINE: State Board of Registration for Professional Engineers, State House, Station 92, Augusta 04333. Secretary, Telephone: (207) 289-3236.

MARYLAND: State Board of Registration for Professional Engineers, 501 St. Paul Place, Room 902, Baltimore 21202. Executive Secretary, Telephone: (410) 659-6322.

MASSACHUSETTS: State Board of Registration of Professional Engineers and of Land Surveyors, Room 1512, Leverett Saltonstall Building, 100 Cambridge Street, Boston 02202. Secretary, Ph: (617) 727-9957.

MICHIGAN: Board of Professional Engineers, P. O. Box 30018, (611 West Ottawa), Lansing 48909. Administrative Secretary, Telephone: (517) 335-1669.

MINNESOTA: State Board of Registration for Architects, Engineers, Land Surveyors and Landscape Architects, 133 7th St. E., St. Paul 55101. Executive Secretary, Ph: (612) 296-2388.

MISSISSIPPI: State Board of Registration for Professional Engineers and Land Surveyors P. 0. Box 3, Jackson 39205. Executive Director, Telephone: (601) 359-6160.

MISSOURI: Board of Architects, Professional Engineers and Land Surveyors, P. O. Box 184, Jefferson City 65102. Executive Director, Telephone: (314) 751-0047.

MONTANA: State Board of Professional Engineers and Land Surveyors, Department of Commerce, 111 N. Jackson, POB 200513, Helena 59620-0513. Administrative Secretary, Telephone: (406) 444-4285.

NEBRASKA: State Board of Examiners for Professional Engineers and Architects, P. 0. Box 94751, Lincoln 68509. Executive Director, Telephone: (402) 471-2021.

NEVADA: State Board of Registered Professional Engineers and Land Surveyors, 1755 East Plum Lane, Ste. 135, Reno 89502. Executive Secretary, Telephone: (702) 688-1231.

NEW HAMPSHIRE: State Board of Professional Engineers, 57 Regional Drive, Concord 03301. Executive Secretary, Telephone: (603) 271-2219.

NEW JERSEY: State Board of Professional Engineers and Land Surveyors, P.O. Box 45015, Newark 07101. Executive Secretary-Director, Telephone: (201) 504-6460.

NEW MEXICO: State Board of Registration for Professional Engineers and Land Surveyors, 1010 Marquez Pl., Santa Fe 87501. Secretary, Telephone: (505) 827-7561.

NEW YORK: State Board for Engineering and Land Surveying, The State Education Department, Cultural Education Center, Madison Avenue, Albany 12230. Executive Secretary, Telephone: (518) 474-3846.

NORTH CAROLINA: State Board of Professional Engineers and Land Surveyors, 3620 Six Forks Rd., Suite 300, Raleigh 27609. Executive Secretary, Telephone (919) 781-9499

NORTH DAKOTA: State Board of Registration for Professional Engineers and Land Surveyors, P. 0. Box 1357, Bismarck 58502. Executive Secretary, Telephone: (701) 258-0786.

OHIO: State Board of Registration for Professional Engineers and Surveyors, 77 S. High St., 16th Fl., Columbus 43266-0314. Executive Secretary, Telephone: (614) 466-8999.

OKLAHOMA: State Board of Registration for Professional Engineers and Land Surveyors, Oklahoma Engineering Center, Room 120, 201 N.E. 27th Street, Oklahoma City, 73105. Executive Secretary, Telephone: (405) 521-2874.

OREGON: State Board of Engineering Examiners, Department of Commerce, 750 Front St., NE, Suite 240, Salem 97310. Executive Secretary, Telephone: (503) 378-4180.

PENNSYLVANIA: State Registration Board for Professional Engineers, P. O. Box 2649, Harrisburg 17120-264g. Administrative Assistant, Telephone: (800) 877-3926.

PUERTO RICO: Board of Examiners of Engineers, Architects, and Surveyors, Box 3271, San Juan 00904. Director, Examining Boards, Telephone: (809) 722-2122.

RHODE ISLAND: State Board of Registration for Professional Engineers and Land Surveyors, 10 Orms St., Suite 324, Providence 02904. Administrative Assistant, Telephone: (401) 277-2565.

SOUTH CAROLINA: State Board of Registration for Professional Engineers and Land Surveyors, P.O. Drawer 50408, Columbia 29250. Agency Director, Ph: (803) 734-9166.

SOUTH DAKOTA: State Commission of Engineering, Architectural Examiners and Land Surveying, 2040 West Main Street, Suite 304, Rapid City 57702-2447. Executive Secretary, Telephone: (605) 394-2510.

TENNESSEE: State Board of Architectural and Engineering Examiners, 3rd Fl. Volunteer Plaza, 500 James Robertson Pkwy, Nashville 37243-1142. Administrator, Telephone: (615) 741-3221.

TEXAS: State Board of Registration for Professional Engineers, P.O. Drawer 18329, Austin 78760-8329. Executive Director, Telephone: (512) 440-7723.

UTAH: Division of Occupational and Professional Licensing, P.O. Box 45805, Salt Lake City 84145-0805. Director, Ph: (801) 530-6628.

VERMONT: State Board of Registration for Professional Engineers, Division of Licensing and Registration, Pavilion Building, Montpelier 05602. Executive Secretary, Telephone: (802) 828-2875.

VIRGINIA: State Board of Architects, Professional Engineers, Land Surveyors and Certified Landscape Architects, 3600 West Broad St., Seaboard Building, 5th Floor, Richmond 23230-4917. Assistant Director, Telephone: (804) 367-8506.

VIRGIN ISLANDS: Board for Architects, Engineers and Land Surveyors, Bldg 1, Sub-Base, Rm 205, St. Thomas 00801. Secretary, Telephone: (809) 774-3130.

WASHINGTON: State Board of Registration for Professional Engineers and Land Surveyors, P.O. Box 9649, Olympia 98504. Executive Secretary, Telephone: (206) 753-3634.

WEST VIRGINIA: State Board of Registration for Professional Engineers, 608 Union Building, Charleston 25301. Executive Director, Telephone: (304) 348-3554.

WISCONSIN: State Examining Board of Professional Engineers, P.O. Box 8935, Madison 53708. Administrator, Telephone: (608) 266-1397.

WYOMING: State Board of Examining Engineers, Herschler Building, Room 4135, Cheyenne 82002. Secretary-Accountant, Telephone: (307) 777-6156.

English and SI Units

The following tables presenting SI (Systems International) units and the conversion of English units to SI units conclude this introduction.

SI Prefixes

Multiplication Factor	Prefix	Symbol
10^{15}	peta	P
10^{12}	tera	T
10^{9}	giga	G
10^{6}	mega	M
10^{3}	kilo	k
10^{-1}	deci	d
10^{-2}	centi	c
10^{-3}	mili	m
10^{-6}	micro	μ
10^{-9}	nano	n
10^{-12}	pico	p
10^{-15}	femto	f

SI base units

Quantity	Name	Symbol
length	meter	m
mass	kilogram	kg
time	second	s
electric current	ampere	A
temperature	kelvin	K
amount of substance	mole	mol
luminous intensity	candela	cd

SI Derived Units

Quantity	Name	Symbol	In Terms of Other Units
area	square meter		m^2
volume	cubic meter		m^3
velocity	meter per second		m/s
acceleration	meter per second squared		m/s^2
density	kilogram per cubic meter		kg/m^3
specific volume	cubic meter per kilogram		m^3/kg
frequency	hertz	Hz	s^{-1}
force	newton	N	$m \cdot kg/s^2$
pressure, stress	pascal	Pa	$kg/(m \cdot s^2)$
energy, work, heat	joule	J	$N \cdot m$
power	watt	W	J/s
electric charge	coulomb	C	$A \cdot s$
electric potential	volt	V	W/A
capacitance	farad	F	C/V
electric resistance	ohm	Ω	V/A
conductance	siemens	S	A/V
magnetic flux	weber	Wb	$V \cdot s$
inductance	henry	H	Wb/A
viscosity	pascal second		$Pa \cdot s$
moment (torque)	meter newton		$N \cdot m$
heat flux	watt per square meter		W/m^2
entropy	joule per kelvin		J/K
specific heat	joule per kilogram-kelvin		$J/(kg \cdot K)$
conductivity	watt per meter-kelvin		$W/(m \cdot K)$

Conversion Factors to SI Units

English	SI	SI Symbol	To Convert from English to SI Multiply by
Area			
square inch	square centimeter	cm^2	6.452
square foot	square meter	m^2	0.09290
acre	hectare	ha	0.4047
Length			
inch	centimeter	cm	2.54
foot	meter	m	0.3048
mile	kilometer	km	1.6093
Volume			
cubic inch	cubic centimeter	cm^3	16.387
cubic foot	cubic meter	m^3	0.02832
gallon	cubic meter	m^3	0.003785
gallon	liter	L	3.785
Mass			
pound mass	kilogram	kg	0.4536
slug	kilogram	kg	14.59
Force			
pound	newton	N	4.448
kip (1000 lb)	newton	N	4448
Density			
pound/cubic foot	kilogram/cubic meter	kg/m^3	16.02
pound/cubic foot	grams/liter	g/L	16.02
Work, Energy, Heat			
foot-pound	joule	J	1.356
Btu	joule	J	1055
Btu	kilowatt-hour	kWh	0.000293
therm	kilowatt-hour	kWh	29.3

Conversion Factors to SI Units (continued)

English	SI	SI Symbol	To Convert from English to SI Multiply by
Power, Heat, Rate			
horsepower	watt	W	745.7
foot pound/sec	watt	W	1.356
Btu/hour	watt	W	0.2931
Btu/hour-ft^2-°F	watt/meter squared-°C	W/m^2·°C	5.678
tons of refrig.	kilowatts	kW	3.517
Pressure			
pound/square inch	kilopascal	kPa	6.895
pound/square foot	kilopascal	kPa	0.04788
inches of H$_2$O	kilopascal	kPa	0.2486
inches of Hg	kilopascal	kPa	3.374
one atmosphere	kilopascal	kPa	101.3
Temperature			
Fahrenheit	Celsius	°C	5 (°F-32)/9
Fahrenheit	kelvin	K	5 (°F+460)/9
Velocity			
foot/second	meter/second	m/s	0.3048
mile/hour	meter/second	m/s	0.4470
mile/hour	kilometer/hour	km/h	1.609
Acceleration			
foot/second squared	meter/second squared	m/s^2	0.3048
Torque			
pound-foot	newton-meter	N·m	1.356
pound-inch	newton-meter	N·m	0.1130
Viscosity, Kinematic Viscosity			
pound-sec/square foot	newton-sec/square meter	N·s/m^2	47.88
square foot/second	square meter/second	m^2/s	0.09290
Flow Rate			
cubic foot/minute	cubic meter/second	m^3/s	0.0004719
cubic foot/minute	liter/second	L/s	0.4719
Frequency			
cycles/second	hertz	Hz	1.00

Conversion Factors

Length

1 cm	= 0.3937 in
1 m	= 3.281 ft
1 yd	= 3 ft
1 mi	= 5280 ft
1 mi	= 1760 yd
1 km	= 3281 ft

Area

1 cm^2	= 0.155 in^2
1 m^2	= 10.76 ft^2
1 ha	= 10^4 m^2
1 acre	= 100 m^2
1 acre	= 4047 m^2
1 acre	= 43,560 ft^2

Volume

1 ft^3	= 28.32 L
1 L	= 0.03531 ft^3
1 L	= 0.2642 gal
1 m^3	= 264.2 gal
1 ft^3	= 7.481 gal
1 m^3	= 35.31 ft^3
1 acre-ft	= 43,560 ft^3
1 m^3	= 1000 L

Velocity

1 m/s	= 3.281 ft/s
1 mph	= 1.467 ft/s
1 mph	= 0.8684 knot
1 knot	= 1.688 ft/s
1 km/h	= 0.2778 m/s
1 km/h	= 0.6214 mph

Force

1 lb	= 4.448 x 10^5 dyne
1 lb	= 32.17 pdl
1 lb	= 0.4536 kg
1 N	= 10^5 dyne
1 N	= 0.2248 lb
1 kip	= 1000 lb

Mass

1 oz	= 28.35 g
1 lb	= 0.4536 kg
1 kg	= 2.205 lb
1 slug	= 14.59 kg
1 slug	= 32.17 lb

Work and Heat

1 Btu	= 778.2 ft-lb
1 Btu	= 1055 J
1 Cal	= 3.088 ft-lb
1 J	= 10^7 ergs
1 kJ	= 0.9478 ft-lb
1 Btu	= 0.2929 W·hr
1 ton	= 12,000 Btu/hr
1 kWh	= 3414 Btu
1 quad	= 10^{15} Btu
1 therm	= 10^5 Btu

Power

1 Hp	= 550 ft-lb/s
1 HP	= 33,000 ft-lb/min
1 Hp	= 0.7067 Btu/s
1 Hp	= 2545 Btu/hr
1 Hp	= 745.7 W
1 W	= 3.414 Btu/hr
1 kW	= 1.341 Hp

Volume Flow Rate

1 cfm	= 7.481 gal/min
1 cfm	= 0.4719 L/s
1 m^3/s	= 35.31 ft^3/s
1 m^3/s	= 2119 cfm
1 gal/min	= 0.1337 cfm

Torque

1 N·m	= 10^7 dyne·cm
1 N·m	= 0.7376 lb-ft
1 N·m	= 10 197 g·cm
1 lb-ft	= 1.356 N·m

Viscosity

1 lb-s/ft^2	= 478 poise
1 poise	= 1 g/cm·s
1 N·s/m^2	= 0.02089 lb-s/ft^2

Pressure

1 atm	= 14.7 psi
1 atm	= 29.92 in Hg
1 atm	= 33.93 ft H$_2$0
1 atm	= 1.013 bar
1 atm	= 1.033 kg/cm^2
1 atm	= 101.3 kPa
1 psi	= 2.036 in Hg
1 psi	= 6.895 kPa
1 psi	= 68 950 dyne/cm^2
1 ft H$_2$0	= 0.4331 psi
1 kPa	= 0.145 psi

1. Mathematics

by Merle C. Potter

The engineer uses mathematics as a tool to help solve the problems encountered in the analysis and design of physical systems. We will review those parts of mathematics that are used fairly often by the engineer when solving engineering design problems. The topics include: algebra, trigonometry, analytic geo-metry, linear algebra (matrices), calculus and differential equations. The review here is intended to be brief and not exhaustive. The majority of the problems on the exam can be solved using the material included in this chapter. There may be a few problems, however, that will require information not covered here; to cover all possible points would not be in the spirit of an efficient review.

1.1 Algebra

It is assumed that the reader is familiar with most of the laws of algebra as applied to both real and complex numbers. We will review some of the more important of these and illustrate several with examples.

1.1.1 Exponents

Laws of exponents are used in many manipulations. For positive x and y we use

$$\begin{aligned} x^{-a} &= \frac{1}{x^a} \\ x^a x^b &= x^{a+b} \\ (xy)^a &= x^a y^a \\ x^{ab} &= \left(x^a\right)^b \end{aligned} \qquad (1.1.1)$$

1.1.2 Logarithms

Logarithms are actually exponents. For example if $b^x = y$ then $x = \log_b y$; that is, the exponent x is equal to the logarithm of y to the base b. Most engineering applications involve common logs which have a base of 10, written as $\log y$, or natural logs which have a base of e ($e = 2.7183\cdots$), written as $\ln y$.

Remember, logarithms of numbers less than one are negative, the logarithm of one is zero, and logarithms of numbers greater that one are positive. The following identities are often useful when manipulating logarithms:

$$\ln x^a = a \ln x$$

$$\ln(xy) = \ln x + \ln y$$

$$\ln(x/y) = \ln x - \ln y$$

$$\ln x = 2.303 \log x$$

$$\log_b b = 1 \tag{1.1.2}$$

$$\ln 1 = 0$$

$$\ln e^a = a$$

1.1.3 The Quadratic Formula and the Binomial Theorem

We often encounter the quadratic equation $ax^2 + bx + c = 0$ when solving engineering problems. The *quadratic formula* provides its solution; it is

$$x = \frac{-b \pm \sqrt{b^2 - 4ac}}{2a}. \tag{1.1.3}$$

If $b^2 < 4ac$, the two roots are complex numbers. Cubic and higher order equations are most often solved by trial and error.

The *binomial theorem* is used to expand an algebraic expression of the form $(a + x)^n$. It is

$$(a+x)^n = a^n + na^{n-1}x + \frac{n(n-1)}{2!}a^{n-2}x^2 + \cdots. \tag{1.1.4}$$

If n is a positive integer, the expansion contains $(n + 1)$ terms. If it is a negative integer or a fraction, an infinite series expansion results.

1.1.4 Partial Fractions

A rational fraction $P(x)/Q(x)$, where $P(x)$ and $Q(x)$ are polynomials, can be resolved into partial fractions for the following cases.

Case 1: $Q(x)$ factors into n different linear terms,

$$Q(x) = (x - a_1)(x - a_2)\ldots(x - a_n).$$

Then

$$\frac{P(x)}{Q(x)} = \sum_{i=1}^{n} \frac{A_i}{x - a_i}. \tag{1.1.5}$$

Case 2: $Q(x)$ factors into n identical terms,

$$Q(x) = (x - a)^n.$$

Then

$$\frac{P(x)}{Q(x)} = \sum_{i=1}^{n} \frac{A_i}{(x - a)^i}. \tag{1.1.6}$$

Case 3: $Q(x)$ factors into n different quadratic terms,

$$Q(x) = \left(x^2 + a_1 x + b_1\right)\left(x^2 + a_2 x + b_2\right)\ldots\left(x^2 + a_n x + b_n\right).$$

Then

$$\frac{P(x)}{Q(x)} = \sum_{i=1}^{n} \frac{A_i x + B_i}{x^2 + a_i x + b_i}. \tag{1.1.7}$$

Case 4: $Q(x)$ factors into n identical quadratic terms,

$$Q(x) = \left(x^2 + ax + b\right)^n.$$

Then

$$\frac{P(x)}{Q(x)} = \sum_{i=1}^{n} \frac{A_i x + B_i}{\left(x^2 + a x + b\right)^i}. \tag{1.1.8}$$

Case 5: $Q(x)$ factors into a combination of the above. The partial fractions are the obvious ones from the appropriate expansions above.

──────── **EXAMPLE 1.1** ────────

The temperature at a point in a body is given by $T(t) = 100 e^{-0.02t}$. At what value of t does $T = 20$?

Solution. The equation takes the form

$$20 = 100 e^{-0.02t}$$

$$0.2 = e^{-0.02t}.$$

Take the natural logarithm of both sides and obtain

$$\ln 0.2 = \ln e^{-0.02t}.$$

Using a calculator, we find

$$-1.6094 = -0.02 t.$$

$$\therefore t = 80.47.$$

──────── **EXAMPLE 1.2** ────────

Find an expansion for $(9 + x)^{1/2}$.

Solution. Using the binomial theorem, Eq. 1.1.4, we have

$$(9 + x)^{1/2} = 3\left(1 + \frac{x}{9}\right)^{1/2}$$

$$= 3\left[1 + \frac{1}{2}\left(\frac{x}{9}\right) + \frac{1/2(-1/2)}{2}\left(\frac{x}{9}\right)^2 + \frac{1/2(-1/2)(-3/2)}{3 \cdot 2}\left(\frac{x}{9}\right)^3 + \cdots\right]$$

$$= 3 + \frac{x}{6} - \frac{x^2}{216} + \frac{x^3}{3888} + \cdots.$$

Note: We factored out $9^{1/2} = 3$ so that in Eq. 1.1.4 $a = 1$; this simplifies the expansion.

EXAMPLE 1.3

Resolve $\dfrac{x^2 + 2}{x^4 + 4x^3 + x^2}$ into partial fractions.

Solution. The denominator is factored into

$$x^4 + 4x^3 + x^2 = x^2(x^2 + 4x + 1).$$

Using Cases 2 and 3 there results

$$\frac{x^2 + 2}{x^4 + 4x^3 + x^2} = \frac{A_1}{x} + \frac{A_2}{x^2} + \frac{A_3 x + B_3}{x^2 + 4x + 1}.$$

This can be written as

$$\frac{x^2 + 2}{x^4 + 4x^3 + x^2} = \frac{A_1 x(x^2 + 4x + 1) + A_2(x^2 + 4x + 1) + (A_3 x + B_3)x^2}{x^2(x^2 + 4x + 1)}$$

$$= \frac{(A_1 + A_3)x^3 + (4A_1 + A_2 + B_3)x^2 + (A_1 + 4A_2)x + A_2}{x^2(x^2 + 4x + 1)}.$$

The numerators on both sides must be equal. Equating the coefficients of the various powers of x provides us with the four equations:

$$A_1 + A_3 = 0$$

$$4A_1 + A_2 + B_3 = 1$$

$$A_1 + 4A_2 = 0$$

$$A_2 = 2.$$

These are solved quite easily to give $A_2 = 2$, $A_1 = -8$, $A_3 = 8$, $B_3 = 31$. Finally,

$$\frac{x^2 + 2}{x^4 + 4x^3 + x^2} = -\frac{8}{x} + \frac{2}{x^2} + \frac{8x + 31}{x^2 + 4x + 1}.$$

1.2 Trigonometry

The primary functions in trigonometry involve the ratios between the sides of a right triangle. Referring to the right triangle in Fig. 1.1, the functions are defined by

$$\sin\theta = \frac{y}{r}, \qquad \cos\theta = \frac{x}{r}, \qquad \tan\theta = \frac{y}{x}. \tag{1.2.1}$$

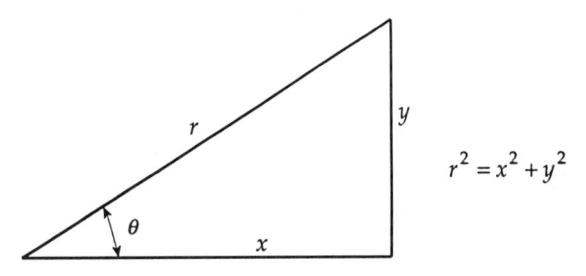

Figure 1.1 A right triangle.

In addition, there are three other functions that find occasional use, namely,

$$\cot\theta = \frac{x}{y}, \qquad \sec\theta = \frac{r}{x}, \qquad \csc\theta = \frac{r}{y}. \tag{1.2.2}$$

The trig functions $\sin\theta$ and $\cos\theta$ are periodic functions with a period of 2π. Fig. 1.2 shows a plot of the three primary functions.

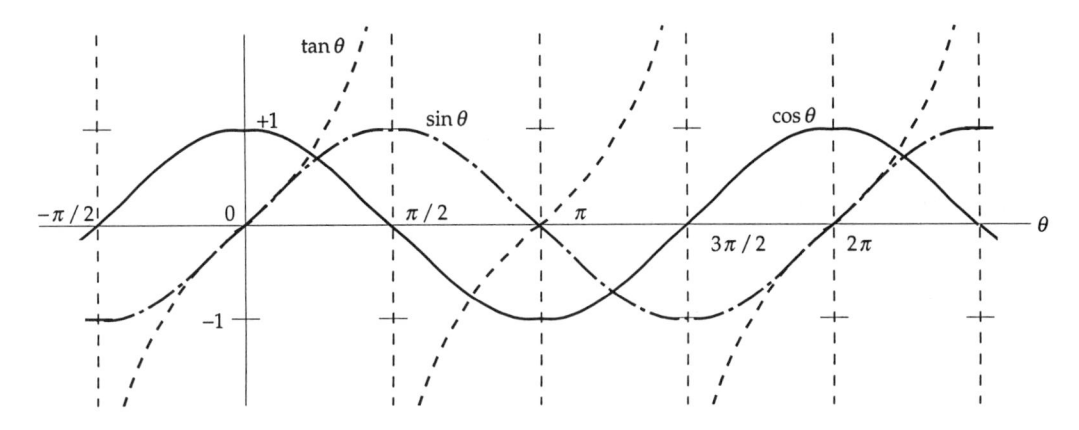

Figure 1.2 The Trig Functions.

In the above relationships, the angle θ is usually given in radians for mathematical equations. It is possible, however, to express the angle in degrees; if that is done it may be necessary to relate degrees to radians. This can be done by remembering that there are 2π radians in 360°. Hence, we multiply radians by $(180/\pi)$ to obtain degrees, or multiply degrees by $(\pi/180)$ to obtain radians. A calculator may use either degrees or radians for an input angle.

Most problems involving trigonometry can be solved using a few fundamental identities. They are

$$\sin^2\theta + \cos^2\theta = 1 \tag{1.2.3}$$

$$\sin 2\theta = 2\sin\theta\cos\theta \tag{1.2.4}$$

$$\cos 2\theta = \cos^2\theta - \sin^2\theta \tag{1.2.5}$$

$$\sin(\alpha \pm \beta) = \sin\alpha\cos\beta \pm \sin\beta\cos\alpha \tag{1.2.6}$$

$$\cos(\alpha \pm \beta) = \cos\alpha\cos\beta \mp \sin\alpha\sin\beta. \tag{1.2.7}$$

A general triangle may be encountered such as that shown in Fig. 1.3. For this triangle we may use the following equations:

law of sines:
$$\frac{\sin\alpha}{a} = \frac{\sin\beta}{b} = \frac{\sin\gamma}{c} \tag{1.2.8}$$

law of cosines: $$a^2 = b^2 + c^2 - 2bc\cos\alpha \qquad (1.2.9)$$

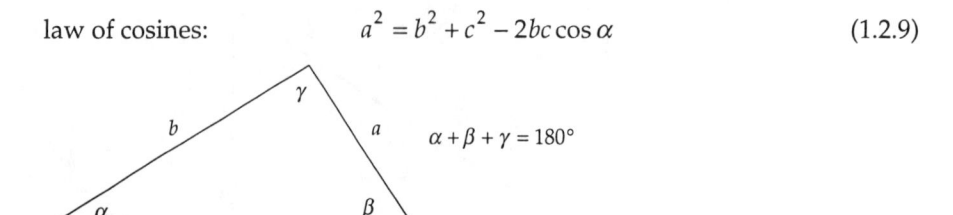

$$\alpha + \beta + \gamma = 180°$$

Figure 1.3 A general triangle.

Note that if $\gamma = 90°$, the law of cosines becomes the *Pythagorean Theorem*

$$c^2 = a^2 + b^2. \qquad (1.2.10)$$

The hyperbolic trig functions also find occasional use. They are defined by

$$\sinh x = \frac{e^x - e^{-x}}{2}, \qquad \cosh x = \frac{e^x + e^{-x}}{2}, \qquad \tanh x = \frac{\sinh x}{\cosh x}. \qquad (1.2.11)$$

Useful identities follow:

$$\cosh^2 x - \sinh^2 x = 1 \qquad (1.2.12)$$

$$\sinh(x \pm y) = \sinh x \cosh y \pm \cosh x \sinh y \qquad (1.2.13)$$

$$\cosh(x \pm y) = \cosh x \cosh y \pm \sinh x \sinh y \qquad (1.2.14)$$

The values of the primary trig functions of certain angles are listed in Table 1.1.

TABLE 1.1 Functions of Certain Angles.

	0	30°	45°	60°	90°	135°	180°	270°	360°
$\sin\theta$	0	1/2	$\sqrt{2}/2$	$\sqrt{3}/2$	1	$\sqrt{2}/2$	0	-1	0
$\cos\theta$	1	$\sqrt{3}/2$	$\sqrt{2}/2$	1/2	0	$-\sqrt{2}/2$	-1	0	1
$\tan\theta$	0	$1/\sqrt{3}$	1	$\sqrt{3}$	∞	-1	0	$-\infty$	0

EXAMPLE 1.4

Express $\cos^2\theta$ as a function of $\cos 2\theta$.

Solution. Substitute Eq. 1.2.3 into Eq. 1.2.5 and obtain

$$\cos 2\theta = \cos^2\theta - (1 - \cos^2\theta)$$
$$= 2\cos^2\theta - 1.$$

There results

$$\cos^2\theta = \frac{1}{2}(1 + \cos 2\theta).$$

If $\sin\theta = x$, what is $\tan\theta$?

Solution. Think of $x = x/1$. Thus, the hypotenuse of an imaginary triangle is of length unity and the side opposite θ is of length x. The adjacent side is of length $\sqrt{1-x^2}$. Hence,

$$\tan\theta = \frac{x}{\sqrt{1-x^2}}.$$

EXAMPLE 1.6

An airplane leaves Lansing flying due southwest at 300 km/hr, and a second leaves Lansing at the same time flying due west at 500 km/hr. How far apart are the airplanes after 2 hours?

Solution. After 2 hours, the respective distances from Lansing are 600 km and 1000 km. A sketch is quite helpful. The distance d that the two airplanes are apart is found using the law of cosines:

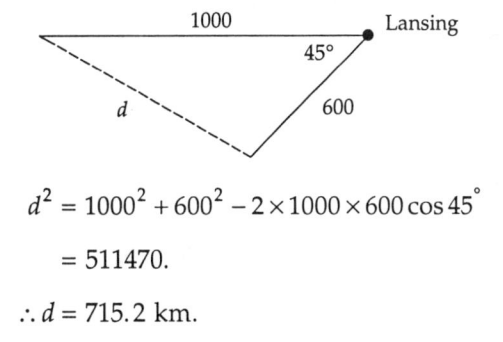

$$d^2 = 1000^2 + 600^2 - 2 \times 1000 \times 600 \cos 45°$$
$$= 511470.$$
$$\therefore d = 715.2 \text{ km}.$$

1.3 Geometry

A regular polygon with n sides has a vertex angle (the central angle subtended by one side) of $2\pi/n$. The included angle between two successive sides is given by $\pi(n-2)/n$.

Some common geometric shapes are displayed in Fig. 1.4:

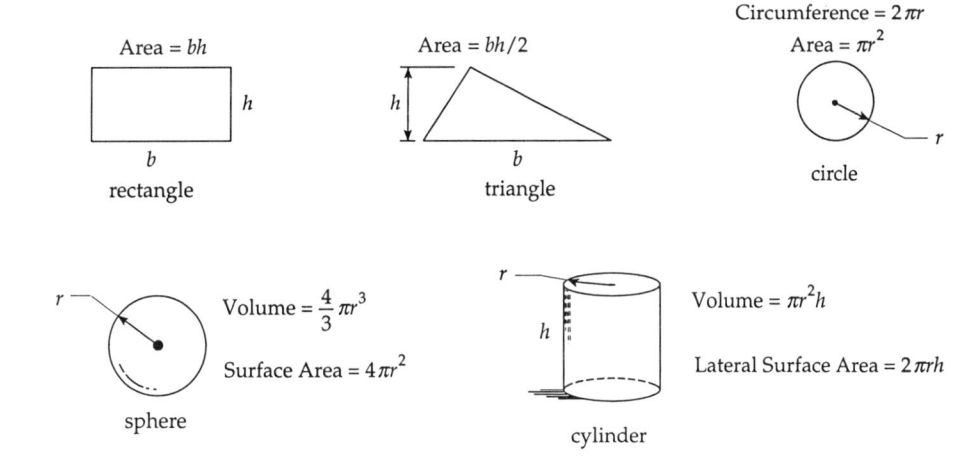

Figure 1.4 Common geometric shapes.

The equation of a straight line can be written in the general form

$$Ax + By + C = 0. \tag{1.3.1}$$

There are three particular forms that this equation can take. They are:

Point-slope: $\quad y - y_1 = m(x - x_1) \tag{1.3.2}$

Slope-intercept: $\quad y = mx + b \tag{1.3.3}$

Two-intercept: $$\frac{x}{a}+\frac{y}{b} = 1. \tag{1.3.4}$$

In the above equations m is the slope, (x_1, y_1) a point on the line, "a" the x-intercept, and "b" the y-intercept (see Fig. 1.5). The perpendicular distance d from the point (x_3, y_3) to the line $Ax + By + C = 0$ is given by (see Fig. 1.5)

$$d = \frac{|Ax_3 + By_3 + C|}{\sqrt{A^2 + B^2}}. \tag{1.3.5}$$

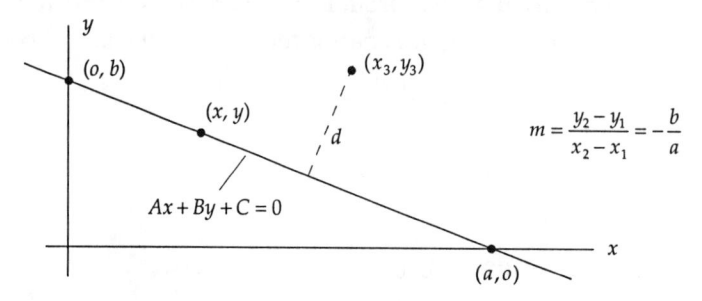

Figure 1.5 A straight line.

The equation of a plane surface is given as

$$Ax + By + Cz + D = 0. \tag{1.3.6}$$

The general equation of second degree

$$Ax^2 + 2Bxy + Cy^2 + 2Dx + 2Ey + F = 0 \tag{1.3.7}$$

represents a set of geometric shapes called *conic sections*. They are classified as follows:

ellipse: $B^2 - AC < 0$ (circle: $B = 0$, $A = C$)

parabola: $B^2 - AC = 0$ (1.3.8)

hyperbola: $B^2 - AC > 0$

If $A = B = C = 0$, the equation represents a line in the xy-plane, not a parabola. Let's consider each in detail.

Circle: The circle is a special case of an ellipse with $A = C$. Its general form can be expressed as

$$(x-a)^2 + (y-b)^2 = r^2 \tag{1.3.9}$$

where its center is at (a, b) and r is the radius.

Ellipse: The sum of the distances from the two foci, F, to any point on an ellipse is a constant. For an ellipse centered at the origin

$$\frac{x^2}{a^2} + \frac{y^2}{b^2} = 1 \tag{1.3.10}$$

where a and b are the semi-major and semi-minor axes. The foci are at $(\pm c, 0)$ where $c^2 = a^2 - b^2$. See Fig. 1.6a.

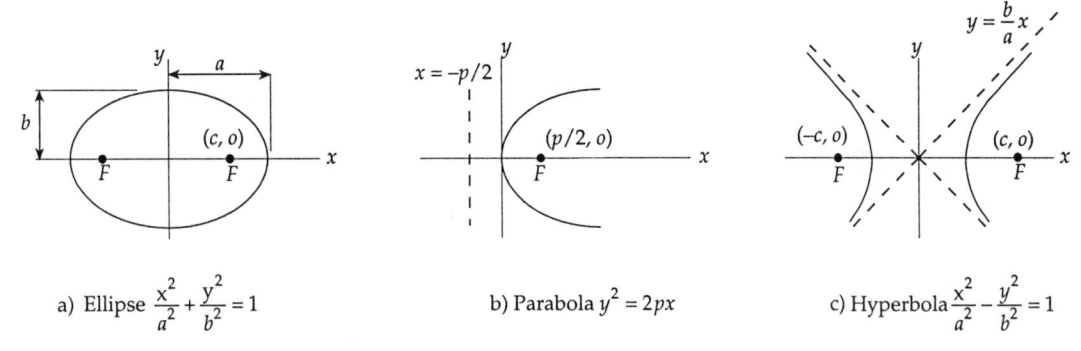

a) Ellipse $\dfrac{x^2}{a^2}+\dfrac{y^2}{b^2}=1$ b) Parabola $y^2 = 2px$ c) Hyperbola $\dfrac{x^2}{a^2}-\dfrac{y^2}{b^2}=1$

Figure 1.6 The three conic sections.

Parabola: The locus of points on a parabola are equidistant from the focus and a line (the directrix). If the vertex is at the origin and the parabola opens to the right, it is written as

$$y^2 = 2px \tag{1.3.11}$$

where the focus is at $(p/2, 0)$ and the directrix is at $x = -p/2$. See Fig. 1.6b. For a parabola opening to the left, simply change the sign of p. For a parabola opening upward or downward, interchange x and y.

Hyperbola: The difference of the distances from the foci to any point on a hyperbola is a constant. For a hyperbola centered at the origin opening left and right, the equation can be written as

$$\frac{x^2}{a^2} - \frac{y^2}{b^2} = 1. \tag{1.3.12}$$

The lines to which the hyperbola is asymptotic are asymptotes:

$$y = \pm \frac{b}{a} x. \tag{1.3.13}$$

If the asymptotes are perpendicular, a rectangular hyperbola results. If the asymptotes are the x and y axes, the equation can be written as

$$xy = \pm k^2. \tag{1.3.14}$$

Finally, in our review of geometry, we will present three other coordinate systems often used in engineering analysis. They are the polar (r, θ) coordinate system, the cylindrical (r, θ, z) coordinate system, and the spherical (r, θ, ϕ) coordinate system. The polar coordinate system is restricted to a plane:

$$x = r \cos \theta, \qquad y = r \sin \theta. \tag{1.3.15}$$

a) polar   b) cylindrical   c) spherical

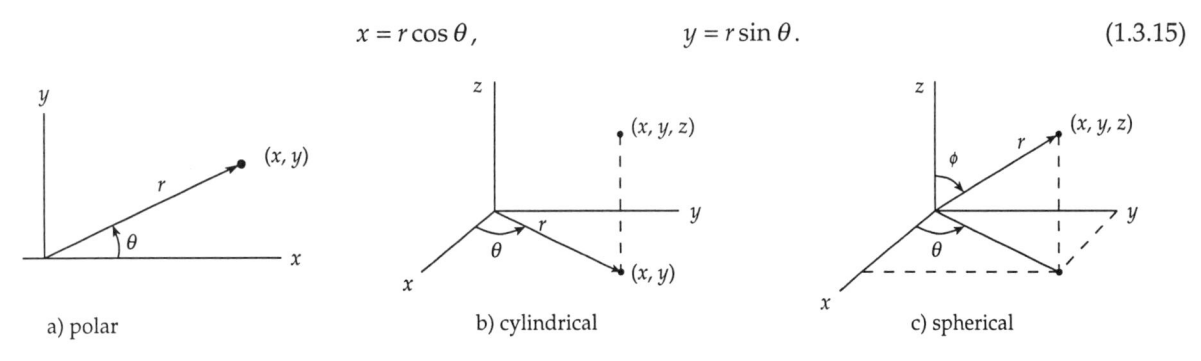

Figure 1.7 The polar, cylindrical and spherical coordinate systems.

For the cylindrical coordinate system

$$x = r \cos \theta, \qquad y = r \sin \theta, \qquad z = z. \tag{1.3.16}$$

And, for the spherical coordinate system

$$x = r\sin\phi\cos\theta, \qquad y = r\sin\phi\sin\theta, \qquad z = r\cos\phi. \qquad (1.3.17)$$

EXAMPLE 1.7

What conic section is represented by $2x^2 - 4xy + 5x = 10$?

Solution. Comparing this with the general form Eq. 1.3.7, we see that

$$A = 2, \qquad B = -2, \qquad C = 0.$$

Thus, $B^2 - AC = 4$, which is greater than zero. Hence, the conic section is a hyperbola.

EXAMPLE 1.8

Write the general form of the equation of a parabola, vertex at (2, 4), opening upward, with directrix at $y = 2$.

Solution. The equation of the parabola (see Eq. 1.3.11) can be written as

$$(x - x_1)^2 = 2p(y - y_1)$$

where we have interchanged x and y so that the parabola opens upward. For this example, $x_1 = 2$, $y_1 = 4$ and $p = 4$ ($p/2$ is the distance from the vertex to the directrix). Hence, the equation is

$$(x - 2)^2 = 2(4)(y - 4)$$

or, in general form,

$$x^2 - 4x - 8y + 36 = 0.$$

EXAMPLE 1.9

Express the rectangular coordinates (3, 4, 5) in cylindrical coordinates and spherical coordinates.

Solution. In cylindrical coordinates

$$r = \sqrt{x^2 + y^2}$$
$$= \sqrt{3^2 + 4^2} = 5,$$
$$\theta = \tan^{-1} y/x$$
$$= \tan^{-1} 4/3 = 0.927 \text{ rad}.$$

Thus, in cylindrical coordinates, the point is located at (5, 0.927, 5).

In spherical coordinates

$$r = \sqrt{x^2 + y^2 + z^2}$$
$$= \sqrt{3^2 + 4^2 + 5^2} = 7.071,$$
$$\phi = \cos^{-1} z/r$$

$$= \cos^{-1} 5/7.071 = 0.785 \text{ rad},$$

$$\theta = \tan^{-1} y/x$$

$$= \tan^{-1} 4/3 = 0.927 \text{ rad}.$$

Finally, in spherical coordinates, the point is located at (7.071, 0.927, 0.785).

1.4 Complex Numbers

A complex number consists of a real part x and an imaginary part y, written as $x + iy$, where $i = \sqrt{-1}$. (In electrical engineering, however, it is common to let $j = \sqrt{-1}$ since i represents current.) In real number theory, the square root of a negative number does not exist; in complex number theory, we would write $\sqrt{-4} = \sqrt{4(-1)} = 2i$. The complex number may be plotted using the real x-axis and the imaginary y-axis, as shown in Fig. 1.8.

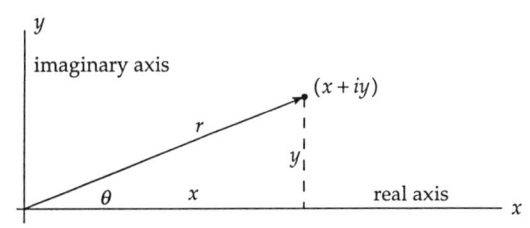

Figure 1.8 The complex number.

It is often useful to express a complex number in polar form as

$$x + iy = re^{i\theta} \tag{1.4.1}$$

where we use *Euler's equation*

$$e^{i\theta} = \cos\theta + i\sin\theta \tag{1.4.2}$$

to verify the relations

$$x = r\cos\theta, \quad y = r\sin\theta. \tag{1.4.3}$$

Using Euler's equation we can show that

$$\sin\theta = \frac{e^{i\theta} - e^{-i\theta}}{2i}, \quad \cos\theta = \frac{e^{i\theta} + e^{-i\theta}}{2}. \tag{1.4.4}$$

It is usually easier to find powers and roots of complex numbers using the polar form.

―――― **EXAMPLE 1.10** ――――

Divide $(3 + 4i)$ by $(4 + 3i)$.

Solution. We perform the division as follows:

$$\frac{3+4i}{4+3i} = \frac{3+4i}{4+3i} \cdot \frac{4-3i}{4-3i} = \frac{12+16i-9i+12}{16+9} = \frac{24+7i}{25} = 0.96 + 0.28i.$$

Note that we multiplied the numerator and the denominator by the *complex conjugate* of the denominator. A complex conjugate is formed simply by changing the sign of the imaginary part.

EXAMPLE 1.11

Find $(2+3i)^4$.

Solution. This can be done using the polar form. Hence,

$$r = \sqrt{2^2 + 3^2} = \sqrt{13}, \qquad \theta = \tan^{-1} 3/2 = 0.9828 \text{ rad}.$$

We normally express θ in radians. The complex number, in polar form, is

$$2 + 3i = \sqrt{13}\, e^{0.9828i}.$$

Thus,

$$(2+3i)^4 = \left(\sqrt{13}\right)^4 e^{4(0.9828i)}$$

$$= 169\, e^{3.9312i}.$$

Converting back to rectangular form we have

$$169\, e^{3.9312i} = 169(\cos 3.9312 + i \sin 3.9312)$$

$$= 169(-0.7041 - 0.7101i)$$

$$= -119 - 120i.$$

EXAMPLE 1.12

Find the three one-third roots of $2 + 3i$.

Solution. We express the complex number (see Example 1.11) in polar form as

$$2 + 3i = \sqrt{13}\, e^{0.9828i}.$$

Since the trig functions are periodic, we know that

$$\sin\theta = \sin(\theta + 2\pi) = \sin(\theta + 4\pi)$$

$$\cos\theta = \cos(\theta + 2\pi) = \cos(\theta + 4\pi).$$

Thus, in addition to the first form, we have

$$2 + 3i = \sqrt{13}\, e^{(0.9828 + 2\pi)i} = \sqrt{13}\, e^{(0.9828 + 4\pi)i}.$$

Taking the one-third root of each form, we find the three roots to be

$$(2+3i)^{1/3} = \left(\sqrt{13}\right)^{1/3} e^{0.3276i}$$

$$= 1.533(0.9468 + 0.3218i) = 1.452 + 0.4935i.$$

$$(2+3i)^{1/3} = \left(\sqrt{13}\right)^{1/3} e^{2.422i}$$

$$= 1.533(-0.7521 + 0.6591i) = -1.153 + 0.010i.$$

$$(2+3i)^{1/3} = \left(\sqrt{13}\right)^{1/3} e^{4.516i}$$

$$= 1.533(-0.1951 - 0.9808i) = -0.2991 - 1.504i.$$

If we added 6π to the angle we would be repeating the first root, so obviously this is not done. If we were finding the one-fourth root, four roots would result.

1.5 Linear Algebra

The primary objective in linear algebra is to find the solution to a set of n linear algebraic equations for n unknowns. To do this we must learn how to manipulate a matrix, a rectangular array of quantities arranged into rows and columns.

An $m \times n$ matrix has m rows (the horizontal lines) and n columns (the vertical lines). We are primarily interested in square matrices since we usually have the same number of equations as unknowns, such as

$$\begin{aligned} a_{11}x_1 + a_{12}x_2 + a_{13}x_3 + a_{14}x_4 &= r_1 \\ a_{21}x_1 + a_{22}x_2 + a_{23}x_3 + a_{24}x_4 &= r_2 \\ a_{31}x_1 + a_{32}x_2 + a_{33}x_3 + a_{34}x_4 &= r_3 \\ a_{41}x_1 + a_{42}x_2 + a_{43}x_3 + a_{44}x_4 &= r_4. \end{aligned} \quad (1.5.1)$$

In matrix form this can be written as

$$[a_{ij}][x_j] = [r_i] \quad (1.5.2)$$

where $[x_j]$ and $[r_i]$ are column matrices. (A column matrix is often referred to as a *vector*.) The coefficient matrix $[a_{ij}]$ and the column matrix $[r_i]$ are assumed to be known quantities. The solution $[x_j]$ is expressed as

$$[x_j] = [a_{ij}]^{-1}[r_i] \quad (1.5.3)$$

where $[a_{ij}]^{-1}$ is the *inverse* matrix of $[a_{ij}]$. It is defined as

$$[a_{ij}]^{-1} = \frac{[a_{ij}]^+}{|a_{ij}|} \quad (1.5.4)$$

where $[a_{ij}]^+$ is the *adjoint* matrix and $|a_{ij}|$ is the *determinant* of $[a_{ij}]$. Let us review how the determinant and the adjoint are evaluated.

In general, the determinant may be found using the *cofactor* A_{ij} of the element a_{ij}. The cofactor is defined to be $(-1)^{i+j}$ times the determinant obtained by deleting the ith row and the jth column. The determinant is then

$$|a_{ij}| = \sum_{j=1}^{n} a_{ij} A_{ij}, \quad (1.5.5)$$

where i is any value from 1 to n. Recall that the third-order determinant can be evaluated by writing the first two columns after the determinant and then summing the products of the elements of the diagonals, using negative signs with the diagonals sloping upward.

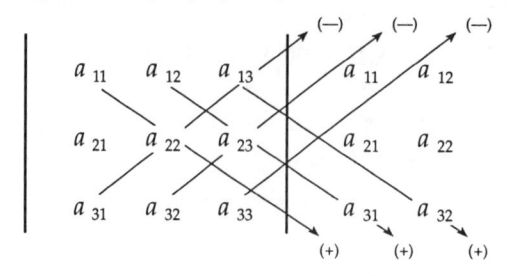

$$\begin{vmatrix} a_{11} & a_{12} & a_{13} \\ a_{21} & a_{22} & a_{23} \\ a_{31} & a_{32} & a_{33} \end{vmatrix} \begin{matrix} a_{11} & a_{12} \\ a_{21} & a_{22} \\ a_{31} & a_{32} \end{matrix}$$

The elements of the adjoint $[a_{ij}]^+$ are the cofactors A_{ij} of the elements a_{ij}; for the matrix $[a_{ij}]$ of Eq. 1.5.1 we have

$$[a_{ij}]^+ = \begin{bmatrix} A_{11} & A_{21} & A_{31} & A_{41} \\ A_{12} & A_{22} & A_{32} & A_{42} \\ A_{13} & A_{23} & A_{33} & A_{43} \\ A_{14} & A_{24} & A_{34} & A_{44} \end{bmatrix} \qquad (1.5.6)$$

Note that A_{ij} takes the position of a_{ji}.

Finally, the solution $[x_j]$ results if we multiply a square matrix by a column matrix. In general, we multiply the elements in each left-hand matrix row by the elements in each right-hand matrix column, add the products, and place the sum at the location where the row and column intersect. The following examples will illustrate.

Before we work some examples though, we should point out that the above matrix presentation can also be presented as *Cramer's rule*, which states that the solution element x_n can be expressed as

$$x_n = \frac{|b_{ij}|}{|a_{ij}|} \qquad (1.5.7)$$

where $|b_{ij}|$ is formed by replacing the nth column of $|a_{ij}|$ with the elements of the column matrix $[r_i]$.

Notes: If the system of equations is homogeneous, i.e., $r_i = 0$, a solution may exist if $|a_{ij}| = 0$.
If the determinant of a matrix is zero, that matrix is *singular*.

EXAMPLE 1.13

Calculate the determinants of $\begin{bmatrix} 2 & -3 \\ 1 & 4 \end{bmatrix}$ and $\begin{bmatrix} 2 & 3 & 0 \\ 1 & 4 & -2 \\ 0 & 3 & 5 \end{bmatrix}$.

Solution. For the first matrix we have

$$\begin{bmatrix} 2 & -3 \\ 1 & 4 \end{bmatrix} = 2 \times 4 - 1(-3) = 11.$$

The second matrix is set up as follows:

$$\begin{vmatrix} 2 & 3 & 0 \\ 1 & 4 & -2 \\ 0 & 3 & 5 \end{vmatrix} \begin{matrix} 2 & 3 \\ 1 & 4 \\ 0 & 3 \end{matrix} = 40 + 0 + 0 - 0 - (-12) - 15 = 37.$$

EXAMPLE 1.14

Expanding with cofactors, evaluate the determinant $D = \begin{vmatrix} 1 & 0 & -2 \\ -1 & 2 & 0 \\ 1 & 2 & 1 \end{vmatrix}$.

Solution. Choosing the first row ($i = 1$ in Eq. 1.5.5) we have

$$D = 1 \begin{vmatrix} 2 & 0 \\ 2 & 1 \end{vmatrix} (-1)^2 - 0 \begin{vmatrix} -1 & 0 \\ 1 & 1 \end{vmatrix} (-1)^3 + (-2) \begin{vmatrix} -1 & 2 \\ 1 & 2 \end{vmatrix} (-1)^4$$
$$= 2 + 0 - 2(-4) = 10.$$

Note: If any two columns (or rows) are a multiple of each other, a determinant is zero.

EXAMPLE 1.15

Find the adjoint of the matrix $[a_{ij}] = \begin{bmatrix} 1 & 0 & -2 \\ -1 & 2 & 0 \\ 1 & 2 & 1 \end{bmatrix}$.

Solution. The cofactor of each element of the matrix must be determined. The cofactor is found by multiplying $(-1)^{i+j}$ times the determinant formed by deleting the ith row and the jth column. They are found to be

$$A_{11} = 2, \quad A_{12} = 1, \quad A_{13} = -4$$
$$A_{21} = -4, \quad A_{22} = 3, \quad A_{23} = -2$$
$$A_{31} = 4, \quad A_{32} = 2, \quad A_{33} = 2.$$

The adjoint is then

$$[a_{ij}]^+ = [A_{ji}] = \begin{bmatrix} 2 & -4 & 4 \\ 1 & 3 & 2 \\ -4 & -2 & 2 \end{bmatrix}.$$

Note: The matrix $[A_{ji}]$ is called the *transpose* of $[A_{ij}]$, i.e., $[A_{ji}] = [A_{ij}]^T$.

EXAMPLE 1.16

Find the inverse of the matrix $[a_{ij}] = \begin{bmatrix} 1 & 0 & -2 \\ -1 & 2 & 0 \\ 1 & 2 & 1 \end{bmatrix}$.

Solution. The inverse is defined to be the adjoint matrix divided by the determinant $|a_{ij}|$. Hence, the inverse is (see Examples 1.14 and 1.15)

$$|a_{ij}|^{-1} = \frac{1}{10} \begin{bmatrix} 2 & -4 & 4 \\ 1 & 3 & 2 \\ -4 & -2 & 2 \end{bmatrix} = \begin{bmatrix} 0.2 & -0.4 & 0.4 \\ 0.1 & 0.3 & 0.2 \\ -0.4 & -0.2 & 0.2 \end{bmatrix}.$$

EXAMPLE 1.17

Find the solution to

$$x_1 \quad\quad - 2x_3 = 2$$
$$-x_1 + 2x_2 \quad\quad = 0$$
$$x_1 + 2x_2 + x_3 = -4.$$

Solution. The solution matrix is (see Example 1.14 and 1.15)

$$[x_j] = [a_{ij}]^{-1}[r_i] = \frac{[a_{ij}]^+}{|a_{ij}|}[r_i]$$

$$= \frac{1}{10} \begin{bmatrix} 2 & -4 & 4 \\ 1 & 3 & 2 \\ -4 & -2 & 2 \end{bmatrix} \begin{bmatrix} 2 \\ 0 \\ -4 \end{bmatrix}.$$

First, let's multiply the two matrices; they are multiplied row by column as follows:

$$2 \cdot 2 + (-4) \cdot 0 + 4 \cdot (-4) = -12$$
$$1 \cdot 2 + 3 \cdot 0 + 2 \cdot (-4) = -6$$
$$-4 \cdot 2 - 2 \cdot 0 + 2 \cdot (-4) = -16$$

The solution vector is then

$$[x_i] = \frac{1}{10} \begin{bmatrix} -12 \\ -6 \\ -16 \end{bmatrix} = \begin{bmatrix} -1.2 \\ -0.6 \\ -1.6 \end{bmatrix}.$$

In component form, the solution is

$$x_1 = -1.2, \quad x_2 = -0.6, \quad x_3 = -1.6.$$

EXAMPLE 1.18

Use Cramer's rule and solve

$$x_1 \quad\quad - 2x_3 = 2$$
$$-x_1 + 2x_2 \quad\quad = 0$$
$$x_1 + 2x_2 + x_3 = -4.$$

Solution. The solution is found (see example 1.14) by evaluating the ratios as follows:

$$x_1 = \frac{\begin{vmatrix} 2 & 0 & -2 \\ 0 & 2 & 0 \\ -4 & 2 & 1 \end{vmatrix}}{D} = \frac{-12}{10} = -1.2 \qquad x_2 = \frac{\begin{vmatrix} 1 & 2 & -2 \\ -1 & 0 & 0 \\ 1 & -4 & 1 \end{vmatrix}}{D} = \frac{-6}{10} = -0.6$$

$$x_3 = \frac{\begin{vmatrix} 1 & 0 & 2 \\ -1 & 2 & 0 \\ 1 & 2 & -4 \end{vmatrix}}{D} = \frac{-16}{10} = -1.6$$

where

$$D = \begin{vmatrix} 1 & 0 & -2 \\ -1 & 2 & 0 \\ 1 & 2 & 1 \end{vmatrix} = 10.$$

Note that the numerator is the determinant formed by replacing the ith column with right-hand side elements r_i when solving for x_i.

1.6 Calculus

1.6.1 Differentiation

The slope of a curve $y = f(x)$ is the ratio of the change in y to the change in x as the change in x becomes infinitesimally small. This is the first derivative, written as

$$\frac{dy}{dx} = \lim_{\Delta x \to 0} \frac{\Delta y}{\Delta x}. \qquad (1.6.1)$$

This may be written using abbreviated notation as

$$\frac{dy}{dx} = Dy = y' = \dot{y}. \qquad (1.6.2)$$

The second derivative is written as

$$\frac{d^2 y}{dx^2} = D^2 y = y'' = \ddot{y}, \qquad (1.6.3)$$

and is defined by

$$\frac{d^2 y}{dx^2} = \lim_{\Delta x \to 0} \frac{\Delta y'}{\Delta x}. \qquad (1.6.4)$$

Some derivative formulas, where f and g are functions of x, and k is constant, are given below.

$$\frac{dk}{dx} = 0$$

$$\frac{d(k x^n)}{dx} = k n x^{n-1}$$

$$\frac{d}{dx}(f+g) = f' + g'$$

$$\frac{df^n}{dx} = n f^{n-1} f'$$

$$\frac{d}{dx}(fg) = fg' + gf' \qquad (1.6.5)$$

$$\frac{d}{dx}(\ln x) = \frac{1}{x}$$

$$\frac{d}{dx}\left(e^{kx}\right) = k e^{kx}$$

$$\frac{d}{dx}(\sin x) = \cos x$$

$$\frac{d}{dx}(\cos x) = -\sin x$$

If a function f depends on more than one variable, partial derivatives are used. If $z = f(x, y)$, then $\partial z / \partial x$ is the derivative of z with respect to x holding y constant. It would represent the slope of a line tangent to the surface in a plane of constant y.

1.6.2 Maxima and Minima

Derivatives are used to locate points of inflection, maxima, and minima. Note the following:

$$f'(x) = 0 \text{ at a maximum or a minimum.}$$
$$f''(x) = 0 \text{ at an inflection point.}$$
$$f''(x) > 0 \text{ at a minimum.}$$
$$f''(x) < 0 \text{ at a maximum.}$$

An inflection point always exists between a maximum and a minimum.

1.6.3 L'Hospital's Rule

Differentiation is also useful in establishing the limit of $f(x)/g(x)$ as $x \to x_0$ if $f(x_0)$ and $g(x_0)$ are both zero or $\pm\infty$. L'Hospital's rule is used in such cases and is as follows:

$$\lim_{x \to x_0} \frac{f(x)}{g(x)} = \lim_{x \to x_0} \frac{f'(x)}{g'(x)}. \qquad (1.6.6)$$

1.6.4 Taylor's Series

Derivatives are used to expand a continuous function as a power series around $x = a$. Taylor's series is as follows:

$$f(x) = f(a) + (x-a)f'(a) + (x-a)^2 f''(a)/2! + \cdots. \qquad (1.6.7)$$

This series is often used to express a function as a polynomial near a point $x = a$ providing the series can be truncated after a few terms. Using Taylor's series we can show that (expanding about $a = 0$):

$$\sin x = x - x^3/3! + x^5/5! - \cdots$$

$$\cos x = 1 - x^2/2! + x^4/4! - \cdots$$

$$\ln(1+x) = x - x^2/2 + x^3/3 - \cdots \tag{1.6.8}$$

$$\frac{1}{1-x} = 1 + x + x^2 + \cdots$$

$$e^x = 1 + x + x^2/2! + x^3/3! + \cdots$$

1.6.5 Integration

The inverse of differentiation is the process called integration. If a curve is given by $y = f(x)$, then the area under the curve from $x = a$ to $x = b$ is given by

$$A = \int_a^b y\,dx. \tag{1.6.9}$$

The length of the curve between the two points is expressed as

$$L = \int_a^b \left(1 + y'^2\right)^{1/2} dx. \tag{1.6.10}$$

Volumes of various objects are also found by an appropriate integration.

If the integral has limits, it is a *definite integral*; if it does not have limits, it is an *indefinite integral* and a constant is always added. Some common indefinite integrals follow:

$$\int dx = x + C.$$

$$\int cy\,dx = c\int y\,dx$$

$$\int x^n dx = \frac{x^{n+1}}{n+1} + C \quad n \neq -1$$

$$\int x^{-1} dx = \ln x + C$$

$$\int e^{ax} dx = \frac{1}{a} e^{ax} + C \tag{1.6.11}$$

$$\int \sin x\, dx = -\cos x + C$$

$$\int \cos x\, dx = \sin x + C$$

$$\int \cos^2 x\, dx = \frac{x}{2} + \frac{1}{4} \sin 2x + C$$

$$\int u\, dv = uv - \int v\, du.$$

This last integral is often referred to as "integration by parts." If the integrand (the coefficients of the differential) is not one of the above, then in the last integral, $\int v\,du$ may in fact be integrable. An example will illustrate.

―――― **EXAMPLE 1.19** ――――

Find the slope of $y = x^2 + \sin x$ at $x = 0.5$.

Solution. The derivative is the slope:

$$y'(x) = 2x + \cos x.$$

At $x = 0.5$ the slope is

$$y'(0.5) = 2 \cdot 0.5 + \cos 0.5 = 1.878.$$

―――― **EXAMPLE 1.20** ――――

Find $\frac{d}{dx}(\tan x)$.

Solution. Writing $\tan x = \sin x / \cos x = f(x) \cdot g(x)$ we find

$$\frac{d}{dx}(\tan x) = \frac{1}{\cos x}\frac{d}{dx}(\sin x) + \sin x \frac{d}{dx}(\cos x)^{-1}$$

$$= \frac{\cos x}{\cos x} + \frac{\sin^2 x}{\cos^2 x} = 1 + \tan^2 x$$

$$= \frac{\cos^2 x + \sin^2 x}{\cos^2 x} = \frac{1}{\cos^2 x} = \sec^2 x.$$

Either expression is acceptable although the latter is usually used.

―――― **EXAMPLE 1.21** ――――

Locate the maximum and minimum points of the function $y(x) = x^3 - 12x - 9$ and evaluate y at those points.

Solution. The derivative is

$$y'(x) = 3x^2 - 12.$$

The points at which $y'(x) = 0$ are at

$$x = 2, -2.$$

At these two points the extrema are:

$$y_{min} = (2)^3 - 12 \cdot 2 - 9 = -25$$

$$y_{max} = (-2)^3 - 12(-2) - 9 = 7.$$

Let us check the second derivative. At the two points we have:

$$y''(2) = 6 \cdot 2 = 12$$

$$y''(-2) = 6 \cdot (-2) = -12$$

Obviously, the point $x = 2$ is a minimum since its second derivative is positive there.

EXAMPLE 1.22

Find the limit as $x \to 0$ of $\sin x / x$.

Solution. If we let $x = 0$ we are faced with the ratio of 0/0, an indeterminate quantity. Hence, we use L'Hospital's rule and differentiate both numerator and denominator to obtain

$$\lim_{x \to 0} \frac{\sin x}{x} = \lim_{x \to 0} \frac{\cos x}{1}.$$

Now, we let $x = 0$ and find

$$\lim_{x \to 0} \frac{\sin x}{x} = \frac{1}{1} = 1.$$

EXAMPLE 1.23

Verify that $\sin x = x - x^3/3! + x^5/5! - \cdots$.

Solution. We expand in a Taylor's series about $x = 0$:

$$f(x) = f(0) + x f'(0) + x^2 f''(0)/2! + \cdots.$$

Letting $f(x) = \sin x$ so that $f' = \cos x$, $f'' = -\sin x$, etc., there results

$$\sin x = 0 + x(1) + x^2(0/2!) + x^3(-1)/3! + \cdots$$

$$= = x - x^3/3! + x^5/5! - \cdots.$$

EXAMPLE 1.24

Find the area of the shaded area in the figure.

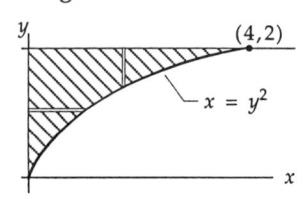

Solution. We can find this area by using either a horizontal strip or a vertical strip. We will use both. First, for a horizontal strip:

$$A = \int_0^2 x \, dy$$

$$= \int_0^2 y^2 \, dy = \left. \frac{y^3}{3} \right|_0^2 = 8/3.$$

Using a vertical strip we have

$$A = \int_0^4 (2 - y) \, dx$$

$$= \int_0^4 \left(2 - x^{1/2}\right) dx = \left[2x - \frac{2}{3} x^{3/2} \right]_0^4 = 8 - \frac{2}{3} 8 = 8/3.$$

Either technique is acceptable. The first appears to be the simpler one.

EXAMPLE 1.25

Find the volume enclosed by rotating the shaded area of Example 1.24 about the y-axis.

Solution. If we rotate the horizontal strip about the y-axis we will obtain a disc with volume

$$dV = \pi x^2 dy.$$

This can be integrated to give the volume, which is

$$V = \int_0^2 \pi x^2 dy$$

$$= \pi \int_0^2 y^4 dy = \frac{\pi y^5}{5}\bigg|_0^2 = \frac{32\pi}{5}.$$

Now, let us rotate the vertical strip about the y-axis to form a cylinder with volume

$$dV = 2\pi x(2-y)dx.$$

This can be integrated to yield

$$V = \int_0^4 2\pi x(2-y)dx$$

$$= \int_0^4 2\pi x\left(2 - x^{1/2}\right)dx = 2\pi\left[x^2 - \frac{2x^{5/2}}{5}\right]_0^4 = \frac{32\pi}{5}.$$

Again, the horizontal strip is simpler.

EXAMPLE 1.26

Show that $\int xe^x dx = (x-1)e^x + C$.

Solution. Let's attempt the last integral of (1.6.11). Define the following:

$$u = x, \quad dv = e^x dx$$

Then,

$$du = dx, \qquad v = \int e^x dx = e^x$$

and we find that

$$\int xe^x dx = xe^x - \int e^x dx$$

$$= xe^x - e^x + C = (x-1)e^x + C.$$

1.7 Vectors

There are two vector multiplications. The first, the *scalar product*, or *dot product*, is the scalar defined by

$$\vec{A} \cdot \vec{B} = AB \cos \theta \tag{1.7.1}$$

where θ is the angle between the two vectors, as shown in Fig. 1.9, and A and B are the magnitudes of the two vectors. In a rectangular coordinate system the dot product becomes

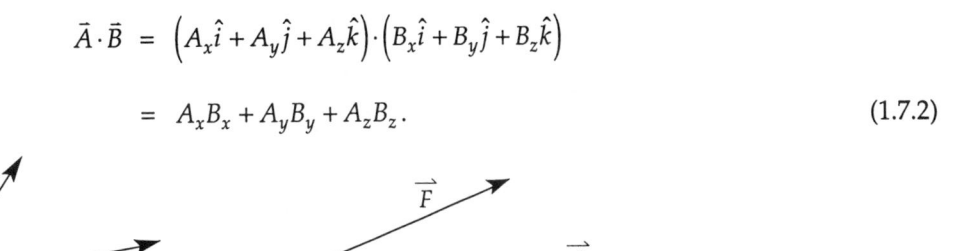

$$\begin{aligned}\vec{A} \cdot \vec{B} &= \left(A_x \hat{i} + A_y \hat{j} + A_z \hat{k}\right) \cdot \left(B_x \hat{i} + B_y \hat{j} + B_z \hat{k}\right) \\ &= A_x B_x + A_y B_y + A_z B_z.\end{aligned} \tag{1.7.2}$$

Figure 1.9 Vectors.

The scalar quantity *work* can be defined using the dot product:

$$w = \vec{F} \cdot \vec{d} \tag{1.7.3}$$

where $\vec{d}$ is the directed distance moved by the force.

The second, the *vector product*, or *cross product*, of the two vectors $\vec{A}$ and $\vec{B}$ is a vector defined by

$$\vec{C} = \vec{A} \times \vec{B} \tag{1.7.4}$$

where the magnitude of $\vec{C}$ is given by

$$C = AB \sin \theta. \tag{1.7.5}$$

The vector $\vec{C}$ acts in a direction perpendicular to the plane of $\vec{A}$ and $\vec{B}$ so that the three vectors form a right-handed set of vectors. (If the fingers curl $\vec{A}$ into $\vec{B}$, the thumb points in the direction of $\vec{C}$.) In a rectangular coordinate system the cross product is

$$\vec{A} \times \vec{B} = \begin{vmatrix} \hat{i} & \hat{j} & \hat{k} \\ A_x & A_y & A_z \\ B_x & B_y & B_z \end{vmatrix} = \left(A_y B_z - A_z B_y\right)\hat{i} + \left(A_z B_x - A_x B_z\right)\hat{j} + \left(A_x B_y - A_y B_x\right)\hat{k}. \tag{1.7.6}$$

The magnitude of $\vec{C}$ is the area of the parallelogram with sides $\vec{A}$ and $\vec{B}$.

The volume of the parallelepiped with sides $\vec{A}$, $\vec{B}$, and $\vec{C}$ is the scalar triple product given by

$$\vec{A} \times \vec{B} \cdot \vec{C} = \begin{vmatrix} A_x & A_y & A_z \\ B_x & B_y & B_z \\ C_x & C_y & C_z \end{vmatrix}. \tag{1.7.7}$$

EXAMPLE 1.27

Find the projection of $\vec{A}$ on $\vec{B}$ if $\vec{A} = 12\hat{i} - 18\hat{j} + 6\hat{k}$ and $\vec{B} = 2\hat{i} - 4\hat{j} + 4\hat{k}$.

Solution. Find the unit vector $\hat{i}_B$ in the direction of $\vec{B}$:

$$\hat{i}_B = \frac{\vec{B}}{B} = \frac{2\hat{i} - 4\hat{j} + 4\hat{k}}{\sqrt{2^2 + 4^2 + 4^2}} = \frac{1}{6}\left(2\hat{i} - 4\hat{j} + 4\hat{k}\right).$$

The projection of $\vec{A}$ on $\vec{B}$ is then

$$\vec{A} \cdot \hat{i}_B = \left(12\hat{i} - 18\hat{j} + 6\hat{k}\right) \cdot \frac{1}{6}\left(2\hat{i} - 4\hat{j} + 4\hat{k}\right)$$

$$= 4 + 12 + 4 = 20.$$

EXAMPLE 1.28

Find the area of a parallelogram with two sides identified by vectors from the origin to the points (3,4) and (8,0).

Solution. The two vectors are represented by

$$\vec{A} = 3\hat{i} + 4\hat{j}, \qquad \vec{B} = 8\hat{i}.$$

The area of the parallelogram is then

$$\left|\vec{A} \times \vec{B}\right| = \left|\left(3\hat{i} + 4\hat{j}\right) \times 8\hat{i}\right| = 32$$

since $\hat{i} \times \hat{i} = 0$ and $\hat{j} \times \hat{i} = -\hat{k}$.

EXAMPLE 1.29

Find a unit vector perpendicular to the plane that contains both $\vec{A} = \hat{i} - 2\hat{j} + 3\hat{k}$ and $\vec{B} = \hat{i} + 2\hat{j} - \hat{k}$.

Solution. The vector $\vec{C} = \vec{A} \times \vec{B}$ is perpendicular to the plane of $\vec{A}$ and $\vec{B}$. Using Eq. 1.8.6,

$$\vec{C} = \left[-2(-1) - 3 \times 2\right]\hat{i} + \left[3 \times 1 - 1(-1)\right]\hat{j} + \left[1 \times 2 - (-2) \times 1\right]\hat{k}$$

$$= -4\hat{i} + 4\hat{j} + 4\hat{k}.$$

A unit vector in the direction of $\vec{C}$ is the desired unit vector:

$$\hat{i}_c = \frac{-4\hat{i} + 4\hat{j} + 4\hat{k}}{\sqrt{4^2 + 4^2 + 4^2}} = \frac{1}{\sqrt{3}}\left(-\hat{i} + \hat{j} + \hat{k}\right).$$

Practice Problems

ALGEBRA

1.1 A growth curve is given by $A = 10 e^{2t}$. At what value of t is $A = 100$?
    a) 5.261   b) 3.070   c) 1.151   d) 0.726   e) 0.531

1.2 If $\ln x = 3.2$, what is x?
    a) 18.65   b) 24.53   c) 31.83   d) 64.58   e) 126.7

1.3 If $\ln_5 x = -1.8$, find x.
    a) 0.00483   b) 0.0169   c) 0.0552   d) 0.0783   e) 0.1786

1.4 One root of the equation $3x^2 - 2x - 2 = 0$ is
    a) 1.215   b) 1.064   c) 0.937   d) 0.826   e) 0.549

1.5 $\sqrt{4+x}$ can be written as the series
    a) $2 - x/4 + x^2/64 + \cdots$ d) $2 + x^2/8 + x^4/128 + \cdots$
    b) $2 + x/8 - x^2/128 + \cdots$ e) $2 + x/4 - x^2/64 + \cdots$
    c) $2 - x^2/4 - x^4/64 + \cdots$

1.6 Resolve $\dfrac{2}{x(x^2 - 3x + 2)}$ into partial fractions.
    a) $1/x + 1/(x-2) - 2/(x-1)$ d) $-1/x + 2/(x-2) + 1/(x-1)$
    b) $1/x - 2/(x-2) + 1/(x-1)$ e) $-1/x - 2/(x-2) + 1/(x-1)$
    c) $2/x - 1/(x-2) - 2/(x-1)$

1.7 Express $\dfrac{4}{x^2(x^2 - 4x + 4)}$ as the sum of fractions.
    a) $1/x + 1/(x-2)^2 - 1/(x-2)$
    b) $1/x + 1/x^2 - 1/(x-2) + 1/(x-2)^2$
    c) $1/x^2 + 1/(x-2)^2$
    d) $1/x + 1/x^2 + 1/(x-2) + 1/(x-2)^2$
    e) $1/x^2 - 1/(x-2) + 1/(x-2)^2$

1.8 A germ population has a growth curve of $Ae^{0.4t}$. At what value of t does its original value double?
    a) 9.682   b) 7.733   c) 4.672   d) 1.733   e) 0.5641

TRIGONOMETRY

1.9 If $\sin \theta = 0.7$, what is $\tan \theta$?
    a) 0.98   b) 0.94   c) 0.88   d) 0.85   e) 0.81

1.10 If the short leg of a right triangle is 5 units long and the long leg is 7 units long, find the angle opposite the short leg, in degrees.
   a) 26.3   b) 28.9   c) 31.2   d) 33.8   e) 35.5

1.11 The expression $\tan\theta \sec\theta(1-\sin^2\theta)/\cos\theta$ simplifies to
   a) $\sin\theta$ b) $\cos\theta$ c) $\tan\theta$ d) $\sec\theta$ e) $\csc\theta$

1.12 A triangle has sides of length 2, 3 and 4. What angle, in radians, is opposite the side of length 3?
   a) 0.55   b) 0.61   c) 0.76   d) 0.81   e) 0.95

1.13 The length of a lake is to be determined. A distance of 850 m is measured from one end to a point x on the shore. A distance of 732 m is measured from x to the other end. If an angle of 154° is measured between the two lines connecting x, what is the length of the lake?
   a) 1542   b) 1421   c) 1368   d) 1261   e) 1050

1.14 Express $2\sin^2\theta$ as a function of $\cos 2\theta$.
   a) $\cos 2\theta - 1$ b) $\cos 2\theta + 1$ c) $\cos 2\theta + 2$ d) $2 - \cos 2\theta$ e) $1 - \cos 2\theta$

GEOMETRY

1.15 The included angle between two successive sides of a regular eight-sided polygon is
   a) 150°   b) 135°   c) 120°   d) 75°   e) 45°

1.16 A large 15-m-dia cylindrical tank that sits on the ground is to be painted. If one liter of paint covers 10 m², how many liters are required if it is 10 m high? (Include the top.)
   a) 65   b) 53   c) 47   d) 38   e) 29

1.17 The equation of a line that has a slope of –2 and intercepts the x-axis at $x = 2$ is
   a) $y + 2x = 4$
   b) $y - 2x = 4$
   c) $y + 2x = -4$
   d) $2y + x = 2$
   e) $2y - x = -2$

1.18 The equation of a line that intercepts the x-axis at $x = 4$ and the y-axis at $y = -6$ is
   a) $2x - 3y = 12$
   b) $3x - 2y = 12$
   c) $2x + 3y = 12$
   d) $3x + 2y = 12$
   e) $3y - 2x = 12$

1.19 The shortest distance from the line $3x - 4y = 3$ to the point (6, 8) is
   a) 4.8   b) 4.2   c) 3.8   d) 3.4   e) 2.6

1.20 The equation $x^2 + 4xy + 4y^2 + 2x = 10$ represents which conic section?
   a) circle   b) ellipse   c) parabola   d) hyperbola   e) plane

1.21 The x- and y-axes are the asymptotes of a hyperbola that passes through the point (2, 2). Its equation is
   a) $x^2 - y^2 = 0$ b) $xy = 4$ c) $y^2 - x^2 = 0$ d) $x^2 + y^2 = 4$ e) $x^2y = 8$

1.22 A 100-m-long track is to be built 50 m wide. If it is to be elliptical, what equation could describe it?
   a) $50x^2 + 100y^2 = 1000$
   b) $2x^2 + y^2 = 250$
   c) $4x^2 + y^2 = 2500$
   d) $x^2 + 2y^2 = 250$
   e) $x^2 + 4y^2 = 10000$

1.23 The cylindrical coordinates (5, 30°, 12) are expressed in spherical coordinates as
   a) (13, 30°, 67.4°)
   b) (13, 30°, 22.6°)
   c) (15, 52.6°, 22.6°)
   d) (15, 52.6°, −22.6°)
   e) (13, 30°, −67.4°)

1.24 The equation of a 4-m-radius sphere using cylindrical coordinates is
   a) $x^2 + y^2 + z^2 = 16$
   b) $r^2 = 16$
   c) $r^2 + z^2 = 16$
   d) $x^2 + y^2 = 16$
   e) $r^2 + y^2 = 16$

COMPLEX NUMBERS

1.25 Divide $3 - i$ by $1 + i$.
   a) $1 - 2i$
   b) $1 + 2i$
   c) $2 - i$
   d) $2 + i$
   e) $2 + 2i$

1.26 Find $(1 + i)^6$.
   a) $1 + i$
   b) $1 - i$
   c) $8i$
   d) $-8i$
   e) $-1 - i$

1.27 Find the first root of $(1+i)^{1/5}$.
   a) $0.17 + 1.07i$
   b) $1.07 + 0.17i$
   c) $1.07 - 0.17i$
   d) $0.17 - 1.07i$
   e) $-1.07 - 0.17i$

1.28 Express $(3 + 2i)\,e^{2it} + (3 - 2i)\,e^{-2it}$ in terms of trigonometric functions.
   a) $3\cos 2t - 4\sin 2t$
   b) $3\cos 2t - 2\sin 2t$
   c) $6\cos 2t - 4\sin 2t$
   d) $3\sin 2t + 2\sin 2t$
   e) $6\cos 2t + 4\sin 2t$

1.29 Subtract $5e^{0.2i}$ from $6e^{2.3i}$.
   a) $-8.90 + 5.48i$
   b) $-0.90 + 3.48i$
   c) $-0.90 - 3.48i$
   d) $8.90 - 5.48i$
   e) $-8.90 + 3.48i$

LINEAR ALGEBRA

1.30 Find the value of the determinant $\begin{vmatrix} 3 & 2 & 1 \\ 0 & -1 & -1 \\ 2 & 0 & 2 \end{vmatrix}$.
   a) 8
   b) 4
   c) 0
   d) −4
   e) −8

1.31 Evaluate the determinant $\begin{vmatrix} 1 & 0 & 1 & 1 \\ 2 & -1 & 0 & 1 \\ 0 & 0 & 2 & 0 \\ 3 & 2 & 1 & 1 \end{vmatrix}$.

    a) 8   b) 4   c) 0   d) –4   e) –8

1.32 The cofactor A_{21} of the determinant of Prob. 1.30 is

    a) –5   b) –4   c) 3   d) 4   e) 5

1.33 The cofactor A_{34} of the determinant of Prob. 1.31 is

    a) 4   b) 6   c) 0   d) –4   e) –6

1.34 Find the adjoint matrix of $\begin{vmatrix} 1 & -4 \\ 0 & 2 \end{vmatrix}$.

    a) $\begin{bmatrix} 4 & 2 \\ 0 & 1 \end{bmatrix}$ b) $\begin{bmatrix} 1 & 0 \\ 4 & 2 \end{bmatrix}$ c) $\begin{bmatrix} 2 & 4 \\ 1 & 0 \end{bmatrix}$ d) $\begin{bmatrix} 2 & 4 \\ 0 & 1 \end{bmatrix}$ e) $\begin{bmatrix} 1 & 4 \\ 0 & 2 \end{bmatrix}$

1.35 The inverse matrix of $\begin{bmatrix} 2 & 3 \\ 1 & 1 \end{bmatrix}$ is

    a) $\begin{bmatrix} -1 & 3 \\ 1 & -2 \end{bmatrix}$ b) $\begin{bmatrix} 1 & -1 \\ -3 & 2 \end{bmatrix}$ c) $\begin{bmatrix} -1 & 1 \\ -3 & 2 \end{bmatrix}$ d) $\begin{bmatrix} -2 & 3 \\ 1 & -1 \end{bmatrix}$ e) $\begin{bmatrix} 2 & 3 \\ -1 & -1 \end{bmatrix}$

1.36 Calculate $\begin{bmatrix} 2 & -1 \\ 3 & 2 \end{bmatrix} \begin{bmatrix} 2 \\ 1 \end{bmatrix}$.

    a) $\begin{bmatrix} 8 \\ 3 \end{bmatrix}$ b) $\begin{bmatrix} 3 \\ 8 \end{bmatrix}$ c) $\begin{bmatrix} -3 \\ -8 \end{bmatrix}$ d) [3,8]   e) [8,3]

1.37 Determine $\begin{bmatrix} 1 & 2 \\ 2 & 1 \end{bmatrix} \begin{bmatrix} -1 & 0 \\ 1 & 2 \end{bmatrix}$.

    a) $\begin{bmatrix} 1 & 4 \\ -1 & 2 \end{bmatrix}$ b) $\begin{bmatrix} 1 & -1 \\ 4 & 2 \end{bmatrix}$ c) $\begin{bmatrix} 1 \\ -1 \end{bmatrix}$ d) $\begin{bmatrix} 4 \\ 2 \end{bmatrix}$ e) [1,4]

1.38 Solve for $[x_i]$. $\begin{aligned} 3x_1 + 2x_2 &= -2 \\ x_1 - x_2 + x_3 &= 0 \\ 4x_1 + 2x_3 &= 4 \end{aligned}$

    a) $\begin{bmatrix} 2 \\ 4 \\ -6 \end{bmatrix}$ b) $\begin{bmatrix} -2 \\ 4 \\ 12 \end{bmatrix}$ c) $\begin{bmatrix} 2 \\ 8 \\ 4 \end{bmatrix}$ d) $\begin{bmatrix} -6 \\ 8 \\ 14 \end{bmatrix}$ e) $\begin{bmatrix} 6 \\ 4 \\ 3 \end{bmatrix}$

CALCULUS

1.39 The slope of the curve $y = 2x^3 - 3x$ at $x = 1$ is

    a) 3   b) 5   c) 6   d) 8   e) 9

1.40 If $y = \ln x + e^x \sin x$, find dy/dx at $x = 1$.
   a) 1.23   b) 3.68   c) 4.76   d) 6.12   e) 8.35

1.41 At what value of x does a maximum of $y = x^3 - 3x$ occur?
   a) 2   b) 1   c) 0   d) –1   e) –2

1.42 Where does an inflection point occur for $y = x^3 - 3x$?
   a) 2   b) 1   c) 0   d) –1   e) –2

1.43 Evaluate $\lim\limits_{x \to \infty} \dfrac{2x^2 - x}{x^2 + x}$.
   a) 2   b) 1   c) 0   d) –1   e) –2

1.44 If a quantity η and its derivatives η' and η'' are known at a point, its approximate value at a small distance h is
   a) $\eta + h^2 \eta'' / 2$
   b) $\eta + h\eta / 2 + h^2 \eta''$
   c) $\eta + h\eta' + h^2 \eta'' / 2$
   d) $\eta + h\eta' + h^2 \eta''$
   e) $\eta + h\eta'$

1.45 Find an approximation to $e^x \sin x$ for small x.
   a) $x - x^2 + x^3$
   b) $x + x^2 + x^3/3$
   c) $x - x^2/2 + x^3/6$
   d) $x + x^2 - x^3/6$
   e) $x + x^2 + x^3/2$

1.46 Find the area between the y-axis and $y = x^2$ from $y = 4$ to $y = 9$.
   a) 29/3   b) 32/3   c) 34/3   d) 38/3   e) 43/3

1.47 The area contained between $4x = y^2$ and $4y = x^2$ is
   a) 10/3   b) 11/3   c) 13/3   d) 14/3   e) 16/3

1.48 Rotate the shaded area of Example 1.24 about the x-axis. What volume is formed?
   a) 4π b) 6π c) 8π d) 10π e) 12π

1.49 Evaluate $\int_0^2 (e^x + \sin x)\, dx$.
   a) 7.81   b) 6.21   c) 5.92   d) 5.61   e) 4.21

1.50 Evaluate $2\int_0^1 e^x \sin x\, dx$.
   a) 1.82   b) 1.94   c) 2.05   d) 2.16   e) 2.22

1.51 Derive an expression $\int x \cos x\, dx$.
   a) $x \cos x - \sin x + C$
   b) $-x \cos x + \sin x + C$
   c) $x \sin x - \cos x + C$
   d) $x \cos x + \sin x + C$
   e) $x \sin x + \cos x + C$

VECTOR ANALYSIS

1.52 Given: $\vec{A} = 3\hat{i} - 6\hat{j} + 2\hat{k}$, $\vec{B} = 10\hat{i} + 4\hat{j} - 6\hat{k}$. Find: $\vec{A} \cdot \vec{B}$.
   a) $-6i$ b) 6   c) -6 d) $30\hat{i} - 24\hat{j} - 12\hat{k}$ e) $28\hat{i} + 38\hat{j} + 72\hat{k}$

1.53 Given: $\vec{A} = 2\hat{i} - 5\hat{k}$, $\vec{B} = \hat{j}$. Find: $\vec{A} \times \vec{B}$.
   a) 0   b) $5\hat{i} + 2\hat{k}$ c) -3 d) $-3\hat{k}$ e) 7

1.54 Find the projection of $\vec{A}$ in the direction of $\vec{B}$ if $\vec{A} = 14\hat{i} - 7\hat{j}$ and $\vec{B} = 6\hat{i} + 3\hat{j} - 2\hat{k}$.
   a) -9 b) $12\hat{i} - 3\hat{j}$ c) 0   d) 15   e) 9

1.55 The equation of a plane perpendicular to and passing through the end of the vector $\vec{A} = 2\hat{i} - 4\hat{j} + 6\hat{k}$ is given by $2x - 4y + 6z = k$ where k is
   a) 56   b) 24   c) 0   d) -8 e) -16

1.56 Estimate the area of the parallelogram with sides $\vec{A} = 2\hat{i} + 3\hat{j}$ and $\vec{B} = 4\hat{i} - 6\hat{j} + 5\hat{k}$.
   a) 32   b) 30   c) 26   d) 20   e) 10

Solutions to Problems

1.1 c) $100 = 10e^{2t}$. $\therefore \ln e^{2t} = \ln 10$. $\therefore 2t = 2.303$ $\therefore t = 1.151$

1.2 b) $\ln x = 3.2$. $\therefore e^{3.2} = x$. $\therefore x = 24.53$

1.3 c) $\ln_5 x = -1.8$. $\therefore 5^{-1.8} = x$. $\therefore x = 0.0552$

1.4 a) $x = \dfrac{-(-2) \pm \sqrt{2^2 - 4(3)(-2)}}{3 \cdot 2} = 1.215$

1.5 e) $(4+x)^{1/2} = 4^{1/2} + \dfrac{1}{2}4^{-1/2}x + \dfrac{\frac{1}{2}(1-\frac{1}{2})}{2}4^{-3/2}x^2 + \cdots = 2 + x/4 - x^2/64 + \cdots$

1.6 a) $\dfrac{2}{x(x^2 - 3x + 2)} = \dfrac{A_1}{x} + \dfrac{A_2}{x-2} + \dfrac{A_3}{x-1} = \dfrac{A_1(x^2 - 3x + 2) + A_2(x^2 - x) + A_3(x^2 - 2x)}{x(x^2 - 3x + 2)}$

$\therefore \begin{aligned} A_1 + A_2 + A_3 &= 0 \\ -3A_1 - A_2 - 2A_3 &= 0 \\ 2A_1 &= 2 \end{aligned} \qquad \begin{aligned} A_1 &= 1 \\ A_2 + A_3 &= -1 \\ -A_2 - 2A_3 &= 3 \end{aligned} \qquad \begin{aligned} A_3 &= -2 \\ A_2 &= 1 \end{aligned}$

1.7 b) $\dfrac{4}{x^2(x^2 - 4x + 4)} = \dfrac{A_1}{x} + \dfrac{A_2}{x^2} + \dfrac{A_3}{x-2} + \dfrac{A_4}{(x-2)^2}$

$= \dfrac{A_1(x^3 - 4x^2 + 4x) + A_2(x^2 - 4x + 4) + A_3(x^3 - 2x^2) + A_4 x^2}{x^2(x-2)^2}$

$$\begin{aligned}
\therefore \quad A_1 + A_3 &= 0 \\
-4A_1 + A_2 - 2A_3 + A_4 &= 0 \\
4A_1 - 4A_2 &= 0 \\
4A_2 &= 4
\end{aligned} \qquad \begin{aligned} A_2 &= 1 \\ A_1 &= 1 \\ A_3 &= -1 \\ A_4 &= 1 \end{aligned}$$

1.8 d)   at $t = 0$, population $= A$. $\therefore 2A = Ae^{0.4t}$. $\ln 2 = 0.4t$. $\therefore t = 1.733$

1.9 a) $\sin\theta = 0.7$ $\therefore \theta = 44.43°$. $\tan 44.43° = 0.980$

1.10 e) $\tan\theta = 5/7$. $\therefore \theta = 35.54°$

1.11 c) $\tan\theta \sec\theta (1 - \sin^2\theta)/\cos\theta = \tan\theta \dfrac{1}{\cos\theta} \cos^2\theta \dfrac{1}{\cos\theta} = \tan\theta$

1.12 d) $3^2 = 4^2 + 2^2 - 2\cdot 2\cdot 4\cos\theta$. $\therefore \cos\theta = 0.6875$. $\theta = 46.6°$. $\therefore \text{rad} = 0.813$

1.13 a) $L^2 = 850^2 + 732^2 - 2\cdot 850\cdot 732\cos 154°$. $\therefore L = 1542$ m

1.14 e) $\cos 2\theta = \cos^2\theta - \sin^2\theta = 1 - \sin^2\theta - \sin^2\theta = 1 - 2\sin^2\theta$. $\therefore 2\sin^2\theta = 1 - \cos 2\theta$

1.15 b) $\theta = \pi(n-2)/n = \pi(8-2)/8$ radians. $\dfrac{6\pi}{8} \times \dfrac{180}{\pi} = 135°$

1.16 a)   Area $= \text{Area}_{\text{top}} + \text{Area}_{\text{sides}} = \pi R^2 + \pi DL$
                    $= \pi \times 7.5^2 + \pi \times 15 \times 10 = 648$ m^2. $648 \div 10 \approx 65$

1.17 a) $y = mx + b$. $y = -2x + b$. $0 = -2(2) + b$. $\therefore b = 4$. $\therefore y = -2x + 4$

1.18 b) $y = mx + b$. $0 = 4m + b$. $-6 = b$. $\therefore m = 3/2$. $\therefore y = 3x/2 - 6$ or $2y = 3x - 12$

1.19 d) $3x - 4y - 3 = 0$. $A = 3, B = -4$. $d = \dfrac{|3 \times 6 - 4 \times 8 - 3|}{\sqrt{3^2 + (-4)^2}} = 3.4$

1.20 c) $B^2 - AC = 2^2 - 1 \times 4 = 0$. $\therefore$ parabola

1.21 b) $xy = \pm k^2 = 4$

1.22 c) $\dfrac{x^2}{25^2} + \dfrac{y^2}{50^2} = 1$. $\therefore 4x^2 + y^2 = 2500$

1.23 b) $x = r\cos\theta = 5 \times 0.866 = 4.33$. $y = r\sin\theta = 5 \times 0.5 = 2.5$
         Spherical coordinates: $r = \sqrt{4.33^2 + 2.5^2 + 12^2} = 13$. $\phi = \cos^{-1}\dfrac{z}{r} = \cos^{-1}\dfrac{12}{13} = 22.6°$.
                          $\theta = \tan^{-1}\dfrac{y}{x} = \tan^{-1}\dfrac{2.5}{4.33}$. $\therefore \theta = 30°$.

1.24 c)   (A) is rectangular coordinates. (B) is spherical coordinates.

1.25 a) $\dfrac{3-i}{1+i} = \dfrac{3-i}{1+i} \dfrac{1-i}{1-i} = \dfrac{3-1-4i}{1-(-1)} = \dfrac{1}{2}(2 - 4i) = 1 - 2i$

1.26 d) $1 + i = re^{i\theta}$. $r = \sqrt{1^2 + 1^2} = \sqrt{2}$. $\theta = \tan^{-1}\dfrac{1}{1} = \pi/4$ rad. $\therefore 1 + i = \sqrt{2}e^{i\pi/4}$.

$$\therefore (1+i)^6 = \left(\sqrt{2}\right)^6 e^{i\pi 3/2} = 8\left(\cos\frac{3\pi}{2} + i\sin\frac{3\pi}{2}\right) = -8i$$

1.27 b) $1+i = \sqrt{2}e^{i\pi/4}$. $\therefore (1+i)^{1/5} = 1.414^{1/5} e^{i\pi/20} = 1.072\left(\cos\frac{\pi}{20} + i\sin\frac{\pi}{20}\right) = 1.06 + 0.168i$

1.28 c) $(3+2i)(\cos 2t + i\sin 2t) + (3-2i)(\cos 2t - i\sin 2t) = 6\cos 2t + 4i(i\sin 2t) = 6\cos 2t - 4\sin 2t$

1.29 e) $6(\cos 2.3 + i\sin 2.3) - 5(\cos 0.2 + i\sin 0.2) = 6(-0.666 + 0.746i) - 5(0.98 + 0.199i) = -8.90 + 3.48i$

1.30 e) $\begin{vmatrix} 3 & 2 & 1 \\ 0 & -1 & -1 \\ 2 & 0 & 2 \end{vmatrix} = -6 - 4 + 2 = -8$

1.31 a) $\begin{vmatrix} 1 & 0 & 1 & 1 \\ 2 & -1 & 0 & 1 \\ 0 & 0 & 2 & 0 \\ 3 & 2 & 1 & 1 \end{vmatrix} = 2\begin{vmatrix} 1 & 0 & 1 \\ 2 & -1 & 1 \\ 3 & 2 & 1 \end{vmatrix} = 2(-1 + 4 + 3 - 2) = 8$ Note: Expand using the third row.

1.32 b) $(-1)^3 \begin{vmatrix} 2 & 1 \\ 0 & 2 \end{vmatrix} = -4$

1.33 e) $(-1)^7 \begin{vmatrix} 1 & 0 & 1 \\ 2 & -1 & 0 \\ 3 & 2 & 1 \end{vmatrix} = -(-1 + 4 + 3) = -6$

1.34 d) $[a_{ij}]^+ = \begin{bmatrix} A_{11} & A_{21} \\ A_{12} & A_{22} \end{bmatrix} = \begin{bmatrix} 2 & 4 \\ 0 & 1 \end{bmatrix}$

1.35 a) $[a_{ij}] = \dfrac{[a_{ij}]^+}{|a_{ij}|} = \dfrac{\begin{bmatrix} 1 & -3 \\ -1 & 2 \end{bmatrix}}{-1} = \begin{bmatrix} -1 & 3 \\ 1 & -2 \end{bmatrix}$

1.36 b) $\begin{bmatrix} 2 & -1 \\ 3 & 2 \end{bmatrix}\begin{bmatrix} 2 \\ 1 \end{bmatrix} = \begin{bmatrix} 4-1 \\ 6+2 \end{bmatrix} = \begin{bmatrix} 3 \\ 8 \end{bmatrix}$

1.37 a) $\begin{bmatrix} 1 & 2 \\ 2 & 1 \end{bmatrix}\begin{bmatrix} -1 & 0 \\ 1 & 2 \end{bmatrix} = \begin{bmatrix} 1 & 4 \\ -1 & 2 \end{bmatrix}$

1.38 d) $[a_{ij}] = \begin{bmatrix} 3 & 2 & 0 \\ 1 & -1 & 1 \\ 4 & 0 & 2 \end{bmatrix}$ $[a_{ij}]^+ = \begin{bmatrix} -2 & -4 & 2 \\ 2 & 6 & -3 \\ 4 & 8 & -5 \end{bmatrix}$. $|a_{ij}| = -2$.

$$\therefore [a_{ij}]^{-1} = \dfrac{[a_{ij}]^+}{|a_{ij}|} = \begin{bmatrix} 1 & 2 & -1 \\ -1 & -3 & 3/2 \\ -2 & -4 & 5/2 \end{bmatrix}. \quad [x_j] = [a_{ij}]^{-1}[r_i] = [a_{ij}]^{-1}\begin{bmatrix} -2 \\ 0 \\ 4 \end{bmatrix} = \begin{bmatrix} -6 \\ 8 \\ 14 \end{bmatrix}$$

1.39 a) $\dfrac{dy}{dx} = 6x^2 - 3 = 6(1)^2 - 3 = 3$

1.40 c) $\dfrac{dy}{dx} = \dfrac{1}{x} + e^x \cos x + e^x \sin x = 1 + e\cos 1 + e\sin 1 = 4.76.$ $\quad (\cos 1 = \cos 57.3°)$

1.41 d) $\dfrac{dy}{dx} = 3x^2 - 3 = 0. \quad \therefore x^2 = 1. \quad \therefore x = \pm 1. \quad \dfrac{d^2 y}{dx^2} = 6x. \quad \therefore x = -1$ is a maximum.

1.42 c) $y' = 3x^2 - 3. \quad y'' = 6x. \quad \therefore x = 0$ is inflection.

1.43 a) $\lim\limits_{x\to\infty} \dfrac{2x^2 - x}{x^2 + x} = \lim\limits_{x\to\infty} \dfrac{4x-1}{2x+1} = \lim\limits_{x\to\infty} \dfrac{4}{2} = 2$

1.44 c) $\eta(x+h) = \eta + h\eta' + \dfrac{h^2}{2}\eta''$

1.45 b) $e^x \sin x = \left(1 + x + \dfrac{x^2}{2}\right)\left(x - \dfrac{x^3}{6}\right) = x + x^2 + \dfrac{x^3}{2} - \dfrac{x^3}{6} = x + x^2 + x^3/3$

1.46 d) Area $= \int\limits_4^9 x\,dy = \int\limits_4^9 y^{1/2}\,dy = \dfrac{2}{3}(27 - 8) = 12\dfrac{2}{3}$

1.47 e) Area $= \int\limits_0^4 (x_2 - x_1)\,dy = \int\limits_0^4 \left(2y^{1/2} - \dfrac{y^2}{4}\right)dy$

$= 2 \times \dfrac{2}{3} \times 8 - \dfrac{1}{12} \times 64 = 16/3$

1.48 c) $V = \int\limits_0^2 2\pi y\, x\,dy = 2\pi \int\limits_0^2 y^3\,dy = 2\pi \times \dfrac{2^4}{4} = 8\pi$

1.49 a) $\int\limits_0^2 (e^x + \sin x)\,dx = e^x - \cos x \big|_0^2 = e^2 - 1 - \cos 2 + 1 = 7.81$

1.50 a) $\int\limits_0^1 e^x \sin x\,dx = e^x \sin x \big|_0^1 - \int\limits_0^1 e^x \cos x\,dx \quad$ $u = \sin x \quad dv = e^x dx \quad u = \cos x \quad dv = e^x dx$
$\quad du = \cos x\,dx \quad v = e^x \quad\quad v = e^x \quad du = -\sin x\,dx$
$\underbrace{}_{\text{1st integral}} \quad \underbrace{}_{\text{2nd integral}}$

$\therefore \int\limits_0^1 e^x \sin x\,dx = e\sin 1 - \left[e^x \cos x \big|_0^1 + \int\limits_0^1 e^x \sin x\,dx\right]$

$\therefore 2\int\limits_0^1 e^x \sin x\,dx = e\sin 1 - e\cos 1 + 1 \times 1 = 1.819. \quad \therefore \int\limits_0^1 e^x \sin x\,dx = 0.909$

1.51 e) $\int x \cos x\,dx = x \sin x - \int \sin x\,dx = x \sin x + \cos x + C \quad$ $u = x \quad dv = \cos x\,dx$
$\quad du = dx \quad v = \sin x$
$\underbrace{}_{\text{1st integral}}$

1.52 c) $\vec{A} \cdot \vec{B} = 3 \cdot 10 + (-6) \cdot 4 + 2(-6) = -6$

1.53 b) $\vec{A} \times \vec{B} = (2\hat{i} - 5\hat{k}) \times \hat{j} = 2\hat{i} \times \hat{j} - 5\hat{k} \times \hat{j} = 2\hat{k} - 5(-\hat{i}) = 5\hat{i} + 2\hat{k}$

1.54 e) $\hat{i}_B = \left(6\hat{i} + 3\hat{j} - 2\hat{k}\right)/\sqrt{6^2 + 3^2 + 2^2} = \frac{1}{7}\left(6\hat{i} + 3\hat{j} - 2\hat{k}\right)$
$\vec{A} \cdot \hat{i}_B = [14 \cdot 6 - 7(3)]/7 = 12 - 3 = 9$

1.55 a) $\left[\left(x\hat{i} + y\hat{j} + z\hat{k}\right) - \left(2\hat{i} - 4\hat{j} + 6\hat{k}\right)\right] \cdot \left(2\hat{i} - 4\hat{j} + 6\hat{k}\right) = 0.$
$2(x-2) - 4(y+4) + 6(z-6) = 0. \quad \therefore 2x - 4y + 6z = 56$

1.56 b) $\left|\vec{A} \times \vec{B}\right| = \left|15\hat{i} - 10\hat{j} - 24\hat{k}\right| = \sqrt{901} \cong 30$

2. Statics

by Merle C. Potter

Statics is concerned primarily with the equilibrium of bodies subjected to force systems. Also, traditionally in engineering statics we consider centroids, center of gravity and moments of inertia.

2.1 Forces, Moments and Resultants

A force is the manifestation of the action of one body upon another. Forces arise from the direct action of two bodies in contact with one another, or from the "action at a distance" of one body upon another as occurs with gravitational and magnetic forces. We classify forces as either *body forces* which act (and are distributed) throughout the volume of the body, or as *surface forces*, which act over a surface portion (either external or internal) of the body. If the surface over which the force system acts is very small we usually assume localization at a specific point in the surface and speak of a *concentrated force* at that point.

Mathematically, forces are represented by *vectors*. Geometrically, a vector is a directed line segment having a head and a tail, i.e., an arrow. The length of the arrow corresponds to the magnitude of the force, its orientation defines the line of action, and the direction of the arrow (tail to head) gives the sense of the force.

Systems of concentrated forces are listed as *concurrent* when all of the forces act, or could act, at a single point; otherwise they are termed *non-concurrent* systems. Parallel force systems are in this second group. Also, force systems are often described as two-dimensional (acting in a single plane), or three-dimensional (spatial systems).

In addition to the "push or pull" effect on the point at which it acts, a force creates a *moment* about other points of the body. Conceptually, a moment may be thought of as a torque. Mathematically, the moment of the force $\vec{F}$ with respect to point A when the force acts at point B is defined by the vector cross product

$$\vec{M}_A = \vec{r}_{B/A} \times \vec{F} \tag{2.1.1}$$

where $\vec{r}_{B/A}$ is the position vector of B relative to A as shown by Fig. 2.1. Actually, it is easily shown that

$$\vec{M}_A = \vec{r}_{Q/A} \times \vec{F} = \vec{r}_{B/A} \times \vec{F} \tag{2.1.2}$$

where Q is any point on the line of action of $\vec{F}$, as shown in Fig. 2.1.

A moment must always be designated with respect to a specific point, and is represented by a vector perpendicular to the plane of $\vec{r}$ and $\vec{F}$. Moment vectors are denoted by double-headed arrows to distinguish them from force vectors. The component of the moment vector in the direction of any axis (line) passing through A is said to be the moment of the force about that axis. If the direction of the axis is defined by the unit vector $\hat{\lambda}$, the moment component, a scalar along that axis, is given by

$$M_\lambda = \vec{r} \times \vec{F} \cdot \hat{\lambda}. \tag{2.1.3}$$

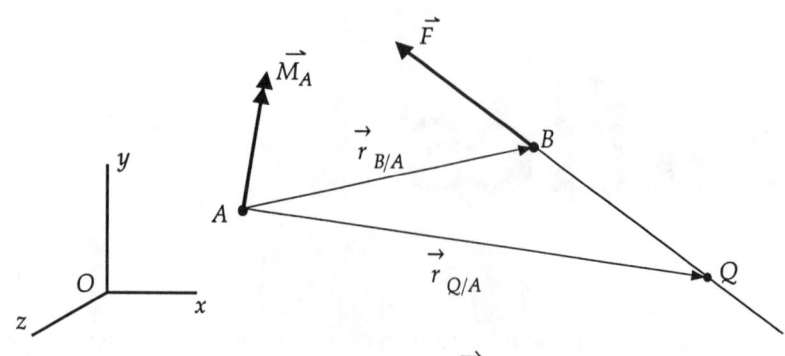

Figure 2.1 Moment of $\vec{F}$ about A.

The "turning effect" about the hinge axis of a door due to the application of a force to the doorknob is an example. If $\hat{\lambda} = \hat{k}$ in Eq. 2.1.3, the moment component is labeled M_z, the scalar moment about the z-axis, and we find that

$$M_z = xF_y - yF_x \tag{2.1.4}$$

where x and y are the components of $\vec{r}_{B/A}$ of Fig. 2.1. Similarly, for $\hat{\lambda} = \hat{j}$ and $\hat{\lambda} = \hat{i}$, respectively,

$$M_y = zF_x - xF_z \tag{2.1.5}$$

$$M_x = yF_z - zF_y. \tag{2.1.6}$$

Furthermore, it may be shown from Eq. 2.1.4 that the scalar moment about any point A in the plane of $\vec{r}$ and $\vec{F}$ is given by

$$M_A = Fd \tag{2.1.7}$$

where d is the perpendicular distance from A to the line of action of $\vec{F}$.

The *resultant* of a system of forces is the equivalent force and moment of the total system at any point. Thus, the resultant of a concurrent force system is a single force acting at the point of concurrency, and being the vector sum of the individual forces. In contrast, the resultant of a non-concurrent system depends upon the point at which it is determined, and in general consists of a resultant force and resultant moment. Actually, the resultant force (the vector sum of the individual forces) is the same at all points, but the resultant moment will vary (in magnitude and direction) from point to point.

The resultant of a pair of equal, but oppositely directed parallel forces, known as a *couple*, is simply a moment having a magnitude Fd where F is the magnitude of the forces and d the perpendicular distance between their lines of action. Figure 2.2 illustrates equivalent couples and the curly symbol often used.

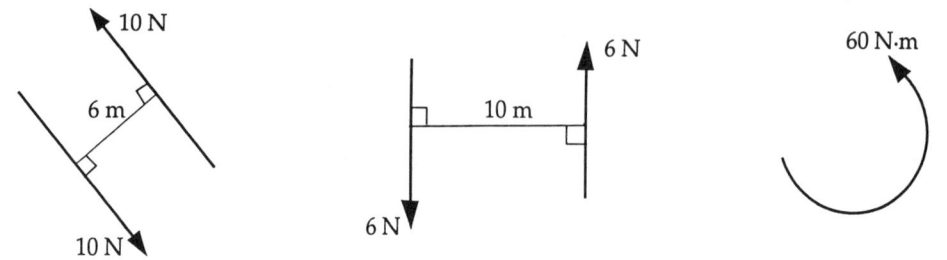

Figure 2.2 Equivalent couples.

EXAMPLE 2.1

Determine the resultant force for the (a) plane, and (b) space concurrent systems shown.

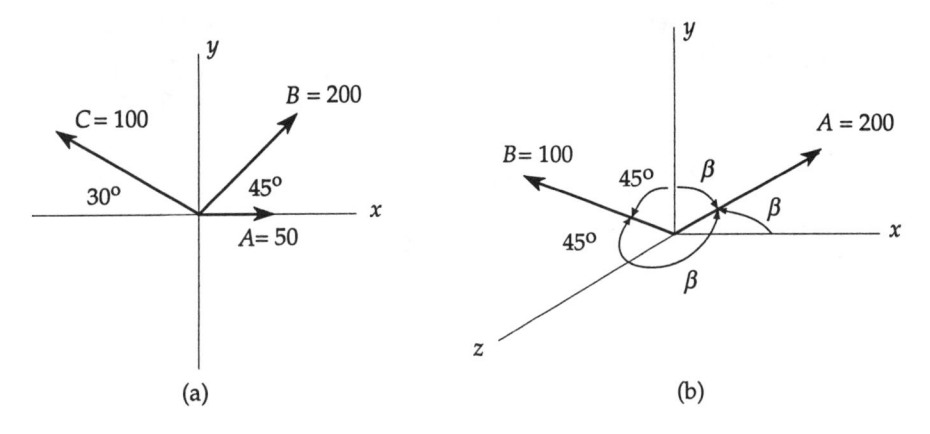

Solution. (a) $\vec{A} = 50\hat{i}$, $\quad \vec{B} = 141.4\hat{i} + 141.4\hat{j}$ $\quad \vec{C} = -86.6\hat{i} + 50\hat{j}$

$$\vec{R} = \vec{A} + \vec{B} + \vec{C} = 104.8\hat{i} + 191.4\hat{j}$$

(b) $\vec{A} = 200(\hat{i} + \hat{j} + \hat{k})/\sqrt{3}$, $\quad \vec{B} = 70.7\hat{j} + 70.7\hat{k}$

$$\vec{R} = 115.47\hat{i} + 186.17\hat{j} + 186.17\hat{k}$$

EXAMPLE 2.2

Determine the moment of force $\vec{F}$

(a) with respect to the origin.
(b) with respect to point Q.
(c) about the axis OQ.
(d) about the x-axis.
(e) about the y-axis.

Solution. $\vec{F} = 0.8(200)\hat{i} - 0.6(200)\hat{k} = 160\hat{i} - 120\hat{k}$.

(a) $\vec{M}_o = 4\hat{i} \times (160\hat{i} - 120\hat{k}) = 3\hat{k} \times (160\hat{i} - 120\hat{k}) = 480\hat{j}$.

(b) $\vec{M}_Q = (3\hat{i} - \hat{j}) \times (160\hat{i} - 120\hat{k}) = (-\hat{i} - \hat{j} + 3\hat{k}) \times (160\hat{i} - 120\hat{k}) = 120\hat{i} + 360\hat{j} + 160\hat{k}$.

(c) $M_{OQ} = (120\hat{i} + 360\hat{j} + 160\hat{k}) \cdot (\hat{i} + \hat{j})/\sqrt{2}$
$= (120 + 360)/\sqrt{2} = 480/\sqrt{2} = 339.5$.

(d) $M_x = 0$ (force $\vec{F}$ intersects the x-axis).
(e) $M_y = Fd = 200(3 \cos \beta) = 200(2.4) = 480 \text{ N} \cdot \text{m}$.

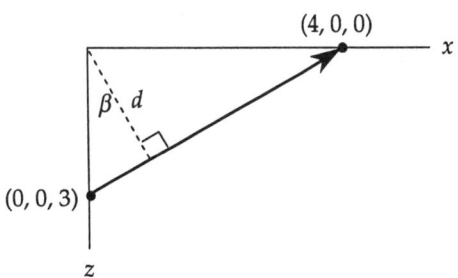

2.2 Equilibrium

If the system of forces acting on a body is one whose resultant is absolutely zero (vector sum of all forces is zero, and the resultant moment of the forces about every point is zero) the body is in *equilibrium*. Mathematically, equilibrium requires the equations

$$\sum \vec{F} = 0, \quad \sum \vec{M}_A = 0 \tag{2.2.1}$$

to be simultaneously satisfied, with A arbitrary. These two vector equations are equivalent to the six scalar equations

$$\begin{aligned} \sum F_x &= 0 & \sum M_x &= 0 \\ \sum F_y &= 0 & \sum M_y &= 0 \\ \sum F_z &= 0 & \sum M_z &= 0 \end{aligned} \tag{2.2.2}$$

which must hold at every point A for any orientation of the xyz-axes. The moment components in Eq. 2.2.2 are the coordinate axes components of $\vec{M}_A$. If the forces are concurrent and their vector sum is zero, the sum of moments about every point will be satisfied automatically.

If all the forces act in a single plane, say the xy-plane, one of the above force equations, and two of the moment equations are satisfied identically, so that equilibrium requires only

$$\sum F_x = 0, \quad \sum F_y = 0, \quad \sum M_z = 0. \tag{2.2.3}$$

In this case we can solve for only three unknowns instead of six as when Eqs. 2.2.2 are required.

The solution for unknown forces and moments in equilibrium problems rests firmly upon the construction of a good *free body diagram*, abbreviated FBD, from which the detailed equations 2.2.2 or 2.2.3 may be obtained. A free body diagram is a neat sketch of the body (or of any appropriate portion of it) showing all forces and moments acting on the body, together with all important linear and angular dimensions.

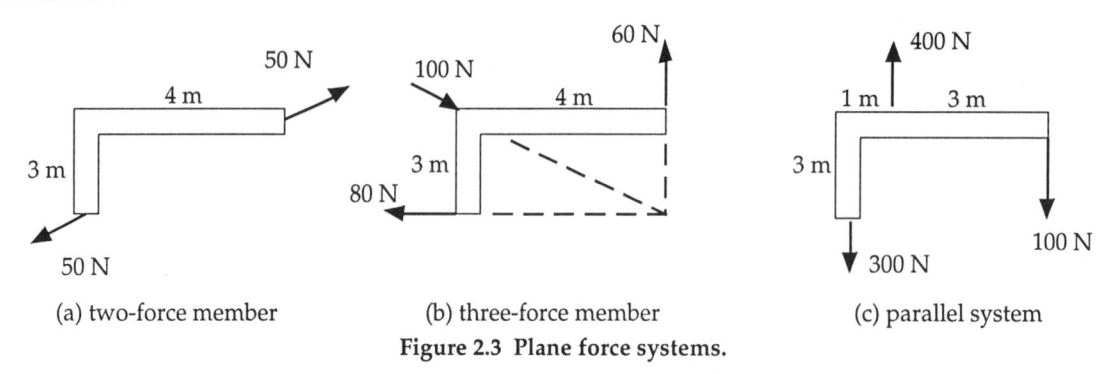

(a) two-force member   (b) three-force member   (c) parallel system

Figure 2.3 Plane force systems.

A body in equilibrium under the action of two forces only is called a *two-force member*, and the two forces must be equal in magnitude and oppositely directed along the line joining their points of application. If a body is in equilibrium under the action of three forces (a *three-force member*) those forces must be coplanar, and concurrent (unless they form a parallel system). Examples are shown in Fig. 2.3.

A knowledge of the possible reaction forces and moments at various supports is essential in preparing a correct free body diagram. Several of the basic reactions are illustrated in Fig. 2.4 showing a block of concrete subjected to a horizontal pull P, Fig. 2.4a, and a cantilever beam carrying both a distributed and concentrated load, Fig. 2.4b. The correct FBDs are to the right of the sketches.

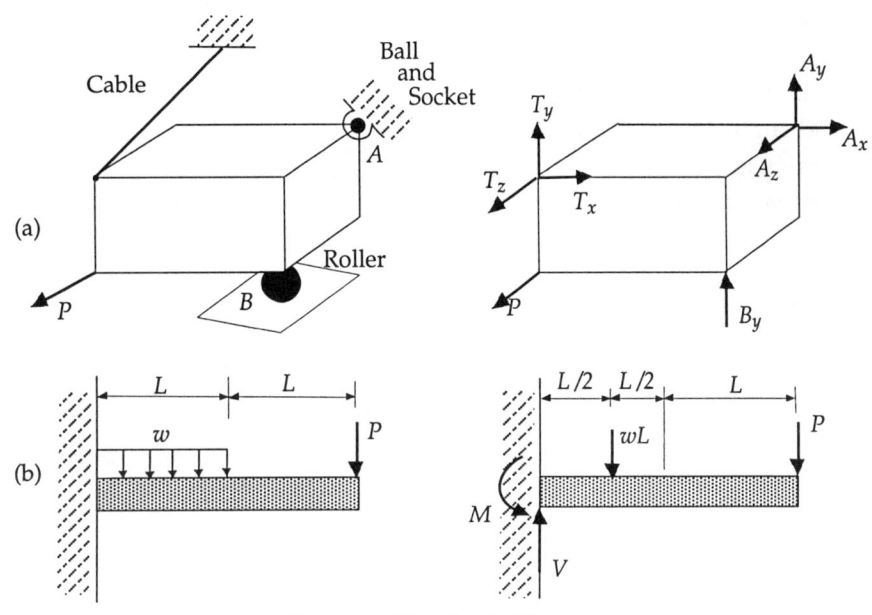

Figure 2.4 Free body diagrams.

EXAMPLE 2.3

Determine the tension in the two cables supporting the 700 N block.

Solution. Construct the FBD of junction A of the cables. Sum forces in x and y directions:

$$\sum F_x = -0.707 T_{AB} + 0.8 T_{AC} = 0$$

$$\sum F_y = 0.707 T_{AB} + 0.6 T_{AC} - 700 = 0.$$

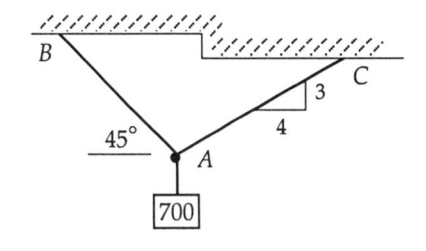

Solve, simultaneously, and find

$$T_{AC} = 700/1.4 = 500 \text{ N}$$

$$T_{AB} = 0.8(500)/0.707 = 565.8 \text{ N}.$$

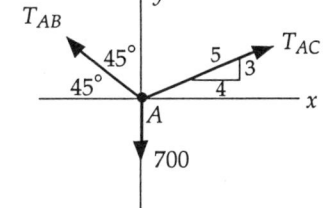

Alternative solution. Draw the force polygon (vector sum of forces) which must close for equilibrium. Determine angles, and use law of sines:

$$\frac{700}{\sin 81.87°} = \frac{T_{AB}}{\sin 53.1°} = \frac{T_{AC}}{\sin 45°}$$

$$T_{AC} = \frac{700(0.707)}{(0.99)} = 500 \text{ N}$$

$$T_{AB} = \frac{700(0.8)}{(0.99)} = 565.8 \text{ N}.$$

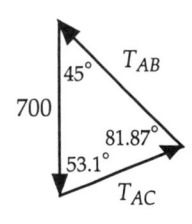

EXAMPLE 2.4

A 12 m bar weighing 140 N is hinged to a vertical wall at A, and supported by the cable BC. Determine the tension in the cable together with the horizontal and vertical components of the force reaction at A.

Solution. Construct the FBD showing force components at A and B. Write the equilibrium equations and solve:

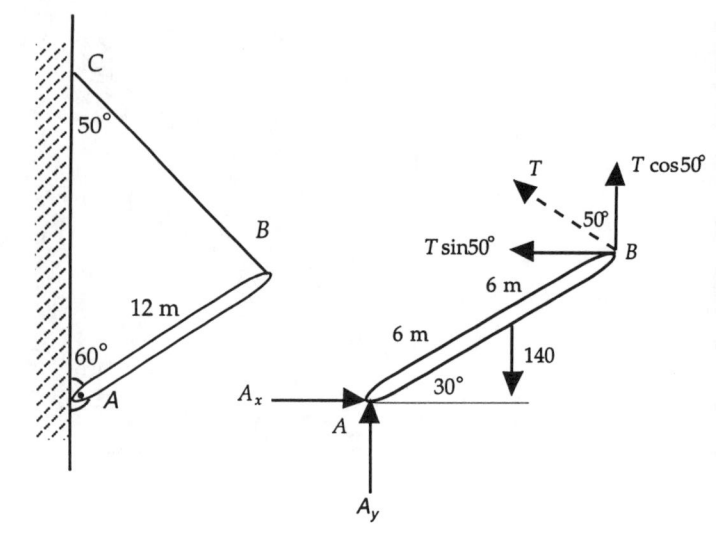

$$\sum M_A = 6T\sin 50° + 6\sqrt{3}T\cos 50° - 140(6)\sqrt{3}/2 = 0$$

$$\therefore T = 64.5 \text{ N}$$

$$\sum F_x = A_x - T\sin 50° = A_x - 64.5(0.766) = 0$$

$$\therefore A_x = 49.4 \text{ N}$$

$$\sum F_y = A_y + T\cos 50° - 140 = A_y + 64.5(0.643) - 140 = 0$$

$$\therefore A_y = 98.5 \text{ N}.$$

2.3 Trusses and Frames

Simple pin-connected trusses and plane frames provide us with elementary examples of structures that may be solved by the equilibrium concepts of statics.

The classic truss problem resembles the one-lane country bridge, as shown schematically in Fig. 2.5a. All members are assumed to be two-force members and are therefore in simple (axial) tension or compression. All loads are assumed to act at the joints (labeled A, B, C, etc.) where the members are pinned together. External reactions such as A_x, A_y and E_y may be determined as a non-concurrent force problem from a FBD of the entire truss. Following that, the internal forces in the members themselves

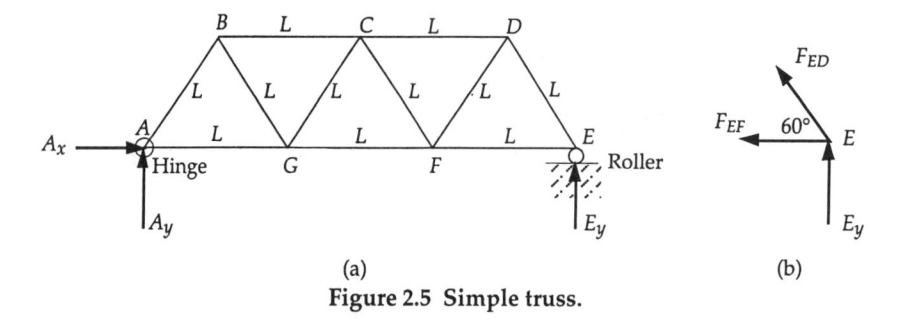

Figure 2.5 Simple truss.

may be determined from a FBD of each joint in turn (method of joints) starting, for example, with joint E of the truss as shown in Fig. 2.5b. Thus, we solve a sequence of concurrent force problems at successive joints having only two unknowns. As noted by Fig. 2.5b we may assume the unknown internal forces such as F_{ED}, F_{EF}, etc., to be tension. A negative result indicates compression.

EXAMPLE 2.5

Determine the forces in the members of the pin-connected truss loaded as shown below. All members have length L.

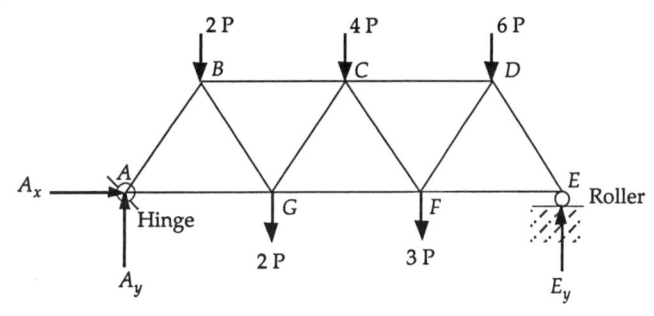

Solution. Summing moments about pin A we solve for the reaction at roller E (counterclockwise moments are positive):

$$\sum M_A = 3LE_y - 2PL - 3P(2L) - 2P(L/2) - 4P(3L/2) - 6P(5L/2) = 0.$$

$$\therefore E_y = 10P.$$

Also,

$$\sum M_E = 3LA_y + 3P(L) + 2P(2L) + 6P(L/2) + 4P(3L/2) + 2P(5L/2) = 0.$$

$$\therefore A_y = 7P.$$

Summing Horizontal forces,

$$\sum F_x = A_x = 0.$$

Consider joint E (see FBD at right),

$$\sum F_y = T_{ED} \sin 60° + 10P = 0$$

$$\therefore T_{ED} = -11.55P \text{ (comp)}.$$

$$\sum F_x = -T_{EF} - T_{ED} \cos 60° = 0$$

$$\therefore T_{EF} = 5.775P \text{ (tens)}.$$

Consider next joint D (see FBD at right),

$$\sum F_y = -6P - 0.866F_{DF} + 0.866(11.55P) = 0.$$

$$\therefore F_{DF} = 4.62P \text{ (tens)}.$$

$$\sum F_x = -F_{DC} - 0.5(11.55P) - 0.5(4.62P) = 0$$

$$\therefore F_{DC} = -8.085P \text{ (comp)}.$$

Consider next joint F (see FBD at right),

$$\sum F_y = -3P + 0.866(4.62P) + 0.866F_{FC} = 0.$$

$$\therefore F_{FC} = -1.156P \text{ (comp)}.$$

$$\sum F_x = -F_{FG} - 0.5(-1.156P) + 0.5(4.62P) + 5.775P = 0$$

$$\therefore F_{FG} = 8.663P \text{ (tens)}.$$

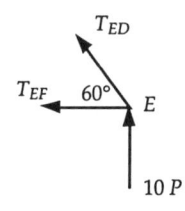

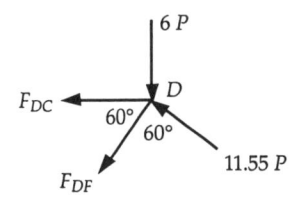

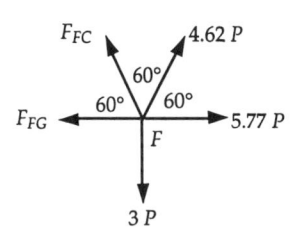

Note: The student should complete this example by considering joint C next, then joint G, and so on.

If the internal forces in only a few selected members are required, the *method of sections* may be used. For example, to obtain only the forces in members CD, CF and GF of the above truss, we "section" it into two portions by cutting across those members as shown in Fig. 2.6. Each portion becomes a sub-truss, and the internal forces of the sectioned members become external reactions of the two sub-trusses. Both force $\left(\sum \vec{F} = 0\right)$ and moment $\left(\sum \vec{M}_A = 0\right)$ equations are useful in this method.

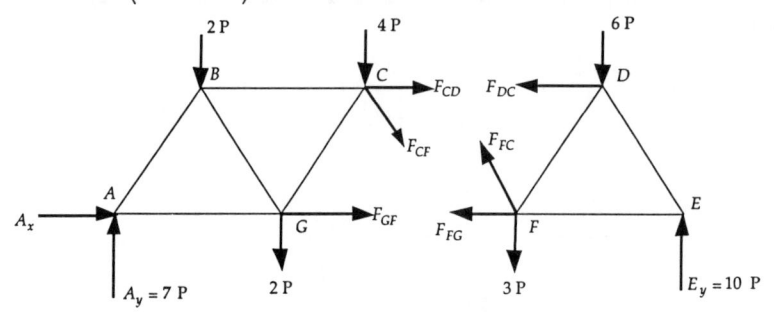

Figure 2.6 Sectioned truss.

EXAMPLE 2.6

Using the right sub-truss of Fig. 2.6 determine the forces in members CD, CF and FG.

Solution. Summing moments about pin F of the FBD of the right sub-truss,

$$\sum M_F = F_{DC}(\sqrt{3}L/2) + 10P(L) - 6P(L/2) = 0.$$

$$\therefore F_{DC} = -8.085P \text{ (comp)}.$$

Summing vertical forces on the sub-truss,

$$\sum F_y = 10P - 3P - 6P + 0.866 F_{FC} = 0.$$

$$\therefore F_{FC} = -1.156P \text{ (comp)}.$$

Summing horizontal forces,

$$\sum F_x = -F_{FG} + 8.085P + 1.156P/2 = 0.$$

$$\therefore F_{FG} = 8.663P \text{ (tens)}.$$

(Note: These results agree with those determined by the method of joints in Example 2.5)

A plane frame is a structure that consists of both two-force and three-force members, or even four-force members, etc. Loads may act at any location on the frame. The problem is to determine the components of the reactions at all pins of the frame. This usually requires not only a FBD of the entire frame, but also a FBD of each member. We illustrate by an example.

EXAMPLE 2.7

For the frame shown, determine the horizontal and vertical components of the reactions at all pins.

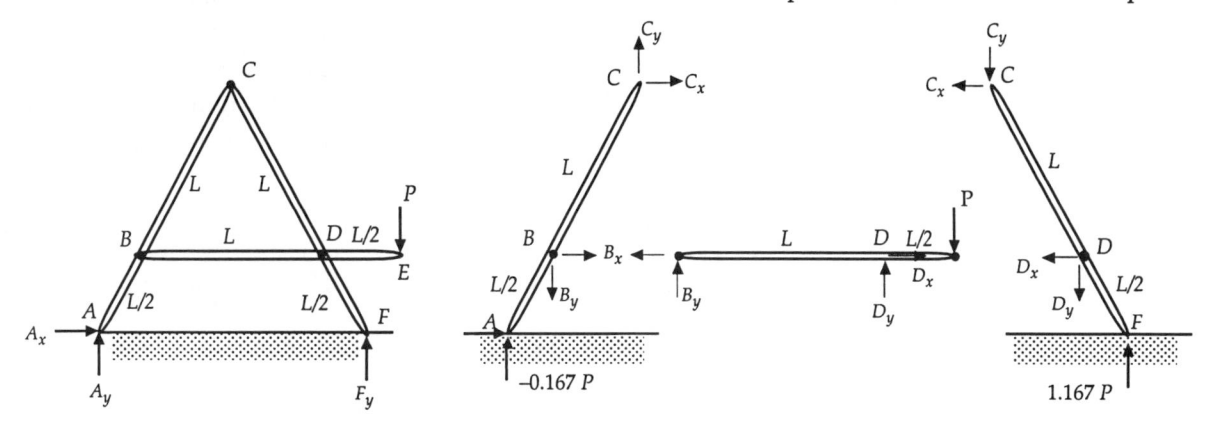

Solution. From a FBD of entire frame,

$\sum F_x = A_x = 0.$ $\therefore A_x = 0.$

$\sum M_A = (1.5L)F_y - (1.75L)P = 0.$ $\therefore F_y = 1.167P$

$\sum F_y = A_y + 1.167P - P = 0.$ $\therefore A_y = -0.167P$

From a FBD of a member BDE,

$\sum M_D = -(L)B_y - (L/2)P = 0.$ $\therefore B_y = -0.5P.$

$\sum F_y = -0.5P + D_y - P = 0.$ $\therefore D_y = 1.5P.$

Transfer the vertical components at B and D of members BDE to members ABC and CDF by changing directions (action and reaction principle). From a *FBD* of member ABC,

$\sum F_y = -0.167P + 0.5P + C_y = 0.$ $\therefore C_y = -0.333P$

$\sum M_B = -(\sqrt{3}L/2)C_x - (L/2)0.333P + (L/4)0.167P = 0.$ $\therefore C_x = -0.144P$

$\sum F_x = B_x - 0.144P = 0.$ $\therefore B_x = 0.144P$

Transfer horizontal components B_x and C_x to members BDE and CDF. From a FBD of member BDE,

$\sum F_x = -0.144P + D_x = 0.$ $\therefore D_x = 0.144P.$

From a FBD of member CDF,

$\sum F_y = 0.333P + D_y + 1.167P = 0.$ $\therefore D_y = -1.5P.$

Completed FBDs of all members are shown below.

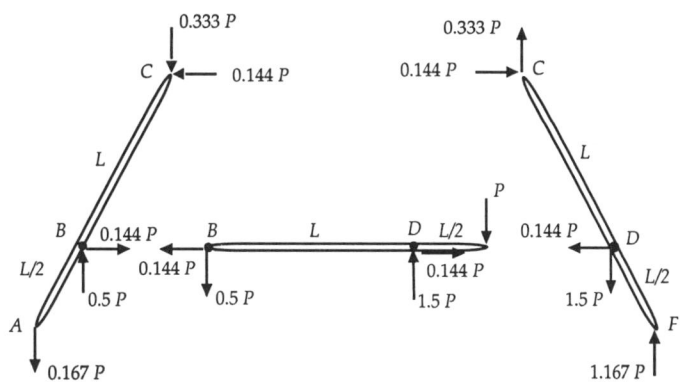

2.4 Friction

Consider a block of weight W at rest on a dry, rough, horizontal plane, Fig. 2.7a. Let a horizontal force P act to the right on the block, Fig. 2.7b. A *friction force* F is developed at the surface of contact between the block and plane, and maintains equilibrium ($F = P$) as long as $P < \mu_s N = \mu_s W$, where μ_s is the *static coefficient of friction*. If P is increased until $P = P^* = \mu_s N = \mu_s W$ (Fig. 2.7c), the equilibrium state is on the verge of collapse, and motion to the right is impending. In summary, dry *Coulomb friction* described here is governed by

$$F \leq \mu_s N. \tag{2.4.1}$$

If relative motion between the block and the plane occurs (Fig. 2.7d), the friction force is given by

$$F = \mu_k N \tag{2.4.2}$$

where μ_k is the *kinetic coefficient of friction*, slightly less in value than μ_s, with both coefficients having a range $0 \leq \mu \leq 1$.

At impending motion, the equality holds in Eq. 2.4.1 and the resultant R of F and N makes an angle ϕ with the normal N, called the *friction angle*. From Fig. 2.8a it is clear that

$$\phi = \tan^{-1} \mu_s. \tag{2.4.3}$$

Also, it may be readily shown that a block will remain at rest (in equilibrium) on a rough inclined plane, Fig. 2.8b, as long as $\beta < \phi$ for the surfaces of contact.

Figure 2.7 Friction forces.

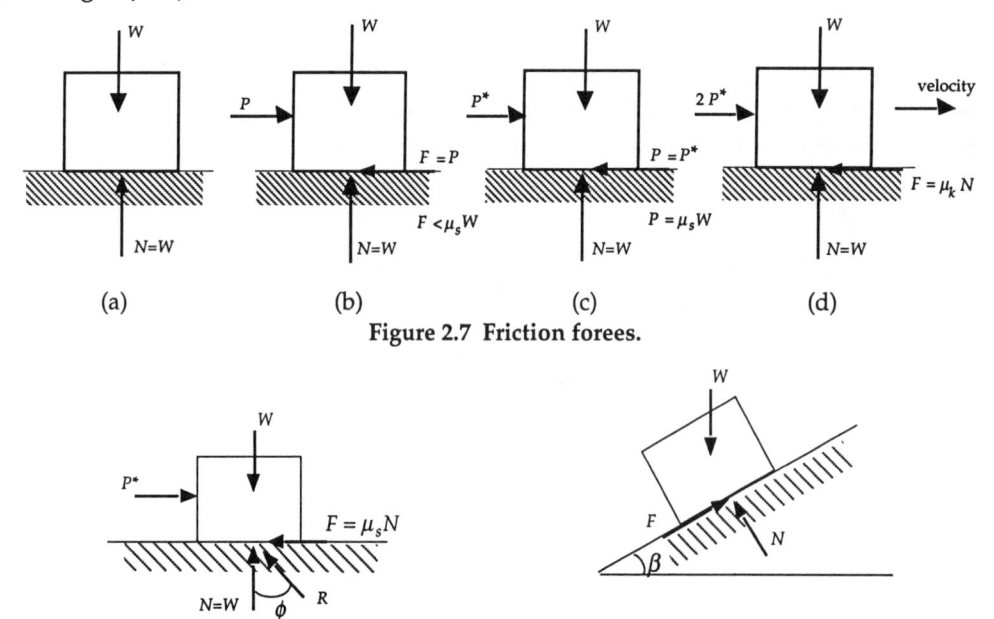

Figure 2.8 Friction angle.

EXAMPLE 2.8

Determine the horizontal force P required to cause impending motion of the 50 kg block, (a) up the plane, (b) down the plane, if $\mu_s = 0.6$ between the block and plane.

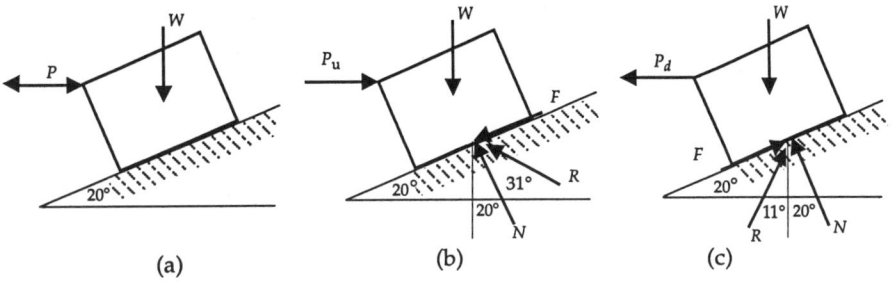

Solution. Note that, $\phi = \tan^{-1}(0.6) = 31°$ so that the block would remain in place if undisturbed.

(a) At impending motion up the plane, Fig. b, the resultant R of the friction force F and normal force N makes an angle of 51° to the right of the vertical and from the FBD of the block,

$$\sum F_x = P_u - R\sin 51° = 0$$

$$\sum F_y = R\cos 51° - 50 \times 9.81 = 0.$$

Solving these equations,

$$R = 779 \text{ N}, \qquad P_u = 606 \text{ N}.$$

(b) At impending motion down the plane, R makes an angle of 11° to the left of the vertical, Fig. c, so that now

$$\sum F_x = -P_d + R\sin 11° = 0$$

$$\sum F_y = R\cos 11° - 50 \times 9.81 = 0.$$

Therefore

$$R = 499 \text{ N}, \qquad P_d = 95.2 \text{ N}.$$

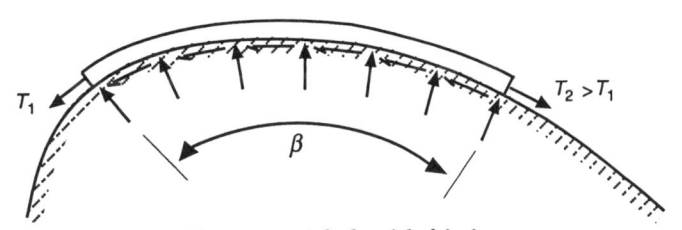

Figure 2.9 A belt with friction.

If a belt, or rope, is pressed firmly against some portion of a rough stationary curved surface, and pulled in one direction or the other, the tension in the belt will increase in the direction of pull due to the frictional resistance between the belt and surface, as shown in Fig. 2.9. It may be shown that

$$T_2 = T_1 e^{\mu_s \beta} \qquad T_2 > T_1 \qquad (2.4.4)$$

where β is the angle of contact, in radians, and μ_s is the static coefficient of friction.

EXAMPLE 2.9

A 100 kg block rests on a 30° rough inclined plane ($\mu_s = 0.4$) and is attached by a rope to a mass m in the arrangement shown. If the static coefficient of friction between the rope and the circular support is 0.25, determine the maximum m that can be supported without slipping.

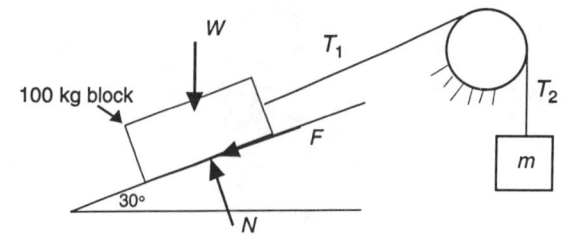

Solution. Summing forces perpendicular to the plane we determine

$$N = 981\cos 30° = 849.5 \text{ N}.$$

Thus,

$$F = 0.4(849.5) = 339.8 \text{ N}$$

at impending motion, and by summing forces along the plane

$$T_1 = 339.8 + 981\sin 30° = 830.3 \text{ N}.$$

For the circular support we have,

$$T_2 = T_1 e^{(0.25)(120\times\pi/180)} = 830.3(1.69) = 1403 \text{ N}.$$

and

$$m = 1403/9.81 = 143 \text{ kg}.$$

2.5 Properties of Plane Areas

Associated with every plane area A (in the xy-plane, for example) there is a point C, known as the *centroid*, whose coordinates $\bar{x}$ and $\bar{y}$ are defined by the integrals

$$\bar{x} = \frac{\int_A x\,dA}{\int_A dA} \qquad \bar{y} = \frac{\int_A y\,dA}{\int_A dA} \qquad (2.5.1)$$

where dA is differential element of area having coordinates x and y as shown in Fig. 2.10a.

Although the integrals in Eq. 2.5.1 may be evaluated by a double integration using $dA = dxdy$, in practice it is often advantageous to calculate centroidal coordinates by a single integration using either a horizontal or vertical strip for dA (Fig. 2.10b and Fig. 2.10c) for which Eq. 2.5.1 may be expressed in the form

$$\bar{x} = \frac{\int_A x_e\,dA}{\int_A dA} \qquad \bar{y} = \frac{\int_A y_e\,dA}{\int_A dA} \qquad (2.5.2)$$

where x_e and y_e are the coordinates of the centroids of the strip elements as shown.

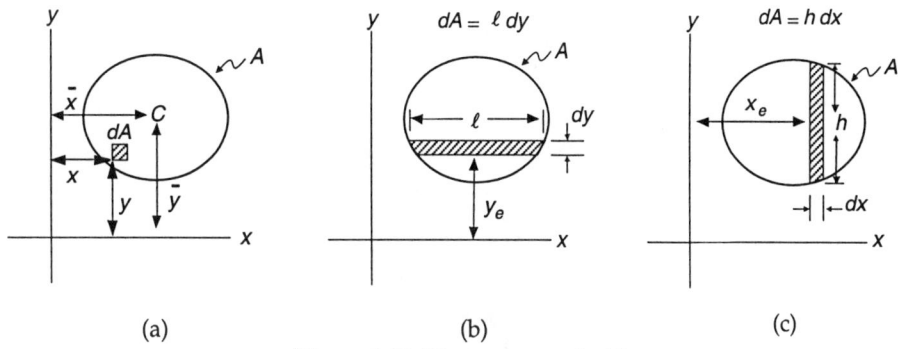

Figure 2.10 Plane area centroid.

The *plane moments of inertia* I_x and I_y of A with respect to the x and y axes, respectively, are defined by

$$I_x = \int_A y^2 dA, \qquad I_y = \int_A x^2 dA \qquad (2.5.3)$$

where in integrating for I_x we use the horizontal strip for dA, and for I_y the vertical strip. The *polar moment of inertia* with respect to the origin O is defined by

$$J_O = \int_A r^2 dA = \int_A (x^2 + y^2) dA = I_x + I_y. \qquad (2.5.4)$$

All moments of inertia may be expressed in terms of their respective *radii of gyration*. Thus,

$$I_x = k_x^2 A, \qquad I_y = k_y^2 A, \qquad J_O = k_O^2 A. \qquad (2.5.5)$$

The *product of inertia* of A is defined with respect to a pair of perpendicular axes. For the coordinate axes we have

$$I_{xy} = \int_A xy\, dA \qquad (2.5.6)$$

which normally must be evaluated by a double integration over A. If either one (or both) of the reference axes is an axis of symmetry the product of inertia is zero relative to that pair of axes.

The *parallel-axis theorem* establishes a relationship between the moment of inertia about an arbitrary axis and the moment of inertia about a parallel axis passing through C. Thus

$$I_P = I_C + Ad^2, \qquad J_P = J_C + Ad^2 \qquad (2.5.7)$$

where the subscript C indicates the centroidal moment of inertia, subscript P indicates the moment of inertia about the parallel axis, A is the area, and d the distance separating the two axes. Similarly, for products of inertia,

$$I_{x_P y_P} = I_{x_C y_C} + A x_1 y_1 \qquad (2.5.8)$$

where x_1 and y_1 are the distances between axes x and x_C, and y and y_C, respectively.

EXAMPLE 2.10

For the shaded area shown by the sketch, determine $\bar{x}$, $\bar{y}$, I_x, I_y, J_O, I_{xy}, $(I_x)_C$, $(I_y)_C$, J_C and $I_{x_C y_C}$. Units are mm.

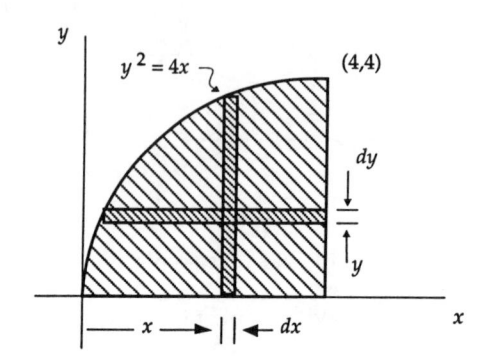

Solution. First we calculate A using the vertical strip, $dA = ydx = 2\sqrt{x}\,dx$:

$$A = \int_0^4 2\sqrt{x}\,dx = 32/3 \text{ mm}^2.$$

Thus, from Eq. 2.5.1, with $dA = 2\sqrt{x}\,dx$,

$$\bar{x} = \frac{3}{32}\int_0^4 x(2\sqrt{x})dx = 2.4 \text{ mm}$$

and, using the horizontal strip $dA = (4 - y^2/4)dy$,

$$\bar{y} = \frac{3}{32}\int_0^4 y(4 - y^2/4)dy = 1.5 \text{ mm}.$$

From Eq. 2.5.3, with $dA = (4 - y^2/4)dy$,

$$I_x = \int_0^4 y^2(4 - y^2/4)dy = 34.13 \text{ mm}^4$$

and, using $dA = 2\sqrt{x}\,dx$,

$$I_y = \int_0^4 x^2(2\sqrt{x})dx = 73.14 \text{ mm}^4.$$

From Eq. 2.5.4

$$J_O = I_x + I_y = 34.13 + 73.14 = 107.27 \text{ mm}^4.$$

From Eq. 2.5.6, with $dA = dxdy$,

$$I_{xy} = \int_0^4 \int_0^{2\sqrt{x}} xy\,dydx = \int_0^4 \left[\frac{xy^2}{2}\right]_0^{2\sqrt{x}} dx = \int_0^4 2x^2 dx = 42.67 \text{ mm}^4.$$

From Eq. 2.5.7,

$$(I_x)_C = I_x - Ad^2 = 34.13 - 10.67(1.5)^2 = 10.13 \text{ mm}^4$$

$$(I_y)_C = I_y - Ad^2 = 73.14 - 10.67(2.4)^2 = 11.68 \text{ mm}^4$$

$$J_C = J_O - Ad^2 = 107.27 - 10.67\left[(1.5)^2 + (2.4)^2\right] = 21.80 \text{ mm}^4.$$

Note that moments of inertia are always minimum about a centroidal axis. Finally, from Eq. 2.5.8,

$$I_{x_C y_C} = I_{xy} - A(-1.5)(-2.4) = 42.67 - 38.40 = 4.27 \text{ mm}^4$$

EXAMPLE 2.11

Determine I_x and J_O for the circular sector shown below.

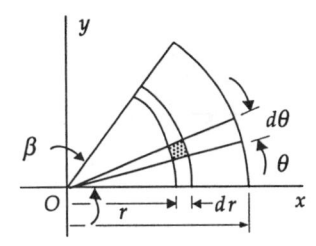

Solution. Using polar coordinates with $dA = r\, dr\, d\theta$ and $y = r\sin\theta$ in Eq. 2.5.3:

$$I_x = \int_0^\beta \int_0^a (r\sin\theta)^2 r\, dr\, d\theta = \frac{a^4}{4}\left[\frac{\beta}{2} - \frac{\sin 2\beta}{4}\right].$$

When $\beta = \pi/2$,

$$I_x = \pi a^4/16.$$

From Eq. 2.5.4,

$$J_O = \int_0^\beta \int_0^a r^3\, dr\, d\theta = \frac{a^4 \beta}{4}.$$

When $\beta = \pi/2$,

$$J_O = \pi a^4/8.$$

The properties of common areas may be determined by integration. A brief list is given in Table 2.1. Using data from Table 2.1, we may calculate centroids and moments of inertia of *composite areas* made up of combinations of two or more (including cutouts) of the common areas. Thus,

$$\bar{x} = \frac{\sum_{i=1}^N x_i A_i}{\sum_{i=1}^N A_i}, \qquad \bar{y} = \frac{\sum_{i=1}^N y_i A_i}{\sum_{i=1}^N A_i} \qquad (2.5.9)$$

where N is the number of areas, and x_i is the centroidal distance for area A_i. Likewise, for moments of inertia

$$I = \sum_{i=1}^N I_i = I_1 + I_2 + \ldots + I_N. \qquad (2.5.10)$$

An example illustrates.

TABLE 2.1. Properties of Areas.

Shape	Dimensions	Centroid	Inertia
Rectangle		$\bar{x} = b/2$ $\bar{y} = h/2$	$I_C = bh^3/12$ $I_x = bh^3/3$ $I_y = hb^3/3$
Triangle		$\bar{y} = h/3$	$I_C = bh^3/36$ $I_x = bh^3/12$
Circle		$\bar{x} = 0$ $\bar{y} = 0$	$I_x = \pi r^4/4$ $J = \pi d^4/32$
Quarter Circle		$\bar{y} = 4r/3\pi$	$I_x = \pi r^4/16$
Half Circle		$\bar{y} = 4r/3\pi$	$I_x = \pi r^4/8$

EXAMPLE 2.12

Determine the centroidal coordinates, and I_x and I_y for the composite area shown.

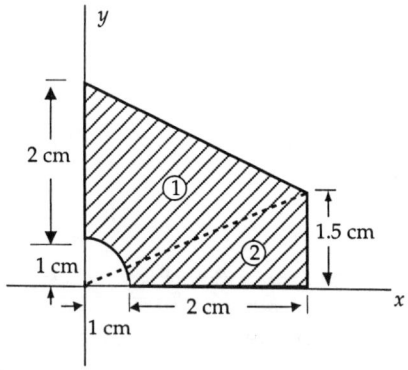

Solution. Decompose the composite into two triangular areas 1 and 2, and the negative quarter circular area 3:

$$A = A_1 + A_2 + A_3 = 4.5 + 2.25 - \pi/4 = 5.97 \text{ cm}^2.$$

$$A\bar{x} = x_1 A_1 + x_2 A_2 + x_3 A_3$$
$$= (1)(4.5) + (2)(2.25) + (4/3\pi)(-\pi/4) = 8.67.$$
$$\therefore \bar{x} = 8.67/5.97 = 1.45 \text{ cm}.$$

$$A\bar{y} = y_1 \cdot A_1 + y_2 A_2 + y_3 A_3$$
$$= (1.5)(4.5) + (0.5)(2.25) + (4/3\pi)(-\pi/4) = 7.54.$$
$$\therefore \bar{y} = 7.54/5.97 = 1.26 \text{ cm}.$$

$$I_x = I_{1x} + I_{2x} + I_{3x}$$
$$= \left\{ 2\left[3(1.5)^3/36\right] + (4.5)(1.5)^2 \right\} + 3(1.5)^3/12 - \pi/16 = 11.33 \text{ cm}^4,$$

where we have used the parallel-axis theorem to obtain I_{1x}. Finally,

$$I_y = I_{1y} + I_{2y} + I_{3y}$$
$$= 3(3)^3/12 + 1.5(3)^3/4 - \pi/16 = 16.68 \text{ cm}^4.$$

2.6 Properties of Masses and Volumes

The coordinates of the *center of gravity* G of an arbitrary mass m occupying a volume V of space are defined by

$$x_G = \frac{\int_V x\rho dV}{\int_V \rho dV}, \qquad y_G = \frac{\int_V y\rho dV}{\int_V \rho dV}, \qquad z_G = \frac{\int_V z\rho dV}{\int_V \rho dV} \qquad (2.6.1)$$

where ρ is the mass density, dV is the differential element of volume, and x, y and z are the coordinates of dV, as shown in Fig. 2.11.

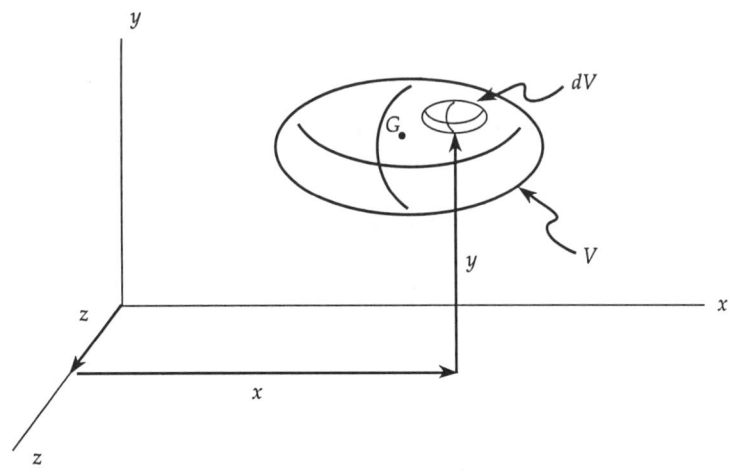

Figure 2.11 Center of gravity of an arbitrary mass.

The density may be a function of the space variables, $\rho = \rho(x,y,z)$. If the density is constant throughout the volume, the integrals in Eq. 2.6.1 reduce to

$$x_C = \frac{\int_V x dV}{V}, \qquad y_C = \frac{\int_V y dV}{V}, \qquad z_C = \frac{\int_V z dV}{V} \qquad (2.6.2)$$

which defines the coordinates of the centroid C of volume V. If ρ is constant, G and C coincide for a given body. As with areas, if an axis of symmetry exists for the volume, C is on that axis. The coordinates of G and C are readily calculated for geometries having an axis of revolution.

EXAMPLE 2.13

Let the area of Example 2.10 be rotated about the x-axis to form a solid (volume) of revolution. Determine (a) G for the solid if $\rho = \rho_o x$ where ρ_o is a constant, (b) C for the volume.

Solution. Since there is symmetry about the x-axis, and since ρ is at most a function of x, we have

$$y_G = z_G = y_C = z_C = 0$$

(a) Let the element of mass be a thin disk for which $\rho dV = \rho_o x \pi y^2 dx$. Therefore,

$$x_G = \frac{\int_0^4 \rho_o \pi x^2 (4x) dx}{\int_0^4 \rho_o \pi x (4x) dx} = \frac{4 \times 4^4/4}{4 \times 4^3/3} = 3.00 \text{ mm.}$$

(b) For the same element with $\rho = \rho_o$

$$x_C = \frac{\int_0^4 \rho_o \pi x (4x) dx}{\int_0^4 \rho_o \pi (4x) dx} = \frac{4 \times 4^3/3}{4 \times 4^2/2} = 2.67 \text{ mm.}$$

Because the topic of *mass moments of inertia* properly belongs in the subject of dynamics, and is not a factor in statics, we present only a brief comment or two here for comparison with area moments of inertia. The mass moment of inertia of the three-dimensional body in Fig. 2.11 about any axis is defined by the integral

$$I = \int_V r^2 \rho dV = \int_V r^2 dm \tag{2.6.3}$$

where r is the perpendicular distance of the mass element dm from the axis. With respect to the coordinate axes, Eq. 2.6.3 may be specialized to yield

$$I_x = \int_V (y^2 + z^2) dm \tag{2.6.4}$$

$$I_y = \int_V (z^2 + x^2) dm \tag{2.6.5}$$

$$I_z = \int_V (x^2 + y^2) dm. \tag{2.6.6}$$

There is also a parallel-axis theorem for mass moments of inertia, namely,

$$I_P = I_G + md^2 \tag{2.6.7}$$

with d being the distance between the center of gravity axis and the parallel axis of interest. Several mass moments of inertia are presented in Table 2.2.

TABLE 2.2 Mass Moments of Inertia.

Shape	Dimensions	Moment of Inertia
Slender rod		$I_y = mL^2/12$ $I_{y'} = mL^2/3$
Circular cylinder		$I_x = mr^2/2$ $I_y = m(L^2 + 3r^2)/12$
Disk		$I_x = mr^2/2$ $I_y = mr^2/4$
Rectangular parallelpiped		$I_x = m(a^2 + b^2)/12$ $I_y = m(L^2 + b^2)/12$ $I_z = m(L^2 + a^2)/12$ $I_{y'} = m(4L^2 + b^2)/12$
Sphere		$I_x = 2mr^2/5$

Practice Problems

GENERAL

2.1 Find the component of the vector $\vec{A} = 15\hat{i} - 9\hat{j} + 15\hat{k}$ in the direction of $\vec{B} = \hat{i} - 2\hat{j} - 2\hat{k}$.

  a) 1   b) 3   c) 5   d) 7   e) 9

2.2 Find the magnitude of the resultant of $\vec{A} = 2\hat{i} + 5\hat{j}$, $\vec{B} = 6\hat{i} - 7\hat{k}$, and $\vec{C} = 2\hat{i} - 6\hat{j} + 10\hat{k}$.

  a) 8.2   b) 9.3   c) 10.5   d) 11.7   e) 12.8

2.3 Determine the moment about the y-axis of the force $\vec{F} = 200\hat{i} + 400\hat{j}$ acting at (4, –6, 4).

  a) 0   b) 200   c) 400   d) 600   e) 800

2.4 What total moment do the two forces $\vec{F_1} = 50\hat{i} - 40\hat{k}$ and $\vec{F_2} = 60\hat{j} + 80\hat{k}$ acting at (2, 0, –4) and (–4, 2, 0), respectively, produce about the x-axis?

  a) 0   b) 80   c) 160   d) 240   e) 320

2.5 The force system shown may be referred to as being

  a) concurrent
  b) coplanar
  c) parallel
  d) two-dimensional
  e) non-concurrent, non-coplanar

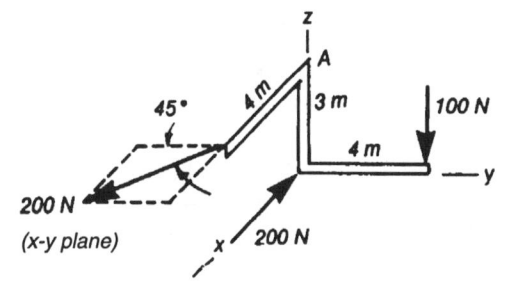

EQUILIBRIUM

2.6 If equilibrium exists due to a rigid support at A in the figure of Prob. 3.5, what reactive force must exist at A?

  a) $-59\hat{i} - 141\hat{j} + 10\hat{k}$
  b) $59\hat{i} + 141\hat{j} + 100\hat{k}$
  c) $341\hat{i} - 141\hat{j} - 100\hat{k}$
  d) $341\hat{i} + 141\hat{j} - 100\hat{k}$
  e) $59\hat{i} - 141\hat{j} + 100\hat{k}$

2.7 If equilibrium exists on the object in Prob. 3.5, what reactive moment must exist at the rigid support A?

  a) $600\hat{i} + 400\hat{j} + 564\hat{k}$
  b) $400\hat{i} + 564\hat{k}$
  c) $400\hat{i} - 600\hat{j} + 564\hat{k}$
  d) $400\hat{i} - 600\hat{j}$
  e) $-600\hat{j} + 564\hat{j}$

2.8 If three nonparallel forces hold a rigid body in equilibrium, they must

  a) be equal in magnitude.
  b) be concurrent.
  c) be non-concurrent.
  d) form an equilateral triangle.
  e) be colinear.

2.9 A truss member

  a) is a two-force body.
  b) is a three-force body.
  c) resists forces in compression only.
  d) may resist three concurrent forces.
  e) resists lateral forces.

2.10 Find the magnitude of the reaction at support B.
   a) 400   b) 500   c) 600
   d) 700   e) 800

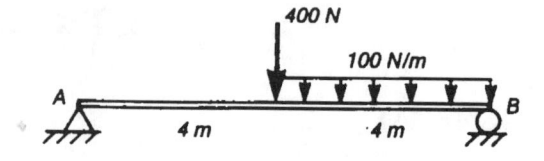

2.11 What moment M exists at support A?
   a) 5600   b) 5000   c) 4400
   d) 4000   e) 3600

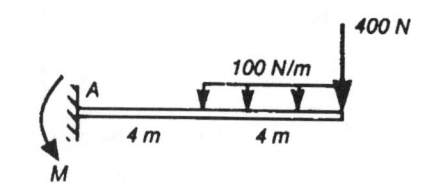

2.12 Calculate the reactive force at support A.
   a) 250   b) 350   c) 450
   d) 550   e) 650

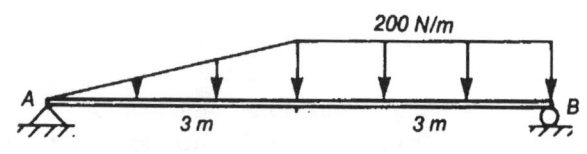

2.13 Find the support moment at A.
   a) 66   b) 77   c) 88
   d) 99   e) 111

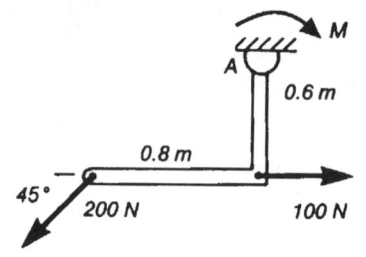

2.14 To ensure equilibrium, what couple must be applied to this member?
   a) 283 cw   b) 283 ccw   c) 400 cw
   d) 400 ccw   e) 0

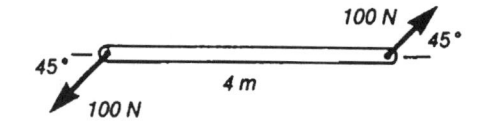

2.15 Calculate the magnitude of the equilibrating force at A for the three-force body shown.
   a) 217   b) 287   c) 343
   d) 385   e) 492

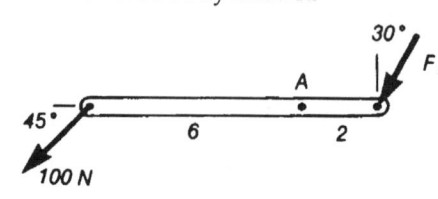

2.16 Find the magnitude of the equilibrating force at point A.
   a) 187   b) 142   c) 114
   d) 99   e) 84

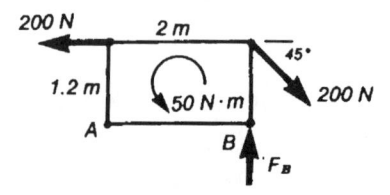

TRUSSES AND FRAMES

2.17 Find F_{DE} if all angles are equal.

    a) 121   b) 142   c) 163

    d) 176   e) 189

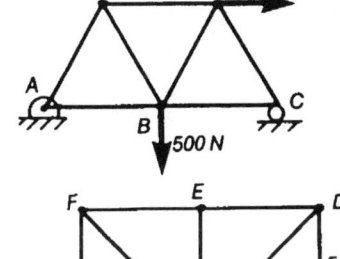

2.18 Find F_{DE}.

    a) 0   b) 1000   c) 2000

    d) 2500   e) 5000

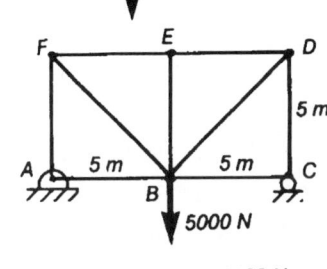

2.19 What is the force in member DE?

    a) 1532   b) 1768   c) 1895

    d) 1946   e) 2231

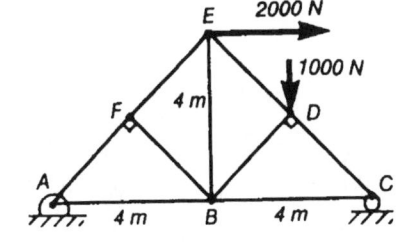

2.20 Calculate F_{FB} in the truss of Problem 2.19.

    a) 0   b) 932   c) 1561   d) 1732   e) 1887

2.21 Find the force in member IC.

    a) 0   b) 1000   c) 1250

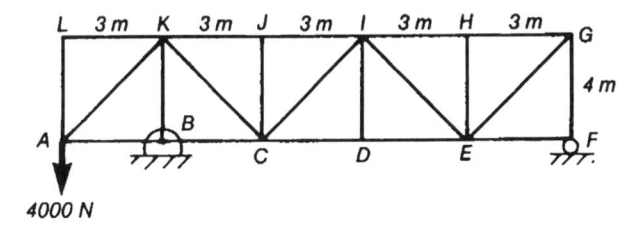

    d) 1500   e) 2000

2.22 What force exists in member BC in the truss of Prob. 2.21?

    a) 0   b) 1000   c) 2000   d) 3000   e) 4000

2.23 Find the force in member FC.

    a) 5320   b) 3420   c) 2560

    d) 936   e) 0

2.24 Determine the force in member BC in the truss of Prob. 2.23.

    a) 3560   b) 4230   c) 4960   d) 5230   e) 5820

Statics

2.25 Find the magnitude of the reactive force at support A.

a) 1400 b) 1300 c) 1200
d) 1100 e) 1000

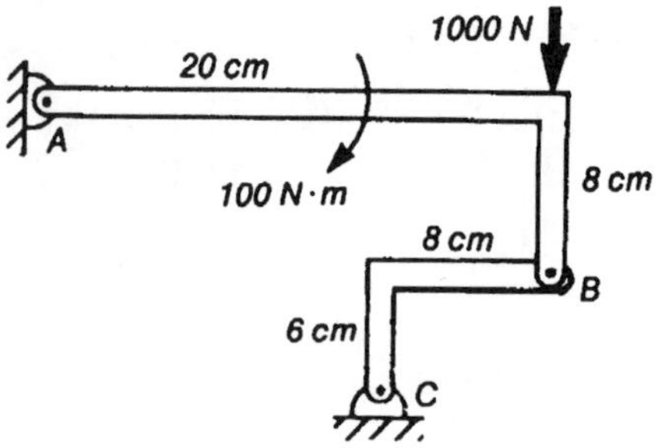

2.26 Determine the distributed force intensity w, in N/m, for equilibrium to exist.

a) 2000 b) 4000 c) 6000
d) 8000 e) 10 000

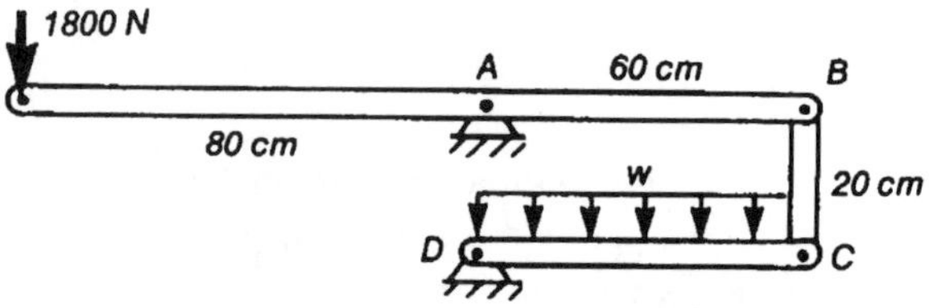

2.27 Find the magnitude of the reactive force at support A.

a) 2580 b) 2670 c) 2790
d) 2880 e) 2960

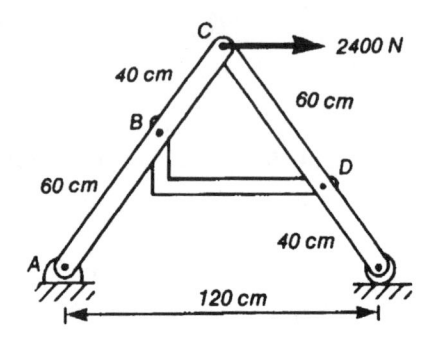

2.28 Calculate the magnitude of the force in member BD of Prob. 2.27.

a) 2590 b) 2670 c) 2790 d) 2880 e) 2960

FRICTION

2.29 What force, in Newtons, will cause impending motion up the plane?

a) 731 b) 821 c) 973
d) 1102 e) 1245

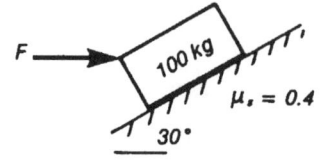

2.30 What is the maximum force F, in Newtons, that can be applied without causing motion to impend?

a) 184 b) 294 c) 316
d) 346 e) 392

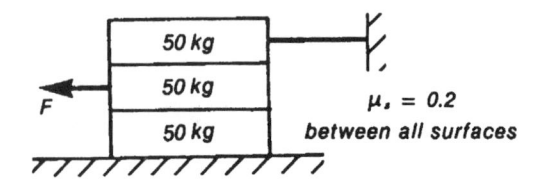

Statics

2.31 Only the rear wheels provide braking. At what angle θ will the car slide if $\mu_s = 0.6$?
- a) 10
- b) 12
- c) 16
- d) 20
- e) 24

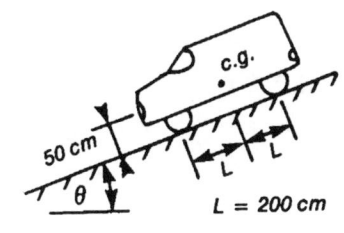

2.32 Find the minimum h value at which tipping will occur.
- a) 8
- b) 10
- c) 12
- d) 14
- e) 16

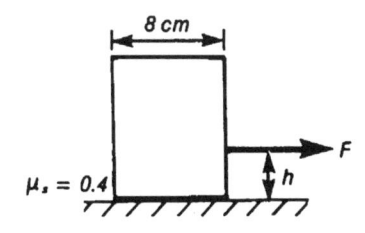

2.33 What force F, in Newtons, will cause impending motion?
- a) 240
- b) 260
- c) 280
- d) 300
- e) 320

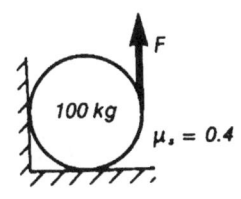

2.34 The angle θ at which the ladder is about to slip is
- a) 50
- b) 46
- c) 42
- d) 38
- e) 34

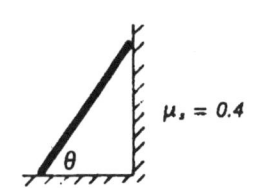

2.35 A boy and his dad put a rope around a tree and stand side by side. What force by the boy can resist a force of 800 N by his dad? Use $\mu_s = 0.5$.
- a) 166
- b) 192
- c) 231
- d) 246
- e) 297

2.36 What moment, in N·m, will cause impending motion?
- a) 88
- b) 99
- c) 110
- d) 121
- e) 146

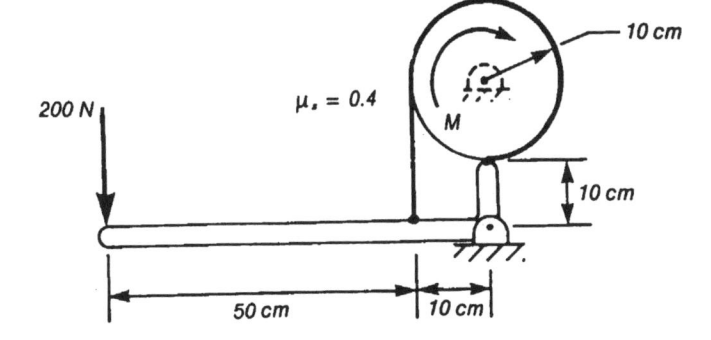

2.37 A 12-m-long rope is draped over a horizontal cylinder of 1.2-m-diameter so that both ends hang free. What is the length of the longer end at impending motion? Use $\mu_s = 0.5$.
- a) 6.98
- b) 7.65
- c) 7.92
- d) 8.37
- e) 8.83

CENTROIDS AND MOMENTS OF INERTIA

2.38 Find the x-coordinate of the centroid of the area bounded by the x-axis, the line $x = 3$, and the parabola $y = x^2$.

    a) 2.0   b) 2.15   c) 2.20   d) 2.25   e) 2.30

2.39 What is the y-coordinate of the centroid of the area of Prob. 2.38?

    a) 2.70   b) 2.65   c) 2.60   d) 2.55   e) 2.50

2.40 Calculate the x-coordinate of the centroid of the area enclosed by the parabolas $y = x^2$ and $x = y^2$.

    a) 0.43   b) 0.44   c) 0.45   d) 0.46   e) 0.47

2.41 Find the y-component of the centroid of the area shown.

    a) 3.35   b) 3.40   c) 3.45
    d) 3.50   e) 3.55

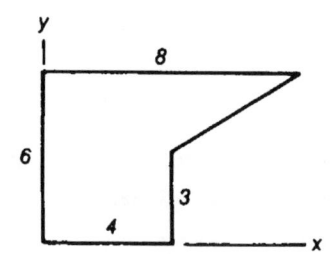

2.42 Calculate the y-component of the centroid of the area shown.

    a) 3.52   b) 3.56   c) 3.60
    d) 3.64   e) 3.68

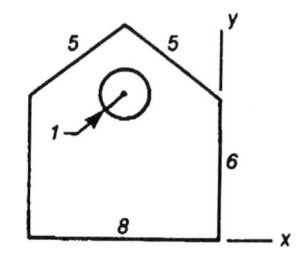

2.43 Find the x-component of the center of gravity of the three objects.

    a) 2.33
    b) 2.42
    c) 2.84
    d) 3.22
    e) 3.64

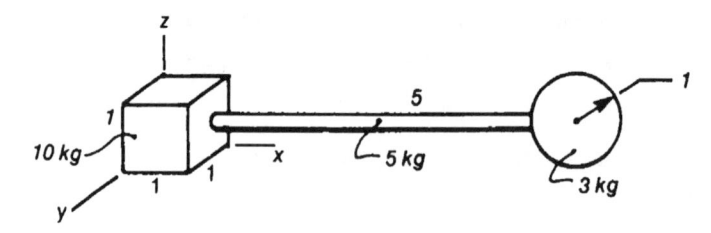

2.44 Calculate the moment of inertia about the x-axis of the area of Prob. 2.38.

    a) 94   b) 104   c) 112   d) 124   e) 132

2.45 What is I_x for the area of Prob. 2.42?

    a) 736   b) 842   c) 936   d) 982   e) 1056

Statics

2.46 Find I_y for the symmetrical area shown.

   a) 4267   b) 4036   c) 3827
   d) 3652   e) 3421

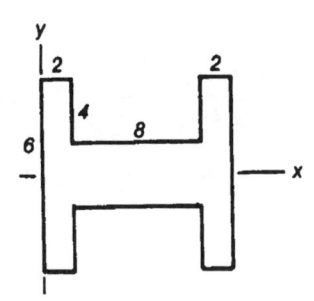

2.47 Determine the mass moment of inertia of a cube with edges of length b, about an axis passing through an edge.

   a) $2mb^2/3$ b) $mb^2/6$ c) $3mb^2/2$ d) $mb^2/2$ e) $3mb^2/4$

2.48 Find the mass moment of inertia about the x-axis if the mass of the rods per unit length is 1.0 kg/m.

   a) 224   b) 268   c) 336
   d) 386   e) 432

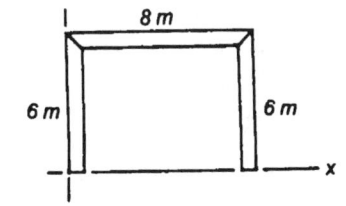

Solutions to Problems

2.1 a) $\hat{i}_B = \dfrac{\hat{i} - 2\hat{j} - 2\hat{k}}{\sqrt{1+4+4}} = \dfrac{1}{3}(\hat{i} - 2\hat{j} - 2\hat{k})$

$\vec{A} \cdot \hat{i}_B = (15\hat{i} - 9\hat{j} + 15\hat{k}) \cdot \dfrac{1}{3}(\hat{i} - 2\hat{j} - 2\hat{k})$

$= 5 + 6 - 10 = 1$

2.2 c) $\vec{A} + \vec{B} + \vec{C} = (2\hat{i} + 5\hat{j}) + (6\hat{i} - 7\hat{k}) + (2\hat{i} - 6\hat{j} + 10\hat{k})$

$= 10\hat{i} - \hat{j} + 3\hat{k}$

magnitude $= \sqrt{10^2 + 1^2 + 3^2} = 10.49$

2.3 e) $\vec{M} = \vec{r} \times \vec{F} = (4\hat{i} - 6\hat{j} + 4\hat{k}) \times (200\hat{i} + 400\hat{j})$

$M_y = 4 \times 200 = 800$ since $\hat{k} \times \hat{i} = \hat{j}$

2.4 c) $\vec{M} = \vec{r_1} \times \vec{F_1} + \vec{r_2} \times \vec{F_2}$

$= (2\hat{i} - 4\hat{k}) \times (50\hat{i} - 40\hat{k}) + (-4\hat{i} + 2\hat{j}) \times (60\hat{j} + 80\hat{k})$

$M_x = 2 \times 80 = 160$ since $\hat{j} \times \hat{k} = \hat{i}$

2.5 e) Concurrent $\Rightarrow$ all pass through a point.

Coplanar $\Rightarrow$ all in the same plane.

The forces are three-dimensional.

2.6 b) $\sum \vec{F} = 0.$ $\therefore \vec{R} + 141\hat{i} - 141\hat{j} - 200\hat{i} - 100\hat{k} = 0$

$\therefore \vec{R} = 59\hat{i} + 141\hat{j} + 100\hat{k}$

2.7 c) $\sum \vec{M} = 0.$ $\therefore \vec{M}_A + (4\hat{j} - 3\hat{k}) \times (-100\hat{k}) - 3\hat{k} \times (-200\hat{i}) + 4\hat{i} \times (141\hat{i} - 141\hat{j}) = 0$

$\therefore \vec{M}_A = 400\hat{i} - 600\hat{j} + 564\hat{k}$

2.8 b) They must be concurrent, otherwise a resultant moment would occur.

2.9 a) It is a two-force body.

2.10 b) $\sum M_A = 0.$ $F_B \times 8 = 400 \times 4 + 400 \times 6.$ $\therefore F_B = 500$ N

2.11 a) $M_A = 400 \times 8 + 400 \times 6 = 5600$ N·m

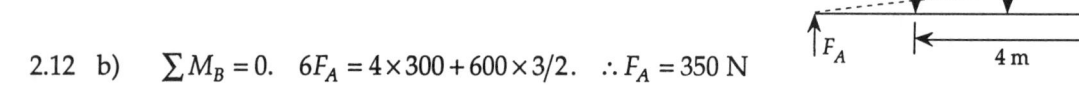

2.12 b) $\sum M_B = 0.$ $6F_A = 4 \times 300 + 600 \times 3/2.$ $\therefore F_A = 350$ N

2.13 c) $M_A = 0.6 \times 100 - 141 \times 0.6 + 141 \times 0.8 = 88.2$

Statics

2.14 a) $M = 100 \sin 45° \times 4 = 282.8$ cw

2.15 c) $\sum M_A = 0.$ $\therefore 6 \times 70.7 = 2 \times 0.866 F_1.$ $\therefore F_1 = 245$

$\sum F_x = 0.$ $\therefore -70.7 - 245 \times 0.5 + F_{Ax} = 0.$ $\therefore F_{Ax} = 193$

$\sum F_y = 0.$ $\therefore -70.7 - 245 \times 0.866 + F_{Ay} = 0.$ $\therefore F_{Ay} = 283$

$\therefore F_A = \sqrt{F_{Ax}^2 + F_{Ay}^2} = \sqrt{193^2 + 283^2} = 343$

2.16 e) $\sum M_A = 0.$ $\therefore 2F_B + 1.2 \times 200 - 141.4 \times 2 - 141.4 \times 1.2 + 50 = 0.$ $\therefore F_B = 81.2$

$\sum F_x = 0.$ $\therefore F_{Ax} - 200 + 141.4 = 0.$ $\therefore F_{Ax} = 58.6$

$\sum F_y = 0.$ $\therefore F_{Ay} + 81.2 - 141.4 = 0.$ $\therefore F_{Ay} = 60.2$

$\therefore F_A = \sqrt{F_{Ax}^2 + F_{Ay}^2} = \sqrt{58.6^2 + 60.2^2} = 84.0$

2.17 e) $\sum M_A = 0.$ $\therefore 500\ell + 200 \times 0.866\ell - F_C \times 2\ell = 0.$ $\therefore F_C = 337$

$0.866 F_{DC} = 337.$ $\therefore F_{DC} = 389$

$0.866 \times 389 = 0.866 F_{BD}.$ $\therefore F_{BD} = 389$

$-F_{DE} + 200 - 389 \times 0.5 - 389 \times 0.5 = 0.$ $\therefore F_{DE} = -189$

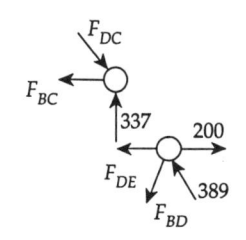

2.18 d) $\sum M_A = 0.$ $\therefore 5 \times 5000 = 10 \times F_C.$ $\therefore F_C = 2500$ $\therefore F_{DC} = 2500$

$0.707 F_{BD} = 2500.$ $\therefore F_{BD} = 3536$

$0.707 \times 3536 = F_{DE}.$ $\therefore F_{DE} = 2500$

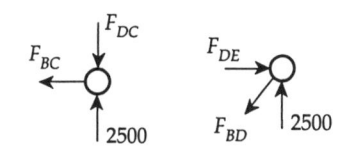

2.19 b) $\sum M_A = 0.$ $\therefore 4 \times 2000 + 6 \times 1000 = 8 F_C.$ $\therefore F_C = 1750$

$0.707 F_{DC} = 1750.$ $\therefore F_{DC} = 2475$

Sum forces in dir. of F_{DE}: $F_{DE} - 2475 + 1000 \times 0.707 = 0.$ $\therefore F_{DE} = 1768$

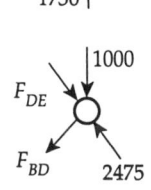

2.20 a) Sum forces in dir. of F_{FB} at F. $F_{FB} = 0$.

2.21 c) $\sum M_B = 0.$ $\therefore 12 F_F = 3 \times 4000.$ $\therefore F_F = 1000 \downarrow$

$\sum F_y = 0.$ $\therefore 0.8 F_{IC} = 1000.$ $\therefore F_{IC} = 1250$

2.22 d) Cut vertically through link KA. Then $F_{KA} = 5000$.

Obviously, $F_{AL} = 0$. $\therefore F_{AB} = 3000$. $\therefore F_{BC} = 3000$

2.23 e) At E we see that $F_{EC} = 0$. $\therefore$ At C, $F_{FC} = 0$

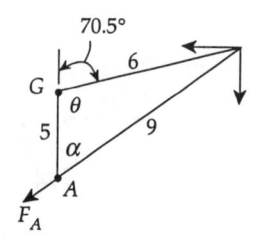

2.24 e) $9^2 = 6^2 + 5^2 - 2 \times 5 \times 6 \cos\theta$. $\therefore \theta = 109.5°$

$6^2 = 9^2 + 5^2 - 2 \times 9 \times 5 \cos\alpha$. $\therefore \alpha = 38.9°$

From pts E, C, F, B we see that $F_{EC} = F_{FC} = F_{FB} = F_{GB} = 0$.

Also, $F_A = F_{BC}$. $\sum M_G = 0$. $\therefore 5 \times F_A \sin 38.9° + 5000 \times 6 \sin 70.5° = 5000 \times 6 \cos 70.5°$. $\therefore F_A = -5817$

2.25 b) Recognize that link BC is a two-force member. $\sum M_A = 0$.

$\therefore 0.2 \times 1000 + 100 = 0.08 \times F_{BC} \times 0.8 + 0.2 \times F_{BC} \times 0.6$

$\therefore F_{BC} = 1630$. $A_x = 1630 \times 0.8 = 1304$. $A_y = 1630 \times 0.6 - 1000 = -22$

$\therefore F_A = \sqrt{1304^2 + 22^2} = 1304$ N

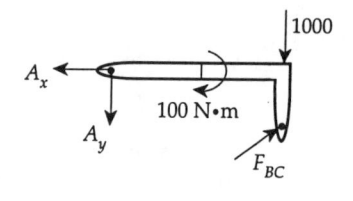

2.26 d) $1800 \times 0.8 = 0.6 F_{BC}$. $\therefore F_{BC} = 2400$

$0.3 \times 0.6w = 0.6 \times 2400$. $\therefore w = 8000$

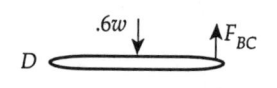

2.27 d) $\sum M_A = 0$. $\therefore 1.2 F_E = 0.8 \times 2400$. $\therefore F_E = 1600$

$\therefore A_x = 2400$. $A_y = 1600$

$\therefore F_A = \sqrt{2400^2 + 1600^2} = 2884$ N

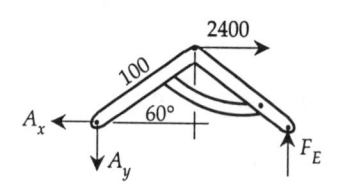

2.28 a) Link BD is a two-force member. $\therefore$ the force acts from D to B. Hence, the angles are found.

$120^2 = 100^2 + 100^2 - 2 \times 100 \times 100 \cos\beta$. $\therefore \beta = 73.7°$

$\overline{BD}^2 = 60^2 + 40^2 - 2 \times 60 \times 40 \cos 73.7°$. $\therefore BD = 62.1$

$\dfrac{62.1}{\sin 73.7°} = \dfrac{40}{\sin\phi}$. $\therefore \phi = 38.2°$. $\alpha = (180 - 73.7)/2 = 53.2°$

$\sum M_C = 0$. $1600 \times 100 \cos 53.2° = 60 \times F_{BD} \sin 38.2°$. $\therefore F_{BD} = 2587$

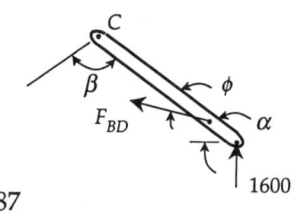

2.29 e) $\sum F_y = 0$. $N \times 0.866 - 980 - 0.4N \times 0.5 = 0$. $\therefore N = 1471$

$\sum F_x = 0$. $F = 1471 \times 0.5 + 0.4 \times 1471 \times 0.866 = 1245$

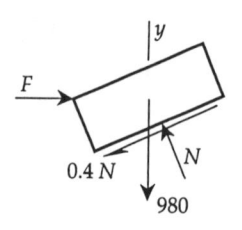

2.30 b) $N_1 = 490$. $N_2 = 980$. $\therefore F = 0.2(490 + 980) = 294$

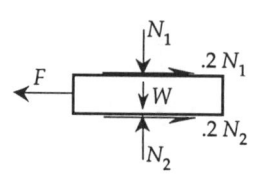

2.31 c) $\sum M_{\text{front wheel}} = 0$. $\therefore 400 N_2 - W \cos\theta \times 200 + W \sin\theta \times 50 = 0$

$\sum F_x = 0$. $\therefore 0.6 N_2 = W \sin\theta$. $\therefore 400(W \sin\theta)/0.6 + 50W \sin\theta = 200W \cos\theta$

$\therefore \dfrac{\sin\theta}{\cos\theta} = \dfrac{200}{716.7} = \tan\theta$. $\therefore \theta = 15.6°$

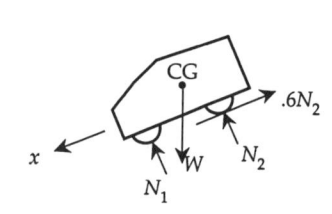

Statics

2.32 b) If $h < h_{min}$ then sliding occurs, and $F_f = 0.4N$.
If $h > h_{min}$ tipping occurs and $F_f < 0.4N$.
When $h = h_{min}$, $F_f = 0.4N = 0.4W = F$.
$\sum M_A = 0$. $\therefore 4W = hF = h \times 0.4W$. $\therefore h = 10$ cm

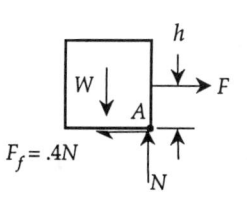

2.33 e) $\sum F_x = 0$. $\therefore N_2 = 0.4N_1$. Also, $W = 980$

$\sum M_A = 0$. $\therefore W \cdot r = (N_1 + 0.4N_1 + 2 \times 0.4N_2)r$. $\therefore N_1 = 0.5814W = 570$

$\sum F_y = 0$. $\therefore F = 980 - 570 - 0.16 \times 570 = 319$

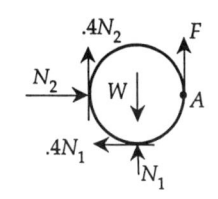

2.34 b) $\sum F_x = 0$. $\therefore N_2 = 0.4N_1$. $\sum F_y = 0$. $\therefore N_1 + 0.4N_2 = W$. $\therefore N_2 = 0.345W$

$\sum M_A = 0$. $\therefore \frac{L}{2} \times W \cos\theta = N_2 \times L\sin\theta + 0.4N_2 \times L\cos\theta$.

This gives $\tan\theta = 1.049$. $\therefore \theta = 46.4°$

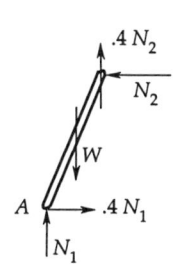

2.35 a) $F_B = F_D e^{-\mu\theta} = 800 e^{-0.5\pi} = 166$

2.36 a) $\sum M_A = 0$. $\therefore 200 \times 0.6 = 0.1 \times T_1 + 0.1 \times T_2$.

$T_1 = T_2 e^{0.4 \times 3\pi/2} = 6.59 T_2$. Thus, $T_2 = 158$ and $T_1 = 1042$.

$\sum M_{center} = 0$. $\therefore M = 0.1 \times (1042 - 158) = 88.4$ N·m

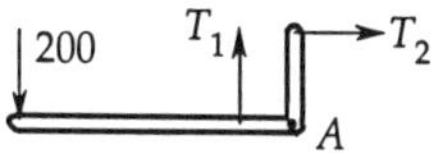

2.37 d) Let h = long end. m = mass/unit length. Then, $(12 - 1.88 - h)mg e^{0.5\pi} = hmg$. $\therefore h = 8.38$ m

2.38 d) $\bar{x} = \dfrac{\int_0^3 xy\,dx}{\int_0^3 y\,dx} = \dfrac{\int_0^3 x^3\,dx}{\int_0^3 x^2\,dx} = \dfrac{3^4/4}{3^3/3} = 2.25$

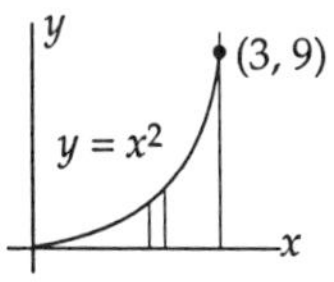

2.39 a) $\bar{y} = \dfrac{\int_0^3 \frac{y}{2} y\,dx}{\int_0^3 y\,dx} = \dfrac{\frac{1}{2}\int_0^3 x^4\,dx}{\int_0^3 x^2\,dx} = \dfrac{3^5/10}{3^3/3} = 2.7$

2.40 c) $\bar{x} = \dfrac{\int_0^1 (\sqrt{x} - x^2)x\,dx}{\int_0^1 (\sqrt{x} - x^2)\,dx} = \dfrac{\frac{1}{5/2} - \frac{1}{4}}{\frac{1}{3/2} - \frac{1}{3}} = 0.45$

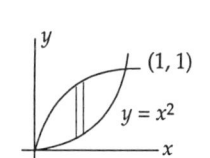

2.41 b) $\bar{y} = \dfrac{24 \times 3 + 6 \times 5}{6 \times 4 + 4 \times 3/2} = 3.4$

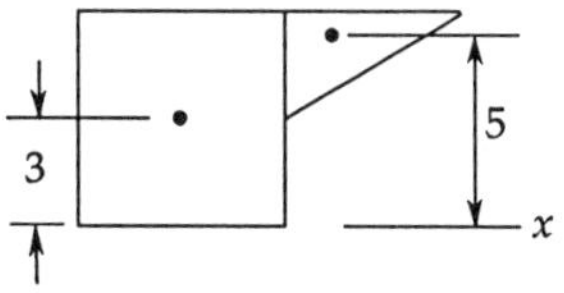

2.42 e) $\bar{y} = \dfrac{48 \times 3 + 12 \times 7 - \pi \times 6}{8 \times 6 + 3 \times 4 - \pi} = 3.68$

2.43 b) $\bar{x} = \dfrac{10 \times \frac{1}{2} + 5 \times 3.5 + 3 \times 7}{10 + 5 + 3} = 2.42$

2.44 b) $I_x = \int_0^3 y^3\, dx/3 = \int_0^3 x^6\, dx/3 = 3^7/21 = 104.1.$

With a horizontal strip: $I_x = \int_0^9 y^2(3-x)dy = \int_0^9 y^2(3-\sqrt{y})dy = 9^3 - \dfrac{9^{7/2}}{7/2} = 104.1$

2.45 e) $I_x = 8 \times 6^3/3 + \left(8 \times 3^3/36 + 12 \times 7^2\right) - \left(\pi \times 1^4/4 + \pi \times 6^2\right) = 1056$

2.46 a) $I_y = 12 \times 12^3/3 - \left(8 \times 8^3/12 + 64 \times 6^2\right) = 4267.$ Or, alternatively:

$I_y = 8 \times 2^3/3 + 4 \times 12^3/3 + 8 \times 2^3/12 + 16 \times 11^2 = 4267$

2.47 a) $I_{edge} = I_{c.g.} + Md^2 = \dfrac{1}{12}M(b^2 + b^2) + M\dfrac{b^2}{2} = \dfrac{2}{3}Mb^2$

2.48 e) $I_x = \dfrac{1}{3}(6m) \times 6^2 \times 2 + 8m \times 6^2 = 432$ with $m = 1$

3. Mechanics of Materials

by Merle C. Potter

The mechanics of materials is one of a number of names given to the study of deformable solids subjected to applied forces and moments. The foundations of this subject reside in three basic topics:

1. internal equilibrium (stress concepts)
2. geometry of deformation (strain concepts)
3. mechanical and thermal properties (by which stress and strain are related)

Additionally, we assume homogeneity (properties are independent of position) and isotropy (absence of directional properties) in the materials considered.

3.1 Stress and Strain, Elastic Behavior

Consider a prismatic bar of length L and a cross sectional area A situated along the x-axis as shown in Fig. 3.1a. Let the bar be subjected to a constant axial force P applied at the centroids of the end faces so as to stretch the bar by an amount δ, Fig. 3.1b.

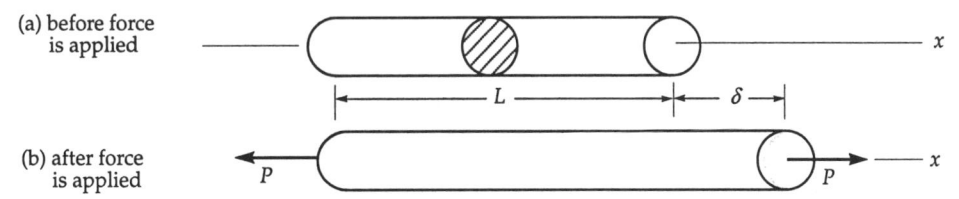

Figure 3.1 Normal stress and strain.

We define the *longitudinal*, or *normal strain* ε_x by the ratio

$$\varepsilon_x = \delta/L. \tag{3.1.1}$$

Strain is dimensionless, having the units of m/m, or in/in, etc. Normal strains are positive if due to elongation, negative if the result is shortening. The stress

$$\sigma_x = P/A \tag{3.1.2}$$

is the *normal*, or *axial stress* in the bar. Stress has units of N/m^2, lbs/in^2 (psi) or $kips/in^2$ (ksi). One Newton per square meter is called a *Pascal*, abbreviated Pa. Note that the stress and strain defined here are averages, constant over the length of the bar, and uniform over its cross section. By contrast, for a tapered bar hanging from the ceiling under its own weight, the stress and strain would vary along the bar. Also, it is natural that a positive longitudinal strain will be accompanied by a negative lateral strain. Indeed, this ratio

$$\nu = -\frac{\text{lateral strain}}{\text{longitudinal strain}} \qquad (3.1.3a)$$

is called *Poisson's ratio*, an important property of a given material. Thus, for the bar in Fig. 3.1,

$$\varepsilon_y = \varepsilon_z = -\nu\varepsilon_x. \qquad (3.1.3b)$$

If the bar in Fig. 3.1 is made of a *linear elastic* material, its axial stress and strain are related by the formula, often called *Hooke's law*,

$$\sigma_x = E\varepsilon_x \qquad (3.1.4)$$

where E is a material constant called *Young's modulus*, or the *modulus of elasticity*. The units of E are the same as those of stress. By inserting Eqs. 3.1.1 and 3.1.2 into Eq. 3.1.4 and solving for δ we obtain the useful formula

$$\delta = \frac{PL}{AE}. \qquad (3.1.5)$$

Stress is always accompanied by strain, but strain may occur without stress. In particular, a temperature change in an unconstrained bar will cause it to expand (or shrink) inducing a thermal deformation

$$\delta_t = \alpha L(T - T_o) \qquad (3.1.6)$$

where α is the *coefficient of thermal expansion*, and $(T - T_o)$ the temperature change. Typical units of α are meters per meter per degree Celsius (°C^{-1}). Important properties of several materials are listed in Table 3.1.

Table 3.1 Average Material Properties.

	Modulus of Elasticity E		Shear Modulus G		Poisson's Ratio ν	Density ρ		Coefficient of Thermal Expansion α × 10	
	× 10^6 kPa	× 10^6 psi	× 10^6 kPa	× 10^6 psi		kg/m^3	lb/ft^3	°C^{-1}	°F^{-1}
steel	210	30	83	12	.28	7850	490	11.7	6.5
aluminum	70	10	27	3.9	.33	2770	173	23.0	12.8
magnesium	45	6.5	17	2.4	.35	1790	112	26.1	14.5
cast iron	140	20	55	8	.27	7080	442	10.1	5.6
titanium	106	15.4	40	6	.34	4520	282	8.8	4.9
brass	100	15	40	6	.33	8410	525	21.2	11.8
concrete	20	3	—	—	—	2400	150	11.2	6.2

Next consider a material cube subjected to a pair of equilibrating couples acting in the plane of the faces of the cube as shown pictorially in Fig. 3.2a, and schematically in Fig. 3.2b. For a cube whose faces have an area A we define the *shear stress* in a plane parallel to those on which the forces act as

$$\tau_{xy} = F/A. \qquad (3.1.7)$$

If the material of the cube is linearly elastic, the top will be displaced relative to the bottom as shown in Fig. 3.2c. The angle γ_{xy} measures the *shear strain* of the cube, and since for elastic behavior this angle is very small, we define the shear strain as

$$\gamma_{xy} \approx \tan \gamma_{xy} = \Delta x/h. \qquad (3.1.8)$$

Also, for elastic behavior,

$$\tau_{xy} = G\gamma_{xy} \qquad (3.1.9)$$

where G is the *shear modulus*, or *modulus of rigidity*, having the units of Pa or psi. It is related to E and ν by

$$G = \frac{E}{2(1+\nu)}. \qquad (3.1.10)$$

There are only two independent material properties in an isotropic, elastic solid. These properties may depend on position as in a heat-treated steel.

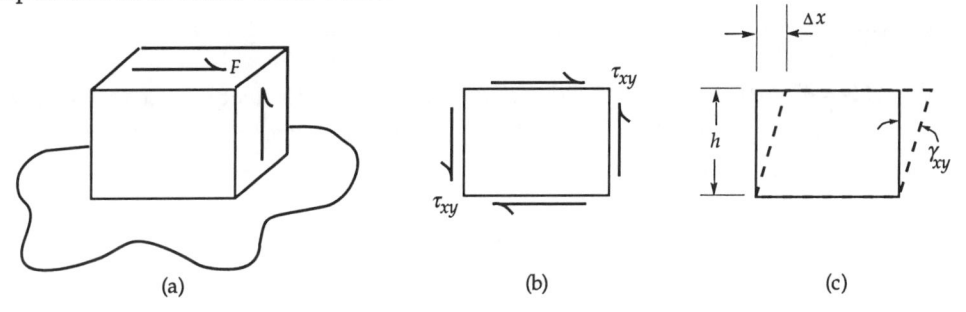

Figure 3.2 Shear stress and strain.

--- EXAMPLE 3.1 ---

A 2 cm × 2 cm square aluminum bar AB supported by 1.25 cm diameter steel cable BC carries a 7000 N load in the arrangement shown. Determine the stresses in the steel and in the aluminum. Also, calculate the elongation of the cable and the shortening of the bar.

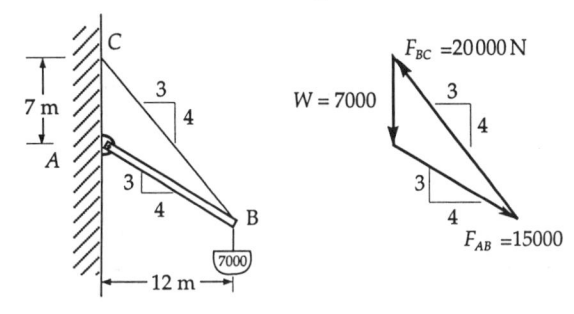

Solution. As shown by the force polygon for the equilibrium, $F_{AB} = -15\,000$ N and $F_{BC} = 20\,000$ N. Thus there results

$$\sigma_{al} = \frac{P}{A} = -\frac{15\,000}{0.02 \times 0.02} = -37.5 \times 10^6 \text{ Pa} \quad \text{or} \quad -37.5 \text{ MPa}.$$

$$\sigma_{st} = \frac{P}{A} = \frac{20\,000}{\pi(0.0125)^2/4} = 163 \times 10^6 \text{ Pa} \quad \text{or} \quad 163 \text{ MPa}.$$

From Table 3.1 and Eq. 3.1.5,

$$\delta_{al} = \frac{PL}{AE} = \frac{-15\,000(15)}{0.02 \times 0.02 \times 70 \times 10^9} = -0.00804 \text{ m} \quad \text{or} \quad -8.04 \text{ mm}.$$

$$\delta_{st} = \frac{PL}{AE} = -\frac{20\,000(20)}{(\pi \times 0.0125^2/4) \times 210 \times 10^9} = 0.0155 \text{ m} \quad \text{or} \quad 15.5 \text{ mm}.$$

--- EXAMPLE 3.2 ---

A 5-cm-dia, 80-cm-long steel bar is restrained from moving. If its temperature is increased $100°C$, what compressive stress is induced?

Solution. The strain can be calculated using the deformation of Eq. 3.1.6:

$$\varepsilon = \frac{\delta_t}{L} = \alpha(T - T_o)$$
$$= 11.7 \times 10^{-6} \times 100 = 11.7 \times 10^{-4}.$$

Hence, the induced stress is

$$\sigma = \varepsilon E = 11.7 \times 10^{-4} \times 210 \times 10^{6} = 246\ 000 \text{ kPa}.$$

EXAMPLE 3.3

The steel block ($G = 83 \times 10^6$ kPa) is welded securely to a horizontal platen and subjected to 1000 kN horizontal force as shown. Determine the shear stress in a typical horizontal plane of the block, and the horizontal displacement of the top edge AB.

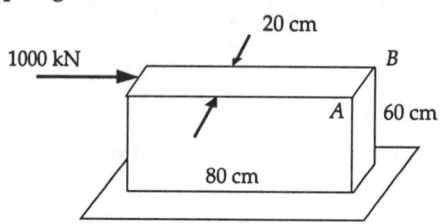

Solution. From Eq. 3.1.7

$$\tau = \frac{F}{A} = \frac{1000}{(0.8)(0.2)} = 6250 \text{ kPa}.$$

From Eq. 3.1.9

$$\gamma_{xy} = \frac{\tau_{xy}}{G} = \frac{6250}{83 \times 10^6} = 7.53 \times 10^{-5} \text{ rad}.$$

From Eq. 3.1.8 the horizontal displacement is

$$\Delta x = h\gamma_{xy} = 600(7.53 \times 10^{-5}) = 0.0452 \text{ mm}.$$

3.2 Torsion

A straight member of constant circular cross section subjected to a twisting couple at each end is said to be in *torsion*, and such a member is called a *shaft*. For an elastic shaft of length L and radius c subjected to a *torque* T (pair of equilibrium couples), as shown in Fig. 3.3a, the angular displacement of one end relative to the other is given by the angle ϕ (in radians) as

$$\phi = \frac{TL}{JG} \qquad (3.2.1)$$

where $J = \pi a^4/2$ is the *polar moment of inertia* of the circular cross section. Also, the *torsional shear stress* at the radial distance r from the axis of the shaft in a given cross section will be

$$\tau = \frac{Tr}{J} \qquad (3.2.2)$$

which increases linearly as shown in Fig. 3.3b. Thus, the maximum shear stress occurs at $r = a$,

$$\tau_{max} = \frac{Ta}{J}. \qquad (3.2.3)$$

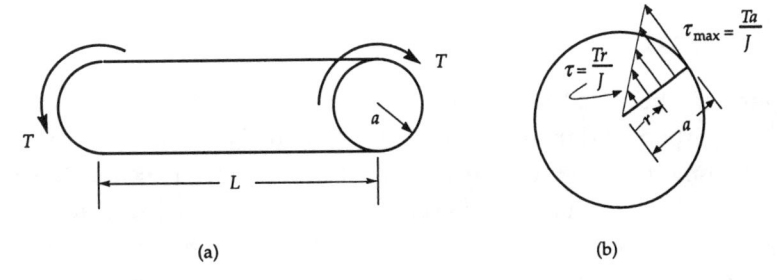

Figure 3.3 Circular shaft subject to a torque.

For a hollow shaft having an inner radius a_i and an outer a_o the above formulas are all valid, but with

$$J = \pi(a_o^4 - a_i^4)/2. \qquad (3.2.4)$$

--- **EXAMPLE 3.4** ---

A 6 cm diameter, 2 m long magnesium ($G = 17 \times 10^9$ Pa) shaft is welded to a hollow ($c_o = 3$ cm and $c_i = 1.5$ cm) aluminum ($G = 27 \times 10^9$ Pa) shaft 1.2 m long. A moment of 2000 m·N is applied at end A. Determine the maximum torsional stress in each material and the angle of twist of end A relative to fixed end B.

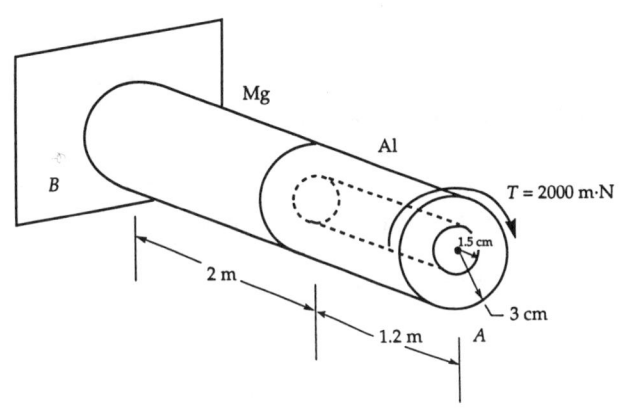

Solution. The polar moments of inertia are

$$J_{mg} = \pi a^4/2 = \pi(0.03)^4/2 = 1.272 \times 10^{-6} \text{ m}^4.$$

$$J_{al} = \pi(a_o^4 - a_i^4)/2 = \pi(0.03^4 - 0.015^4)/2 = 1.193 \times 10^{-6} \text{ m}^4.$$

From Eq. 3.2.3

$$\tau_{mg} = \frac{Ta}{J} = \frac{2000 \times 0.03}{1.272 \times 10^{-6}} = 47.17 \times 10^6 \text{ Pa}.$$

$$\tau_{al} = \frac{2000 \times 0.03}{1.193 \times 10^{-6}} = 50.29 \times 10^6 \text{ Pa}.$$

From Eq. 3.2.1 the angle of twist is

$$\phi = \phi_{mg} + \phi_{al} = \left(\frac{TL}{JG}\right)_{mg} + \left(\frac{TL}{JG}\right)_{al}$$

$$= \frac{2000 \times 2}{1.272 \times 10^{-6} \times 17 \times 10^9} + \frac{2000 \times 1.2}{1.193 \times 10^{-6} \times 27 \times 10^9}$$

$$= 0.1850 + 0.0745 = 0.2595 \text{ rad}.$$

3.3 Beam Theory

The usual geometry of a beam is that of a member having the length much larger than the depth with the forces applied perpendicular to this long dimension. The beams considered here have a longitudinal plane of symmetry in which the forces act and in which beam deflections occur. To illustrate we consider a T shaped beam having the cross section shown in Fig. 3.4b, supported either as a cantilever beam, Fig. 3.4a, or as a simply-supported beam, Fig. 3.4c. The longitudinal axis of the beam (x-axis here) passes through the centroidal points of all cross sections. The xy-plane is the plane of symmetry. Any combination of concentrated and distributed loads may act on the beam.

(a) cantilever beam   (b) cross section   (c) simply-supported beam

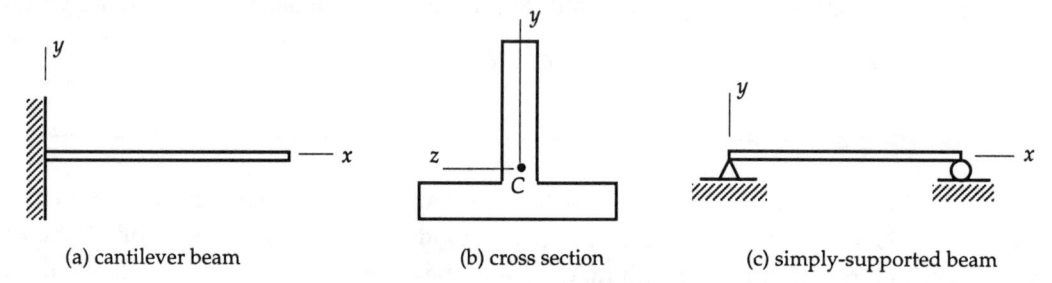

Figure 3.4 Beam geometry.

At the typical cross section of the loaded beam there is an internal force V called the *shear force*, and an internal moment M called the *bending moment*. Both V and M may be determined by a free-body diagram of the left hand portion of the beam, and are, in general, functions of x as shown in Fig. 3.5, where positive values of V and M are displayed. A plot of $V(x)$ is called a *shear diagram* and a plot of $M(x)$ is a *moment diagram*. At a given cross section where the moment has the value M, the (longitudinal) bending stress acting normal to the cross section is

$$\sigma_x = -\frac{My}{I} \tag{3.3.1}$$

where I is the plane moment of inertia of the cross sectional area relative to the centroidal axis. The minus sign is needed to assure a compressive stress for positive y values when the moment M is positive. The stress is a linear function of y as is shown in Fig. 3.6b, with the maximum compression occurring at the top of the beam, and the maximum tension at the bottom for the positive M. The bending stress is zero at $y = 0$, the so-called *neutral axis*.

The stress due to the shear force V is a vertical shear stress

$$\tau_{xy} = \frac{VQ}{Ib} \tag{3.3.2}$$

where b is breadth, or thickness of the beam at the position (y coordinate) at which the shear stress is calculated. The symbol Q stands for the *statical moment* about the neutral axis of the area between the position and the top of the beam. For a rectangular beam τ_{xy} is parabolic. For the T beam it has the shape shown in Fig. 3.6c. In both cases the maximum shear stress occurs at the neutral axis (centroidal position).

$$V = \frac{dM}{dx}$$
$$\Delta M = \int V dx$$

(a) cantilever beam   (b) simply-supported beam

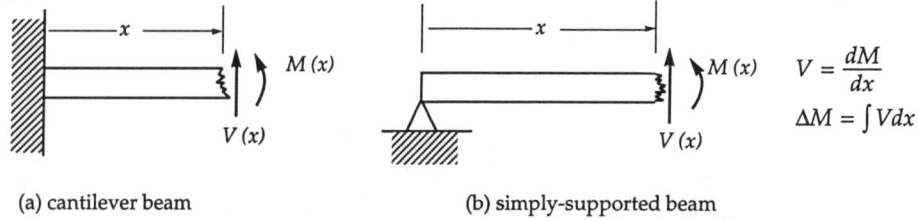

Figure 3.5 Internal shear force and bending moment.

Mechanics of Materials

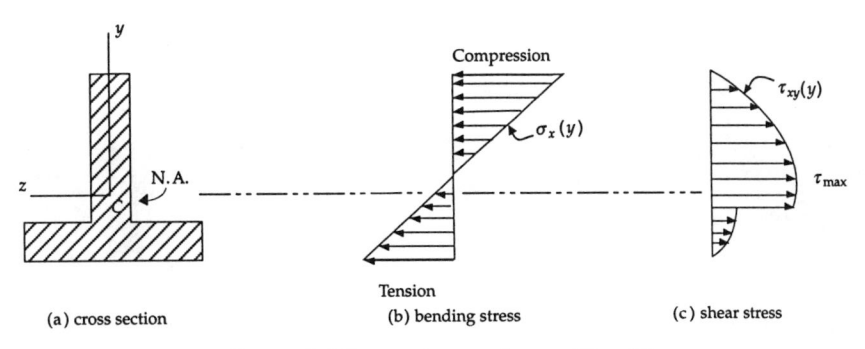

Figure 3.6 Beam stresses for positive M.
(a) cross section (b) bending stress (c) shear stress

The vertical displacement of the x-axis of a loaded beam measures the beam deflection. The curve of this deflection $v = v(x)$ is called the *equation of the elastic line*, shown in Figure 3.7. Also, the slope $\theta = \theta(x) = dv/dx$ of the deflection curve is an important quantity in beam theory. Table 3.2 lists some useful formulas for basic beams.

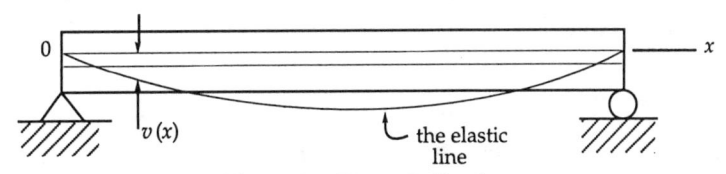

Figure 3.7 Beam deflection.

TABLE 3.2 Beam Formulas.

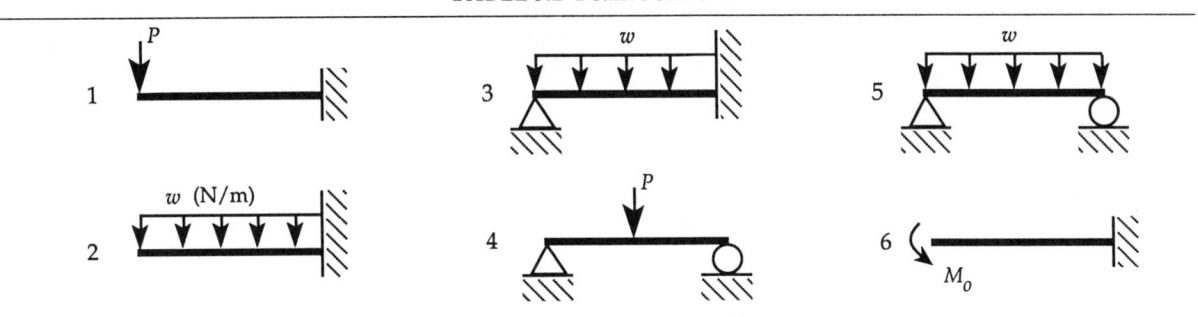

Number	Max Shear	Max Moment	Max Deflection	Max Slope
1	P	PL	$PL^3/3EI$	$PL^2/2EI$
2	wL	$wL^2/2$	$wL^4/8EI$	$wL^3/6EI$
3	$5wL/8$	$wL^2/8$	$wL^4/185EI$	
4	$P/2$	$PL/4$	$PL^3/48EI$	$PL^2/16EI$
5	$wL/2$	$wL^2/8$	$5wL^4/384EI$	$wL^3/24EI$
6		M_o	$M_o L^2/2EI$	$M_o L/EI$

─────── **EXAMPLE 3.5** ───────

Sketch the shear and moment diagrams for the beam shown.

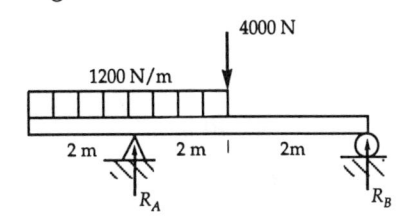

Solution.

First, determine the support reactions R_A and R_B:

$$\Sigma M_B = 4R_A - 4000(2) - 4(1200)4 = 0$$

$$\therefore R_A = 6800 \text{ N}$$

$$\Sigma M_A = 4R_B - 4000(2) = 0$$

$$\therefore R_B = 2000 \text{ N}$$

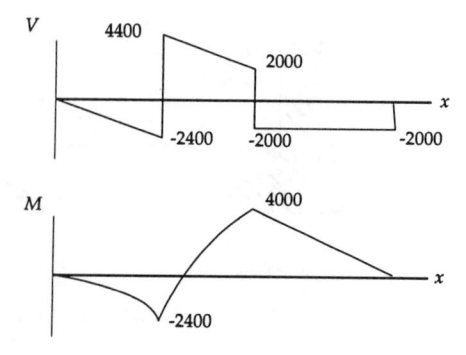

The values of V and M as functions of x are shown in the sketches; V is the resultant of all forces acting on the portion of the beam to the left of the x-location. The change in moment is the area under the shear diagram, $\Delta M = \int V dx$. Note from the diagrams that the maximum positive and negative moments occur at locations where the shear plot crosses the x-axis; the values of -2400 and 4000 are simply the appropriate areas under the shear diagram. Note that $M = 0$ at both ends and $V = 0$ at the left end.

──────── **EXAMPLE 3.6** ────────

If the beam of Example 3.5 has the cross section shown below, determine the maximum tensile and compressive stresses, and the maximum shear stress.

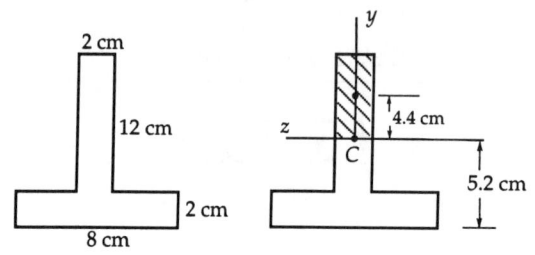

Solution. First, we locate the neutral axis by determining C relative to the bottom of the beam, as shown above:

$$\bar{y} = \frac{\Sigma y_i A_i}{\Sigma A_i} = \frac{16(1) + 24(8)}{16 + 24} = 5.2 \text{ cm}.$$

The moment of inertia is (use $I = \bar{I} + Ad^2$)

$$I = \frac{8(2^3)}{12} + 16(4.2)^2 + \frac{2(12)^3}{12} + 24(2.8)^2$$

$$= 763.7 \text{ cm}^4.$$

The maximum positive M is 4000 N·m at $x = 4$, so the maximum compressive stress is (intuitively, we can visualize compression in the top fibers under the 4000 N force)

$$\left(\sigma_c\right)_{max} = \frac{My}{I} = \frac{4000(0.088)}{764 \times 10^{-8}} = 46.1 \times 10^6 \text{ Pa}.$$

The maximum negative M is 2400 N·m at $x = 2$, so the maximum tensile stress is (intuitively, we know that tension occurs in the top fibers to the left of R_A)

$$\left(\sigma_t\right)_{max} = \frac{My}{I} = \frac{2400(0.088)}{764 \times 10^{-8}} = 27.6 \times 10^6 \text{ Pa}.$$

The maximum V is 4400 N. The moment of the shaded area with respect to the neutral axis is Q. Therefore,

$$Q = \bar{y}A = 0.044 \times (0.02 \times 0.088) = 7.744 \times 10^{-5} \text{ m}^3.$$

Since $b = 0.02$ at $y = 0$, the maximum shear stress is

$$\tau_{max} = \frac{VQ}{Ib} = \frac{4400 \times 7.744 \times 10^{-5}}{764 \times 10^{-8} \times 0.02} = 2.23 \times 10^6 \text{ Pa}.$$

EXAMPLE 3.7

Determine the maximum deflection of a 3 cm × 24 cm rectangular aluminum beam, 5 m long, if a concentrated load of 800 N acts downward at its mid-point.

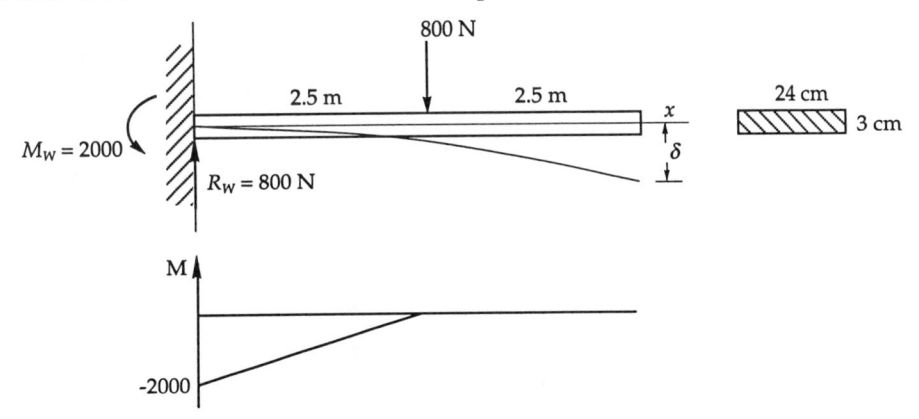

Solution. As the moment diagram shows, M is zero for the right-hand half of the beam. The right-hand half remains straight, but is inclined at the slope of the beam at mid-point. The left hand half is a simple end-loaded cantilever. From Table 3.2 with $L = 2.5$ m,

$$\delta = \delta_{middle} + \theta L = \frac{PL^3}{3EI} + \frac{PL^2}{2EI}(L) = \frac{5PL^3}{6EI}$$

$$= \frac{5(800)(2.5)^3}{6 \times (70 \times 10^9)(0.24)0.03^3 / 12} = 0.276 \text{ m}$$

3.4 Combined Stress

It often happens that structural members are simultaneously subjected to some combination of axial, torsional and bending loads. In such cases the state of stress at points on the surface of the member consists of both normal and shear components, and is called *combined stress*. At any given point of interest on the surface we introduce a local set of coordinate axes and focus attention on the stresses acting on a very small rectangular element of material at the same point P as shown by Fig. 3.8a. For an element aligned with a rotated set of $x'y'$-axes at the same point, located by the c.c.w. angle θ, shown in Fig. 3.8b, the primed stresses will differ from the original unprimed stresses, the relationship being a function of θ. At a certain angle θ_p, with which we associate the axes x^* and y^*, Fig. 3.8c, the normal stresses will reach their maximum (x^*-direction) and minimum (y^*-direction) values, while the shear stresses vanish. These axes are called *principal axes of stress*, and the values σ_1 and σ_2 are called the *principal stresses*. It turns out that

$$\sigma_1 = \sigma_{max} = \frac{\sigma_x + \sigma_y}{2} + \sqrt{\left(\frac{\sigma_x - \sigma_y}{2}\right)^2 + \tau_{xy}^2} \qquad (3.4.1a)$$

$$\sigma_2 = \sigma_{min} = \frac{\sigma_x + \sigma_y}{2} - \sqrt{\left(\frac{\sigma_x - \sigma_y}{2}\right)^2 + \tau_{xy}^2} \qquad (3.4.1b)$$

and the angle θ_p at which they occur is calculated from

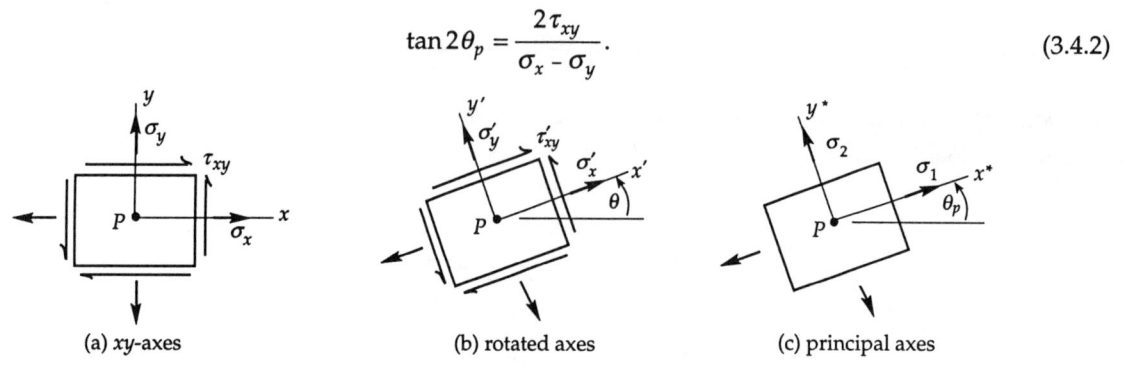

$$\tan 2\theta_p = \frac{2\tau_{xy}}{\sigma_x - \sigma_y}. \qquad (3.4.2)$$

(a) xy-axes (b) rotated axes (c) principal axes

Figure 3.8 State of stress.

As stated, relative to the x^*y^*-axes the shear stresses are zero. The maximum shear stress occurs with respect to axes rotated 45° relative to the principal axes; its value is

$$\tau_{max} = \frac{\sigma_1 - \sigma_2}{2} = \sqrt{\left(\frac{\sigma_x - \sigma_y}{2}\right)^2 + \tau_{xy}^2} \qquad (3.4.3)$$

A graphical method is often used when obtaining stresses on a particular plane; it utilizes *Mohr's circle*. Mohr's circle is sketched by locating both ends of a diameter, whose center is always on the horizontal axis. For the stress state of Fig. 3.8a, which shows positive stresses, we plot the diameter ends as $(\sigma_x, -\tau_{xy})$ and (σ_y, τ_{xy}). The stresses on any plane, oriented at an angle θ with respect to the stresses on any known plane, are then the coordinates of a point on Mohr's circle located an angle 2θ from the known point. Once Mohr's circle is sketched, it is relatively obvious that the maximum shear

(a) Mohr's circle (b) maximum normal stress (c) maximum shear stress

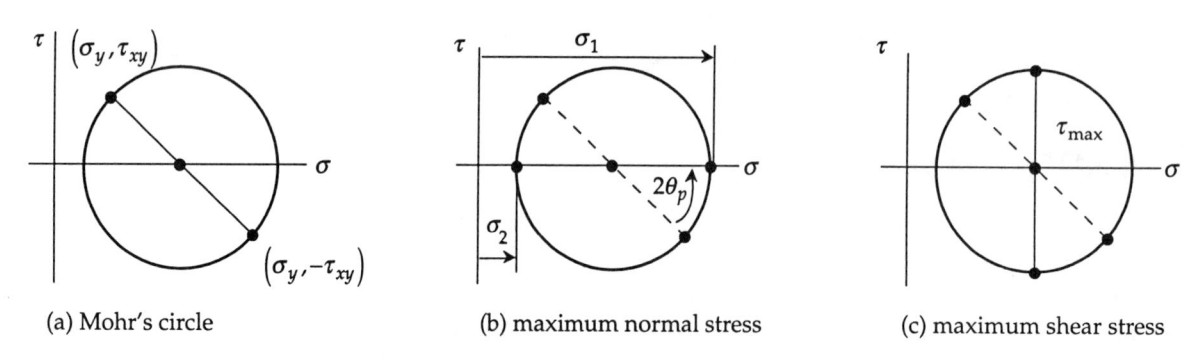

Figure. 3.9 Mohr's circle and maximum stresses.

stress is the circle's radius, and the maximum normal stress is the circle's radius plus $(\sigma_x + \sigma_y)/2$, as observed in Fig. 3.9a. These are, in fact, equivalent to the formulas in the above equations. Rather than refer to the formulas, we can simply sketch Mohr's circle and easily find τ_{max} and σ_1, the quantities often of interest, since they may lead to failure.

EXAMPLE 3.8

A solid circular shaft of radius 5 cm and length 3 m has a 2 m rigid bar welded to end A, and is "built in" to the vertical wall at B. A load of 8 kN acts at end C, and an axial force of 80 kN compresses the shaft as shown. Determine the maximum normal, and maximum shear stress at point P on the top of the shaft, midway between A and B.

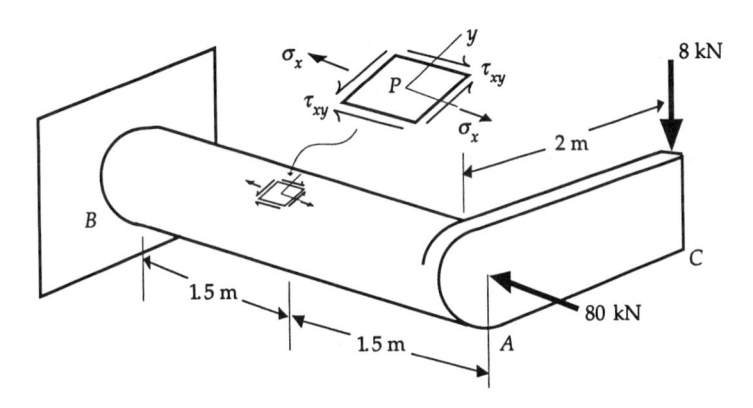

Solution. The torque on the shaft is $T = (8000)(2) = 16\,000$ N·m. The torsional shear stress on the element at P is

$$\tau_{xy} = \frac{Tc}{J} = \frac{16\,000(0.05)}{\pi(0.05)^4/2} = 81.5 \times 10^6 \text{ Pa} \quad \text{or} \quad 81.5 \text{ MPa}.$$

The axial compressive stress is

$$\sigma_x = \frac{F}{A} = \frac{-80\,000}{\pi(0.05)^2} = -10.2 \times 10^6 \text{ Pa} \quad \text{or} \quad -10.2 \text{ MPa}.$$

The tensile bending stress on the element at P (also a σ_x stress) is

$$\sigma_x = \frac{Mc}{I} = \frac{8000(1.5)(0.05)}{\pi(0.05)^4/4} = 122.2 \times 10^6 \text{ Pa} \quad \text{or} \quad 122.2 \text{ MPa}.$$

Thus, the stress components on the element at P (in MPa) are

$$\sigma_x = 122.2 - 10.2 = 112, \quad \sigma_y = 0, \quad \tau_{xy} = 81.5$$

so that from Eq. 3.4.1a the maximum normal stress is

$$\sigma_{max} = \frac{112+0}{2} + \sqrt{\left(\frac{112-0}{2}\right)^2 + (81.5)^2} = 155 \text{ MPa}.$$

From Eq. 3.4.3 the maximum shear stress is

$$\tau_{max} = \sqrt{\left(\frac{112-0}{2}\right)^2 + (81.5)^2} = 98.9 \text{ MPa}.$$

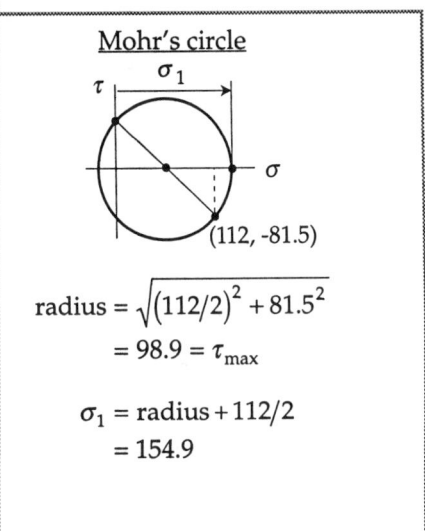

Mohr's circle

(112, −81.5)

$$\text{radius} = \sqrt{(112/2)^2 + 81.5^2}$$
$$= 98.9 = \tau_{max}$$

$$\sigma_1 = \text{radius} + 112/2$$
$$= 154.9$$

As another case illustrating the ideas of combined stress, let us consider a cylindrical vessel of inside diameter D and wall thickness t (with $t/D \ll 0.1$) containing a fluid under a pressure p, and subjected to a torque T as shown in Fig. 3.9. We consider the stresses acting upon a small element of the wall having sides parallel and perpendicular, respectively, to the axis of the cylinder. By sectioning the cylinder perpendicular to its axis at the element, we find from axial equilibrium that the *longitudinal stress*, also called *axial stress*, in the wall is

$$\sigma_a = \frac{pD}{4t}. \tag{3.4.4}$$

Similarly by sectioning lengthwise through the axis, radial equilibrium requires the *circumferential stress*, also called *hoop stress*, to be

$$\sigma_t = \frac{pD}{2t}. \tag{3.4.5}$$

And finally, from torsional equilibrium about the vessel's axis, the shear stress in the wall is

$$\tau = \frac{Tr}{J} \tag{3.4.6}$$

where $J = 2\pi r^3 t$, the approximate polar moment of inertia. From these formulas we may calculate the maximum normal and shear stresses as a problem in combined stress.

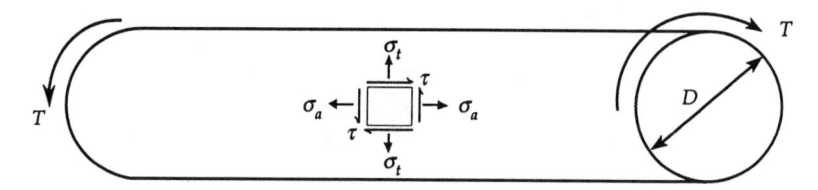

Figure 3.10 Pressurized cylinder under torque.

Finally, we note that in the absence of the torque T in Fig. 3.10, the maximum tensile stress is σ_t, the minimum tensile stress is σ_a, and the maximum shear stress is

$$\tau_{max} = \frac{\sigma_t - \sigma_a}{2}. \tag{3.4.7}$$

Also, for a thin-walled spherical container under pressure p the normal stress in the wall is

$$\sigma = \frac{pD}{4t} \tag{3.4.8}$$

in every direction, and the shear stress in the wall is zero everywhere.

EXAMPLE 3.9

A cylindrical tank of radius 40 cm and wall thickness 3 mm is subjected to an internal pressure of 2 MPa, and a torque of 0.5 MN·m. Determine the maximum normal and shear stresses in the cylinder wall.

Solution. From Eq. 3.4.4 the longitudinal stress is

$$\sigma_\ell = \frac{pr}{2t} = \frac{2(0.4)}{2(0.003)} = 133 \text{ MPa.}$$

From Eq. 3.4.5 the circumferential stress is

$$\sigma_c = \frac{pr}{t} = \frac{(2)(0.4)}{(0.003)} = 267 \text{ MPa.}$$

From Eq. 3.4.6 the shear stress is

$$\tau_{xy} = \frac{Tr}{J} = \frac{(0.5)(0.4)}{2\pi(0.4)^3(0.003)} = 166 \text{ MPa.}$$

Thus, from Eq. 3.4.1a

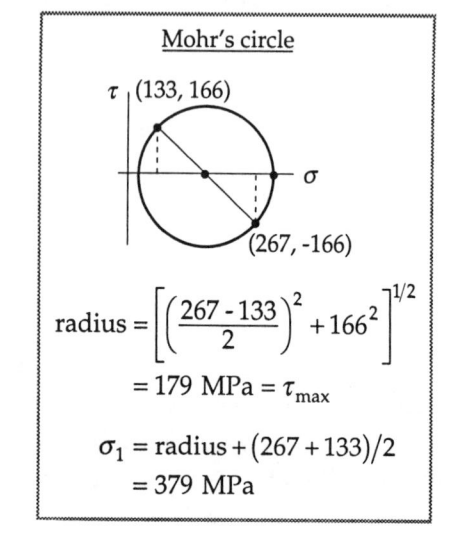

Mohr's circle

τ (133, 166)

(267, -166)

$$\text{radius} = \left[\left(\frac{267 - 133}{2}\right)^2 + 166^2\right]^{1/2}$$
$$= 179 \text{ MPa} = \tau_{max}$$

$$\sigma_1 = \text{radius} + (267 + 133)/2$$
$$= 379 \text{ MPa}$$

$$\sigma_{max} = \frac{133+267}{2} + \sqrt{\left(\frac{133-267}{2}\right)^2 + (166)^2} = 379 \text{ MPa}$$

and from Eq. 3.4.3

$$\tau_{max} = \sqrt{\left(\frac{133-267}{2}\right)^2 + (166)^2} = 179 \text{ MPa}.$$

3.5 Composite Bars and Beams

Consider a member composed of several parallel portions, each of a particular material, securely bonded together and loaded axially. As an example, we show in Fig. 3.11 a composite bar of three materials subjected through rigid and parallel end plates to an axial force P. Let the portion of the bar have cross-sectional areas A_1, A_2 and A_3, as well as moduli of elasticity E_1, E_2, and E_3, respectively. Furthermore, let $E_1 \leq E_2 \leq E_3$ and form the ratios

$$m = E_2/E_1 \quad \text{and} \quad n = E_3/E_1. \tag{3.5.1}$$

Since the axial deformation is the same for each material, Hooke's law requires

$$\sigma_2 = m\sigma_1 \quad \text{and} \quad \sigma_3 = n\sigma_1 \tag{3.5.2}$$

and also that $\sigma_1 = P/A_T$ where A_T is the "transformed area" such that

$$\sigma_1 = \frac{P}{A_T} = \frac{P}{A_1 + mA_2 + nA_3}. \tag{3.5.3}$$

The generalization to a bar of any number of materials is obvious.

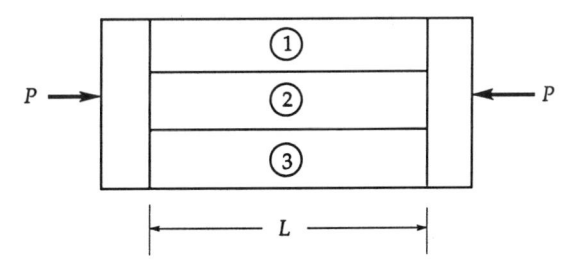

Figure 3.11 Composite bar.

──────── **EXAMPLE 3.10** ────────

Let a composite bar made of aluminum, steel, and brass be subjected to an axial load of 500 kN. Determine the stress in each material if $E_{al} = 70$ GPa, $E_{st} = 210$ GPa, and $E_{br} = 105$ GPa, together with $A_{al} = 0.04 \text{ m}^2$, $A_{st} = 0.006 \text{ m}^2$, and $A_{br} = 0.08 \text{ m}^2$.

Solution. From Eq. 3.5.1, with reference to Fig. 3.10,

$$m = \frac{E_{br}}{E_{al}} = \frac{105}{70} = 1.5.$$

$$n = \frac{E_{st}}{E_{al}} = \frac{210}{70} = 3.0.$$

From Eq. 3.5.3 we have

$$\sigma_{al} = \frac{P}{A_1 + mA_2 + nA_3} = \frac{500 \times 10^3}{0.04 + 1.5(0.08) + 3(0.006)} = 2.8 \text{ MPa}.$$

From Eq. 3.5.2 there results

$$\sigma_{br} = m\sigma_1 = 1.5(2.8) = 4.2 \text{ MPa}.$$

$$\sigma_{st} = n\sigma_1 = 3.0(2.8) = 8.4 \text{ MPa}.$$

For a beam having a composite section and subjected to a bending moment M we again determine the ratios $E_2/E_1 = m$, $E_3/E_1 = n$, etc., and from them construct a "transformed cross section" by multiplying the width of each material by the corresponding ratio. We then determine the centroid of the transformed cross section, and calculate the moment of inertia I_T about the neutral axis of the transformed section. Thus,

$$\sigma_1 = -\frac{My}{I_T} \tag{3.5.4}$$

with $\sigma_2 = m\sigma_1$ and $\sigma_3 = n\sigma_1$, etc. An example illustrates the method.

EXAMPLE 3.11

Let a composite steel and aluminum beam having the cross section shown be subjected to a positive bending moment of 90 N·m. Determine the maximum bending stress in each material.

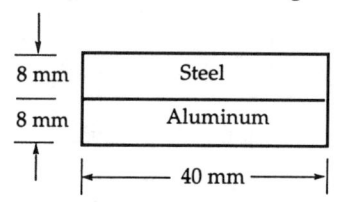

Solution. Here $m = E_s/E_a = 210/70 = 3$ so that the area has the geometry shown below with the neutral axis (N.A.) calculated to be 10 mm from the bottom.

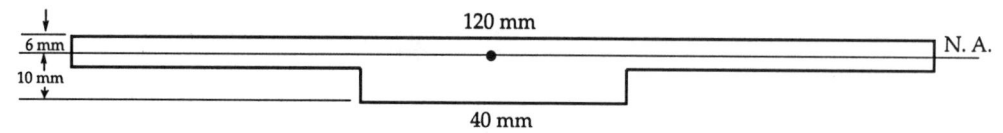

The moment of inertia about the N.A. is

$$I_T = \frac{(0.12)(0.006)^3}{3} + \frac{(0.04)(0.01)^3}{3} + \frac{(0.08)(0.002)^3}{3}$$

$$= 22 \times 10^{-9} \text{ m}^4.$$

Thus, the maximum bending stress in the aluminum is

$$(\sigma_{al})_{max} = -\frac{My}{I_T} = -\frac{90(-0.01)}{22 \times 10^{-9}} = 41 \text{ MPa}$$

and the maximum bending stress in the steel is

$$(\sigma_{st})_{max} = -\frac{My}{I_t} = -3\left(\frac{90(0.006)}{22 \times 10^{-9}}\right) = -74 \text{ MPa}.$$

3.6 Columns

Long slender members loaded axially in compression are referred to as *columns*. Such members frequently fail by *buckling* (excessive lateral deflection) rather than by crushing. Buckling onset depends not only on the material properties but also the geometry and type of end supports of the column. The axial load at the onset of buckling is called the *critical load*.

If the *slenderness ratio* of the column, defined as L/r (length divided by least radius of gyration r where $r = \sqrt{I/A}$), is greater than 120, the critical load for a column is the Euler load

$$P_{cr} = \pi^2 EI/k^2 L^2. \tag{3.6.1}$$

Values of k, with end supports shown in parentheses, are given as:

$$\begin{aligned} k &= 1 && \text{(pinned - pinned)} \\ k &= 0.5 && \text{(fixed - fixed)} \\ k &= 0.7 && \text{(pinned - fixed)} \\ k &= 2 && \text{(free - fixed)} \end{aligned} \tag{3.6.2}$$

Intermediate columns are those whose slenderness ratios are less than 120 but greater than that at which failure occurs by crushing. For these, empirical formulas have been developed to predict buckling.

EXAMPLE 3.12

Determine the critical load for a square steel (E = 210 GPa) strut 8 cm × 8 cm if its length is 6 m under (a) pinned ends, (b) fixed ends.

Solution. The moment of inertia is

$$I = bh^3/12 = (0.08)(0.08)^3/12 = 3.4 \times 10^{-6} \text{ m}^4.$$

a) The critical load for pinned ends is

$$P_{cr} = \frac{\pi^2 EI}{L^2} = \frac{\pi^2 (210 \times 10^9)(3.4 \times 10^{-6})}{6^2} = 195\ 000 \text{ N}.$$

The normal stress, which must not exceed the yield stress, is

$$\sigma = \frac{F}{A} = \frac{195 \times 10^3}{0.0064} = 30.5 \times 10^6 \text{ Pa}.$$

This is substantially less than the yield stress for all steels.

b) The critical load for fixed ends is

$$P_{cr} = \frac{\pi^2 (210 \times 10^9)(3.4 \times 10^{-6})}{0.5^2 \times 6^2} = 780\ 000 \text{ N}.$$

The normal stress for this case is

$$\sigma = \frac{780 \times 10^3}{0.0064} = 122 \times 10^6 \text{ Pa}.$$

Practice Problems

STRESS AND STRAIN

3.1 A structural member with the same material properties in all directions at any particular point is

    a) homogeneous b) isotropic c) isentropic d) holomorphic e) orthotropic

3.2 The amount of lateral strain in a tension member can be calculated using

    a) the bulk modulus.
    b) the moment of inertia.
    c) the yield stress.
    d) Hooke's law.
    e) Poisson's ratio.

3.3 Find the allowable load, in pounds, on a 1/2" dia, 4-ft-long, steel rod if its maximum elongation cannot exceed 0.04 inches.

    a) 9290 b) 6990 c) 5630 d) 4910 e) 3220

3.4 An elevator is suspended by a 1/2" dia, 100-ft-long steel cable. Twenty people, with a total weight of 3500 lbs, enter. How far, in inches, does the elevator drop?

    a) 1.3 b) 1.1 c) 0.9 d) 0.7 e) 0.5

3.5 A hole in a 3-ft-long structural steel member fixed at one end is 1/32" shy of matching another hole for possible connection. What force, in pounds, is necessary to stretch it for connection? The cross section is 1/8" × 1".

    a) 3300 b) 3000 c) 2700 d) 2400 e) 2100

3.6 As the load is applied, edge AB moves 0.0012" to the right. Determine the shear modulus, in psi.

    a) 7.5×10^6 b) 6.2×10^6 c) 5.7×10^6
    d) 5.2×10^6 e) 4.5×10^6

3.7 A 2" dia steel shaft is subjected to an axial tensile force of 150,000 lbs. What is the diameter, in inches, after the force is applied? Use $v = 0.28$.

    a) 1.999 b) 1.998 c) 1.997 d) 1.996 e) 1.995

3.8 An aluminum cylinder carries an axial compressive load of 400,000 lbs. Its diameter measures exactly 5.923" and its height 8.314". What was its original diameter, in inches?

    a) 5.922 b) 5.920 c) 5.918 d) 5.916 e) 5.914

THERMAL STRESS

3.9 A tensile stress of 16,000 psi exists in a 1" dia steel rod that is fastened securely between two rigid walls. If the temperature increases by 50° F, determine the final stress, in psi, in the rod.

    a) 12,200 b) 9400 c) 8600 d) 7400 e) 6200

3.10 A steel bridge span is normally 1000 ft long. What is the difference in length, in inches, between January (–30° F) and August (100° F)?

    a) 10 b) 9 c) 8 d) 7 e) 6

3.11 An aluminum bar at 80° F is inserted between two rigid stationary walls by inducing a compressive stress of 10,000 psi. At what temperature, in ° F, will the bar drop out?

    a) 36   b) 24   c) 10   d) 2   e) –10

3.12 Brass could not be used to reinforce concrete because

    a) it is not sufficiently strong.   d) its coefficient of thermal expansion is not right.

    b) its density is too large.   e) it does not adhere well to concrete.

    c) it is too expensive.

TORSION

3.13 The maximum shearing stress, in psi, that exists in a 2" dia shaft subjected to a 2000 in-lb torque is

    a) 1270   b) 1630   c) 1950   d) 2610   e) 3080

3.14 The shaft of Prob. 3.13 is replaced with a 2" outside diameter, 1.75" inside diameter hollow shaft. What is the maximum shearing stress, in psi?

    a) 1270   b) 1630   c) 1950   d) 2610   e) 3080

3.15 The maximum allowable shear stress in a 4" dia shaft is 20,000 psi. What maximum torque, in ft-lb, can be applied?

    a) 20,900   b) 15,600   c) 11,200   d) 8,600   e) 4,210

3.16 A builder uses an 18" long, 7/8" dia steel drill. If two opposite forces of 160 lbs are applied normal to the shaft, each with a moment arm of 12", what angle of twist, in degrees, occurs in the drill?

    a) 10.3   b) 8.29   c) 6.95   d) 5.73   e) 4.68

BENDING MOMENTS IN BEAMS

3.17 The maximum bending stress at a given cross section of an I-beam occurs

    a) where the shearing stress is maximum.   d) at the neutral axis.

    b) at the outermost fiber.   e) at the joint of the web and the flange.

    c) just below the joint of the web and the flange.

3.18 The moment diagram for a simply supported beam with a load at the midpoint is

    a) a triangle.   b) a parabola.   c) a trapezoid.   d) a rectangle.   e) a semicircle.

3.19 Find the bending moment, in ft-lb, at A.

    a) 7500   b) 7000   c) 6500

    d) 6000   e) 5000

3.20 What is the bending moment, in ft-lb, at A?

    a) 18,000   b) 15,000   c) 12,000

    d) 10,000   e) 8000

STRESSES IN BEAMS

3.21 Find the maximum tensile stress, in psi.

    a) 4360   b) 3960   c) 3240

    d) 2860   e) 2110

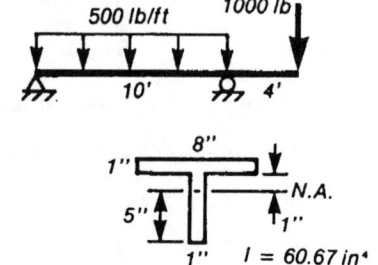

3.22 What is the maximum compressive stress, in psi, in the beam of Prob. 3.21?

    a) 4360   b) 3960   c) 3240   d) 2860   e) 2110

3.23 What is the maximum shearing stress, in psi, in the beam of Prob. 3.21?

    a) 1000   b) 900   c) 800   d) 700   e) 600

3.24 The shearing stress distribution $\tau = VQ/Ib$ on the cross section of the T-beam in Prob. 3.21 most resembles which sketch?

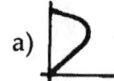

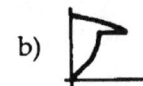

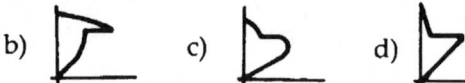

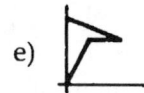

3.25 If the allowable bending stress is 20,000 psi in the beam of Prob. 3.20, calculate the *section modulus* defined by I/y, in in^3.

    a) 11   b) 10   c) 9   d) 8   e) 7

3.26 Find the maximum bending stress, in psi, if the 4" wide beam is 2" deep.

    a) 29,200   b) 23,400   c) 18,600

    d) 15,600   e) 11,700

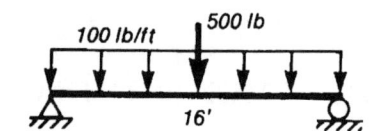

3.27 If the beam of Prob. 3.26 were 2" wide and 4" deep, find the maximum bending stress, in psi.

    a) 29,200   b) 23,400   c) 18,600   d) 15,600   e) 11,700

3.28 Find the maximum shearing stress, in psi, of a simply supported, 20-ft-long beam with a 2"×2" cross section if it has a 500-lb load at the mid point.

    a) 188   b) 152   c) 131   d) 109   e) 94

DEFLECTION OF BEAMS

3.29 What is the maximum deflection, in inches, of a simply supported, 20-ft-long steel beam with a 2"×2" cross section if it has a 500-lb load at the midpoint?

    a) 7.2   b) 6.0   c) 4.8   d) 3.6   e) 2.4

3.30 Find the maximum deflection, in inches, for the steel beam of Prob. 3.26.

    a) 4.22   b) 4.86   c) 3.52   d) 2.98   e) 2.76

3.31 If the deflection of the right end of the 2" dia steel beam is 4", what is the load P, in pounds?

    a) 220   b) 330   c) 440

    d) 550   e) 660

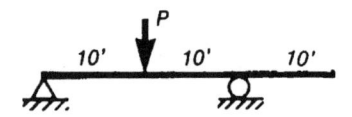

COMBINED STRESSES

3.32 Find the maximum shearing stress, in psi.
- a) 8000
- b) 7000
- c) 6000
- d) 5000
- e) 4000

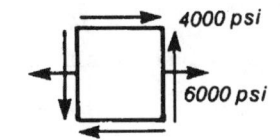

3.33 What is the maximum tensile stress, in psi?
- a) 4000
- b) 3000
- c) 2000
- d) 1000
- e) 0

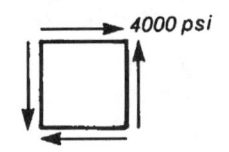

3.34 Determine the maximum shearing stress, in psi.
- a) 8000
- b) 6000
- c) 5000
- d) 4000
- e) 3000

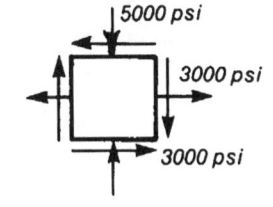

3.35 Find the maximum shearing stress, in psi, in the shaft.
- a) 7340
- b) 6520
- c) 5730
- d) 4140
- e) 3160

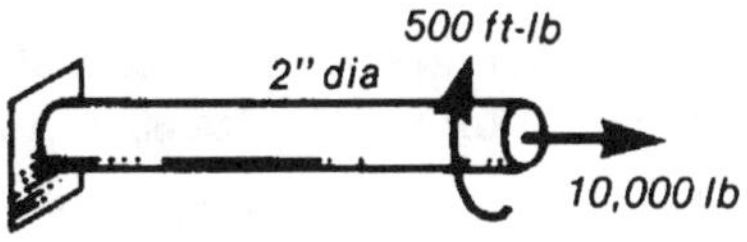

3.36 The maximum normal stress, in psi, in the shaft of Prob. 3.35 is
- a) 7340
- b) 6520
- c) 5730
- d) 4140
- e) 3160

3.37 The normal stress, in psi, at pt. A is
- a) 25,000
- b) 35,000
- c) 41,000
- d) 46,000
- e) 55,000

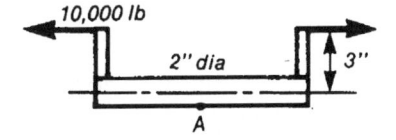

3.38 The maximum shearing stress, in psi, at pt. A in Prob. 3.37 is
- a) 12,500
- b) 15,000
- c) 17,500
- d) 20,500
- e) 23,000

3.39 The maximum shearing stress, in psi, in the circular shaft is
- a) 12,000
- b) 18,000
- c) 24,000
- d) 30,000
- e) 36,000

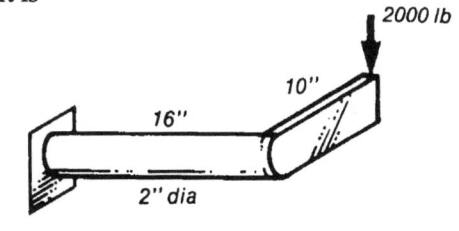

3.40 The maximum tensile stress, in psi, in the circular shaft of Prob. 3.39 is
- a) 11,100
- b) 22,200
- c) 33,300
- d) 44,400
- e) 55,500

THIN-WALLED PRESSURE VESSELS

3.41 The allowable tensile stress for a pressurized cylinder is 24,000 psi. What maximum pressure, in psi, is allowed if the 2-ft-dia cylinder is made of 1/4" thick material?
- a) 250
- b) 500
- c) 1000
- d) 1500
- e) 2000

3.42 The maximum normal stress that can occur in a 4-ft-dia steel sphere is 30,000 psi. If it is to contain a pressure of 2000 psi, what must be the minimum thickness, in inches?

    a) 1.0   b) 0.9   c) 0.8   d) 0.7   e) 0.6

3.43 What is the maximum shearing stress, in psi, in the sphere of Prob. 3.42?

    a) 30,000   b) 20,000   c) 15,000   d) 10,000   e) 0

COMPOSITE SECTIONS

3.44 A compression member, composed of 1/2" thick steel pipe with 10" inside diameter is filled with concrete. Find the stress, in psi, in the steel if the load is 400,000 lbs.

    a) 16,400   b) 14,300   c) 12,600   d) 10,100   e) 8,200

3.45 If the flanges are aluminum and the rib is steel, find the maximum tensile stress, in psi, in the beam.

    a) 2170   b) 1650   c) 1320

    d) 1150   e) 1110

3.46 If the flanges of the I-beam of Prob. 3.45 are steel and the rib is aluminum, what is the maximum tensile stress, in psi?

    a) 2170   b) 1650   c) 1320   d) 1150   e) 1110

COLUMNS

3.47 What is the minimum length, in ft, for which a 4" × 4" wooden post can be considered a long column? Assume a maximum slenderness ratio of 60.

    a) 9.2   b) 7.6   c) 6.8   d) 5.8   e) 4.2

3.48 A free-standing platform, holding 500 lb, is to be supported by a 4" dia vertical aluminum strut. How long, in ft, can it be if a safety factor of 2 is used?

    a) 66   b) 56   c) 46   d) 36   e) 26

3.49 What increase in temperature, in °F, is necessary to cause a 1" dia, 10 ft long, steel rod with fixed ends to buckle? There is no initial stress.

    a) 66   b) 56   c) 46   d) 36   e) 26

3.50 A column with both ends fixed buckles when subjected to a force of 8000 lbs. One end is then allowed to be free. At what force, in pounds, will it buckle?

    a) 100   b) 500   c) 2000   d) 8000   e) 32,000

Solutions to Problems

3.1 b) Isotropic

3.2 e) Poisson's Ratio

3.3 d) $\sigma = E\delta/L = P/\pi r^2$. $\therefore P = E\delta\pi r^2/L = 30\times 10^6 \times 0.04 \times \pi(1/4)^2/48 = 4909$ lb.

3.4 d) $\dfrac{P}{\pi r^2} = E\dfrac{\delta}{L}$. $\therefore \delta = \dfrac{PL}{\pi r^2 E} = \dfrac{3500\times 100\times 12}{\pi\times(1/4^2)\times 30\times 10^6} = 0.71$ in

3.5 a) $\dfrac{P}{A} = E\dfrac{\delta}{L}$. $\therefore P = \dfrac{AE\delta}{L} = \dfrac{1}{8}\times 1\times 30\times 10^6 \times \dfrac{1/32}{36} = 3255$ lb.

3.6 b) $\dfrac{P}{A} = G\dfrac{\delta}{L}$. $\therefore G = \dfrac{PL}{A\delta} = \dfrac{5000\times 6}{8\times(1/2)\times 0.0012} = 6.25\times 10^6$

3.7 a) $\dfrac{P}{A} = E\varepsilon$. $\therefore \varepsilon = \dfrac{150{,}000}{\pi\times 1^2\times 30\times 10^6} = 0.00159$

$\therefore \Delta d = v\varepsilon d = 0.28\times 0.00159\times 2 = 0.00089$. $\therefore d_{after} = 2 - 0.00095 = 1.9991$ in

3.8 b) $\dfrac{P}{A} = E\varepsilon$. $\therefore \varepsilon = \dfrac{400{,}000}{\pi\times(5.923/2)^2\times 10\times 10^6} = 0.00145$

$\Delta d = v\varepsilon d = 0.33\times 0.00145\times 5.923 = 0.00283$.

$\therefore d_f = d - \Delta d = 5.923 - 0.00283 = 5.920$ in

3.9 e) $\sigma = E\delta/L$. $16{,}000 = 30\times 10^6(\delta/L)$. $\therefore \delta = 5.33\times 10^{-4}L$.

$\delta_T = \alpha L \Delta T = 6.5\times 10^{-6}\times 50 L = 3.25\times 10^{-4}L$.

$\therefore \delta_{final} = 2.08\times 10^{-4}L$. $\therefore \sigma = 30\times 10^6 \times 2.08\times 10^{-4} = 6{,}240$ psi

3.10 a) $\delta = \alpha L\Delta T = 6.5\times 10^{-6}\times 1000\times 12\times 130 = 10.1"$

3.11 d) $\dfrac{\delta}{L} = \dfrac{\sigma}{E} = \alpha\Delta T$. $\therefore \dfrac{10{,}000}{10\times 10^6} = 12.8\times 10^{-6}(80-T)$. $\therefore T = 1.88°F$

3.12 d) It expands at a different rate.

3.13 a) $\tau = \dfrac{Ta}{J} = \dfrac{2000\times 1}{\pi\times 1^4/2} = 1{,}273$ psi

3.14 e) $J = \pi(a_1^4 - a_2^4)/2 = \pi(1^4 - .875^4)/2 = 0.650$ in^4

$\tau = \dfrac{Ta}{J} = 2{,}000\times 1/0.650 = 3{,}077$ psi

3.15 a) $T = \dfrac{\tau J}{a} = 20{,}000\times(\pi 2^4/2)/2 = 251{,}300$ in·lb or 20,940 ft·lb

3.16 d) $\theta = \dfrac{TL}{JG} = \dfrac{160\times 24\times 18}{\left[\pi(7/16)^4/2\right]\times 12\times 10^6} = 0.1001$ rad (5.73°)

3.17 b) $\sigma = My/I$. $\therefore \sigma = \sigma_{max}$ at $y = y_{max}$

3.18 a) A triangle.

3.19 a) $\sum M_{right\,end} = 0$. $\therefore 20F_{left} = 1,000 \times 10 + 1,000 \times 5$. $\therefore F_{left} = 750$.

$M_A = 750 \times 10 = 7,500$ lb·ft

3.20 b) $M_A = 1,000 \times 10 + 1,000 \times 5 = 15,000$ ft·lb

3.21 a) $10F_{right} = 5,000 \times 5 + 1,000 \times 14$. $\therefore F_{right} = 3,900$ lb.

$10F_{left} = 5,000 \times 5 - 1,000 \times 4$. $\therefore F_{left} = 2,100$ lb.

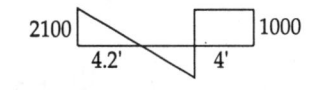

$M_{max} = 2,100 \times (4.2/2) = 4,410$ ft·lb = area under V-diagram.

$\sigma = \dfrac{My}{I} = \dfrac{4,410 \times 12 \times 5}{60.67} = 4,361$ psi

3.22 b) Compression occurs in bottom fibers over right support.

$\sigma = \dfrac{My}{I} = \dfrac{4,000 \times 12 \times 5}{60.67} = 3,956$ psi

3.23 e) $\tau_{max} = \dfrac{VQ}{Ib} = \dfrac{2,900 \times (5 \times 2.5)}{60.67 \times 1} = 597$ psi

3.24 c) τ_{max} occurs on the N.A. with a sudden decrease when b goes from 1" to 8". Also, it is parabolic.

3.25 c) $\sigma = \dfrac{My}{I}$. $\therefore \dfrac{I}{y} = \dfrac{M}{\sigma} = \dfrac{15,000 \times 12}{2,000} = 9$ in³

3.26 b) Area under curve: $M_{max} = 250 \times 8 + 800 \times (8/2) = 5,200$

$\sigma = \dfrac{My}{I} = \dfrac{5,200 \times 12 \times 1}{4 \times 2^3/12} = 23,4000$ psi

3.27 e) $\sigma_{max} = \dfrac{5,200 \times 12 \times 2}{2 \times 4^3/12} = 11,700$ psi

3.28 e) $V_{max} = 250$. $\tau_{max} = \dfrac{VQ}{Ib} = \dfrac{250 \times 2 \times (1/2)}{2 \times (2^3/12) \times 2} = 93.8$

3.29 d) $\delta = PL^3/48EI = 500 \times 240^3/(48 \times 30 \times 10^6 \times 2^4/12) = 3.6"$

3.30 e) $\delta = \dfrac{PL^3}{48EI} + \dfrac{5wL^4}{384EI}$ $I = \dfrac{4 \times 2^3}{12} = 2.67$ in⁴

$= \dfrac{500 \times (16 \times 12)^3}{48 \times 30 \times 10^6 \times 2.67} + \dfrac{5 \times (100/12) \times (16 \times 12)^4}{384 \times 30 \times 10^6 \times 2.67} = 2.76"$

3.31 a) $\delta = \theta L_2 = \dfrac{PL^2}{16EI} \times L_2.$ $\therefore 4 = \dfrac{P \times 240^2 \times 120}{16 \times 30 \times 10^6 \times \pi \times 1^4/4}.$ $\therefore P = 218$ lb

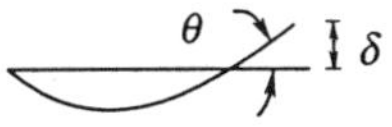

3.32 d) $\tau_{max} = \dfrac{1}{2}\sqrt{(\sigma_x - \sigma_y)^2 + 4\tau^2} = \dfrac{1}{2}\sqrt{6,000^2 + 4 \times 4,000^2} = 5,000$ psi

3.33 a) $\sigma_{max} = \dfrac{1}{2}(\sigma_x + \sigma_y) + \tau_{max} = 0 + 4,000 = 4,000$ psi

3.34 c) $\tau_{max} = \dfrac{1}{2}\sqrt{(3,000+5,000)^2 + 4 \times 3,000^2} = 5,000$ psi

3.35 d) $\tau = Ta/J = 500 \times 12 \times 1/(\pi \times 1^4/2) = 3,820$ psi.
$\sigma = P/A = 10,000/\pi \times 1^2 = 3,180$ psi.
$\tau_{max} = \dfrac{1}{2}\sqrt{3,180^2 + 4 \times 3,820^2} = 4,137$ psi

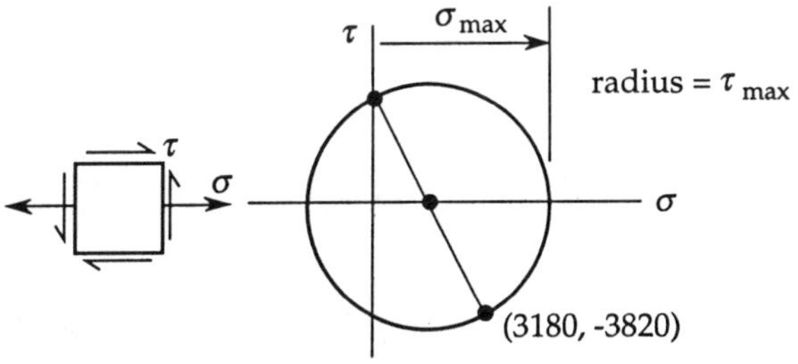

3.36 c) $\sigma_{max} = \dfrac{1}{2} \times 3,180 + 4,137 = 5,727$ psi

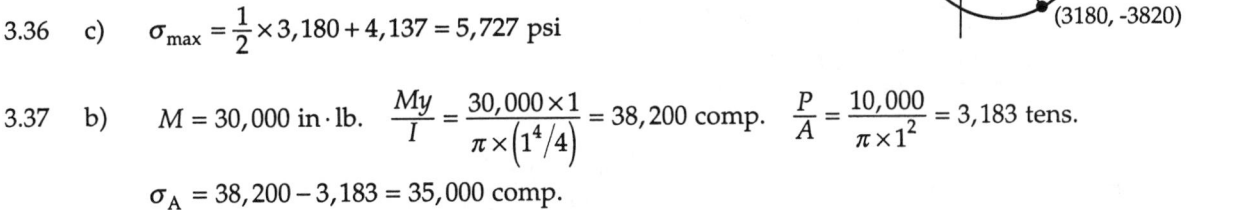

3.37 b) $M = 30,000$ in·lb. $\dfrac{My}{I} = \dfrac{30,000 \times 1}{\pi \times (1^4/4)} = 38,200$ comp. $\dfrac{P}{A} = \dfrac{10,000}{\pi \times 1^2} = 3,183$ tens.
$\sigma_A = 38,200 - 3,183 = 35,000$ comp.

3.38 c) $\tau_{max} = \sigma/2 = 17,500$ psi. $VQ/Ib = 0$ on outer fibers

3.39 c) $\tau = \dfrac{Ta}{J} = \dfrac{2,000 \times 10 \times 1}{\pi \times (1^4/2)} = 12,732$ psi
$\sigma = \dfrac{My}{I} = \dfrac{2,000 \times 16 \times 1}{\pi \times (1^4/2)} = 40,744$ psi
$\tau_{max} = \dfrac{1}{2}\sqrt{40,744^2 + 4 \times 12,732^2} = 24,023$ psi

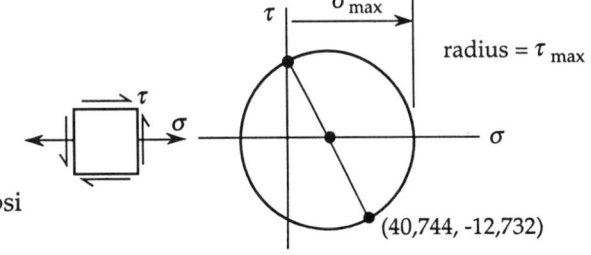

3.40 d) $\sigma_{max} = \dfrac{1}{2} \times 40,744 + 24,023 = 44,395$ psi

3.41 b) $\sigma_t = pD/2t.$ $\therefore p = 24,000 \times 2 \times \dfrac{1}{4}/24 = 500$ psi

3.42 c) $\sigma_a = pD/4t.$ $\therefore t = \dfrac{2000 \times 48}{4 \times 30,000} = 0.800$ in

3.43 e) $\tau_{max} = \dfrac{1}{2}\sqrt{(30,000 - 30,000)^2 + 4 \times 0^2} = 0$

3.44 a) $\left(\dfrac{\Delta L}{L}\right)_s = \left(\dfrac{\Delta L}{L}\right)_c.$ $\therefore \varepsilon_s = \varepsilon_c.$ $\therefore \sigma_s = \dfrac{E_s}{E_c}\sigma_c = 10\,\sigma_c.$
$F_s + F_c = 400,000$ or $A_s\sigma_s + A_c\sigma_c = 400,000.$
$\sigma_s\left[\pi(5.5^2 - 5^2) + \pi \times \dfrac{5^2}{10}\right] = 400,000.$ $\therefore \sigma_s = 16,430$ psi

3.45 e) $n = E/E_{min} = 3$. The area is transformed:

$$I_t = \frac{4 \times 8^3}{12} - \frac{1 \times 6^3}{12} = 152.7 \text{ in}^4.$$

$$\therefore \sigma_{al} = \frac{My}{I} = \frac{2,500 \times 12 \times 4}{152.7} = 786 \text{ psi}$$

3.46 d) $n = 3$. The area is transformed:

$$I_t = \frac{12 \times 8^3}{12} - \frac{11 \times 6^3}{12} = 314 \text{ in}^4.$$

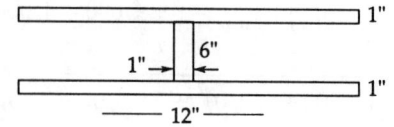

$$\therefore \sigma_s = \frac{nMy}{I} = \frac{3 \times 2,500 \times 12 \times 4}{314} = 1,146 \text{ psi}.$$

3.47 d) $60 = \frac{L}{r} = \frac{L}{\sqrt{I/A}} = \frac{L}{\sqrt{4 \times (4^3/12)/16}}$. $\therefore L = 69.3"$ or $5.77'$

Assume $P = 1,000$ lb using a factor of safety of 2.

3.48 c) $1,000 = \frac{\pi^2 \times 10 \times 10^6 \times \pi \times 4^4/64}{4L^2}$. $\therefore L = 557"$ or $46.4'$

3.49 e) $P_{cr} = 4\pi^2 EI/L^2 = \alpha \Delta TEA$. $\Delta T = \frac{4\pi^2 I}{\alpha AL^2} = \frac{4\pi^2 \times \pi \times 1^4/64}{6.5 \times 10^{-6} \times \pi \times 0.5^2 \times 120^2} = 26.4 \text{ °F}$

3.50 b) $P_{cr} = 4\pi^2 EI/L^2 = 8,000$. $\therefore \pi^2 EI/L^2 = 2,000$. $\therefore P_{cr} = \pi^2 EI/4L^2 = 2,000/4 = 500$ lb.

4. Fluid Mechanics

by Merle C. Potter

Fluid Mechanics deals with the statics, kinematics and dynamics of fluids, including both gases and liquids. Most fluid flows can be assumed to be incompressible (constant density); such flows include liquid flows as well as low speed gas flows (with velocities less than about 100 m/s). In addition, particular flows are either viscous or inviscid. Viscous effects dominate internal flows—such as flow in a pipe—and must be included near the boundaries of external flows (flow near the surface of an airfoil). Viscous flows are laminar if well-behaved, or turbulent if chaotic and highly fluctuating. Inviscid flows occur primarily as external flows outside the boundary layers that contain viscous effects. This review will focus on *Newtonian fluids*, that is, fluids which exhibit linear stress-strain-rate relationships; Newtonian fluids include air, water, oil, gasoline and tar.

4.1 Fluid Properties

Some of the more common fluid properties are defined below and listed in Tables 4.1 and 4.2 for water and air at standard conditions.

$$\text{density} \qquad \rho = \frac{M}{V} \qquad (4.1.1)$$

$$\text{specific weight} \qquad \gamma = \rho g \qquad (4.1.2)$$

$$\text{viscosity} \qquad \mu = \frac{\tau}{du/dy} \qquad (4.1.3)$$

$$\text{kinematic viscosity} \qquad \nu = \frac{\mu}{\rho} \qquad (4.1.4)$$

$$\text{specific gravity} \qquad S = \frac{\rho_x}{\rho_{H_2O}} \qquad (4.1.5)$$

$$\text{bulk modulus} \qquad K = -V \frac{\Delta p}{\Delta V} \qquad (4.1.6)$$

$$\text{speed of sound} \qquad c_{liquid} = \sqrt{K/\rho} \qquad c_{gas} = \sqrt{kRT} \qquad (k_{air} = 1.4) \qquad (4.1.7)$$

EXAMPLE 4.1

A velocity difference of 2.4 m/s is measured between radial points 2 mm apart in a pipe in which 20°C water is flowing. What is the shear stress?

Solution. Using Eq. 4.1.3 we find, with $\mu = 10^{-3}$ N·s/m^2 from Table 4.1,

$$\tau = \mu \frac{du}{dy} \cong \mu \frac{\Delta u}{\Delta y}$$

$$= 10^{-3} \frac{2.4}{0.002} = 1.2 \text{ Pa.}$$

---EXAMPLE 4.2---

Find the speed of sound in air at an elevation of 1000 m.

Solution. From Table 4.2 we find $T = 281.7$ K. Using Eq. 4.1.7, with $R = 287$ J/kg·K, there results

$$c = \sqrt{kRT} = \sqrt{1.4 \times 287 \times 281.7} = 336.4 \text{ m/s}.$$

Note: Temperature must be absolute.

TABLE 4.1 Properties

Property	Symbol	Definition	Water (20°C, 68°F)	Air (STP)
density	ρ	$\dfrac{\text{mass}}{\text{volume}}$	1000 kg/m³ 1.94 slug/ft³	1.23 kg/m³ 0.0023 slug/ft³
viscosity	μ	$\dfrac{\text{shear stress}}{\text{velocity gradient}}$	10^{-3} N·s/m² 2×10^{-5} lb-sec/ft²	2.0×10^{-5} N·s/m² 3.7×10^{-7} lb-sec/ft²
kinematic viscosity	ν	$\dfrac{\text{viscosity}}{\text{density}}$	10^{-6} m²/s 10^{-5} ft²/sec	1.6×10^{-5} m²/s 1.6×10^{-4} ft²/sec
speed of sound	c	velocity of propagation of a small wave	1480 m/s 4900 ft/sec	343 m/s 1130 ft/sec
specific weight	γ	$\dfrac{\text{weight}}{\text{volume}}$	9800 N/m³ 62.4 lb/ft³	12 N/m³ 0.077 lb/ft³
surface tension	σ	stored energy per unit area	0.073 J/m² 0.005 lb/ft	
bulk modulus	K	$-\text{volume} \dfrac{\Delta \text{pressure}}{\Delta \text{volume}}$	220×10^4 kPa 323,000 psi	
vapor pressure	p_v	pressure at which liquid & vapor are in equilibrium	2.45 kPa 0.34 psia	

Notes:
- Kinematic viscosity is used because the ratio μ/ρ occurs frequently.
- Surface tension is used primarily for calculating capillary rise.
- Vapor pressure is used to predict *cavitation* which exists whenever the local pressure falls below the vapor pressure.

4.2 Fluid Statics

Typical problems in fluid statics involve manometers, forces on plane and curved surfaces, and buoyancy. All of these problems are solved by using the pressure distribution derived from summing forces on an infinitesimal element of fluid; in differential form with h positive downward, it is

$$dp = \gamma \, dh. \tag{4.2.1}$$

For constant specific weight, assuming $p = 0$ at $h = 0$, we have

$$p = \gamma h. \tag{4.2.2}$$

TABLE 4.2 Properties of Water and Air

Properties of Water (Metric)

Temperature °C	Density kg/m³	Viscosity N·s/m²	Kinematic Viscosity m²/s	Bulk Modulus kPa	Surface Tension N/m	Vapor Pressure kPa
0	999.9	1.792×10^{-3}	1.792×10^{-6}	204×10^4	7.62×10^{-2}	0.588
5	1000.0	1.519	1.519	206	7.54	0.882
10	999.7	1.308	1.308	211	7.48	1.176
15	999.1	1.140	1.141	214	7.41	1.666
20	998.2	1.005	1.007	220	7.36	2.447
30	995.7	0.801	0.804	223	7.18	4.297
40	992.2	0.656	0.661	227	7.01	7.400
50	988.1	0.549	0.556	230	6.82	12.220
60	983.2	0.469	0.477	228	6.68	19.600
70	977.8	0.406	0.415	225	6.50	30.700
80	971.8	0.357	0.367	221	6.30	46.400
90	965.3	0.317	0.328	216	6.12	68.200
100	958.4	0.284×10^{-3}	0.296×10^{-6}	207×10^4	5.94×10^{-2}	97.500

Properties of Air at Standard Pressure (Metric)

Temperature	Density kg/m³	Specific Weight N/m³	Viscosity N·s/m²	Kinematic Viscosity m²/s
−20°C	1.39	13.6	1.56×10^{-5}	1.13×10^{-5}
−10°C	1.34	13.1	1.62×10^{-5}	1.21×10^{-5}
0°C	1.29	12.6	1.68×10^{-5}	1.30×10^{-5}
10°C	1.25	12.2	1.73×10^{-5}	1.39×10^{-5}
20°C	1.20	11.8	1.80×10^{-5}	1.49×10^{-5}
40°C	1.12	11.0	1.91×10^{-5}	1.70×10^{-5}
60°C	1.06	10.4	2.03×10^{-5}	1.92×10^{-5}
80°C	0.99	9.71	2.15×10^{-5}	2.17×10^{-5}
100°C	0.94	9.24	2.28×10^{-5}	2.45×10^{-5}

Properties of the Atmosphere (Metric)

Altitude m	Temperature K	p/p_0 (p_0=101 kPa)	ρ/ρ_0 (ρ_0=1.23 kg/m³)
0	288.2	1.000	1.000
1 000	281.7	0.8870	0.9075
2 000	275.2	0.7846	0.8217
4 000	262.2	0.6085	0.6689
6 000	249.2	0.4660	0.5389
8 000	236.2	0.3519	0.4292
10 000	223.3	0.2615	0.3376
12 000	216.7	0.1915	0.2546
14 000	216.7	0.1399	0.1860
16 000	216.7	0.1022	0.1359
18 000	216.7	0.07466	0.09930
20 000	216.7	0.05457	0.07258
22 000	218.6	0.03995	0.05266
26 000	222.5	0.02160	0.02797
30 000	226.5	0.01181	0.01503
40 000	250.4	0.2834×10^{-2}	0.3262×10^{-2}
50 000	270.7	0.7874×10^{-3}	0.8383×10^{-3}
60 000	255.8	0.2217×10^{-3}	0.2497×10^{-3}
70 000	219.7	0.5448×10^{-4}	0.7146×10^{-4}
80 000	180.7	0.1023×10^{-4}	0.1632×10^{-4}
90 000	180.7	0.1622×10^{-5}	0.2588×10^{-5}

TABLE 4.2(E) Properties of Water and Air

Properties of Water (English)

Temperature °F	Density slugs/ft^3	Viscosity lb-sec/ft^2	Surface Tension lb/ft	Vapor Pressure lb/in^2	Bulk Modulus lb/in^2
32	1.94	3.75×10^{-5}	0.518×10^{-2}	0.089	293,000
40	1.94	3.23×10^{-5}	0.514×10^{-2}	0.122	294,000
50	1.94	2.74×10^{-5}	0.509×10^{-2}	0.178	305,000
60	1.94	2.36×10^{-5}	0.504×10^{-2}	0.256	311,000
70	1.94	2.05×10^{-5}	0.500×10^{-2}	0.340	320,000
80	1.93	1.80×10^{-5}	0.492×10^{-2}	0.507	322,000
90	1.93	1.60×10^{-5}	0.486×10^{-2}	0.698	323,000
100	1.93	1.42×10^{-5}	0.480×10^{-2}	0.949	327,000
120	1.92	1.17×10^{-5}	0.465×10^{-2}	1.69	333,000
140	1.91	0.98×10^{-5}	0.454×10^{-2}	2.89	330,000
160	1.90	0.84×10^{-5}	0.441×10^{-2}	4.74	326,000
180	1.88	0.73×10^{-5}	0.426×10^{-2}	7.51	318,000
200	1.87	0.64×10^{-5}	0.412×10^{-2}	11.53	308,000
212	1.86	0.59×10^{-5}	0.404×10^{-2}	14.7	300,000

Properties of Air at Standard Pressure (English)

Temperature °F	Density slugs/ft^3	Viscosity lb-sec/ft^2	Kinematic Viscosity ft^2/sec
0	0.00268	3.28×10^{-7}	12.6×10^{-5}
20	0.00257	3.50×10^{-7}	13.6×10^{-5}
40	0.00247	3.62×10^{-7}	14.6×10^{-5}
60	0.00237	3.74×10^{-7}	15.8×10^{-5}
68	0.00233	3.81×10^{-7}	16.0×10^{-5}
80	0.00228	3.85×10^{-7}	16.9×10^{-5}
100	0.00220	3.96×10^{-7}	18.0×10^{-5}
120	0.00215	4.07×10^{-7}	18.9×10^{-5}

Properties of the Atmosphere (English)

Altitude ft	Temperature °F	Pressure lb/ft^2	Density slugs/ft^3	Kinematic Viscosity ft^2/sec	Velocity of Sound ft/sec
0	59.0	2116	0.00237	1.56×10^{-4}	1117
1,000	55.4	2041	0.00231	1.60×10^{-4}	1113
2,000	51.9	1968	0.00224	1.64×10^{-4}	1109
5,000	41.2	1760	0.00205	1.77×10^{-4}	1098
10,000	23.4	1455	0.00176	2.00×10^{-4}	1078
15,000	5.54	1194	0.00150	2.28×10^{-4}	1058
20,000	−12.3	973	0.00127	2.61×10^{-4}	1037
25,000	−30.1	785	0.00107	3.00×10^{-4}	1016
30,000	−48.0	628	0.000890	3.47×10^{-4}	995
35,000	−65.8	498	0.000737	4.04×10^{-4}	973
36,000	−67.6	475	0.000709	4.18×10^{-4}	971
40,000	−67.6	392	0.000586	5.06×10^{-4}	971
50,000	−67.6	242	0.000362	8.18×10^{-4}	971
100,000	−67.6	22.4	3.31×10^{-5}	89.5×10^{-4}	971
110,000	−47.4	13.9	1.97×10^{-5}	1.57×10^{-6}	996
150,000	113.5	3.00	3.05×10^{-6}	13.2×10^{-6}	1174
200,000	160.0	0.665	6.20×10^{-7}	68.4×10^{-6}	1220
260,000	−28	0.0742	1.0×10^{-7}	321×10^{-6}	1019

Equation 4.2.2 can be used to interpret manometer readings directly. By summing forces on elements of a plane surface, we would find the magnitude and location of a force acting on one side (refer to Fig. 4.1) to be

$$F = \gamma \bar{h} A \tag{4.2.3}$$

$$y_p = \bar{y} + \frac{\bar{I}}{\bar{y}A} \tag{4.2.4}$$

where $\bar{y}$ locates the centroid and $\bar{I}$ is the second moment* of the area about the centroidal axis.

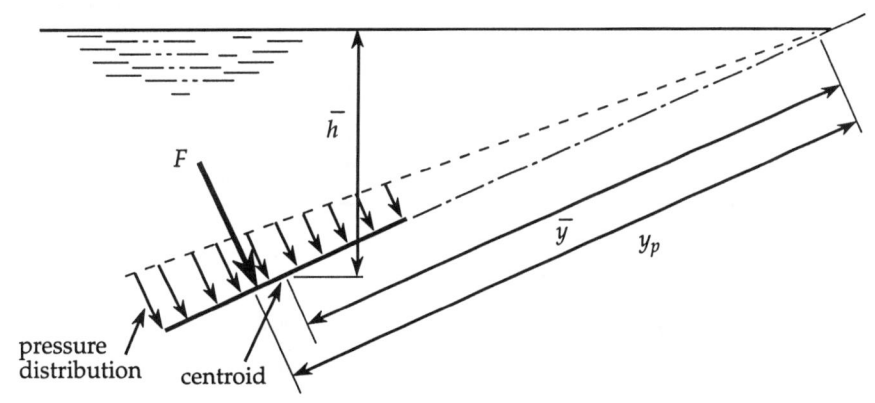

Figure 4.1 Force on a plane surface.

To solve problems involving curved surfaces, we simply draw a free-body diagram of the liquid contained above the curved surface and, using the above formulas, solve the problem.

To solve buoyancy-related problems we us Archimedes' ancient principle which states: the buoyant force on a submerged object is equal to the weight of displaced liquid; that is,

$$F_b = \gamma V_{displaced}. \tag{4.2.5}$$

—— **EXAMPLE 4.3** ——

Find the pressure difference between the air pipe and the water pipe.

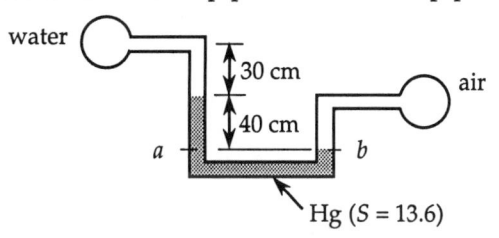

Solution. We first locate points "a" and "b" in the same fluid where $p_a = p_b$; then using Eq. 4.2.2

$$p_{water} + 9800 \times 0.3 + (9800 \times 13.6) \times 0.4 = p_{air} + \overset{\text{neglect}}{\cancel{\gamma_{air} \times 0.4}}$$

$$\therefore p_{air} - p_{water} = 56\,300\,\text{Pa} \quad \text{or} \quad 56.3\,\text{kPa}$$

---

*The second moment $\bar{I}$ (often symbolized by I_C) of three common areas:

$$\bar{I} = \frac{bh^3}{12} \qquad \bar{I} = \frac{bh^3}{36} \qquad \bar{I} = \frac{\pi r^4}{4}$$

EXAMPLE 4.4

What is the pressure in pipe A?

Solution. Locate points "a" and "b" so that $p_a = p_b$. Then, using Eq. 4.2.2 there results

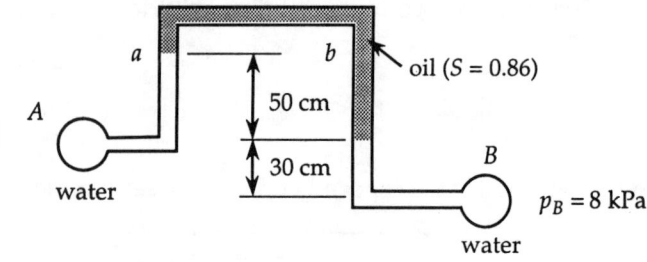

$$p_A - 9800 \times 0.5 = 8000 - 9800 \times 0.3 - (9800 \times 0.86) \times 0.5$$

$$\therefore p_A = 5750 \text{ Pa} \quad \text{or} \quad 5.75 \text{ kPa}.$$

EXAMPLE 4.5

Find the force P needed to hold the 5-m-wide gate closed.

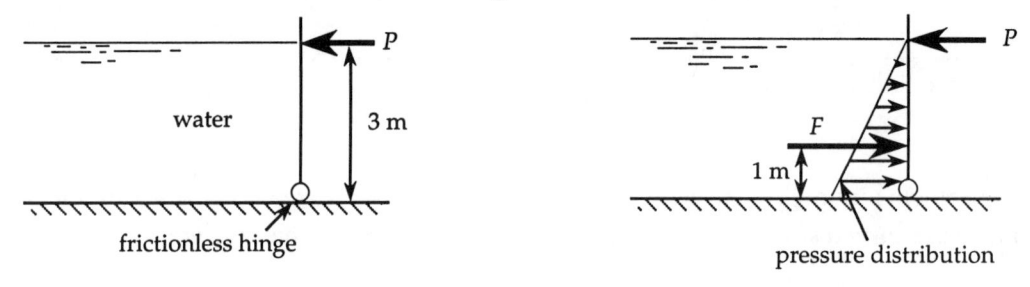

Solution. First, we note that the pressure distribution is triangular, as shown. Hence, the resultant force F acts 1/3 up from the hinge through the centroid of the triangular distribution. Summing moments about the hinge gives

$$F \times 1 = P \times 3$$

$$\therefore P = F/3 = \gamma \bar{h} A / 3$$

$$= 9800 \times \frac{3}{2} \times (5 \times 3)/3 = 73\,500 \text{ N}.$$

Note: If the top of the gate were not at the free surface, we would find y_p using Eq. 4.2.4.

EXAMPLE 4.6

A rectangular 4 m × 20 m vessel has a mass of 40 000 kg. How far will it sink in water when carrying a load of 100 000 kg?

Solution. The total weight of the loaded vessel must equal the weight of the displaced water. This is expressed as

$$W = \gamma V$$

$$(40\,000 + 100\,000) \times 9.8 = 9800 \times 4 \times 20 \times h$$

$$\therefore h = 1.75 \text{ m}.$$

4.3 Dimensionless Parameters and Similitude

Information involving phenomena encountered in fluid mechanics is often presented in terms of dimensionless parameters. For example, the lift force F_L on a streamlined body can be represented by a lift coefficient C_L, a dimensionless parameter. Rather than plotting the lift force as a function of velocity, the lift coefficient could be plotted as a function of the Reynolds number, or the Mach number—two other dimensionless parameters.

To form dimensionless parameters, we first list various quantities encountered in fluid mechanics in Table 4.3. A dimensionless parameter involving several quantities is then formed by combining the quantities so that the combination of quantities is dimensionless. If all units are present in the quantities to be combined, this usually requires four quantities. For example, the four quantities power $\dot{W}$, flow rate Q, specific weight γ, and head H can be arranged as the dimensionless parameter $\dot{W}/\gamma QH$. Many dimensionless parameters have special significance; they are identified as follows:

$$\text{Reynolds number} = \frac{\text{inertial force}}{\text{viscous force}} \qquad Re = \frac{V\ell\rho}{\mu}$$

$$\text{Froude number} = \frac{\text{inertial force}}{\text{gravity force}} \qquad Fr = \frac{V^2}{\ell g}$$

$$\text{Mach number} = \frac{\text{inertial force}}{\text{compressibility force}} \qquad M = \frac{V}{c}$$

$$\text{Weber number} = \frac{\text{inertial force}}{\text{surface tension force}} \qquad We = \frac{V^2 \ell \rho}{\sigma}$$

$$\text{Strouhal number} = \frac{\text{centrifugal force}}{\text{inertial force}} \qquad St = \frac{\ell \omega}{V}$$

$$\text{Pressure coefficient} = \frac{\text{pressure force}}{\text{inertial force}} \qquad C_p = \frac{\Delta p}{\tfrac{1}{2}\rho V^2}$$

$$\text{Drag coefficient} = \frac{\text{drag force}}{\text{inertial force}} \qquad C_D = \frac{\text{drag}}{\tfrac{1}{2}\rho V^2 A}$$

So, rather than writing the drag force on a cylinder as a function of length ℓ, diameter d, velocity V, viscosity μ, and density ρ, i.e.,

$$F_D = f(\ell, d, V, \mu, \rho) \tag{4.3.2}$$

we express the relationship using dimensionless parameters as

$$C_D = f\left(\frac{V\rho d}{\mu}, \frac{\ell}{d}\right). \tag{4.3.3}$$

The subject of similitude is encountered when attempting to use the results of a model study in predicting the performance of a prototype. We always assume *geometric similarity*, that is, the model is constructed to scale with the prototype; the length scale $\ell_p/\ell_m = \lambda$ is usually designated. The primary notion is simply stated: *Dimensionless quantities associated with the model are equal to corresponding dimensionless quantities associated with the prototype.* For example, if viscous effects dominate we would require

$$Re_m = Re_p. \tag{4.3.4}$$

TABLE 4.3 Symbols and Dimensions of Quantities Used in Fluid Mechanics

Quantity	Symbol	Dimensions	Quantity	Symbol	Dimensions
Length	ℓ	L	Pressure	p	M/LT^2
Time	t	T	Stress	τ	M/LT^2
Mass	m	M	Density	ρ	M/L^3
Force	F	ML/T^2	Specific Weight	γ	M/L^2T^2
Velocity	V	L/T	Viscosity	μ	M/LT
Acceleration	a	L/T^2	Kinematic Viscosity	ν	L^2/T
Frequency	ω	T^{-1}	Work	W	ML^2/T^2
Gravity	g	L/T^2	Power	$\dot{W}$	ML^2/T^3
Area	A	L^2	Heat Flux	$\dot{Q}$	ML^2/T^3
Flow Rate	Q	L^3/T	Surface Tension	σ	M/T^2
Mass Flux	$\dot{m}$	M/T	Bulk Modulus	K	M/LT^2

Then if we are interested in, for example, the drag force, we would demand the dimensionless forces to be equal:

$$(F_D)_m^* = (F_D)_p^* \tag{4.3.5}$$

where the asterisk * denotes a dimensionless quantity. Since force is pressure (ρV^2) times area (ℓ^2), the above equation can be expressed in terms of dimensional quantities:

$$\frac{(F_D)_m}{\rho_m \ell_m^2 V_m^2} = \frac{(F_D)_p}{\rho_p \ell_p^2 V_p^2}. \tag{4.3.6}$$

This would allow us to predict the drag force expected on the prototype as

$$(F_D)_p = (F_D)_m \frac{\rho_p \ell_p^2 V_p^2}{\rho_m \ell_m^2 V_m^2}. \tag{4.3.7}$$

The same strategy is used for other quantities of interest.

EXAMPLE 4.7

Combine $\dot{W}$, Q, γ, and H as a dimensionless parameter.

Solution. First, let us note the dimensions on each variable:

$$[\dot{W}] = \frac{ML^2}{T^3} \qquad [Q] = \frac{L^3}{T} \qquad [\gamma] = \frac{M}{L^2T^2} \qquad [H] = L$$

Now, by inspection we simply form the dimensionless parameter. Note that to eliminate the mass unit, $\dot{W}$ and γ must appear as the ratio, $\dot{W}/\gamma$. This puts an extra time unit in the denominator; hence, Q must appear with γ as $\dot{W}/\gamma Q$. Now, we inspect the length unit and find one length unit still in the numerator. This requires H in the denominator giving the dimensionless parameter as

$$\frac{\dot{W}}{\gamma Q H}$$

EXAMPLE 4.8

If a flow rate of 0.2 m³/s is measured over a 9-to-1 scale model of a weir, what flow rate can be expected on the prototype?

Solution. First, we recognize that gravity forces dominate (as they do in all problems involving weirs, dams, ships, and open channels), and demand that

$$Fr_p = Fr_m \quad \text{or} \quad \frac{V_p^2}{\ell_p g_p} = \frac{V_m^2}{\ell_m g_m}.$$

$$\therefore \frac{V_p}{V_m} = \sqrt{\frac{\ell_p}{\ell_m}} = 3.$$

The dimensionless flow rates are now equated:

$$Q_p^* = Q_m^*$$

$$\frac{Q_p}{V_p \ell_p^2} = \frac{Q_m}{V_m \ell_m^2}$$

recognizing that velocity times area ($V \times \ell^2$) give the flow rate. We have

$$Q_p = Q_m \frac{V_p \ell_p^2}{V_m \ell_m^2}$$

$$= 0.2 \times 3 \times 9^2 = 48.6 \text{ m}^3/\text{s}.$$

4.4 Control Volume Equations

When solving problems in fluid dynamics, we are most often interested in volumes into which and from which fluid flows; such volumes are called *control volumes*. The control volume equations include the conservation of mass (the continuity equation), Newton's second law (the momentum equation), and the first law of thermodynamics (the energy equation). We will not derive the equations but simply state them and then apply them to some situations of interest. We will assume *steady, incompressible flow* with *uniform velocity profiles*. The equations take the following forms:

continuity: $$A_1 V_1 = A_2 V_2 \qquad (4.4.1)$$

momentum: $$\Sigma \vec{F} = \rho Q \left(\vec{V}_2 - \vec{V}_1 \right) \qquad (4.4.2)$$

energy: $$-\frac{\dot{W}_s}{\gamma Q} = \frac{V_2^2 - V_1^2}{2g} + \frac{p_2 - p_1}{\gamma} + z_2 - z_1 + h_L \qquad (4.4.3)$$

where

$$Q = AV = \text{flow rate} \qquad (4.4.4)$$

$\dot{W}_s$ = work output (positive for a turbine)

h_L = head loss.

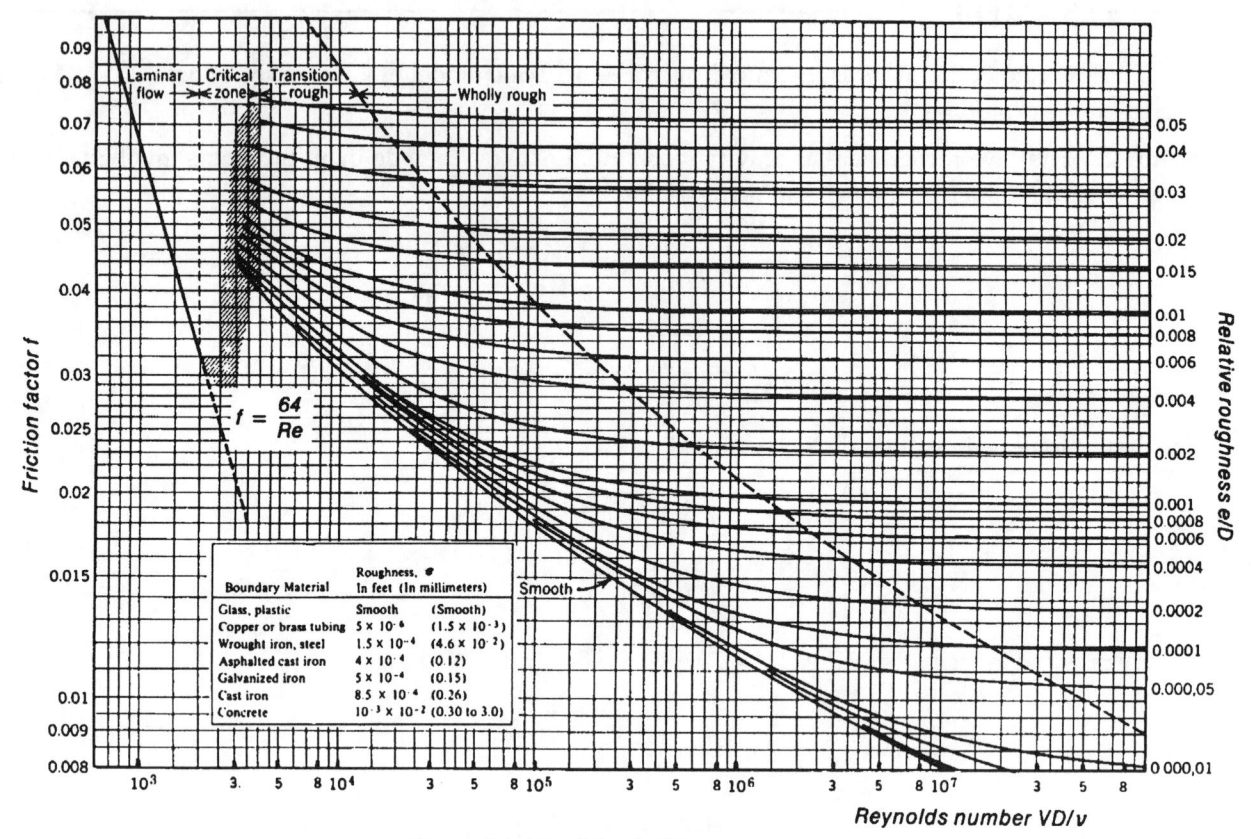

Figure 4.2 The Moody Diagram.

If there is no shaft work term $\dot{W}_s$ (due to a pump or turbine) between the two sections and the losses are zero, then the energy equation reduces to the Bernoulli equation, namely,

$$\frac{V_2^2}{2g} + \frac{p_2}{\gamma} + z_2 = \frac{V_1^2}{2g} + \frac{p_1}{\gamma} + z_1. \tag{4.4.5}$$

For flow in a pipe, the head loss can be related to the friction factor by the Darcy-Weisbach equation,

$$h_L = f \frac{L}{D} \frac{V^2}{2g} \tag{4.4.6}$$

where the friction factor is related to the Reynolds number, $Re = VD/\nu$, and the relative roughness e/D by the Moody diagram, Fig. 4.2; the roughness e is given for various materials. Note that for completely turbulent flows, the friction factor is constant so that the head loss varies with the square of the velocity. For laminar flow the head loss is directly proportional to the velocity.

For sudden geometry changes, such as valves, elbows, and enlargements, the head loss (often called a minor loss) is determined by using a loss coefficient K; that is,

$$h_L = K \frac{V^2}{2g} \tag{4.4.7}$$

where V is the characteristic velocity associated with the device. Typical values are given in Table 4.4.

TABLE 4.4 Loss Coefficients

Geometry	K	Geometry	K
Globe valve (fully open)	6.4	Reentrant entrance	0.8
(half open)	9.5	Well-rounded entrance	0.03++
Angle valve (fully open)	5.0	Pipe exit	1.0
Swing check valve (fully open)	2.5	Sudden contraction (2 to 1)*	0.25++
Gate valve (fully open)	0.2	(5 to 1)*	0.41++
(half open)	5.6	(10 to 1)*	0.46++
(one-quarter open)	24.0	Orifice plate (1.5 to 1)*	0.85
Close return bend	2.2	(2 to 1)*	3.4
Standard tee	1.8	(4 to 1)*	29
Standard elbow	0.9	Sudden enlargement+	$(1 - A_1/A_2)^2$
Medium sweep elbow	0.75	90° miter bend (without vanes)	1.1
Long sweep elbow	0.6	(with vanes)	0.2
45° elbow	0.4	General contraction (30° included angle)	0.02
Square-edged entrance	0.5	(70° included angle)	0.07

*Area ratio   +Based on V_1 ++Based on V_2

In engineering practice, the loss coefficient is often expressed as an *equivalent length* L_e of pipe; if that is done, the equivalent length is expressed as

$$L_e = K \frac{D}{f} \tag{4.4.8}$$

The above analysis, using the Moody diagram and the loss coefficients, can be applied directly to only circular cross-section conduits; if the cross section is non-circular but fairly "open" (rectangular with aspect ratio less than four, oval, or triangular), a good approximation can be obtained by using the hydraulic radius defined by

$$R = A/P \tag{4.4.9}$$

where A is the cross sectional area and P is the wetted perimeter (that perimeter where the fluid is in contact with the solid boundary). Using this formula the diameter of a pipe is $D = 4R$. The Reynolds number then takes the form

$$Re = \frac{4VR}{\nu} \tag{4.4.10}$$

If the shape is not "open," such as flow in an anulus, the error in using the above relationships will be quite significant.

A final note in this article defines the energy grade line (*EGL*) and the hydraulic grade line (*HGL*). The distance $(z + p/\gamma)$ above the datum (the zero elevation line) locates the *HGL*, and the distance $(z + p/\gamma + V^2/2g)$ above the datum locates the *EGL*. These are shown in Fig. 4.3. Note that the pump head H_p is given by

$$H_p = -\frac{\dot{W}_p}{\gamma Q} \tag{4.4.11}$$

The negative sign is necessary since the pump power $\dot{W}_p$ is negative.

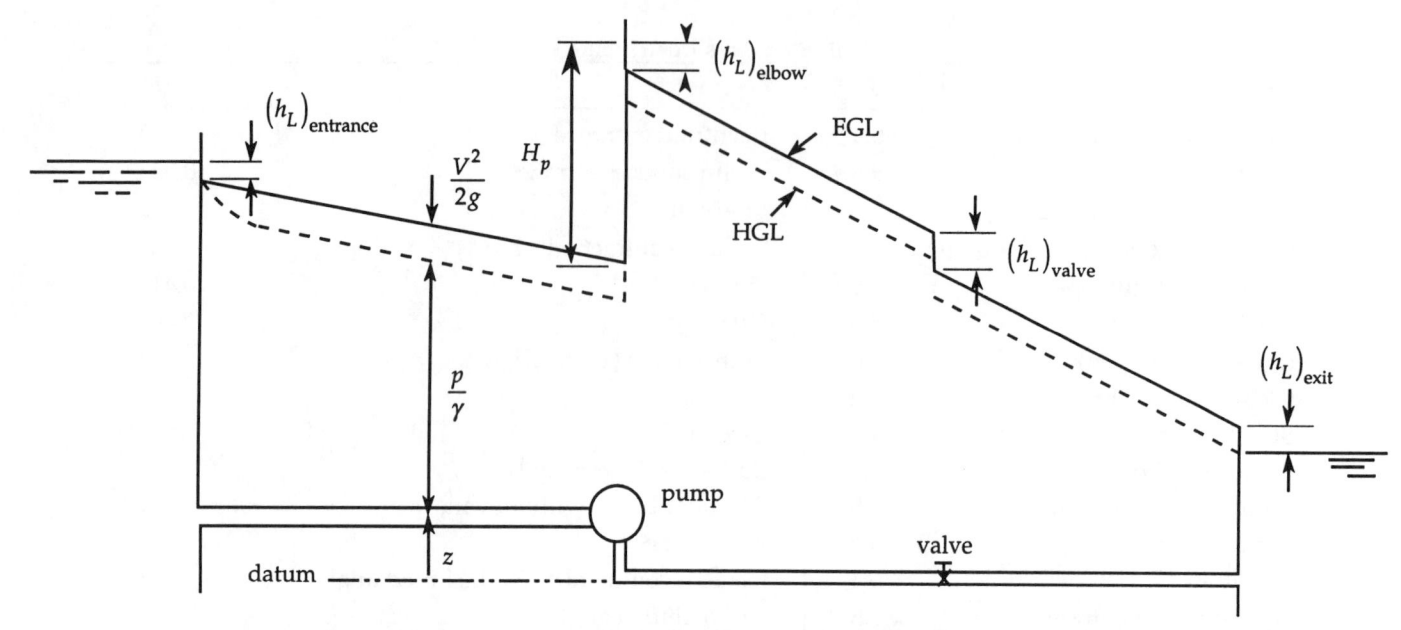

Figure 4.3 The energy grade line (EGL) and the hydraulic grade line (HGL).

EXAMPLE 4.9

The velocity in a 2-cm-dia pipe is 10 m/s. If the pipe enlarges to 4-cm-dia, find the velocity and the flow rate.

Solution. The continuity equation is used as follows:

$$A_1 V_1 = A_2 V_2$$

$$\frac{\pi D_1^2}{4} V_1 = \frac{\pi D_2^2}{4} V_2.$$

$$\therefore V_2 = V_1 \frac{D_1^2}{D_2^2} = 10 \times \frac{2^2}{4^2} = 2.5 \text{ m/s}.$$

The flow rate is

$$Q = A_1 V_1 = \pi \times 0.01^2 \times 10 = 0.00314 \text{ m}^3/\text{s}.$$

EXAMPLE 4.10

What is the force R needed to hold the plate as shown?

Solution. The momentum equation (4.4.2) is a vector equation; applying it in the x-direction results in

$$-R = \rho Q (V_{2x}^{\,0} - V_{1x}).$$

$$\therefore R = \rho A_1 V_1^2$$

$$= 1000 \times \pi \times 0.02^2 \times 20^2 = 503 \text{ N}.$$

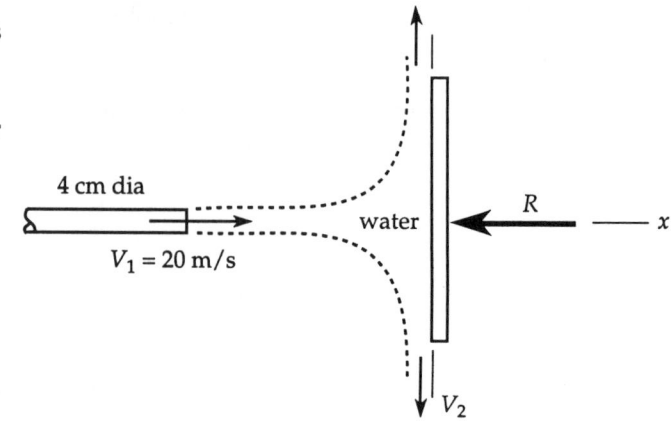

Note: Since the water is open to the atmosphere, $p_2 = p_1$, and if we neglect elevation changes, Bernoulli's equation requires $V_2 = V_1$. However, here $V_{2x} = 0$ so V_2 was not necessary.

EXAMPLE 4.11

What force is exerted on the joint if the flow rate of water is $0.01 \text{ m}^3/\text{s}$?

Solution. The velocities are found to be

$$V_1 = \frac{Q}{A_1} = \frac{0.01}{\pi \times 0.02^2} = 7.96 \text{ m/s}$$

$$V_2 = \frac{Q}{A_2} = \frac{0.01}{\pi \times 0.01^2} = 31.8 \text{ m/s}$$

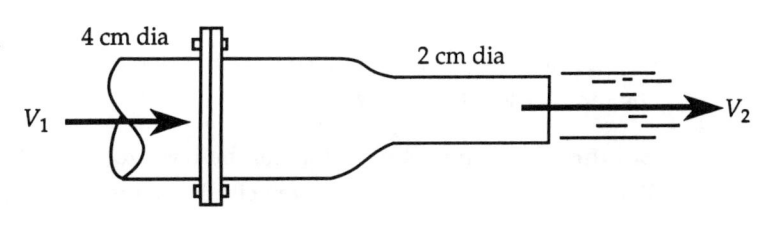

Bernoulli's equation is used to find the pressure at section 1. There results, using $p_2 = 0$ (atmospheric pressure is zero gage),

$$\frac{V_1^2}{2g} + \frac{p_1}{\gamma} = \frac{V_2^2}{2g} + \frac{p_2}{\gamma}$$

$$\frac{7.96^2}{2 \times 9.8} + \frac{p_1}{9800} = \frac{31.8^2}{2 \times 9.8} \qquad \therefore p_1 = 474\,000 \text{ Pa}$$

Now, using the control volume shown, we can apply the momentum equation (4.4.2) in the x-direction:

F_j = the force of the contraction on the water

$$p_1 A_1 - F_j = \rho Q (V_2 - V_1)$$

$$474\,000 \times \pi \times 0.02^2 - F_j = 1000 \times 0.01 (31.8 - 7.96)$$

$$\therefore F_j = 357 \text{ N}$$

Note: Remember, all forces on the control volume must be included. Never forget the pressure force.

EXAMPLE 4.12

What is the pump power needed to increase the pressure by 600 kPa in a 8-cm-dia pipe transporting $0.04 \text{ m}^3/\text{s}$ of water?

Solution. The energy equation (4.4.3) is used:

$$-\frac{\dot{W}_s}{\gamma Q} = \cancel{\frac{V_2^2 - V_1^2}{2g}}^0 + \frac{p_2 - p_1}{\gamma} + \cancel{z_2 - z_1}^0$$

$$-\frac{-\dot{W}_p}{9800 \times 0.04} = \frac{600\,000}{9800} \qquad \therefore \dot{W}_p = 24\,000 \text{ W} \quad \text{or} \quad 24 \text{ kW}.$$

EXAMPLE 4.13

A pitot tube is used to measure the velocity in the pipe. If $V = 15$ m/s, what is H?

Solution. Bernoulli's equation can be used to relate the pressure at pt. 2 to the velocity V. It gives

$$\cancel{\frac{V_2^2}{2g}}^0 + \frac{p_2}{\gamma} + z_2 = \frac{V_1^2}{2g} + \frac{p_1}{\gamma} + z_1$$

$$\therefore p_2 = p_1 + \gamma \frac{V_1^2}{2g}$$

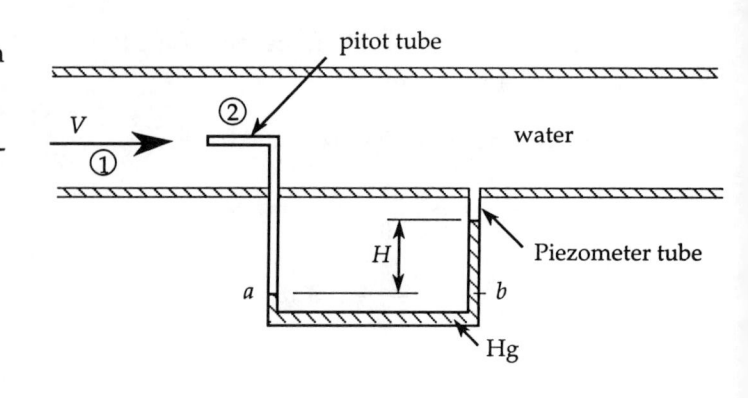

The manometer allows us to write

$$p_a = p_b$$

$$\gamma H + p_2 = \gamma_{Hg} H + p_1$$

Substituting for p_2 we have

$$\gamma H + p_1 + \gamma \frac{V_1^2}{2g} = \gamma_{Hg} H + p_1$$

$$\therefore H = \frac{\gamma}{\gamma_{Hg} - \gamma} \cdot \frac{V_1^2}{2g}$$

$$= \frac{9800}{13.6 \times 9800 - 9800} \cdot \frac{15^2}{2 \times 9.8} = 0.91 \text{ m}$$

Note: The piezometer tube on the right leg measures the pressure p_1 in the pipe.

EXAMPLE 4.14

For a flow rate of 0.02 m³/s, find the turbine output if it is 80% efficient.

Solution. The energy equation (4.4.3) takes the form:

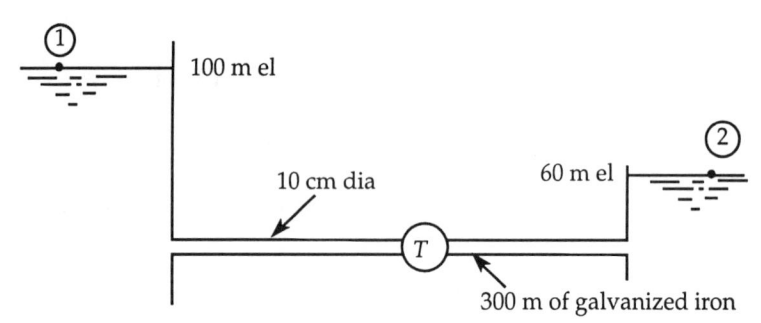

$$-\frac{\dot{W}_T}{\gamma Q} = \cancel{\frac{V_2^2}{2g}}^0 - \cancel{\frac{V_1^2}{2g}}^0 + \cancel{\frac{p_2}{\gamma}}^0 - \cancel{\frac{p_1}{\gamma}}^0 + z_2 - z_1 + \left(K_{entrance} + K_{exit} + f\frac{L}{D}\right)\frac{V^2}{2g}$$

where Eqs. 4.4.6 and 4.4.7 have been used for the head loss. To find f, using the Moody diagram, we need

$$V = \frac{0.02}{\pi \times 0.05^2} = 2.55 \text{ m/s} \qquad Re = \frac{2.55 \times 0.1}{10^{-6}} = 2.55 \times 10^5 \qquad \frac{e}{D} = \frac{0.15}{100} = 0.0015$$

$$\therefore f = 0.022$$

Using the loss coefficients from Table 4.4 we have

$$-\frac{\dot{W}_T/0.8}{9800\times 0.02} = 60-100+\left(0.5+1.0+0.022\frac{300}{0.1}\right)\frac{2.55^2}{2\times 9.8}$$

$$\therefore \dot{W}_T = 2760\text{ W} \quad \text{or} \quad 2.76\text{ kW}$$

EXAMPLE 4.15

The pressure drop over a 4-cm-dia, 300-m-long section of pipe is measured to be 120 kPa. If the elevation drops 25 m over that length of pipe and the flow rate is 0.003 m³/s, calculate the friction factor and the power loss.

Solution. The velocity is found to be

$$V = \frac{Q}{A} = \frac{0.003}{\pi \times 0.02^2} = 2.39 \text{ m/s}.$$

The energy equation (4.4.3) with Eq. 4.4.6 then gives

$$-\frac{\dot{W}_s}{\gamma Q}^{\!0} = \frac{V_2^2 \cancel{-V_1^2}^{\!0}}{2g} + \frac{p_2-p_1}{\gamma} + z_2 - z_1 + f\frac{L}{D}\frac{V^2}{2g}$$

$$0 = -\frac{120\,000}{9800} - 25 + f\frac{300}{0.04}\frac{2.39^2}{2\times 9.8}$$

$$\therefore f = 0.0170, \quad h_L = 37.2 \text{ m}$$

The power loss is

$$\dot{W}_{friction} = h_L \gamma Q$$

$$= 37.2 \times 9800 \times 0.003 = 1095 \text{ W}$$

EXAMPLE 4.16

Estimate the loss coefficient for an orifice plate if $A_1/A_0 = 2$.

Solution. We approximate the flow situation shown as a gradual contraction up to A_c and a sudden enlargement from A_c back to A_1. The loss coefficient for the contraction is very small so it will be neglected. For the enlargement, we need to know A_c; it is

$$A_c = C_c A_0 \qquad C_c = 0.60 + 0.40\left(\frac{A_0}{A_1}\right)^2$$

$$A_c = C_c A_0$$

$$= \left[0.6+0.4\left(\frac{A_0}{A_1}\right)^2\right]A_0 = \left[0.6+0.4\left(\frac{1}{2}\right)^2\right]\frac{A_1}{2} = 0.35 A_1$$

Using the loss coefficient for an enlargement from Table 4.4, there results

$$h_L = K\frac{V_c^2}{2g}$$

$$= \left(1-\frac{A_c}{A_1}\right)^2 \frac{V_c^2}{2g} = (1-0.35)^2 \frac{1}{0.35^2}\frac{V_1^2}{2g} = 3.45\frac{V_1^2}{2g}$$

where the continuity equation $A_c V_c = A_1 V_1$ has been used. The loss coefficient for the orifice plate is thus

$$K = 3.4$$

Note: Two-place accuracy is assumed since C_c is known to only two significant figures.

4.5 Open Channel Flow

If liquid flows down a slope in an open channel at a constant depth, the energy equation (4.4.3) takes the form

$$-\cancel{\frac{\dot{W}_s}{\gamma Q}}^0 = \frac{V_2^2 - \cancel{V_1^2}^0}{2g} + \cancel{\frac{p_2 - p_1}{\gamma}}^0 + z_2 - z_1 + h_L \qquad (4.5.1)$$

which shows that the head loss is given by

$$h_L = z_1 - z_2 = LS \qquad (4.5.2)$$

where L is the length of the channel between the two sections and S is the slope. Since we normally have small angles, we can use $S = \tan\theta = \sin\theta = \theta$ where θ is the angle that the channel makes with the horizontal.

The Chezy-Manning equation is used to relate the flow rate to the slope and the cross section; it is

$$Q = \frac{1.0}{n} A R^{2/3} S^{1/2} \qquad \text{(metric-SI)} \qquad (4.5.3)$$

where R is the hydraulic radius given by Eq. 4.4.9, A is the cross sectional area, and n is the Manning n, given in Table 4.5. The constant 1.0 must be replaced by 1.49 if English units are used. The most efficient cross section occurs when the width is twice the depth for a rectangular section, and when the sides make angles of 60° with the horizontal for a trapezoidal cross section.

TABLE 4.5 Average Values* of the Manning n

Wall Material	Manning n	Wall Material	Manning n
Planed wood	.012	Concrete pipe	.015
Unplaned wood	.013	Riveted steel	.017
Finished concrete	.012	Earth, straight	.022
Unfinished concrete	.014	Corrugated metal flumes	.025
Sewer Pipe	.013	Rubble	.03
Brick	.016	Earth with stones and weeds	.035
Cast iron, wrought iron	.015	Mountain streams	.05

*If $R > 3$ m, increase n by 15%.

EXAMPLE 4.17

A 2-m-dia concrete pipe transports water at a depth of 0.8 m. What is the flow rate if the slope is 0.001?

Solution. Calculate the geometric properties:

$$\alpha = \sin^{-1} \frac{0.2}{1.0} = 11.54°$$

$$\therefore \theta = 156.9°$$

$$A = \pi \times 1^2 \times \frac{156.9}{360} - 0.2 \times \cos 11.54° \times \frac{1}{2} \times 2 = 1.174 \text{ m}^2$$

$$P = \pi \times 2 \times \frac{156.9}{360} = 2.738 \text{ m}$$

For concrete pipe, $n = 0.015$, so

$$Q = \frac{1.0}{n} AR^{2/3} S^{1/2} = \frac{1.0}{0.015} \times 1.174 \times \left(\frac{1.174}{2.738}\right)^{2/3} \times 0.001^{1/2} = 1.41 \text{ m}^3/\text{s}$$

Practice Problems

GENERAL

4.1 A fluid is a substance that
   a) is essentially incompressible.
   b) always moves when subjected to a shearing stress.
   c) has a viscosity that always increases with temperature.
   d) has a viscosity that always decreases with temperature.
   e) expands until it fills its space.

4.2 Viscosity has dimensions of?
   a) FT^2/L b) F/TL^2 c) M/LT^2 d) M/LT e) ML/T

4.3 The viscosity of a fluid varies with
   a) temperature.   d) temperature and pressure.
   b) pressure.   e) temperature, pressure, an density.
   c) density.

4.4 In an isothermal atmosphere the pressure
   a) is constant with elevation.
   b) decreases linearly with elevation.
   c) cannot be related to elevation.
   d) decreases near the surface but approaches a constant value.
   e) decreases exponentially with elevation.

4.5 A torque of 1.2 ft-lb is needed to rotate the cylinder at 1000 rad/sec. Estimate the viscosity $(lb\text{-}sec/ft^2)$.
   a) 8.25×10^{-4} b) 7.16×10^{-4} c) 6.21×10^{-4}
   d) 5.27×10^{-4} e) 4.93×10^{-4}

4.6 A pressure of 80 psi applied to 60 ft^3 of liquid results in a volume change of 0.12 ft^3. The bulk modulus, in psi, is
   a) 20,000   b) 30,000   c) 40,000   d) 50,000   e) 60,000

4.7 Water at 70°F will rise, in a clean 0.04" radius glass tube, a distance, in inches, of
   a) 1.21   b) 0.813   c) 0.577   d) 0.401   e) 0.311

4.8 Water at 70°F flows in a piping system at a low velocity. At what pressure, in psia, will cavitation result?
   a) 0.79   b) 0.68   c) 0.51   d) 0.42   e) 0.34

4.9 A man is observed to strike an object and 1.2 sec later the sound is heard. How far away, in feet, is the man?
   a) 1750   b) 1550   c) 1350   d) 1150   e) 950

4.10 The viscosity of a fluid with specific gravity 1.3 is measured to be 7.2×10^{-5} $lb\text{-}sec/ft^2$. Its kinematic viscosity, in ft^2/sec, is
   a) 2.85×10^{-5} b) 1.67×10^{-5} c) 1.02×10^{-5} d) 9.21×10^{-4} e) 8.32×10^{-4}

FLUID STATICS

4.11 Fresh water 6 ft deep flows over the top of 12 ft of salt water ($S = 1.04$). The pressure at the bottom, in psi, is
a) 8   b) 9   c) 10   d) 11   e) 12

4.12 What pressure, in psi, is equivalent to 28" of Hg?
a) 14.6   b) 14.4   c) 14.2   d) 14.0   e) 13.8

4.13 What pressure, in psi, must be maintained in a diving bell, at a depth of 4000 ft, to keep out ocean water ($S = 1.03$)?
a) 1480   b) 1540   c) 1660   d) 1780   e) 1820

4.14 Predict the pressure, in psi, at an elevation of 6000 ft in an isothermal atmosphere assuming $T = 70°F$. Assume $p_{atm} = 14.7$ psi.
a) 10.7   b) 11.2   c) 11.9   d) 12.3   e) 12.8

4.15 The force F, in lbs, is
a) 15.7   b) 13.5   c) 12.7
d) 11.7   e) 10.2

4.16 A U-tube manometer, attached to an air pipe, measures 10" of Hg. The pressure, in psi, in the air pipe is
a) 4.91   b) 4.42   c) 4.01   d) 3.81   e) 3.12

4.17 The pressure p, in psi, is
a) 8.32   b) 7.51   c) 6.87
d) 6.21   e) 5.46

4.18 A 2-ft dia, 10-ft high, cylindrical water tank is pressurized such that the pressure at the top is 3 psi. The force, in lb, acting on the bottom is
a) 4250   b) 3960   c) 3320   d) 2780   e) 2210

4.19 The force, in lb, acting on one of the 5-ft sides of an open cubical water tank which is full is
a) 4200   b) 3900   c) 3600   d) 3300   e) 3000

4.20 The force P, in lb, for the 10-ft wide gate is
a) 17,000   b) 16,000   c) 15,000
d) 14,000   e) 13,000

4.21 The force P, in lb, to just open the 12-ft wide gate is
a) 88,600   b) 82,500   c) 79,100
d) 73,600   e) 57,100

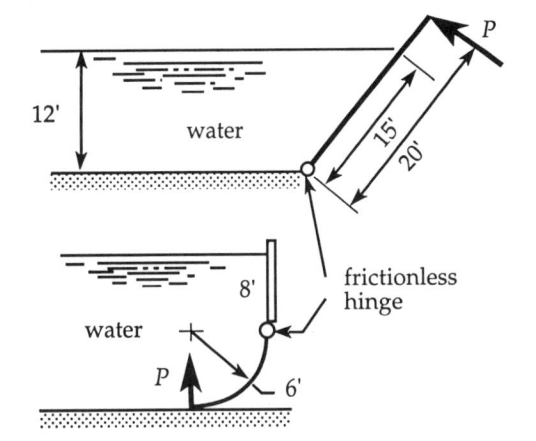

4.22 The force P, in lb, on the 15-ft wide gate is
a) 90,000   b) 78,000   c) 57,000
d) 48,000   e) 32,000

4.23 Four cars, with a mass of 3200 lb each, are loaded on a 20-ft-wide, 40-ft-long small-car ferry. How far, in inches, will it sink in the water?
a) 7   b) 6   c) 5   d) 4   e) 3

4.24 An object weighs 25 lb in air and 6 lb when submerged in water. Its specific gravity is
a) 1.5   b) 1.4   c) 1.3   d) 1.2   e) 1.1

4.25 What pressure differential, in psf, exists at the bottom of a 10-ft-vertical wall if the temperature inside is 70°F and outside is –10°F? Assume equal pressures at the top.
a) 0.478   b) 0.329   c) 0.211   d) 0.133   e) 0.101

DIMENSIONLESS PARAMETERS

4.26 Arrange pressure p, flow rate Q, diameter D, and density ρ into a dimensionless group.
a) $pQ^2/\rho D^4$ b) $p/\rho Q^2 D^4$ c) $pD^4\rho/Q^2$ d) $pD^4/\rho Q^2$ e) $p/\rho Q^2$

4.27 Combine surface tension σ, density ρ, diameter D, and velocity V into a dimensionless para-meter.
a) $\sigma/\rho V^2 D$ b) $\sigma D/\rho V$ c) $\sigma \rho/VD$ d) $\sigma V/\rho D$ e) $\sigma D^2/\rho V$

4.28 The Reynolds number is a ratio of
a) velocity effects to viscous effects.
b) inertial forces to viscous forces.
c) mass flux to viscosity.
d) flow rate to kinematic viscosity.
e) mass flux to kinematic viscosity.

SIMILITUDE

4.29 What flow rate, in ft^3/sec, is needed using a 20:1 scale model of a dam over which 120 ft^3/sec of water flows?
a) 0.20   b) 0.18   c) 0.092   d) 0.067   e) 0.052

4.30 It is proposed to model a submarine moving at 30 ft/sec by testing a 10:1 scale model. what velocity, in ft/sec, would be needed in the model study?
a) 300   b) 400   c) 500   d) 700   e) 900

4.31 The drag force on a 40:1 scale model of a ship is measured to be 2 lb. What force, in lb, is expected on the ship?
a) 128,000   b) 106,000   c) 92,000   d) 80,000   e) 60,000

4.32 The power output of a 10:1 scale model of a water wheel is measured to be 0.06 Hp. The power output, in Hp, expected from the prototype is
a) 130   b) 150   c) 170   d) 190   e) 210

CONTINUITY

4.33 The velocity in a 1"-dia pipe is 60 ft/sec. If the pipe enlarges to 2.5"-dia the velocity, in ft/sec, will be
    a) 24.0   b) 20.0   c) 16.2   d) 12.4   e) 9.6

4.34 A 1"-dia pipe transports water at 60 ft/sec. If it exits out 100 small 0.1"-dia holes, the exiting velocity, in ft/sec, will be
    a) 180   b) 120   c) 90   d) 60   e) 30

4.35 Water flows through a 1"-dia pipe at 60 ft/sec. It then flows radially outward between two discs, 0.1" apart. When it reaches a radius of 20", its velocity, in ft/sec, will be
    a) 30   b) 22.5   c) 15.0   d) 7.50   e) 3.75

BERNOULLI'S EQUATION

4.36 The pressure force, in lb, on the 6"-dia headlight of an automobile traveling at 90 ft/sec is
    a) 1.02   b) 1.83   c) 2.56   d) 3.75   e) 4.16

4.37 The pressure inside a 2"-dia hose is 100 psi. If the water exits through a 1"-dia nozzle, what velocity, in ft/sec, can be expected inside the hose?
    a) 39.8   b) 37.6   c) 35.1   d) 33.7   e) 31.5

4.38 Calculate V, in ft/sec.
    a) 19.7   b) 18.9   c) 17.4
    d) 16.4   e) 15.2

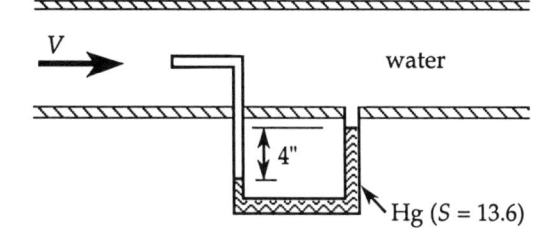

4.39 Water enters a turbine at 150 psi with negligible velocity. What maximum speed, in ft/sec, can it reach before it enters the turbine rotor?
    a) 119   b) 131   c) 156   d) 168   e) 183

MOMENTUM

4.40 If the density of the air is 0.0024 slug/ft^3, find F, in lb.
    a) 0.524   b) 0.711   c) 0.916
    d) 1.17   e) 2.25

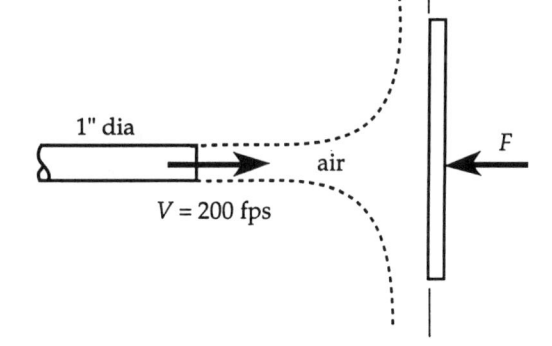

4.41 A rocket exits exhaust gases with $\rho = 0.001 \text{ slug}/\text{ft}^3$ out a 20"-dia nozzle at a velocity of 4000 ft/sec. Estimate the thrust, in lb.
    a) 24,600   b) 30,100   c) 34,900   d) 36,200   e) 41,600

4.42 A high-speed vehicle, traveling at 150 ft/sec, dips a 30"-wide scoop into water and deflects the water 180°. If it dips 2" deep, what force, in lb, is exerted on the scoop?
    a) 36,400   b) 32,100   c) 26,200   d) 22,100   e) 19,900

4.43   What force, in lb, acts on the nozzle?
   a) 1700   b) 1600   c) 1500
   d) 1400   e) 1300

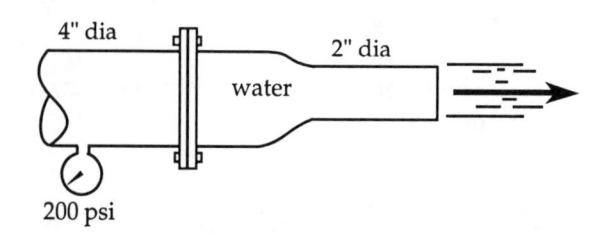

ENERGY

4.44   The locus of elevations that water will rise in a series of pitot tubes is called
   a) the hydraulic grade line.
   b) the energy grade line.
   c) the velocity head.
   d) the pressure head.
   e) the head loss.

4.45   A pressure rise of 75 psi is needed across a pump in a pipe transporting 6 cfs of water. If the pump is 85% efficient, the power needed, in Hp, would be
   a) 140   b) 130   c) 120   d) 110   e) 100

4.46   An 85% efficient turbine accepts 3 cfs of water at a pressure of 90 psi. What is the maximum power output, in Hp?
   a) 100   b) 90   c) 80   d) 70   e) 60

4.47   If the turbine is 88% efficient, the power output, in Hp, is
   a) 60   d) 90
   b) 70   e) 120
   c) 80

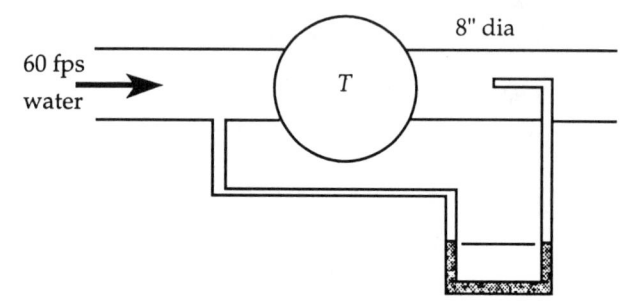

LOSSES

4.48   In a completely turbulent flow the head loss
   a) increases with the velocity.
   b) increases with the velocity squared.
   c) decreases with wall roughness.
   d) increases with diameter.
   e) increases with flow rate.

4.49   The shear stress in a turbulent pipe flow
   a) varies parabolically with the radius.
   b) is constant over the pipe radius.
   c) varies according to the 1/7th power law.
   d) is zero at the center and increases linearly to the wall.
   e) is zero at the wall and increases linearly to the center.

4.50   The velocity distribution in a turbulent flow in a pipe is often assumed to
   a) vary parabolically.
   b) be zero at the wall and increase linearly to the center.
   c) vary according to the 1/7th power law.
   d) be unpredictable and is thus not used.
   e) be maximum at the wall and decrease linearly to the center.

4.51 The Moody diagram is sketched. The friction factor for turbulent flow in a smooth pipe is given by curve?
a) A b) B c) C
d) D e) E

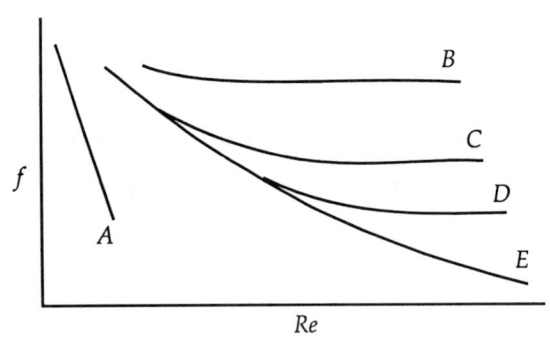

4.52 For the Moody diagram given in Problem 4.51, the completely turbulent flow is best represented by curve
a) A b) B c) C d) D e) E

4.53 The pressure gradient $(\Delta p/\Delta x)$ in a developed turbulent flow in a horizontal pipe
a) is constant.
b) varies linearly with axial distance.
c) is zero.
d) decreases exponentially.
e) varies directly with the average velocity.

4.54 The head loss in a pipe flow can be calculated using
a) the Bernoulli equation.
b) Darcy's law.
c) the Chezy-Manning equation.
d) the momentum equation.
e) the Darcy-Weisbach equation.

4.55 Minor losses in a piping system are
a) less than the friction factor losses, $f \dfrac{L}{D}\dfrac{V^2}{2g}$.
b) due to the viscous stresses.
c) assumed to vary linearly with the velocity.
d) found by using loss coefficients.
e) independent of the flow rate.

4.56 In a turbulent flow in a pipe we know the
a) Reynolds number is greater than 10,000.
b) fluid particles move in straight lines.
c) head loss varies linearly with flow rate.
d) shear stress varies linearly with radius.
e) viscous stresses dominate.

4.57 Water flows through a 4"-dia, 300-ft-long pipe connecting two reservoirs with an elevation difference of 120 ft. The average velocity is 20 fps. Neglecting minor losses, the friction factor is
a) 0.0257 b) 0.0215 c) 0.0197 d) 0.0193 e) 0.0182

4.58 Find the power required, in Hp, by the 85% efficient pump if $Q=0.6$ cfs.
a) 16.1 b) 14.9 c) 13.3
d) 11.2 e) 10.1

4.59 The pressure at section A in a 2"-dia, wrought-iron, horizontal pipe is 70 psi. A fully open globe valve, two elbows, and 150 ft of pipe connect section B. If $Q=0.2$ cfs of water, the pressure p_B, in psi, is
a) 35 b) 40 c) 45 d) 50 e) 55

4.60 Air at 70°F and 14.7 psia is transported through 1500 ft of smooth, horizontal, 40"x15" rectangular duct with a flow rate of 40 cfs. The pressure drop, in psf, is
   a) 2.67   b) 1.55   c) 0.96   d) 0.52   e) 0.17

4.61 Estimate the loss coefficient K in a sudden contraction $A_1/A_2 = 2$ by neglecting the losses up to the vena contracta A_c. Assume that $A_c/A_2 = 0.62 + 0.38\left(A_2/A_1\right)^3$ and $h_L = KV_2^2/2g$.
   a) 0.28   b) 0.27   c) 0.26   d) 0.25   e) 0.24

4.62 An elbow exists in a 4"-dia galvanized iron pipe transporting 0.6 cfs of water. Find the equivalent length of the elbow in ft.
   a) 9.7   b) 10.3   c) 11.0   d) 12.1   e) 13.6

OPEN CHANNEL FLOW

4.63 The depth of water in a 10-ft-wide, rectangular, finished concrete channel is 6 ft. If the slope is 0.001, estimate the flow rate, in cfs.
   a) 460   b) 350   c) 290   d) 280   e) 270

4.64 At what depth, in ft, will 300 cfs of water flow in a 12-ft-wide, rectangular, brick channel if the slope is 0.001?
   a) 7.5   b) 6.5   c) 5.5   d) 4.5   e) 3.5

4.65 A 6-ft-dia, brick storm sewer transports 100 cfs when it's nearly full. Estimate the slope of the sewer.
   a) 0.0008   b) 0.002   c) 0.003   d) 0.004   e) 0.005

Solutions to Problems

4.1 (b) (a) is true for a liquid and low speed gas flows. (c) and (e) are true of gases. (d) is true for a liquid.

4.2 (d) $\tau = \mu\, du/dy$. $\therefore \mu = \tau/du/dy$. $[\mu] = \dfrac{F/L^2}{\frac{L}{T}/L} = \dfrac{FT}{L^2} = \dfrac{(ML/T^2)T}{L^2} = \dfrac{M}{LT}$

4.3 (a) Viscosity μ varies with temperature only.

4.4 (e) $dp = -\gamma dz = -\rho g dz$. $p = \rho RT$ (ideal gas)

$\therefore dp = -\dfrac{p}{RT} g\, dz$ or $\dfrac{dp}{p} = -\dfrac{g}{RT} dz$.

$\int \dfrac{dp}{p} = -\dfrac{g}{RT} \int dz$. $\therefore \ln p = -Cz$. $\therefore p = e^{-Cz}$

4.5 (a) $T = \tau A r = \mu \dfrac{du}{dy} A r = \mu \dfrac{r\omega}{t} 2\pi r L r$.

$1.2 = \dfrac{2/12 \times 1000}{.04/12} \times 2\pi \times \dfrac{2}{12} \times \dfrac{2}{12} \times \dfrac{2}{12} \mu$. $\therefore \mu = 8.25 \times 10^{-4}$

4.6 (c) $K = -\mathcal{V} \dfrac{\Delta p}{\Delta \mathcal{V}} = -60 \dfrac{80}{0.12} = 40{,}000$ psi

4.7 (c) $\sigma 2\pi r = \gamma \pi r^2 L$

$0.005 \times 2\pi \times \dfrac{.04}{12} = 62.4\pi \dfrac{.04^2}{144} L$

$\therefore L = 0.0481'$ or $0.577''$

4.8 (e) Cavitation occurs when the pressure reaches the vapor pressure = 0.34 psi abs. (See Table 4.1.)

4.9 (c) $L = V\Delta t = \sqrt{kRT}\, \Delta t = \sqrt{1.4 \times 53.3 \times 32.2 \times 530} \times 1.2 = 1354'$

Assume $T = 70°F = 530°R$

4.10 (a) $v = \dfrac{\mu}{\rho} = \dfrac{7.2 \times 10^{-5}}{1.3 \times 1.94} = 2.85 \times 10^{-5}$. ρ must be in slug/ft^3.

4.11 (a) $p = \gamma_1 \Delta h_1 + \gamma_2 \Delta h_2 = 62.4 \times 6 + (62.4 \times 1.04) \times 12 = 1153$ psf or 8.01 psi.

4.12 (c) $p = \gamma h = (13.6 \times 62.4) \times 28/12 = 1980$ psf or 13.75 psi.

4.13 (d) $p = \gamma h = (1.03 \times 62.4) \times 4000 = 257{,}000$ psf or 1785 psi.

4.14 (c) $dp = -\gamma dz = -\rho g dz = -p\dfrac{g}{RT}dz$ using $p = \rho RT$

$\therefore \dfrac{dp}{p} = -\dfrac{g}{RT}dz.$ $\int_{14.7}^{p} dp/p = -\dfrac{g}{RT}\int_{0}^{6000} dz.$

$\therefore \ln\dfrac{p}{14.7} = -\dfrac{32.2}{53.3 \times 32.2 \times 530} \times 6000.$ $\therefore p = 11.89$ psia.

4.15 (c) $F = pA.$ $p = \dfrac{200}{\pi \times 2^2} + 0.86 \times 62.4 \times \dfrac{10}{12}\Big/144 = 16.2$ psi.

$\therefore F = 16.2 \times \pi \times (1/2)^2 = 12.74$ lb.

4.16 (a) $p = \gamma h = (13.6 \times 62.4) \times \dfrac{10}{12}\Big/144 = 4.91$ psi.

4.17 (e) $p + 62.4 \times 1 = 13.6 \times 62.4 \times 1.$ $\therefore p = 786$ psf

or $p = 5.46$ psi.

4.18 (c) $F = pA = (3 \times 144 + 62.4 \times 10)\pi \times 1^2 = 3318$ lb.

4.19 (b) $F = \bar{p}A = 62.4 \times \dfrac{5}{2} \times 5^2 = 3900$ lb.

4.20 (d) $20P = \dfrac{15}{3}(62.4 \times 6 \times 150).$ $\therefore P = 14{,}040$ lb.

4.21 (d) All pressures on the curved section pass thru the center. Moments about the hinge give
$P = F_v = \gamma$ Volume $= 62.4 \times 12 \times \pi 6^2/4 + 62.4 \times 6 \times 8 \times 12 = 57{,}000$ lb.

4.22 (b) $y_p = \bar{y} + \dfrac{\bar{I}}{\bar{y}A} = 15 + \dfrac{15 \times 10^3/12}{150 \times 15} = 15.56'$

$10P = 62.4 \times 15 \times 150 \times 5.56.$ $\therefore P = 78{,}000$ lb.

4.23 (e) $W = \gamma \mathcal{V}.$ $4 \times 3200 = 62.4 \times 20 \times 40h$ $\therefore h = .256'$ or $3.08''.$

4.24 (c) $6 = 25 - 62.4\mathcal{V}.$ $\therefore \mathcal{V} = 0.3045$ ft^3. $25 = 62.4 S \times 0.3045.$ $\therefore S = 1.316.$

4.25 (d) $\Delta p = \Delta \gamma \times h = \left(\dfrac{1}{450} - \dfrac{1}{530}\right)\dfrac{14.7 \times 144}{53.3} \times 10 = 0.1332$ psf.

4.26 (d) $[p] = \dfrac{M}{LT^2}$ $[Q] = \dfrac{L^3}{T}$ $[D] = L$ $[\rho] = \dfrac{M}{L^3}.$ First, eliminate M, then T, then L:

$\dfrac{M}{LT^2} \cdot \dfrac{L^3}{M} \cdot \dfrac{T^2}{L^6} \cdot L^4 = p \cdot \dfrac{1}{\rho} \cdot \dfrac{1}{Q^2} \cdot D^4 = \dfrac{pD^4}{\rho Q^2}.$

4.27 (a) $[\sigma] = \dfrac{M}{T^2}$ $[\rho] = \dfrac{M}{L^3}$ $[D] = L$ $[V] = \dfrac{L}{T}.$

$\dfrac{M}{T^2} \cdot \dfrac{T^2}{L^2} \cdot \dfrac{L^3}{M} \cdot \dfrac{1}{L} = \sigma \cdot \dfrac{1}{V^2} \cdot \dfrac{1}{\rho} \cdot \dfrac{1}{D} = \dfrac{\sigma}{\rho D V^2}.$

4.28 (b) Inertial force to viscous forces.

4.29 (d) $(Fr)_m = (Fr)_p$. $\therefore \dfrac{V_m^2}{l_m g} = \dfrac{V_p^2}{l_p g}$. $\therefore \dfrac{V_m^2}{V_p^2} = \dfrac{1}{20}$.

$Q_m^* = Q_p^*$ or $\dfrac{Q_m}{V_m l_m^2} = \dfrac{Q_p}{V_p l_p^2}$. $\therefore Q_m = 120 \times \dfrac{1}{20^2} \times \dfrac{1}{\sqrt{20}} = .0671$ cfs.

4.30 (a) $Re_m = Re_p$. $\left(\dfrac{Vl}{\nu}\right)_m = \left(\dfrac{Vl}{\nu}\right)_p$. $\therefore \dfrac{V_m}{V_p} = \dfrac{l_p}{l_m} = 10$.

$\therefore V_m = 10 V_p = 10 \times 30 = 300$ fps.

4.31 (a) $Fr_m = Fr_p$. $\therefore \left(\dfrac{V^2}{\ell g}\right)_m = \left(\dfrac{V^2}{\ell g}\right)_p$. $\therefore \dfrac{V_p^2}{V_m^2} = \dfrac{\ell_p}{\ell_m}$.

$(F_D)_m^* = (F_D)_p^*$ or $\dfrac{(F_D)_m}{\rho_m V_m^2 \ell_m^2} = \dfrac{(F_D)_p}{\rho_p V_p^2 \ell_p^2}$. $\therefore (F_D)_p = 2 \dfrac{\rho_p}{\rho_m} \dfrac{V_p^2}{V_m^2} \dfrac{\ell_p^2}{\ell_m^2} = 2 \dfrac{\ell_p^3}{\ell_m^3}$.

$\therefore (F_D)_p = 2 \times 40^3 = 128{,}000$ lb.

4.32 (d) $Fr_m = Fr_p$. $\left(\dfrac{V^2}{lg}\right)_m = \left(\dfrac{V^2}{lg}\right)_p$. $\therefore \dfrac{V_p^2}{V_m^2} = 10$.

$\dot{W}_p^* = \dot{W}_m^*$. or $\dfrac{\dot{W}_m}{\rho_m V_m^3 l_m^2} = \dfrac{\dot{W}_p}{\rho_p V_p^3 l_p^2}$. $\therefore \dot{W}_p = .06 \dfrac{V_p^3}{V_m^3} \dfrac{l_p^2}{l_m^2} \dot{W}_m = .06 \times 10\sqrt{10} \times 10^2 = 189.7$ Hp.

4.33 (e) $V_2 = 60\pi \times \left(\dfrac{1}{2}\right)^2 \big/ \pi \times 1.25^2 = 9.6$ fps.

4.34 (d) $V_2 = 60\pi \times \left(\dfrac{1}{2}\right)^2 \big/ 100\pi \times .05^2 = 60$ fps.

4.35 (e) $V_2 \times 0.1 \times 2\pi \times 20 = 60 \times \pi \times \left(\dfrac{1}{2}\right)^2$. $\therefore V_2 = 3.75$ fps.

4.36 (b) $p = \rho V^2 / 2 = .0023 \times 90^2 / 2 = 9.32$ psf.

$F = pA = 9.32 \pi \times 3^2 / 144 = 1.83$ lb.

4.37 (e) $V_2 A_2 = V_1 A_1$. $\therefore V_2 = V_1 \times 2^2 / 1^2 = 4 V_1$.

$\dfrac{p_1}{\rho_1} + \dfrac{V_1^2}{2} = \cancel{\dfrac{p_2}{\rho}}^0 + \dfrac{V_2^2}{2}$. $\dfrac{100 \times 144}{1.94} + \dfrac{V_1^2}{2} = \dfrac{16 V_1^2}{2}$. $\therefore V_1 = 31.5$ fps.

4.38 (d) $p + \rho \dfrac{V^2}{2} + 62.4 \times \dfrac{4}{12} = p + 13.6 \times 62.4 \times \dfrac{4}{12}$.

Using $\rho = 1.94$ slug/ft^3, $V = 16.44$ fps.

4.39 (c) Cavitation results if $p_2 = -14.7$ psi.

$$p_1/\rho + \cancel{V_1^2/2}^{0} = p_2/\rho + V_2^2/2.$$

$$150 \times 144/1.94 = -14.7 \times 144/1.94 + V_2^2/2. \quad \therefore V_2 = 156 \text{ fps}.$$

4.40 (a) $F = \rho A V^2 = .0024 \times \pi \times \dfrac{5^2}{144} \times 200^2 = 0.524$ lb.

4.41 (c) $F = \rho A V^2 = .001 \times \pi \times \dfrac{10^2}{144} \times 4000^2 = 34,900$ lb.

4.42 (a) $-F = \rho A V(-V - V). \quad \therefore F = 2\rho A V^2.$

$$\therefore F = 2 \times 1.94 \times \dfrac{2 \times 30}{144} \times 150^2 = 36,400 \text{ lb}.$$

4.43 (c) $V_2 = 4V_1$. $\dfrac{p_1}{\rho} + \dfrac{V_1^2}{2} = \cancel{\dfrac{p_2}{\rho}}^{0} + \dfrac{V_2^2}{2} = \dfrac{16V_1^2}{2}. \quad \therefore V_1^2 = \dfrac{2p_1}{15\rho}.$

$$\therefore V_1 = \sqrt{\dfrac{2 \times 200 \times 144}{15 \times 1.94}} = 44.5. \quad \therefore V_2 = 178.$$

$$p_1 A_1 - F = \rho A_1 V_1 (V_2 - V_1).$$

$$\therefore F = 200\pi \times 2^2 - 1.94 \times \pi \times \dfrac{2^2}{144} \times 44.5 \times 133.5 = 1508 \text{ lb}.$$

4.44 (b) The energy grade line.

4.45 (a) $\dot{W}_P = \gamma Q \dfrac{\Delta p}{\gamma} \Big/ 0.85 = 6 \times (75 \times 144)/.85 = 76,200$ ft-lb/sec or 139 Hp.

4.46 (e) $\dot{W}_T = \gamma Q \dfrac{\Delta p}{\gamma} \times 0.85 = 3 \times (90 \times 144) \times 0.85 = 33,050$ ft-lb/sec or 60.1 Hp.

4.47 (a) manometer: $p_1 = p_2 + \rho V_2^2/2.$

$$-\dfrac{\dot{W}_T}{\gamma Q} = \cancel{\dfrac{V_2^2}{2g}} + \dfrac{p_2}{\gamma} - \cancel{\dfrac{V_1^2}{2g}} - \dfrac{p_1}{\gamma} \quad (100\% \text{ efficient})$$

$$\therefore \dot{W}_T = Q \dfrac{V_2^2}{2} \rho \eta = \left(60 \times \pi \times \dfrac{4^2}{144}\right) \dfrac{60^2}{2} \times 1.94 \times .88 = 64,400 \text{ ft-lb/sec or } 117 \text{ Hp}.$$

4.48 (b) Increases with the velocity squared.

4.49 (d) Increases linearly to the wall.

4.50 (c) Vary as the 1/7th power law.

4.51 (e) By curve E.

4.52 (b) By curve B.

4.53 (a) Pressure varies linearly $\therefore \dfrac{\Delta p}{\Delta x} = $ Const.

4.54 (e) The Darcy-Weisbach equation.

4.55 (d) Found by using loss coefficients.

4.56 (d) Shear stress varies linearly with radius.

4.57 (b) $h_L = f \dfrac{L}{d} \dfrac{V^2}{2g}$. $\therefore f = 120 \dfrac{4/12}{300} \dfrac{2 \times 32.2}{20^2} = .0215$

4.58 (c) $V = Q/A = \dfrac{0.6}{\pi \times 2^2 / 144} = 6.875$ fps.

$\mathrm{Re} = \dfrac{Vd}{\nu} = 6.875 \times \dfrac{4}{10} / 10^{-5} = 2.3 \times 10^5 \quad \dfrac{e}{d} = \dfrac{.00085}{4/12} = .0025$

From Fig. 4.2 $f = .02$.

$\dot{W}_p = \dfrac{\gamma Q}{\eta}\left(\dfrac{V_2^{\,\cancel{2}\,0}}{2g} + \dfrac{\cancel{p_2}^{\,0}}{\gamma} + z_2 - \dfrac{V_1^{\cancel{2}\,0}}{2g} - \dfrac{\cancel{p_1}^{\,0}}{\gamma} - z_1 + f\dfrac{L}{d}\dfrac{V^2}{2g} + K\dfrac{V^2}{2g}\right)$

$= \dfrac{62.4 \times .6}{.85}\left[200 - 50 + \left(.02\dfrac{300}{4/12} + 1 + .5\right)\dfrac{6.875^2}{64.4}\right] = 73.0$ ft-lb/sec or 13 Hp.

4.59 (e) $V = Q/A = \dfrac{.02}{\pi \times 1^2 / 144} = 9.167$ fps

$\mathrm{Re} = \dfrac{Vd}{\nu} = 9.167 \times \dfrac{2}{12} / 10^{-5} = 1.53 \times 10^5. \quad \dfrac{e}{d} = \dfrac{.00015}{2/12} = .0009$

From Fig. 4.2 $f = .021$. $\quad 0 = \dfrac{p_B - p_A}{\gamma} + f\dfrac{L}{d}\dfrac{V^2}{2g} + K\dfrac{V^2}{2g}$.

$\therefore p_B = 70 - \left(.021\dfrac{150}{2/12} + 6.4 + 2 \times .9\right)\dfrac{9.167^2}{2 \times 32.2} \times 62.4/144 = 54.7$ psi.

4.60 (b) $V = Q/A = 40 / \left(\dfrac{40}{12} \times \dfrac{15}{12}\right) = 9.6$ fps. $\quad R = \dfrac{40 \times 15}{110} = 5.455''$.

$\mathrm{Re} = \dfrac{4 \times 9.6 \times 5.45/12}{1.6 \times 10^{-4}} = 1.10 \times 10^5$. With $\dfrac{e}{d} = 0, \; f = 0.0175$.

$\gamma = p/RT = \dfrac{14.7 \times 144}{53.3 \times 530} = .075$

$\Delta p = f\dfrac{L}{4R}\dfrac{V^2}{2g}\gamma = .0175 \dfrac{1500}{4 \times 5.45/12}\dfrac{9.6^2}{64.4} \times .075 = 1.55$ psf.

4.61 (d) $A_c/A_2 = .62 + .38(.5)^3 = .6675. \quad K_1 = (1 - .6675)^2 = .111.$

$.111\, V_c^2/2g = K\, V_2^2/2g. \quad \therefore K = .111(A_2/A_c)^2 = .111/.6675^2 = .249.$

4.62 (e) $V = Q/A = .6/(\pi \times 2^2/144) = 6.875 \quad e/d = .0005/\dfrac{1}{3} = .0015$

$\mathrm{Re} = \dfrac{6.875 \times 4/12}{10^{-5}} = 2.3 \times 10^5. \quad \therefore f = .022.$

$L_e = Kd/f = .9 \times \dfrac{4}{12} / .022 = 13.6'.$

4.63 (a) $Q = \dfrac{1.49}{n} AR^{2/3} S^{1/2} = \dfrac{1.49}{.012} 60 \times 2.73^{2/3} \times .001^{1/2} = 460$ cfs where $R = 60/22 = 2.73'$.

4.64 (b) $Q = \dfrac{1.49}{n} AR^{2/3} S^{1/2} = \dfrac{1.49}{.016} 12h \left(\dfrac{12h}{12 + 2h}\right)^{2/3} \times .001^{1/2} = 300$ cfs. Trial-and-error: $h = 4.52'$.

4.65 (b) $Q = \dfrac{1.49}{n} AR^{2/3} S^{1/2} = \dfrac{1.49}{.016} \pi \times 3^2 \times 1.5^{2/3} S^{1/2} = 100$, where $R = \dfrac{A}{P} = \dfrac{\pi r^2}{2\pi r} = \dfrac{r}{2} = 1.5'. \quad \therefore S = .00084.$

5. Open Channel Flow

by David C. Wiggert

Open channel flow is commonly encountered in civil engineering practice. Natural occurrences include water waves in oceans and lakes, river discharges, and rainfall-induced flows. Civil engineers may design canals, culverts, drainage ditches, and other devices to convey water. In addition, they may construct dams, weirs, protective levees, or storage ponds to divert or contain water for specific purposes such as flood protection or supply. All of the flow situations mentioned here have a common characteristic: the free surface interface between the air and the upper layer of the water. Generally, the position of the free surface does not remain constant; it may vary from one location to another. Additionally, free surface flows can be three-dimensional, as in estuaries or rivers under flood conditions. However, most flows in rivers and channels are treated as one-dimensional. In this chapter, the principles of one-dimensional, steady, uniform and nonuniform flow in open channels are reviewed, and applied to design situations.

5.1 Basic Concepts

5.1.1 Channel Geometry

The three most common channel cross-sectional shapes are rectangular, trapezoidal, and circular; these are shown in Fig. 5.1. The expressions for the section area A, free surface width B, and wetted perimeter P are given in Table 1.

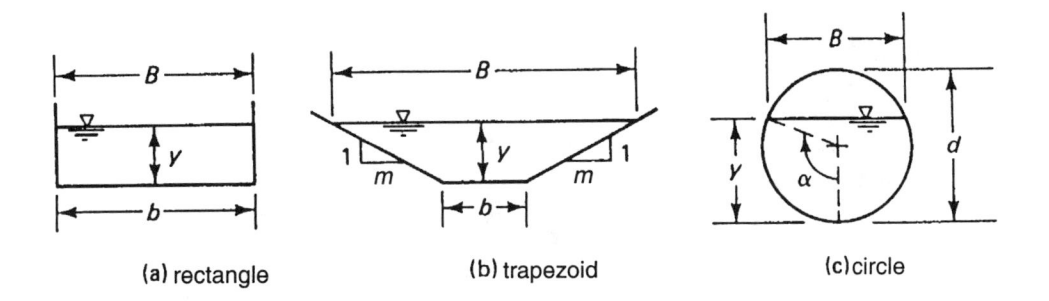

Figure 5.1 Channel cross sections.

TABLE 5.1 Cross-Sectional Formulas

Section	A	B	P
Rectangle	by	b	$b+2y$
Trapezoid	$by + my^2$	$b + 2my$	$b + 2y\sqrt{1+m^2}$
Circle	$\frac{1}{4}(\alpha - \sin\alpha \cos\alpha)d^2$ where $\cos\alpha = 1 - 2\frac{y}{d}$	$d \sin\alpha$	αd

---

EXAMPLE 5.1

A trapezoidal channel has a bottom width of 15 ft, side slope of 1 ft vertical: 4 ft horizontal, and is conveying water at a depth of 6 ft. Compute the flow area, width of the free surface, and the wetted perimeter.

Solution. Using the trapezoidal formulas from Table 5.1 one finds, with $b = 15$, $m = 4$, and $y = 6$,

$A = 15 \times 6 + 4 \times 6^2 = 234 \text{ ft}^2$

$B = 15 + 2 \times 4 \times 6 = 63 \text{ ft}$

$P = 15 + 2 \times 6 \times \sqrt{1 + 4^2} = 64.5 \text{ ft}$

5.1.2 Uniform Flow

Consider the nearly horizontal, parallel channel flow shown in Fig. 5.2. Designate z as the elevation of the channel bottom and y as the depth of flow. One can reasonably assume that the pressure distribution is hydrostatic ($p = \gamma h$), and that the hydraulic grade line, given by $z + y$, coincides with the water surface. Uniform flow occurs when the terminal velocity has been reached in a channel of constant cross section, so that the depth as well as the velocity is invariant along the length of the channel. Here the energy grade line, water surface and channel bottom are all parallel.

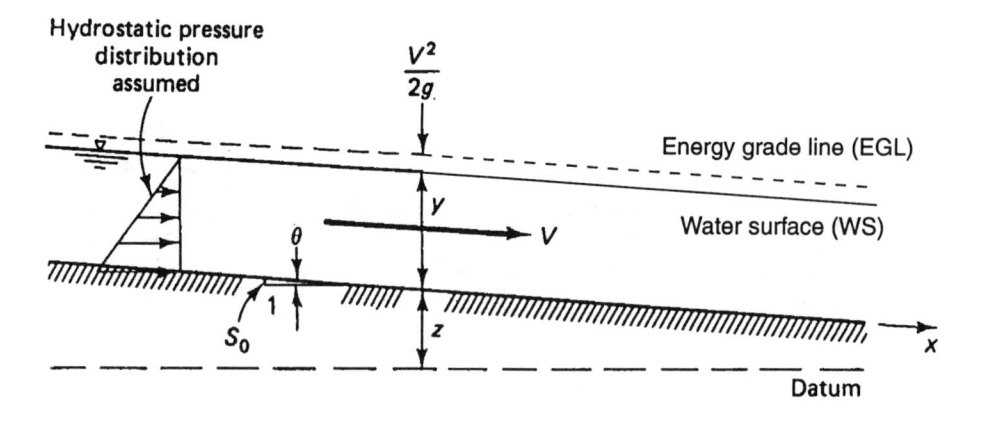

Figure 5.2 Reach of open-channel flow.

Open Channel Flow

The relationship between discharge Q, channel slope S_o, and cross sectional geometry is given by the Chezy-Manning equation:

$$Q = \frac{1.49}{n} AR^{2/3} \sqrt{S_o} \tag{5.1.1}$$

in which the hydraulic radius $R = A/P$, and n is the Manning roughness coefficient. Values of n are given in Table 5.2.

TABLE 5.2 Values of the Manning Roughness Coefficient n

Material	Manning n
Planed wood	0.012
Unplaned wood	0.013
Finished concrete	0.012
Unfinished concrete	0.014
Sewer pipe	0.013
Brick	0.016
Cast iron, wrought iron	0.015
Concrete pipe	0.015
Riveted steel	0.017
Earth, straight	0.022
Corrugated metal flumes	0.025
Rubble	0.030
Earth with stones and weeds	0.035
Mountain streams	0.050

(Note: If $R > 3$ m, increase n by 15%.)

──── EXAMPLE 5.2 ────

A 3-ft-diameter concrete pipe transports water at a depth of 1.3 ft. If the channel slope is 0.003, compute the flow rate.

Solution.
$$\cos \alpha = 1 - \frac{2 \times 1.3}{3} = 0.133$$

$$\therefore \alpha = 82.3°, \text{ or } 1.44 \text{ rad}$$

$$\therefore A = \frac{1}{4} \times (1.44 - \sin 1.44 \times \cos 1.44) \times 3^2 = 2.94 \text{ ft}^2$$

$$P = 1.44 \times 3 = 4.31 \text{ ft}$$

$$R = \frac{2.94}{4.31} = 0.681 \text{ ft}$$

$$Q = \frac{1.49}{0.015} \times 2.94 \times 0.681^{2/3} \times \sqrt{0.003} = 13 \text{ ft}^3/\text{s}$$

Note that only two significant figures are used in the answer since the Manning n is known to only two significant figures.

5.1.3 Energy Principle

The energy equation, taken between any two locations in an open-channel flow, is given by

$$y_1 + \frac{V_1^2}{2g} + z_1 = y_2 + \frac{V_2^2}{2g} + z_2 + h_L \tag{5.1.2}$$

in which V = mean (or average) velocity, and h_L = head loss between locations 1 and 2. Various expressions for h_L will be given in subsequent example problems. Equation 5.1.2 is a general relationship used to analyze both rapidly- and gradually-varied flow.

It is convenient in open-channel flow calculations to make use of *specific energy E*, defined as the sum of the depth of flow and the kinetic energy:

$$E = y + \frac{V^2}{2g} \tag{5.1.3}$$

For a rectangular cross-section, define the *specific discharge q* to be

$$q = \frac{Q}{b} = Vy \tag{5.1.4}$$

Then Eq. 5.1.3 becomes

$$E = y + \frac{q^2}{2gy^2} \tag{5.1.5}$$

which is valid for a rectangular channel only. Equation 5.1.5 is diagrammed in Figure 5.3. Observe the existence of a unique depth at minimum E. That depth is termed *critical depth* y_c and is given by

$$y_c = \left(\frac{q^2}{g}\right)^{1/3} \tag{5.1.6}$$

The *Froude number Fr* is a non dimensional ratio of inertial force to gravity force; it plays a significant role in open channel flows. For a rectangular channel it is defined as

$$Fr = \frac{V}{\sqrt{gy}} = \frac{q}{\sqrt{gy^3}} \tag{5.1.7}$$

On the E-y diagram, note that at a given E, two depths are possible; these depths are termed *alternate depths*. For $y > y_c$, $Fr < 1$ and the flow state is subcritical. Conversely, when $y < y_c$, $Fr > 1$ and the flow is supercritical. Note that at critical flow the minimum energy $E_c = 3y_c/2$ and $Fr = 1$.

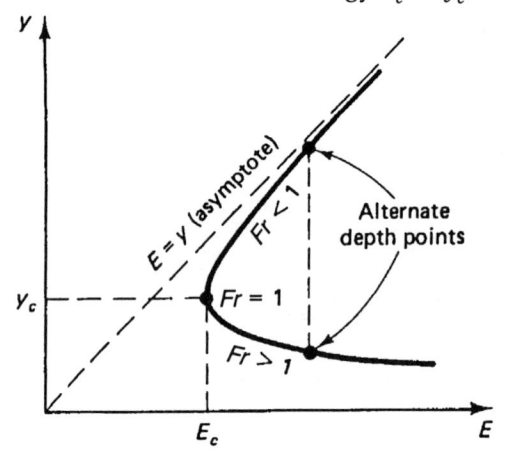

Figure 5.3 Specific energy E versus depth y for a rectangular channel section.

For a non rectangular cross-section, the Froude number is expressed in terms of discharge and channel geometry:

$$Fr = \frac{Q/A}{\sqrt{gA/B}} \tag{5.1.8}$$

Note that the specific discharge q is used only for rectangular channel sections.

EXAMPLE 5.3

Water is flowing in a 10-ft-wide rectangular channel at a discharge of 75 ft³/sec. If the water depth is 3 ft, determine the specific energy, Froude number, critical depth, and alternate depth.

Solution. The flow area and top width are $A = 10 \times 3 = 30$ ft² and $b = 10$ ft. The specific discharge is

$$q = \frac{Q}{b} = \frac{75}{10} = 7.5 \text{ ft}^2/\text{sec}$$

The specific energy, critical depth, and Froude number are computed using Eqs. 5.1.5, 5.1.6 and 5.1.7:

$$E = 3 + \frac{7.5^2}{2 \times 32.2 \times 3^2} = 3.097 \text{ ft}$$

$$Fr = \frac{q}{\sqrt{gy^3}} = \frac{7.5}{\sqrt{32.2 \times 3^3}} = 0.254$$

$$y_c = \left(\frac{q^2}{g}\right)^{1/3} = \left(\frac{7.5^2}{32.2}\right)^{1/3} = 1.205 \text{ ft}$$

The alternate depth is calculated using Eq. 5.1.5 and noting that alternate depths are defined for a constant E (Fig. 5.3):

$$3.097 = y + \frac{7.5^2}{2 \times 32.2 \times y^2} = y + \frac{0.8734}{y^2}$$

Solving by trial and error, the alternate depth $y = 0.59$ ft.

5.1.4 Momentum Principle

The equation of motion is applied to a control volume, Fig. 5.4, with the following assumptions: hydrostatic pressure distribution, negligible frictional forces, and small channel slope. The submerged 2-dimensional obstacle imparts a force F on the control volume directed opposite to the flow. The resulting equation is

$$M_1 - M_2 = \frac{F}{\gamma} \tag{5.1.9}$$

in which the *momentum function M* is

$$M = A\bar{y} + \frac{Q^2}{gA} \tag{5.1.10}$$

for a general cross-section, and

$$M = \frac{by^2}{2} + \frac{bq^2}{gy} \tag{5.1.11}$$

for a rectangular section. In Eqs. 5.1.9 to 5.1.11, γ = specific weight of water, $\bar{y}$ = distance to the centroid of the channel area measured from the free surface (for a rectangular channel $\bar{y} = y/2$). Equation 5.1.11 is sketched in Fig. 5.5; note that two depths, termed *conjugate depths*, exist for a given M and q. The conjugate depths are found by setting $M_1 = M_2$. The upper leg applies to subcritical flow ($y < y_c$), and the lower leg to supercritical flow ($y > y_c$). Under certain circumstances (e.g., hydraulic jump) F may be equal to zero; this is illustrated in Example 5.4.

Open Channel Flow

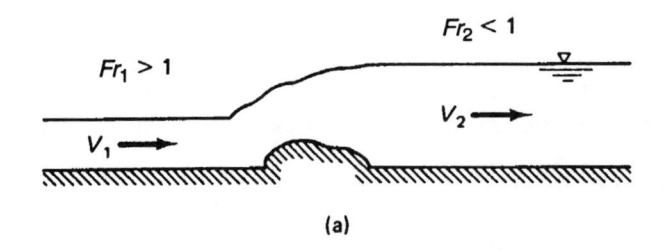

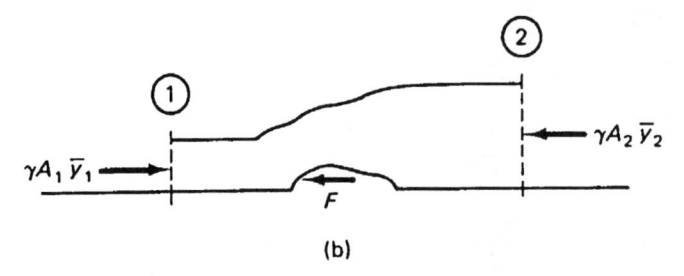

Figure 5.4 Flow past a submerged 2-D obstacle; a) idealized flow; b) control volume.

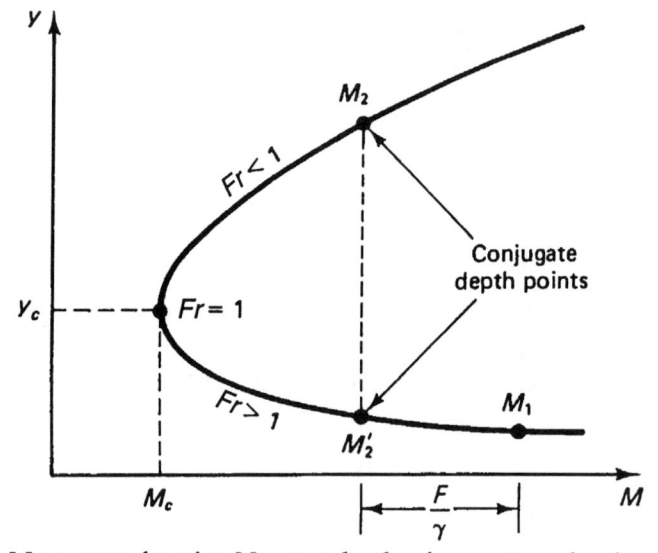

Figure 5.5 Momentum function M versus depth y for a rectangular channel section.

5.1.5 Comment

Equations 5.1.3 to 5.1.7 and 5.1.11 apply only to channels with rectangular cross-sections. Likewise, Figs. 5.3 and 5.5 pertain to rectangular channels; however the general trends of the curves are valid for non rectangular section geometry as well. The various formulas developed in this section are summarized in Table 5.3.

TABLE 5.3 Formulas for Rectangular and General Sections

Section	Fr	y_c	E	M
Rectangular	$\dfrac{q}{\sqrt{gy^3}}$	$\left(\dfrac{q^2}{g}\right)^{1/3}$	$y+\dfrac{q^2}{2gy^2}$	$\dfrac{by^2}{2}+\dfrac{q^2}{gy}$
General	$\dfrac{Q/A}{\sqrt{gA/B}}$	$\dfrac{Q^2 B}{gA^3}=1$	$y+\dfrac{Q^2}{2gA^2}$	$A\bar{y}+\dfrac{Q^2}{gA}$

5.2 Rapidly-Varied Flow; Transitions

Rapidly-varied flow exhibits quick changes in depth and velocity. It may take place over relatively short reaches of a channel, termed a *transition*, where a change in width or bottom elevation may occur. In addition, rapidly-varied flow may also exist with no geometric changes in the channel geometry; for example, boundary conditions create the situation known as a hydraulic jump. The energy and momentum equations are combined in a selective manner with the continuity equation to provide solutions for rapidly-varied flow situations.

5.2.1 Hydraulic Jump

A *hydraulic jump* is a phenomenon in which a supercritical flow undergoes a transition to the subcritical state. An idealized jump in a rectangular channel is shown in Fig. 5.6. Equation 5.1.11 is substituted into Eq. 5.1.9, and with $F = 0$ the result is solved for the ratio y_2/y_1:

$$\frac{y_2}{y_1} = \frac{1}{2}\left(\sqrt{1+8Fr_1^2} - 1\right) \tag{5.2.1}$$

Note that Eq. 5.2.1 is also valid if the subscripts on the depths and Froude number are reversed. In Fig. 5.6 h_j is the energy loss associated with the jump. Evaluation of h_j is shown in Example 5.4.

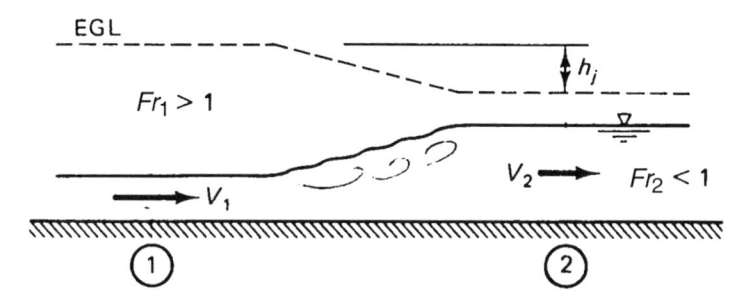

Figure 5.6 A hydraulic jump.

─────── **EXAMPLE 5.4** ───────

A hydraulic jump is situated in a 12-ft-wide rectangular canal. The discharge is 250 ft³/sec, and the depth upstream of the jump is 0.65 ft. Determine the depth downstream of the jump, and the rate of energy dissipated by the jump.

Solution. Compute the specific discharge and upstream Froude number:

$$q = \frac{Q}{b} = \frac{250}{12} = 20.83 \text{ ft}^2/\text{sec}$$

$$Fr_1 = \frac{q}{\sqrt{gy_1^3}} = \frac{20.83}{\sqrt{32.2 \times 0.65^3}} = 7.01$$

Using Eq. 5.2.1 the downstream depth is evaluated to be:

$$y_2 = \frac{y_1}{2}\left(\sqrt{1+8Fr_1^2} - 1\right)$$

$$= \frac{0.65}{2}\left(\sqrt{1+8\times 7.01^2} - 1\right) = 6.13 \text{ ft}$$

The rate of energy dissipation is given by $\dot{W} = \gamma Q h_j$. First compute the head loss:

$$h_j = E_1 - E_2$$
$$= y_1 + \frac{q^2}{2gy_1^2} - y_2 - \frac{q^2}{2gy_2^2}$$
$$= 0.65 + \frac{20.83^2}{2 \times 32.2 \times 0.65^2} - 3.38 - \frac{20.83^2}{2 \times 32.2 \times 3.38^2} = 12.63 \text{ ft}$$

Then compute $\dot{W}$:

$$\dot{W} = \gamma Q h_j$$
$$= 62.4 \times 250 \times 12.63 = 197,500 \text{ ft-lb/sec, or 358 horsepower}$$

5.2.2 Channel Expansions and Contractions

Transitions for subcritical flow conditions in which changes take place in the channel geometry can be analyzed by use of the energy equation. Consider a rectangular canal whose bottom elevation is raised by a magnitude h over a short region, Fig. 5.7. Recognizing that $E = y + V^2/2g$, Eq. 5.1.2, applied over the transition, is

$$E_1 = E_2 + h \qquad (5.2.2)$$

where the head loss has been neglected. The depth y_2 at the end of the transition is visualized on the E-y diagram, Fig. 5.7. Since the flow at location 1 is subcritical, the depth is located on the upper leg as shown. The magnitude of h is sufficiently small so that $y_2 > y_c$ and, as a result, the flow at location 2 is likewise subcritical. As h is increased, y_2 approaches y_c. Once $y_2 = y_c$ the flow is "choked," and any increases in h will result in flow changes no longer localized at the transition. (A narrowing of the channel width will create a situation similar to a raised elevation of the channel bottom.) Either lowering the channel bottom or increasing the channel width will result in $y_2 > y_1$ provided $y_1 > y_c$.

(a) raised channel bottom

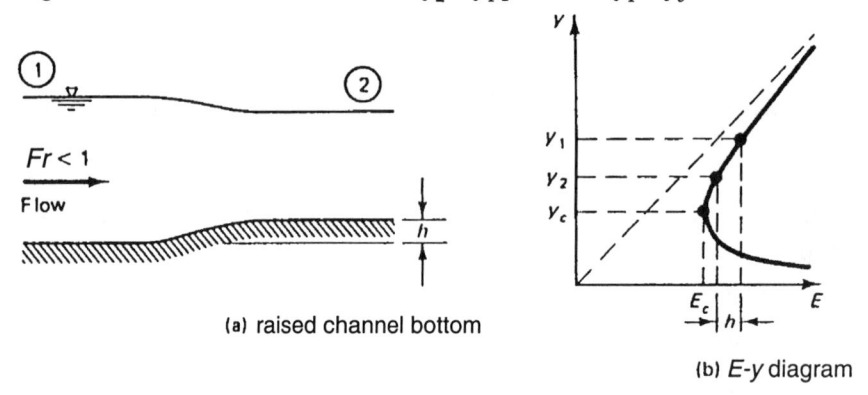

(b) E-y diagram

Figure 5.7 Channel constriction.

Energy losses usually are relatively small for subcritical flow conditions. If desired, losses can be included in Eq. 5.2.2; however, they will not significantly alter the results. Supercritical flows in transitions exhibit standing waves; proper design requires an understanding of wave mechanics, which is beyond the scope of this presentation.

EXAMPLE 5.5

A rectangular channel 10 ft wide carries water at a depth of 5 ft and a velocity of 6 ft/sec. A transition exists as shown in Fig. 5.7, where the rise in bottom elevation is 0.67 ft. Neglecting losses, (a) compute the depth and velocity at location 2, and (b) the change in bottom elevation for choking to occur.

Solution. (a) The specific discharge is computed from Eq. 5.1.4:

$$q = V_1 y_1 = 6 \times 5 = 30 \text{ ft}^2/\text{sec}$$

The Froude number at location 1 is

$$Fr = \frac{q}{\sqrt{gy_1^3}} = \frac{30}{\sqrt{32.2 \times 5^3}} = 0.47$$

which is less than one, hence the flow at location 1 is subcritical. The specific energy at location 2 is found using Eq. 5.2.2:

$$E_2 = y_1 + \frac{q^2}{2gy_1^2} - h$$

$$= 5 + \frac{30^2}{2 \times 32.2 \times 5^2} - 0.67 = 5.60 - 0.67 = 4.89 \text{ ft}$$

If $E_2 > E_c$ it is possible to determine y_2. Hence, compute E_c:

$$E_c = \frac{3}{2} y_c = \frac{3}{2} \left(\frac{q^2}{g} \right)^{1/3} = \frac{3}{2} \left(\frac{30^2}{32.2} \right)^{1/3} = 4.55 \text{ ft}$$

Since $E_2 > E_c$ the depth y_2 at location 2 is subcritical (i.e., $y_2 > y_c$) and can be evaluated by substituting known values into Eq. 5.2.2:

$$4.89 = y_2 + \frac{30^2}{2 \times 32.2 \times y_2^2} = y_2 + \frac{14}{y_2^2}$$

Solving by trial and error yields $y_2 = 4.0$ ft and $V_2 = q/y_2 = 30/4 = 7.5$ ft/sec.

(b) The value of h for critical flow to occur at location 2 is found by setting $E_2 = E_c$ in Eq. 5.2.2:

$$h = E_1 - E_c$$
$$= 5.60 - 4.55 = 1.05 \text{ ft}$$

Similar analyses prevail for other types of channel constrictions or expansions. The general procedure is as follows:

1. Write the energy equation between the two sections in question, including any changes in channel geometry, such as elevation change of channel bottom or constriction or expansion of channel width.

2. Introduce known data into energy equation.

3. Solve for the unknown parameter (depth, channel elevation, or channel width).

5.2.3 Measurement of Discharge; Weirs

Flow is commonly measured using a weir; two types of weirs are shown in Fig. 5.8. Under design conditions, the flow upstream of the weir will be subcritical, and the downstream overflow exhibits a nappe, which flows freely into the atmosphere. The broad-crested weir, Fig. 5.8a, has sufficient elevation above

(a) broad-crested weir   (b) sharp-crested weir

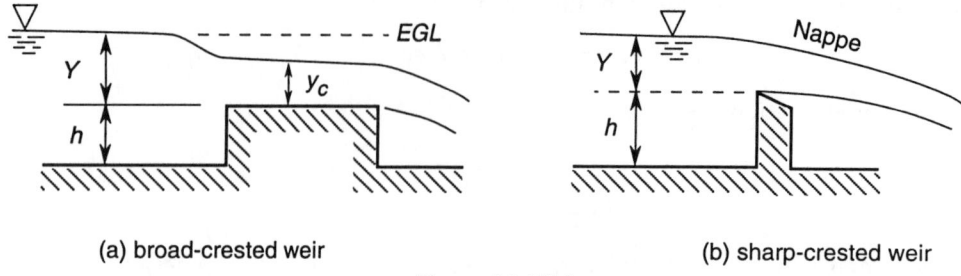

Figure 5.8 Weirs.

the bottom of the channel, and is long enough to cause critical flow to occur at some position on the top. For a nearly horizontal weir of height h and width b, application of the energy equation, assuming negligible losses, will provide the following expression for the discharge:

$$Q = \left(\frac{2}{3}\sqrt{\frac{2}{3}g}\right)bY^{3/2}$$
$$= 3.087bY^{3/2}$$
(5.2.3)

The parameter Y is the vertical distance from the top of the weir to the free surface at a location upstream of the weir where the flow is relatively undisturbed, Fig. 5.8a. Equation 5.2.3 is accurate to within several percent of the actual flow for a well-designed broad-crested weir.

A rectangular sharp-crested weir, Fig. 5.8b, has a horizontal sharp-edge crest placed normal to the flow with the nappe behaving as a free jet. The energy equation can be applied to provide an estimate for the discharge. However, unlike the broad-crested weir, it is necessary to include a discharge coefficient C_d to account for the effect of contractions and losses that occur over the crest. The resulting expression is

$$Q = C_d\left(\frac{2}{3}\sqrt{2g}\right)bY^{3/2}$$
$$= 5.35 C_d bY^{3/2}$$
(5.2.4)

For $Y/h < 0.1$, use $C_d = 0.61$; otherwise use $C_d = 0.61 + 0.08 Y/h$.

───── **EXAMPLE 5.6** ─────

Determine the discharge of water over a rectangular sharp-crested weir, $b = 6.0$ ft, $Y = 1.25$ ft, $h = 3.7$ ft.

Solution. Since $Y/h = 1.25/3.7 = 0.34$, the discharge coefficient is

$$C_d = 0.61 + 0.08 \times 0.34 = 0.64$$

Hence the discharge is

$$Q = 5.35 \times 0.64 \times 6 \times 1.25^{3/2} = 28.7 \text{ ft}^3/\text{sec}$$

5.3 Nonuniform, Gradually Varied Flow

In contrast to rapidly varied flow, gradually varied flow consists of long reaches of open channel where the depth and velocity do not experience sudden changes, but instead vary slowly. The shear stress is the primary mechanism that resists the flow.

5.3.1 Water Surface Profiles and Controls

Water surface profiles can be classified according to the position of y relative to uniform depth y_o (i.e., the depth at which uniform flow occurs, see Section 5.1.2) and critical depth y_c. Note that y_o and y_c are uniquely determined once the channel properties and discharge are known. Table 5.4 gives a classification for three commonly encountered types of profiles: mild (M), steep (S), and horizontal (H). The existence of the profiles shown depend upon the boundary conditions that are specified at given locations in the channel. Very often a control will define that condition. A *control* occurs when a known depth-discharge relation occurs at a section. The profiles shown in Table 5.4 are all influenced by controls; examples of controls include sluice gates, weirs, dams, and constricting flumes, all of which may force a rapidly-varied flow situation, with critical flow occurring within the transition.

Appropriate identification of controls and their interaction with the corresponding profiles of Table 5.4 is necessary for correct design and analysis procedures. Note that controls are basically the short reach transitions, and that the rapidly-varied flow procedures presented in Section 5.2 will determine the required depth-discharge relation. Once the controls have been established, the profiles of Table 5.5 can be selected and the range of influence of those controls can be identified. Two general rules regarding controls that may be observed are the following:

1. When the channel profile is in the subcritical range (M1, M2, S1, H2), that profile will be influenced by a control at the downstream end of the reach.
2. When the channel profile is in the supercritical range (M3, S2, S3, H3), that profile will be influenced by a control at the upstream end of the reach.

TABLE 5.4 Water Surface Profiles

Channel slope	Profile type	Depth range	Fr	dy/dx	dE/dx	
Mild	M1	$y > y_o > y_c$	<1	>0	>0	
	M2	$y_o > y_c > y$	<1	<0	<0	
	M3	$y_o > y_c > y$	>1	>0	<0	
Steep	S1	$y > y_c > y_o$	<1	>0	>0	
	S2	$y_c > y > y_o$	>1	<0	>0	
	S3	$y_c > y_o > y$	>1	>0	<0	
Horizontal	H2	$y > y_c$	<1	<0	<0	
	H3	$y_c > y$	>1	>0	<0	

EXAMPLE 5.8

Water is flowing in a long rectangular channel ($b = 12$ ft, $n = 0.012$, $S_o = 0.0013$). For a discharge of 1000 ft³/sec, the depth at the upstream end of the channel is 2.0 ft, and at the downstream end the flow is critical. Determine the nature of the water surface and the energy grade line.

Solution. First the critical and uniform depths are computed.

$$y_c = \left(\frac{q^2}{g}\right)^{1/3} = \left(\frac{(1000/12)^2}{32.2}\right)^{1/3} = 5.99 \text{ ft}$$

Uniform depth is computed by trial and error using Eq. 5.1.1:

$$\frac{Qn}{1.49\sqrt{S_o}} = AR^{2/3}$$

$$\frac{1000 \times 0.012}{1.49 \times \sqrt{0.0013}} = 12y_o \left(\frac{12y_o}{12 + 2y_o}\right)^{2/3}$$

$$18.61 = y_o \left(\frac{12y_o}{12 + 2y_o}\right)^{2/3}$$

Solving by trial and error, $y_o = 8.15$ ft. Since $y_o > y_c$, the profile is mild. Furthermore, the depth at the entrance is less than critical depth, hence an M3 profile will exist downstream of the entrance. An M2 profile occurs upstream from the exit. The M3 profile is in the supercritical range, thus it will join the subcritical M2 profile by means of a hydraulic jump somewhere within the channel. The exact location of the jump can only be determined by means of a numerical analysis, as demonstrated in the next section.

5.3.2 Equation for Gradually Varied Flow

The energy equation is applied from location 1 to location 2 in an incremental reach of gradually varied flow, Fig. 5.9. The resulting differential equation is

$$\frac{dE}{dx} = S_o - S \tag{5.3.1}$$

in which S_o = slope of channel bottom, x = distance parallel to channel bottom, and S = slope of energy grade line. The objective of numerical analysis is to predict the variation in depth and energy grade line with respect to the distance. Equation 5.3.1 is written in finite-difference form between two locations x_i and x_{i+1}:

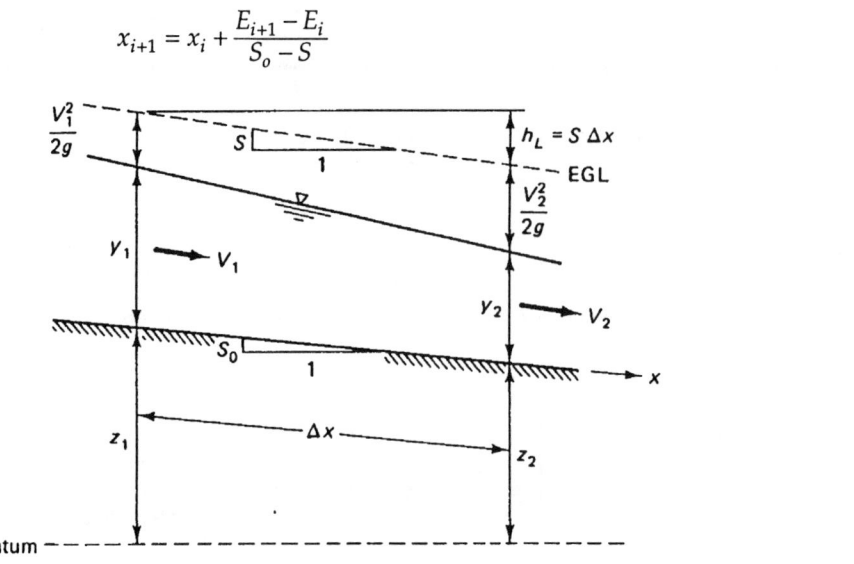

$$x_{i+1} = x_i + \frac{E_{i+1} - E_i}{S_o - S} \tag{5.3.2}$$

Figure 5.9 Gradually varied flow.

The variable S is determined from the Chezy-Manning equation:

$$S = \left(\frac{Qn}{1.49}\right)^2 \frac{1}{A^2 R^{4/3}} \tag{5.3.3}$$

Note that S is dependent upon the depth as well as Q and n.

The calculations take place stepwise, starting at some location where the depth is known. For example, consider generating values of x_i, y_i, and E_i for $i = 1,2,...,k$, where k is the location at the opposite side of the reach. The evaluation of conditions from location i, where y_i and E_i are known, to location $i+1$ proceeds as follows:

1. Choose y_{i+1}.
2. Evaluate S with Eq. 5..3.3, using the mean value $y_m = (y_{i+1} + y_i)/2$ when calculating R and A.
3. Compute E_i from Eq. 5.1.3 or Eq. 5.1.5 (rectangular channel).
4. Compute x_i from Eq. 5.3.2.
5. The conjugate depth can be calculated, if it is required to determine the possible location of a hydraulic jump.

―――― **EXAMPLE 5.9** ――――

Water is flowing at a discharge of 750 ft³/sec in a rectangular channel whose width is 30 ft and length is 3000 ft. Beginning at the downstream end, where the depth is 5.3 ft, compute the water surface and energy grade line over the length of the channel. Use $n = 0.015$, and $S_o = 0.0006$.

Solution. The critical and normal depths are computed to be $y_c = 2.69$ ft and $y_o = 4.50$ ft. The water surface is an M1 profile, with depth control at the downstream end of the channel. The solution is tabulated below.

i	y_i (ft)	V_i (ft/sec)	E_i (ft)	y_m (ft)	$R(y_m)$ (ft)	$S(y_m)$	x_i (ft)
1	5.30	4.72	5.65				3000
2	5.15	4.85	5.52	5.23	3.88	0.00038	2407
3	5.00	5.00	5.39	5.08	3.79	0.00042	1713
4	4.85	5.15	5.26	4.93	3.71	0.00046	845
5	4.70	5.32	5.14	4.78	3.62	0.00050	-383

Starting with $y_1 = 5.3$ ft, subsequent depths are reduced by 0.15 ft, and the corresponding values of x_i are computed until the last value (x_5) is negative.

―――― **EXAMPLE 5.10** ――――

With the data provided in Example 5.8, compute the water surface profile and energy grade line, and determine the location of the hydraulic jump for a channel length of 1500 ft.

Solution. The normal and critical depths were found to be 8.15 and 5.99 ft, respectively. It was concluded that the upper portion of the channel contained an M3 curve, terminating in a hydraulic jump, followed by an M2 profile in the downstream reach. The calculations using the standard step method are tabulated below in two parts. First the M3 profile is computed until the depth approaches critical conditions.

i	y_i (ft)	E_i (ft)	y_m (ft)	$R(y_m)$ (ft)	$S(y_m)$	x_i (ft)
1	2.00	28.96				0
2	3.00	14.98	2.50	1.76	0.03380	430
3	4.00	10.74	3.50	2.21	0.01277	800
4	5.00	9.31	4.50	2.57	0.00632	1084
5	6.00	9.00	5.50	2.87	0.00365	1219

Next the M2 profile is calculated beginning at the downstream end, until the until the channel entrance is approached. In addition the conjugate depth y_j is computed from Eq. 5.2.1 and tabulated.

i	y_i (ft)	E_i (ft)	y_m (ft)	$R(y_m)$ (ft)	$S(y_m)$	x_i (ft)	y_j (ft)
1	6.00	9.00				1500	
2	6.50	9.05	6.25	3.06	0.00259	1456	5.52
3	7.00	9.20	6.75	3.18	0.00212	1275	5.09
4	7.50	9.42	7.25	3.28	0.00176	801	4.71

To locate the hydraulic jump, one can plot the curves of y_j and y_i versus x_i; their intersection is the position of the jump. Alternately, one can observe from the tables that the jump will occur between locations 3 and 4 on the M1 profile, and between locations 3 and 4 on the M2 profile. The intersection of the two lines between the pairs of points will show that the jump occurs at $x \cong 1060$ ft.

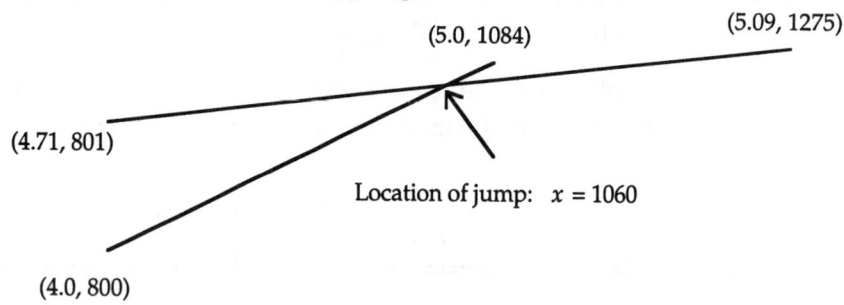

# Practice Problems (Essay)

5.1 Determine the uniform depth y, area A, and wetted perimeter P for a rectangular channel with the following conditions: b = 25 ft, Q = 500 ft3/sec, n = 0.020, and So = 0.0004.

5.2 Water is flowing in a rectangular channel at a velocity of 10 ft/sec and a depth of 8 ft. Determine the changes in the water surface elevation for an increase (upward step) of 8 inches in the channel bottom.

5.3 Steep, that is, critical flow conditions exist at the outlet from a reservoir into an open canal. The canal width is not yet fixed, but the required depth at the entrance is 3 ft and the required discharge is 635 ft3/sec. Find the necessary bottom widths of the canal for the following geometry: (a) rectangular section, and (b) trapezoidal section, m =3.

5.4 Rapidly-varied flow takes place over a rectangular sill, with free outfall conditions at location C. Throughout the reach losses can be neglected. Determine the following: (a) discharge, and (b) depths at locations A, B, and C.

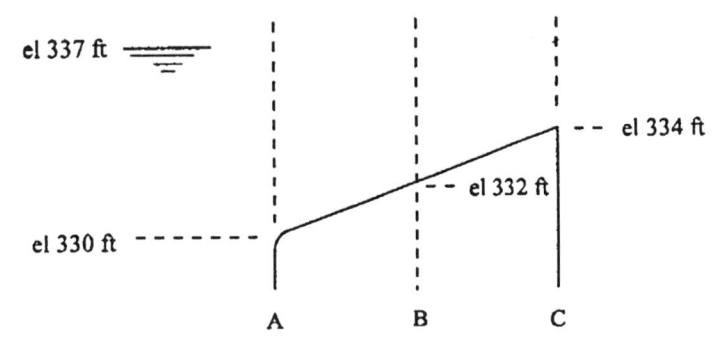

5.5 Water is flowing in a rectangular channel, b = 20 ft. A hydraulic jump occurs, with the downstream depth y_2 = 4.5 ft. The discharge Q = 400 ft³/sec. Compute the upstream depth y_1 and the power dissipated by the jump.

5.6 A flow of 300 ft³/sec occurs in a long rectangular channel 10 ft wide with the normal depth y_o = 5 ft. There is a smooth constriction in the channel to 6 ft wide. Determine (a) the depths within and immediately upstream of the constriction, and (b) the nature of the gradually varied flow upstream of the constriction.

5.7 Water is flowing at a discharge of Q = 88 ft³/sec in a 1600-ft-long circular conduit with d = 8 ft, S_o = 0.001, and n = 0.015. At the inlet the water depth is 1.3 ft and at the outlet critical flow conditions occur. The critical depth is computed to be y_c = 2.31 ft. Identify the nature of the water surface in the conduit.

5.8 A sluice gate is placed in a long rectangular channel, b = 15 ft, n = 0.014, and S_o = 0.0008. The depth upstream of the gate is y_1 = 6.1 ft and downstream of the gate the depth is y_2 = 1.1 ft. Identify the water surface profiles on either side of the gate if the discharge is Q = 300 ft³/sec.

5.9 Gradually-varied flow takes place in reach of rectangular channel with n = 0.013, S_o = 0.005, and b = 8 ft. At location 1 the depth is 3.5 ft, and at location 2 the depth is 4.0 ft. If the two locations are 200 ft apart, find the discharge. Use one reach.

5.10 Water is flowing at a discharge of 1200 ft³/sec in a 40-ft-wide rectangular channel, whose bottom slope is 0.0008 and Manning coefficient is 0.015. It passes over a small dam, creating a "backwater" curve. Immediately upstream of the dam the water depth is 8.5 ft, and at a distance upstream of the dam the flow is at normal conditions. The normal depth is y_o = 4.5 ft. Determine (a) the profile type, and (b) the distance upstream from the dam to where the normal depth occurs. Use four reaches.

Practice Problems (Multiple Choice)

Questions 5.11-5.16

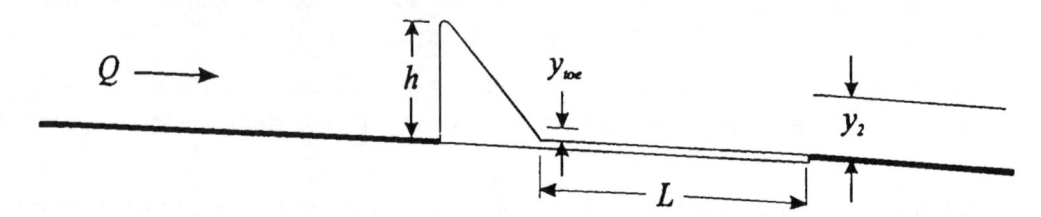

A design flow of $Q = 680$ ft³/sec occurs in a river that can be approximated as a wide rectangular channel with the following properties: $b = 60$ ft, $S_o = 0.0005$, $n = 0.016$. A small dam is to be placed in the channel as part of a flood control plan. The height of the dam is $h = 20$ ft, and the depth of the flow at the toe of the dam is $y_{toe} = 0.35$ ft. A concrete spillway apron ($n = 0.014$) is to be situated downstream of the dam. The purpose of the apron is to keep any hydraulic jump that may occur downstream of the dam off of the river bed for the given design conditions.

5.11 Critical flow conditions prevail over the crest of the dam. The critical depth y_c, in feet, is

   a) 21.58   b) 0.35   c) 1.58   d) 2.87   e) 0.93

5.12 The normal depth in the river y_o, in feet, is

   a) 2.87   b) 2.50   c) 1.58   d) 21.58   e) 0.93

5.13 If the depth behind the jump at the end of the apron is $y_2 = 2.5$ ft, then the depth y_1, in feet immediately ahead of the jump, is

   a) 0.64   b) 1.58   c) 0.35   d) 0.93   e) 2.87

5.14 The power $\dot{W}$, in horsepower, dissipated by the hydraulic jump, is

   a) 17,280   b) 0.42   c) 324   d) 1,540   e) 32.4

5.15 The length of the apron is the distance from the toe to downstream of the jump. If the length of the jump is 6 y_2, and the depth immediately upstream of the jump is $y_1 = 0.9$ ft, then the length of the apron L, in feet, is approximately

   a) 15   b) 100   c) 180   d) 200   e) 165

5.16 The two gradually-varied-flow profiles upstream of the dam, and between the toe and the hydraulic jump, respectively, are

   a) M1, M2   b) M1, M3   c) S1, M3   d) S1, S3   e) M3, M1

Questions 17-20

A trapezoidal canal has the following properties: $b = 10$ ft, side slopes 1 vertical to 2 horizontal ($m = 2$), $n = 0.012$, and $S_o = 0.0002$. Water is flowing in the canal at a normal depth of $y_o = 3$ ft. Two circular concrete conduits ($n = 0.013$) placed at the same bottom slope as the canal are to convey the flow underneath a roadway. It is required that the conduits flow one-half full.

5.17 The discharge in the canal and conduits Q, in ft³/sec, is

   a) 125   b) 80   c) 167   d) 152   e) 136

5.18 Assuming normal flow conditions have been attained in the conduits, then the required diameter d of the conduits, d, to the nearest ft, is

   a) 8   b) 7   c) 10   d) 16   e) 6

5.19 If the slope in the trapezoidal canal is changed to $S_o = 0.0003$, the altered normal depth y_o, in feet, is

   a) 3.0   b) 3.8   c) 2.7   d) 4.1   e) 5.2

5.20 Assume that a rectangular 25-ft-wide broad-crested weir is located at some location in the canal. Upstream of the weir, the height of the water surface relative to the top of the weir, in feet, will be

   a) 0.97   b) 2.34   c) 3.10   d) 1.46   e) 2.15

Solutions to Problems

5.1 Using the Chezy-Manning equation, substitute in the appropriate expressions for A and $R = A/P$:

$$Q = \frac{1.49}{n} AR^{2/3} \sqrt{S_o} = \frac{1.49}{n} \frac{A^{5/3}}{P^{2/3}} \sqrt{S_o} = \frac{1.49}{n} \frac{(by)^{5/3}}{(b+2y)^{2/3}} \sqrt{S_o}$$

Substitute in known data and reduce the resulting equation:

$$500 = \frac{1.49(25y)^{5/3}\sqrt{0.0004}}{0.020(25+2y)^{2/3}}, \quad \text{or} \quad 1.568 = \frac{y^{5/3}}{(25+2y)^{2/3}}$$

Solving by trial and error, we find $y = 5.49$ ft. Now compute A and P:
$A = 25 \times 5.49 = 137$ ft², and $P = 25 + 2 \times 5.49 = 36$ ft

5.2 Given: $V_1 = 10$ ft/sec, $y_1 = 8$ ft, and $h = 8/12$ ft. Compute the unit discharge, energy at the upstream location 1, and the critical depth:

$$q = V_1 y_1 = 10 \times 8 = 80 \text{ ft}^2/\text{sec}, \quad E_1 = y_1 + \frac{V_1^2}{2g} = 8 + \frac{10^2}{2 \times 32.2} = 9.55 \text{ ft},$$

$$y_c = \sqrt[3]{\frac{q^2}{g}} = \sqrt[3]{\frac{80^2}{32.2}} = 5.83 \text{ ft, and } E_c = \frac{3}{2} y_c = \frac{3}{2} \times 5.83 = 8.74 \text{ ft}$$

The energy equation taken from location 1 upstream of the transition to location 2 in the transition is $E_2 = E_1 - h = 9.55 - 8/12 = 8.88$ ft. Since $E_2 > E_c$, the depth at location 2 is subcritical. Therefore

$$E_2 = y_2 + \frac{q^2}{2gy_2^2}, \quad \text{or} \quad 8.88 = y_2 + \frac{80^2}{2 \times 32.2 \times y_2^2} = y_2 + \frac{99.38}{y_2^2}$$

Solving by trial and error yields $y_2 = 6.60$ ft. Hence the change in the water surface is
$y_2 + h - y_1 = 6.6 + 0.67 - 8 = -0.73$ ft

5.3 At the canal entrance, the condition of critical flow is $Fr^2 = Q^2B/(gA^3) = 1$, to be solved for the unknown width b.

  a) Rectangular channel: $B = b$, $A = by_c$, and

$$\therefore \frac{Q^2 b}{g(by_c)^3} = \frac{Q^2}{gb^2 y_c^3} = 1, \text{ or } b = \frac{Q}{\sqrt{gy_c^3}} = \frac{635}{\sqrt{32.2 \times 3^3}} = 21.5 \text{ ft}$$

  b) Trapezoidal channel:
  $B = b + 2my_c = b + 2 \times 3 \times 3 = b + 18$, and
  $A = by_c + my_c^2 = b \times 3 = 3 \times 3^2 = 3b + 27$

$$\therefore \frac{Q^2 B}{gA^3} = \frac{635^2 (b+18)}{32.2 \times (3b+27)^3} = 1, \text{ or } \frac{(b+9)^3}{(b+18)} = 464$$

  Solving by trial and error gives us $b = 16.1$ ft.

5.4 a) Write the energy equation from the reservoir to section C, where critical conditions exist:

  $y_R + z_R = E_C + z_C$, or $E_C = y_R + z_R - z_C = 337 - 334 = 3$ ft.

  Now the critical depth and specific discharge can be computed:

$$y_C = \tfrac{2}{3} E_C = \tfrac{2}{3} \times 3 = 2 \text{ ft, and } q = \sqrt{gy_C^3} = \sqrt{32.2 \times 2^3} = 16.05 \text{ ft}^2/\text{sec}$$

  b) The energy equation from the reservoir to section A is:

$$337 = y_A + \frac{16.05^2}{2 \times 32.2 \times y_A^2} + 330, \text{ or } 7 = y_B + \frac{4}{y_B^2}$$

  Solving by trial and error, $y_A = 6.92$ ft. In a similar fashion, the energy equation is written from the reservoir to section B, and the depth is computed to be $y_B = 4.83$ ft.

5.5 Compute the specific discharge and the downstream Froude number:

$$q = \frac{Q}{b} = \frac{400}{20} = 20 \text{ ft}^2/\text{sec}, \quad Fr_2 = \frac{q}{\sqrt{gy_2^3}} = \frac{20}{\sqrt{32.2 \times 4.5^3}} = 0.37$$

  The upstream depth is

$$y_1 = \frac{y_2}{2}\left(\sqrt{1 + 8Fr_2^2} - 1\right) = \frac{4.5}{2}\left(\sqrt{1 + 8 \times 0.37^2} - 1\right) = 1.00 \text{ ft}$$

  Compute the head loss across the jump:

$$h_j = y_1 + \frac{q^2}{2gy_1^2} - y_2 - \frac{q^2}{2gy_2^2} = 1.00 + \frac{400}{2 \times 32.2 \times 1.00^2} - 4.5 - \frac{400}{2 \times 32.2 \times 4.5^2} = 2.40 \text{ ft}$$

  Hence the rate of energy dissipation is

$$\dot{W} = \gamma Q h_j = 62.4 \times 400 \times 2.4 = 60{,}000 \text{ ft-lb/sec, or 109 horsepower}$$

5.6 Compute the specific discharges and critical depths at location 1, upstream of the transition, and at location 2 in the transition:

$$q_1 = \frac{300}{10} = 30 \text{ ft}^2/\text{sec}, \quad (y_c)_1 = \sqrt[3]{\frac{30^2}{32.2}} = 3.03 \text{ ft},$$

$$q_2 = \frac{300}{6} = 50 \text{ ft}^2/\text{sec}, \quad (y_c)_2 = \sqrt[3]{\frac{50^2}{32.2}} = 4.26 \text{ ft}$$

  The critical energy at location 2 is $(E_c)_2 = 1.5(y_c)_2 = 1.5 \times 4.26 = 6.39$ ft. The specific energy at normal flow conditions is

$$E_o = y_o + \frac{q_1^2}{2gy_o^2} = 5 + \frac{30^2}{2 \times 32.2 \times 5^2} = 5.56 \text{ ft}$$

Since $E_o < (E_c)_2$, choking (a critical condition) occurs at location 2, and the depth at location 1 is greater than y_o. To compute y_1, set $E_1 = (E_c)_2$:

$$y_1 + \frac{q_1^2}{2gy_1^2} = y_1 + \frac{30^2}{2 \times 32.2 y_1^2} = y_1 + \frac{14}{y_1^2} = 6.39$$

A trial and error solution yields $y_1 = 6.0$ ft. Since $y_o > (y_c)_1$, a mild slope condition exists, and an M1 curve occurs upstream of the constriction.

5.7 First we need to compute the normal depth:

$$\frac{Qn}{1.49\sqrt{S_o}} = \frac{\left[d^2/4(\alpha - \sin\alpha \cos\alpha)\right]^{5/3}}{(\alpha d)^{2/3}}$$

Substituting in the given data and reducing we have

$$1.10 = \frac{(\alpha - \sin\alpha \cos\alpha)^{5/3}}{(\alpha)^{2/3}}$$

Solving by trial and error one finds $\alpha = 1.385$ radians, or 79 degrees. Hence the normal depth is

$$y_o = \frac{d}{2}(1 - \cos\alpha) = \frac{8}{2} \times (1 - \cos 1.385) = 3.26 \text{ ft}$$

Since $y_o > y_c$, a mild slope condition exists. The water surface consists of an M3 curve beginning at the inlet, followed by a hydraulic jump to an M2 curve, which terminates at critical depth at the outlet.

5.8 Compute the specific discharge and critical depth:

$$q = 300/15 = 20 \text{ ft}^2/\text{sec}, \quad \text{and} \quad y_c = \sqrt[3]{20^2/32.2} = 2.31 \text{ ft}$$

Compute the normal depth using the Chezy-Manning equation:

$$\frac{Qn}{1.49\sqrt{S_o}} = \frac{300 \times 0.014}{1.49 \times \sqrt{0.0008}} = 99.7 = \frac{(15y_o)^{5/3}}{(15 + 2y_o)^{2/3}}$$

Solving by trial and error, $y_o = 3.65$ ft. Since $y_o > y_c$, mild channel conditions exist. Upstream of the gate there will be an M1 profile, with an M3 profile downstream of the gate terminating in a hydraulic jump to normal flow conditions.

5.9 Evaluate Q using the varied flow equation. First evaluate the specific energies and the slope of the energy grade line:

$$E_1 = y_1 + \frac{Q^2}{2gA_1^2} = 3.5 + \frac{Q^2}{2 \times 32.2 \times (8 \times 3.5)^2} = 3.5 + \frac{Q^2}{50490},$$

$$E_2 = y_2 + \frac{Q^2}{2gA_2^2} = 4 + \frac{Q^2}{2 \times 32.2 \times (8 \times 4)^2} = 4 + \frac{Q^2}{65950},$$

$$y_m = \frac{1}{2}(y_1 + y_2) = \frac{1}{2}(3.5 + 4) = 3.75 \text{ ft,}$$

$$S = \left(\frac{Qn}{1.49}\right)^2 \frac{(b + 2y_m)^{4/3}}{(by_m)^{10/3}} = \frac{Q^2 \times 0.013^2 \times (8 + 2 \times 3.75)^{4/3}}{1.49^2 (8 \times 3.75)^{10/3}} = 3.507 \times 10^{-8} Q^2$$

Substitute into the varied-flow equation $\Delta x(S_o - S) = E_2 - E_1$:

$$200\left(0.005 - 3.507 \times 10^{-8} Q^2\right) = 4 + \frac{Q^2}{65950} - 3.5 - \frac{Q^2}{50490}, \text{ or}$$

$$2.37 \times 10^{-6} Q^2 = 0.5$$

Hence $Q = 459$ ft³/sec.

5.10   a) Compute specific discharge and critical depth:

$$q = 1200/40 = 30 \text{ ft}^2/\text{sec}, \text{ and } y_c = \sqrt[3]{30^2/32.2} = 3.03 \text{ ft}$$

Since $y_o > y_c$, mild flow conditions prevail. From Table 5.4, an M1 profile exists upstream of the dam.

b) Use the step method to evaluate the M1 profile. First set up the functions S and E:

$$S = \left(\frac{Qn}{1.49}\right)^2 \frac{1}{A^2 R^{4/3}} = \frac{145.9}{A^2 R^{4/3}}, \quad A = 40y, \quad R = \frac{40y}{40 + 2y}$$

$$E = y + \frac{q^2}{2gy^2} = y + \frac{30^2}{2 \times 32.2 \times y^2} = y + \frac{14}{y^2}$$

Then use the relation $x_{i+1} = x_i + (E_{i+1} - E_i)/(S_o - S) = x_i + (E_{i+1} - E_i)/(0.0008 - S)$ to complete the M1 profile evaluation. The results are tabulated below:

i	y_i	E_i	y_m	$A(y_m)$	$R(y_m)$	$S(y_m)$	x_i
1	8.50	8.69					0
2	7.50	7.75	8.00	320	5.71	0.000140	-1431
3	6.50	6.83	7.00	280	5.19	0.000207	-2979
4	5.50	5.96	6.00	240	4.62	0.000330	-4826
5	4.50	5.19	5.00	200	4.00	0.000575	-8251

Note that since we are computing the M1 profile upstream from the starting depth, the x-values are negative. The approximate location where normal depth is reached is 8250 ft upstream of the dam.

5.11   c) The critical depth is $y_c = \sqrt[3]{(680/60)^2/32.2} = 1.58$ ft.

5.12   a) Set up the Chezy-Manning equation:

$$680 = \frac{1.49}{0.016}(60y_o)\left(\frac{60y_o}{60+2y_o}\right)^{2/3} \sqrt{0.0005} = 1917 \times \frac{(y_o)^{5/3}}{(60+2y_o)^{2/3}}$$

Solving by trial and error, $y_o = 2.87$ ft.

5.13   d) Use the hydraulic jump equation to evaluate y_1. The downstream Froude number is

$$V_2 = \frac{680}{2.5 \times 60} = 4.53 \text{ ft/sec}, \quad Fr_2 = \frac{4.53}{\sqrt{32.2 \times 2.5}} = 0.505$$

$$\therefore y_1 = \frac{2.5}{2}\left(\sqrt{1 + 8 \times 0.505^2} - 1\right) = 0.93 \text{ ft}.$$

5.14   e) First find the energy loss across the jump:

$$q = \frac{680}{60} = 11.33 \text{ ft}^2/\text{sec}$$

$$h_j = E_1 - E_2 = 0.93 + \frac{11.33^2}{2 \times 32.2 \times 0.93^2} - 2.5 - \frac{11.33^2}{2 \times 32.2 \times 2.5^2} = 0.42 \text{ ft}$$

The power dissipated is $\dot{W} = \gamma Q h_j / 550 = 62.4 \times 680 \times 0.42 / 550 = 32.4$ Hp.

5.15 d) Use Eq. 5.3.2 to estimate the distance on the apron from the toe of the dam to immediately upstream of the jump (location 1):

$$E_{toe} = 0.35 + \frac{11.33^2}{2 \times 32.2 \times 0.35^2} = 16.62 \text{ ft}, \quad E_1 = 0.9 + \frac{11.33^2}{2 \times 32.2 \times 0.9^2} = 3.36 \text{ ft},$$

$$y_m = \frac{1}{2}(0.35 + 0.9) = 0.625 \text{ ft}, \quad A = 60 \times 0.625 = 37.5 \text{ ft}^2,$$

$$P = 60 + 2 \times .625 = 61.2 \text{ ft}, \quad S = \left(\frac{680 \times 0.014}{1.49 \times 37.5}\right)^2 \frac{1}{(37.5/61.2)^{4/3}} = 0.0558,$$

$$\therefore \text{ Profile length} = \frac{3.36 - 16.2}{0.0005 - 0.0558} = 232 \text{ ft}$$

The length of the jump is $6 \times 2.5 = 15$ ft. Hence the required length of the apron is $L = 232 + 15 = 247$ ft, or about 250 ft. (Normal design practice would require additional length as a safety factor.)

5.16 b) In a manner similar to Problem 12, the normal depth on the apron is $y_o = 2.65$ ft. Since y_o is greater than y_c in both the river and apron ($y_o = 2.87$ ft in river from Prob. 12), mild slope conditions prevail. Ahead of the dam, the depth is $h + y_c = 20 + 1.6 = 21.6$ ft, hence the profile there is an M1 curve. On the apron, between the toe and the hydraulic jump, an M3 curve exists.

5.17 e) Use the Chezy-Manning equation to determine the discharge:

$$A = 10 \times 3 + 2 \times 3^2 = 48 \text{ ft}^2, \quad P = 10 + 2 \times 3 \times \sqrt{1 + 2^2} = 23.4 \text{ ft}, \quad R = \frac{48}{23.4} = 2.05 \text{ ft}$$

$$\therefore Q = \frac{1.49}{0.012} \times 48 \times 2.05^{2/3} \times \sqrt{0.0002} = 136 \text{ ft}^3/\text{sec}$$

5.18 a) The conduits are flowing half full. For each conduit the area and hydraulic radius are $A = \pi d^2 / 8$, and $R = d/4$. Apply the Chezy-Manning equation to one of the conduits using one-half of the discharge:

$$68 = \frac{1.49}{0.013} \times \frac{\pi d^2}{8} \times \left(\frac{d}{4}\right)^{2/3} \times \sqrt{0.0002} = 0.252 d^{8/3}, \quad \therefore d = 8.2 \text{ ft}$$

The diameter to the nearest foot is $d = 8$ ft.

5.19 c) Use the Chezy-Manning equation to evaluate the normal depth:

$$136 = \frac{1.49}{0.012} \frac{A^{5/3}}{P^{2/3}} \sqrt{0.0003}, \quad \text{which reduces to} \quad 63.2 = \frac{\left(10 y_o + 2 y_o^2\right)^{5/3}}{\left(10 + 2\sqrt{5} y_o\right)^{2/3}}$$

A trial and error solution yields $y_o = 2.7$ ft.

5.20 e) Use Eq. 5.2.3 (assume that there are rectangular side walls in the vicinity of the weir):

$$136 = 3.087 \times 25 \times Y^{3/2}, \quad \therefore Y = 1.46 \text{ ft}.$$

6. Flow in Piping Systems

by David C. Wiggert

The design and analysis of piping systems and associated turbomachinery are common occurrences for civil and environmental engineers. The complexity of piping ranges from a single pipeline draining water from a reservoir to an extensive network supplying potable water to a metropolitan area. Piping networks are analyzed using computer algorithms, and they will not be considered in this chapter. We will focus on simpler systems that can be solved by hand calculation: pipe elements either operating singly, or joined to form series, parallel, or branching configurations. The analysis of a single pipeline is presented first, followed by the more complex systems that are treated using ad hoc solutions suitable for use with calculators. The basic idea behind the ad hoc method of solution presented herein is to identify all of the unknowns and write an equivalent number of independent equations to be solved. Subsequently, the system is simplified by eliminating as many unknowns as possible and reducing the problem to a series of single pipe problems. It is possible that the resulting equation or grouping of equations will be nonlinear, so that an iterative solution will be required; either a trial-and-error or numerical procedure could be utilized. An alternate method would be use of a graphical procedure. A brief introduction to pumps is provided, focusing primarily on centrifugal, or radial flow pumps operating in piping systems.

6.1 Losses in Piping

Losses are divided into two categories: (1) those due to wall friction in the piping, and (2) those created by piping components such as valves, bends, orifices, etc. The former are distributed along the length of the pipe, and the latter, termed "minor losses", are concentrated at prescribed locations in the system. They are defined separately and then introduced into the system energy equation, allowing one to determine the discharge, pressure drop, or pipe diameter.

6.1.1 Frictional Losses

Frictional pipe loss is normally expressed using either the Darcy-Weisbach or the Hazen-Williams relation. The *Darcy-Weisbach* equation is

$$h_L = \frac{fL}{D}\frac{V^2}{2g} = \frac{8fL}{g\pi^2 D^5}Q^2 \qquad (6.1.1)$$

in which h_L = head loss, f = friction factor, L = pipe length, D = pipe diameter, V = mean velocity, and Q = discharge. The behavior of the friction factor is described by the Moody diagram, shown in Fig. 4.2. Generally, f is a function of the Reynolds number $Re = VD/\nu$ and the relative roughness $\varepsilon = e/D$, where ν = kinematic viscosity of the liquid, and e = absolute roughness of the pipe wall. An empirical formula for f that accurately reproduces the turbulent portion of the Moody diagram is:

$$f = 1.325\left[\ln(0.27\varepsilon + 5.74Re^{-0.9})\right]^{-2} \tag{6.1.2}$$

Either Eq. 6.1.2 or the Moody diagram can be used to determine the friction factor. Equation 6.1.2 is valid over the ranges $0.01 > \varepsilon > 10^{-6}$, and $10^8 > Re > 3000$. For values of Re and ε outside of those ranges, the Moody diagram must be used directly. Numerical values of e are provided in Table 6.1.

In many instances it is acceptable to assume that the friction factor is independent of the Reynolds number. It can be observed in Eq. 6.1.2 that the second term in brackets becomes relatively smaller as Re increases, so that it can be used in the form $f = 1.325[\ln 0.27\varepsilon]^{-2}$, either to estimate f as a constant for a given problem, or to obtain an estimate for an iterative solution in which initially Re is unknown. In many engineering situations this simplified form is acceptable.

TABLE 6.1 Values of the Average Wall Roughness Height e

Type of pipe	e (ft)
Riveted steel	~0.01
Concrete	~0.001-0.01
Wood	~0.001
Cast iron	0.00085
Galvanized iron	0.0005
Wrought iron	0.00015
Drawn tubing	0.000005

The *Hazen-Williams* formula applies only to water as the flowing liquid. It normally expresses the head loss as a function of the discharge Q:

$$h_L = \frac{4.72L}{C^{1.85}D^{4.87}} Q^{1.85} \tag{6.1.3}$$

in which C is the *Hazen-Williams loss coefficient*. Values of C are given in Table 6.2.

TABLE 6.2 Values of the Hazen-Williams Coefficient C

Type of pipe	C
Extremely smooth; asbestos-cement	140
New or smooth cast iron; concrete	130
Wood stave; newly welded steel	120
Average cast iron; newly riveted steel	110
Cast iron or riveted steel after some years of use	95-100
Deteriorated old pipes	60-80

Note that in the form given the Hazen-Williams formula is applicable only with English units. It is neither as universally applicable, nor as accurate, as is the Darcy-Weisbach equation.

EXAMPLE 6.1

Water is flowing at $Q = 20$ gal/min in a 500-ft-long, 2-inch-diameter cast iron steel pipe. The water temperature is 50° F. Determine the head loss h_L using (a) the Darcy-Weisbach formula and (b) the Hazen-Williams formula.

Solution. From a table of properties of water, the kinematic viscosity of water at the given temperature is found to be $v = 1.41 \times 10^{-5}$ ft^2/sec. Using Table 6.1, one finds the roughness height to be $e = 0.00085$ ft. The diameter is $D = 2/12 = 0.167$ ft and the pipe length is $L = 500$ ft. The pipe area, discharge and velocity are:

$$A = (\pi/4) \times (2/12)^2 = 0.0218 \text{ ft}^2$$
$$Q = (20 \text{ gal/min}) \times (1 \text{ ft}^3/7.48 \text{ gal}) \times (1 \text{ min}/60 \text{ sec}) = 0.0446 \text{ ft}^3/\text{sec}$$
$$V = Q/A = 0.0446/0.0218 = 2.05 \text{ ft/sec}$$

The Reynolds number is $Re = 2.05 \times 0.167 / 1.41 \times 10^{-5} = 24,300$, and $\varepsilon = 0.00085/0.167 = 0.00509$. Substitute the data into Eq. 6.1.2 to determine the friction factor:

$$f = 1.325 \left[\ln(0.27 \times 0.00509 + 5.74 \times 24300^{-0.9}) \right]^{-2} = 0.034$$

a) The head loss is computed with Eq. 6.1.1:

$$h_L = 0.034 \times \frac{500}{0.167} \times \frac{2.05^2}{2 \times 32.2} = 6.64 \text{ ft}$$

b) From Table 6.2, the Hazen-Williams coefficient is $C = 110$. Application of Eq. 6.1.3 yields the head loss:

$$h_L = \frac{4.72 \times 500}{110^{1.85} \times 0.167^{4.87}} \times 0.0446^{1.85} = 7.64 \text{ ft}$$

These two results, although numerically different, are sufficiently close for most design problems.

6.1.2 Minor losses

Minor losses are expressed proportional to the kinetic energy in the manner

$$h_L = K \frac{V^2}{2g} = \frac{8K}{g\pi^2 D^4} Q^2 \qquad (6.1.4)$$

in which K is the loss coefficient. Representative values of K can be found in Table 4.4. Application of Eq. 6.1.4 is relatively straight-forward and will be shown in subsequent examples.

It is often convenient to express a loss coefficient as an equivalent length L_e of pipe. This is done by equating Eq. 6.1.4 to Eq. 6.1.1 and eliminating like terms to produce the relation

$$L_e = K \frac{D}{f} \qquad (6.1.5)$$

Collectively, frictional losses and minor losses make up the head loss term. Equations 6.1.1 to 6.1.5 are used in the following sections to express h_L in the energy equation.

6.2 Energy and Continuity Equations

The two fundamental control volume equations for piping problems are presented here without derivation. We assume steady, incompressible flow with uniform velocity profiles and constant pressures at the control surfaces, Fig. 6.1. Between two control surfaces 1 and 2 the equations take the form:

Continuity: $\quad Q = A_1 V_1 = A_2 V_2 \quad$ (6.2.1)

Energy: $\quad \dfrac{p_1}{\gamma} + \dfrac{V_1^2}{2g} + z_1 + H_P = \dfrac{p_2}{\gamma} + \dfrac{V_2^2}{2g} + z_2 + H_T + h_L \quad$ (6.2.2)

in which p = pressure, γ = specific weight of the fluid, z = elevation, H_P = head rise across a pump and H_T = head drop across a turbine. If

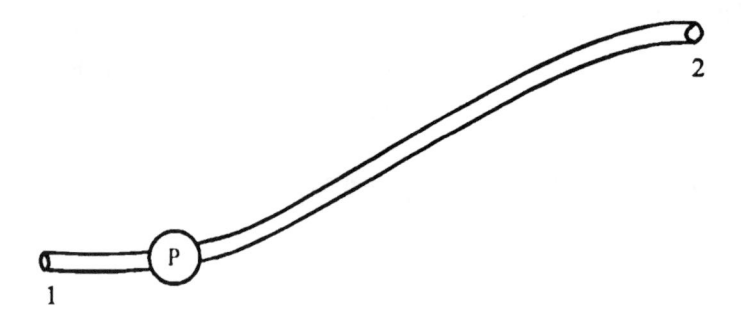

Figure 6.1 Control volume for a piping system.

no pump is present, then $H_P = 0$. It is convenient in many applications to make use of the hydraulic grade line (HGL), defined in the manner

$$H = \frac{p}{\gamma} + z \quad (6.2.3)$$

in which H is the distance from a selected datum to the HGL. If the kinetic energy terms $V^2/2g$ in Eq. 6.2.2 are relatively small, they can be neglected, and the energy equation becomes

$$H_1 + H_P = H_2 + h_L \quad (6.2.4)$$

In this form, the energy equation is useful for analyzing parallel and branching systems.

The power $\dot{W}_f$ delivered to the fluid by a pump is designated as

$$\dot{W}_f = \gamma Q H_P \quad (6.2.5)$$

and the power $\dot{W}_P$ delivered to the pump, called the brake power, is

$$\dot{W}_P = \frac{\dot{W}_f}{\eta} = \frac{\gamma Q H_P}{\eta} \quad (6.2.6a)$$

where η = the pump efficiency. The units for $\dot{W}_f$ and $\dot{W}_P$ are (ft-lb/sec); to convert to horsepower, divide the expressions by 550 ft-lb/sec. Equations 6.2.5 and 6.2.6a are used in conjunction with the energy and continuity equations when a pump is present in the system. IF a turbine is present we use

$$\dot{W}_T = \eta \gamma Q H_T \quad (6.2.6b)$$

6.2.1 Single Pipe Systems

A single pipe system consists of one pipe connected to one or more reservoirs, Fig. 6.2. The system may include a pump, or the flow may be induced by gravity forces. There are three categories of single pipe problems:

Category 1: Determining the pressure drop, or head loss
Category 2: Determining the discharge
Category 3: Determining the pipe diameter

In conjunction with the Darcy-Weisbach formulation, the first category can be solved directly by substi-

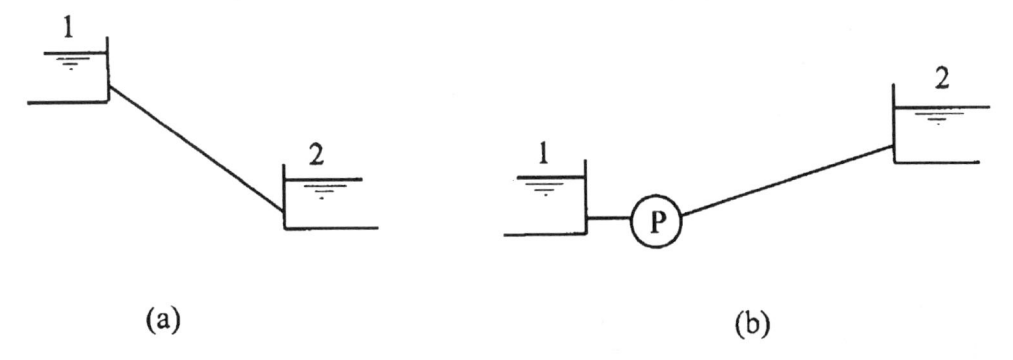

Figure 6.2 Single pipe systems: (a) gravity flow; (b) pump-driven flow.

tution into the energy equation, and the second and third categories require either trial and error solutions using the Moody diagram, or use of empirically derived formulas. Table 6.3 lists the various alternative solutions.

TABLE 6.3 Solutions for Single Pipe Problems using the Darcy-Weisbach Formulation

Category	Variables	Moody diagram $f=f(Re,\varepsilon)$	Empirical formula*
1	Known: Q, D, e, v Unknown: h_L	Direct solution	$h_L = 1.07 \dfrac{Q^2 L}{gD^5} \left\{ \left[\ln \dfrac{e}{3.7D} + 4.62 \left(\dfrac{vD}{Q} \right)^{0.9} \right] \right\}^{-2}$ Range: $10^{-6} < \varepsilon < 10^{-2}$, $3000 < Re < 3 \times 10^8$
2	Known: D, e, v, h_L Unknown: Q	Trial and error: fQ^2 = Constant	$Q = -0.965 \left(\dfrac{gD^5 h_L}{L} \right)^{0.5} \ln \left[\dfrac{e}{3.7D} + \left(\dfrac{3.17 v^2 L}{gD^3 h_L} \right)^{0.5} \right]$ Range: $Re > 2000$
3	Known: Q, e, v, h_L Unknown: D	Trial and error: $\dfrac{f}{D^5}$ = Constant	$D = 0.66 \left[e^{1.25} \left(\dfrac{LQ^2}{gh_L} \right)^{4.75} + vQ^{9.4} \left(\dfrac{L}{gh_L} \right)^{5.2} \right]^{0.04}$ Range: $10^{-6} < \varepsilon < 10^{-2}$, $5000 < Re < 3 \times 10^6$

*No pumps or components with minor losses are incorporated in the empirical formulas.

EXAMPLE 6.2

A pumping scheme requires that water be transported from a river at elevation $z_1 = 610$ ft to a reservoir at elevation $z_2 = 795$ ft. The discharge $Q = 1,030$ gallons per minute, the pipe has length $L = 1,500$ ft, and diameter $D = 6$ inches. The pipe is made of concrete and contains a gate valve half open, two standard elbows, and a pipe exit loss. Determine the following: (a) pump head, (b) fluid power, and (c) brake power, if the pump efficiency is $\eta = 0.75$. Use the Hazen-Williams formula to estimate pipe frictional loss.

Solution. This is a category 1 problem. The discharge, pipe area and velocity are:

$$Q = (1,030 \text{ gal/min})/(7.48 \text{ gal/ft}^3 \times 60 \text{ sec/min}) = 2.30 \text{ ft}^3/\text{sec}$$
$$A = (\pi/4) \times 0.5^2 = 0.196 \text{ ft}^2$$
$$V = 2.30/0.196 = 11.7 \text{ ft/sec}$$

From Table 6.2, $C = 130$, and the sum of the minor loss coefficients from Table 4.4 is $5.6 + 2 \times 0.9 + 1 = 8.4$. The system head loss is

$$h_L = \frac{4.72 \times 1,500}{130^{1.85} \times 0.5^{4.87}} \times 2.30^{1.85} + 8.4 \times \frac{11.7^2}{2 \times 32.2} = 118.7 + 17.9 = 137 \text{ ft}$$

The energy equation, Eq. 6.2.2, is solved for H_P, noting that the kinetic energy terms $V^2/2g$ and the pressure terms p/γ at the two reservoirs are zero:

(a) $H_P = z_2 - z_1 + h_L = 795 - 610 + 137 = 322$ ft

(b) $\dot{W}_f = \gamma Q H_P$
$\quad\quad = 62.4 \times 2.30 \times 322 = 46,200$ ft-lb/sec, or 84 horsepower

(c) $\dot{W}_P = \dfrac{\dot{W}_f}{\eta} = \dfrac{46,200}{0.75} = 61,600$ ft-lb/sec, or 112 horsepower

Note: The information given does not merit more than two significant numbers. However, it is common practice to provide three significant numbers.

It is customary to neglect minor loss terms if the pipe length is sufficiently large: one "rule-of-thumb" is to ignore them when the ratio $L/D > 1,000$. Caution is urged using such a rule; in the previous example, the minor losses, primarily dominated by the gate valve, amounted to 13 percent of the head loss and 6 percent of the required brake power. It is recommended to include the minor losses when sufficient information is provided to estimate the terms.

EXAMPLE 6.3

A pump is situated between sections 1 and 2 of a horizontal pipeline. The upstream section has dimensions $D_1 = 2$ in, with $p_1 = 50$ lb/in², and the downstream section has dimensions $D_2 = 3$ in, with $p_2 = 110$ lb/in². Determine the required fluid power if $Q = 25$ gal/min, $h_L = 20$ ft, and $\gamma = 62.4$ lb/ft³.

Solution. This is also a category 1 problem. Substitute $V = Q/A$ and $H_P = \dot{W}_f/(\gamma Q)$ into the energy relation, Eq. 6.2.2, noting that $z_1 = z_2$,

$$\frac{p_1}{\gamma} + \frac{Q^2}{2gA_1^2} + \frac{\dot{W}_f}{\gamma Q} = \frac{p_2}{\gamma} + \frac{Q^2}{2gA_2^2} + h_L$$

Substitute $Q = 25/(7.48 \times 60) = 0.0557$ ft^3/sec, $A_1 = (\pi/4) \times (2/12)^2 = 0.0218$ ft^2, $A_2 = (\pi/4) \times (3/12)^2 = 0.0491$ ft^2, $p_1/\gamma = 50 \times 144/62.4 = 115.4$ ft, and $p_2/\gamma = 110 \times 144/62.4 = 253.8$ ft into the relation:

$$115.4 + \frac{(0.0557)^2}{2 \times 32.2 \times 0.0218^2} + \frac{\dot{W}_f}{62.4 \times 0.0557} = 253.8 + \frac{(0.0557)^2}{2 \times 32.2 \times 0.0491^2} + 20$$

Reducing and solving for the fluid power, one finds $\dot{W}_f = 551$ ft-lb/sec, or approximately 1 horsepower. The kinetic energy terms are quite small relative to the head loss term, and could justifiably be neglected.

EXAMPLE 6.4

A liquid with a specific gravity of 0.68 is pumped from a storage tank to a discharge station through a 1,500-ft-long, 8-in-diameter pipe. The pump provides 15 horsepower of useful power to the liquid. Determine the discharge in the system. Assume a constant friction factor of $f = 0.015$ and include minor losses.

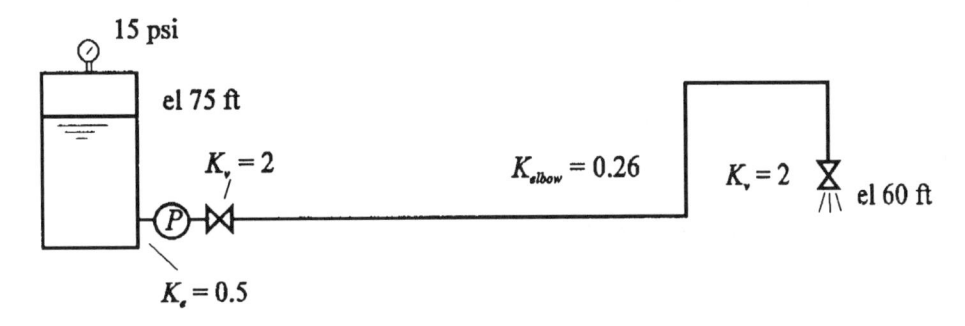

Solution. This is a category 2 problem. Write the energy equation from the liquid surface in the tank (location 1) to the valve exit (location 2):

$$\frac{p_1}{\gamma} + z_1 + \frac{\dot{W}_f}{\gamma Q} = \left(\Sigma K + \frac{fL}{D} + 1 \right) \frac{Q^2}{2gA^2} + z_2$$

Rearrange and substitute in known data:

$$\frac{p_1}{\gamma} + z_1 - z_2 + \left(\frac{\dot{W}_f}{\gamma} \right) \frac{1}{Q} - \left(\Sigma K + \frac{fL}{D} + 1 \right) \frac{Q^2}{2gA^2} = 0$$

$$\frac{15 \times 144}{0.68 \times 62.4} + 75 - 60 + \frac{15 \times 550}{0.68 \times 62.4} \times \frac{1}{Q} - \frac{(0.5 + 3 \times 0.25 + 2 + 0.015 \times 1500/0.667 + 1) \times Q^2}{2 \times 32.2 \times (\pi/4)^2 \times 0.667^4} = 0$$

The equation reduces to $F(Q) = 65.9 + 194.4/Q - 4.8Q^2 = 0$. A trial and error solution (Newton's method) is used to solve for Q, and is tabulated below. The first value of Q is selected by noting that $F(4)$ is positive and $F(5)$ is negative, so that the root lies between those two values.

Q (ft³/sec)	$F = 65.9 + 194.4/Q - 4.8Q^2$	$F' = -194.4/Q^2 - 9.6Q$	$\Delta Q = -F/F'$
4	+37.7	-50.5	+0.75
4.75	-1.47	-54.2	-0.03
4.72	+0.15	-54.0	+0.003

Hence, $Q \cong 4.7$ ft³/sec. (We could have avoided Newton's method and simply guessed values of Q until $F = 0$.)

EXAMPLE 6.5

Select a pipe to convey 50° F water delivering at least 250 gallons per minute from a higher reservoir to a lower one. The distance between the reservoirs is 750 ft, and the difference in water surface elevations is 20 ft. The pipe is made of concrete, with $e = 0.001$ ft, and minor losses can be neglected. The flow is gravity driven.

Solution. This is a category 3 problem, where we need to find the diameter. The following data is provided: $Q = 250/(7.48 \times 60) = 0.557$ ft³/sec, $v = 2.74 \times 10^{-5}$ ft²/sec, $z_1 - z_2 = 20$ ft, and $L = 750$ ft. Substitute the data into the energy equation and reduce:

$$z_1 - z_2 = \frac{8fLQ^2}{g\pi^2 D^5}, \quad 20 = \left(\frac{8 \times 750 \times 0.557^2}{32.2 \times \pi^2}\right)\frac{f}{D^5}, \quad \text{or} \quad D = 0.782 f^{1/5} \quad \text{(A)}$$

Equation 6.1.2 is expressed in terms of f and D, recognizing that $Re = 4Q/(\pi v D)$:

$$f = 1.325\left\{\ln\left[\frac{0.27 \times 0.001}{D} + 5.74 \times \left(\frac{4 \times 0.557}{\pi \times 2.74 \times 10^{-5}}\right)^{-0.9} D^{0.9}\right]\right\}^{-2}$$

$$= 1.325\left\{\ln\left[\frac{2.7 \times 10^{-4}}{D} + 6.13 \times 10^{-4} D^{0.9}\right]\right\}^{-2} \quad \text{(B)}$$

A trial and error solution will yield the diameter. To begin the iteration, guess a value of f, say, 0.02 (this is an intermediate value that is commonly used), and then solve for D with Eq. A and f with Eq. B. The results are tabulated as follows:

f (Eq B)	0.02 (guess)	0.028	0.026	0.0275
D (Eq A)	0.358	0.382	0.378	0.381

With $D = 0.381$ ft, the computed friction factor is $f = 0.0275$. Hence a diameter of $0.381 \times 12 = 4.57$ inches is to be used. Likely, a 6-in-diameter pipe available "off the shelf" would be specified. This problem can also be solved using Eq. A and the Moody diagram, Fig. 4.2, or solved directly with the empirical equation given in Table 6.3, using $h_L = z_1 - z_2 = 20$ ft..

6.2.2 Series and Parallel Piping Systems

Multiple pipe systems can be generalized by considering combinations of series and parallel elements, Fig. 6.3. If several pipe segments are connected end to end, then the system is labeled series piping. The discharge is the same in each element, and the head loss is the sum of the loss in each element:

$$\begin{aligned}&\text{(a)} \quad Q_{sys} = Q_1 = Q_2 = Q_3 = \cdots \\ &\text{(b)} \quad (h_L)_{sys} = (h_L)_1 + (h_L)_2 + (h_L)_3 + \cdots = \Sigma(h_L)_i\end{aligned} \qquad (6.2.7)$$

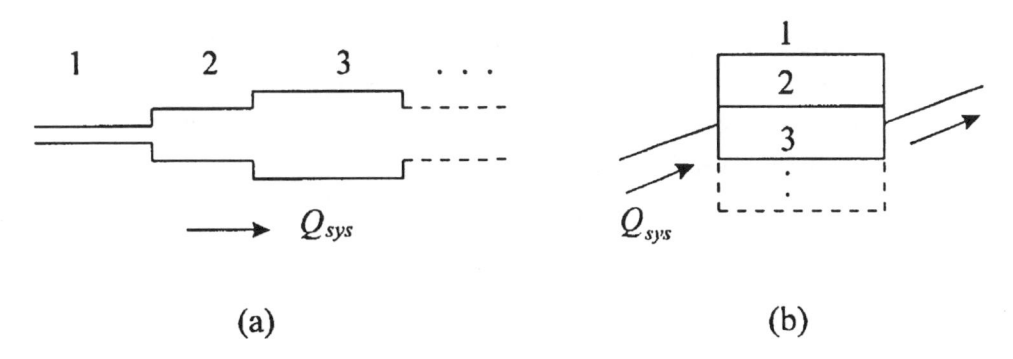

Figure 6.3 Two piping system configurations: (a) series; (b) parallel.

For pipes in parallel, Fig. 6.3b, the system discharge is cumulative, and the system head loss is the same for each element:

$$\begin{aligned}&\text{(a)} \quad Q_{sys} = Q_1 + Q_2 + Q_3 + \cdots = \Sigma Q_i \\ &\text{(b)} \quad (h_L)_{sys} = (h_L)_1 = (h_L)_2 = (h_L)_3 = \cdots = (h_L)_i\end{aligned} \qquad (6.2.8)$$

The standard problem associated with parallel piping is category 2, i.e., determining the discharge for a given head loss across the system. The methodology for a trial and error solution is as follows. The head loss for any pipe i can be expressed as

$$Q_i = \sqrt{\frac{(h_L)_{sys}}{R_i}} \qquad (6.2.9)$$

$$R_i = \frac{1}{2gA_i^2}\left[\frac{fL}{D} + \Sigma K\right]_i = \left[\frac{8f(L+L_e)}{g\pi^2 D^5}\right]_i \qquad (6.2.10)$$

where L_e comes from Eq. 6.1.5. Combining Eqs. 6.2.9 and 6.2.8b, and eliminating Q_i results in

$$(h_L)_{sys} = \left[\frac{Q_{sys}}{\Sigma \sqrt{1/R_i}}\right]^2 \qquad (6.2.11)$$

An iterative procedure to solve for the discharges Q_i and system head loss $(h_L)_{sys}$, given the system discharge Q_{sys} is as follows:

Step 1: Estimate or guess the initial friction factor in each element, provided they are not given.

Step 2: Compute R_i for each element with Eq. 6.2.10 and estimate $(h_L)_{sys}$ with Eq. 6.2.11.

Step 3: Compute Q_i in each element using Eq. 6.2.9.

Step 4: With the current values of Q_i, update estimates of f for each element using Eq. 6.1.2 or the Moody diagram.

Step 5: Repeat steps 2 to 4 until $(h_L)_{sys}$ and Q_i do not vary significantly.

Quite often in problems of this nature the friction factor is known. In that case R_i is found using Eq. 6.2.10, steps 4 and 5 are unnecessary, and a solution results with the first iteration.

EXAMPLE 6.6

Three pipes in series deliver flow between two reservoirs under the action of gravity. The upper reservoir elevation is 820 ft, and the elevation at the lower reservoir is 351 ft. Minor losses are proportional to the square of the discharge, and the Hazen-Williams formula is used to account for frictional losses. Determine the system discharge Q.

Pipe	L (ft)	D (in)	K	C
1	600	8	2	100
2	300	10	3	120
3	900	12	0	90

Solution. The energy equation between the two reservoirs is

$$H_A - H_B = (h_L)_1 + (h_L)_2 + (h_L)_3, \quad \text{with} \quad h_L = \frac{4.72L}{C^{1.85}D^{4.87}}Q^{1.85} + \frac{\Sigma K}{2gA^2}Q^2.$$

Substituting in known data and reducing the equations results in

$$820 - 351 = (4.07 + 0.49 + 1.03)Q^{1.85} + (0.255 + 0.157)Q^2, \quad \text{or} \quad 469 = 5.59Q^{1.85} + 0.41Q^2.$$

To compute Q it is necessary to employ a numerical solution. Newton's method is chosen and the solution is tabulated below. For the first iteration, neglect the minor losses, so that $Q = (469/5.59)^{1/1.85} = 10.96$.

Iteration	Q (cfs)	$F = 5.59Q^{1.85} + 0.41Q^2 - 469$	$F' = 10.34Q^{0.85} + 0.82Q$	$\Delta Q = -F/F'$
1	10.96	49.26	88.13	-0.56
2	10.40	1.083	84.23	-0.013
3	10.39	0.0004	84.14	-0.000005

Hence the discharge is approximately 10.4 ft³/sec.

EXAMPLE 6.7

Determine the flow distribution of water in a three-pipe parallel system, as shown in Fig. 6.3b. The system flow is $Q_{sys} = 0.35$ ft³/sec, and the friction factors are specified.

Pipe	L (ft)	D (in)	f	K
1	90	2	0.020	3
2	120	3	0.025	5
3	180	2.5	0.022	1

Solution. For each pipe we first compute the equivalent lengths and the coefficients R_i:

Pipe 1: $L_e = \dfrac{2/12}{0.02} \times 3 = 25$, $R_1 = \dfrac{8 \times 0.02 \times 115}{32.2 \times \pi^2 \times (2/12)^5} = 450.2$

Pipe 2: $L_e = \dfrac{3/12}{0.025} \times 5 = 50$, $R_2 = \dfrac{8 \times 0.025 \times 170}{32.2 \times \pi^2 \times (3/12)^5} = 109.6$

Pipe 3: $L_e = \dfrac{2.5/12}{0.022} \times 1 = 9.5$, $R_3 = \dfrac{8 \times 0.022 \times 189.5}{32.2 \times \pi^2 \times (2.5/12)^5} = 267.4$

Since the friction factors are constant, no iterations are required. The solution for the system head loss and the discharges are now computed:

$$(h_L)_{sys} = \dfrac{0.35^2}{\left[1/\sqrt{450.2} + 1/\sqrt{109.6} + 1/\sqrt{267.4}\right]^2} = 2.95 \text{ ft}$$

$Q_1 = \sqrt{2.95/450.2} = 0.081 \text{ ft}^3/\text{sec}$

$Q_2 = \sqrt{2.95/109.6} = 0.164 \text{ ft}^3/\text{sec}$

$Q_3 = \sqrt{2.95/267.4} = 0.105 \text{ ft}^3/\text{sec}$

A check of the sum of the individual discharges reveals that $Q_{sys} = 0.35 \text{ ft}^3/\text{sec}$.

6.2.3 Branching Pipe Systems

Consider the branching pipe system illustrated in Fig. 6.4a, which is made up of three pipe elements, each connected to a reservoir, and joined at a single junction. As in the previous section, the most common situation is to determine the flow distribution, a category 2 problem. First assume a flow direction in each pipe as shown; then, neglecting the kinetic energy terms, the energy equations can be written in the manner

$$H_A - H_B = (h_L)_1, \quad H_B - H_C = (h_L)_2, \quad H_B - H_D = (h_L)_3 \tag{6.2.12}$$

and the continuity balance at location B is

$$\Sigma Q = Q_1 - Q_2 - Q_3 = 0 \tag{6.2.13}$$

The hydraulic grade lines at A, C, and D (that is, the reservoir elevations) are considered known, and the unknowns are H_B, Q_1, Q_2, and Q_3. A convenient ad hoc solution is outlined below.

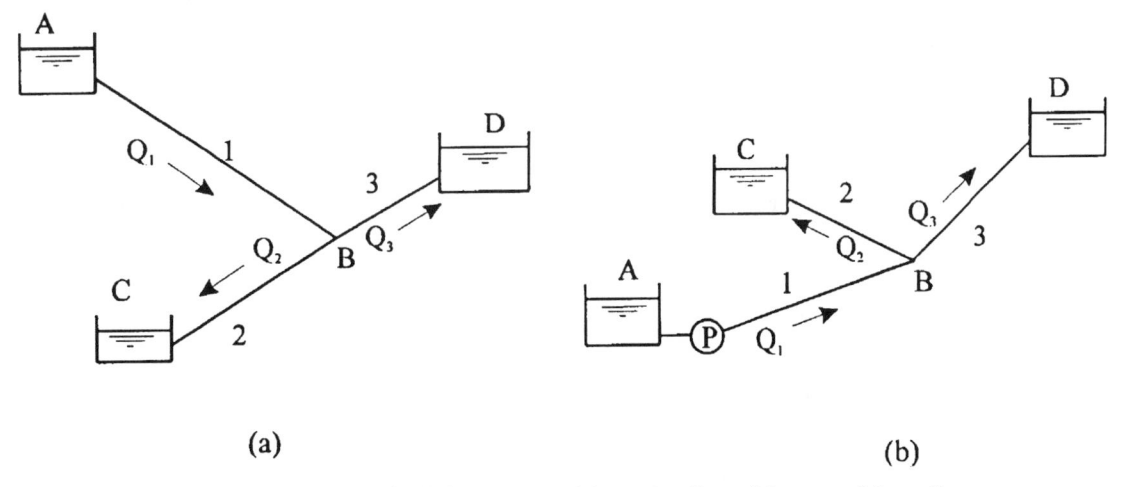

Figure 6.4 Branch piping system: (a) gravity flow; (b) pump-driven flow.

Step 1: Assume Q_1 and compute the hydraulic grade line H_B at location B.

Step 2: Compute Q_2 and Q_3 in branches 2 and 3.

Step 3: Substitute Q_1, Q_2, and Q_3 into the continuity equation. Generally there will be a flow imbalance, i.e., $\Sigma Q \neq 0$.

Step 4: Adjust the flow Q_1 and repeat steps 2 and 3 until ΣQ is within desired limits.

If a pump exists in pipe one (see Figure 6.4b) the energy equation for that pipe is altered as

$$H_A - H_B + H_P = (h_L)_1 \qquad (6.2.14)$$

An additional unknown H_P is introduced. Another necessary relationship is the pump characteristic curve. It is often convenient to follow the solution graphically by plotting the assumed discharge in pipe 1 versus either the hydraulic grade line at B or Q at B. Such a procedure is helpful to determine in what manner the discharge in pipe 1 should be altered for the next iteration.

EXAMPLE 6.8

Consider the system shown in Figure 6.4b, with the data tabulated below. Determine the water distribution and the piezometric head at the junction. Assume constant friction factors and a pump characteristic curve given by $H_P = 55 - 0.1Q^2$, with units of H_P in ft and Q in ft³/sec. The reservoir elevations are $H_A = 20$ ft, $H_C = 50$ ft, and $H_D = 45$ ft.

Pipe	L (ft)	D (in).	f	K
1	100	10	0.020	2
2	200	8	0.015	0
3	300	6	0.025	0

Solution. The losses for each pipe are expressed using Eqs. 6.1.5 and 6.1.1:

Pipe 1: $L_e = \dfrac{10/12}{0.02} \times 2 = 83.3$, $(h_L)_1 = \dfrac{8 \times 0.02 \times 183.3}{32.2 \times \pi^2 \times (10/12)^5} Q_1^2 = 0.23 Q_1^2$

Pipe 2: $(h_L)_2 = \dfrac{8 \times 0.015 \times 200}{32.2 \times \pi^2 \times (8/12)^5} Q_2^2 = 0.57 Q_2^2$

Pipe 3: $(h_L)_3 = \dfrac{8 \times 0.025 \times 300}{32.2 \times \pi^2 \times (6/12)^5} Q_3^2 = 6.04 Q_3^2$

The energy relations for the system are Eq. 6.2.14 for pipe 1 and the latter two of Eq. 6.2.12 for pipes 2 and 3:

Pipe 1: $20 + 55 - 0.1 Q_1^2 = H_B + 0.23 Q_1^2$, or $H_B = 75 - 0.33 Q_1^2$

Pipe 2: $H_B = 50 + 0.57 Q_2^2$, or $Q_2 = \sqrt{(H_B - 50)/0.57}$

Pipe 3: $H_B = 45 + 6.04 Q_3^2$, or $Q_3 = \sqrt{(H_B - 45)/6.04}$

The flow balance at the junction is $\Sigma Q = Q_1 - Q_2 - Q_3$. It is assumed that the hydraulic grade line at B lies below the water surfaces at C and D. If during the iterations, it becomes apparent that the flow direction will reverse in either pipe 2 or pipe 3, it will be necessary to reformulate the appropriate energy equation and the flow balance. The solution is tabulated below. The first estimate of Q_1 is based on assuming that $H_B = 65$ ft, so that $Q_1 = \sqrt{(75-65)/0.33} = 5.5$.

Q_1 (ft³/sec)	H_B(ft)	Q_2 (ft³/sec)	Q_3 (ft³/sec)	ΣQ
5.5	65	5.13	1.82	-1.45
6	63.12	4.80	1.73	-0.53
6.5	61.10	4.41	1.63	+0.46
6.27*	62.03	4.59	1.68	0.00

*Estimated by linear interpolation between the two previous values of Q_1 and Q.

Hence the solution is $H_B = 62$ ft, $Q_1 = 6.3$ ft³/sec, $Q_2 = 4.6$ ft³/sec, and $Q_3 = 1.7$ ft³/sec.

6.3 Use of Centrifugal Pumps in Piping Systems

Thus far, we have included pumps only by making simplified assumptions regarding their behavior. Let us now focus our attention on centrifugal pumps, which include radial-flow, mixed-flow and axial-flow pumps. If a centrifugal pump is included in the piping system and the flow rate or useful power is specified, the solution is straight-forward, employing techniques we have already developed. On the other hand, if the flow rate is not specified, a situation that is often the case (category 2 problem), a trial-and-error solution is required. The reason is that the head produced by a centrifugal pump depends on the discharge in a unique relationship, termed the *pump characteristic curve*. Sometimes the characteristic curve is approximated by a polynomial equation (refer to Example 6.8), but it is more appropriate and accurate to make use of one provided by the pump manufacturer. The characteristic curve provides a relationship between the discharge Q and the pump head H_P; a representative one is shown in Fig. 6.5 along with the pump efficiency. The pump characteristic curve and the appropriate energy equation(s) for the piping system are solved simultaneously to determine the desired discharge.

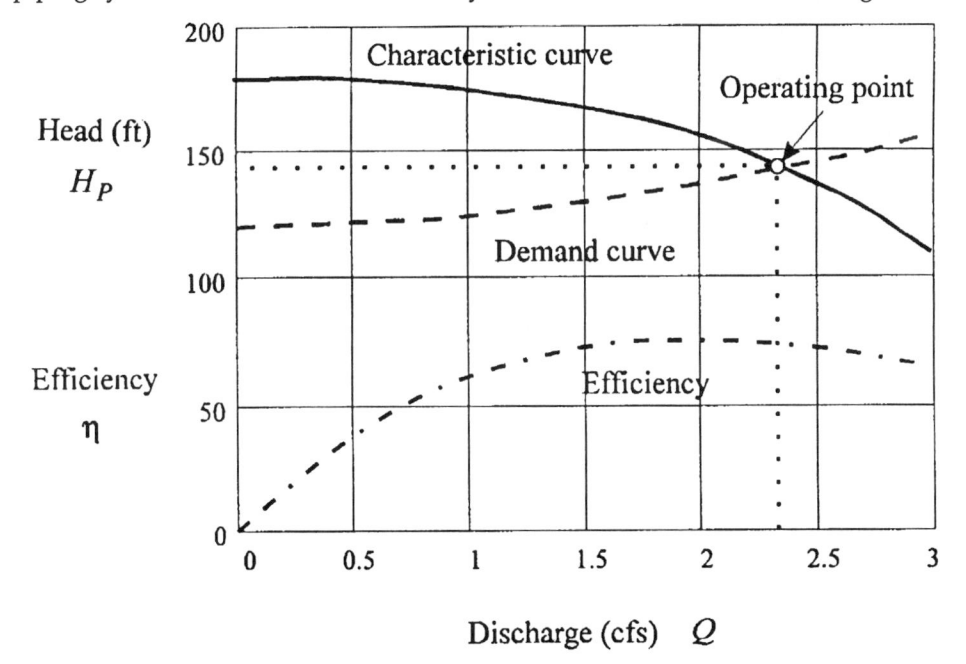

Figure 6.5 Pump characteristic curve and system demand curve.

6.3.1 Matching pump to system demand

Consider a single pipeline that contains a pump to deliver fluid between two reservoirs, Fig. 6.2b. The energy equation for the system is

$$H_P = z_2 - z_1 + \left(\frac{fL}{D} + \Sigma K\right)\frac{Q^2}{2gA^2} \qquad (6.3.1)$$

and is termed the *system demand curve*. Note that z_1 can be less than z_2 and that the friction factor can vary with the discharge as shown earlier. Equation 6.3.1 is represented by the dashed curve in Fig. 6.5; on the right-hand-side, $z_1 - z_2$ is the static head, and the remaining term is the head loss. The steep nature of the demand curve is dependent on the magnitude of the losses: as the losses increase, the pumping head required for a given discharge increases. Piping systems may experience short-term changes in the demand such as throttling of valves, and long-term changes due to aging of pipes. The intersection of the characteristic curve with the demand curve is termed the *operating point*; it provides the unknown head and discharge. When selecting a pump for a given application, it is desirable to have the operating point occur as close as possible to the point of maximum efficiency.

EXAMPLE 6.9

The piping system shown in Fig. 6.2b has the following data: $z_1 = 100$ ft, $z_2 = 220$ ft, $D = 8$ in (wrought iron pipe), $L = 1{,}500$ ft. The entrance and exit loss coefficients are $K_{ent} = 0.5$, and $K_{exit} = 1.0$. Using the characteristic pump curve shown in Fig. 6.5, determine the operating head, system discharge, and required horsepower.

Solution. To generate the system demand curve, we assume that the Reynolds number is sufficiently large, so that it can be neglected when evaluating the friction factor with Eq. 6.1.2. From Table 6.1, $e = 0.00015$ ft, and $\varepsilon = 0.00015/(8/12) = 0.000225$. Then the friction factor is computed to be $f = 1.325[\ln(0.27 \times 0.000225)]^{-2} = 0.014$. The system demand curve is thus

$$H_P = z_2 - z_1 + \left(\frac{fL}{D} + \Sigma K\right)\frac{Q^2}{2gA^2}$$

$$= 220 - 100 + \left[\frac{0.014 \times 1500}{8/12} + 1.5\right]\frac{Q^2}{2 \times 32.2 \times (\pi/4)^2 \times (8/12)^4}$$

$$= 120 + 4Q^2$$

The demand curve is plotted in Fig. 6.5. At the operating point the desired head, discharge, and efficiency are observed to be $H_P \cong 140$ ft, $Q \cong 2.3$ ft^3/sec, or 1030 gal/min, and $\cong 70\%$. The required input power is

$$\dot{W}_P = \gamma Q H_P / \eta = 62.4 \times 2.3 \times 140 / 0.70 = 28{,}700 \text{ ft-lb/sec, or 52 horsepower}$$

For more complex systems, one can work the pump characteristic curve into the system equations in an ad hoc fashion. One example is the branching system presented in Section 6.2.3, where in Eq. 6.2.14 the pump head H_P is considered a function of the discharge Q_1, with values provided by the characteristic curve.

6.3.2 Cavitation and Net Positive Suction Head

Cavitation refers to conditions within a pump where the local pressures reduce to vapor pressure. Consequently, vapor cavities are formed, and when transported into regions of greater pressure they collapse, creating high localized pressures near solid boundaries. After repeated collapse cycles, the pump impeller and other components may become materially damaged. The proper selection of a pump for a given piping system requires that a calculation be made to eliminate the possibility of cavitation at the design flow condition.

A convenient parameter used to designate the potential for cavitation is the *net positive suction head*, NPSH. Consider a pump operating as shown in Figure 6.6; location 1 is at the liquid surface on the inlet, or suction, reservoir, and location 2 is the center of the pump. The NPSH design requirement for the pump is established by the inequality

$$NPSH \leq \frac{p_{atm} - p_{vap}}{\gamma} - \Delta z - h_L \tag{6.3.2}$$

in which p_{atm} = atmospheric pressure, p_{vap} = vapor pressure of the liquid, $\Delta z = z_2 - z_1$, the difference in elevation between locations 1 and 2, and h_L = head loss on the suction pipe between locations 1 and 2. The performance data provided by the pump manufacturer usually include the NPSH curve. Application of Eq. 6.3.2 is demonstrated in the following example.

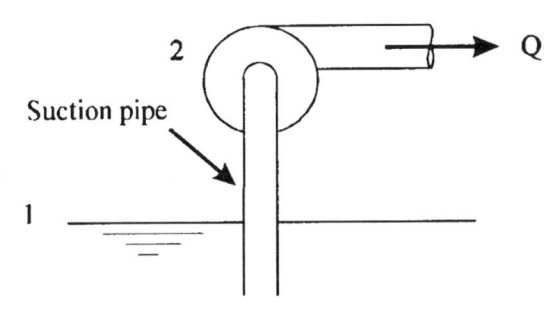

Figure 6.6 Cavitation setting for a pump.

---- **EXAMPLE 6.10** ----

Water at 80° F is being pumped through a 20-ft-long, 4-inch-diameter suction pipe at a rate of 500 gal/min. The suction pipe includes two vaned bends ($K=0.2$) and a reentrant entrance ($K = 0.8$). At the given flow rate the friction factor is 0.015. If the allowable NPSH at that flow rate is 20 ft, determine the height z above the inlet water surface that the pump can be located to operate without cavitating.

Solution. Assume p_{atm} = 14.7 psi. From a table of properties of water one finds that at a temperature of 80° F, p_{vap} = 0.507 psi, and γ = 62.1 lb/ft³. The suction pipe head loss is

$$h_L = \left(\frac{fL}{D} + \Sigma K\right)\frac{Q^2}{2gA^2}$$

$$= \left(\frac{0.015 \times 20}{4/12} + 2 \times 0.2 + 0.8\right)\frac{500/(7.48 \times 60)}{2 \times 32.2 \times (\pi/4)^2 \times (4/12)^4} = 4.77 \text{ ft}$$

Substituting the accumulated data into Eq. 6.3.2 results in

$$20 \leq \frac{(14.7 - 0.507) \times 144}{62.1} - \Delta z - 4.77, \text{ or } \Delta z \leq 8.1 \text{ ft}$$

Hence, the pump should be placed no higher than about 8 ft above the suction reservoir surface.

6.3.3 Pump Similarity Rules

Following the principles of similitude presented in Section 4.3, relationships for two pumps that are geometrically similar can be developed. Let $\dot{W}$ = pump power, N = pump speed, and D = impeller diameter, with subscripts 1 and 2 designating the two pumps. Then the pump similarity relationships are:

$$\frac{H_2}{H_1} = \left(\frac{N_2}{N_1}\right)^2 \left(\frac{D_2}{D_1}\right)^2 \tag{6.3.3}$$

$$\frac{Q_2}{Q_1} = \frac{N_2}{N_1}\left(\frac{D_2}{D_1}\right)^3 \tag{6.3.4}$$

$$\frac{\dot{W}_2}{\dot{W}_1} = \left(\frac{N_2}{N_1}\right)^3 \left(\frac{D_2}{D_1}\right)^5 \tag{6.3.5}$$

These relations are used to design or select a pump from a family of geometrically similar units. Another use of them is to examine the effects of changing speed on a given unit.

EXAMPLE 6.11

A pump is designed to operate at 1,500 rev/min with an efficiency of 0.70 when delivering water at 600 gal/min against a head of 65 ft. If the same pump is now required to deliver water at a head of 100 ft at the same efficiency, determine the required rotational speed, the discharge and power delivered to the pump.

Solution. The given data are the following:

$$H_1 = 65 \text{ ft}; \quad Q_1 = 600 \text{ gal/min, or } 1.34 \text{ ft}^3/\text{sec}; \quad N_1 = 1,500 \text{ rev/min};$$
$$\dot{W}_1 = 62.4 \times 1.34 \times 65 / 0.7 = 7,760 \text{ ft-lb/sec, or } 14.1 \text{ horsepower}; \quad H_2 = 100 \text{ ft}$$

Solve for N_2, Q_2, and $\dot{W}_2$ using Eqs. 6.3.3, 6.3.4, and 6.3.5, respectively:

$$N_2 = \sqrt{\frac{H_2}{H_1}}\left(\frac{D_2}{D_1}\right) N_1 = \sqrt{\frac{100}{65}} \times 1 \times 1,500 = 1,860 \text{ rev/min}$$

$$Q_2 = \frac{N_2}{N_1}\left(\frac{D_2}{D_1}\right)^3 Q_1 = \frac{1,860}{1,500} \times 1^3 \times 600 = 744 \text{ gal/min}$$

$$\dot{W}_2 = \left(\frac{N_2}{N_1}\right)^3 \left(\frac{D_2}{D_1}\right)^5 \dot{W}_1 = \left(\frac{1,860}{1,500}\right)^3 \times 1^5 \times 14.1 = 26.9 \text{ horsepower}$$

Note that for this problem the ratio $D_2/D_1 = 1$ since the same pump is used for both flow conditions.

Practice Problems (Essay)

6.1 Water at 60° F flows in a 1.5-in-diameter cast iron pipe at flow rate of 27 gal/min. Determine the head loss in a 350-ft-section of pipe using (a) the Darcy-Weisbach relation, and (b) the Hazen-Williams formula.

6.2 A pressure drop of 5 psi is not to be exceeded over a 600-ft length of horizontal 4-ft-diameter concrete pipe transporting water at 60° F. What flow rate can be accommodated?

6.3 Find the size of 1,000 ft of plastic tubing that can transport 20 gal/min of water at 80° F if the maximum allowable drop in the hydraulic grade line is 210 ft.

6.4 Determine the flow rate in the pipe shown. Discuss the nature of the hydraulic grade line.

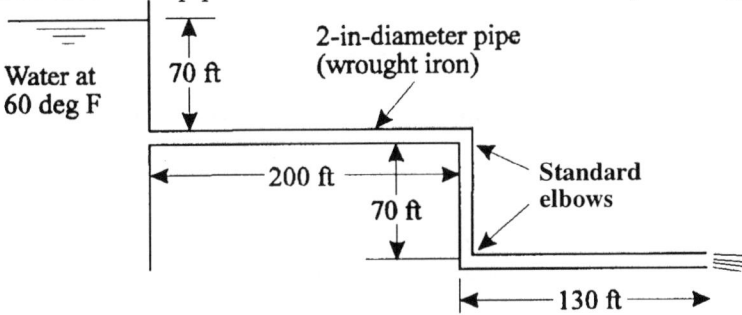

6.5 Water at 60° F is being pumped through the three pipes in series as shown. The power delivered to the pump is 2,500 horsepower, and the pump efficiency is 0.82. Compute the discharge. Assume constant friction factors.

Pipe	L (ft)	D (ft)	f	K
1	600	5	0.018	2
2	1,000	3	0.020	0
3	400	4	0.019	10

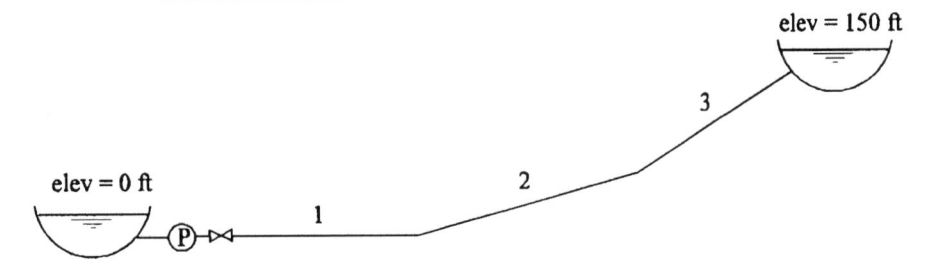

6.6 The system discharge for the four-pipe parallel system is $Q_{sys} = 5$ ft^3/sec. Compute the flow distribution. Assume $f = 0.020$ for all four pipes.

Pipe	L (ft)	D (ft)	K
1	200	0.333	2
2	300	0.500	0
3	180	0.167	4
4	260	0.333	1

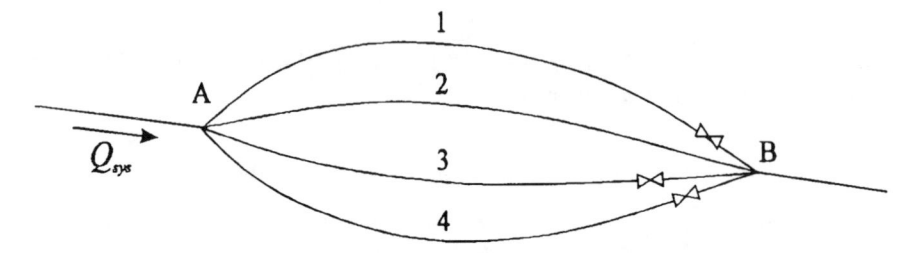

6.7 Consider the three branch piping system shown in Fig. 6.4a with the data tabulated below. The reservoir elevations are $z_A = 65$ ft, $z_C = 5$ ft, and $z_D = 43$ ft. Find the water flow distribution assuming constant friction factors.

Pipe	L (ft)	D (ft)	f	K
1	2,500	0.500	0.02	2
2	1,600	0.333	0.025	3
3	3,300	0.500	0.018	7

6.8 An oil ($\gamma = 53$ lb/ft^3) is pumped through a single pipe system with the following constraints: $L = 8,000$ ft, $D = 8$ in, $f = 0.02$, $K = 12.5$. The upper reservoir is 105 ft higher than the lower one. Using the pump characteristic curve shown in Fig. 6.5, determine (a) the oil discharge in the pipe, and (b) the power requirement.

6.9 During a test on a water pump, no cavitation is detected when the atmospheric pressure is 15 psi, the water temperature is 80° F, and the discharge is 13 gal/sec. The inlet galvanized pipe diameter is 6 in, and the pump is situated 16 ft above the suction reservoir.

(a) Compute the *NPSH* at standard atmospheric pressure conditions. (b) If the pump is to produce the same head and discharge at a location where the atmospheric pressure is 12 psi, what is the required change in elevation of the pump relative to the inlet reservoir to avoid cavitation?

6.10 A pump at design conditions delivers 230 ft^3/sec of water at a head of 38 ft. The pump speed is 970 rev/min, and its efficiency is 0.85. A geometrically similar pump that is one-half the size is to run at a speed of 1,200 rev/min. Find the resulting head, discharge, and required power.

Practice Problems (Multiple Choice)

A pump, whose performance curve is given below, has been chosen to deliver water in a piping system consisting of four pipes arranged as shown. The pumping system consists of four pipes arranged as shown. Water at 60° F is being pumped from reservoir A and exits at either reservoir B or at location D. The atmospheric pressure is $p_{atm} = 14.7$ psi, the vapor pressure is $p_{vap} = 0.256$ psi, and the specific weight is $\gamma = 62.4$ lb/ft³. The pipe characteristics are shown in the accompanying table. All of the pipe diameters are 4 inches, and the friction factor in each pipe is assumed to be constant, $f = 0.02$.

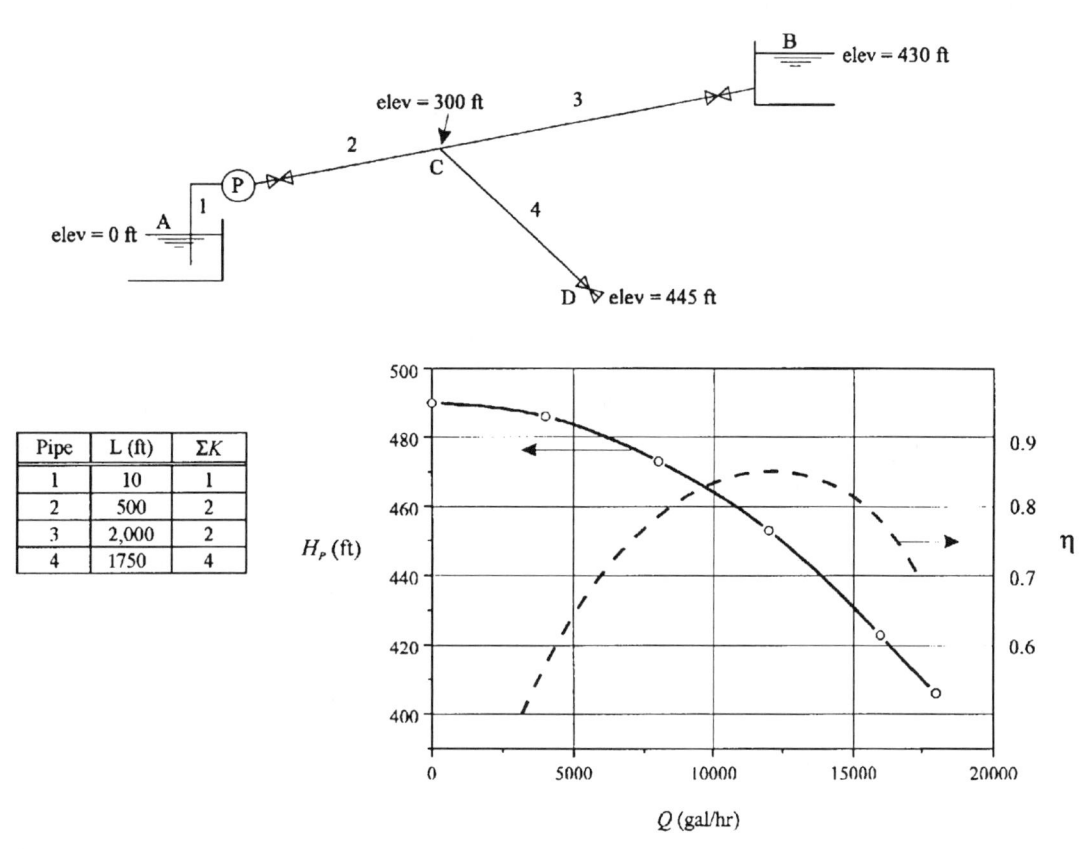

Pipe	L (ft)	ΣK
1	10	1
2	500	2
3	2,000	2
4	1750	4

Questions 6.11 - 6.20

6.11 If the discharge through the pump is 5,000 gal/hr, then the head loss across pipe 2, in feet, is
    a) 11.2   b) 300   c) 485   d) 2.3   e) 8.6

6.12 Assume that the valve at location D is closed, then the discharge in the system is, in gal/hr
    a) 5,000   b) 9,000   c) 11,000   d) 7,000   e) 13,000

6.13 If the valve at D is open and the discharge through the pump is 11,000 gal/hr, then the discharge in pipe 4, in gal/hr, is
    a) 3,600   b) 11,000   c) 7,400   d) 6,000   e) 2,400

6.14 Assume that the discharge through the pump is 8,000 gal/hr. Then the required power for the pump, in horsepower, is approximately
    a) 148   b) 16   c) 118   d) 20   e) 27

6.15 At a given operating state, the *NPSH* for the pump is 25 ft, and the discharge is 10,000 gal/hr. The elevation of the pump above the water surface at location A, in ft, is about
    a) 33   b) 25   c) 8   d) 0.4   e) -8

6.16 Assume that the pump delivers 12,000 gal/hr. The pressure at location C, in pounds per square inch, is
    a) -2   b) 61   c) 195   d) -63   e) 130

6.17 With the same assumption as Problem 16, if the valve at location D is closed, the gage pressure at that location, in pounds per square inch, is approximately
    a) 195   b) -2   c) 130   d) 61   e) -63

6.18 If the valve at location D is open and the head rise across the pump is 460 ft, then the discharge in pipe 3, in gal/hr, is
    a) 3,600   b) 11,000   c) 2,400   d) 7,400   e) 6,000

6.19 If the valve at location B were closed, and pipe 4 was open, the head across the pump, in ft, would be about
    a) 485   b) 450   c) 470   d) 490   e) 435

6.20 Assume that the pump is delivering 12,000 gal/hr to the system at a rotational speed of 900 rev/min. If a geometrically similar pump 2/3 the original size were used to deliver the same discharge, the speed of the new pump, in rev/min, would be
    a) 2,020   b) 276   c) 400   d) 900   e) 3,040

Solutions to Problems

6.1 This is a category 1 problem.

a) The kinematic viscosity of 60° F water is 1.22×10^{-5} ft²/sec. From Table 6.1 we find $e = 0.00085$ ft. Compute the velocity, Reynolds number, and relative roughness:

$$D = 1.5/12 = 0.125 \text{ ft}, \quad V = \frac{27/(7.48 \times 60)}{\pi \times 0.125^2 /4} = 4.9 \text{ ft/sec},$$

$$Re = \frac{4.9 \times 0.125}{1.22 \times 10^{-5}} = 3.2 \times 10^6, \quad \varepsilon = \frac{0.00085}{0.125} = 0.0068$$

Compute the friction factor:

$$f = 1.325\left[\ln\left(0.27 \times 0.0068 + 5.74 \times (3.2 \times 10^6)^{-0.9}\right)\right]^{-2} = 0.033$$

(One could also use the Moody Diagram to estimate the friction factor.) Substitute into the Darcy-Weisbach equation:

$$h_L = \frac{0.033 \times 350 \times 4.9^2}{0.125 \times 2 \times 32.2} = 34.4 \text{ ft}, \quad \text{or} \quad 34.4 \times \frac{62.4}{144} = 14.9 \text{ psi}$$

b) From Table 6.2, assume $C = 110$. Then substitute known data into the Hazen-Williams formula:

$$h_L = \frac{4.72 \times 350 \times [27/(7.48 \times 60)]^{1.85}}{110^{1.85} \times 0.125^{4.87}} = 38.1 \text{ ft}, \text{ or } 38.1 \times \frac{62.4}{144} = 16.5 \text{ psi}$$

Note the difference in the two results; both are acceptable answers for most problems. Solution (a), however, is more accurate.

6.2 This is a category 2 problem. The head loss due to friction is $h_L = 5 \times 144/62.4 = 11.5$ ft. Substitute into the Darcy-Weisbach equation:

$$11.5 = f \frac{600}{4} \times \frac{V^2}{2 \times 32.2}, \text{ or } fV^2 = 4.94$$

The Moody diagram is used to estimate f. The kinematic viscosity is $v = 1.22 \times 10^{-5}$ ft²/sec, and the relative roughness is $\varepsilon = 0.01/4 = 0.0025$. First assume $f = 0.03$. Then $V = \sqrt{4.94/0.03} = 12.8$ ft/sec, $Re = 12.8 \times 4/1.22 \times 10^{-5} = 4.2 \times 10^6$, and from the Moody diagram, we find the friction factor to be $f = 0.025$. Hence the new estimate of the velocity is $V = 14.1$ ft/sec, and $Re = 14.1 \times 4/1.22 \times 10^{-5} = 4.6 \times 10^6$. The friction factor remains at 0.025, so that the final velocity is $V = 14.1$ ft/sec, and the discharge is $Q = 14.1 \times \pi \times 4^2/4 = 177$ ft³/sec.

As an alternative to the trial and error approach, the empirical formula for a category 2 problem, Table 6.3, can be used since no pump is involved:

$$Q = -0.965 \times \left(\frac{32.2 \times 4^5 \times 11.5}{600}\right)^{0.5} \ln\left[\frac{0.01}{3.7 \times 4} + \left(\frac{3.17 \times (1.22 \times 10^{-5})^2 \times 600}{32.2 \times 4^3 \times 11.5}\right)^{0.5}\right]$$

$$= 177 \text{ ft}^3/\text{sec}$$

The empirical formula gives an answer that is as precise as the trial and error solution.

6.3 This is a category 3 pipe problem. The following data are given: $L = 1000$ ft, $Q = 20/(7.48 \times 60) = 0.0446$ ft³/sec, and $h_L = 210$ ft. Substitute into the Darcy-Weisbach equation and simplify:

$$210 = \frac{8 \times f \times 1000 \times 0.0446^2}{32.2 \times \pi^2 \times D^5}, \text{ or } D = 0.189 f^{1/5}$$

This relation will be used in conjunction with the Moody diagram to estimate the friction factor. For drawn tubing, $e = 5 \times 10^{-6}$ ft from Table 6.1, and the kinematic viscosity for water at 80° F is $v = 0.93 \times 10^{-5}$ ft²/sec. The two parameters required for using the diagram are

$$\varepsilon = \frac{e}{D} = \frac{5 \times 10^{-6}}{D}, \text{ and } Re = \frac{4Q}{\pi D v} = \frac{4 \times 0.0446}{\pi \times D \times 0.93 \times 10^{-5}} = \frac{6110}{D}$$

Begin the iterative solution by assuming that $f = 0.015$. Then

$$D = 0.189 \times 0.015^{1/5} = 0.0816 \text{ ft}, \quad \varepsilon = \frac{5 \times 10^{-6}}{0.0816} = 6 \times 10^{-5}, \quad Re = \frac{6110}{0.0816} = 75,000$$

The updated friction factor read from the Moody diagram is $f = 0.019$. Repeating the calculations one finds that $D = 0.0855$ ft, $\varepsilon = 6 \times 10^{-5}$, and $Re = 71,000$. From the Moody diagram, the friction factor again is $f = 0.019$, unchanged from the previous estimate. Hence $D = 0.0855 \times 12 = 1.03$ in. One could use either a 1-inch-diameter pipe or a 2-inch-diameter pipe. A 1-inch-diameter pipe will yield a head loss of $h_L = 237$ ft, and a 2-inch-diameter pipe will yield $h_L = 74$ ft.

The solution using the formula for a category 3 problem in Table 6.3 is

$$D = 0.66\left[(5\times10^{-6})^{1.25}\left(\frac{1{,}000\times 0.0446^2}{32.2\times 210}\right)^{4.75} + 0.93\times 10^{-5}\times 0.0446^{9.4}\left(\frac{1{,}000}{32.2\times 210}\right)^{5.2}\right]^{0.04}$$

$$= 0.0868 \text{ ft, or } 1.04 \text{ in}$$

Note that the solution is as accurate as using the Moody diagram with the trial and error procedure.

6.4 The kinematic viscosity is $v = 1.22\times 10^{-5}$ ft²/sec, the relative roughness is $\varepsilon = 0.00015/0.167 = 0.0009$, and the pipe diameter is $D = 2/12 = 0.167$ ft. Assume a completely turbulent flow regime, so that the friction factor can be approximated by $f = 1.325[\ln(0.27\times 0.0009)]^{-2} = 0.019$. Next, write the energy equation from the reservoir surface to the pipe outlet:

$$140 = \frac{V^2}{2g} + \left[\frac{fL}{D} + 2K_{elbow} + K_{entrance}\right]\frac{V^2}{2g}$$

Assume standard elbows so that $K_{elbow} = 0.9$, and with $K_{entrance} = 0.5$ from Table 4.4, substitute known data into the energy equation:

$$140 = \left[1 + \frac{0.019\times 400}{0.167} + 2\times 0.9 + 0.5\right]\frac{V^2}{2\times 32.2} \quad \therefore\ V = 13.6 \text{ ft/sec}$$

We perform one more iteration to determine a better estimate of the velocity:

$$Re = \frac{13.6\times 0.167}{1.22\times 10^{-5}} = 1.9\times 10^5,$$

$$f = 1.325\left[\ln\left(0.27\times 0.0009 + 5.74(1.9\times 10^5)^{-0.9}\right)\right]^{-2} = 0.021$$

Substituting the new value of f into the energy equation and solving we find that $V = 13.0$ ft/sec. Hence $Q = 13.0\times \pi \times 0.167^2/4 = 0.285$ ft³/sec. The hydraulic grade line has sudden drops at the entrance and elbows, and a linear drop over the length of the pipe.

6.5 The energy equation for the system is $H_P = (h_L)_{sys} + 150$, or

$$\frac{\dot{W}_P \eta}{\gamma Q_{sys}} = (R_1 + R_2 + R_3)Q_{sys}^2 + 150, \text{ in which } R_i = \frac{1}{2gA_i^2}\left(\frac{fL}{D} + \Sigma K\right)_i$$

Compute the R-values:

$A_1 = \frac{\pi}{4}\times 5^2 = 19.6$ ft², $R_1 = \frac{1}{2\times 32.2\times 19.6^2}\left(\frac{0.018\times 600}{5} + 2\right) = 0.00017$

$A_2 = \frac{\pi}{4}\times 3^2 = 7.1$ ft², $R_2 = \frac{1}{2\times 32.2\times 7.1^2}\left(\frac{0.020\times 1000}{3} + 0\right) = 0.00205$

$A_3 = \frac{\pi}{4}\times 4^2 = 12.6$ ft², $R_3 = \frac{1}{2\times 32.2\times 12.6^2}\left(\frac{0.019\times 400}{4} + 10\right) = 0.00116$

Substituting the known data into the energy equation we have

$$\frac{2500\times 550\times 0.82}{62.4 Q_{sys}} = (0.00017 + 0.00205 + 0.00116)Q_{sys}^2 + 150$$

which reduces to $18{,}070 = 0.00338 Q_{sys}^3 + 150 Q_{sys}$. Solving the equation by trial and error gives

$$Q_{sys} = 99 \text{ ft}^3/\text{sec.}$$

6.6 Compute the R-values:

Pipe 1: $L_e = \dfrac{2 \times 0.333}{0.02} = 33$, $\quad R_1 = \dfrac{8 \times 0.02 \times 233}{32.2 \times \pi^2 \times 0.333^5} = 29$

Pipe 2: $L_e = 0$, $\quad\quad\quad\quad\quad\; R_2 = \dfrac{8 \times 0.02 \times 300}{32.2 \times \pi^2 \times 0.5^5} = 5$

Pipe 3: $L_e = \dfrac{2 \times 0.167}{0.02} = 33$, $\quad R_3 = \dfrac{8 \times 0.02 \times 213}{32.2 \times \pi^2 \times 0.167^5} = 826$

Pipe 4: $L_e = \dfrac{1 \times 0.333}{0.02} = 17$, $\quad R_4 = \dfrac{8 \times 0.02 \times 277}{32.2 \times \pi^2 \times 0.333^5} = 34$

We next find the system head loss, and then the discharge in each pipe:

$$(h_L)_{sys} = \dfrac{5^2}{\left(1/\sqrt{29} + 1/\sqrt{5} + 1/\sqrt{826} + 1/\sqrt{34}\right)^2} = 35.5 \text{ ft}$$

$Q_1 = \sqrt{35.5/29} = 1.11 \text{ ft}^3/\text{sec}$, $\quad Q_2 = \sqrt{35.5/5} = 2.66 \text{ ft}^3/\text{sec}$

$Q_3 = \sqrt{35.5/826} = 0.21 \text{ ft}^3/\text{sec}$, $\quad Q_4 = \sqrt{35.5/34} = 1.02 \text{ ft}^3/\text{sec}$

The sum of the four discharges is equal to the system discharge.

6.7 Compute the loss terms for each pipe:

Pipe 1: $L_e = \dfrac{2 \times 0.5}{0.02} = 50$, $\quad (h_L)_1 = \dfrac{8 \times 0.02 \times 2550}{32.2 \times \pi^2 \times 0.5^5} Q_1^2 = 41 Q_1^2$

Pipe 2: $L_e = \dfrac{3 \times 0.333}{0.025} = 40$, $\quad (h_L)_2 = \dfrac{8 \times 0.025 \times 1640}{32.2 \times \pi^2 \times 0.333^5} Q_2^2 = 252 Q_2^2$

Pipe 3: $L_e = \dfrac{7 \times 0.5}{0.018} = 190$, $\quad (h_L)_3 = \dfrac{8 \times 0.018 \times 3490}{32.2 \times \pi^2 \times 0.5^5} Q_3^2 = 51 Q_3^2$

Assume the flow directions shown in Fig. 6.4a, and write the energy relations for each pipe:

Pipe 1: $H_B = 65 - 41 Q_1^2$, or $Q_1 = \sqrt{(65 - H_B)/41}$

Pipe 2: $H_B = 5 + 252 Q_2^2$, or $Q_2 = \sqrt{(H_B - 5)/252}$

Pipe 3: $H_B = 43 + 51 Q_3^2$, or $Q_3 = \sqrt{(H_B - 43)/51}$

The solution is found by iterating values of H_B, computing Q_1, Q_2, and Q_3, and checking the continuity balance at the junction: $\Sigma Q = Q_1 - Q_2 - Q_3$. The results are tabulated below; five iterations are performed until the continuity balance is satisfied within a tolerable limit.

Iteration	H_B	Q_1	Q_2	Q_3	ΣQ
1	43	0.732	0.388	0	+0.344
2	45	0.698	0.398	0.198	+0.102
3	46	0.681	0.403	0.242	+0.036
4	47	0.663	0.408	0.280	-0.025
5	46.6*	0.670	0.406	0.266	-0.002

* Estimated by linear interpolation between the two previous values of H_B and Q.

Hence the discharges are $Q_1 = 0.67$ ft^3/sec, $Q_2 = 0.40$ ft^3/sec, and $Q_3 = 0.27$ ft^3/sec.

$$H_P = 105 + \left(\frac{0.02 \times 8{,}000}{0.667} + 12.5\right) \frac{Q^2}{2 \times 32.2 \times \left(\pi \times 0.667^2 / 4\right)^2}$$

$$= 105 + 32 Q^2$$

a) Assume $Q = 1.0$ ft^3/sec, then $H_p = 105 + 32 \times 1^2 = 137$ ft. On the pump characteristic curve of Fig. 6.5, $H_p \cong 175$ ft at $Q = 1.0$ ft^3/sec. Try another discharge, say $Q = 1.5$ ft^3/sec; then from the demand curve $H_p = 177$ ft, and from the pump curve $H_p \cong 170$ ft. Finally, with $Q = 1.4$ ft^3/sec, from the demand curve $H_p = 168$ ft, and from the pump curve $H_p \cong 170$ ft. Thus $Q = 1.4$ ft^3/sec and $H_p = 170$ ft.

b) From Fig 6.4, the pump efficiency at $Q = 1.4$ ft^3/sec is approximately 0.7. Hence, the required power input to the pump is $\dot{W}_P = 53 \times 1.4 \times 170 / 0.7 = 18{,}000$ ft-lb/sec, or 32.8 horsepower.

6.9 First compute the head loss in the suction pipe. Neglecting minor losses (no information is given) we have, see Eq. 6.1.1 and use Fig. 4.2,

$$h_L = \frac{8 \times 0.02 \times 16 \times (13/7.48)^2}{32.2 \times \pi^2 \times 0.5^5} = 0.78 \text{ ft}$$

a) For water at 80° F, $\gamma = 62.2$ lb/ft^3, and $p_{vap} = 0.507$ psi. Substituting into Eq. 6.3.2, the condition for imminent cavitation is

$$NPSH = \frac{(15 - 0.507) \times 144}{62.2} - 16 - 0.78 = 16.8 \text{ ft}$$

b) With $p_{atm} = 12$ psi and the same flow conditions for part a), use Eq. 6.3.2 to determine the required z:

$$\Delta z = \frac{p_{atm} - p_{vap}}{\gamma} - h_L - NPSH = \frac{(12 - 0.507) \times 144}{62.2} - 0.78 - 16.8 = 9.0 \text{ ft}$$

Hence the change in elevation is $16 - 9 = 7$ ft.

6.10 The following parameters are given: $N_1 = 970$ rpm, $N_2 = 1{,}200$ rpm, $H_1 = 38$ ft, $Q_1 = 230$ ft^3/sec, and $D_2 : D_1 = 1 : 2$. Using the similitude relations, Eqs. 6.3.4 and 6.3.5, compute Q_2 and H_2:

$$H_2 = 38 \times \left(\frac{1{,}200}{970}\right)^2 \times \left(\frac{1}{2}\right)^2 = 14.5 \text{ ft}$$

$$Q_2 = 230 \times \frac{1{,}200}{970} \times \left(\frac{1}{2}\right)^3 = 35.6 \text{ ft}^3/\text{sec}$$

The power is $\dot{W}_2 = 62.4 \times 35.6 \times 14.5 / 0.85 = 37{,}900$ ft-lb/sec, or 70 horsepower.

Questions 11-20

Before answering the specific questions, it is expedient to compute the R-values for each pipe (recall that $h_L = RQ^2$):

$$\text{Pipe 1: } Le = \frac{1 \times 0.333}{0.02} = 17 \text{ ft}, \quad R_1 = \frac{8 \times 0.02 \times 27}{32.2 \times \pi^2 \times 0.333^5} = 3$$

$$\text{Pipe 2: } Le = \frac{2 \times 0.333}{0.02} = 33 \text{ ft}, \quad R_2 = \frac{8 \times 0.02 \times 533}{32.2 \times \pi^2 \times 0.333^5} = 66$$

$$\text{Pipe 3: } Le = \frac{2 \times 0.333}{0.02} = 33 \text{ ft}, \quad R_3 = \frac{8 \times 0.02 \times 2,033}{32.2 \times \pi^2 \times 0.333^5} = 250$$

$$\text{Pipe 4: } Le = \frac{4 \times 0.333}{0.02} = 67 \text{ ft}, \quad R_4 = \frac{8 \times 0.02 \times 1,817}{32.2 \times \pi^2 \times 0.333^5} = 224$$

6.11 d) Compute the discharge and head loss in pipe 2:

$$Q_2 = 5,000 \frac{\text{gal}}{\text{hr}} \times \frac{1 \text{ ft}^3}{7.48 \text{ gal}} \times \frac{1 \text{ hr}}{3600 \text{ sec}} = 0.186 \frac{\text{ft}^3}{\text{sec}},$$

$$(h_L)_2 = R_2 Q_2^2 = 66 \times 0.186^2 = 2.28 \text{ ft}$$

6.12 b) The flow path is pipe 1 - pipe 2 - pipe 3; there is no flow in pipe 4. The demand curve from A to B is

$$H_P = z_B - z_A + (h_L)_1 + (h_L)_2 + (h_L)_3 = 430 + (3 + 66 + 250)Q^2 = 430 + 319Q^2$$

A trial and error solution is employed: assume Q, compute H_P from the demand curve, and compare it with H_P from the pump characteristic curve.

Trial 1: $Q = 10,000$ gal/hr $= 0.371$ ft^3/sec,
H_P (demand) $= 430 + 319 \times 0.371^2 = 474$ ft,
H_P (pump) $= 465$ ft

Trial 2: $Q = 9,000$ gal/hr $= 0.334$ ft^3/sec,
H_P (demand) $= 430 + 319 \times 0.334^2 = 467$ ft,
H_P (pump) $= 466$ ft

6.13 a) $Q_2 = 11,000$ gal/min $= 0.409$ ft^3/sec, and from the pump curve, $H_P = 460$ ft. Write the energy equation from A to C:

$$H_C = H_P - (R_1 + R_2)Q_2^2 = 460 - 69 \times 0.409^2 = 449 \text{ ft}$$

The energy equation from C to D is $H_C = z_D + R_4 Q_4^2$, where H_C is the hydraulic grade line at location C. Therefore the discharge in pipe 4 is

$$Q_4 = \sqrt{\frac{H_C - z_D}{R_4}} = \sqrt{\frac{449 - 445}{224}} = 0.134 \text{ ft}^3/\text{sec, or 3,600 gal/hr.}$$

6.14 d) The discharge is $Q = 8,000/(7.48 \times 3600) = 0.297$ ft^3/sec. From the pump characteristic curve,

$H_P = 470$ ft, and $\eta = 0.8$. Therefore the required power is

$$\dot{W}_P = \gamma Q H_P / \eta = 62.4 \times 0.297 \times 470 / 0.8 = 10,900 \text{ ft-lb/sec, or } 19.8 \text{ horsepower.}$$

6.15 c) First determine the head loss in the suction pipe:

$$Q = \frac{10,000}{7.48 \times 3600} = 0.371 \text{ ft}^3/\text{sec}, \quad (h_L)_1 = 3 \times 0.371^2 = 0.41 \text{ ft}$$

Using Eq. 6.3.2, the required pump height is:

$$\Delta z = \frac{p_{atm} - p_{vap}}{\gamma} - (h_L)_1 - NPSH = \frac{(14.7 - 0.256) \times 144}{62.4} - 0.41 - 25 = 7.9 \text{ ft}$$

6.16 b) The discharge through the pump is $Q = 12,000/(7.48 \times 3600) = 0.446 \text{ ft}^3/\text{sec}$, and from the characteristic curve the pump head is $H_P = 454$ ft. The energy equation from A to C is $H_P = z_C + p_C/\gamma - z_A + (h_L)_1 + (h_L)_2$. Therefore

$$\frac{p_C}{\gamma} = 454 - 300 - (3 + 66) \times 0.446^2 = 140 \text{ ft}, \quad \text{and} \quad p_C = \frac{62.4 \times 140}{144} = 61 \text{ lb/in}^2$$

6.17 b) Write the energy equation from C to D and solve for the pressure at location D:

$p_C/\gamma + z_C = p_D/\gamma + z_D$. (Since there is no flow in pipe 4, the energy equation reduces to the law of hydrostatics.) From Problem 16, the pressure at location C is 61 lb/in². Therefore,

$$p_D = 61 + 62.4 \times \frac{300 - 445}{144} = -1.83 \text{ lb/in}^2$$

6.18 d) From the pump curve, for $H_P = 460$ ft, $Q = 11,000$ gal/hr, or $0.409 \text{ ft}^3/\text{sec}$. Write the energy equation from A to C and evaluate the hydraulic grade line:

$$H_C = H_P - (h_L)_1 - (h_L)_2 = 460 - 69 \times 0.409^2 = 449 \text{ ft}$$

Thus the discharge in pipe 3, between location C and location B is

$$Q_3 = \sqrt{\frac{(H_C - H_B)}{R_3}} = \sqrt{\frac{449 - 430}{250}} = 0.276 \text{ ft}^3/\text{sec}, \text{ or } 7,400 \text{ gal/hr}$$

6.19 c) The flow path is pipe 1 - pipe 2 - pipe 4; there is no flow in pipe 3. Write the energy equation from location A to location D:

$$H_P = z_D - z_A + (h_L)_1 + (h_L)_2 + (h_L)_4 = 445 + (3 + 66 + 224)Q^2 = 445 + 293Q^2$$

A trial and error solution is employed: assume Q, compute H_P from the demand curve, and compare it with H_P from the pump characteristic curve.

Trial 1: $Q = 10,000$ gal/hr $= 0.371 \text{ ft}^3/\text{sec}$,
H_P (demand) $= 445 + 293 \times 0.371^2 = 485$ ft,
H_P (pump) $= 465$ ft

Trial 2: $Q = 7,500$ gal/hr $= 0.278 \text{ ft}^3/\text{sec}$,
H_P (demand) $= 445 + 293 \times 0.278^2 = 468$ ft,
H_P (pump) $= 475$ ft

Trial 3: $Q = 8{,}000 \text{ gal/hr} = 0.297 \text{ ft}^3/\text{sec}$,
$H_P \text{ (demand)} = 445 + 293 \times 0.297^2 = 471 \text{ ft}$,
$H_P \text{ (pump)} = 472 \text{ ft}$

6.20 e) Given data are $Q_1 = Q_2$, $N_1 = 900$ rev/min. $D_1 : D_2 = 2 : 3$. Use the similitude relation, Eq. 6.3.4 to compute the new pump speed N_2:

$$N_2 = N_1 \left(\frac{Q_2}{Q_1}\right)\left(\frac{D_1}{D_2}\right)^3 = 900 \times 1 \times \left(\frac{3}{2}\right)^3 = 3{,}040 \text{ rev/min}$$

7. Hydrology

by David A. Hamilton

Hydrology is the study of the occurrence and distribution of water, both on and under the earth's surface. Through hydrologic analysis, engineers are able to quantify the flow of water under a variety of circumstances, allowing them to safely locate and design structures in or adjacent to waterways. Hydrologic analysis is also used to study water supply, design remedial environmental cleanup, and develop environmental protection measures. This chapter presents a brief review of hydrologic engineering concepts and common design techniques. It is not intended to be an exhaustive development of hydrologic theory. More detailed development and explanation of limitations of the methodologies may be found in various texts.

The distribution of water varies greatly geographically as well as with time. Precipitation patterns vary seasonally and from year to year. The variation is accounted for through probabilistic analysis. Design criteria are frequently expressed in terms of a probability, such as the 1% chance or once-in-a-hundred year flood.

Precipitation is the starting point for the hydrologic cycle (Fig. 7.1). As rain falls, some is intercepted by plants and buildings, and never reaches the ground. Some infiltrates into the ground, and some runs off over the ground surface. A large portion of the infiltrated water is taken up by plants and evapotranspirated back into the atmosphere. The remainder enters the saturated groundwater and eventually flows back to the surface water system. Some of the runoff flows into depressions, where it either infiltrates to the groundwater, or evaporates into the atmosphere. A portion of the original precipitation runs off into the stream network. The hydrologic cycle eventually closes when water is evaporated from the oceans and lakes and delivered again through precipitation by weather systems.

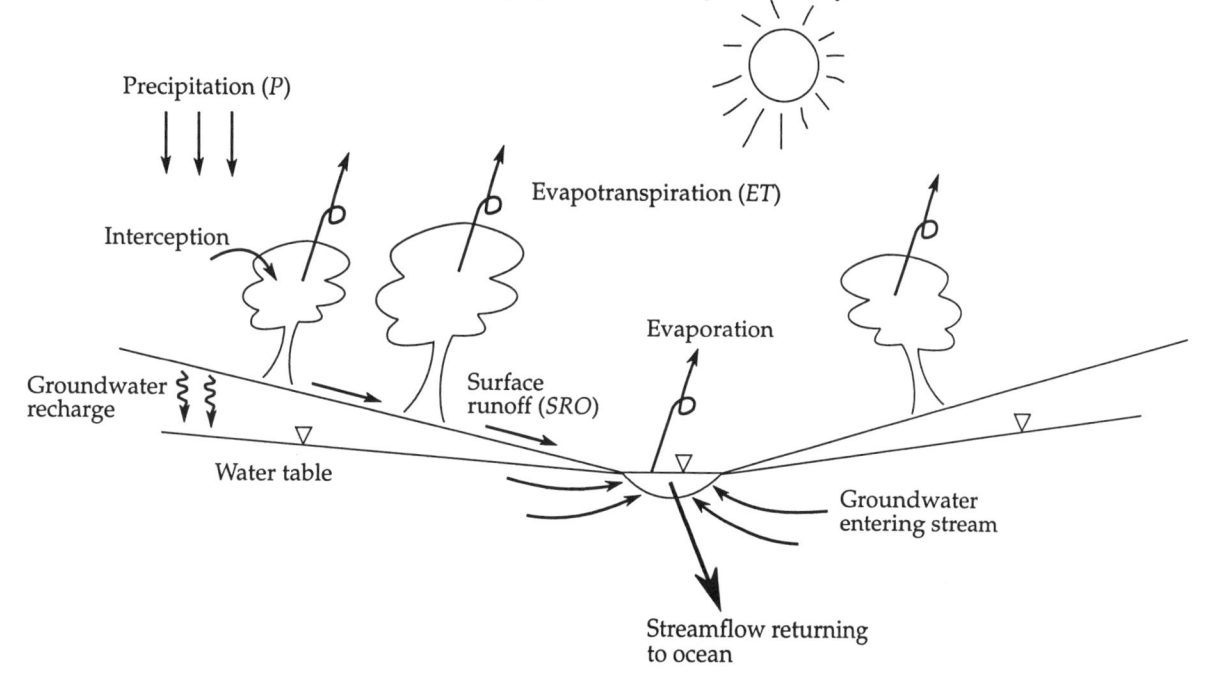

Figure 7.1. Hydrologic cycle.

7.1 Water Balance

It is important to consider water movement as part of a system. Where has the water come from, and what are the factors influencing its movement? Fig. 7.2 illustrates the system approach through a series of simple "tank" models. To quantify the movement of water, we define the system in terms of a mass balance. Simply stated, the difference between the inflows and outflows equals the change in storage. Generally, the mass balance can be written over the hydrologic system, including both surface and groundwater components, as a simple water budget equation:

$$P - Q - G - ET = \Delta S \qquad (7.1.1)$$

where
- P = precipitation over the period of interest
- Q = net surface water outflow (flow out – flow in)
- G = net groundwater outflow (flow out – flow in)
- ET = evapotranspiration
- ΔS = net increase in both surface and groundwater storage over the period of interest

All units are in volume per time. English units are usually ft^3/sec (cfs). It is also conventional to divide all terms by the area so that the units are length per time (e.g., in./yr).

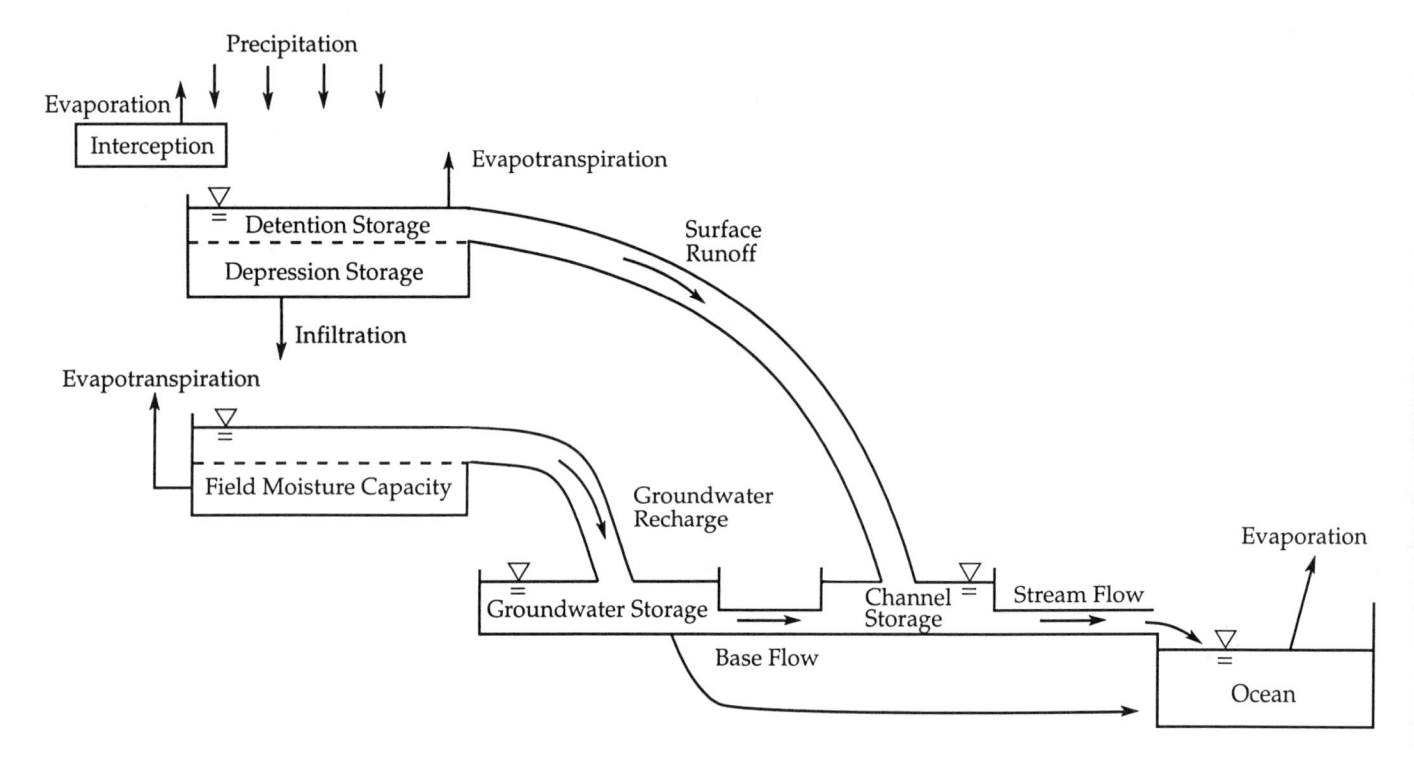

Figure 7.2. Hydrologic cycle illustrated by tank models.

EXAMPLE 7.1

The average discharge recorded at a stream gage in the Eastern U.S. is 400 cfs. The drainage area is 250 mi² with an average annual precipitation of 32 inches. Estimate the average annual evapotranspiration loss from the watershed in inches per year.

Solution. The change in storage from year to year is usually a small number, near zero. Since the question deals with long-term average values, we can assume that there is no net change in storage over the long term; therefore, $\Delta S = 0$. Since the watershed is also relatively large, and located in the humid region of the U.S., we can assume the vast majority of water that recharged aquifers in the watershed has discharged back into the river and has been recorded as streamflow at the gage. Therefore, the gage represents essentially all the outflow from the watershed ($Q + G$). The remaining unknown, ET, can be found using Eq. 7.1.1 and averaging over the drainage area:

$$ET = P - Q - G - \Delta S$$

$$= 32 \tfrac{\text{in}}{\text{yr}} - \frac{400 \tfrac{\text{ft}^3}{\text{sec}} \times 3600 \tfrac{\text{sec}}{\text{hr}} \times 24 \tfrac{\text{hr}}{\text{day}} \times 365 \tfrac{\text{day}}{\text{yr}} \times 12 \tfrac{\text{in}}{\text{ft}}}{250 \, \text{mi}^2 \times 5280^2 \tfrac{\text{ft}^2}{\text{mi}^2}}$$

$$= 10.3 \tfrac{\text{in}}{\text{yr}}$$

7.2 Precipitation

Precipitation is the major input of the hydrologic cycle. A hydrologic analysis focuses on rainfall. The characteristics needed to evaluate the hydrologic impact of rainfall are

- total amount
- duration
- intensity
- recurrence interval
- areal distribution

Most designs are based on a hypothetical or design storm. Actual precipitation patterns are studied to determine the most reasonable characteristics to be assigned to the design storm. The most commonly used nationwide study, frequently referred to as "TP-40," for determining total precipitation and recurrence interval used in design storms is by D. Hershfield of the U.S. Weather Bureau (1961). It is slowly being superseded by regional studies that incorporate the decades of additional precipitation data. Fig. 7.3 contains the maps for the 24-hour duration, 100-year and 10-year recurrence interval design storms.

The design rainfall from Fig. 7.3 is considered to represent "point rainfall" because it was derived from analyzing rainfall data at discrete rain gages. These values may be used for watersheds of up to 10 square miles. However, as the watershed becomes larger, it is less likely that the rainfall will be uniform over the entire area, and an areal adjustment is required. The ratios for areal adjustment of point rainfalls are included in Table 7.1.

TABLE 7.1 Ratios for areal adjustment of 24 hour point rainfalls.

Area (mi²)	Ratio
10 or less	1.00
15	.978
20	.969
25	.964
30	.960
35	.957
40	.953

The flood response to a storm depends largely on the rainfall intensity (inches per hour) and duration. A 0.5 inch rainfall spread out over a day will probably not result in much runoff, but if it occurred in a few minutes, it could cause temporary flooding in parking lots and streets. The longer a storm lasts, the greater its impact will be. Short storms, even if they are intense, will only affect small watersheds. As the storm duration increases, so does its potential impact on larger watersheds.

While snowfall can be significant in some parts of the country, it is not hydrologically significant until it melts and provides runoff to the streams and recharge to the groundwater.

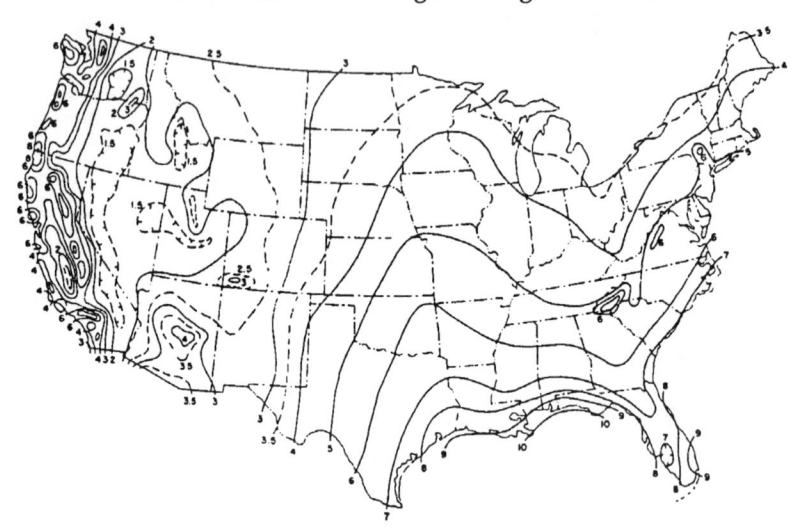

Ten-year 24-hr rainfall (inches).

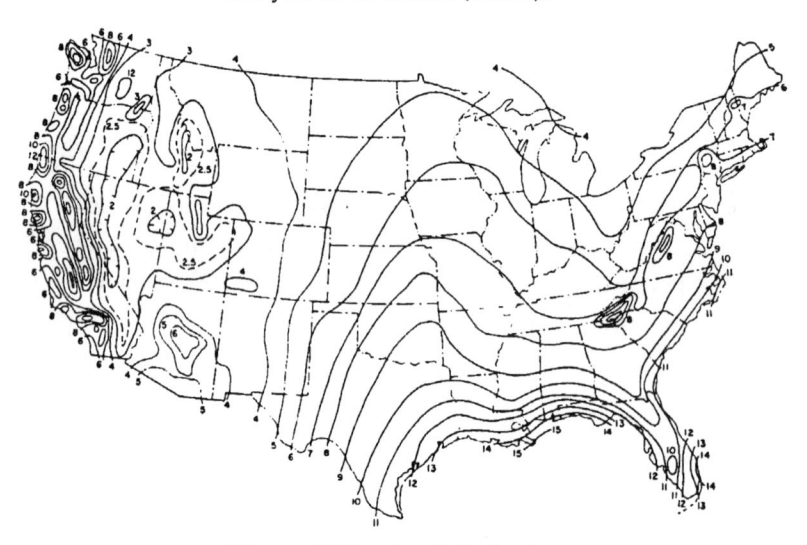

100-year 24-hr rainfall (inches).

Figure 7.3 Design precipitation (U.S. Weather Bureau).

EXAMPLE 7.2

Determine the design rainfall for a 24-hour, 10-year flood, on a 25 mi² watershed near Chicago, Illinois.

Solution. Using Fig. 7.3, the point precipitation is 4 inches. Since the watershed is larger than 10 mi², the point rainfall must be adjusted. From Table 7.1, the adjustment is 0.964. Therefore, the design rainfall is

$$4 \times 0.964 = 3.86 \text{ in}.$$

7.3 Precipitation Losses

Not all of the precipitation runs off and contributes to flows. Some never reaches the ground. A small amount, on the order of 0.1 inches, is intercepted by vegetation and buildings. More significant is the amount of *infiltration*, which is the water entering the soil. The rate of infiltration is influenced by several factors:

- soil (structure, hydraulic conductivity, porosity)
- soil moisture content
- vegetation
- rainfall intensity
- soil surface conditions (such as crusting)

Some of these factors change with time during a rainstorm. Typically the infiltration rate is highest at the beginning of a storm. The soil is driest and most receptive to infiltration. The rain pounding on the soil surface may cause it to seal, and as the soil moisture increases during a storm, infiltration rates drop. At some point the rainfall rate becomes greater than the infiltration rate, ponding on the surface occurs, and runoff starts. The total amount of water intercepted, evaporated and infiltrated before surface runoff begins is called the *initial abstraction* (I_a). The soil has a continuing infiltration capacity after the initial abstraction has been met. The difference between this and the rainfall volume is the surface runoff (*SRO*). These quantities are illustrated in Fig. 7.4. There are several empirical methods used to calculate the infiltration runoff and rates. The SCS method is frequently used and is explained in section 7.4.

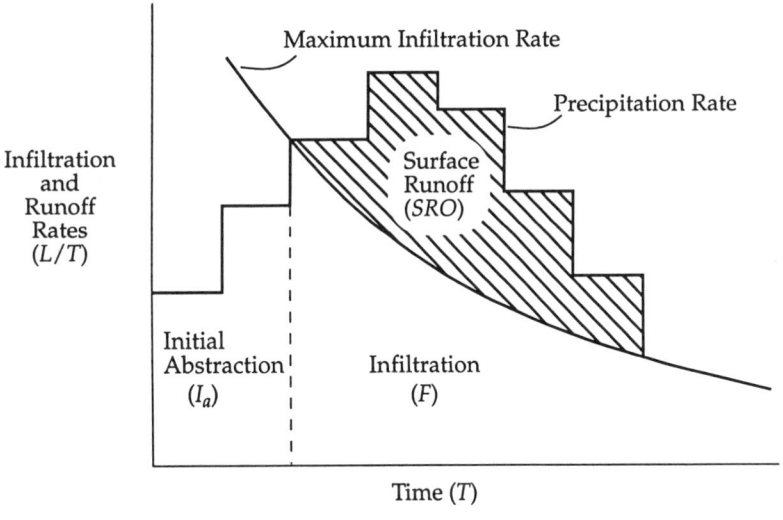

Figure 7.4. Infiltration–Runoff Relationship.

7.4 Runoff

The total precipitation volume (*P*) is distributed as

$$P = SRO + I_a + F \tag{7.4.1}$$

The *SCS* method assumes the ratio of the actual water infiltrated (*F*) to the maximum possible amount of water retained (*S*) (maximum infiltrated plus initial abstraction) is equal to the ratio of the actual surface runoff (*SRO*) to the maximum possible runoff ($P - I_a$). This is expressed as

$$\frac{F}{S} = \frac{SRO}{P - I_a} \tag{7.4.2}$$

Rearranging Eq. 7.4.1 in terms of *F*, substituting it into Eq. 7.4.2 and solving for *SRO* results in

$$SRO = \frac{(P-I_a)^2}{P-I_a+S} \tag{7.4.3}$$

The *SCS* method empirically determined that, in general, $I_a = 0.2S$, therefore, Eq. 7.4.3 becomes

$$SRO = \frac{(P-0.2S)^2}{P+0.8S} \tag{7.4.4}$$

The maximum water retained (S) is a function of the soil and land use. The *SCS* method defined a curve number (*CN*) that reflects the runoff characteristics represented by the combination of soil type and land use. The equation relating S and *CN* is

$$S = \frac{1000}{CN} - 10 \tag{7.4.5}$$

The curve number is a function of the soil, land use and antecedent moisture content. Soils are classified into one of four Hydrologic Soil Groups:

Soil A. Soils having high infiltration rates even when thoroughly wetted and consisting chiefly of deep, well to excessively drained sands or gravels. These soils have a high rate of water transmission. (Low runoff potential.)

Soil B. Soils having moderate infiltration rates when thoroughly wetted and consisting of moderately deep to deep, moderately well to well drained soils with moderately fine to moderately coarse textures. These soils have a moderate rate of water transmission.

Soil C. Soils having slow infiltration rates when thoroughly wetted and consisting chiefly of soils with a layer that impedes the downward movement of water or soils with moderately fine to fine texture. These soils have a low rate of water transmission.

Soil D. Soils having very slow infiltration rates when thoroughly wetted and consisting chiefly of clay soils with a high swelling potential, soils with a permanent high water table, soils with a claypan or clay layer at or near the surface, and shallow soils over nearly impervious material. These soils have a very low rate of water transmission. (High runoff potential.)

The combination of a hydrologic soil group and a land use is a "hydrologic soil-cover complex." Each combination is assigned a *CN* which is an index to its runoff potential on soil that is not frozen. A list of these values is shown in Table 7.2. The tabulated *CN* values are for normal soil moisture conditions referred to as Antecedent Moisture Condition II (AMC-II). AMC-I has the lowest runoff potential since the watershed soils are dry. AMC-III has the highest runoff potential as the watershed is practically saturated from antecedent rainfall or snowmelt. The AMC can be estimated from the 5-day antecedent rainfall total by using Table 7.3. The limits for "dormant" season apply when the soils are not frozen and there is no snow on the ground. Although the *CN* in Table 7.2 is for AMC-II conditions, an analysis of a specific storm event may require an equivalent *CN* for AMC-I or AMC-III. They are computed by the following equations:

$$CN(\text{I}) = \frac{4.2 \times CN(\text{II})}{10 - 0.058 \times CN(\text{II})} \tag{7.4.6}$$

$$CN(\text{III}) = \frac{23 \times CN(\text{II})}{10 + 0.13 \times CN(\text{II})} \tag{7.4.7}$$

TABLE 7.2 Runoff curve numbers (*CN*) for hydrologic soil-cover complexes (AMC-II conditions).

Land Use	Treatment or Practice	Hydrologic Condition	A	B	C	D
Fallow	Straight row		77	86	91	94
Row crops	Straight row	Poor	72	81	88	91
	Straight row	Good	67	78	85	89
	Contoured	Poor	70	79	84	88
	Contoured	Good	65	75	82	86
	Contoured & terraced	Poor	66	74	80	82
	Contoured & terraced	Good	62	71	78	81
Small grain	Straight row	Poor	65	76	84	88
	Straight row	Good	63	75	83	87
	Contoured	Poor	63	74	82	85
	Contoured	Good	61	73	81	84
	Contoured & terraced	Poor	61	72	79	82
	Contoured & terraced	Good	59	70	78	81
Close-seeded legumes or rotation meadow	Straight row	Poor	66	77	85	89
	Straight row	Good	58	72	81	85
	Contoured	Poor	64	75	83	85
	Contoured	Good	55	69	78	83
	Contoured & terraced	Poor	63	73	80	83
	Contoured & terraced	Good	51	67	76	80
Pasture or range		Poor	68	79	86	89
		Fair	49	69	79	84
		Good	39	61	74	80
	Contoured	Poor	47	67	81	88
	Contoured	Fair	25	59	75	83
	Contoured	Good	6	35	70	79
Meadow			30	58	71	78
Woods		Poor	45	66	77	83
		Fair	36	60	73	79
		Good	25	55	70	77
Residential	1/8 acre or less lot size		77	85	90	92
	1/4 acre		61	75	83	87
	1/3 acre		57	72	81	86
	1/2 acre		54	70	80	85
	1 acre		51	68	79	84
Open spaces	(parks, golf courses, cemeteries, etc.)					
	Good condition: Grass cover > 75% of area		39	61	74	80
	Fair condition: Grass cover > 50–75% of area		49	69	79	84
Commercial or business area (85% impervious)			89	92	94	95
Industrial district (72% impervious)			81	88	91	93
Farmsteads			59	74	82	86
Paved areas (roads, driveways, parking lots, roofs)			98	98	98	98
Water surfaces (lakes, ponds, reservoirs, etc.)			100	100	100	100
Swamp	At least 1/3 is open water		85	85	85	85
Swamp	Vegetated		78	78	78	78

TABLE 7.3 Definition of Antecedent Moisture Conditions by season and 5-day antecedent rainfall total (inches).

AMC Group	Dormant Season	Growing Season
I	< 0.5	< 1.4
II	0.5 – 1.1	1.4 – 2.1
III	> 1.1	> 2.1

EXAMPLE 7.3

Determine the CN for a small suburban watershed with the following land uses and hydrologic soil groups:

50 acre subdivision with 1/4 acre lots on B soils,
100 acre subdivision with 1/2 acre lots on A soils,
20 acre park on C soils.

Also, what would the CN be if it had not rained in two weeks?

Solution. Design conditions are usually for AMC-II, so find CN in Table 7.2 for each combination of land use and soil group. The composite CN, representative for the entire watershed, can be determined by multiplying the individual land use CN by its area, summing the products, then dividing by the total area. We assume "good" grass cover in the park.

Land Use	Soil Group	CN	Area (ac)	Product
1/4 ac lots	B	75	50	3750
1/2 ac lots	A	54	100	5400
park	C	74	20	1480
		Total	170	10,630

Therefore, the composite CN for the watershed is 10,630/170 = 62.5.

Note: This method of calculating the composite CN is accurate as long as no more than 20% of the watershed has CN < 45. If the area with CN < 45 is over 20%, the incremental runoff can be calculated for each land use, summed for the total SRO; CN is then calculated using Eqs. 7.4.4 and 7.4.5.

If it had not rained for two weeks, referring to Table 7.3, AMC-I would be selected, and using Eq. 7.4.6

$$CN(I) = (4.2 \times 62.5)/(10 - 0.058 \times 62.5) = 41$$

The surface runoff is simply the difference between the precipitation and the losses (interception and infiltration). It is easily calculated with the methods outlined above. The relationship is also depicted in Fig. 7.5 which provides a quick estimate if the equations are not used.

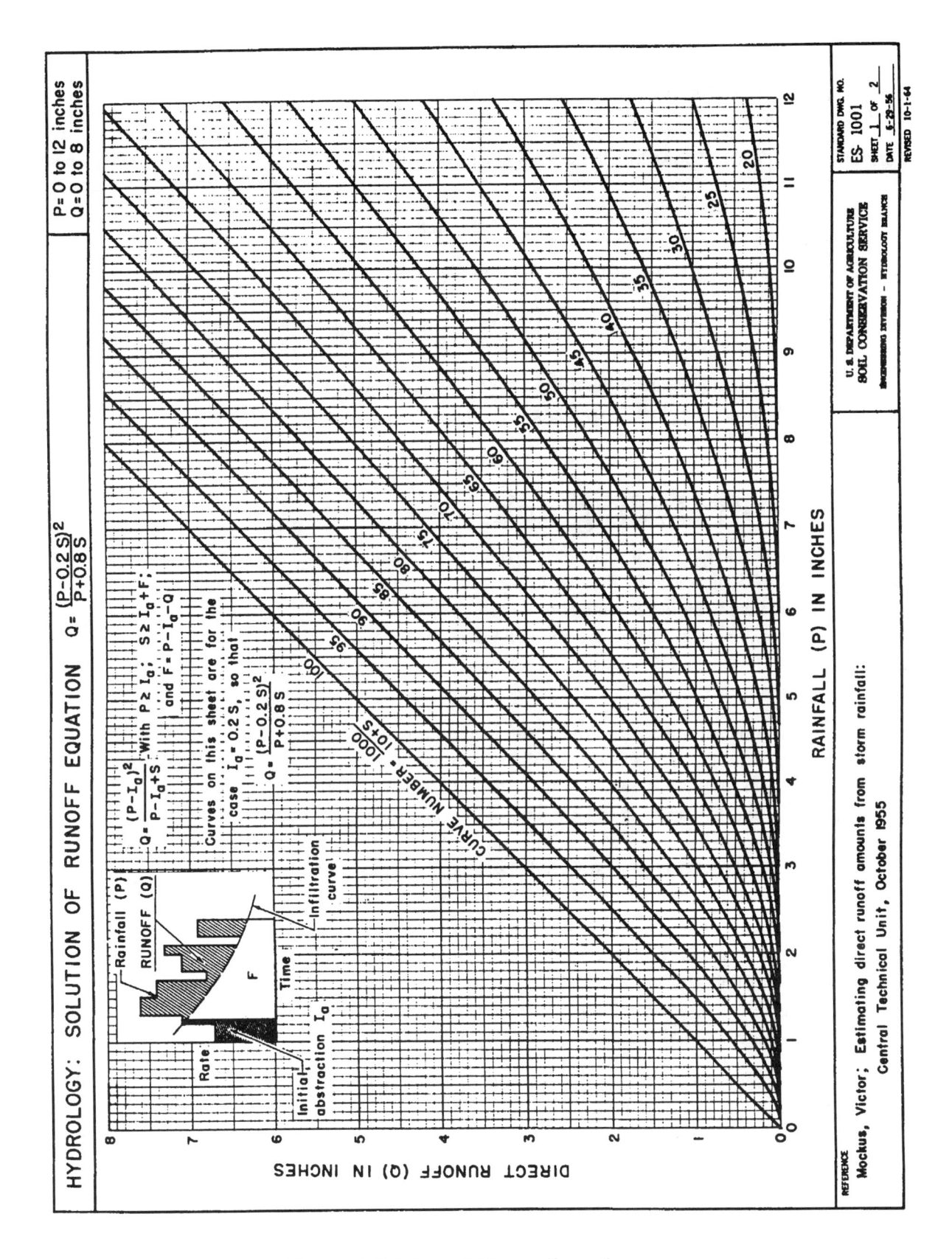

Figure 7.5. Solution of SCS runoff equation.

EXAMPLE 7.4

Compare the surface runoff in Example 7.3 under the AMC-I and AMC-II conditions for a design rainfall of 4 inches.

Solution. Using $CN = 63$ for AMC-II, Eq. 7.4.5 gives $S = (1000/63) - 10 = 5.87$ in. Eq. 7.4.4 then provides

$$SRO = \frac{(P - 0.2S)^2}{P + 0.8S} = \frac{(4 - 0.2 \times 5.87)^2}{4 + 0.8 \times 5.87} = 0.92 \text{ in.}$$

For the AMC-I situation $CN = 41$ so that Eq. 7.4.5 and Eq. 7.4.4 give $S = (1000/41) - 10 = 14.4$ in. and

$$SRO = \frac{(4 - 0.2 \times 14.4)^2}{4 + 0.8 \times 14.4} = 0.08 \text{ in.}$$

There is obviously a significant difference in runoff depending on the Antecedent Moisture Condition. These can also be quickly estimated by using Fig. 7.5.

7.5 Runoff Hydrograph

The next step in a hydrologic analysis is to determine the relationship between surface runoff volume and the ensuing flood, especially the flood peak. Stream discharge is depicted graphically with hydrographs. Each watershed will respond differently to precipitation events. The volume of runoff is determined by the combination of land use and soil characteristics. The flood peak and shape of the runoff hydrograph are determined by watershed and stream channel characteristics such as slope, shape, size, and channel conveyance.

If there is a constant intensity rainfall uniformly falling over a watershed, the runoff will peak and remain constant until the rainfall stops. This concept is illustrated in Fig. 7.6. As the storm starts, the flow in the channel responds slowly. At first, the only runoff is from precipitation directly on the channel. After the initial abstraction is met and the infiltration rate exceeded, then surface runoff begins. The areas immediately adjacent to the channel quickly contribute the excess precipitation as runoff. The hydrograph increases rapidly as more area in the watershed begins to contribute runoff. At t_1 the majority of the watershed is contributing runoff, and the rate of streamflow increase begins to slow. This is the point of inflection on the rising limb of the hydrograph. At t_c, the *time of concentration*, the whole watershed is contributing runoff. This represents the time it takes the most hydraulically distant location in the watershed to contribute runoff to the design location. Theoretically, the basin reaches a point where all the storage capacity is filled and the whole watershed contributes to the runoff. The runoff outflow is then constant. The hydrograph begins to drop at t_2, when the rainfall stops.

The basic assumption of the Rational Formula, to be presented in Section 7.8, is: if the rainfall is uniform and it lasts longer than the time of concentration for the watershed, then the runoff peak is directly proportional to the rainfall intensity.

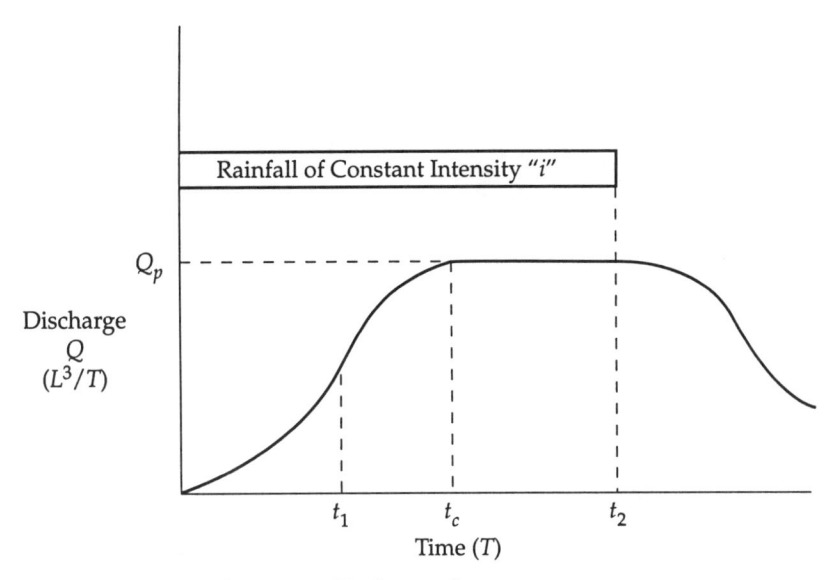

Figure 7.6. Hydrograph response to a storm.

A watershed will have runoff patterns that reflect constant watershed characteristics, such as size, shape, slope and degree of channel development. This concept leads to the *unit hydrograph*. A characteristic runoff hydrograph can be developed for each watershed. A runoff hydrograph of unit volume (typically one inch or one cm) is developed to represent the basin response to runoff events. This is illustrated in Fig. 7.7. The time to peak (T_p) is measured from the beginning of precipitation excess (when surface runoff begins) to the peak runoff rate. The lag time (T_ℓ) is measured from the centroid of the precipitation excess to the peak runoff rate. The lag time is empirically related to the time of concentration by the relationship: $T_\ell = 0.6 \times T_c$. In actual hydrographs, T_c can be determined by measuring from the centroid of rainfall excess to the point of inflection on the recession limb of the hydrograph.

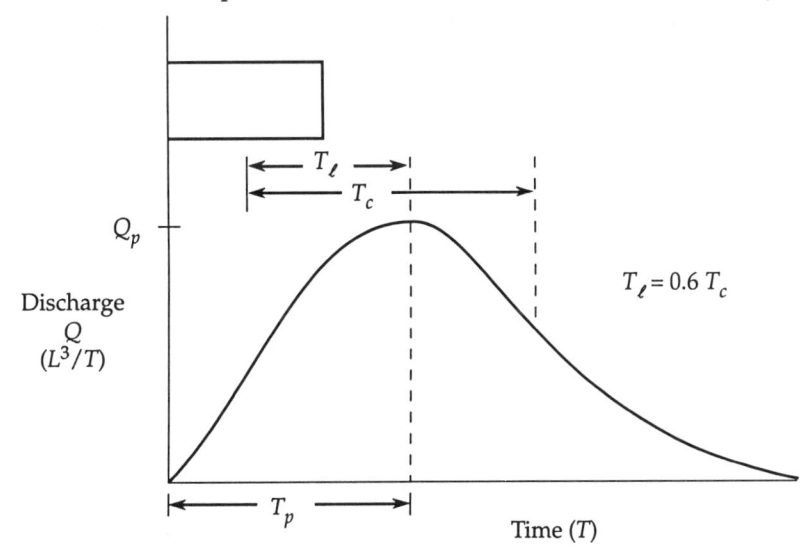

Figure 7.7. Unit hydrograph, definition of terms.

A storm with 0.5 inch of runoff will have runoff rates half the values of the unit hydrograph and a storm with 2.0 inches of runoff will have twice the values. This is illustrated in Fig. 7.8. Note the timing of the runoff response is the same for each event, it is only the magnitude of the response that varies linearly with runoff volume.

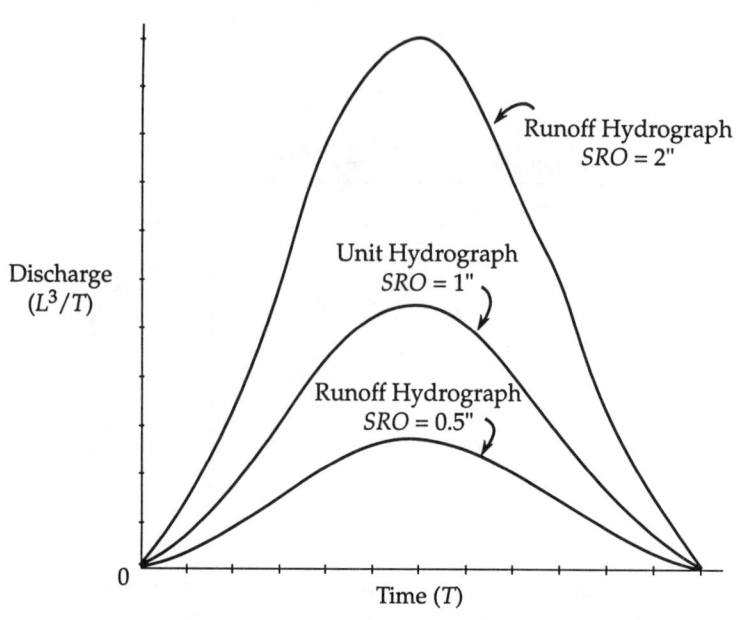

Figure 7.8. Application of different SRO's to unit hydrograph.

A runoff hydrograph only includes the portion of stream discharge that is derived from surface runoff. The baseflow (groundwater contribution) must be added to reflect the total hydrograph observed in a river.

7.6 Effects of Storage

Storage influences the timing and volume of surface runoff. Natural storage of runoff occurs in depressions and wetland areas. These areas must fill before any surface runoff leaves the vicinity. The surface runoff volume is decreased by the amount of this depression storage. There are also temporary storage areas in the floodplains adjacent to rivers and streams. As the flood waters rise, some of the water enters the floodplains and slows down or even becomes stagnant. This water is temporarily stored or delayed. It will reenter the stream and continue downstream as the flood recedes. The net impact of storage is to lower the flood peak and delay the timing of the flood peak. In many areas of the U.S. there are large natural storage areas.

As the U.S. developed, many of the natural wetland storage areas were drained and placed into intensive agricultural production. Large scale urban development and suburban sprawl filled natural wetland and floodplain storage areas. Urbanization also greatly increases the impervious area, which increases runoff volume. Development alters channel characteristics to more easily transport increased flood flows from the urban area. The net impact of development is to increase runoff volumes, increase channel conveyance, decrease the time of concentration, and therefore increase flood peaks. These factors can significantly increase the flooding problems for downstream areas. This is illustrated in Fig. 7.9.

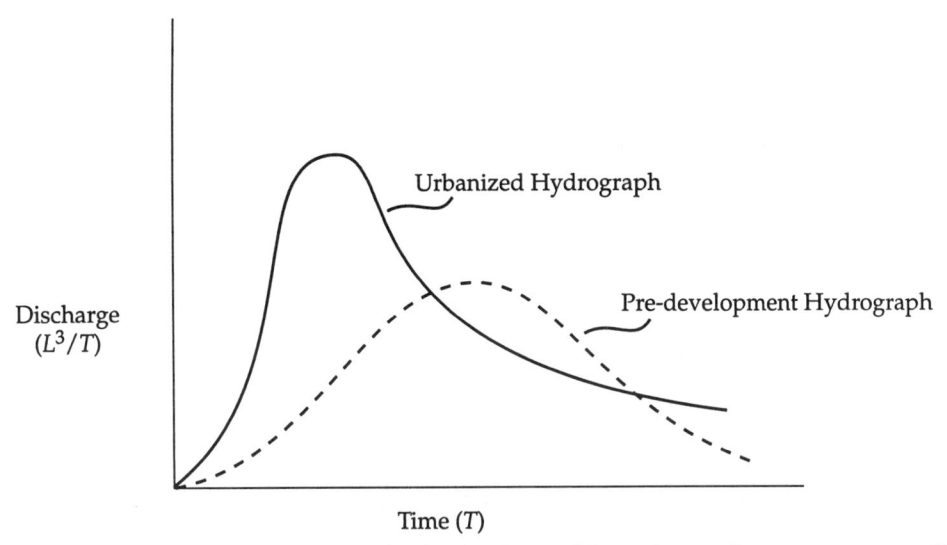

Figure 7.9. Generalized impact of urbanization and loss of natural storage on a runoff hydrograph.

The increased flooding problems associated with development can be counteracted by creating artificial storage areas. These are usually called detention or retention ponds. Proper design can protect communities from flooding problems and help protect the water quality of urban streams. Hydrologic engineers are increasingly called upon to design these structures. Techniques described in the following sections are used to design detention ponds.

7.7 Flood Probability

Flood frequency and flood probability are closely related terms. *Flood frequency*, or *recurrence interval*, is more commonly used. It is the frequency with which a flood of a given size may be expected to occur at a site, expressed as an interval of years. A "T" year flood will, over a long period of time, be equaled or exceeded on the average once every T years. Therefore, a 10-year flood would be expected to occur approximately 10 times in a 100-year period. *Flood probability* is the probability of a flood of a given size being equaled or exceeded in a given period, expressed in percent. It is inversely related to the flood frequency:

$$p = \frac{1}{T} \times 100 \qquad (7.7.1)$$

where
- p = flood probability (in percent)
- T = flood frequency or recurrence interval (years)

Consequently, a one percent chance flood is the same as a 100-year flood.

A flood probability must be determined for any project. A primary consideration is the level of protection desired, relative to the cost – benefit considerations. For example, if a large dam is built in the mountains upstream of a city, then its failure would have catastrophic consequences. It would be designed to withstand a "probable maximum flood." A common design criterion for siting a subdivision along a water course is that it should be placed above the 1% chance flood (100-year flood) elevation. Storm sewers are frequently designed to handle a 10% chance flood (10-year flood) without surcharging. The actual design criteria vary with jurisdiction and application, depending on local ordinances, state statutes and federal regulations.

The probability (in percent) of a T-year event occurring sometime in a period of n years is

$$p(T-\text{year event in } n \text{ years}) = \left[1-\left(1-\frac{1}{T}\right)^n\right]\times 100. \qquad (7.7.2)$$

The same concepts can also be applied to rainfall. Eqs. 7.7.1 and 7.7.2 can also be used to express rainfall probability.

EXAMPLE 7.5

A facility is designed for a useful life of 50 years. If it is located in the 100-year flood plain, what is the probability that a 100-year flood will occur during its useful life?

Solution. Using Eq. 7.7.2 with $T = 100$, and $n = 50$, there results

$$p(100\text{-year event in } 50 \text{ years}) = \left[1-\left(1-\frac{1}{100}\right)^{50}\right]\times 100$$

$$= 39\%$$

7.8 Flood Flow Calculation Techniques

7.8.1 Rational Method

The *Rational Formula* is a commonly used empirical equation to estimate peak runoff rates from small urban watersheds (usually less than a few hundred acres). It is

$$Q = CiA \qquad (7.8.1)$$

where
- Q = peak discharge (cfs)
- A = area (acres)
- i = rainfall intensity (inches/hour)
- C = a runoff coefficient that incorporates all rainfall losses (see Table 7.4)

A dimensional analysis of the equation reveals that Q should have dimensions of acre-inches/hour. The conversion factor from this to cubic feet per second (cfs) is 1.008, so Q is considered to be in cfs.

The rational formula is based on a number of assumptions:
- The rainfall intensity is uniform with time and over the entire watershed.
- The rainfall lasts for a period longer than the time of concentration for the watershed.
- The flood probability is equal to the rainfall probability.
- The coefficient C accounts for all rainfall losses including interception, infiltration and storage, and considered to be constant throughout the storm duration, and uniform across the watershed.

These assumptions severely limit the accuracy and applicability of the method, therefore it is generally limited to small watersheds, usually in urban areas.

Drainage area is determined from topographic maps. Surface runoff is determined using the C coefficient. Table 7.4 contains C coefficients for various land uses. The remaining unknown is the rainfall intensity, which must be related to the time of concentration (T_c). Since T_c is the time it takes water to travel from the hydraulically most distant part of the watershed to the design point, it can be calculated using hydraulic equations such as Manning's equation (discussed in the Hydraulics Chapter). There are

TABLE 7.4 Values of Runoff Coefficient C. (Source: Chow, 1964.)

Type of drainage area	Runoff coefficient, C
Lawns:	
Sandy soil, flat, 2%	0.05–0.10
Sandy soil, average, 2–7%	0.10–0.15
Sandy soil, steep, 7%	0.15–0.20
Heavy soil, flat, 2%	0.13–0.17
Heavy soil, average, 2–7%	0.18–0.22
Heavy soil, steep, 7%	0.25–0.35
Business:	
Downtown areas	0.70–0.95
Neighborhood areas	0.50–0.70
Residential:	
Single-family areas	0.30–0.50
Multi units, detached	0.40–0.60
Multi units, attached	0.60–0.75
Suburban	0.25–0.40
Apartment dwelling areas	0.50–0.70
Industrial:	
Light areas	0.50–0.80
Heavy areas	0.60–0.90
Parks, cemeteries	0.10–0.25
Playgrounds	0.20–0.35
Railroad yard areas	0.20–0.40
Unimproved areas	0.10–0.30
Streets:	
Asphalt	0.70–0.95
Concrete	0.80–0.95
Brick	0.70–0.85
Drives and walks	0.75–0.85
Roofs	0.75–0.95

also a number of empirical methods that have been developed to estimate T_c based on easily obtainable parameters, such as slope and generalized stream channel characteristics. One very simplified equation is

$$T_c = 0.00013 \times L^{0.77} S_0^{-0.385} \qquad (7.8.2)$$

where T_c = time of concentration (hours)
L = maximum length of flow (feet)
S_0 = average watershed gradient (foot/foot)

This equation was developed from SCS data for rural basins in Tennessee with well-defined channels and steep slopes (3 – 10%). If it is used for overland flow on concrete or asphalt surfaces, multiply T_c by 0.4; for concrete channels, multiply by 0.2; no adjustments are needed for overland flow on bare soil or flow in roadside ditches.

The rainfall duration T_d is set equal to T_c; the corresponding rainfall intensity must now be determined. Research has related rainfall intensity to the duration and frequency (an IDF curve); it is

$$i = \frac{c}{T_d^e + f} \qquad (7.8.3)$$

where T_d is the duration (min), and c, e, and f are coefficients that vary with location and return period. Table 7.5 contains coefficients for a 10-year return period at several locations across the U.S.. A sample IDF curve is shown for Chicago in Fig. 7.10.

TABLE 7.5 Constants for rainfall Eq. 7.8.3 for 10-year return period storm intensities at various locations.

Location	c	e	f
Atlanta	97.5	0.83	6.88
Chicago	94.9	0.88	9.04
Cleveland	73.7	0.86	8.25
Denver	96.6	0.97	13.90
Houston	97.4	0.77	4.80
Los Angeles	20.3	0.63	2.06
Miami	124.2	0.81	6.19
New York	78.1	0.82	6.57
Santa Fe	62.5	0.89	9.10
St. Louis	104.7	0.89	9.44

Constants correspond to i in inches per hour and T_d in minutes. Source: Wenzel, 1982.

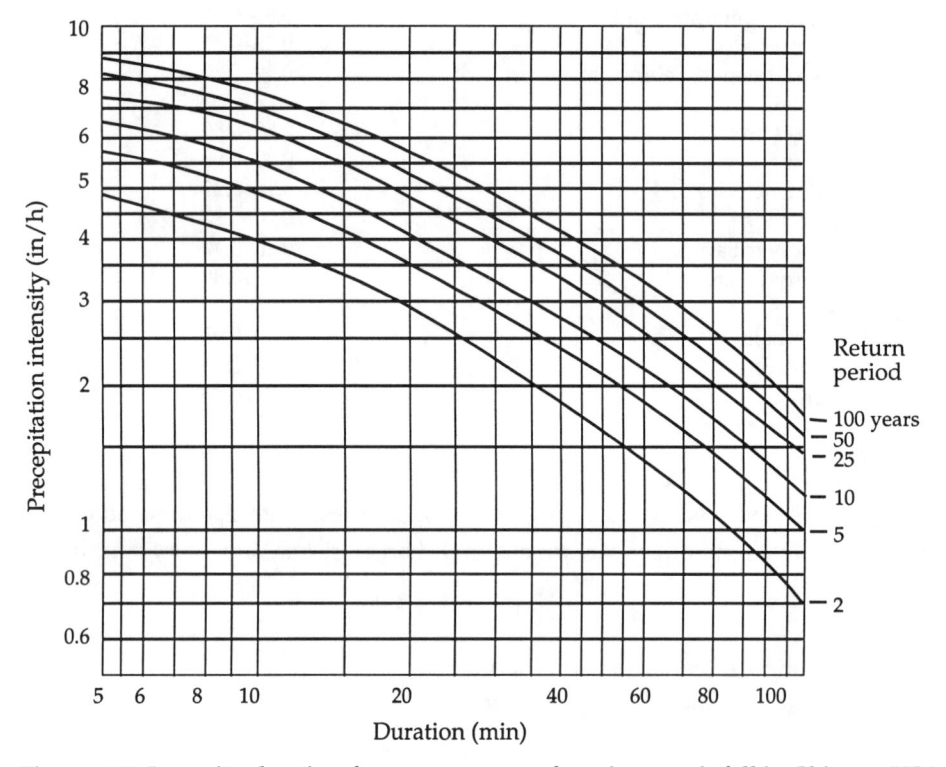

Figure 7.10 Intensity-duration-frequency curves of maximum rainfall in Chicago, USA.

EXAMPLE 7.6

Determine the 10-year, 20-minute design rainfall intensity for Chicago. Compare the results using Eq. 7.8.3 and Fig. 7.10.

Solution. Table 7.5 provides $c = 94.9$, $e = 0.88$, and $f = 9.04$. Equation 7.8.3 then gives

$$i = \frac{94.9}{20^{0.88} + 9.04}$$

$$= 4.1 \text{ in/hr}$$

From Fig. 7.10 we obtain 4.0 in./hr, essentially the same as above.

Once the design discharge entering a storm sewer has been calculated, the diameter D_p of the pipe is determined. Manning's equation is rearranged to provide D_p:

$$D_p = \left(2.16 Q n / S_0^{0.5}\right)^{3/8} \qquad (7.8.4)$$

where
- D_p = pipe diameter (feet)
- Q = design discharge (cfs)
- n = Manning's coefficient for the pipe (dimensionless)
- S_0 = bed slope of the pipe (ft/ft)

The following assumptions are incorporated into Eq. 7.8.4:
- the pipe is flowing full under gravity, and is not pressurized
- friction slope is set equal to the bed slope of the pipe

The rational formula can be used to perform a composite analysis of watersheds made up of several subcatchment areas. The peak runoff is computed using the following form of the rational formula:

$$Q = i \times \sum \left(C_j \times A_j\right) \qquad (7.8.5)$$

where the summation is from 1 to m, m being the number of subcatchment areas draining to the design point; i should be based on the largest time of concentration.

Summary of steps in the Rational Method:

1. Determine the watershed drainage area.
2. Determine T_c, which in turn sets the design rainfall duration. This is the sum of all surface flow travel times, including overland flow, open channel flow, and/or storm sewer flow, from the design point to the hydraulically most distant point in the watershed.
3. Select a design storm frequency, and a rainfall intensity (i) using a rainfall duration equal to T_c.
4. Select a value of C.
5. Calculate Q.
6. Use hydraulic techniques to determine storm sewer or open channel sizes to carry design peak flows.

—————— EXAMPLE 7.7 ——————

Determine the peak runoff rate for a commercial, 100-acre watershed, located near Chicago for a 10-year storm, using the Rational Method. The length of the flow path is 2700 feet, with an average slope of 0.1%.

Solution. The rational formula is $Q = CiA$. A is given. C can be estimated from Table 7.4 as 0.75 by assuming the commercial area is similar to a "Downtown Business" area. T_c must be determined before estimating i. Using Eq. 7.8.2, and adjusting by 0.4 for paved overland flow, we obtain

$$T_c = 0.00013 L^{0.77} S_0^{-0.385} \times 0.4$$

$$= 0.00013 \times (2700)^{0.77} \times (0.001)^{-0.385} \times 0.4$$

$$= 0.33 \text{ hr} = 20 \text{ min}$$

Then using Fig. 7.10, the IDF for Chicago (or use Table 7.5 and Eq. 7.8.3 by setting $T_d = T_c$) we find $i = 4$ in./hr.

Therefore,

$$Q = CiA = 0.75 \times 4 \times 100 = 300 \text{ cfs}$$

7.8.2 Unit Hydrograph Method

The Unit Hydrograph Method provides a way of not only calculating the flood peak, but also the entire runoff hydrograph. It also provides a means of explicitly accounting for the variation within a watershed that influences runoff characteristics. The method has two basic steps. The design rainfall must be converted to a runoff volume, and the runoff volume must be converted to a flood peak or runoff hydrograph.

There are a number of unit hydrograph techniques. Once the unit hydrograph has been developed, it is used to calculate the runoff hydrograph. To illustrate the development of a unit hydrograph, we will focus on the SCS triangular unit hydrograph shown in Fig. 7.11. As with all unit hydrograph methods, there are a number of assumptions:

1. The excess rainfall that becomes the surface runoff (SRO) has a constant intensity over a time period equal to the unit hydrograph duration. This requires storms of relatively short duration.

2. Excess rainfall (SRO) is uniformly distributed over the entire watershed.

3. The time ordinates for a runoff hydrograph do not change for a given duration unit hydrograph event.

4. The ordinates of the unit hydrograph are directly proportional to the amount of SRO. The principles of superposition and proportionality are assumed and allow the extension of the method to evaluate any storm event.

5. The unit hydrograph reflects the unchanging characteristics of the watershed. It is time invariant, so it does not change from one storm to another.

6. The flood probability is equal to the rainfall probability.

The unit hydrograph can be approximated by the triangular shape illustrated in Fig. 7.11; it is the response to a uniform excess rainfall of duration T_d. SCS determined that the *time of recession* for the hydrograph is $T_r = 1.67 \times T_p$. The "recession" is the falling limb of the hydrograph, after the time to peak. The ordinate of the triangular hydrograph is flow per drainage area ($q = Q/A$) expressed in units of in./hr, and q_p is the unit hydrograph peak. The area under the unit hydrograph is the unit volume of SRO. This may be expressed by simply calculating the area under the triangular hydrograph:

$$SRO = \frac{1}{2}(q_p \times T_p) + \frac{1}{2}(q_p \times T_r) \tag{7.8.6}$$

Solving for q_p:

$$q_p = 2 \times SRO/(T_p + T_r)$$
$$= 2 \times SRO/(T_p + 1.67 \times T_p) \quad (7.8.7)$$
$$= 0.75 \times SRO/T_p$$

The total hydrograph peak flow rate is q_p multiplied by the contributing drainage area ($Q_p = q_p \times A_d$). The conversion factor from mi^2 in./hr to cfs is 645. The equation for the peak flow based on the SCS triangular unit hydrograph is then

$$Q_p = (A_d \times 0.75 \times SRO/T_p) \times 645$$
$$= 484 \times A_d \times SRO/T_p \quad (7.8.8)$$

where the units are: SRO (inches)

T_p, T_r (hours)

q_p (in/hour)

Q_p (cfs)

The drainage area A_d is determined from topographic maps. Surface runoff is determined using the curve number procedure. The remaining unknown is the time to peak. This is related to the time of concentration, which is the time it takes water to travel from the hydraulically most distant part of the watershed to the design point. Equation 7.8.8 can be rewritten in terms of T_c since $T_p = T_l + T_d/2$ and $T_l = 0.6 T_c$:

$$Q_p = \frac{484 \times A_d \times SRO}{0.6 T_c + T_d/2} \quad (7.8.9)$$

where T_d is the uniform excess rainfall duration (hours) and T_c is the time of concentration (hours).

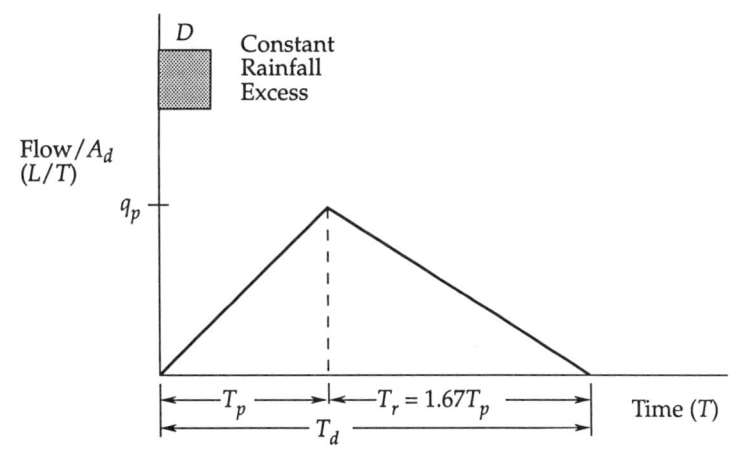

Figure 7.11. SCS triangular unit hydrograph.

EXAMPLE 7.8

Determine the unit hydrograph peak flow (in units of cfs/mi²in) for a watershed with $T_c = 8$ hours for 6, 12 and 24-hour duration storms.

Solution. Using Eq 7.8.9 for a 6-hour storm:

$$Q_p/(A_d \times SRO) = 484/(0.6T_c + T_d/2)$$

$$= 484/(0.6 \times 8 + 6/2)$$

$$= 62 \text{ cfs / mi}^2\text{in}$$

For a 12-hour storm: $Q_p/(A_d \times SRO) = 45$ cfs / mi²in

and for a 24-hour storm: $Q_p/(A_d \times SRO) = 29$ cfs / mi²in.

Note these are based on uniform rainfall excess over the time period T_d. The hydrograph response becomes proportionally less as the duration increases.

Design rainfall distributions are often used. A common one, recommended by SCS for most of the continental U.S., is the Type II rainfall distribution for a 24-hour rainfall. The rainfall intensity changes during the storm. A unit hydrograph peak can be developed that takes into account the rainfall pattern. Fig. 7.12 illustrates the relationship between $Q_p/(A_d \times SRO)$ (cfs / mi²in) and T_c for a 24-hour Type II rainfall.

The time of concentration T_c can be calculated using hydraulic equations such as Manning's equation (also discussed in the Hydraulics Chapter). There are also a number of empirical methods that have been developed to estimate the travel time based on easily obtainable parameters, such as slope and generalized stream channel characteristics. An example is Eq. 7.8.2, which can also be used with this method.

It is more accurate to calculate T_c from an estimate of the velocity through the various components of the stream network where the characteristics are relatively uniform. Another simple method that allows better definition of the stream network expresses velocity as

$$V = K \times S_0^{0.5} \tag{7.8.10}$$

where the coefficient K depends on the type of flow and is given in Table 7.6, S_0 is the slope of the flow path in percent, and V is the velocity in feet per second.

Three flow types are used based on their designation on U.S. Geological Survey topographic maps.

- **Small tributary:** Permanent or intermittent streams which appear as solid or dashed blue lines on the topo maps. This also applies to a swamp that has a defined stream channel.

- **Waterway:** Any overland route which is a well-defined swale by elevation contours but does not have a blue line denoting a defined channel. This also applies to a swamp that does not have a defined channel flowing through it.

- **Sheet flow:** Any overland flow path which does not conform to the waterway definition.

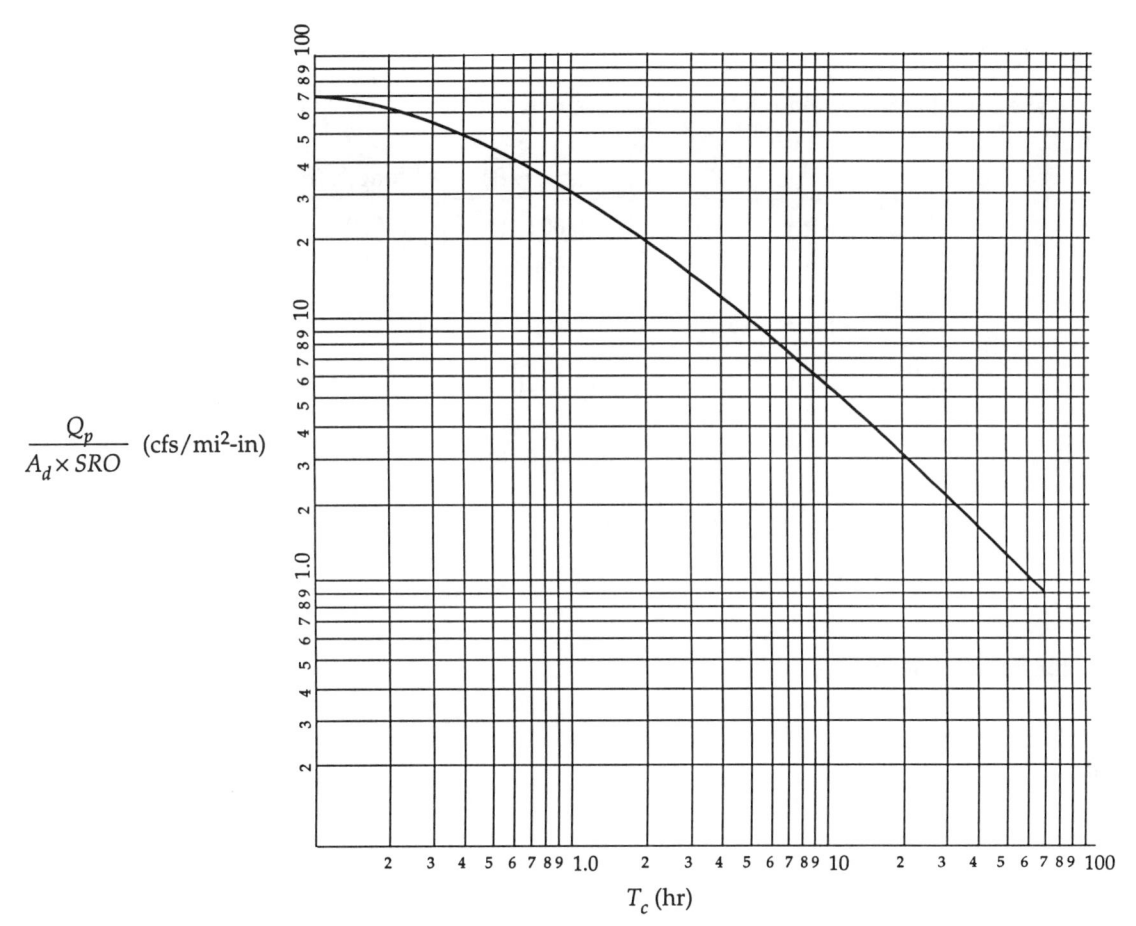

Figure 7.12 $\dfrac{Q_p}{A_d \times SRO}$ versus T_c for a 24-hour Type II rainfall. (Source: Michigan Dept of Natural Resources.)

The coefficients in Table 7.6 are a means of estimating velocities when detailed stream hydraulic data are unavailable.

Once the velocity is determined, time of concentration can be computed as

$$T_c = \frac{L}{3600V} \tag{7.8.11}$$

where L is the length in feet of the particular flow path and the factor 3600 converts T_c from seconds to hours.

TABLE 7.6 Coefficients K (ft/sec) for calculating velocity V.

Flow Type	K
Small tributary	2.1
Waterway	1.2
Sheet flow	0.48

In most watersheds, all three flow types will be present. Starting at the basin divide, the runoff could proceed from sheet flow to waterway, back to sheet flow, then waterway again, and then small tributary. The T_c for each segment should be computed and then summed to give the total T_c. It is important that each length used to compute T_c has a uniform slope.

EXAMPLE 7.9

A watershed has been divided into the following reaches based on channel characteristics and slope. Calculate T_c.

	S_0 (ft/ft)	length (ft)
Small trib	0.001	5000
Waterway	0.0007	1500

Solution. Use Eq. 7.8.10 to calculate velocity. For the small trib portion,

$$V = 2.1 \times (0.1)^{0.5} = 0.66 \text{ fps}$$

where S_0 is in %. Then using Eq. 7.8.11, the concentration time is

$$T_c = \frac{L}{3600V}$$

$$= \frac{5000}{3600 \times 0.66} = 2.1 \text{ hr}$$

Similarly, for the waterway portion,

$$V = 1.2 \times (0.07)^{0.5} = 0.32 \text{ fps}$$

$$T_c = \frac{1500}{3600 \times 0.32} = 1.3 \text{ hr}$$

The total T_c is the sum of the time of the two reaches:

$$T_c = 2.1 + 1.3 = 3.4 \text{ hr}$$

Peak flows determined by unit hydrograph methods assume that the topography is such that surface flow into ditches, drains, and streams is approximately uniform. The presence of the depressions and wetlands in a watershed tend to temporarily store water and slow down the runoff process. Adjustments can be made to the Unit Hydrograph Method to reflect the effects of storage on flood peaks. Table 7.7 provides adjustment factors to determine this reduction based on the ratio of ponding or swampy area to the total drainage area for a range of flood frequencies. The three sections of this table provide different adjustment factors depending on where the ponding occurs in the watershed relative th the design point where we wish to calculate the flood peak. These values were determined by SCS from experimental watersheds of less than 2000 acres. In some cases, it is appropriate to apply the ponding adjustment more than once.

EXAMPLE 7.10

Assume a watershed has two per cent ponding scattered throughout with a lake that is one per cent of the drainage area located at the design point. What factor should multiply the peak flow rate?

Solution. If the 100-year frequency flood is being determined, Table 7.7 suggests that the peak flow should be multiplied by 0.87 for the scattered ponding and further reduced by 0.89 for the lake. This produces a total reduction factor of $0.87 \times 0.89 = 0.77$ which would multiply the Q_p calculated by the Unit Hydrograph Method.

TABLE 7.7 Adjustment factors to peak runoff rates for swamps and ponding.

Percentage of ponding and swampy area	Storm frequency (years)					
	2	5	10	25	50	100
a) Ponding occurs in central parts of the watershed or is spread throughout						
0.2	.94	.95	.96	.97	.98	.99
0.5	.88	.89	.90	.91	.92	.94
1.0	.83	.84	.86	.87	.88	.90
2.0	.78	.79	.81	.83	.85	.87
2.5	.73	.74	.76	.78	.81	.84
3.3	.69	.70	.71	.74	.77	.81
5.0	.65	.66	.68	.72	.75	.78
6.7	.62	.63	.65	.69	.72	.75
10	.58	.59	.61	.65	.68	.71
20	.53	.54	.56	.60	.63	.68
b) Ponding occurs only in upper reaches of watershed						
0.2	.96	.97	.98	.98	.99	.99
0.5	.93	.94	.94	.95	.96	.97
1.0	.90	.91	.92	.93	.94	.95
2.0	.87	.88	.88	.90	.91	.93
2.5	.85	.85	.86	.88	.89	.91
3.3	.82	.83	.84	.86	.88	.89
5.0	.80	.81	.82	.84	.86	.88
6.7	.78	.79	.80	.82	.84	.86
10	.77	.77	.78	.80	.82	.84
20	.74	.75	.76	.78	.80	.82
c) Ponding occurs at the design point						
0.2	.92	.94	.95	.96	.97	.98
0.5	.86	.87	.88	.90	.92	.93
1.0	.80	.81	.83	.85	.87	.89
2.0	.74	.75	.76	.79	.82	.86
2.5	.69	.70	.72	.75	.78	.82
3.3	.64	.65	.67	.71	.75	.78
5.0	.59	.61	.63	.67	.71	.75
6.7	.57	.58	.60	.64	.67	.71
10	.53	.54	.56	.60	.63	.68
20	.48	.49	.51	.55	.59	.64

EXAMPLE 7.11

A bridge is scheduled for replacement in the southern tip of Illinois. The watershed area above the bridge is 4 mi^2; it is all in row crops with B type soils. About 1% of the watershed consists of small depressions that form ponds during wet periods. The time of concentration is 5 hours. Determine the 1%- chance flood peak based on a Type II 24-hour rainfall event.

Solution. The drainage area is given as 4 mi^2. The design rainfall is determined from Fig. 7.3 where we find the 24-hour precipitation for a 100-year (1% chance) event is 7.0 inches for southern Illinois. No adjustment of the rainfall is needed since the drainage area is less then 10 mi^2. The curve number is found from Table 7.3, as $CN = 78$, for B soils with straight row crops and assuming "good" hydrologic condition. The runoff is determined using Fig. 7.5, as 4.5 inches for $CN = 78$ and precipitation = 7.0 inches. The total runoff volume for the watershed is 4.5 in $\times$ 4 mi^2 = 18 mi^{2}in. Since T_c is given as 5 hours, the unit hydrograph peak for a 24-hour Type II storm is determined from Fig. 7.12 as 95 cfs/mi²in. The adjustment to the peak flow for 1% ponding throughout the watershed is 0.9 from Table 7.7. Therefore,

$$Q_p = 95 \ \frac{\text{cfs}}{\text{mi}^2\text{in.}} \times 18 \ \text{mi}^2\text{in.} \times 0.9 = 1540 \ \text{cfs}$$

Summary of steps in the Unit Hydrograph Method:

1. Determine the drainage area of the watershed.
2. Determine the design rainfall, both total amount and rainfall distribution. Adjust if necessary for drainage area.
3. Determine the surface runoff volume. This can be done by determining the runoff Curve Number (CN) through evaluation of soils and land use in the watershed. Apply appropriate Antecedent Moisture Condition (AMC) adjustment, if necessary.
4. Determine T_c.
5. Determine the unit hydrograph.
6. Determine adjustments for storage.
7. Calculate the flood peak or runoff hydrograph.

7.9 Flood Routing

Engineers are frequently asked to determine the effects of natural or artificial storage areas on flood peaks. Often the engineer must design the storage area as part of a stormwater management system. To properly evaluate the effect of these storage features and structures, it is necessary to develop the runoff hydrograph and route the hydrograph through the storage area. A simplified approach, hydrologic or reservoir routing, is based on the continuity equation. The change in reservoir storage is equal to inflow minus outflow:

$$\frac{\Delta S}{\Delta t} = I(t) - O(t) \qquad (7.9.1)$$

where
- S = reservoir storage (L^3)
- I = inflow (L^3/T)
- O = outflow (L^3/T)

The following assumptions are made:

1. The reservoir pool is level, so that storage (S) is a single valued function of depth.
2. The reservoir outflow is a single valued function of depth, i.e., the same elevation always produces the same outflow. Therefore, it is assumed that the gates or other outflow controls do not change during the flood event.

Using a small time step Δt, the continuity equation can be written as

$$\frac{(S_2 - S_1)}{\Delta t} = (I_1 + I_2)/2 - (O_1 + O_2)/2 \qquad (7.9.2)$$

This equation is then rearranged to

$$(S_2/\Delta t + O_2/2) = (I_1 + I_2)/2 + (S_1/\Delta t + O_1/2) - O_1 \qquad (7.9.3)$$

I_1 and I_2 are known from the inflow hydrograph, and S_1 and O_1 are known for the current time step. The only unknowns are S_2 and O_2. Three relationships are needed to solve this equation: discharge vs. elevation; storage vs. elevation; and $(S/\Delta t + O/2)$ vs. O. The first is derived from hydraulic equations for outlets, the second is determined from topographic information, and the third is calculated. A small Δt must be selected. A good initial approximation is $\Delta t = T_p/5$. If $\Delta t > 2S_2/O_2$, negative outflows can occur in the calculation and the time increment must then be reduced.

EXAMPLE 7.12

Determine the peak outflow from a stormwater detention pond for a design storm where the $SRO = 3.5$ inches, and the peak inflow is 190 cfs. The watershed area is 120 acres, and the inflow hydrograph is shown below. The stage-storage-discharge curves are also provided.

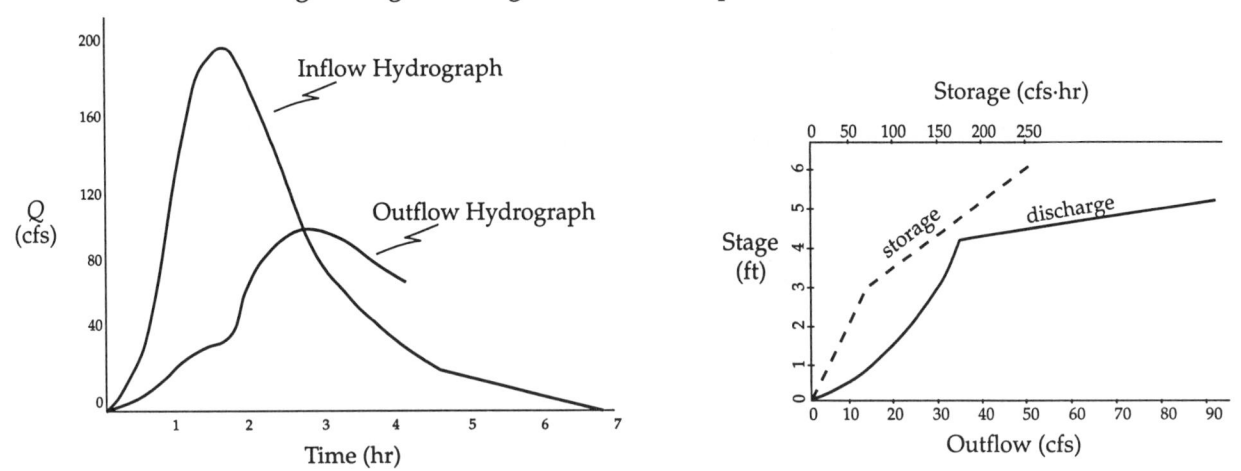

Flood routing example: Inflow and outflow hydrographs.   Flood routing example: Stage-Discharge-Storage curves.

Solution. Use the flood routing procedure, Eq. 7.9.3. Reviewing the rising limb of the hydrograph indicates Δt should be less than 0.3 hr; we will use $\Delta t = 0.2$ hr. Develop the curve O vs. $(S/\Delta t + O/2)$ using the stage-storage-discharge curves. Notice the break in the discharge curve at about 35 cfs, where there is a shift in the outlet hydraulic control. Choose O, look up the corresponding S for the same stage, and calculate $(S/\Delta t - O/2)$. The numbers should be as follows:

O (cfs)	S (cfs-hr)	$(S/\Delta t - O/2)$
5	3	18
10	10	55
15	17	93
20	30	160
30	70	365
35	130	668
40	140	720
60	165	855
90	200	1045

Plot this curve. At $t = 0$, $I = O$, and $(S/\Delta t - O/2)$ is determined from the curve for the initial conditions. I is always known from the inflow hydrograph, so $(I_1 + I_2)/2$ can always be calculated. Using Eq. 7.9.3, $(S/\Delta t - O/2)$ can be calculated from the previous time step. Then O is determined from the plotted curve. A solution table is then constructed. The outlet hydrograph is plotted along with the inflow hydrograph.

The Solution Table

t (hr)	I (cfs)	$(I_1+I_2)/2$	$(S/\Delta t - O/2)$	O (cfs)
0	0	5	0	0
.2	10	16	5	1
.4	22	35	20	5
.6	48	69	50	10
.8	90	111	109	16
1.0	132	150	204	23
1.2	168	176	331	29
1.4	184	187	478	33
1.6	190	186	632	35
1.8	182	174	783	49
2.0	165	155	908	69
2.2	145	135	994	82
2.4	125	115	1047	90
2.6	105	96	1072	93
2.8	87	79	1075	94
3.0	71	67	1060	92
3.2	63	58	1035	88
3.4	53	50	1005	83
3.6	46	43	972	78
3.8	39	36	937	72
4.0	32		901	67

The peak outflow is 94 cfs. Notice how the peak is flattened and delayed by the artificial storage provided by the detention pond.

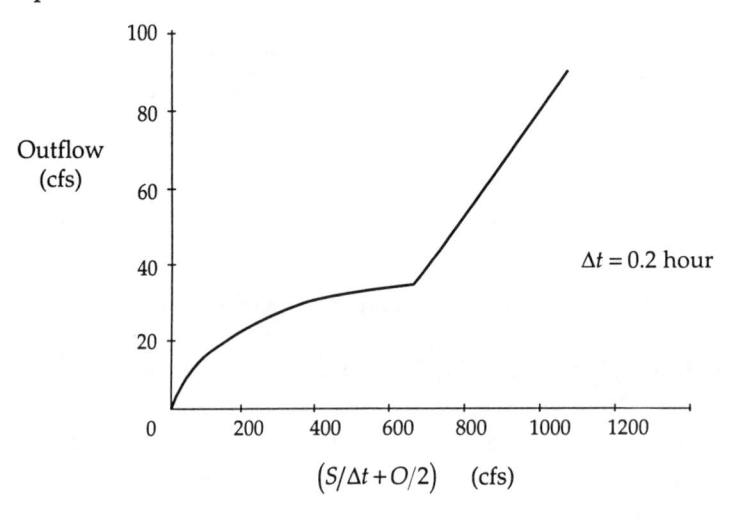

Flood routing example: O vs. $(S/\Delta t + O/2)$.

7.10 Groundwater

Groundwater provides an important water supply. Description of *aquifers*, saturated underground formations, are developed from data obtained from wells. Aquifers are often complex, varying greatly over short distances. Well information is essentially point data that must be interpreted and extrapolated to estimate the aquifer properties.

Information about the aquifer's physical properties is obtained by examining the cuttings as a well is drilled. Wells also provide the piezometric heads (water level in the well) and water quality through well water samples. From this information hydrologists are able to quantify an aquifer's yield and estimate it's response to management alternatives.

The basic empirical law that governs groundwater flow is Darcy's Law,

$$Q = K I A \tag{7.10.1}$$

where
- Q = total flow rate through a cross section of the aquifer (L^3/T)
- K = aquifer hydraulic conductivity (L/T)
- I = gradient (L/L)
- A = cross sectional area normal to the gradient (L^2)

Sometimes it is important to know the velocity of the groundwater. The Darcy velocity V is simply Q/A. However, this represents the flow averaged over the entire cross sectional area. Much of the area is occupied by the aquifer material; the water actually only flows through the void space between particles. To determine the actual velocity of the water, or *seepage velocity*, the Darcy velocity must be divided by the *effective porosity* (the proportion of the porous media volume that will allow the transport of water). The seepage velocity is

$$V_s = \frac{K I}{n_e} \tag{7.10.2}$$

where n_e is the effective porosity (L^3/L^3).

The effective porosity can be approximated by the *specific yield* (S_y) in a water table aquifer. The specific yield is approximately the same as the storativity coefficient (S) determined in a non-equilibrium pump test of a granular water table aquifer.

Transmissivity is frequently used to express the ability of an aquifer to transmit water. It is the hydraulic conductivity multiplied by the saturated aquifer thickness:

$$T = K \times b \tag{7.10.3}$$

where
- T = transmissivity (L^2/T)
- K = hydraulic conductivity (L/T)
- b = aquifer saturated thickness (L)

Note that b in a confined aquifer is the difference between the elevations of the confining units that define the top of the aquifer and the bottom of the aquifer. In a water table or phreatic aquifer, it is the difference between the static water elevation and the aquifer bottom. This is illustrated in Fig. 7.13.

Flow direction is normal to the lines of constant piezometric head. These can be contoured from piezometric head data supplied by a number of observation wells, or it is determined through triangulation from any three wells.

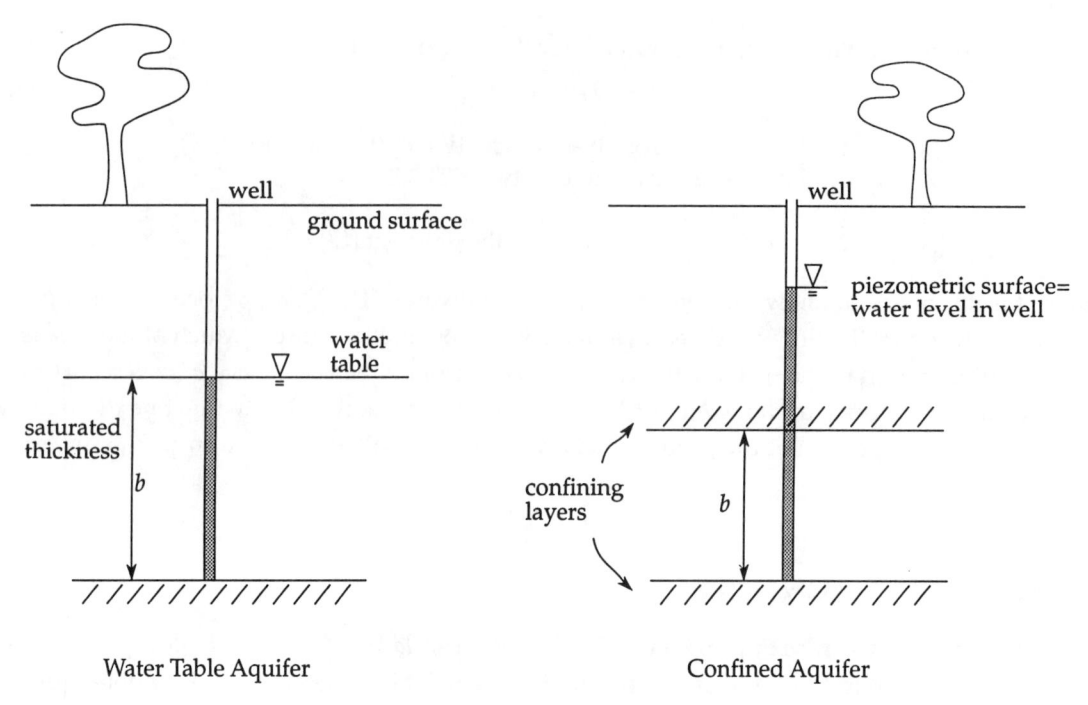

Figure 7.13 Aquifer saturated thickness.

EXAMPLE 7.13

Three monitoring wells were drilled into a water table aquifer with measured static water elevations as shown below. The saturated thickness at the wells averages 50 ft, hydraulic conductivity is estimated to be 100 ft/day, and the effective porosity is estimated to be 0.25. Determine the flow direction and seepage velocity.

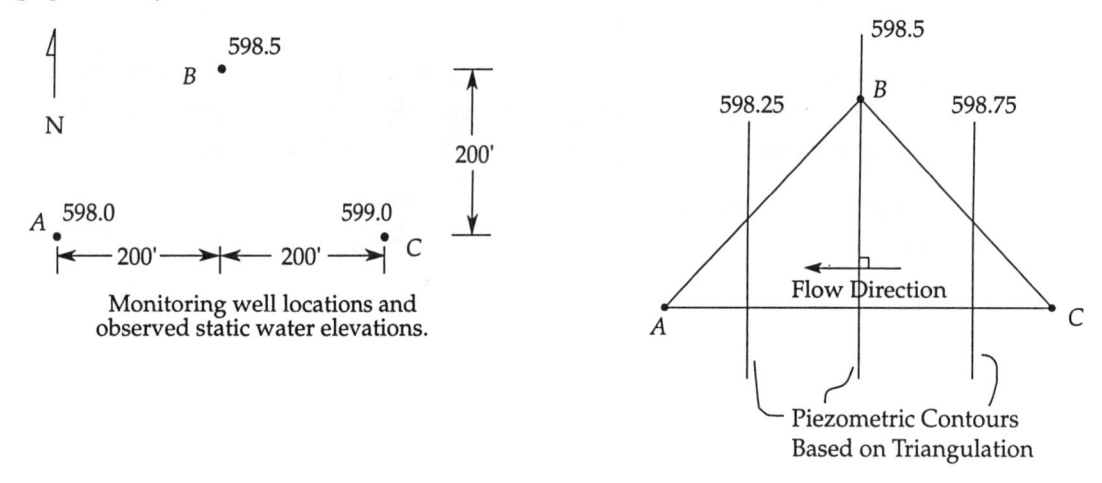

Solution. Contour lines are drawn between points of equal piezometric head. Interpolate between the wells to find the location of the head values 598.25, 598.5 and 598.75, and draw lines connecting those points, as shown above. The 598.5 contour line passes exactly half way between wells A and C. From the sketch, A–C is on an east-west orientation, and flow is toward A, the low point. Therefore, flow direction is west.

From the sketch, the gradient along the flow direction between A and C is $I = 1 \text{ ft}/400 \text{ ft} = 0.0025$. The seepage velocity is then

$$V_s = KI/n_e$$
$$= (100 \text{ ft/day} \times 0.0025)/0.25$$
$$= 1 \text{ ft/day}$$

The hydraulic conductivity can be estimated from the type of aquifer material. For example, a clean course sand will have a much higher hydraulic conductivity than a silty fine sand. But it is more accurate to estimate hydraulic conductivity for the site aquifer by stressing the wells and monitoring the change in head. There are a number of aquifer testing techniques and ways of analyzing the data, such as the Theis method for evaluating nonequilibrium pump tests. These are beyond the scope of this review.

An example of an equilibrium (steady state) relationship for well hydraulics is based on the Dupuit approximation; it applies to a water table aquifer. It is assumed that:

- well drawdown is small compared to the total saturated thickness
- flow is horizontal and uniform over depth
- the well is fully penetrating
- the aquifer bottom is horizontal
- there is no areal recharge
- the aquifer is homogeneous and isotropic

While the assumptions are restrictive, the method is quite useful for estimating aquifer properties and responses.

With the above assumptions, water approaches a pumping well uniformly from all horizontal radial directions. The flow rate from the well can be written in terms of conditions observed at two monitoring wells at distances r_1 and r_2 from the well pumping at rate Q:

$$Q = \pi K \frac{h_2^2 - h_1^2}{\ln(r_2/r_1)} \qquad (7.10.4)$$

where h is the observed water head elevation above the aquifer bottom in the respective observation wells.

EXAMPLE 7.14

Estimate the hydraulic conductivity of an aquifer near a well that is pumping 100 gpm. Observation well 1 is located 50 ft away and has a head of 725 ft. Observation well 2 is located 100 ft away with a head of 727 ft. The heads in the observation wells have been observed to be steady. The aquifer bottom is relatively flat and has an elevation of about 700 ft.

Solution. Rearrange Eq. 7.10.4 and solve for K:

$$K = \frac{Q \ln(r_2/r_1)}{\pi (h_2^2 - h_1^2)}$$

$$= \frac{100 \text{ gpm} \times 192.5 \text{ ft}^3/\text{day/gpm} \times \ln(100/50)}{\pi (27^2 - 25^2)} = 40.8 \text{ ft/day}.$$

Practice Problems (Multiple Choice)

7.1 With all site conditions held constant, such as soils and slope, which land use would produce the least volume of runoff?
   a) cornfield   b) meadow   c) subdivision   d) golf course   e) parking lot

7.2 Which set of circumstances will probably produce the largest flood response for a given watershed from a very intense 3-hour rain storm, followed by below freezing temperatures?
   a) late summer growing season
   b) late winter, bare frozen ground
   c) late winter, deep snow pack
   d) late fall after growing season
   e) all of the above will produce the same flood

7.3 A 5% chance flood can be expected to occur how many times during a forty-year period.
   a) 1   b) 2   c) 3   d) 4   e) 5

7.4 Natural depressions tend to have which of the following impacts on streamflow?
   a) decrease the amount of runoff volume
   b) decrease the flood peak
   c) increase the base flow
   d) increase the time to the flood peak
   e) all of the above

7.5 The recurrence interval of a 2% chance flood is:
   a) 2 years   b) 20 years   c) 50 years   d) 200 years   e) none of these

7.6 A watershed has a fairly complex flow path, with varying slopes and channel characteristics. The most reliable way of estimating the time of concentration (T_c) is:
   a) Use the average slope for the watershed to estimate T_c.
   b) Determine reaches with relatively uniform slope and channel characteristics, calculate the travel time through each reach, and sum for the total watershed T_c.
   c) Use a regression equation that relates peak flow and drainage area, thereby completely bypassing the need to calculate T_c.
   d) Estimate T_c from a hydrograph on a nearby stream.
   e) All of the above yield equivalent estimates.

7.7 Which of the following must be considered when designing a stormwater detention basin?
   a) design storm recurrence interval
   b) peak outflow from the basin
   c) storage volume of the basin
   d) all of the above
   e) none of the above

7.8 A stormwater detention pond tends to:
   a) increase the time to peak downstream of the structure
   b) decrease the time to peak downstream of the structure
   c) not affect the timing of the downstream flood peak
   d) increase the flood volume
   e) decrease the base flow

Hydrology

7.9 The most effective way of containing and managing a groundwater contamination plume is:
   a) to limit recharge
   b) to control the groundwater gradient
   c) to limit the hydraulic conductivity around the plume
   d) excavate all the contaminated material
   e) all of the above are equally effective

Practice Problems (Essay)

7.10 An irrigation district is being formed to provide water to 5000 acres of cropland. The estimated average evapotranspiration (ET) for the crop is 24 inches for the growing season. The average precipitation for the same period is 11 inches. The soil has a water holding capacity of 0.1 inch water/inch soil

   A) Estimate the average amount of irrigation water required for a growing season.
   B) Discuss other considerations in designing the system.

7.11 A community in southern Ohio requires developers to store the entire runoff of a 10-year, 24-hour storm on site.

   A) What detention pond volume will be required for a 100-acre subdivision, on C soils with 1/2 acre lots?
   B) What would the detention pond volume be if the development were located on a nearby A-soil parcel?

7.12 Half of a 3-mi² watershed is in row crops on C soils, and the other half is a subdivision with 1/4-acre lots, equally divided between A and B soils. The watershed is located in the middle of Missouri and has a unit hydrograph peak of 75 cfs/mi²in. for a 24-hour, Type II rainfall. Estimate the peak runoff rate for a 100-year, 24-hour storm.

7.13 Use the Rational Formula to estimate the peak outflow, and size the storm sewer pipes A-B and B-C for the 14-acre watershed illustrated in the schematic below. The watershed is located near Atlanta, Georgia, it is for a commercial development, and local ordinance requires that storm sewers have capacity for a ten-year storm.

Sub watershed	Area (ac)	T_c (min)	C
I	2	5	0.75
II	3	7	0.7
III	4	10	0.6
IV	5	15	0.6

	Ground Elevation (ft)
A	704.7
B	701.0
C	698.3

The distances between points A and B and points B and C are each 500 feet.

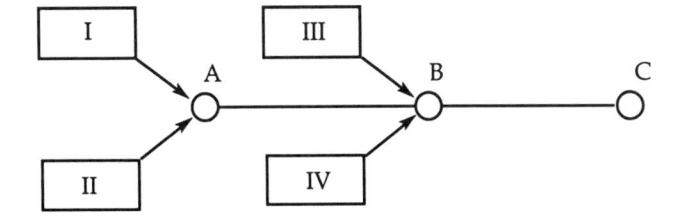

7.14 Calculate T_c for a watershed with reaches based on the following slope and channel characteristics:

	S (ft/ft)	length (ft)
small trib	0.001	5000
small trib	0.0015	2000
waterway	0.0007	1500
small trib	0.0009	3000
sheet flow	0.003	500
sheet flow	0.01	200

7.15 Given a design 6-hour rainfall of 4.5 inches, a composite CN for the watershed of 75, T_c calculated in Problem 14 above, and a drainage area of 3 square miles, determine the SCS 6-hour duration Unit Hydrograph peak and the peak design outflow.

7.16 A subdivision is planned for development in southwestern Michigan. It is adjacent to a stream with a drainage area of 15 square miles and T_c = 12 hours. About 10% of the watershed is covered by wetlands, 60% is in agricultural crops, 20% is forested, and the remaining 10% is in a 1/4-acre-lot suburban development. The soils are very sandy. State regulation requires the subdivision be built outside the 100-year floodplain. Determine the 100-year flood peak using a 24-hour, Type II rainfall.

7.17 Determine the probability of a 1% chance flood occurring during a 100-year period.

7.18 Answer the following questions using the observation well data and sketch below. From observation wells it was determined that the aquifer is an average of 100 feet thick. Pump test data yielded: Transmissivity T = 10,000 ft²/day and Storativity coefficient S = 0.25.

A) What direction is the groundwater flow?

B) How much water will discharge through a 1000 foot wide portion of the aquifer normal to the flow direction?

C) Estimate how long it will take a water particle to travel between Road II and Road I.

Observation well	Ground surface elev. (ft)	Depth to water (ft)
A	925	25
B	932	37
C	908	18
D	927	32

7.19 Determine the runoff hydrograph given the following:

Drainage area = 1 square mile

composite CN = 80

T_c = 1 hour

6 - hour design rainfall = 4 inches

7.20 What other factors, besides the storage volume and outlet structure, should be considered in the design of a stormwater detention pond?

Solutions to Problems

7.1 b) Use Table 7.2 to compare CN; meadow is consistently less than the other possibilities.

7.2 b) Late winter with bare frozen ground will provide the most rapid and intense response to the rainfall. Almost all of the precipitation will immediately run off. If the deep snow pack melts rapidly with a rainstorm, it would significantly increase the runoff volume and then increase the flood peak. As described, it is likely the snowpack would absorb much of the rainfall, and then hold it with the return of sub-freezing temperatures. Late summer has the highest infiltration rate, and more vegetation for interception, so much of the rainfall would not runoff.

7.3 b) There is a 5% chance of this flood flow being equaled or exceeded each year. Over a period of time we would expect it to occur in approximately 5% of the years (i.e., $40 \times 0.05 = 2$).

7.4 e) Base flow tends to be increased because there is greater recharge to groundwater from depressions, so more groundwater tends to be available to sustain base flows. Depressions remove water from surface runoff, which then decreases both volume and flood peak.

7.5 c) Using Eq. 7.7.1: $2 = 100/T$; therefore $T = 50$ years.

7.6 b) Answer a) can introduce significant error by neglecting steep reaches with very short travel times; answer c) assumes implicitly average characteristics that may be significantly different form the watershed in question; answer d) assumes the nearby stream has the same T_c and it may be very different.

7.7 d) a)–c) are all integral components of the design. The design storm is chosen based on risk, governmental regulation, or economics. The peak outflow depends on current or projected downstream uses and channel capacity, and governmental regulation. And storage volume is the biggest factor determining the effectiveness of the structure in reducing flood peaks, it is also the major factor in determining the cost for land acquisition and liability associated with deep ponds. Once any two of these factors are determined, the third is automatically fixed.

7.8 a) The effect of storage is to slow down flood peaks. Introducing detention ponds tends to slow down the flood peak, so it reaches downstream areas later. One caution, slowing the timing of the flood peak may cause it to coincide with the peak on a downstream tributary. This has the potential of actually increasing the flood peak below that confluence.

7.,9 b) Wells can be used to control the groundwater gradient to contain and capture a plume. Once contamination is in an aquifer, limiting recharge will not contain it. And even if it is cost effective to isolate a plume by artificially constructing a low hydraulic conductivity barrier around it, the gradient must still be controlled or contamination could still slowly move through the barrier.

7.10 A) First we must consider that there is an additional source of water. The water balance equation must be modified to include I for irrigation as a positive input, along with precipitation. Assume the irrigation system is properly managed so there is no surface runoff or groundwater recharge from precipitation events. Assume the root zone is three feet deep (a reasonable estimate for a problem like this) and on average the soil moisture will be maintained at half its water holding capacity. Assume the soil profile is saturated to its field capacity at the beginning of the season.

From these assumption, $Q = 0$, $G = 0$, and ΔS is the amount of water drawn from storage out of the soil profile. This is half the water holding capacity for a 3 foot thickness of the root zone as assumed. The groundwater storage is

$$\Delta S = 0.5 \times (0.1 \text{ in/in} \times 3 \text{ ft} \times 12 \text{ in/ft}) = 1.8 \text{ in}$$

Inserting I in the water budget equation (7.1.1) allows us to write

$$I = \Delta S + ET - P$$
$$= 1.8 + 24 - 11 = 14.8 \text{ in}$$

The average volume of water needed is:

$$\frac{14.8 \text{ in}}{12 \text{ in/ft} \times 5000 \text{ ac}} = 6167 \text{ ac-ft}$$

B) Other design considerations are:
- risk analysis: what level of protection against drought is desired
- cost effectiveness
- peak flow demand
- economic and environmental impacts of withdrawing this volume from the source water body
- environmental impacts of irrigation return flow on receiving water bodies
- public acceptance of the project.

7.11 A) From Fig. 7.3, the design rainfall is approximately 4 inches. The CN is determined from Table 7.2 to be 80 for the first parcel, and 54 for the alternative site. The runoff is 2.05 in or 17 ac-ft for the first parcel. The runoff is determined from either Fig. 7.5, or Eqs. 7.4.4 and 7.4.5.

B) The runoff for the second site is 0.5 in, or 4 ac-ft. The volume of the detention pond will be significantly less on the A soils.

7.12 The design rainfall is found from Fig. 7.3 as 7.5 in. The CN are found in Table 7.2, and the runoff is determined from either Fig. 7.5, or Eqs. 7.4.4 and 7.4.5. The runoff can be calculated for each land use and summed for the watershed total.

Land Use	Soil Type	CN	SRO (in)	Area (mi^2)	SRO (mi^{2}in)
row crop	C	85	5.75	1.5	8.63
1/4 sub	A	61	3.1	0.75	2.33
1/4 sub	B	75	4.6	0.75	3.45
			Total:	3.0	14.4

$$Q_p = q_p \times SRO$$
$$= 75 \text{ cfs/mi}^2\text{in} \times 14.4 \text{ mi}^2\text{in} = 1,080 \text{ cfs}$$

7.13 Pipe A–B will carry the combined peak runoff from areas I and II. The longest T_c for the combined area is 7 minutes. Using the IDF equation (Eq. 7.8.3) and the parameters from Table 7.5 for Atlanta, the rainfall intensity is

$$i = \frac{c}{T_d^e + f}$$
$$= \frac{97.5}{7^{0.83} + 6.88} = 8.19 \text{ in/hr}$$

Therefore, the flow rate is $Q = i \times \sum CA = 8.19 \times (0.75 \times 2 + 0.7 \times 3) = 29.5 \text{ cfs}$.

Using the ground elevation as a guide for the slope of the pipe,
$$S_0 = (704.7 - 701.0)/500 = 0.0074 \text{ ft/ft}$$

Assume $n = 0.015$ for the pipe. Then from Eq. 7.8.4,
$$D_p = \left(2.16 Q n / S_0^{0.5}\right)^{3/8}$$
$$= \left(2.16 \times 29.5 \times 0.015 / 0.0074^{0.5}\right)^{3/8} = 2.47 \text{ ft}$$

Hydrology

Round up to the nearest commercial size, 2.5 ft or 30 in.

Calculate the travel time:

$$V = \frac{Q}{A} = \frac{29.5}{\pi 2.5^2/4} = 6.0 \text{ ft/sec}$$

Therefore, travel time $= \frac{L}{V} = 500/6.1 = 83$ sec or 1.4 min.

Pipe B–C will add the runoff from Areas III and IV to pipe A–B. The T_c from Areas I and II plus travel time is $7 + 1.4 = 8.4$ min. Compare this with T_c of 15 min for Area IV. Therefore, 15 min will be used to size Pipe B–C.

All four areas contribute flow. The rainfall intensity is $i = \dfrac{97.5}{15^{0.83} + 6.88} = 5.96$ in/hr.

The flow rate is then $Q = 5.96[3.6 + 0.6 \times 4 + 0.6 \times 5] = 54$ cfs.

The slope is $S_0 = \dfrac{701.0 - 698.3}{500} = 0.0054$ ft/ft.

And the diameter of the B–C pipe is $D_p = \left(\dfrac{2.16 \times 54 \times 0.015}{0.0054^{0.5}}\right)^{3/8} = 3.28$.

Round up to the nearest commercial size, 3.5 ft or 42 in.

7.14 Use Eq. 7.8.10 to calculate velocity, and Eq. 7.8.11 to calculate the travel time for each reach. Sum all the travel times for the total T_c. The calculations are summarized below. Use $T_c = 6$ hr.

	S (ft/ft)	length (ft)	V (ft/sec)	T_c (hr)
small trib	0.001	5000	.66	2.09
small trib	0.0015	2000	.81	0.68
waterway	0.0007	1500	.32	1.31
small trib	0.0009	3000	.63	1.32
sheet flow	0.003	500	.26	0.53
sheet flow	0.01	200	.48	0.12

Total = 6.05 hr

7.15 Use Eq. 7.8.9 to determine the unit hydrograph. We have

$$\frac{Q_p}{A_d \times SRO} = \frac{484}{T_d/2 + 0.6T_c}$$

$$= \frac{484}{6/2 + 0.6 \times 6} = 73 \text{ cfs/mi}^2\text{in}$$

Use Fig. 7.5 or Eqs. 7.4.4 and 7.4.5 to find $SRO = 2.05$ in.

Therefore the peak design flood flow from this watershed is

$$Q_p = 73\frac{\text{cfs}}{\text{mi}^2\text{in}} \times 3 \text{ mi}^2 \times 2.05 \text{ in} = 450 \text{ cfs}.$$

7.16 The drainage area is given as 15 square miles. The point rainfall is 5 inches (from Fig. 7.3), which must be adjusted since the drainage area is more than 10 square miles. The point rainfall adjustment from Table 7.1 is 0.978, therefore the design rainfall is 4.89 in. The soils are described as very sandy; assume they are hydrologic soil group A. Use Table 7.2 to determine CN for each land use, and calculate the composite CN for the watershed:

Land Use	Soil Group	CN	Area (mi²)	Product
crops	A	65	9	585
forest	A	45	3	135
1/4 ac sub	A	61	1.5	91.5
wetland	A	85	1.5	127.5
		Total:	15	939

The composite CN is then 939/15 = 63, and from Fig. 7.5,

$$SRO = 1.5 \text{ in.}$$

T_c is given as 12 hours, the unit hydrograph peak from Fig. 7.12 is

$$q_p = 47 \text{ cfs/mi}^2\text{in}.$$

The adjustment for 10% wetland throughout the watershed is 0.71 (from Table 7.7). Therefore, the design peak runoff is

$$Q_p = 47 \text{ cfs/mi}^2\text{in} \times 15 \text{ mi}^2 \times 1.5 \text{ in} \times 0.71 = 750 \text{ cfs}.$$

7.17 In Eq. 7.7.1 $p = 1\%$ so that

$$1\% = 100 \times 1/T. \quad \therefore T = 100 \text{ yr}.$$

Then using Eq. 7.7.2

$$p(T \text{ event in } n \text{ years}) = 1 - (1 - 1/T)^n \times 100$$

$$= 1 - (1 - 1/100)^{100} \times 100 = 63\%.$$

7.18 First calculate the static water elevations (SWE), ground elevation − depth to water:

Observation well	Ground elev. (ft)	Depth to water (ft)	SWE (ft)
A	925	25	900
B	932	37	895
C	908	18	890
D	927	32	895

A) Plotting the SWE on the sketch and using triangulation, the flow direction is north.

B) Rearranging Eq. 7.10.3 and solving for K:

$$K = T/b$$

$$= \frac{10{,}000 \text{ ft}^2/\text{day}}{100 \text{ ft}} = 100 \text{ ft/day}.$$

The gradient can be calculated between A and C since they are in line with the flow direction:

$$I = \frac{10'}{1000'} = 0.01.$$

The total flow through a 1000 foot wide section can be calculated using Eq. 7.10.1:

$$Q = KIA$$

$$= 100 \text{ ft/day} \times 0.01 \times (1000' \times 100') = 100{,}000 \text{ ft}^3/\text{day}.$$

C) Estimate $n_e = S = 0.25$. Using Eq. 7.10.2 the seepage velocity is

$$V_s = \frac{KI}{n_e} = \frac{100 \text{ ft/day} \times 0.01}{0.25} = 4 \text{ ft/day}.$$

Finally,

$$\text{travel time} = \frac{\text{distance}}{\text{seepage velocity}} = \frac{1000 \text{ ft}}{4 \text{ ft/day}} = 250 \text{ days}.$$

7.19 Use the SCS triangular unit hydrograph to develop the runoff hydrograph. Note the Rational Formula only estimates peak flows, not the runoff hydrograph. Since the design rainfall is a six hour duration, develop a six hour unit hydrograph. With $T_d = 6$ hr and $T_c = 1$ hr, and since $T_l = 0.6 T_c$ and $T_p = T_l + (T_d/2)$ (see Fig. 7.7), therefore

$$T_p = 0.6 \times 1 + (6/2) = 3.6 \text{ hr}.$$

The time of recession is then (see Fig. 7.11)

$$T_r = 1.67 \times T_p = 1.67 \times 3.6 = 6 \text{ hours}.$$

The base of the hydrograph is then $6 + 3.6 = 9.6$ hours. From Eq. 7.8.7

$$q_p = \frac{0.75 \times SRO}{T_p}$$

For a unit hydrograph $SRO = 1$ in so that $q_p = 0.75/3.6 = 0.21$ in/hr.

Use Fig. 7.11 as a guide. The runoff from the 6-hour storm (using $CN = 80$, and Fig. 7.5) is $SRO = 2.05$ in. Multiply q_p by this, and 1 square mile and 645 to convert to cfs will yield the maximum runoff rate:

$$Q_p = A_d \times SRO \times q_p$$
$$= 1 \times 2.05 \times 0.21 \times 645 = 275 \text{ cfs}.$$

The hydrograph follows:

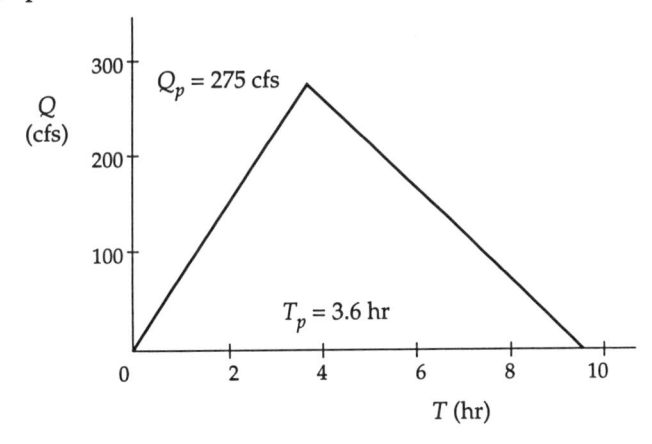

7.20 Other factors to be considered in the design:

1. Evaluate downstream impacts. The change in flood peak timing may cause an increase in flooding downstream.
2. Evaluate the impact of extreme flood events on the structure. For example, if the design is to control a 25-year flood peak, what will happen during a 100-year flood? An emergency spillway may be needed to protect the integrity of the structure and downstream people and property.
3. Have an effective sediment and erosion control program.
4. Develop operation and maintenance guidelines and schedule. Non-maintenance of detention ponds is the major cause of their failure; make sure maintenance needs are clearly understood, documented and can be done.
5. Determine cost effectiveness.

8. Structural Steel

by Richard W. Furlong

Basic concepts in this chapter for the design and analysis of structural steel are described in the context of principles for Load and Resistance Factor Design (LRFD) by the American Institute of Steel Construction, Chicago, 1986, after which similar concepts applicable to the rules for Allowable Stress Design (ASD), Ninth Edition, American Institute of Steel Construction, Chicago, 1989, are presented. The Allowable Stress Design method has been the basic technique for structural steel design in North America for more than 70 years, and it will remain an acceptable technology into the foreseeable future. The Load and Resistance Factor Design system has been in use since 1986, with a second edition expected in 1994. It was developed as a rational system to produce a degree of reliability among structural components more consistent than that which is possible with the ASD procedure. The stress limits of the ASD procedure are more readily understood after the corresponding LRFD principles and applications are understood. Problems can be solved by either methodology, and answers from either system should be accepted in professional engineering examinations. Since the LRFD system is becoming the only system taught by major universities, the methodology for both systems are included in this review summary. For every example problem, a solution with LRFD rules as well as those of ASD is presented.

The AISC manual for the design system being used is a necessary accessory for the solution of problems in steel design. The manuals contain tables with dimensions and properties of all structural shapes available in North America, and each has design aid tables and charts that help with applications of complex rules from each specification.

8.1 Safety Concept

The reliable strength of a structural system must be adequate to support any required load. The service load cases involve Dead Load (D), Live Load (L), Live Load on Roof (L_r), Snow Load (S) or Rain Load (R), Wind Load (W), or Earthquake Load (E). Coefficients related to the predictability and the possibility that combinations of load might occur simultaneously are multiplied by the magnitude of probable service load to produce required limit loads for which strength must be adequate. Six load combination cases are specified in LRFD Section A4.1 for every structure designed in accordance with LRFD rules.

The nominal strength or resistance R_n of structural elements can be predicted with confidence on the basis of knowledge about material behavior, structural form, and the accuracy of fabrication and erection. Some margin of over-strength is used to insure satisfactory minimum performance. The margin of over-strength is largest for conditions of sudden or unanticipated failure, and they are smallest for conditions that involve some warning of structural distress such as a sagging beam. Reliability coefficients are applied to estimates of required strength in order to insure adequate over-strength capacity. The fundamental relationship for expressing structural safety requires that the product of reliability factor ϕ and required resistance R_n exceed the sum of products of load factor LF_i and force P_i:

For LRFD, $\quad\quad\quad\quad\quad\quad \phi R_n \geq \sum LF_i P_i \quad$ (each i represents a load type) $\quad\quad$ (8.1.1)

LRFD Section A4.1 lists 6 required combinations of load factors and load types. In most cases for design, only 1 of the 6 load cases applies to a given structural component.

The design requirements for ASD give results similar to those from LRFD. The logic for ASD differs from LRFD in that all required loads are simply added together to determine a total serviceability loading F_n for any probable combination of load types. Stresses must be estimated on the basis of elastic behavior of structural components that support the serviceability loading. ASD rules specify maximum allowable values of stress f. Allowable stresses are material stress limits divided by a safety factor FS:

For ASD, $\quad\quad\quad\quad\quad\quad$ maximum $f \leq F_n / FS \quad\quad\quad\quad\quad\quad\quad\quad\quad\quad$ (8.1.2)

In perfectly analogous applications, the ASD safety factor FS would be the same as the LRFD ratio LF/ϕ. However, the amount of load factor LF differs for each type of load, and perfectly analogous applications rarely exist.

Material stress limits and member strength limits are based on the characteristic behavior of structural steel as revealed by steel bars pulled in tension. The tension stress P divided by the cross section area A is the tension stress f, and the change in length of specimen ΔL divided by the original (unloaded) length L of specimen is the tension strain E. A typical stress-strain diagram for structural steel is shown in Fig. 8.1.

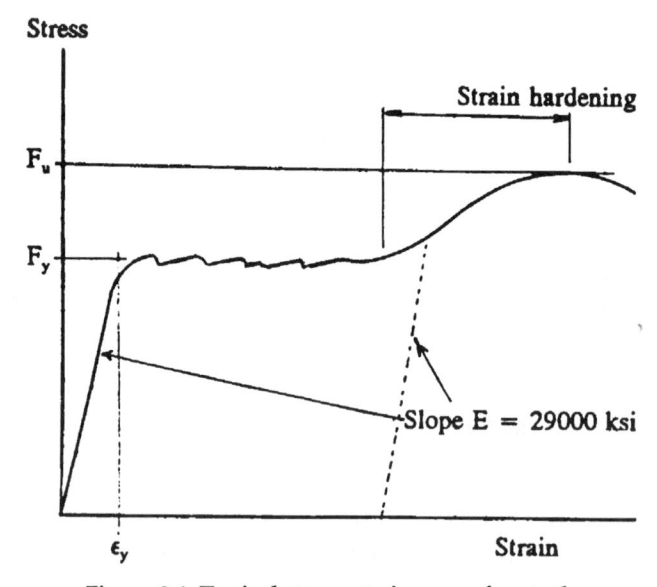

Figure 8.1 Typical stress-strain curve for steel.

The slope of the initial, straight line portion of the diagram is the material stiffness, called the *modulus of elasticity*, E. The straight line ends as stress approaches the yield strength F_y, beyond which the graph is almost a horizontal line at the stress F_y. The strain at which F_y is reached is called the yield strain ε_y, which is $\varepsilon_y = F_y/E$. As the bar stretches after yielding, shear slip lines oriented at 45° from the direction of the tension stress continue to form making the irregular shape shown, with no tensile tearing or fracture of the bar. This material behavior called *ductility* is a very desirable engineering property. Steel can deform extensively before tearing. At a strain 8 to 15 times ε_y, the graph indicates increasing resistance to continued stretching. If load were removed after the yielding had begun, the unloading graph (shown as a dashed line) is parallel to the initial slope E. Subsequent reloading would produce a graph back along the unloading line, possibly to an apparent yield stress higher than the original value. This process can be used to increase the apparent yield strength (at the expense of total ductile stretching before fracture). The process is called *cold working* or *strain hardening* the material. As strain hardening occurs, the weakest region of a bar will begin to stretch more than other regions, and the area of the bar diminishes correspondingly, appearing to neck down at the weakest zone. The highest stress value on the graph is called the material ultimate stress F_u.

8.2 Tension Members

Two strength limits must be satisfied in the design of steel tension members. A *yield strength limit* is simply the product of yield stress F_y and gross cross section area A_g. A *rupture strength limit* is the product of ultimate stress F_u and minimum cross section effective net area A_e. A yield strength limit involves unacceptable overall stretching of the member, and the ultimate strength limit occurs when the member ruptures. The area A_e is the effective cross section remaining to resist force after all interruptions (such as holes) have been deducted from the area, and the effect of shear lag has been considered. When a nominal strength limit involves stretching without rupture (yielding), LRFD rules specify a reliability factor $\phi_{ty} = 0.90$. In contrast, when a strength limit involves rupture, a lower reliability factor $\phi_{tr} = 0.75$ is specified by LRFD. Similarly, with ASD, the nominal factor of safety against yield stretching without rupture can be made smaller than the larger safety factor against rupture.

LRFD rules require that the yield strength limit $\phi_{ty} P_n$ be greater than the required force P_u. The relationship is expressed as

$$P_u \le \phi_{ty} P_n = 0.90 F_y A_g \tag{8.2.1a}$$

The rupture strength limit reduced by the resistance factor ϕ_{tu} likewise must be greater than the required force P_u as expressed as

$$P_u \le \phi_{tr} P_n = 0.75 F_u A_e \tag{8.2.1b}$$

Values of F_y and F_u for all commercial grades of structural steel are given in Table 1, Page 1–8 of the LRFD Manual. Currently, the most common grade of structural steel has a label A36 for which $F_y = 36$ ksi and $F_u = 58$ ksi.

ASD rules employ a factor of safety 1.67 against yielding of the gross area. The allowable tension stress f against yielding is

$$f = \frac{P}{A_g} \le \frac{F_y}{1.67} \tag{8.2.2a}$$

The allowable stress for effective net area fracture employs a factor of safety 2.0 against rupture of the effective net area, and the allowable tension stress against rupture is

$$f = \frac{P}{A_e} \le \frac{F_u}{2.0} \tag{8.2.2b}$$

8.2.1 Effective Net Area

Effective net area is the effective cross section area remaining after areas (such as holes) have been removed as interruptions to the straight "flow" of stress, and an allowance has been made for shear lag, a term used for the local distribution of connection forces to all components of the structural shape. An *area reduction coefficient U* must be taken as low as 0.75 for connections with only 2 bolts per line of bolts in the direction of tension stress. The U factor can be taken as 0.85 when 3 or more bolts are used in each line of bolts at the connection. If all elements of a shape are bolted, $U = 1$.

Net area is the gross area minus the area of material removed for holes. Effective net area is the net area multiplied by the U factor defined above. In determining net width b_n, the width of a bolt hole is taken to be 1/8 inch plus the bolt diameter. If a rupture line is zigzag across a staggered pattern of bolt holes, the quantity $s^2/4g$ should be added to the net width for each diagonal component of the zigzag line. The value s is the hole spacing in the direction of the force, and the value g is the hole spacing perpendicular to the direction of the force.

8.2.2 Tension Member Example Problems

Every tension member must be adequate to resist overall yielding (Eq. 8.2.1a or 8.2.2a) as well as localized tension fracture (Eq. 8.2.1b or 8.2.2b). In the majority of design circumstances, a trial size for a cross section can be based on the yield criterion. The gross area values are given in the steel manuals. The detailed computation of effective net area then can be made for the trial size. If the rupture strength is found to be inadequate, a larger size member must be tried. The following example problems illustrate the design procedure.

——— EXAMPLE 8.1 ———

Select a tension strap (plate) of A36 steel for required tension forces $D = 32$ k and $L = 76$ k if six 7/8 in bolts are used to connect the ends of the strap.

Solution. LRFD Method

Step 1—Load factor combinations pages 6-25 show that Case (A4-2) will produce the largest values for the 2 loads given for this problem.

Compute required tension force:

$$P_u = 1.2D + 1.6L = 1.2(32) + 1.6(76) = 160 \text{ k}.$$

Step 2—Compute required gross area using Eq. 8.2.1a and $F_y = 36$ ksi:

$$A_g \geq \frac{P_u}{\phi_{ty} F_y} = \frac{160}{0.90(36)} = 4.94 \text{ in}^2$$

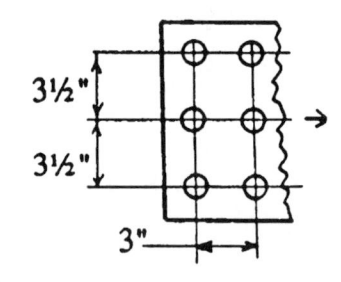

Step 3—Compute required effective net area using Eq. 8.2.1b and $F_u = 58$ ksi:

$$A_e \geq \frac{P_u}{\phi_{tr} F_u} = \frac{160}{0.75(58)} = 3.68 \text{ in}^2$$

Step 4—Sketch and check a trial design with 10-in × 0.5-in strap:

$A_g = b \times t = 10 \times 0.5 = 5.00 \text{ in}^2$, which is greater than 4.94 in². OK.

$A_e = Ub_n t = 1[10 - 3(7/8 + 1/8)]0.50 = 3.50 \text{ in}^2 < 3.68 \text{ in}^2$. Not OK.

Note that $U = 1$ as there is only one component to the shape.

Step 5—Modify the trial design in order to satisfy required $A_e = 3.68$ in². Try strap 10 × 0.625-in:

$A_e = Ub_n t = 1[10 - 3(7/8 + 1/8)]0.625 = 4.38 \text{ in}^2 > 3.68 \text{ in}^2$. OK.

Or, use staggered bolt pattern to make b_n adequate with 0.5 in. strap:

$$b_n = A_e/t = b - 3 \text{ holes} + 2s^2/4g$$

$$4.38/(5/8) = 10 - 3(7/8 + 1/8) + s^2/(4 \times 3.5)$$

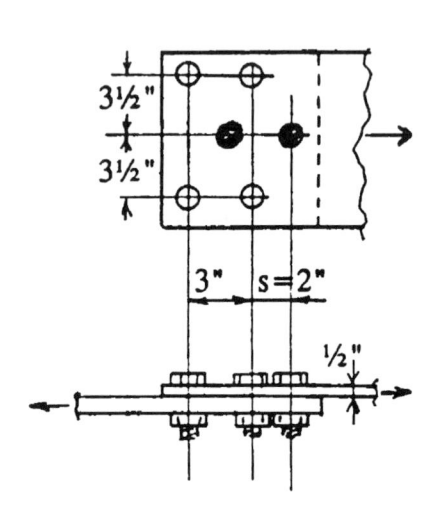

Solve for s:
$$s = \sqrt{0.36(14/2)} = 1.59 \text{ in}$$

Use $s = 2$ in.

ASD Method

Step 1—Compute design service load, $P = 32 + 76 = 108$ k.

Step 2—Compute required gross area using Eq. 8.2.2a:
$$A_g \geq \frac{P}{0.6F_y} = \frac{108}{0.6(36)} = 5.00 \text{ in}^2$$

Step 3—Compute required effective net area using Eq. 8.2.2b:
$$A_e \geq \frac{P}{0.50F_u} = \frac{108}{0.50(58)} = 3.72 \text{ in}^2$$

Step 4—Check actual A_g and A_e provided by $10 \times \frac{1}{2}$-in strap:
$$A_g = b \times t = 10 \times 0.5 = 5.00 \text{ in}^2, \text{ as required}$$
$$A_e = Ub_n t = 1[10 - 3(7/8+1/8)]0.50 = 3.50 \text{ in}^2 < 3.72 \text{ in}^2, \text{ as required. Therefore, revise.}$$

Use staggered line of holes as shown above for LRFD solution.
$$A_e = 10 - 3(7/8+1/8) + \frac{2(2^2)}{4(3.5)} = 7.57 \text{ in.}$$
$$A_e = Ub_n t = 1(7.57)0.50 = 3.88 \text{ in}^2 > 3.72 \text{ in}^2. \text{ OK.}$$

EXAMPLE 8.2

Select a pair of angles (called a double angle member) of A441 Grade 50 steel for a truss member with tension forces $D = 18$ k, $L_r = 44$ k, and $W = 28$ k. Connections are to be made with 3 bolts A325 3/4-in diameter in double shear (see Section 8.5.2) as shown in the sketch.

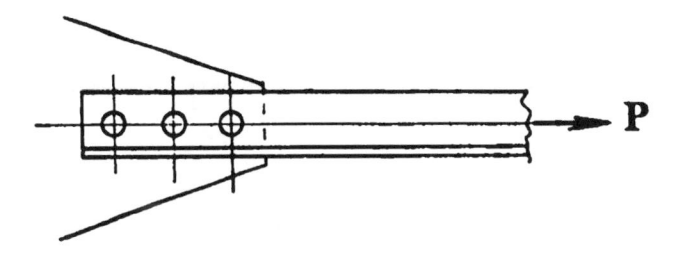

Solution. LRFD Method

Step 1—Refer to LRFD Table 1 to obtain $F_y = 50$ ksi and $F_u = 70$ ksi for A441 steel.

Step 2—Refer to LRFD Section A4, Page 6-25 to determine governing load combination:

$$P_u = 1.2D + 1.6L_r \qquad = 1.2(18) + 1.6(44) \qquad = 92 \text{ k}$$
$$\text{or} \quad P_u = 1.2D + 1.6L_r + 0.8W \quad = 1.2(18) + 1.6(44) + 0.8(28) \quad = 114 \text{ k}$$
$$\text{or} \quad P_u = 1.2D + 1.3W + 0.5L_r \quad = 1.2(18) + 1.3(28) + 0.5(44) \quad = 80 \text{ k}$$

The highest value $P_u = 114$ k governs design.

Step 3—Compute required gross area using Eq. 8.2.1a:

$$A_g \geq \frac{P_u}{\phi_{ty}F_y} = \frac{114}{0.90(50)} = 2.53 \text{ in}^2$$

Step 4—Compute required effective net area using Eq. 8.2.1b:

$$A_e \geq \frac{P_u}{\phi_{tr}F_u} = \frac{114}{0.75(70)} = 2.17 \text{ in}^2$$

Step 5—Select trial member using LRFD Table on pg 1-90. Find a member with gross area greater than 2.53 in². Try 2 angles 3× 2½ ×5/16 for which A_g = 3.24 in².

Step 6—Check effective net area:

$$A_e = U(A_g - A_{holes}); \quad U = 0.85, \text{ since there are 3 or more bolts in line of force.}$$
$$A_e = 0.85[3.24 - 2(3/4 + 1/8)0.3125] = 2.29 \text{ in}^2$$

Since 2.29 in² > required A_e = 2.17 in², use 2 angles 3× 2½ ×5/16.

ASD Method

Step 1—Compute design service load: $P = 18 + 44 \qquad = 62 \text{ k}$
   or $P = 18 + 44 + 28 = 90 \text{ k}$

Since wind acts with other loads, allowable stress can be increased 33%. Increase allowable stress by reducing required P (with wind) to 75% of 90 = 67.5 k.

Step 2—Compute required gross area using Eq. 8.2.2a and the larger service load = 67.5 k:

$$A_g \geq \frac{P}{0.6F_y} = \frac{67.5}{0.60(50)} = 2.25 \text{ in}^2$$

Step 3—Compute required effective net area using Eq. 8.2.2b:

$$A_e \geq \frac{P}{0.50F_u} = \frac{67.5}{0.50(70)} = 1.93 \text{ in}^2$$

Step 4—Select trial member using ASD Table on pg 1-81. Find member with A_g > 2.25 in². Try 2 angles 3× 2½ ×1/4 for which A_g = 2.63 in².

Step 5—Check effective net area: $A_e = U(A_g - A_{holes})$
   $A_e = 0.85[2.63 - 2(3/4 + 1/8)0.25] = 1.86 \text{ in}^2$.

Step 6—Since 1.86 in² < 1.93 in² required A_e, revise trial size to 3× 2½ ×5/16 in order to provide adequate A_e.

EXAMPLE 8.3

Select a WT shape for a tension brace of A572 Grade 50 steel adequate for a wind force of 95 k if connections are to be made with 1/4-in fillet welds to the flange. No other forces cause tension force in the brace.

Solution. LRFD Method

Step 1—Compute the required limit force: $P_u = 1.2D + 1.6L + 1.3W$
$= 0 + 0 + 1.3(95) = 124 \text{ k}$

Step 2—Compute required gross area: $A_g = \frac{P_u}{\phi_{ty}F_y} = \frac{124}{0.9(50)} = 2.76 \text{ in}^2$

Step 3—Use LRFD Page 1-74 to find WT 5 × 11 with A_g = 3.24 in², which is larger than the required A_g = 2.76 in².

Step 4—Compute capacity to resist rupture of effective net area. $A_e = UA_n$. See LRFD B3a, pg 6-30 to find that for this case, U = 0.90. With no holes, $A_n = A_g$.

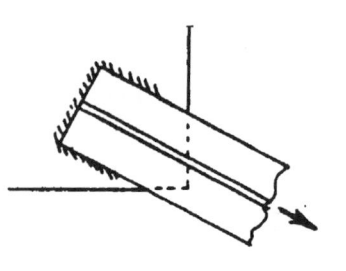

Page 1-8 shows for A572 Grade 50 that $F_u = 65$ ksi:

$$\phi_{tr} P_n = \phi_{tr} F_u U A_n = 0.75(65)0.90(3.24) = 142 \text{ k} > 124 \text{ k}$$

WT 5×11 is acceptable.

ASD Method

Step 1—Required service load is from wind. Therefore, the allowable stresses can be increased by 33%. For design, frequently it is more convenient to consider 3/4 of the force caused by wind and no stress increase instead of using 4/3 times the allowable stress. Use

$$P = 0.75(95) = 71 \text{ k}$$

Step 2—Compute required gross area using Eq. 8.2.2a:

$$A_g = \frac{P}{0.60 F_y} = \frac{71}{0.60(50)} = 2.37 \text{ in}^2$$

Step 3—Use ASD Page 1-66 to find WT 5×8.5 with $A_g = 2.50$ in^2 which is greater than the required $A_g = 2.37$ in^2.

Step 4—Compute allowable strength to resist tension fracture of effective net area. See ASD B3a, pg 5-34 to find that for this case $U = 0.90$. With no holes, $A_n = A_g$, and $A_e = U A_n$. Then

$$\text{Allowable } P = 0.50 F_u U A_n = 0.50(65)0.90(2.50) = 73.1 \text{ k} > \text{required } 71 \text{ k}.$$

Therefore the WT 5×8.5 is OK.

8.3 Compression Members

The stress-strain response of steel in compression is very much the same as the response of steel in tension as described in Fig. 8.1. The major difference in response occurs with yielding and strain hardening. Whereas the tension specimen will experience decreasing cross section area as it lengthens, eventually necking down to a fracture condition, the compression specimen will experience an increasing cross section area as its length shortens, eventually leading to a barrel-shaped specimen that simply 'squashes' outward, not fracturing at all, but "failing" by deforming excessively. If the compression specimen were longer than 2 or 3 diameters, lateral deformations under applied force could prohibit development of the 'squashed barrel' limit mode, and failure would occur as the compression specimen bends away from the line of the compressive forces at each end. The tendency for a compressed bar to 'bend away' from the line of the applied force is called compression buckling.

Compression buckling will occur at stress levels that become smaller as the effective length of the member increases. The effective length of a member is the distance between points of contraflexure (points at which curvature changes direction) as the column buckles. *Slenderness* is a measure of the effective length together with the cross section size. A typical limit stress versus slenderness curve for steel columns is shown in Fig. 8.2 with compression failure stress as the vertical axis and slenderness as the horizontal axis. Slenderness is the ratio kL/r between the effective length kL and the cross section radius of gyration r. If the column is stocky (not slender), the limit stress is the yield strength of the material. If the column is very, very slender, it will buckle "elastically" at a *critical limit stress* F_E (also called the *Euler buckling stress*) related to the material stiffness and the slenderness ratio kL/r:

$$F_E = \frac{\pi^2 E}{(kL/r)^2} \qquad (8.3.1)$$

The graph of Eq. 8.3.1 is shown in Fig. 8.2 as the solid heavy line for high slenderness ratios kl/r, and as a dashed heavy line for smaller slenderness ratios. The observed behavior of real columns rarely involves absolutely straight members, totally homogeneous steel, or perfectly concentric forces. Consequently, there is an intermediate zone of slenderness for which the ordinary solid line LRFD design curves of Fig. 8.2 illustrate actual behavior. The ASD allowable stress graphs are shown as the gray shaded line of Fig. 8.2. These graphs show that the compressive strength of columns is the same as the material yield stress when the slenderness kL/r is less than 20. The strength decreases as slenderness increases.

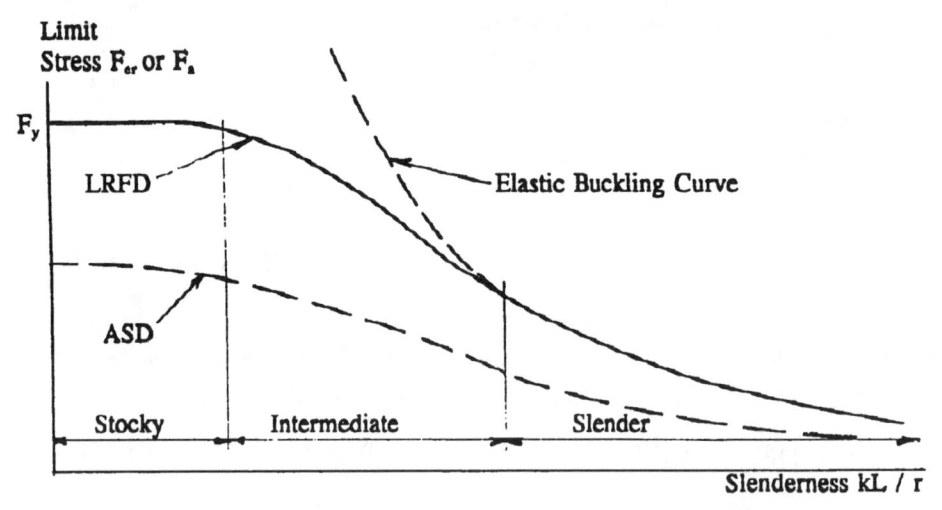

Figure 8.2 Limit compression stress vs. slenderness for compression members.

The actual length L of a column between connections to beams can be multiplied by an effective length factor k to produce the effective length kL. The value of k will be less than 1 and always greater than 1/2 for compression members on which the ends do not deflect laterally during loading. If the ends of a compression member can change lateral position due to loading, such as the top of a flagpole, the effective length factor must be greater than 1. For the case of a compression force at the top of a flagpole the value of k is 2.0. The LRFD Manual Commentary Table C-C2.1 together with Fig. C-C2.1, shown here as Fig. 8.3, gives recommended values for the effective length of compression members with a variety of end framing conditions. It is recommended for the flagpole case (e) in Fig. 8.3 that k be taken as 2.1. The effective length factor k is taken as 1 for most columns in braced structures in which beams are attached to columns with clip angles, simple seats, or shear plates, none of which resist significant amounts of rotation at the connection, represented as Case (d) in Fig. 8.3.

The radius of gyration r of a cross section is the square root of the moment of inertia I divided by the area A, $r = \sqrt{I/A}$. There are two important values of r and I for any non-circular cross section. A maximum value of r is perpendicular to the major axis and a minimum value of r is perpendicular to the minor axis. Lateral supports may provide support in the plane of either the major axis or the minor axis. Any determination of compression strength requires the use of the largest value of kL/r.

Structural Steel

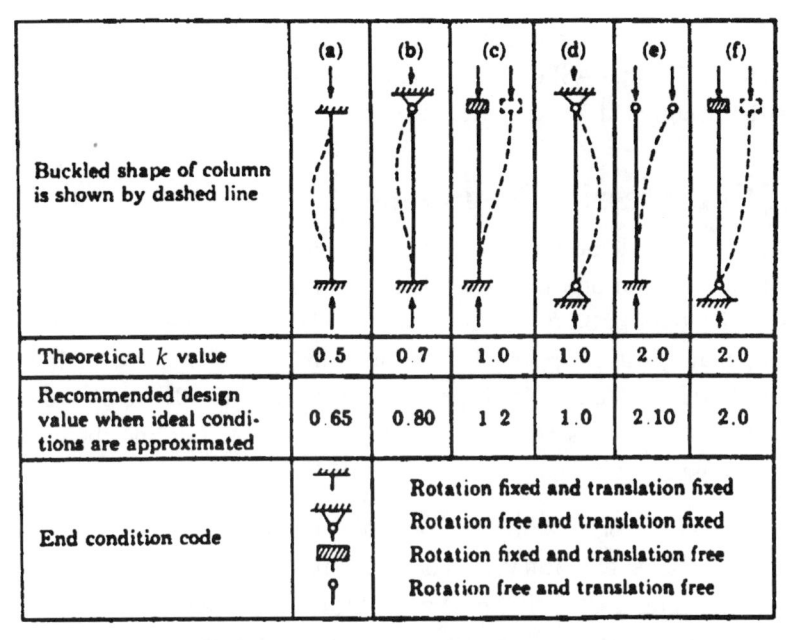

Buckled shape of column is shown by dashed line	(a)	(b)	(c)	(d)	(e)	(f)
Theoretical k value	0.5	0.7	1.0	1.0	2.0	2.0
Recommended design value when ideal conditions are approximated	0.65	0.80	1.2	1.0	2.10	2.0
End condition code		Rotation fixed and translation fixed				
		Rotation free and translation fixed				
		Rotation fixed and translation free				
		Rotation free and translation free				

Figure 8.3 Effective length factors k for compression members.

8.3.1 LRFD Compression Strength Determination

The nominal compression strength P_n of members is expressed as a product of the nominal limit stress F_{cr} and the cross section gross area A_g. Limit compression stress F_{cr} is a function of material yield strength F_y and slenderness parameter $\lambda_c^2 = (kL/r\pi)^2 (F_y/E)$:

$$F_{cr} = \left(0.658^{\lambda_c^2}\right) F_y \qquad \text{for } \lambda_c \leq 1.5 \qquad (8.3.2)$$

$$= \left(0.877/\lambda_c^2\right) F_y \qquad \text{for } \lambda_c > 1.5 \qquad (8.3.3)$$

Note that the parameter λ_c^2 is the ratio between the material yield stress F_y and the Euler buckling stress for the specific slenderness of the column being evaluated.

LRFD rules require that the product of nominal strength P_n and a capacity reduction factor ϕ_c be greater than the required compression force P_u. The value of ϕ_c is 0.85 for compression members. Tables 3-36 and 3-50 on Pages 6-125 and 6-126 of the LRFD Manual show values of $\phi_c F_{cr}$ for kL/r values from 1 to 200. Column load tables are provided on pg 2-16 through 2-45 in the LRFD Manual for standard column shapes. Compression capacity is shown for buckling about the minor axis of the cross section.

8.3.2 ASD Allowable Compression Strength

ASD requirements specify an allowable compression stress F_a which is a function of the slenderness ratio kL/r, the material *yield strength index* C_c, and a factor of safety FS:

$$C_c = \sqrt{2\pi^2 E/F_y}$$

$$FS = \frac{5}{3} + \frac{3(kL/rC_c)}{8} - \frac{(kL/rC_c)^3}{8}$$

The allowable compression stress F_a is

$$F_a = \left[1 - \frac{(kL/r)^2}{C_c^2}\right] F_y/FS \qquad \text{for } kl/r \leq C_c \qquad (8.3.4)$$

$$= \frac{12\pi^2 E}{23(kL/r)^2} \qquad \text{for } kL/r > C_c \qquad (8.3.5)$$

The service load stress P/A_g must be less than F_a. Allowable stress values are provided in Tables C36 and C50, Pages 3-16 and 3-17 of the ASD Manual. Column load tables for standard column shapes are contained on Pages 3-18 through 3-105 of the ASD Manual.

8.3.3 Design of Compression Members

The design process requires demonstration that the member selected for a given force and unsupported length has a radius of gyration r and gross area A_g large enough to support the force within the limits specified by LRFD Eq. 8.3.2 or 8.3.3 or by ASD allowable stress values in Eq. 8.3.4 or 8.3.5. A short, stocky column (low kL/r ratio) can support a load at or near its yield strength, whereas a long, slender column (high kL/r ratio) will buckle elastically at loads that are only 20% to 30% of the yield capacity. Some practice with design and analysis helps develop judgment in estimating the effect slenderness will have on strength or allowable stress.

The most convenient method for selecting compression members involves the use of column load tables in the LRFD Manual or the ASD Manual. Given a required force and an unsupported length kL, the tables can be searched at the length kL until the tabulated force is somewhat larger than that required. If the tabulated force is significantly larger than that required, the column will be larger than a more economical smaller shape, and a smaller, lighter section should be sought. Steel is sold by weight, and efficient design forbids excess overdesign. In the absence of design tables, an iterative design process must be used.

An iterative process is followed. It involves the 1) estimation of limit stress or allowable stress appropriate for the estimated slenderness of a column, followed by 2) calculation of required gross area as a basis for the selection of a structural shape based on area, and then 3) computing the actual slenderness ratio kL/r for the selected shape, and finally 4) comparing the capacity or actual stress with that required. The selection of a shape on the basis of an estimated area is convenient since the unit weight of steel shapes is directly proportional with gross area, (wgt = $3.40\,A_g$). If the selected shape has too much capacity or an allowable stress significantly higher than required, a smaller section should be tried, as the iteration is repeated.

Since the radius of gyration of a cross section squared is the ratio between moment of inertia I and the area A, $r^2 = I/A$, the most efficient cross sections for compression members have their area concentrated around the edges of the section, as for pipes or hollow tubes. Steel W-shapes and I-beams are efficient against buckling about their strong axis, but less efficient for buckling about their weak axis. Lateral bracing frequently is used to reduce effective length for buckling about the weak axis of W-shapes. Thin walled tubing can be quite efficient for compression members. However, the thickness of tube walls or the thickness of flanges for W-shapes must not be too thin.

8.3.4 Local Buckling and Width-to-Thickness Ratios

As any page of a book wrinkles easily when compressed in the plane of the page, it is apparent that wide, thin plate components can buckle locally when compressed in the plane of the plate. Plates that are held in line along two edges (the webs of W-shapes) are less likely to buckle than plates held in line along only one edge (the flanges of W-shapes). Table B5.1 in both the LRFD Manual and in the ASD Manual gives ratios of width to thickness below which local buckling will not occur before the yield strength of the element is attained in compression. For example, the limiting width to thickness ratio b/t for flanges of W-shapes is given as $95/\sqrt{F_y}$ with F_y expressed in ksi units. Thus, when F_y = 36 ksi, cross sections have flanges less than 15.8 times their thickness, there is no hazard of local buckling prior to the development of full compression strength. If F_y were 50 ksi, the limiting width to thickness ratio for flanges would be 13.4. Very few structural shapes are made with flange width-to-thickness ratios that are greater than 10. Consequently, local buckling may be ignored for most shapes of steel sections with $F_y = 50$ ksi.

Similarly, a limit width-to-thickness ratio for webs of W-shapes (sometimes called wide flange shapes) is given as $760/\sqrt{F_y}$. That limit ratio would be 127 for $F_y = 36$ ksi and the limit is 107 for $F_y = 50$ ksi. None of the rolled structural shapes listed have a ratio d/t_w as high as 100. Web buckling in compression members is of concern only for higher grade steels or for special, built-up column cross sections.

The selection of angles for compression members may be the most common condition for which width-to-thickness values of Table B5.1 (limit $b/t = 76/\sqrt{F_y}$) need be considered. That limit of 12.7 for Grade 36 steel would require a thickness of 5/16 in for a 4-in leg of an angle used for a column. An angle $4 \times 4 \times 1/4$ should not be considered. If $F_y = 50$ ksi, were to be used, a $4 \times 4 \times 3/8$ would be the thinnest appropriate size angle for a compression member. If flanges, webs, or legs of structural shapes cannot satisfy width-to-thickness maxima of Table B5.1, the shape should not be used. Both LRFD and ASD regulations (Appendix B, Section B5) contain special rules for the use of segments that might experience local buckling. The rules are based on applications with cold formed metals, and their application is beyond the scope of this summary of fundamental concepts in steel structures.

Subscripts will be used to identify the axis of the column cross section about which dimensions are relevant. For example, kL_x is the distance between supports that restrain deflection perpendicular to the x-axis (the major axis of a W-shape), and r_x is the radius of gyration about the x-axis. The quantity kL_x/r_x or more simply kL/r_x would refer to the slenderness ratio for buckling about the x-axis. In the following example, lateral restraints to displacement perpendicular to the y-axis (the minor axis) are spaced at distances less than the distance between displacement restraints perpendicular to the x-axis. Slenderness ratios must be considered for both possible directions of column buckling.

─── EXAMPLE 8.4 ───

Select a pair of angles (symbol $2\angle^s$) for compression forces $D = 38$ k and $L = 57$ k if the unsupported length $kL = 12' - 8"$. Use A572 Grade 50 steel.

Solution. LRFD Method

Step 1—Use Load Factor combinations LRFD Manual pg 6-25 to see that Case (A4-2) will govern for the 2 loads of this problem. Compute required load: $\quad P_u = 1.2D + 1.6L$
$$= 1.2(38) + 1.6(57) = 137 \text{ k}$$

Step 2—Use LRFD Manual pg 2-60 to 2-65 for angles with long legs back-to-back. From LRFD Manual, pg 2-60, interpolate between values listed for $2\angle^s$, $5 \times 3\frac{1}{2} \times \frac{1}{2}$, to find that the angles can support 148 k for buckling about the Y-Y axis of the pair and 173 k for buckling about the X-X axis. No other pair of angles weighing less than the 20.8 lb/ft of this pair can support 137 k.

ASD Method

Step 1—Compute the design service load: $P = D + L = 38 + 57 = 95$ k.
Step 2—From ASD Manual pg 3-69, interpolate between values listed for $2L^s 5 \times 3\frac{1}{2} \times \frac{1}{2}$ to find that these angles are allowed 108 k for buckling about the Y-Y axis and 125 k for buckling about the X-X axis. No other pair of angles weighing less than these are allowed a load greater than 95 k.

Note: Please take time to examine the load tables on LRFD pg 2-71 to 2-77 and ASD pg 3-76 to 3-84 which give much smaller load values for bending about the X-X axis when pairs of angles have short legs back-to-back. When desired $kL/r_x = kL/r_y$ pairs of angles always are most efficient with long legs back-to-back such that r_x more nearly equals r_y.

EXAMPLE 8.5

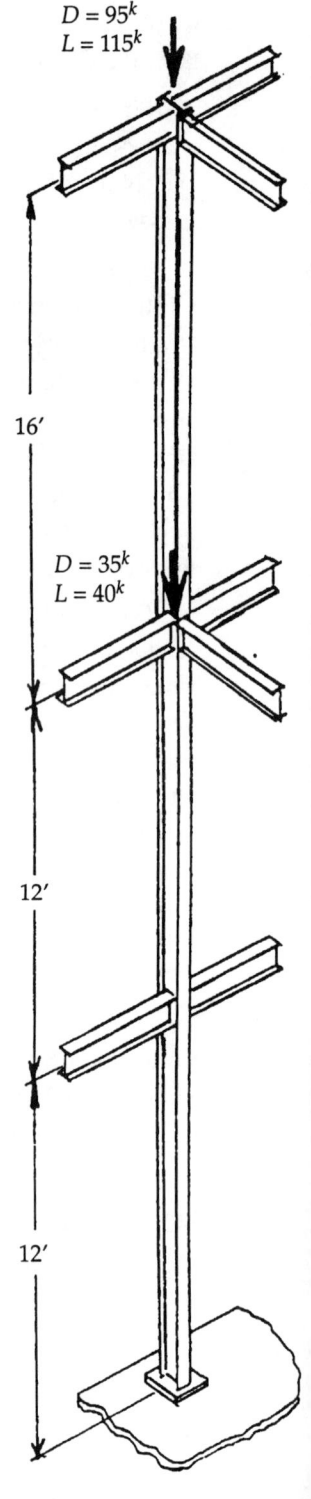

Select a W-shape for the column and loads shown.

Solution. LRFD Method

Step 1—Compute required loads with $U = 1.2D + 1.6L$:

Top 16 ft, $P_u = 1.2(95) + 1.6(115) = 298$ k
$kL_x = kL_y = 16$ ft

Lower 24 ft, $P_u = 1.2(95+35) + 1.6(115+40) = 404$ k
$kL_y = 12$ ft, $kL_x = 24$ ft

Step 2—Assume that buckling about Y-Y axis in the lower 24 ft portion will govern. LRFD Manual pg 2-27, for W10×54 and $kL/r_y = 12$ ft, read $\phi P_n = 409$ k, which is > required $P_u = 404$ k.

Step 3—Check capacity for $kL_x = 24$ ft by using equivalent $kL_y = (kL_x)/(r_x/r_y)$. Ratios r_x/r_y are given at bottom of pg 2-27:

Equivalent $kL_y = 24/(1.71) = 14$ ft for which $\phi P_n = 385$ k.

Since this value is < 404 k, a larger section must be found.

Step 2—LRFD Manual pg 2-25, for W12 × 58 with $kL/r_y = 12$ ft, $\phi P_n = 437$ k, which is > required $P_u = 404$ k

Step 3—Check capacity for $kL_x = 24$ ft:

Equivalent $kL_y = kL_x/(r_x/r_y) = 24/2.10 = 11.4$ ft for which $\phi P_n = 445$ k, which is > required $P_u = 404$ k

Step 4—Check capacity in top portion with $kL_x = kL_y = 16$ ft. Table gives
$\phi P_n = 382$ k which is > $P_u = 298$ k

Selection of W12 × 58 is acceptable for this column.

ASD Method

Step 1—Compute service loads:

Top 16 ft, $P = 115 + 95 = 210$ k
$kL_x = kL_y = 16$ ft

Lower 24 ft, $P = 210 + 40 + 35 = 285$ k
$kL_y = 24$ ft, $kL_x = 12$ ft

Step 2—Assume that buckling about Y-Y axis in the lower 24 ft portion will govern. ASD Manual pg 3-28, for W12×58 with $kL/r_y = 12$ ft.

Allowable $P = 301$ k, which is > 285 k needed

Step 3—Check allowable P for $kL/r_x = 24$ ft. Use equivalent $kL_y = kL_x/(r_x/r_y)$. Ratios r_x/r_y are given in lower part of Table on pg 3-28:

Equivalent $kL_y = 24/(2.10) = 11.4$ ft for which allowable $P = 305$ k, which is > 285 k needed

Step 4—Check capacity in top portion with $kL_x = kL_y = 16$ ft:

Allowable $P = 268$ k, which is > 210 k needed

Selection of W12×58 is acceptable for this column.

EXAMPLE 8.6

Select a shape with A572 Grade 42 steel to support axial loads $D = 185$ k and $L = 225$ k if $kL_x = kL_y = 14$ ft. (Manuals have no charts for $F_y = 42$ ksi, so design charts will not be used.)

Solution. LRFD Method

Step 1—Compute
$$P_u = 1.2D + 1.6L = 1.2(185) + 1.6(225) = 582 \text{ k}.$$

Step 2—For the unsupported length $kL = 14$ ft, assume F_{cr} in the intermediate range of slenderness in Fig. 8.2. Estimate $F_{cr} \approx 0.8F_y = 0.8(42) = 34$ ksi. If $F_{cr} = 34$ ksi,
$$A = P_u/F_{cr} = 582/34 = 17.0 \text{ in}^2$$

Step 3—Look for a shape with an area near 17 in², and values r_y as large as possible. Try a W 10×60 for which $r_y = 2.57$ in and $A = 17.6$ in².

Step 4—Compute $kL/r_y = 14(12)/2.57 = 65.4$, and Section 8.3.1:
$$\lambda_c^2 = \left(\frac{kL}{r_y\pi}\right)^2 (F_y/E) = \left(\frac{14(12)}{2.57\pi}\right)^2 (42/29\,000) = 0.627$$

Step 5—Use Eq. 8.3.2 and compute
$$F_{cr} = (0.658)^{\lambda_c^2}(F_y) = (0.658)^{0.627}(42) = 32.3 \text{ ksi}$$

Then,
$$P_{cr} = AF_{cr} = 17.6(32.3) = 568 \text{ k}$$

This value is less than the required value $P_u = 582$ k. Try a larger shape. Repeat Steps 4 and 5.

Step 6—Try W 12×65 for which $r_y = 3.02$ in and $A = 19.1$ in². Repeat Step 4, $kL/r_y = 14(12)/3.02 = 55.6$:
$$\lambda_c^2 = \left(\frac{kL}{r}\right)^2 (F_y/E) = \left(\frac{14(12)}{3.02\pi}\right)^2 (42/29\,000) = 0.454$$

Step 7—Repeat Step 5,
$$F_{cr} = (0.658)^{0.454}(F_y) = (0.658)^{0.454}(42) = 34.7 \text{ ksi}.$$

Compute
$$\phi P_n = AF_{cr} = 19.1(34.7) = 663 \text{ k}, > P_u = 582 \text{ k}$$

Therefore, W12×65 is selected.

ASD Method

Step 1—Compute Service load
$$P = D + L = 185 + 225 = 410 \text{ k}.$$

Step 2—For the unsupported length $kL = 14$ ft, assume allowable compression stress F_a in the intermediate range $\approx 0.5F_y$:

If $F_a = 0.5(42) = 21$ ksi, required $A = P/F_a = 410/21 = 19.5$ in².

Try W12×65 for which $r_y = 3.02$ in and $A = 19.1$ in².

Step 3—From Section 8.3.2, compute
$$C_c = \sqrt{2\pi^2 E/F_y} = \sqrt{2\pi^2(29\,000)/42} = 116.7$$

$$\frac{kL}{rC_c} = \frac{14(12)}{3.02(116.7)} = 0.477$$

When this value is < 1, Eq 8.3.4 applies:

$$FS = \frac{5}{3} + \frac{3}{8}\left(\frac{kL}{rC_c}\right) - \frac{1}{8}\left(\frac{kL}{rC_c}\right)^3$$

$$= \frac{5}{3} + 0.375(0.477) - 0.125(0.477)^3 = 1.83$$

Eq. 8.3.4 gives

$$F_a = \left[1 - \tfrac{1}{2}(kL/rC_c)^2\right]\left(F_y/FS\right) = \left[1 - \tfrac{1}{2}(0.477)^2\right](42/1.83) = 20.4 \text{ ksi}$$

Step 4—Compute allowable

$$P = A_g F_a = 19.1(20.4) = 390 \text{ k}$$

This is less than the service load value of 410 k. A larger section must be used.

Step 5—Repeat Step 3 and 4 with W12×72 for which r_y = 3.04 in and A = 21.1 in². C_c = 116.7 as before:

$$kL/rC_c = 14(12)/[3.04(116.7)] = 0.474$$

$$FS = 5/3 + 0.375(0.474) - 0.125(0.474)^3 = 1.83$$

$$F_a = \left[1 - \tfrac{1}{2}(0.477)^2\right][42/1.83] = 20.4 \text{ ksi}$$

Compute allowable $P = AF_a = 21.1(20.4) = 430$ k, which is greater than the design service load of 410 k. Therefore, the W12×72 is a satisfactory shape for this application.

8.4 Beams

8.4.1 Flexure of Beams

Beams support forces that are applied perpendicular or transverse to the longitudinal axis of the member. Internally, every cross section along the beam sustains a shear force shown in Fig 8.4 as V and a force couple shown as a compression force C and a tension force T in Fig 8.4. The top portion of the beam is shown to be compressed while the bottom portion is stretched by the transverse forces. There is a "neutral axis" line along which there is neither compression nor tension.

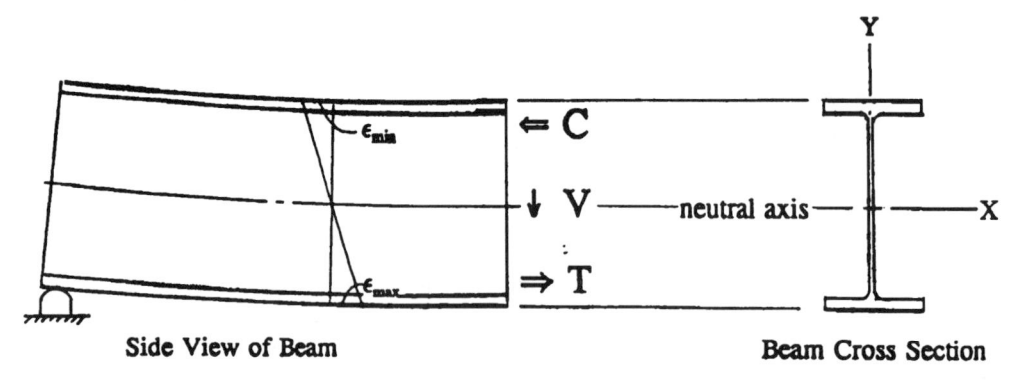

Figure 8.4 Flexure of a beam.

The amount of compression or stretching varies linearly with the distance y from any point to the neutral axis. The exact amount of stretching or compression per unit of beam length is the strain at any position y. Maximum stress f_{max} occurs at the greatest value of y, shown as c in Fig. 8.5. All other stresses are proportional with the ratio y/c. Since stress is proportional to strain while beam material is strained less than its elastic limit ε_y (Fig. 8.1), any increment of force ΔP will be equal to the increment of area ΔA

times the stress f acting at the increment of area located a distance y from the neutral axis. The total moment on the section is the sum of increments of force times their distance y from the neutral axis. When all stresses are less than F_y,

$$M = \sum y \Delta P = \frac{f_{max}}{c} \sum y^2 \Delta A$$

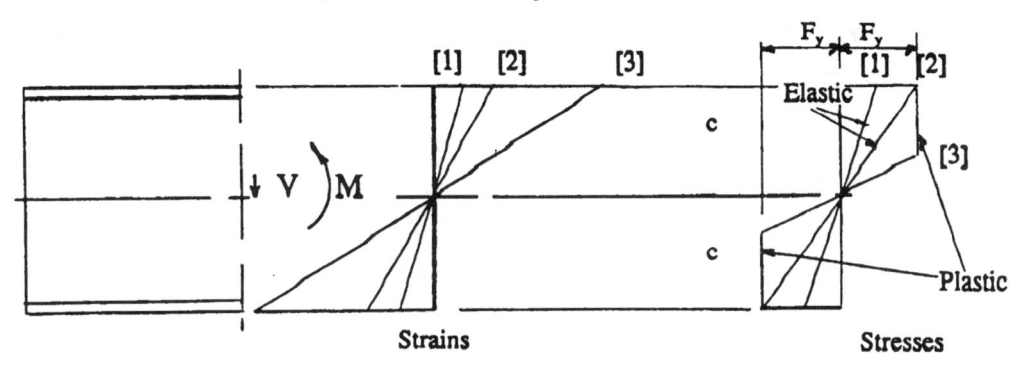

Figure 8.5 Flexural strains and flexural stresses.

The term $\sum y^2 \Delta A$ is called the moment of inertia, a shape factor with the symbol I. Thus for elastic conditions, defined here to mean that all stresses are less than F_y,

$$M = f_{max} I/c \quad \text{or} \quad f_{max} = Mc/I = M/S \tag{8.4.1}$$

The symbol $S = I/c$ is called the section modulus, a geometric shape factor for stress.

Cross sections can be bent through curvatures large enough to make extreme fiber strains larger than the yield strain ε_y of the material. As suggested by line [2] of Fig. 8.5, each fiber for which the strains exceed ε_y will maintain a stress of F_y. Continued increase in curvature eventually will create surface strains large enough for the steel to reach the strain hardening condition. An upper limit to bending resistance of steel beams is taken as the moment for which all fibers of a cross section are yielded, half the fibers in compression and half the fibers in tension. The moment associated with such a stress state is called the *plastic moment M_p* of a section.

Components of cross sections (flanges, webs, T-stems, and angle legs) must be compact or non-slender in order that the components can be compressed to their strain hardening state before they buckle under compression stress. Table B5.1 of the LRFD Manual gives ratios λ_p above which a component element of a section cannot qualify as a part of a compact section. The ratio for flanges of rolled shapes $\frac{1}{2} b_f / t_f = \lambda$ must be less than $65/\sqrt{F_y}$, and the ratio for webs $h_c / t_w = \lambda$ must be less than $640/\sqrt{F_y}$. The LRFD tables and ASD tables of properties of W-shapes provide "Compact Section Criteria" for every rolled shape. For $F_y = 36$ ksi, these λ_r ratios are 9.8 for flanges and 107 for webs. With $F_y = 50$ ksi, the ratios are 9.2 for flanges and 90 for webs. Most rolled shapes satisfy these limits.

Two cross sections with plastic moment stresses are illustrated in Fig. 8.6. The plastic neutral axis is located at the center of area (the area in compression equals the area in tension), not the geometric centroid of the shape. The plastic moment shape factor or plastic section modulus Z is the moment of areas about the plastic neutral axis. Calculations for the value of Z are shown in Fig. 8.6. The plastic moment M_p is the product of F_y and Z. For compact sections,

$$M_p = Z F_y \tag{8.4.2}$$

$$Z = 2(6 \times 5.5 + 5 \times 2.5) = 91 \text{ in}^3 \quad Z = 6 \times 2.5 + 2 \times 1 + 8 \times 4 = 49 \text{ in}^3$$

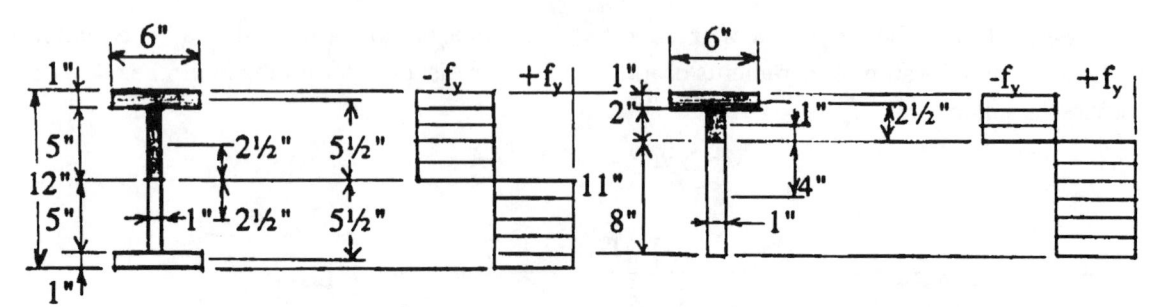

Figure 8.6 Plastic stress conditions on beam sections.

8.4.2 Lateral Stability of Beams

Beams must be restrained from lateral displacement, since one half of a beam cross section is in compression, and an unbraced compression flange can buckle laterally as would any other compression member. Simply supported beams that support floors or roofs have floor slabs, joists or roof purlins that hold the compression flange from lateral displacement. Such beams, usually uniformly loaded, are completely laterally supported. In contrast, beams that are continuous across supporting columns or that extend beyond columns as cantilevers generally have no lateral support along the bottom flange, which is in compression due to negative moment. Frequently the beams that support equipment in industrial structures have regions without lateral restraint on compression flanges. When lateral bracing is not designed to provide restraint for compression flanges of beams, the capacity of laterally unbraced beams must be reduced such that the compression part of the beam will be safe from lateral buckling.

The very large strains that must occur in order to permit a full plastic moment M_p to develop require that lateral bracing be available at the compression flange. The maximum laterally unsupported length of beam L_b that will remain stable until M_p develops is proportional to the weak axis radius of gyration r_y for the beam section. The LRFD rule specifies:

$$L_b = \frac{300 r_y}{\sqrt{F_y}} \tag{8.4.3}$$

If the compression flange of a beam is supported laterally at distances greater than L_b, the LRFD Manual provides equations for computing the resisting moment M_r that can be developed before lateral buckling occurs. The values of M_r are determined as the product of the elastic section modulus S and a compression stress limit F_r.

A typical graph of moment capacity ϕM_n as a function of unbraced length L_u is shown in Fig. 8.7 for a W21×62 steel shape with both F_y = 36 ksi and F_y = 50 ksi. It should be noted that the plastic moment M_p may be used only if the unbraced compression flange is less than 6 ft long when F_y = 50 ksi, but M_p may be used with an unbraced compression flange 8 ft long when F_y = 36 ksi. Moment capacity is reduced as unbraced length increases, and the capacity becomes independent of F_y for unbraced lengths greater than 22 ft.

The ASD Manual similarly limits allowable moment on laterally unsupported beams by specifying allowable stresses that decrease as the unsupported length L_b of compression flange increases. The dotted line graphs of Fig. 8.7 indicate allowable moments on the W21×62 shape when F_y = 36 ksi and when F_y = 50 ksi. When the unbraced length L_u exceeds 20 ft., flexural strength is limited by material stiffness E instead of yield strength F_y, as both curves become the same for L_u greater than 20 ft.

Both the LRFD Manual and the ASD Manual contain complete sets of charts with graphs similar to those of Fig. 8.7 for most of the rolled sections available for use as beams. Representative charts from each Manual are shown in Fig. 8.8. These charts will be used for Flexure example problems. The charts are derived as if the moments were constant across the unsupported length L_b. The charts are accurate for that condition as well as conditions in which the largest moment occurs within the L_b space, but may be smaller at each end of the unbraced space. If the maximum moment occurs at the edge of the unbraced space, moment values can be increased by a

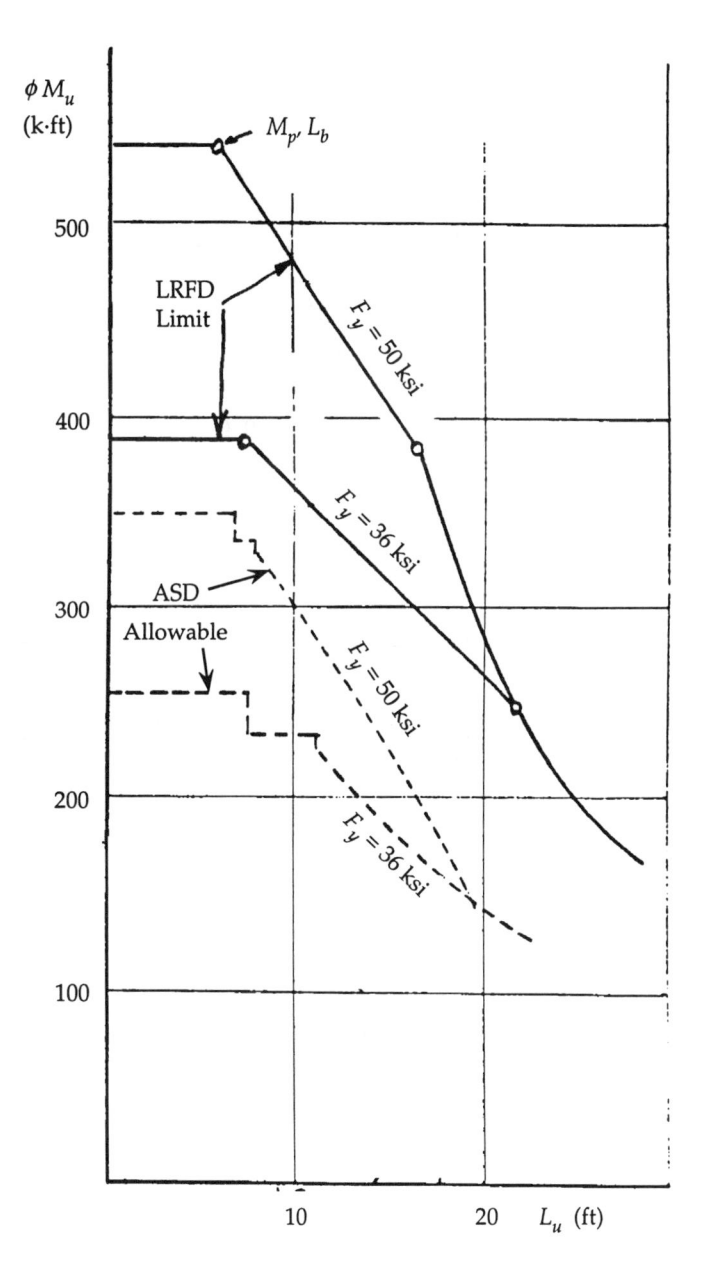

Figure 8.7 Moments vs. L_u for W21×62 beam.

factor C_b, but the absolute upper limit value M_p always represents limit capacity for a section in bending.

$$C_b = 1.75 + 1.05(M_1/M_2) + 0.3(M_1/M_2)^2 \leq 2.3 \qquad 8.4.4$$

in which M_1 = the smaller of the moments at the ends of the space L_b, and M_2 = the larger of the moments at the ends of the space L_b. The algebraic sign for moment is positive if the moment acts clockwise at the end of the space L_b. Perhaps the most common condition for which beams have segments with moments of unequal magnitude between the ends of the segments are beams that support one or more concentrated loads. For example, a simply supported beam with one concentrated load and lateral support at the concentrated load will have a moment of zero at each support and a design moment at the point of the concentrated load. The ratio $M_1/M_2 = 0$ for that case, and $C_b = 1.75$. If the beam were not supported laterally at the point of the concentrated load, C_b must be taken as 1.0, the value which applies for all cases with a moment between supported ends larger than the moments at the ends.

Moment values from the charts shown in Fig. 8.8 are multiplied by C_b to obtain ultimate or allowable moments on laterally braced beams. However, values of $C_b M$ may never exceed values for unbraced lengths less than L_p (Eq 8.4.3 for LRFD). Lengths analogous with L_p for ASD charts are the lengths beyond which the maximum allowable moments must be decreased to stress levels below $0.66F_y$ (or $0.60F_y$ in non-compact shapes).

8.4.3 Beam Shear

Shear forces and internal shear stresses act in the plane of the flexural forces but shear forces are perpendicular to the longitudinal axis. Shear forces V are resisted by the webs of W-shapes, since the beam web is in the vertical plane. Shear stress f_v is greatest at the flexural neutral axis, and it becomes smaller near the top and near the bottom of the web. The effective area of webs is the product of depth d and thickness t_w, $A_w = t_w d$. The average shear stress $f_v = V/A_w$ serves adequately as an indication of strength. The resistance factor for shear $\phi_v = 0.90$.

The limit shear stress for most metals can be taken as $F_y/\sqrt{3}$ or approximately $0.6F_y$. If the web is very high and thin, it is possible that the web will wrinkle or buckle from the action of shear stress. If slender webs can buckle, their shear strength is reduced significantly. Web stiffeners can be used to stabilize webs by restraining buckling, thereby increasing shear capacity. Designating the height h of the web

Limit Moment vs. Unbraced
Length (LRFD)

Allowable Moment vs. Unbraced
Length (ASD)

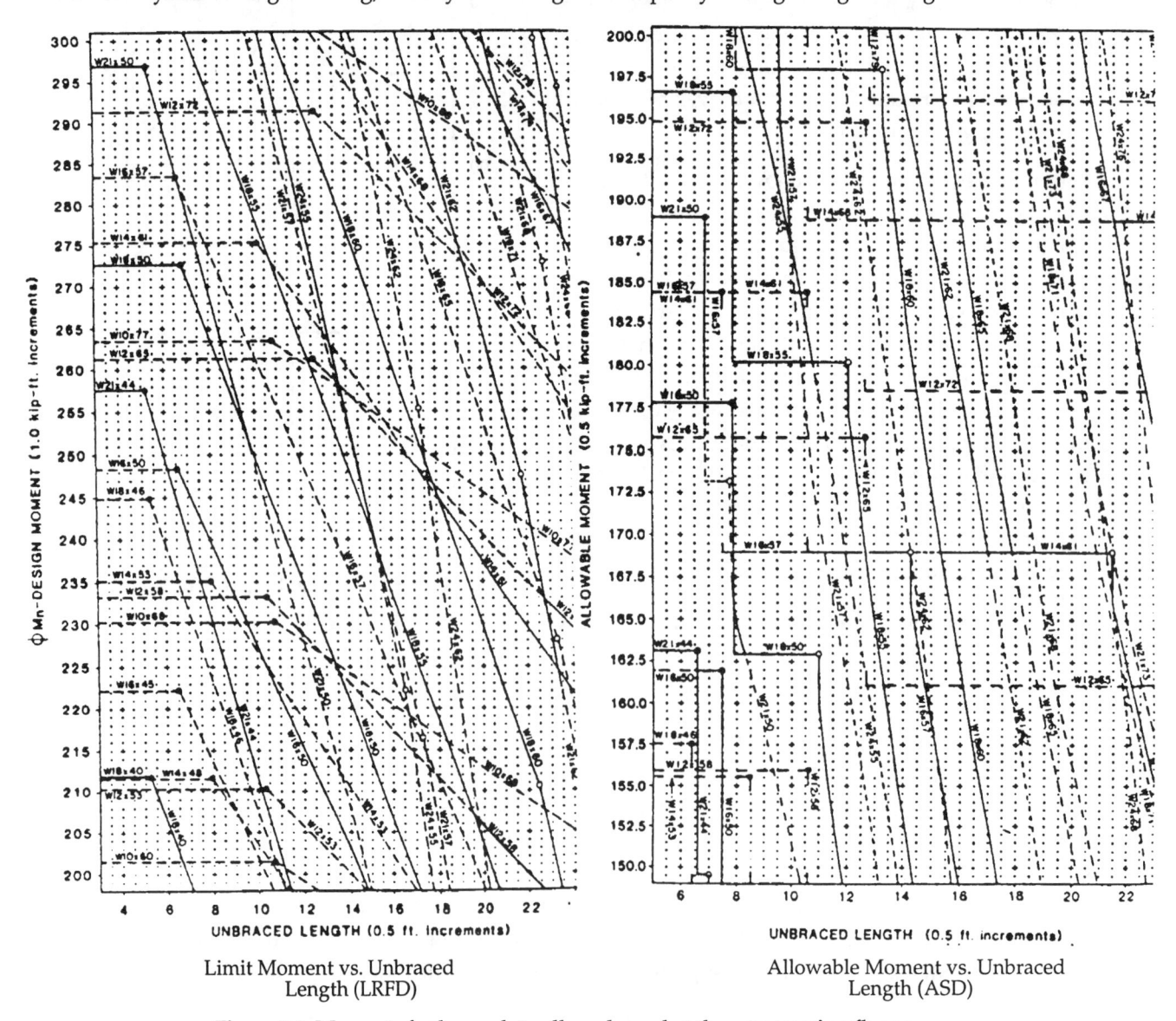

Figure 8.8 Moments for beams laterally unbraced at the compression flange.

between flanges as $h = d - 2t_f$, and the distance between web stiffeners as a, the web plate buckling parameter k is defined as $k = 5 + \dfrac{5}{(a/h)^2}$. By LRFD rules, take F_{yw} as F_y for the web, nominal shear strength V_n is given by

$$V_n = 0.6 F_{yw} A_w \qquad \text{for } h/t_w < 187\sqrt{k/F_{yw}} \qquad (8.4.5a)$$

$$= 0.6 F_{yw} A_w \frac{187\sqrt{kF_{yw}}}{h/t_w} \qquad \text{for } 187\sqrt{k/F_{yw}} \leq h/t_w \leq 234\sqrt{k/F_{yw}} \qquad (8.4.5b)$$

$$= A_w \frac{26\,400}{(h/t_w)^2} \qquad \text{for } h/t_w > 234\sqrt{k/F_{yw}} \qquad (8.4.5c)$$

Furthermore, k shall be taken as 5 if a/h exceeds 3.0 or $[260/(h/t_w)]^2$.

Transverse stiffeners are not required when $h/t_w \leq 418/\sqrt{F_{yw}}$ or when required shear capacity V_u is less than ϕV_n for $k = 5$. When transverse stiffeners are required, they must have a moment of inertia j about an axis through the center of the web which shall be greater than

$$j > \frac{2.5}{(a/h)^2} - 2 \geq 0.5 \qquad (8.4.6)$$

Stiffeners must be connected to web plates with fillet welds spaced not more than 10 in nor 16 times the web thickness apart, or if bolted, the bolts must be spaced no more than 12 in apart.

The ASD rules for allowable shear stress F_v follow much the same logic as that for LRFD, but the terminology differs. Allowable shear stress F_v according to the ASD specification is

$$F_v = 0.40 F_{yw} \qquad \text{for } h/t_w < 380/\sqrt{F_{yw}} \qquad (8.4.7a)$$

$$= F_y (C_v) \leq 0.40 F_{yw} \qquad \text{for } h/t_w > 380/\sqrt{F_{yw}} \qquad (8.4.7b)$$

where

$$C_v = \frac{45\,000 k_v}{F_y (h/t_w)^2} \qquad \text{when } C_v < 0.8 \qquad (8.4.8a)$$

$$= \frac{190}{(h/t_w)^2} \sqrt{\frac{k_v}{F_{yw}}} \qquad \text{when } C_v \geq 0.8 \qquad (8.4.8b)$$

and

$$k_v = 4.00 + \frac{5.34}{(a/h)^2} \qquad \text{when } a/h < 1.0 \qquad (8.4.9a)$$

$$= 5.34 + \frac{4.00}{(a/h)^2} \qquad \text{when } a/h > 1.0 \qquad (8.4.9b)$$

Most of the rolled shapes commonly used for joists and beams have ratios h/t_w less than those which require a reduction of shear strength or allowable stress below the maximum values from Eq. 8.4.5a or 8.4.5b. Also, the flexural demands on uniformly loaded beams generally require that the rolled shape be selected for flexure, and there is substantial extra capacity for the accompanying shear demand. In the presence of large concentrated loads or for heavy uniform loads on beams with a span-to-depth ratio less than 6 to 8, web shear should be reviewed separately.

EXAMPLE 8.7

Select a rolled shape of A36 steel for a simply supported beam with a span of 26 ft. The design dead load is 890 plf (pounds per linear ft) and the design live load is 1400 plf. Assume the top flange is laterally braced throughout the length of the beam.

Solution. LRFD Method

Step 1—Refer LRFD pg 6-25 for governing load combination (A4-2). It is $U = 1.2D + 1.6L$

Compute required load with beam self-weight = 40 plf:

$$w_u = 1.2w_D + 1.6w_L$$
$$= 1.2(890 + 40) + 1.6(1400) \quad = 3360 \text{ plf}$$

Step 2—Compute required moment:
$$M_u = \frac{w_u L^2}{8} = \frac{3360(26)^2}{8} = 284,000 \text{ ft-lb}$$

Step 3—Use Page 3-15 to find a beam for which $\phi M_p \geq 284 \text{ ft} \cdot \text{k}$.

Note that a value $\phi M_p = 284$ for a W16×67 and values ϕM_p are greater than 284 ft·k for all other shapes listed above the W16×57. The **lightest** shape with a value ϕM_p greater than other shapes in each category has its ϕM_p value printed in **bold** type. Select the W21×50 with 297 ft·k shown in bold type.

(Alternate Step 3: use LRFD Chart, pg 3-70. Along the vertical axis at the left side of the page, seek the first **solid** line above the required capacity of 283 ft·k. The W21×50 is the shape indicated.)

Note: In this example, the actual self-weight of the beam is 50 plf, which is greater than the 40 plf that was assumed at the time the required loads were computed. The actual weight will be 10 plf greater than the assumed weight. The real value of w_u will change by an amount $1.2(10) = 12$ plf, less than 0.4% of 3360 plf, the value of w_u used for design. The correction in design load is too trivial to require refiguring.

Step 4—Check shear capacity:
$$V_u = 0.5w_u L = (0.50)3360(26) = 43,700 \text{ lb}$$

Properties of this beam shape are found on pg 1-25 and 1-26 of the LRFD Manual. The ratio $h/t_w = 49.4$ is listed on pg 1-25.

Start with Eq. 8.4.5a, compute $187\sqrt{k/F_y} = 187\sqrt{5/36} = 70$, which is greater than 49.4, so Eq. 8.4.5a applies:

$$\phi V_n = \phi(0.6F_y t_w d) = 0.90(0.6)36,000(20.83)0.380 = 153,900 \text{ lb} \gg 43,700 \text{ lb}$$

Perhaps the Beams Chart on LRFD Page 3-31 is the easier reference for shear capacity. That chart shows for the W21×50 a value $\phi V_n = 154$ k. This beam shape is much more than adequate in shear, as is common when $L/d > 8$.

ASD Method

Step 1—Compute service load moment, again assuming 40 plf self-weight for the beam:

$$M = \frac{wL^2}{8} = \frac{(890 + 40 + 1400)26^2}{8} = 197,000 \text{ ft-lb}$$

Step 2—Use Page 2-11 to find (in bold type), or use pg 2-166 to find that the W24×55 has a value M_R of 226 ft·k, which is more than enough for the design $M = 197$ k·ft.

Step 3—Compute maximum shear = $0.50wL = 0.5(2330)26 = 30,300$ lb. Beam properties are on pg 1-22.

$$\text{Web shear stress} = V/(t_w d) = 30{,}300/(17.89 \times 0.355) = 4800 \text{ psi}$$

Use pg 1-18 and 1-19 for section properties of W24×55. Note $d/t_w = 59.7$

Start with Eq. 8.4.7a:
$$380/\sqrt{F_y} = 380/\sqrt{36} = 63 > 59.7 \, .$$

Thus, allowable shear $F_v = 0.4F_y = 0.4(36{,}000) = 14{,}400$ psi. Allowable shear stress is much higher than actual shear stress, as is common for all beams with $L/d > 8$.

Analogous with LRFD, the ASD Manual pg 2-61 is the more direct reference for beam load values. That table gives an allowable load value of $V = 92$ k, more than 3 times the service load amount.

EXAMPLE 8.8

Select a shape of A36 steel for the beam shown if lateral support to the top flange is provided only at the ends of the beam and at the points of the concentrated loads.

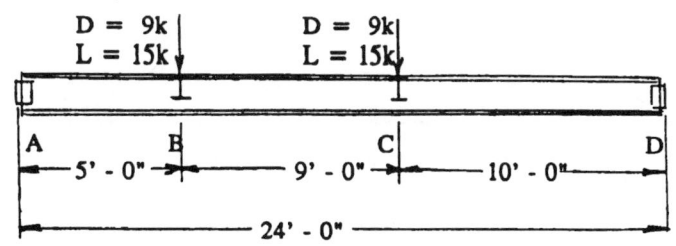

Solution. The compression flange is laterally unsupported for the distances of 5 ft, 9 ft, and 10 ft. Since the values of C_b will be smaller through the 9 ft region, select a shape for that region, and then check beam capacity for the 10 ft region and a value C_b greater than that for the 9 ft region.

LRFD Method

Step 1—Determine required moments. LRFD Load Case (A4-2), pg 6-25 applies:
$$P_u = 1.2D + 1.6L = 1.2(9) + 1.6(15) = 34.8 \text{ k}$$

Right reaction: $R_D = [34.8\,(5) + 34.8\,(14)]/24 = 27.55$ k

Left reaction: $R_A = [34.8(19) + 34.8(10)]/24 = 42.05$ k

Required M_u diagram:

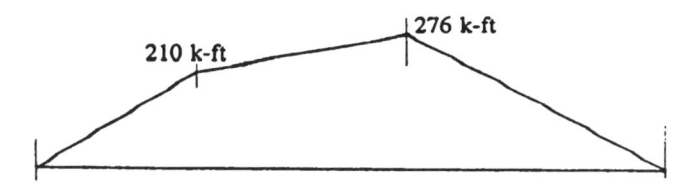

Step 2—For $L_b = 9$ ft, compute C_b with $M_1 = 210$ k·ft and $M_2 = 276$ k·ft. Clause F1.3, pg 6-43 defines the ratio M_1/M_2 as positive when the moments cause reverse curvature. Thus $M_1/M_2 = -(210/276) = -0.76$ for this case:
$$C_b = 1.75 + 1.05(M_1/M_2) + 0.3(M_1/M_2)^2 = 1.75 + (0.76) + (0.76)^2 = 1.13$$

Step 3—For $L_b = 9$ ft, and pg 3-70 chart values for $C_b = 1$, search for a shape with capacity greater than $276/C_b = 276/1.13 = 244$ k·ft.

Both the W18×50 and the W21×50 satisfy the search criteria, but only the W21×50 has a moment capacity adequate for the required $M_u = 276$ ft·k at the point of load. Select W21×50.

Step 4—For $L_b = 10$ ft in the right portion of the beam, the value of $M_1 = 0$ and M_2 remains 276 k·ft. Compute $C_b = 1.75 + 1.05(0) - 0.3(0) = 1.75$.

Since the moment capacity for the W21×50 with $L_u = 10$ ft is 250 k·ft while $C_b = 1$, it would be $1.75 \times 250 = 437$ k·ft if its plastic moment limit strength did not restrict the beam to a limit moment of 297 k·ft.

ASD Method

Step 1—Compute required moments. Compute value of force

$$P = 9 + 15 = 24 \text{ k}$$

Right reaction: $R_D = [24(5) + 24(14)]/24 = 19$ k
Left reaction: $R_A = [24(19) + 24(10)]/24 = 29$ k

Service Load Moment diagram:

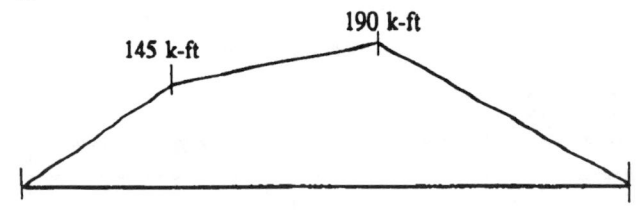

Step 2—For $L_b = 9$ ft, compute C_b with $M_1 = 145$ k·ft and $M_2 = 190$ k·ft.

Clause F1.3, pg 5-47 defines the ratio M_1/M_2 positive when the moments create reverse curvature. Thus $M_1/M_2 = -145/190 = -0.76$, and

$$C_b = 1.75 + 1.05(M_1/M_2) + 0.3(M_1/M_2)^2 = 1.75 + (0.76) + (0.76)^2 = 1.13$$

Step 3—For $L_u = 9$ ft and pg 2-168 chart values for $C_b = 1$, search for a shape with allowable moment greater than $M_2/C_b = 190/1.13 = 168$ k·ft.

Both the W18×55 and W24×55 satisfy the search criteria, but the W18×55 is limited to a moment of 180 k·ft ($0.60F_y$ stress) when its unbraced length exceeds 7.5 ft. Select W24×55.

Step 4—For $L_b = 10$ ft in the right portion of the beam, the value of $M_1 = 0$ and $M_2 = 190$ k·ft. Thus

$$C_b = 1.75 + 1.05(0) - 0.3(0) = 1.75$$

Since the allowable moment for the W24×55 = 187 k·ft when $C_b = 1$ and L_b is 10 ft, the allowable moment would be

$$C_b M_2 = 1.75 \times 187 = 327 \text{ k·ft}$$

if it were not restricted to 209 k·ft by the maximum allowable stress of $0.6F_y = 22$ ksi. The W24×55 is acceptable.

EXAMPLE 8.9

Select a shape of A441 Grade 50 steel for the beam illustrated in the figure. The beam is one of 2 beams that support a large tank. Note that the exterior ends of the beams are braced laterally to one another to stabilize torsional bending in the cantilever regions outside the supports of the beams.

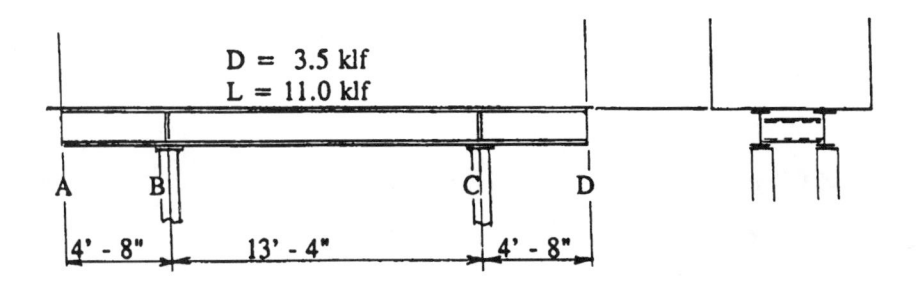

Solution. Step 1—Refer LRFD pg 6-25 for governing Load Case (A4-2). $U = 1.2D + 1.6L$. Compute required ultimate load w_u. Construct required shear and moment diagrams.

$$w_u = 1.2w_D + 1.6w_L = 1.2(3.5) + 1.6(11.0) = 22.0 \text{ klf}$$

Total load on beam $= w_u(4.67 + 13.33 + 4.67) = 22(22.67) = 498.6$ k
Beam reactions $= 1/2$ total load $= 1/2 (498.6) = 249.3$ k
Shear left of B $= -4.67 (w_u) = -4.67 (22.0) = -102.7$ k
Shear right of B $= -102.7 + 249.3 = 146.7$ k
Moment at B $= -102.7 (4.67)/2 = -240$ k·ft
Moment at Center $= -240 + 146.7 (6.67)/2 = 249$ k·ft

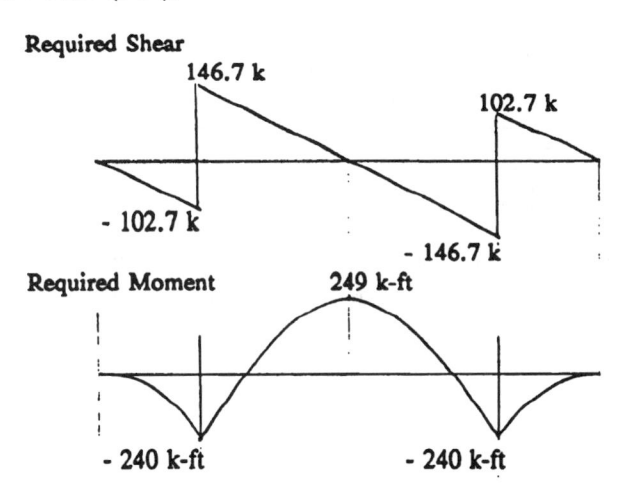

Step 2—Use the LRFD Manual Load Factor Design Selection Table, pg 3-16 to find a shape for which $\phi_b M_p \geq 249$ k·ft when steel has a yield strength of 50 ksi. Use the Chart on pg 3-70 to observe that the laterally unbraced bottom flange at the cantilever could be longer than 4.3 ft without any reduction in moment capacity even with $C_b = 1$. $C_b = 1.75$ since $M_1 = 0$ at the braced end of the cantilevered portion. Try W18×35.

Step 3—Check available shear strength. See W-Shapes Table pg 1-26, 1-27 for section properties. The value for $h/t_w = 53.5$. Start with Eq. 8.4.5a:

$$187\sqrt{k/F_{yw}} = 187\sqrt{5/50} = 59.1 > 53.5 = h/t_w$$

Therefore,

$$\phi_v V_n = \phi_v (0.6) F_{yw} t_w d = 0.90(0.6)50(17.70)0.300 = 143 \text{ k}$$

(The same value could have been read directly from the Beams Chart pg 3-41.)

The available shear strength of 143 k is less than the required value of 146.7 k. A shape with more shear strength must be used. The Table on pg 3-16 indicates that either the W16×40 or W18×40 might be considered. However, the Table 3-42 shows for the W16×40 a shear strength of only 132 k.

The shear strength for the W18×40 is given as 152 k on pg 3-41, and the laterally unbraced length L_b = 4.5 ft within which the plastic hinging moment is permitted. Select the W18×40.

ASD Method

Step 1—Construct service load shear and moment diagrams for uniform w = 3.5 + 11 = 14.5 klf.

Total load on beam = $W = w(4.67 + 13.33 + 4.67) = 14.5(22.67) = 328.7$ k
Each reaction = $1/2\ W = (1/2)328.7 = 164.4$ k
Shear left of B = $4.67w = -4.67(14.5) = -67.7$ k
Shear right of B = $-67.7 + 164.4 = 96.7$ k
Moment at B = $-67.7(4.67)/2 = -158$ k·ft
Moment at mid span = $-158 + 96.7(6.67)/2 = 164.5$ k·ft
Shear diagram:

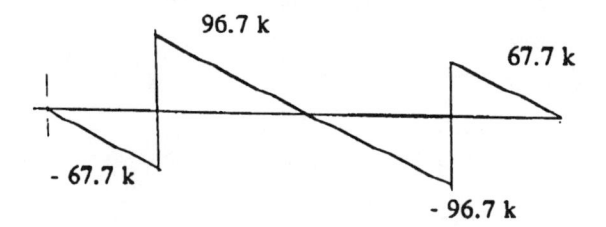

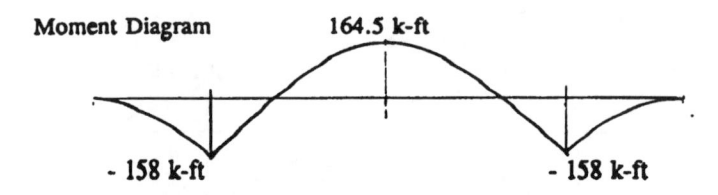

```
                    96.7 k
                    |\
                    | \           67.7 k
                    |  \          |\
    |\              |   \         | \
    | \             |    \        |  \
    |  _____|_____|_____
    |                      \              \
    -67.7 k                 \              \
                             -96.7 k
```

Moment Diagram

```
                         164.5 k-ft
                            /\
                           /  \
        ___         _____/    _____         ___
           \       /                   \       /
            \     /                     \     /
             \___/                       \___/
          -158 k-ft                    -158 k-ft
```

Step 2—Use WSD Allowable Stress Design Selection Table pg 2-11 to find with F_y = 50 ksi a shape with $M_R \geq 164.5$ k·ft. Try W16×40. Refer to pg 2-204 to see that since the laterally unbraced bottom flange is less than 6 ft long, allowable moment is 178 k·ft.

Step 3—Check allowable shear strength. Section properties are given on pg 1-22. Use Eq. 8.4.7a and compute

$$380/\sqrt{F_{yw}} = 380/\sqrt{50} = 53.7$$

Since d/t_w = 52.5, F_v = 0.40F_y. Then

$$\text{Allowable } V = F_v\, dt_w = 0.40(50)16.01(0.305) = 97.7 \text{ k}$$

which is greater than the largest value needed = 96.7 k. Therefore, W16×40 is acceptable. (The allowable shear strength could have been read directly from the Beams Table, pg 2-115. This Table also shows the maximum length of unsupported compression flange as L_c = 6.3 ft before moments must be reduced for lateral torsional buckling.)

8.5 Combined Axial Force Plus Flexure

8.5.1 Cross Section Strength

Beams and columns frequently must resist bending moments in addition to axial force. For example, any transverse force which acts between the joints of truss members creates moments that must be accommodated in addition to the axial force caused by truss action. Columns in rigid frames resist axial forces plus moments associated with continuity and compatibility of rotation at joints.

Member response to axial force was described in Sections 8.2 and 8.3, and response to flexural force has been described in Section 8.4. Specifications of LRFD use an interaction function to describe the limit capacity of sections subjected to both axial force and moments about both the X and the Y axis. It is less confusing to consider only the uniaxial condition here:

$$\frac{P_u}{\phi P_n} + \frac{8}{9}\left(\frac{M_u}{\phi_b M_n}\right) \leq 1.0 \qquad \frac{P_u}{\phi P_n} \geq 0.2 \qquad (8.5.1a)$$

$$\frac{P_u}{2\phi P_n} + \left(\frac{M_u}{\phi_b M_n}\right) \leq 1.0 \qquad \frac{P_u}{\phi P_n} \leq 0.2 \qquad (8.5.1b)$$

Graphs for Eqs. 8.5.1a and 8.5.1b are shown in Fig. 8.9. Any combination of load ratios $P_u/\phi P_n$ and $M_u/\phi_b M_n$ inside the solid lines of the graph represent a safe load condition. The strength limit P_n includes the considerations of unbraced length kL_u as well as material strength F_y, and the strength limit M_n likewise must include full consideration of any laterally unbraced length L_b of flange in compression due to flexure.

When a tension force acts through a member, the tension force will try to straighten the bent member, bringing the member closer to the line of action of the tension force as illustrated in Fig. 8.10(a). Tension straightening will never create between the ends of the member moment values higher than the moments at the ends of the member. Consequently, the largest value of moment M for which a member need be designed will not be affected by an axial tension force. In contrast, when a compression force acts along a member, the member bends away from the line of action of the compression force. The axial force will increase the amount by which the member bends away from the line of action of the force as illustrated in Fig. 8.10(b).

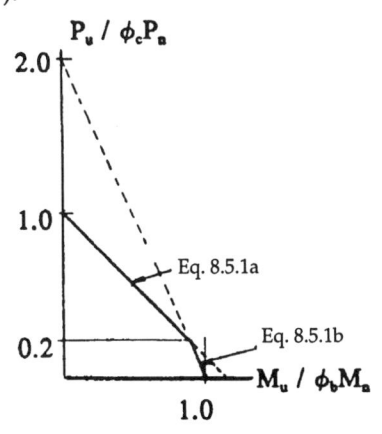

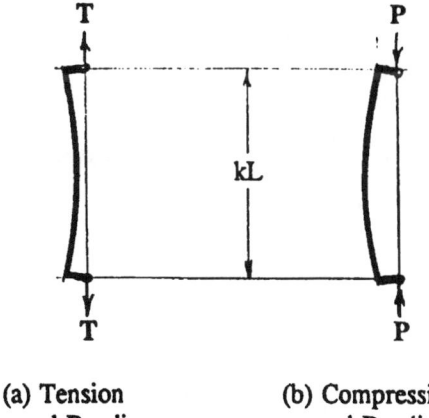

Figure 8.9 P_u and M_u strength limits.

Figure 8.10 Axial forces with moments.

(a) Tension and Bending

(b) Compression and Bending

8.5.2 Moment Magnification

At some distance within a member and away from the end of the member the moment may be larger than the moment at the end of the member. If the member is bent by equal amounts of moment applied at each end as in Fig. 8.10b, the ratio between moment at midheight and moment at the end equals the secant of the angle $\frac{1}{2}kL\sqrt{P/EI}$ expressed in radians. Another form for expressing the ratio of midheight moment to end moment is $\sec\left[\frac{1}{2}\pi\sqrt{P/P_e}\right]$, where P_e is the critical (Euler) buckling load of a column of effective length kL and flexural stiffness EI. The secant function can be considered as a magnification factor B which is approximated as

$$B = \sec\left[\tfrac{1}{2}\pi\sqrt{P/P_e}\right] \approx \frac{1}{(1-P/P_e)} \tag{8.5.2}$$

Compression forces that act through a column with end moments M_1 smaller than M_2 will cause the column to bend away from the line of action of the compression force, but if the column is stiff enough, the member deformation due to the force P will not make the maximum moment between the ends of the member bigger than the larger of the moments at either end. A coefficient $C_m = 0.6 - 0.4M_1/M_2$ is used with Eq. 8.5.2 to provide a magnification factor B_1 in LRFD rules for which the load P is taken as the required compression load P_u. Values of M_1 and M_2 are taken as positive when a moment acts clockwise at the end of the column:

$$B_1 = \frac{C_m}{1 - P_u/P_e} \geq 1.0 \tag{8.5.3}$$

Values of the denominator $1 - P_u/P_e$ larger than C_m would imply a "magnifier" less than one, but obviously the end moment cannot be reduced, and value of B_1 must be at least 1. The diagrams of Fig. 8.11 illustrate several column end conditions and values of C_m. The moment diagrams without the influence of the compression force are shown as solid lines. Deflected shapes for each column with intermediate slenderness are indicated with dashed lines. Dotted lines represent moment diagrams for very slender columns. Dashed line moment diagrams represent moments for columns of intermediate slenderness.

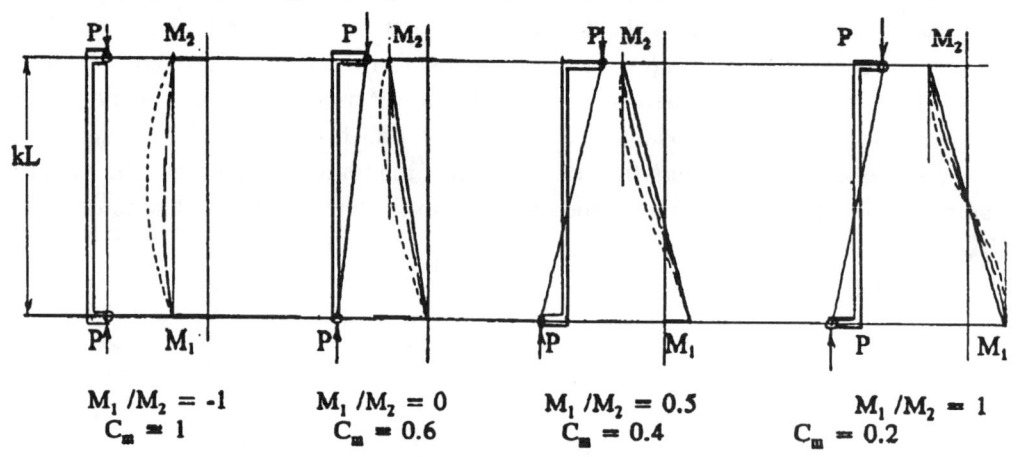

Figure 8.11 Compression force and moments on columns.

The magnitude of the slenderness index force P_e is defined in LRFD rules as (see Section 8.3.1)

$$P_e = A_g F_y / \lambda_c^2 \tag{8.5.4}$$

The effective length kL should be taken as the length of member between connection points at the ends of the member.

LRFD rules require that the moment M_u used in Eq. 8.5.1a or 8.5.1b be taken as the product $B_1 M_u$, the required (factored) moment increased by the magnifying factor B_1.

8.5.3 ASD Allowable Combined Stress

The ASD rules use 2 strength interaction relationships, one applying to maximum stress caused by forces P and M at the ends of the member and the other applying to the allowable stresses including slenderness effects between the ends of the compression member. When slenderness effects are included, a moment magnification function similar to Eq. 8.5.3 is part of the interaction equation. Superposition of stresses at the end of a member produce the requirement:

$$\frac{f_a}{0.6F_y} + \frac{f_b}{F_b} \leq 1.0 \tag{8.5.5}$$

Values for F_b will be $0.66F_y$ unless the $b/2t_f$ ratio is greater than $65/\sqrt{F_y}$ for the compression flange, in which case the bending stress is restricted to $0.6F_y$.

The ASD stress requirement with slenderness effects becomes

$$\frac{f_a}{F_a} + \frac{C_m f_b}{F_b(1 - f_a/F_e')} \leq 1.0 \tag{8.5.6}$$

The value of effective Euler buckling stress is defined as $F_e' = 12\pi^2 E/[23(kL/r)^2]$. The radius of gyration r is taken in the direction of the plane of the bending moment for determining F_e'. That need not be the same as the direction for the radius of gyration in determining F_a. Due to bending, if there is an unbraced length L_b of compression flange, the value of F_b must include such effects as described in Section 8.4.3. The ratio between service load moment and allowable moment will be the same as the ratio f_b/F_b when charts such as Fig. 8.8 are used.

In order to select a structural shape for combined loading in flexure and axial force, the simple estimate of required P_y or area A_g or M_p or section modulus S_x is not as straightforward as it was for the selection of a tension member, a beam or a column. An estimate can be made that the ratio for the axial force component of the interaction relationship will be 50% as will be the ratio for the flexural component. A search of the Manuals for a shape with the estimated area and flexural strength quickly reveals the type revision that may be needed in axial influence or flexural influence for the interaction equations. Examples will illustrate the procedure.

EXAMPLE 8.10

Select a W-shape of A36 steel for the tension chord of a truss. The tension force is 120 k due to Dead Load and 172 k due to live load. There is no bending stress from dead load, but the moment due to Live Load can be 45 k·ft. Assume that bending is about the minor axis of the shape.

Solution. LRFD Method

Step 1—Compute required forces:
$$T_u = 1.2T_D + 1.6T_L = 1.2(120) + 1.6(172) = 419 \text{ k}$$
$$M_u = 1.2M_D + 1.6M_L = 1.2(0) + 1.6(45) = 72 \text{ k·ft}$$

Step 2—Assume that Eq. 8.5.1a governs, with estimated ratio $P_u/\phi P_n = 0.5$:

$$A_g = P_u/(0.5\phi_t F_y) = 419/[0.5(0.9)36] = 25.9 \text{ in}^2$$
$$Z_y = \frac{8 M_u}{9(1-0.5)\phi_b F_y} = \frac{8(72)(12)}{9(0.5)0.9(36)} = 47.4 \text{ in}^3$$

Step 3—Search for a shape with $A_g \approx 25.9 \text{ in}^2$ and $Z_y \approx 47.4 \text{ in}^3$.

Since all shapes with an area as large as 25.9 in² have values Z_y higher than 47.4 in³, the ratio $P_u/\phi P_n$ will be larger than 0.5, and the ratio $M_u/\phi_b M_n$ will be smaller than 0.5.

Try W12×79 with A_g = 23.2 in² and Z_y = 54.3 in³.

Step 4—Check Eq. 8.5.1a for the W12×65 shape:

$$\frac{P_u}{\phi P_n} + \frac{8}{9}\left(\frac{M_u}{\phi_b M_n}\right) = \frac{419}{0.9(23.2)36} + \frac{8}{9}\left(\frac{72(12)}{0.9(54.3)36}\right) = 0.99 \; < \; 1.0$$

W12×79 is acceptable.

ASD Method

Step 1—Compute design axial force P = 120 + 172 = 292 k, and design moment M = 45 k·ft.

Step 2—Equation 8.5.5 will govern for axial tension forces. Estimate $f_a \approx 0.3 F_y$, and $f_b \approx 0.33 F_y$. Then seek a shape with

$$A_g = P/0.3F_y = 292/(0.3 \times 36) = 27.0 \text{ in}^2$$
$$Z_y = M/0.33F_y = 45 \times 12/(0.33 \times 36) = 45.5 \text{ in}^3$$

Step 3—Note that a W12×79 has an area somewhat less than 27.0 in² and Z_y larger than 45.5 in³. Try the W12×79 with A_g = 23.2 in² and Z_y = 54.3 in³.

Step 4—Check Eq. 8.5.5:

$$\frac{P/A_g}{0.6F_y} + \frac{M/Z_y}{0.66F_y} = \frac{(292/23.2)}{0.6(36)} + \frac{45 \times 12/54.3}{0.6(36)} = 1.00$$

The W12×79 is just adequate.

EXAMPLE 8.11

Check the W18×40 shape of A36 steel for the vertical forces at P, Dead Load P = 60 k, Roof Load P = 50 k, and Live Load P = 65 k. Lateral loading from wall beams (girts) due to Wind w = 0.90 k/ft. (The wall beams prevent displacement perpendicular to the web of the column, effectively bracing the compression flange and prohibiting weak axis buckling of the column. Column buckling only in the plane of the web must be checked.)

Solution. LRFD Method

Step 1—Compute required loads. See LRFD Manual pg 6-25 for required load combinations.

Moment due to Wind = $wL^2/8$ = 0.90 (26)²/8 = 76 k·ft.

Load Case (A4-3): U = $1.2D + 1.6L_r + 0.5L + 0.8W$
                              P_u = 1.2(60) + 1.6(50) + 0.5(65)   = 184.5 k
                              M_u = 0.8(76)   = 61 k·ft

Load Case (A4-4): U = $1.2D + 0.5(L_r + 0.5L + 1.3W$
                              P_u = 1.2(60) + 0.5(50) + 0.5(65)   = 129.5 k
                              M_u = 1.3(76)   = 99 k·ft

Step 2—LRFD Manual pg 1-26 & 27 give section properties for the W18×40:

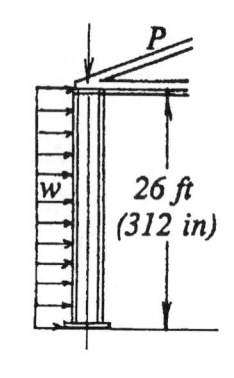

$$A_g = 11.8 \text{ in}^2, \quad r_x = 7.21 \text{ in}, \quad Z_x = 78.4 \text{ in}^3$$

Compute

$kL/r_x = 312/7.21 = 43.3$, and use pg 6-124 to find $\phi_c F_{cr} = 27.7$ ksi.

$$\phi_c P_n = \phi_c A_g F_{cr} = 27.7 (11.8) = 327 \text{ k}$$
$$P_e = \pi^2 E A r_x^2/(kL)^2 = \pi^2(29000)11.8(7.21)^2/(312)^2 = 1804 \text{ k}$$
$$\phi_b M_n = \phi_b Z_x F_y = 0.9(78.4)36/12 = 212 \text{ ft·k}$$

Step 3—Check Load Case (A4-3). Compute magnified M_u:

$$M_u = C_m/(1 - P_u/P_e) = 61(1)/(1 - 184.5/1804) = 68 \text{ k·ft}$$

The value of $C_m = 1$, since the moment at midheight exceeds the moment at each end. Eq. 8.5.1a gives

$$P_u/(\phi_c P_n) + 8M_u/(9\phi_b M_n) = 184.5/327 + 8(68)/(9 \times 212) = 0.85$$

The W18×40 is acceptable for this Load Case.

Step 4—Check Load Case (A4-4). Compute magnified M_u:

$$M_u = C_m/(1 - P_u/P_e) = 99(1)/(1 - 129.5/1804) = 107 \text{ k·ft}$$

Eq. 8.5.1a gives

$$P_u/(\phi_c P_n) + 8M_u/(9\phi_b M_n) = 129.5/327 + 8(107)/(9 \times 212) = 0.85$$

The W18×40 is O.K. for both Load Cases.

ASD Method

Step 1—Compute service loads:

$$P = 60 + 50 + 65 = 175 \text{ k}$$
$$M = wL^2/8 = 0.9(26)^2/8 = 76 \text{ k·ft}$$

Step 2—Check allowable P without Wind. ASD Manual pg 1-22 & 23 give properties of the W18×40 as

$$A_g = 11.8 \text{ in}^2, \quad S_x = 68.4 \text{ in}^3, \quad r_x = 7.21 \text{ in}$$

Compute

$$kL/r_x = 312/7.21 = 43.3$$

Use ASD Manual pg 5-120 Table 4 to find for $F_y = 36$ ksi, $C_c = 126.1$. Then compute

$$kL/(r_x C_c) = 43.3/126.1 = 0.343$$

Use pg 5-114 Table 3 to find $C_B = 0.526$. Compute $F_a = C_B F_y = (0.526)36 = 18.9$ ksi, and $f_a = P/A_g = 175/11.8 = 14.8$ ksi. Since allowable $F_a = 18.9$ ksi is greater than 14.8 ksi, W18×40 is OK for this Load Case.

Step 3—Check Eq (8.22) for midheight region with $P = 175$ k and $M = 76$ k·ft due to Wind. When Wind acts with other forces, allowable stresses can be increased 33%:

$$F'_e = \frac{12\pi^2 E}{23(kL/r_x)^2} = \frac{12\pi^2(29000)}{23(43.3)^2} = 79.6 \text{ ksi}$$

Compute $f_b = M/S_x = 76 \times 12/68.4 = 13.3$ ksi. $F_b = 0.66 F_y = 24$ ksi. Eq 8.5.6 with $C_m = 1$, since moment at midheight is greater than moment at either end, is

$$\frac{f_a}{1.33F_a} + \frac{f_b}{1.33F_b}\left[\frac{C_m}{1-f_a/F'_e}\right] = \frac{14.8}{1.33(18.9)} + \frac{13.3}{1.33(24)}\left[\frac{1}{1-14.8/79.6}\right] = 1.10$$

The W18×40 is NOT adequate for this load case.

EXAMPLE 8.12

Select an A36 shape for the column shown. Loads P_1 are 45 k Dead Load and 64 k Roof Load L_r. Loads P_2 are 5 k Dead Load and 50 k Live Load L. Lateral support exists only at the top and at the bottom of the column. Use $kL = 28$ ft for $\phi_c P_n$ even though total P_u acts only on the lower 20 ft. Use $L_u = 28$ ft and $C_b = 1$ for ϕM_n as neither flange is braced laterally and bracket force causes displacements only to the left of column centerline.

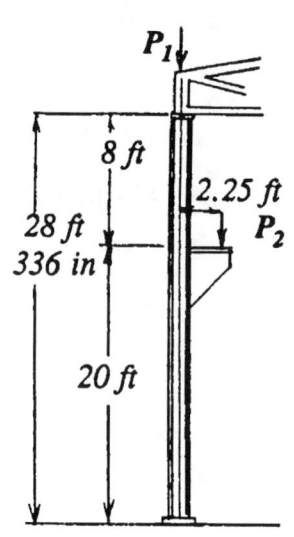

Solution. LRFD Method

Step 1—Compute required loading. Case (A4-2) will govern:

$$U = 1.2D + 1.6L + 0.5L_r$$

Below Crane bracket: $P_u = 1.2(45 + 5) + 1.6(50) + 0.5(64) = 172$ k
Horizontal reactions equal: $2.25P_2/28 = [1.2(5) + 1.6(50)]2.25/28 = 6.91$ k
At Crane bracket: $M_u = 6.91(20) = 138$ k·ft (3317 k-in)

Step 2—Assume $P_u/\phi_c P_n = 0.5$. With $P_u = 0.5\phi_c F_{cr}A_g$, estimate $\phi_c F_{cr} = F_y/3$ in order to compute

$$A_g \approx P_u/(0.5F_y/3) = 172/(0.5 \times 36/3) = 28.7 \text{ in}^2$$

Steel weighs 3.40 lb/sq in/ft, therefore weight = 28.7(3.4) = 98 plf. With $(8/9)M_u/\phi_b M_n = 0.5$, estimate $B_1 = 1.2$, so $M_u = 1.2M_u = 1.2(138) = 166$ k·ft.

Use LRFD pg 3-70 with $L_u = 28$ ft to seek a 98-lb shape for which

$$\phi M_n \approx 2 \times 8 \times 166/9 = 295 \text{ k·ft}$$

Try W18×86 noting on pg 3-68 that $\phi M_n = 372$ ft·k when $L_u = 28$ ft.

Step 3—Obtain section properties on LRFD pgs 1-26 & 27:

$$A_g = 25.3 \text{ in}^2, \quad r_x = 7.77 \text{ in} \quad \text{and} \quad r_y = 2.63 \text{ in}$$

Compute

$kL/r_y = 336/2.63 = 128$, and Table 3-36, pg 6-124 gives $\phi_c F_{cr} = 12.92$ ksi
$kL/r_x = 336/7.77 = 43.2$, and Table 9, pg 6-131 gives $F_e = P_e/A_g = 153$ ksi
$\phi_c P_n = A_g \phi_c F_{cr} = 25.3(12.92) = 327$ k
$P_e = A_g F_e = 25.3(153) = 3870$ k

Compute moment magnifier B_1. Take $C_m = 1$ since the moments at the crane bracket are larger than the moment at either end of the column:

$$B_1 = 1/(1 - P_u/P_e) = 1/(1 - 172/3870) = 1.05 \quad \text{(Estimate in Step 2 was too high!)}$$

Compute magnified M_u:

$$M_u = B_1 M_u = 1.05(166) = 174 \text{ ft·k}$$

Step 4—Check Eq. 8.5.1a,

$$P_u/\phi P_n + 8M_u/(9\phi M_n) = 172/327 + 8(174)/(9 \times 372) = 0.94$$

The W18×86 shape is adequate for the required loading.

A subsequent trial with a W14×82 shape will show the lighter shape to be inadequate largely due to it's lower value of F_a. The flanges of the W14×82 are 2 inches narrower than those of the W18×86.

ASD Method

Step 1—Compute design service loads:

    Below Crane bracket: $P = 45 + 64 + 5 + 50 = 164$ k

    Horizontal reactions equal: $2.25\, P_2/28 = 2.25(5+50)/28 = 4.42$ k

    At crane bracket: $M = 4.42\,(20) = 88.4$ k·ft (1060 k-in)

Step 2—Eq 8.5.6 will govern. Assume $f_a/F_a = 0.5$, and guess $f_a = F_y/4$ in order to compute

$$A_g \approx P/(0.5 F_y/4) = 164/(0.5 \times 36/4) = 36 \text{ in}^2$$

Steel weighs 3.40 lbs/sq in/ft, therefore section weight $= 36(3.4) = 122$ plf. Assume the magnification factor will be 1.2, such that $1.2 f_b/F_b = 0.5 = 1.2\, M/M_{all}$. Use ASD pg 2-164. When $L_u = 28$ ft, find a 122-lb shape with

$$M_{all} > 1.2 M/0.5 = 2.4(88.4) = 212 \text{ k·ft}$$

Try W14×90, noting that $M_{all} = 262$ k·ft (3144 k-in) when $L_u = 28$ ft.

Step 3—Obtain section properties on pg 1-26 & 27. $A_g = 26.5$ in², $r_x = 6.14$ in, $S_x = 143$ in³. Use Column Table pg 3-23 to find for $kL = 28$ ft, $P_{all} = 374$ k. Compute

$$F_a = P_{all}/A_g = 374/26.5 = 14.1 \text{ ksi}$$

$$f_a = P/A_g = 164/26.5 = 6.19 \text{ ksi}$$

$$f_b = M/S_x = 1060/143 = 7.41 \text{ ksi}$$

$$F_b = M_{all}/S_x = 3144/143 = 22 \text{ ksi}$$

Compute $kL/r_x = 336/6.14 = 54.7$ for which Table 8, pg 5-122 gives $F'_e = 49.9$ ksi.

Step 4—Check Eq 8.5.6 with $C_m = 1$, since moment at crane bracket is larger than moments at either end:

$$\frac{f_a}{F_a} + \frac{f_b}{F_b}\left[\frac{C_m}{1-f_a/F'_e}\right] = \frac{6.19}{14.1} + \frac{7.41}{22}\left[\frac{1}{1-6.19/49.9}\right] = 0.82$$

(A subsequent trial with a W18×86 shape will show the lighter shape to be inadequate largely due to it's lower value of F_a. The flanges of the W18×86 are 2 inches less wide than those of the W14×90.)

8.6 Connections

The perfect structure ideally might be fabricated completely in a factory, such that all components fit precisely, attachments among parts are machine controlled, and alignment is perfect. That ideal is quite impossible. Large structures must be shipped in smaller components. Connections allow individual components to be shipped and then assembled at a jobsite. Connections must permit some adjustment to alignment and precision of fit among components while providing adequate strength for load transfer among the structural components. The best connections are simple in detail, easily assembled, and quick to complete at the jobsite.

Structural steel components are connected together by welding or by bolting. In general, the quality of welding is best when performed in a factory environment with all the machine-controlled automation that might be available. Certainly good welding can be achieved at a jobsite, but it is achieved at a cost significantly higher than in the factory environment. Welded components usually require more precise (and expensive) controls on overall dimensions than those required with bolting. Consequently, some structures are designed with "shop" connections welded and "field" connections bolted. Other systems may be completely bolted in the shop as well as the field, or totally welded both in the shop and in the field.

Most erectors desire only one type of connection at a jobsite, all bolting or all welding. However, even an all welded system requires "fit-up" bolted connections to hold components in good alignment until welding is complete. Every field-bolted project eventually seems to require some field welding to accommodate adjustments to fit-up or alignment.

8.6.1 Welded Connections

Welding is the fusing of metal components into one continuous material. The fusing is accomplished with heat adequate to melt the metals in the immediate vicinity of the welding operation. The molten metals fuse together and then cool to a continuous solid state. Heat for the welding operation is produced by material resistance to a strong current of electricity. An electric arc is produced at the end of a steel electrode that heats and melts edges of base components and itself becomes part of the fused metal in the welded connection. Base material adjacent to the weld is effectively heat treated and has a yield strength slightly greater than its original strength. Material for the weld electrode is selected always to be stronger than the base materials being joined.

Cross sections through common weld types are sketched in Fig. 8.12. The butt joint with groove weld is appropriate for joining together the edges of plates or shapes. Please note that the edges must fit precisely at the point of the welds, and the edges may require special preparation of grooves. The full penetration groove weld has essentially the same strength as the base materials that are welded. The fillet weld, as its name implies, produces a fillet at the right angle juncture between connected parts. Forces that pass through the fillet weld create shear stresses which are highest through the throat of the weld.

(a) Complete Penetration Groove Weld   (b) Partial Penetration Groove Weld   (c) Corner Fillet Weld   (d) Gusset Fillet Weld

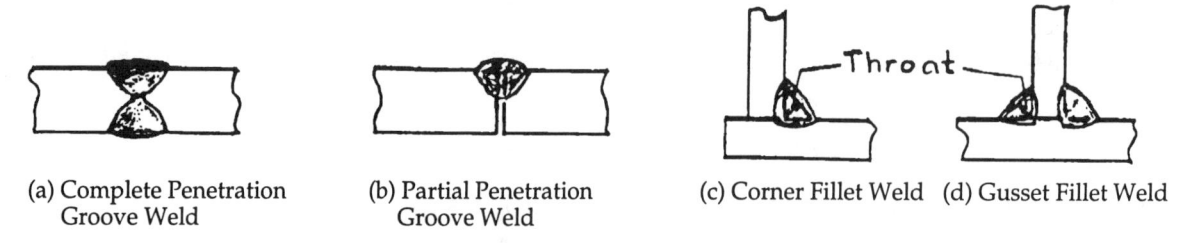

Figure 8.12 Typical structural welds.

The strength of weld material (electrode) is selected always to be stronger than the strength of the base material. Thus, a complete penetration groove weld has a strength in tension, compression, or in shear at least as great as the strength to resist the same forces on the base material. Weld electrode strength is indicated by a symbol E60, E70, E_{xx} to designate the nominal yield capacity of the electrode material. For such welds LRFD rules specify capacity reduction values $\phi = 0.90$, the same as those specified for the base material subjected to tension or compression forces and ductile behavior before rupture. The transfer of shear stress through full penetration welds may be less ductile than through base material, and LRFD rules specify a factor $\phi = 0.80$ instead of 0.90 for shear strength of full penetration welds. The shear strength of weld material is taken as $0.6E_{xx}$, the same as $0.60F_y$ for base material.

In many practical cases, it is not necessary to develop through a connection the full strength of the base materials being connected. In such cases it is reasonable to use less expensive partial penetration welds, or intermittent discontinuous pieces of full penetration groove welds. When partial or intermittent welds are used, the stress concentrations which occur at each discontinuous edge reduce effective ductility. Therefore LRFD rules require a strength reduction factor $\phi = 0.75$ for shear strength parallel with the axis of discontinuous and partial penetration welds.

The effect of lower capacity reduction factors ϕ for weld material can be accommodated most readily by specifying an electrode strength at least 20% greater than F_y of the base material.

Fillet weld strength is limited by the shear strength of weld material at the "throat" of the weld. Weld throat dimensions are indicated in Fig. 8.12 (c) and (d). The throat dimension can be taken safely as 70% of the weld size dimension. Shear forces parallel with the axis of the fillet weld obviously create shear stress which is largest at the throat of the weld. Tension forces perpendicular to the axis of a fillet weld must be transferred through the weld with combined tension and shear stress that reaches a maximum value through the throat of the weld. It is convenient and safe to use the shear strength of fillet welds as the limit strength of the weld whether forces are primarily shear parallel with the weld or tension perpendicular to the weld. With LRFD rules in LRFD Table J3.2, call the size of fillet weld D_w, and the shear strength W_u through the throat of the weld can be expressed as

$$W_u = \phi(0.70)0.60(E_{xx})D_w = 0.75(0.70)0.60\, D_w E_{xx} = 0.325\, D_w E_{xx} \quad \text{per in of weld} \tag{8.6.1a}$$

When S is the number of sixteenths in a weld size, $D_w = S/16$, and the value of $W = 0.02SE_{xx}$ per inch of weld. Thus a 1/4-in E60 electrode could develop $(4)0.02(60) = 4.8$ k per inch of fillet weld.

ASD Table J2.5 of ASD regulations gives similar allowable stress values through the throat of fillet welds, specifying the allowable stress level as 30% of the nominal ultimate strength of the weld material. The allowable force W_a per inch of fillet weld can be expressed as:

$$W_a = 0.30(0.60)E_{xx}(D_w) = 0.18\, D_w\, E_{xx} \quad \text{per in of weld} \tag{8.6.1b}$$

When S is the number of sixteenths in a weld size, $D_w = S/16$, and the value of $W = 0.011SE_{xx}$ per inch of weld. Thus a 1/4-in E60 electrode could be allowed $4(0.11)60 = 2.64$ k per inch of weld.

Welds should be deposited at connections such that the centroid of the welds coincides with the centroid of the connected element. Applied to one leg of an angle, a rule-of-thumb places twice as much weld along the heel as along the toe of the leg. Intermittent or discontinuous welds create stress concentrations at the ends of each weld segment. For fillet welds near the end of a plate, flange or angle, it is good practice to continue welds along the edge to a corner and then continue a "return" around the corner for a distance at least 4 times the weld size. The size of fillet weld is generally 1/8-in less than the thickness of the plate, flange or angle, being welded, but it is permitted to have the same thickness. Along the rolled toe of an angle, welds of the same thickness as the angle must be built up to accommodate the rounded edge of the rolled angle shape.

Applied to thick plates, welds from a small electrode would lose heat too rapidly into the thick plates. Rapid cooling of the welded material produces a brittle and unacceptable material. Tables J2.3 and J2.4 in both LRFD and ASD specifications limit the minimum weld size for various thicknesses of connected parts. When it is necessary to use a small size electrode with a thick plate, (joining a thin plate to a thick plate, for example), the thicker plate can be preheated before the welding is performed thus extending the time for fused materials to cool in the vicinity of the weld.

Standard welding symbols for fabricators are illustrated and defined on LRFD pg 5-179 and on ASD pg 4-155.

8.6.2 Bolted Connections

Bolts consist of a bolt head and a shank that is threaded to receive a nut. The shank is put through holes in connected parts before the nut is tightened to create a pressure between the connected parts. Frequently, a washer is used between the nut and the connected part, both to facilitate rotation of the nut for tightening and to spread the clamping pressure contact area near the bolt. Special nuts are available for such purposes as resisting loosening or indicating limit contact pressures. Holes for bolts are made with a diameter slightly greater than the bolt shank.

TABLE 8.1 Bolt strength.

Bolt Type ASTM	Material, F_y ksi	Limit Strength Tens. ksi	Limit Strength Shear ksi	Allowable Str. Tens. ksi	Allowable Str. Shear ksi
A307	45	33.8	16.2	20	10
A325	90	67.5	46.8	44	20
A490	112.5	84.4	58.5	54	40

The fundamental characteristic of a bolt is the capacity of the bolt shank. Bolt strength is identified on the basis of tensile strength of the shank. When the shank is loaded in tension, the weakest portion of shank is the threaded portion where the cross section area is less than the cross section area of the unthreaded portion. The same principles from Section 8.2 apply for bolt tension. The tension fracture strength of the threaded shank is $0.75F_y A_e$, and the yield limit of the unthreaded shank is $0.90F_y A_g$. Commercial grade bolts are either ASTM A307 machine bolts or high strength bolts designated as ASTM A325 and ASTM A490. Table 8.1 displays bolt strength as determined by stress values for tension alone on the bolt and for shear alone on the unthreaded part of the bolt shank. Limit values for LRFD are shown as are allowable values for ASD.

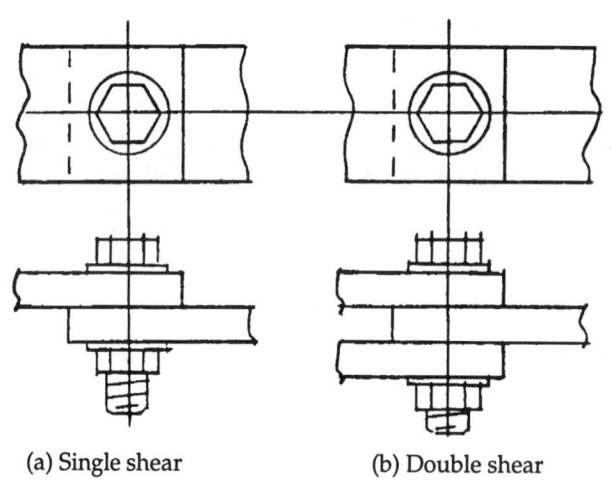

(a) Single shear   (b) Double shear

Figure 8.13 Shear transfer through a bolt.

The diagrams of Fig. 8.13 indicate the mechanism by which forces are transmitted through shear on bearing-type bolts. Any clamping pressure from the bolt against connected parts is ignored, and forces from one plate are assumed to press against one side of the bolt shank while forces from the connected plate press against the opposite side of the bolt shank. The bolt shank is subjected primarily to shear stress across the shank at the plane between connected plates. The bolt shank is said to be in single shear when only one plane occurs between connected elements indicated in Fig. 8.13. The tendency for the bolt shank to rotate is balanced by unsymmetric bearing pressures under the head and the nut of the bolt. The bolt shank is said to be in double shear when two shear planes occur. One connected plate is sandwiched between two connecting plates.

Bolts are said to "fail" when the distortion of connected parts becomes excessive, generally in the order of 1/4 in at a bolt. Bolts rarely fail by fracture with complete separation between connected components. The three basic failure modes for bearing-type bolts are illustrated in Fig. 8.14.

The bearing strength R_n in the direction of the applied force for bolts is specified in LRFD Section J3.6. A capacity reduction factor $\phi = 0.75$ must be used, and in the direction of the force, hole spacing must be at least 3 times the diameter d of the bolts. Plate thickness is designated as t, and the ultimate material strength is F_u. For bolts in standard round holes,

Nominal strength $\qquad R_n = 2.4\, d\, t\, F_u \qquad$ (8.6.2a)

Allowable strength $\qquad R = 1.2\, d\, t\, F_u \qquad$ (8.6.2b)

Limit loads on bolts in bearing are given in LRFD Table I-E, pg 5-7. Allowable loads on bolts in bearing are given in ASD Table I-E, pg 4-6.

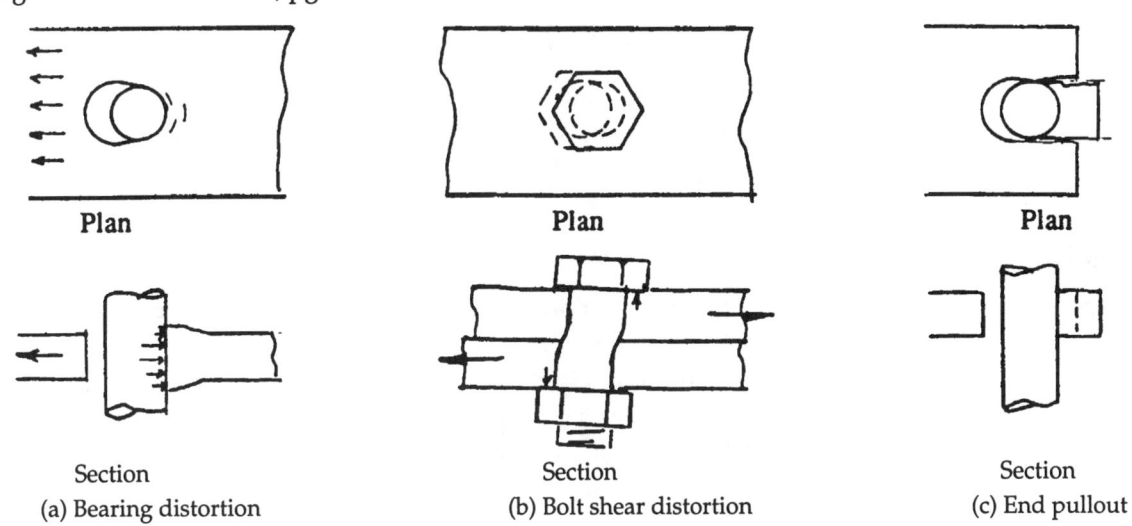

Figure 8.14 Failure modes at bolted connections.
(a) Bearing distortion (b) Bolt shear distortion (c) End pullout

Bolts can be prepared with a washer if necessary such that connection plates will create the shear plane(s) on the bolt shank only in the unthreaded region of the shank. Shear strength R_v of bolts is taken as the bolt unthreaded shank area A_b times 60% of the nominal tensile ultimate strength F_u for the bolt. Bolts referred to as "high strength" bolts (A325 with $F_u = 90$ ksi and A490 with $F_u = 112$ ksi) require a capacity reduction factor $\phi = 0.65$. The ϕ factor is reduced to 0.6 with A307 bolts, and it must be assumed that shear planes are in threaded regions of the shank of A307 bolts for which $F_y = 45$ ksi:

For A307 bolts, $\qquad R_v = 0.60(0.6)45\, A_b = 16.2\, A_b \qquad$ (8.6.3a)

For A325 bolts, $\qquad R_v = 0.65(0.6)120\, A_b = 46.8\, A_b \qquad$ (8.6.3b)

If the shear plane were in a threaded portion of the bolt shank, limit shear strength must be reduced to 75% of the value for the unthreaded shank. LRFD Table I-D, pg 5-5 gives shear capacity for bolts of available diameters from 5/8 in to 1 1/2 in.

ASD Allowable shear stress under service loads on bolt shanks is specified in a similar manner in ASD Table J3.2, pg 5-73. A307 bolts are allowed a shear stress of 10 ksi. For shear on an unthreaded part of the shank, A325 bolts are allowed a stress of 30 ksi, and A490 bolts are permitted a shear stress of 40 ksi. These allowable stress values must be reduced 70% when shear planes occur in a threaded portion of a bolt shank. ASD allowable stress values produce forces about 60% of the limit strength values from the LRFD rules. ASD Table I-D provides a tabulation of allowable shear forces per bolt. The pullout strength of bolts is limited by the shear strength at the edge of a plate. In order for a bolt to push out the material between the bolt and the end of the plate, the

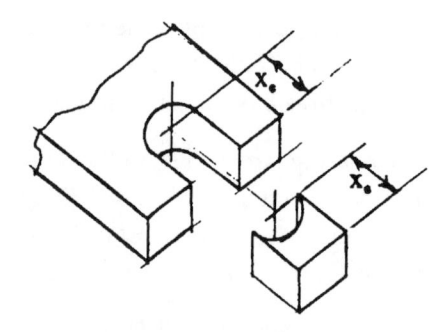

Figure 8.15 End pullout.

material must fail in shear along the two sides of the displaced material as sketched in Fig. 8.15. Call x_e the edge x_e distance from bolt centerline to the plate edge and equate the shear limit strength to the plate bearing capacity of Eq. 8.24) to obtain

$$2x_e \times t(0.6F_u) = 2.4d \times t \times F_u \quad \text{or} \quad x_e = 2d \qquad (8.6.4)$$

Thus, in the direction of force, edge distance should be at least $2d$ unless the capacity of the bolt closer to an edge is to be reduced to a value less than the bearing strength of the plate. The same logic for ASD leads to a recommended minimum edge distance of 2 bolt diameters.

8.6.3 Block Shear

The strength of connected components can be limited by a failure mode related to end pullout. The phenomenon is called Block Shear. Block shear strength is analyzed as a detail in the immediate vicinity of the attachments in a connection. The diagrams of Fig. 8.16 illustrate two common types of critical block shear regions. The regions consist of a beam web or a gusset plate between two angles. Let it be assumed that each of the 2 angle legs have a thickness about the same as the thickness of the connected plate or web such that block shear strength obviously is more critical for the web or plate than for the angle legs. Shear strength and tension strength combine to resist forces through the plate as indicated. The block shear capacity reduction factor is $\phi = 0.75$ both for yield and for fracture.

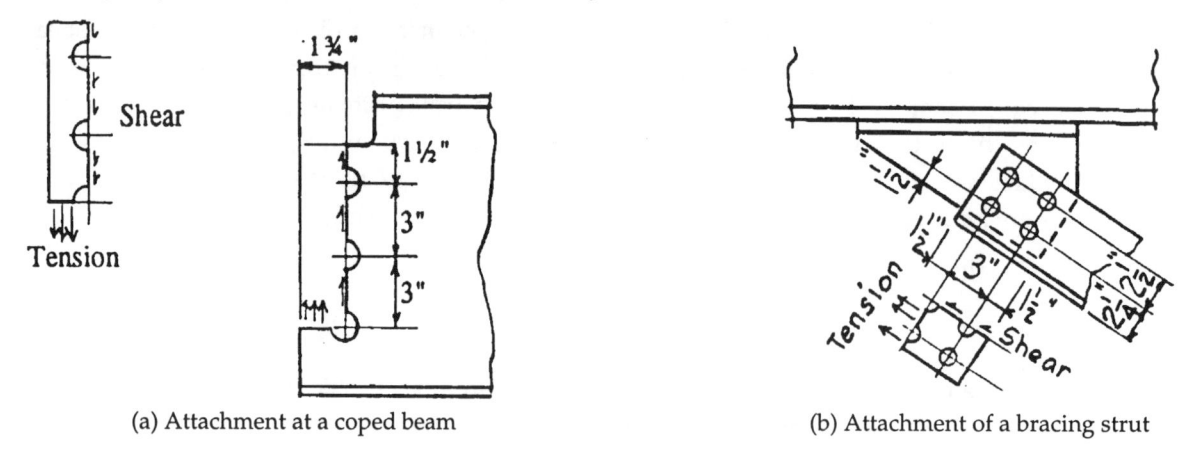

(a) Attachment at a coped beam   (b) Attachment of a bracing strut

Figure 8.16 Block shear failure modes.

Block shear capacity of the connection material is taken to be the larger sum of a) tension yield strength $0.75A_gF_y$ plus shear fracture strength $0.75A_n(0.6F_u)$ or b) tension fracture strength $0.75A_nF_u$ plus shear yield strength $0.75A_g(0.6F_y)$. Values of A_n cannot include plate segments occupied by holes. Hole diameters should be taken as 1/8-in greater than bolt size. For Fig. 8.16(a), the gross tension area of web is $(1.75)(0.40) = 0.70$ in^2. The net tension area is $(1.75 - 7/16)(0.40) = 0.525$ in^2. The gross shear area is $(7.50)(0.40) = 3.00$ in^2. The net shear area is $(7.50 - 2.5 \times 7/8)(0.40) = 2.125$ in^2. For condition a) the capacity is 0.75(0.70)(36) + 0.75(2.125)(0.6)58 = 74.4 k. For condition b) the capacity is 0.75(0.525)58 + 0.75(3.00)0.60(36) = 71.4 k. The larger value of 74.4 k is the block shear strength.

8.6.4 Combined Shear and Tension on Bolts

Bolts in some connections are required to resist tension forces in addition to shear forces. One such connection is illustrated in Fig. 8.17. The bolts must resist a tension force of $T\cos\alpha$ and a shear force of $T\sin\alpha$. Bolts can sustain some shear force in addition to total tension capacity, and bolts can sustain some tension force in addition to shear capacity. However, when significant combinations of the two types of force act simultaneously, both tensile and shear limits must be reduced from the full strength values. AISC rules use interaction functions to define the limit for combined shear and tension strength. Graphs for limit strength values in Table J3.3, LRFD pg 6-68 and ASD pg 5-74 are shown in terms of stress in Fig. 8.18. LRFD values are shown with solid lines and ASD values with dashed lines. Combined stress values within the graphs represent acceptable values.

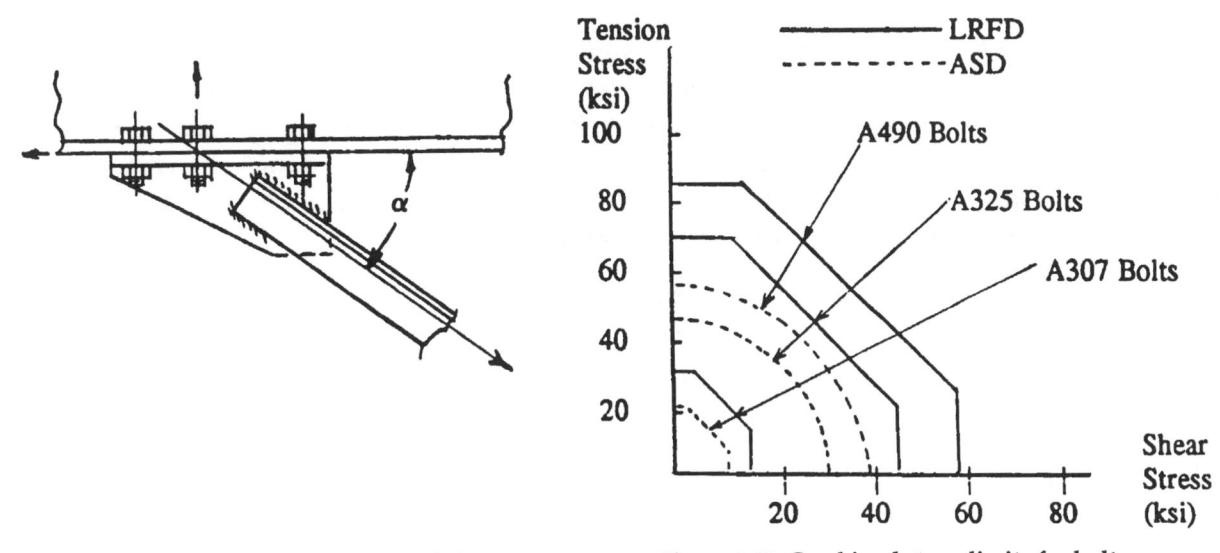

Figure 8.17 Bolts in tension and shear.   Figure 8.18 Combined stress limits for bolts.

The designer must approximate the predominance of either shear or tension force when estimating a strength value for use in proportioning connections that involve combined tension and shear on bolts. It should be noted that the limit combined stress equations can be satisfied when either the tension limit alone or the shear limit alone would be excessive. All 3 limits must be satisfied.

Bolts are not compressed. Combined compression and shear does not occur through bolts.

8.6.5 Slip-Critical Connections

There are structural applications for which no slip can be tolerated at connections. Minute amounts of slip occur ordinarily when bearing surfaces are brought into contact with bearing type bolts. Slip occurs after forces overcome the shear friction resistance of clamping pressure that tightened nuts create between connected surfaces. In slip-critical applications, special precautions are exercised to insure that bolts are tightened until every bolt clamping force is essentially the same as the yield strength of the bolt shank. The design of slip-critical connections is based on serviceability conditions of force, as the service load condition governs non-slip criteria.

Only high strength bolts are acceptable for slip-critical connections. The nominal shear strength limit for A325 bolts is 17 ksi, and for A490 bolts, the shear limit stress is 21 ksi. These shear limit stress values must be reduced by the factor $(1 - T/T_b)$ when slip-critical bolts are subjected to tension force, as the tension force relieves the clamping pressure. The bolt tension force is T, and T_b is 70% of the tensile capacity values that have been given in Table 8.1.

8.6.6 Eccentric Shear on Connections

The arrangement of bolts and welds in standard connections involves a rather tight cluster of connectors for which the centroid of the cluster is coincident with the line of action of the forces to be connected. Connected parts will deform negligible amounts until all bolts or welds share equally in resisting the ultimate or failure force. If the line of action of an applied force does not act through the centroid of a bolt group or a weld arrangement, some components of the connection can reach a failure state before other components. Connection plates and shapes can be assumed to be infinitely stiff while bolts or welds remain elastic, and principles of elastic analysis can be applied to estimate the maximum stress at any point in a connection. Safe designs can be taken as the condition for which the largest elastic stress in a connector reaches its failure state.

A more accurate estimate of "true" strength would require analysis of inelastic distortions. Such analysis is significantly more complex than is the elastic analysis of a connection. Both the LRFD and ASD Manuals contain tabulated values derived on the basis of inelastic response for acceptable strength levels in standard connectors such as those for shear at the end of beams, seated connection brackets, and eccentric forces on connector plates. In the absence of such tables, and for connection types that are not included in any table, the elastic analysis of the connection is recommended.

Elastic analysis of a bolt group or a weld arrangement is an application of the familiar stress superposition relationships for a stress at a point $x = a$ and $y = b$ from the centroid of the arrangement:

$$f_x = \frac{P_x}{A} + \frac{Mb}{I_x} \quad \text{and} \quad f_y = \frac{P_y}{A} + \frac{Ma}{I_y} \tag{8.6.5a}$$

where
- f_x and P_x are components in the x direction
- f_y and P_y are components in the y direction
- I_x is the second moment of area $\left(\sum \Delta A y^2\right)$ about the x axis
- I_y is the second moment of area $\left(\sum \Delta A x^2\right)$ about the y axis

At a radius $r^2 = a^2 + b^2$, the radial-directed stress f_r is the resultant of f_x and f_y,

$$f_r = \sqrt{f_x^2 + f_y^2} = Mr/J \quad \text{where } J = I_x + I_y \tag{8.6.5b}$$

The highest stress or most highly loaded component is the component farthest from the centroid and nearest the line of action of the force. Frequently, it is convenient to evaluate A as inches of weld or number of bolts while evaluating I and J in terms of in^2-inches if weld or in^2-bolts. Units for stress f then become force per inch of weld or force per bolt.

8.6.7 Other Connectors and Special Considerations

Rivets—Many years ago, rivets were the major connector used in steel construction. It is impossible to find riveting equipment today in many of the nation's large cities. The mechanics of behavior for design or analysis of rivets are identical with those described for bearing type bolts. LRFD and ASD equations for combined strength limits reflect the differences in base material used for the rivets.

Beam Bearing Plates—Steel plate bearings are used to distribute reaction forces from steel members onto masonry or concrete supports. Reaction pressures against the steel bearings are assumed to vary linearly across the surface, generally as a uniform reaction pressure on the entire bearing surface. The plate itself must be proportioned such that the beam reaction can be transmitted from the web to the bearing without causing the web itself to yield and fail locally. That concern generally governs the required width of bearing. The plate must be thick enough that maximum bending stress never reaches the yield limit of the plate. The sketch of Fig. 8.19 is for a beam bearing plate. The critical point for determining bending stress is taken at the edge of flange fillet, where the flexural stiffness of the steel shape increases sharply. LRFD Section K3 requires that the stress on the web segment at the start of the web fillet as shown be less than the yield strength of the beam web. The capacity reduction factor for such web strength is to be taken as 1.0. Bearing pressure under factored limit loads against concrete are limited to $1.7\phi f_c'$, where $\phi = 0.6$ and f_c' is the design strength of standard concrete cylinders. Bearing pressure under factored limit loads against masonry should not exceed 500 psi. Comparable service load allowable bearing pressures become $0.70 f_c'$ against concrete and 300 psi against masonry. The local strength and vertical stability of the beam web also must be checked as described in the following paragraphs.

Web Crippling—LRFD Section K4 sets $\phi = 0.75$ and defines reaction capacity R_n when a concentrated load is applied at a distance less than $d/2$ from the end of the beam:

$$R_n = 68 t_w^2 \left\{ 1 + 3(N/d)(t_w/t_f)^{1.5} \right\} \sqrt{F_{yw} t_f / t_w}$$

Replace the coefficient 68 by a value of 135 when a concentrated load is applied at a distance greater than $d/2$ from the end.

ASD Section K4 uses an equation the same as Eq. 8.6.6 with a coefficient of 34 substituted for the coefficient 68, and a coefficient of 67.5 is substituted when a concentrated service load is applied at a distance greater than $d/2$ from the end of the beam.

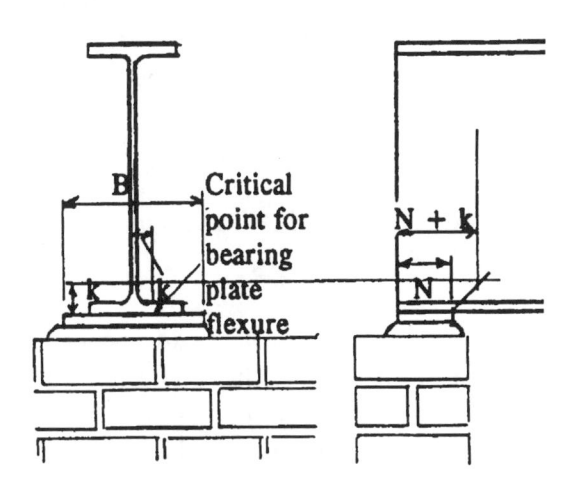

Figure 8.19 Bearing plate for a beam.

Sidesway Web Buckling—LRFD Section K5 sets a value $\phi = 0.85$, and limits the end reaction R_n at reactions for beams with the top flange unrestrained from lateral movement (by web stiffeners or lateral bracing):

$$R_n = \frac{12\,000 t_w^3}{h} \left\{ 0.4 \left(\frac{d_c / t_w}{\ell / b_f} \right)^3 \right\} \qquad (8.6.7)$$

where ℓ = largest laterally unsupported length along either flange from point of load, in
$d_c = d - 2k$, the depth of web clear of fillets

and the quantity $(d_c / t_w)/(\ell / b_f)$ is less than 1.7. When that quantity is greater than 1.7 but less than 2.3, the value of the terms in the brackets may be increased by 1.0. If $(d_c / t_w)/(\ell / b_f)$ is greater than 2.3, sidesway web buckling will not occur.

The allowable web buckling strength under service loads is expressed in an equation identical with Eq. 8.6.7 after a value of 6800 is substituted for 12000. When value of $(d_c / t_w)/(\ell / b_f)$ is greater than 1.7, the term in brackets may be increased by 1. When the value of $(d_c / t_w)/(\ell / b_f)$ is greater than 2.3, there is no concern for no web buckling.

Compression Buckling of Web—The LRFD rules specify in Section K1.6 a limit for any concentrated force R_n that a web can sustain without stiffeners to help transmit the force. A capacity reduction factor

$\phi = 0.90$ applies, and

$$R_n = \left(4100 t_w^3 / d_c\right)\sqrt{F_{yw}} \tag{8.6.8}$$

A web stiffener must be used to help transmit forces larger than $0.90R_n$ from Eq. 8.6.8. The same requirement for stiffeners is required by the ASD Section K1.6 with the service load R substituted for the limit force R_n. Concentrated forces occur when flanges of shape bear against flanges of beams and columns in rigid joints of monolithic frames.

Column Base Plates—A design procedure for column base plates is described in the LRFD Manual beginning on pg 2-101 and in the ASD Manual beginning on pg 3-106. Base plates must be proportioned to distribute the column load to the supporting concrete surface within bearing stress levels that might cause the concrete to fail. A capacity reduction factor $\phi = 0.6$ applies for the bearing strength F_p of concrete. The symbol f'_c represents the design or specified strength (usually in ksi units) for concrete control cylinders. Limit bearing strength under LRFD factored column load P_u can be taken as $1.7 f'_c$ if the concrete bearing surface has an area A_c at least 4 times the area A_{pl} of the bearing plate. If the area A_c is less than the area A_{pl} (such as the condition at the top of a drilled shaft) the bearing strength F_p can be computed as

$$\text{LRFD,} \qquad F_p = 0.85 f'_c \sqrt{A_c / A_{pl}} \leq 1.7 f'_c \tag{8.6.9a}$$

Allowable bearing strength under ASD service loads from columns can be taken as

$$\text{ASD,} \qquad F_p = 0.35 f'_c \sqrt{A_c / A_{pl}} \leq 0.70 f'_c \tag{8.6.9b}$$

Column base plates must be made at least large enough to cover the entire cross section of the column shape, a minimum base plate area can be computed as $(d + 1)(b + 1)$, where d is the depth of the shape, and b is the flange width of the shape. A required bearing area is the column load divided by the appropriate F_p value. If the required bearing area is larger than the minimum area, the base plate must be made thick enough not to yield in flexure from the F_p pressure on cantilevered edge lengths as defined in Fig. 8.20(a). If a plate must be made larger than the required bearing area simply to cover the entire column cross section, an effective bearing area shown shaded in Fig. 8.20(b) is used for determining plate thickness t_p.

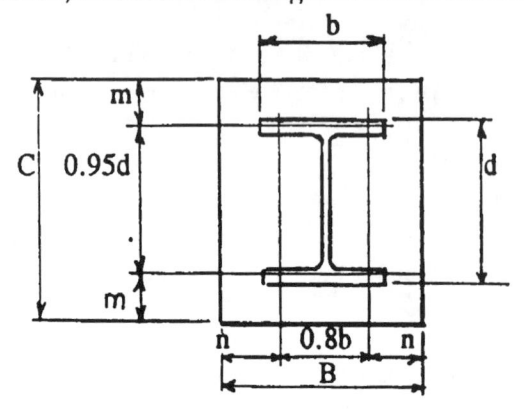

$f_{pl} = 3F_p m^2 / t_p^2$ or $f_{pl} = 3F_p n^2 / t_p^2$
LRFD: $f_{pl} \leq 0.9 F_y$ ASD: $f_{pl} \leq 0.66 F_y$
(a) Base plate larger than column section

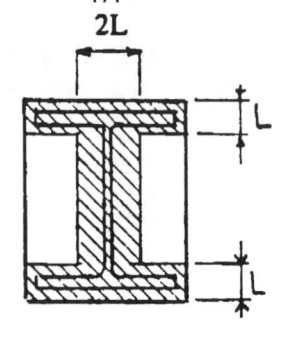

Required $A_{pl} = 2(d + b - 2L)L$
Solve for L
$f_{pl} = 3F_p L^2 / t_p^2$
LRFD: $f_{pl} \leq 0.9 F_y$ ASD: $f_{pl} \leq 0.66 F_y$
(b) Base plate sized to cover column section

Figure 8.20 Column base plate strength analysis.

8.7 Steel-Concrete Composite Members

Concrete floor slabs that are supported by steel beams can be connected to the steel beams to become a composite beam in which the concrete slab resists flexural compression stress and the steel shape resists flexural tension stress. The weight of steel in a composite beam can be reduced 20 to 30 percent less than the weight required if the slab and steel beam were not connected together to form a composite beam.

Diagrams of Fig. 8.5 showed how flexural strains increase in proportion with distance from a neutral axis. The same linear strain relationship occurs in composite beams if connectors bind the slab to the top flange of the steel beam. However, steel is much stiffer than concrete since E_s = 29,000 ksi and E_c is only 3000 to 5000 ksi. Also, concrete has no significant tension strength. Therefore, for a compression strain value of E the stress in steel is $F_s = EE_s$ and the stress in concrete is $f_c = EE_c$. At the same amount of compression strain, steel resists $n = E_s/E_c$ times as much stress as concrete, or at the same strain E, concrete resists only $1/n$ times as much stress as the stress in steel. Fig. 8.21 shows a composite cross section with strain values and corresponding stress behavior as solid lines before any allowable stresses are reached and then as dashed lines the strain and stress conditions at the "plastic" limit state of bending strength.

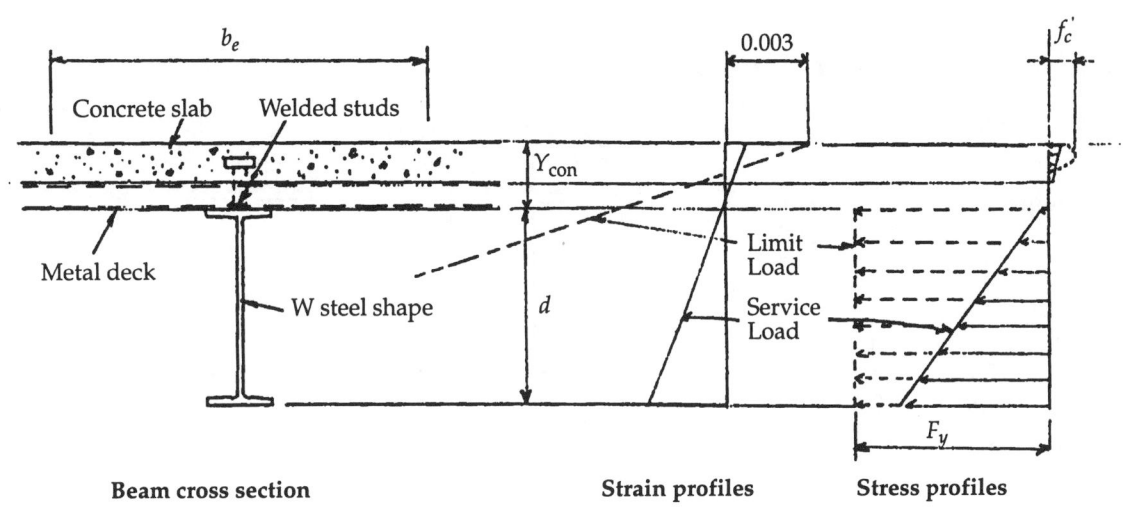

Fig. 8.21 Composite beam cross section in flexure.

The value Y_{con} of Fig. 8.21 is the distance from the top of the steel beam to the top of the concrete slab. Metal decking is attached to steel beams by shear connectors as indicated. Then concrete is cast into the metal deck. After concrete hardens, the metal deck serves as reinforcement for the concrete, and all subsequent loads are resisted by the composite slab and the composite beam cross section. The effective slab width shown as b_e is the smaller of a) the average distance to adjacent beams or b) one-fourth of the span of the beam. At limit bending strength, the maximum tension force on the cross section is simply the tensile capacity of the entire steel shape

$$T_u = A_s F_y \tag{8.7.1}$$

The equilibrating compression force on concrete will be the product of concrete area $b_e a$ times concrete strength $0.85 f_c'$ or

$$C_u = 0.85 f_c' b_e a \tag{8.7.2}$$

Thus,

$$a = A_s F_y / (0.85 f_c' b_e) \tag{8.7.3}$$

and for $a < h_s$,

$$M_n = A_s F_y (0.5d + Y_{con} - 0.5a) \tag{8.7.4}$$

Eq. 8.7.4 is valid only if $a < h_s$, the slab thickness. If a value computed for a were greater than h_s it would signify that the theoretical neutral axis at the plastic limit state is somewhere in the steel part of the cross section, probably in the top flange. The specific location must be determined such that the slab compressive strength plus the compression yield strength of a portion of the beam will equal the tensile yield strength of the remaining portion of the steel beam. A value for M_n then can be computed as the sum of moments of internal forces on the cross section. When $a > h_s$ a safe (underestimated) value for M_n can be made simply by using the relationship

$$M_n \approx 0.85 f_c' b_e h_s (0.5d + Y_{con} - 0.5h_s) \qquad a > h_s \qquad (8.7.5)$$

The connection of the slab to the steel beam must be adequate to transfer between the slab and the steel all of the flexural force Q required for M_n. The force Q, which would shear the slab along the steel flange, is resisted with shear connectors, the most popular of which are headed studs that are welded along the flange. Fig. 8.22 shows a shear stud on which a weld flux at the bottom of the stud is fused to the flange with a welding "gun" made for the specific purpose. Each stud can resist a shear force Q_n limited by the shear strength of the stud shank or by the splitting resistance of the concrete as listed in Table 8.2.

Table 8.2 Stud strength in shear (k per stud)

f_c' (ksi)	Stud diameter (in)		
	0.500	0.625	0.750
Hard rock aggregate concrete			
3.0	9.4	14.6	21.0
4.0	11.5	17.9	25.8
5.0	11.5	17.9	25.8
Lightweight aggregate concrete			
3.0	7.9	12.3	17.7
4.0	9.8	15.2	21.9

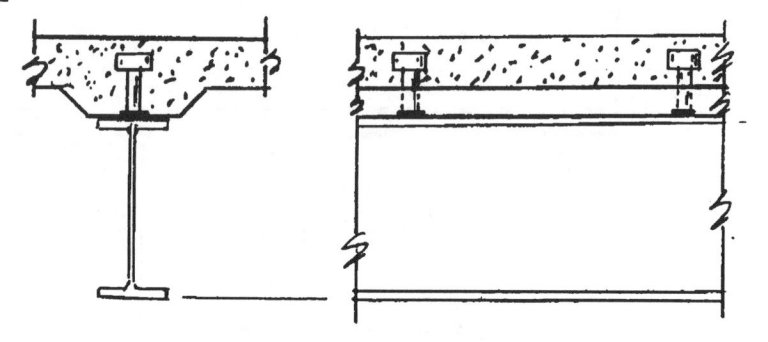

Fig. 8.22 Stud welded shear connector.

The total force to be connected will be the steel capacity $Q = A_s F_y$ for the usual case in which $a < h_s$, but if $a > h_s$, then the capacity Q of concrete is

$$Q = 0.85 f_c' b_e h_s \qquad (8.7.6)$$

The total force Q must be developed along the length of the beam in each direction from the point of maximum moment to the point of zero moment. The studs ordinarily are spaced at equal intetrvals. In each direction from the point of maximum moment, the total number N_s required for full composite action is Q/N_s.

It is desirable that composite beams be constructed without supplementary shoring to help the steel shape support the weight of forms plus concrete during casting of the slab. Consequently, the steel shape may be chosen on the basis of strength to support the construction loading of forms and concrete without composite action. After concrete hardens, composite action resists all subsequent loading. Theoretically, there will be some "locked-in" stresses due to construction forces at the time concrete hardens. Subsequent loads will change the initial "locked-in" stresses. The tendancy of concrete to creep and shrink as it hardens and gains strength will invalidate stress estimates determined simply as a superposition of casting stress on steel alone plus subsequent stress on the composite section. Specific distribution of stress actually is irrelevant, as the plastic stress condition always can develop before flexural failure if there are adequate shear connectors between the concrete and the steel. If without composite action, the steel shape alone cannot support forms and wet concrete, shoring must be used to support the steel shape during casting of the slab.

The force Q and the moment M_n as described above represent forces at which materials reach their strength limit. Research indicates that these limit strength analysis procedures produce accurate estimates of composite beam capacity. A reliability resistance factor $\phi = 0.85$ must be applied for composite beam calculations. Required moments M_u must be less than ϕM_n.

Composite beams can be designed also on the basis of limiting allowable stresses computed under the action of service loads on the cross section, although it is recognized that *actual* stress values differ from those computed. Without temporary shoring, the steel shape must support its own weight plus the weight of the concrete and formwork. After concrete hardens, the composite section resists additional loads. The calculation of stress on the composite section requires first the determination of section modulus values I/c_c and I/c_s for the composite section. In order to determine a moment of inertia I, the effective area of composite concrete must be reduced by the modular ratio $1/n = E_c/E_s$. ASD rules give an expression $E_c = (w^{1.5})\sqrt{f_c'}$ with units of ksi for f_c'. Tension stress due to construction load moment resisted only by the steel shape must be added to tension stress from additional dead load plus all live load. The allowable total stress in steel is $0.90F_y$, and the maximum compression stress in concrete is $0.45f_c'$. The details for computing stress values are shown in Example 8.13.

The steel portion of a composite beam must have a web that is adequate to support all of the shear forces for the composite beam The steel webs of composite beams are connected to supporting columns or girders in the same way that non-composite beams are connected to supports.

Deck reinforced concrete slabs represent a special form of steel-concrete composite member. The steel deck is selected to be stiff and strong enough to support wet concrete during casting, and the deck thereafter serves as reinforcement for the composite slab. The steel decking is formed with folds, dents and protrusions that hold hardened concrete and deck tight enough to provide shear connection for the composite system. Deck manufacturers publish load tables for various configurations of decking and concrete slabs.

EXAMPLE 8.13

Select an A36 steel shape to support a deck-reinforced concrete slab. The 1.5 in. high ribs of the decking support a 3 in. thick slab with $f_c' = 3.0$ ksi. Composite beams are 6 ft center to center, and each beam spans 26 ft between supporting girders. Including an allowance for the self weight of the steel shape, design service loads are 50 psf dead load and 70 psf live load.

Solution.   **LRFD Method**

Select a W shape to support the construction loading, and take the full dead load of 50 psf as the construction load. This is an instance for which the load factor case (A4-1) applies.

Step 1—Required load = 1.4D. The required construction load w_c per ft of beam is

$$w_c = 1.4(50)(6.0) = 420 \text{ plf}$$

for which

$$M_u = w_c L^2/8 = 420(26)^2/8 = 35{,}500 \text{ ft-lb}$$

Required plastic section modulus $Z = M_u/\phi F_y = 35{,}500 \times 12/(0.9 \times 36{,}000) = 13.2$ in^3. Try a W12x14 shape for which $Z_x = 17.4$ and $A_s = 4.16$ in^2.

Step 2—Compute the required total load on the composite section for load condition (A4-2):

$$w_u = 1.2D + 1.6L = [1.2(50) + 1.6(70)]6.0 = 1032 \text{ plf}$$

for which

$$M_u = w_u L^2 / 8 = 1032(26)^2 / 8 = 87{,}200 \text{ ft-lb} = 87.2 \text{ k-ft}$$

and

$$V_u = w_u L/2 = 1032(26)/2 = 13{,}400 \text{ lb} = 13.4 \text{ k}$$

Since $Z = 17.4$ in^3, which is greater than required plastic section modulus = 13.2 in^3, this section can support the constrction load. The deck plates provide lateral support for the top flange, as they will be attached to the W12x14 by shear studs before the wet concrete is placed.

Step 3—Compute ϕM_n for the composite section. The effective width b_e of concrete slab will be the smaller of L of the span or the spacing of beams. Beams are 6 ft apart and $L/4 = 26/4 = 6.5$ ft. The effective width $b_e = 72$ in.

Compute

$$a = A_s F_y / [0.85(f_c') b_e] = 4.16(36) / [0.85(3.00)72] = 0.82 \text{ in}$$

This value is less than the slab thickness h_s, therefore it is valid. Then

$$M_n = A_s F_y (Y_n + d/2 - a/2) = 4.16(36)(4.50 + 11.91/2 - 0.82/2) = 1504 \text{ k-in}$$

$$\phi M_n = 0.85(1504)/12 = 106 \text{ k-ft}.$$

This value is greater than the required value $M_u = 87.2$ k-ft, and the section may be used.

Step 4—Check shear strength by computing

$$\phi V_n = \phi d t_w (0.6 F_y) = 0.85(11.91)0.20(0.6)36 = 46.5 \text{ k}$$

This value is far in excess of the required value $V_u = 13.2$ k

Step 5—Determine shear studs required between the concrete slab and the steel beam.

Compute

$$Q = A_s F_y = 4.17(36) = 150 \text{ k}.$$

Use 5/8-in studs 3.5 in high. In Table 8.2, the shear strength is given as 14.6 k/stud. The number required $N = 150/14.6 = 11$ studs each side of the centerline of the W12x14 beam. The 22 studs should be spaced from each end at about 12 in centers with only one stud per trough of the steel decking. Near midspan the spacing may be increased to 18 in.

Solution.   **ASD Method**

Step 1—Compute required construction load moment

$$M_{constr} = w_{constr} L^2 / 8$$

$$= 50(26)^2 / 8 = 25{,}400 \text{ ft-lb} = 25.4 \text{ k-ft}$$

Select a shape for which construction load stress $\approx 0.5 F_y = 0.5(36) = 18$ k/in^2. Approximate section modulus $S = M/F_s = 25.4(12)/18 = 16.9$ in^3. Try W12x16 for which $S = 17.1$ in^3, $A_s = 4.71$ in^2, and $I_s = 103$ in^4.

Step 2—Compute required service load moment for load in addition to construction loading:

$$M_L = w_L L^2 / 8 = 6(70)26^2 / 8 = 35{,}500 \text{ ft-lb} = 35.5 \text{ k-ft}$$

Step 3—Compute

$$E_c = (w^{1.5})\sqrt{f_c'} = (145^{1.5})\sqrt{3.0} = 3020 \text{ k/in}^2$$

Modular ratio $n = E_s/E_c = 29000/3020 = 9.6$

Effective width of slab = beam spacing or $L/4$
= 72 in

Transformed width of slab $b_{tr} = b_e/n = 72/9.6 = 7.50$ in.

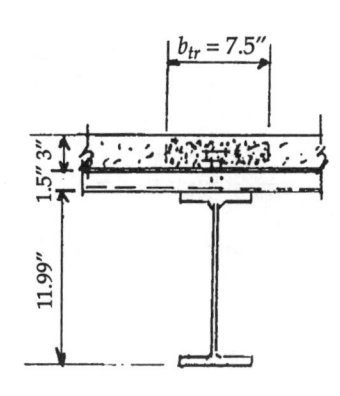

Step 4—Compute moment of inertia I and section modulii.

Locate neutral axis at x from the top of the slab by using sum of moments of areas for $x < h_s$:

$$7.5x(0.5x) = 4.71(4.5 + 11.99/2 - x) \quad \therefore \quad x = 3.06 \text{ in}$$

Since $x = 3.06 > h_s = 3.0$ in, use equation for $x > h_s$:

$$7.5(3)(x - 1.5) = 4.71(4.5 + 11.99/2 - x)$$

for which $x = 3.06$ in. as before.

Then

$$I = b_{tr}(h_s)^3/12 + b_{tr}(h_s)x - h_s/2)^2 + I_s + A_s(4.5 + 11.99/2 - x)^2$$
$$= 7.5(3.0)^3/12 + 7.5(3.0)(3.06 - 1.5)^2 + 103 + 4.71(10.495 - 3.06)^2 = 435 \text{ in}^4$$

For bottom fiber, $S_{bot} = 435/(4.5 + 11.99 - 3.06) = 35.0 \text{ in}^3$. For top fiber of concrete, $S_{top} = 435/3.06 = 142 \text{ in}^3$

Step 5—Check maximum tension stress at bottom:

$$f_{max} = M_{constr}/S + M_L/S_{bot}$$
$$= 25.4(12)/17.1 + 35.5(12)/35.0 = 30.0 \text{ k/in}^2$$

Allowable $f_s = 0.90F_y = 0.9(36) = 32.4 \text{ k/in}^2$, so steel section is OK.

Check maximum concrete stress

$$f_{max} = [M_L/S_{top}](1/n) = [35.5(12)/142](1/9.6) = 0.31 \text{ k/in}^2$$

Allowable concrete stress = $0.45f_c' = 0.45(3.0) = 1.35 \text{ k/in}^2$. The composite section is satisfactory.

Step 6—Determine shear connectors required along the top of the W12x16.

Required force

$$Q = A_s F_y = 4.71(36) = 170 \text{ k}$$

Use 5/8 in studs for which the strength in Table 8.2 is given as 14.6 k/stud. Required number $N = 170/14.6 = 12$ studs each side of midspan. Locate studs at about 12 in. spacing, one stud per trough in deck plate, from each end toward center, increasing spacing to 18 in. near midspan.

Note: Steel beams should be cambered with an unloaded upward displacement equal to the downward midspan deflection of the steel beams under the construction loading. If beams are not cambered, the thickness and weight of concrete will increase by the amount of the deflection as concrete is finished to a level surface.

Practice Problems

8.1 Design fillet welds necessary to connect a $6 \times \frac{5}{8}$ in A36 strap to the back of a 612×25. The weld must be strong enough to develop the full tensile strength of the strap.

8.2 Determine the welds required at the end of a pair of A36 steel angles $3 \times 2\frac{1}{2} \times \frac{5}{16}$ connected to the web of an ST9×18 chord for a truss for a service dead load = 32 k and live load = 58 k. Use E70 electrode and 1/4-in fillet welds.

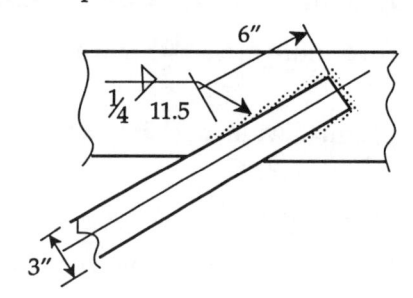

8.3 Compare the strength of the tension hanger connection shown if A) 3/4 in A307 bolts are used and B) 3/4 in A325X bolts are used.

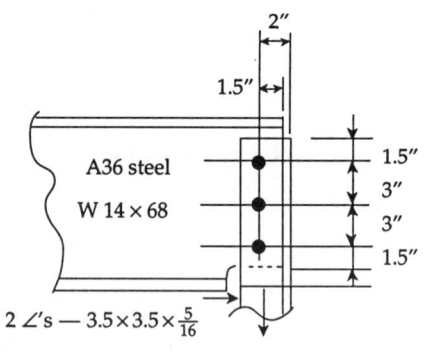

8.4 Determine the capacity of the tension hanger connection shown. Strength is governed by bolt tension limit.

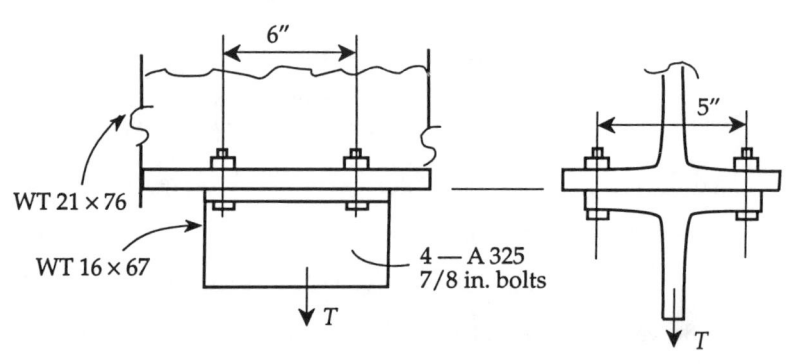

8.5 What load P can be carried by the 6 A325X 3/4-in diam bolts for the connection shown?

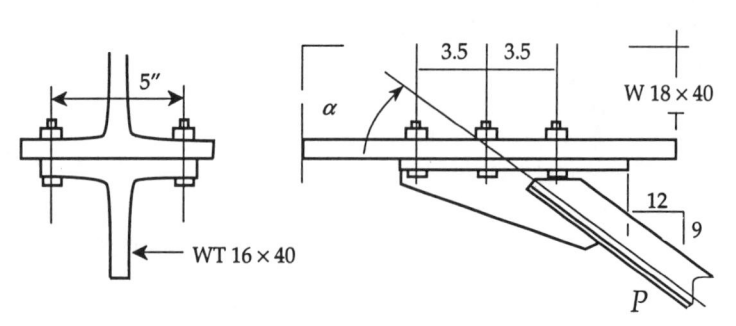

8.6 Using the same connection shown in problem #5, what load P can be allowed by 6 A325X bolts if this were a slip-critical connection?

8.7 Find the maximum force on any of the bolts in the connection.

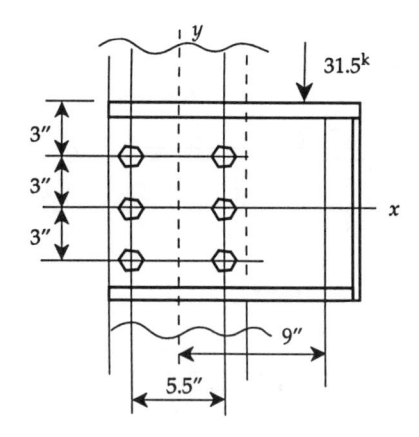

8.8 Find the maximum force on any of the bolts in the connection.

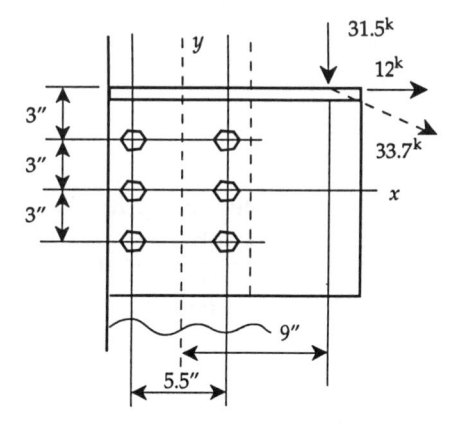

8.9 Find the maximum force per inch of weld for this connection.

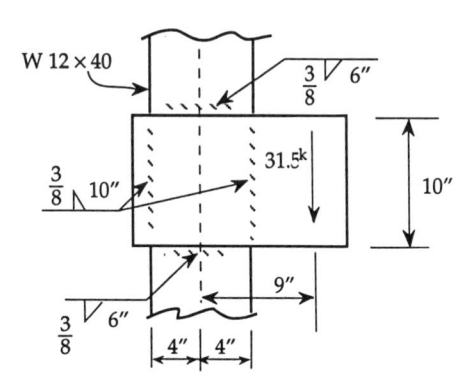

8.10 Find the maximum force per inch of weld for this connection.

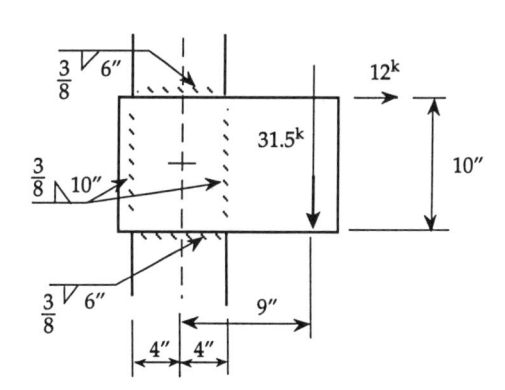

8.11 Design an A36 bearing plate for a reaction of $P_D = 28$ k and $P_L = 40$ k for a W21×50 at an 8 in concrete block wall for which allowable bearing pressure is 600 psi.

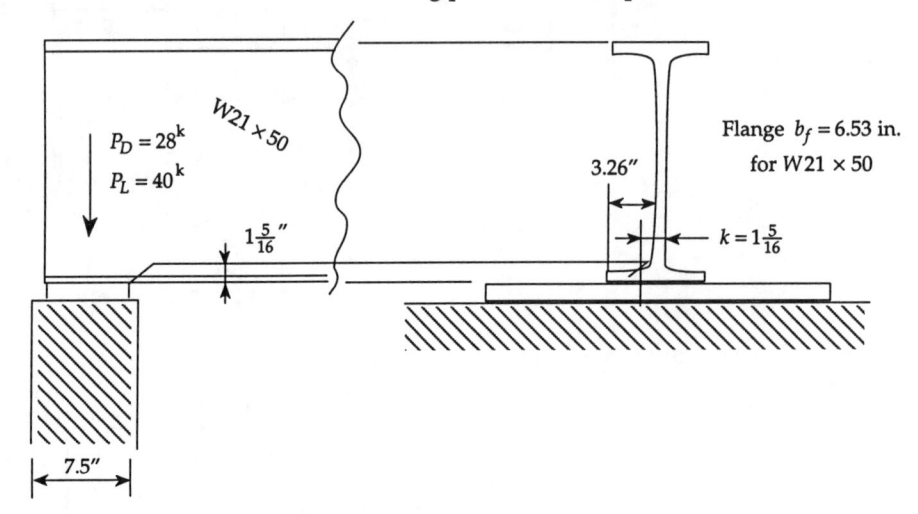

Required bearing plate can be a maximum width of 7.0 in.

Required area = $\dfrac{(28+40)1000}{600} = 113$ in². Required length = $\dfrac{113}{7} = 17$ in. Choose a plate 7 in×17 in. Determine required thickness.

8.12 Design an A36 base plate for a W12×40 column reaction of 72 k DL and 66 k LL. The column is bearing on a large footing of concrete with $f_c' = 3000$ psi. Determine base plate area

$= \dfrac{P}{0.7 f_c'} = \dfrac{(72+66)}{2.1} = 66$ in². In order to cover the W12×40 profile, use a plate 10×14=140 in².

Determine required thickness.

8.13 Design a base plate for a W10×77 column that bears against the top of a 30-in diameter drilled concrete pier with $f_c' = 3$ ksi. Loads on the column are $P_D = 227$ k, $P_L = 185$ k, and $M_D = M_L = 0$. Use A36 steel plate.

Practice Problems
Multiple Choice (PE Format)

The schematic represents a concentrically braced simple frame. Webs of the wideflange beam (AB) and columns (AC and BD) are in the plane of the structure. The slender braces (AD and BC) are ineffective in compression; they buckle elastically at very low loads. All steel is A572 Grade 50. Bolted connections are shear-bearing, with threads of the bolts included in shear planes. Bolts are 7/8 inch diameter, Grade A325, in standard holes. For welded connections, the shielded-metal arc welding process is used, with E70 electrodes.

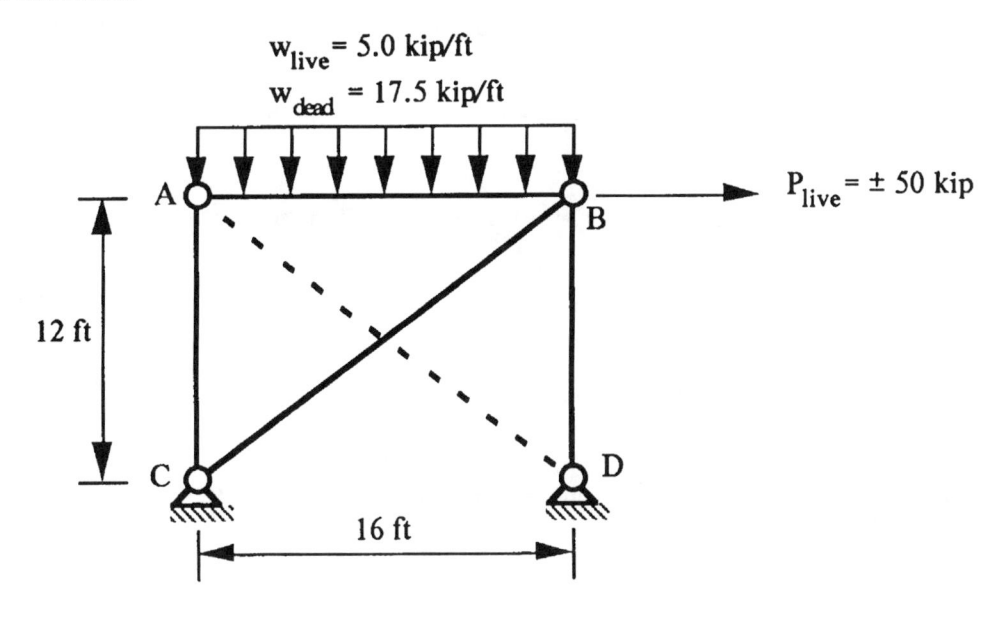

8.14 Select the lightest safe equal leg double angle for brace BC. Connections are fillet welded and the shear lag reduction coefficient $U = 0.85$. Neglect second order effects.

LRFD
(a) 2L2×2×5/16 (b) 2L2×2×3/8 (c) 2L2.5×2.5×3/16 (d) 2L3×3×3/16 (e) 2L3×3×1/4

ASD
(a) 2L2×2×3/8 (b) 2L2×2×5/16 (c) 2L2.5×2.5×3/16 (d) 2L3×3×1/4 (e) 2L3×3×3/16

8.15 Select the lightest safe wideflange shape for column BD, which is braced at top and bottom. Neglect second order effects.

LRFD
(a) W10×33 (b) W8×35 (c) W8×40 (d) W10×39 (e) W12×40

ASD
(a) W8×35 (b) W10×39 (c) W8×31 (d) W10×45 (e) W12×40

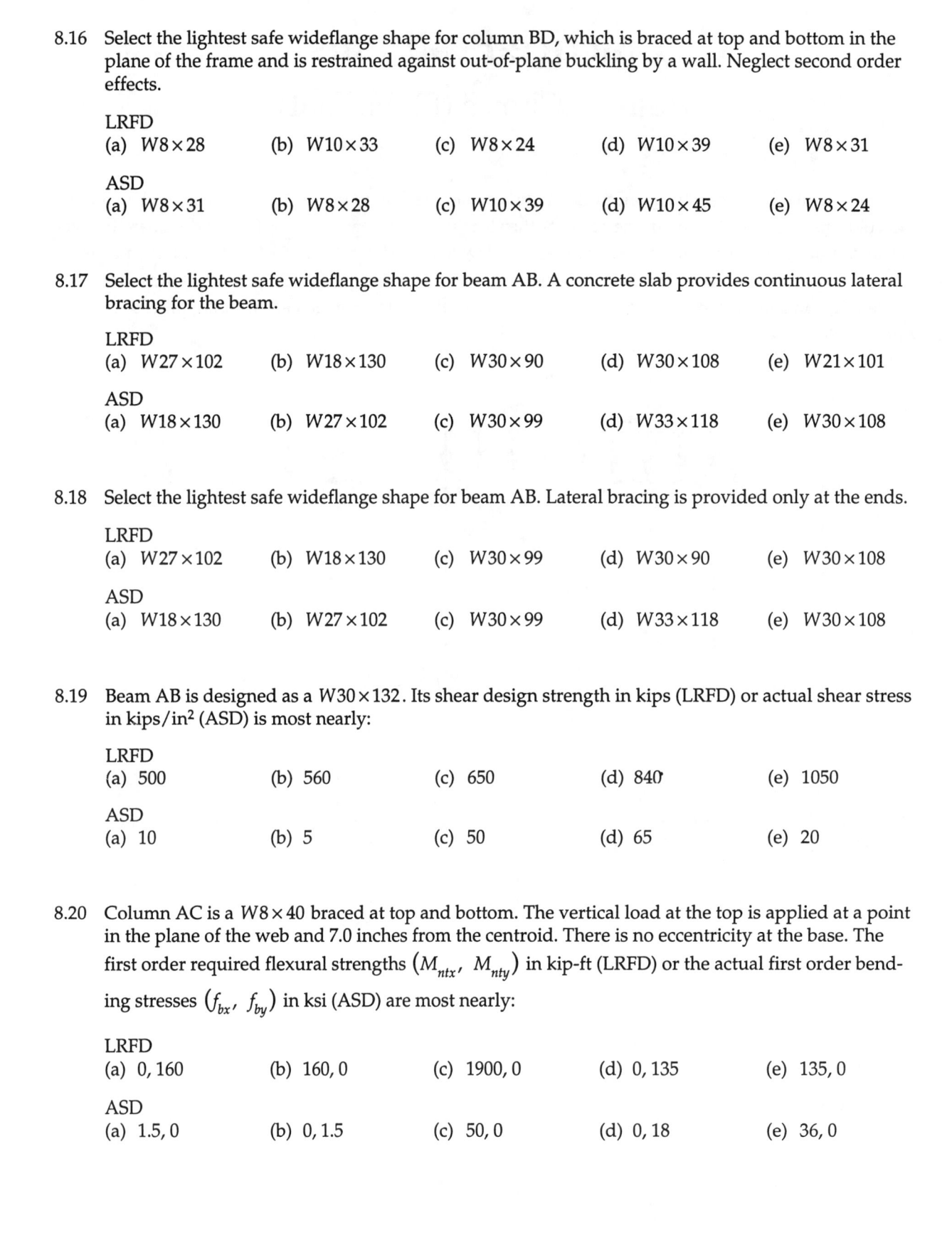

8.16 Select the lightest safe wideflange shape for column BD, which is braced at top and bottom in the plane of the frame and is restrained against out-of-plane buckling by a wall. Neglect second order effects.

LRFD
(a) W8×28   (b) W10×33   (c) W8×24   (d) W10×39   (e) W8×31

ASD
(a) W8×31   (b) W8×28   (c) W10×39   (d) W10×45   (e) W8×24

8.17 Select the lightest safe wideflange shape for beam AB. A concrete slab provides continuous lateral bracing for the beam.

LRFD
(a) W27×102   (b) W18×130   (c) W30×90   (d) W30×108   (e) W21×101

ASD
(a) W18×130   (b) W27×102   (c) W30×99   (d) W33×118   (e) W30×108

8.18 Select the lightest safe wideflange shape for beam AB. Lateral bracing is provided only at the ends.

LRFD
(a) W27×102   (b) W18×130   (c) W30×99   (d) W30×90   (e) W30×108

ASD
(a) W18×130   (b) W27×102   (c) W30×99   (d) W33×118   (e) W30×108

8.19 Beam AB is designed as a W30×132. Its shear design strength in kips (LRFD) or actual shear stress in kips/in² (ASD) is most nearly:

LRFD
(a) 500   (b) 560   (c) 650   (d) 840   (e) 1050

ASD
(a) 10   (b) 5   (c) 50   (d) 65   (e) 20

8.20 Column AC is a W8×40 braced at top and bottom. The vertical load at the top is applied at a point in the plane of the web and 7.0 inches from the centroid. There is no eccentricity at the base. The first order required flexural strengths (M_{ntx}, M_{nty}) in kip-ft (LRFD) or the actual first order bending stresses (f_{bx}, f_{by}) in ksi (ASD) are most nearly:

LRFD
(a) 0, 160   (b) 160, 0   (c) 1900, 0   (d) 0, 135   (e) 135, 0

ASD
(a) 1.5, 0   (b) 0, 1.5   (c) 50, 0   (d) 0, 18   (e) 36, 0

8.21 Beam AB is a $W30 \times 132$ and column AC is a $W8 \times 40$. At A the web of the beam is connected to the flange of the column by an all-bolted double angle shear connection. angles are $6 \times 4 \times 3/8$. Spacing is 3 inches and end distances are 1.5 inches. How many bolts must be used in the connection?

LRFD
(a) 9   (b) 15   (c) 18   (d) 24   (e) 33

ASD
(a) 12   (b) 18   (c) 24   (d) 30   (e) 36

8.22 Assume that the clip angles described in the preceding problem have seven rows of bolts. The shear rupture design strength in kips (LRFD) or the actual shear stress on net area in ksi (ASD) is most nearly:

LRFD
(a) 60   (b) 470   (c) 280   (d) 300   (e) 510

ASD
(a) 25   (b) 11   (c) 20   (d) 17   (e) 8

8.23 Brace BC is a double $L3 \times 3 \times 1/4$. Each end of the brace is connected to a 5/8-inch thick gusset plate by longitudinal fillet welds. The length in inches of 3/16-inch fillet weld needed to connect one end of the double angle is most nearly:

LRFD
(a) 12   (b) 17   (c) 24   (d) 30   (e) 48

ASD
(a) 13   (b) 18   (c) 23   (d) 30   (e) 45

Solutions to Problems

8.1 **LRFD.** Step 1—Compute the tensile strength of the strap.

$$T_u = \phi A_g F_y = 0.90\left(6 \times \tfrac{5}{8}\right)36 = 122 \text{ k}$$

Step 2—Select weld size and electrode. LRFD Table J2.5 requires at least 1/4 in weld size for 5/8 in thick plate. Try 5/16 in fillet weld. Since F_u for A36 is 58 ksi, use electrode E60.

Step 3—Compute weld value using Eq. 8.6.1a.

$$W_u = 0.325 \times (5/16) \times 60 = 6.1 \text{ k/in}$$

Step 4—Compute required length of weld = $T_u/W_u = 122/6.1 = 20$ in.

Step 5—Sketch connection:

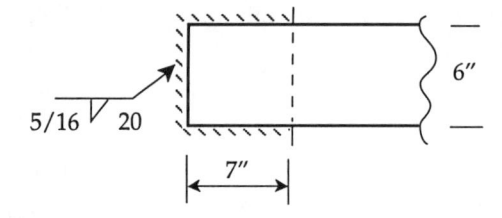

ASD. Step 1—Compute allowable force $T = 0.6 F_y A_g$. $T = 0.6 \times 36 \times 6 \times \tfrac{5}{8} = 81 \text{ k}$.

Step 2—Select weld size and electrode quality. ASD Table J2.4, pg 5-67 requires electrode at least 1/4 in size for plate thickness from 1/2 in to 3/4 in. Select 5/16 in fillet weld. Since F_u for A36 steel is 58 ksi; use E60 electrode.

Step 3—Use Eq. 8.6.1b to compute a weld value $W = 0.18 D_w E_{xx} = 0.18(5/16)60 = 3.37$ k/in.

Step 4—Compute required length of weld = $T/W = 81/3.37 = 24$ in.

Step 5—Sketch connection:

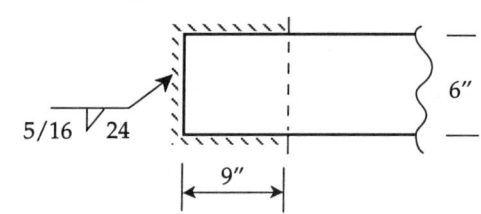

8.2 **LRFD.** Step 1—Compute required $P_u = 1.2D + 1.6L = 1.2(32) + 1.6(58) = 131 \text{ k}$.

Step 2—Use Eq. 8.6.1a to determine weld value $W_u = 0.325 D_w E_{xx} = 0.325(1/4)70 = 5.7$ k/in.

Step 3—Compute length of weld $= \dfrac{P_u}{W_u} = \dfrac{131}{5.7} = 23$ in. Use $11\tfrac{1}{2}$ in of weld on each angle, $2\tfrac{1}{2}$ in at end, 6 in along heel, and 3 in along toe.

ASD. Step 1—Compute required service load = $D + L = 32 + 58 = 90$ k.

Step 2—Use Eq. 8.6.1b to determine weld value $W = 0.18 D_w E_{xx} = 0.18(1/4)70 = 3.15$ k/in.

Step 3—Compute length of weld = $\dfrac{P}{W} = \dfrac{90}{3.15} = 29$ in. Use $14\tfrac{1}{2}$ in of weld on each angle, $2\tfrac{1}{2}$ in at end, 8 in along heel, and 3 in along toe.

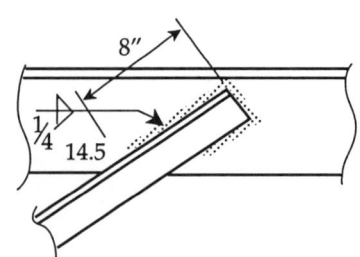

Structural Steel

8.3 **LRFD.** Step 1—Bolts are in double shear. Use Table 1-D, pg 5-5: A307—$T_u = 14.3 \times 3 = 42.9$ k and A325—$T_u = 41.4 \times 3 = 124.2$ k.

Step 2—Determine bolt bearing strength (pg 1-30) for W14×68, $t_w = 0.415$ in. Use Table 1-E (pg 5-7) for 1 in thick $R_b = 91.3$ k. For $t_w = 0.415(91.3)3 = 113.7$ k for both bolts. The thickness of the 2 angles exceeds that of the beam web, so bearing is limited by beam web.

Step 3—Check block shear (independent of bolt quality). Dimensions of shear block are identical for angles and for beam web. The thinner web will restrain the smaller force.
Tension yield and shear fracture:

$$R_{BS} = (1.50)0.415(36) + (7.50 - 2.5 \times 7/8)0.415(0.6)58 = 99 \text{ k}$$

Tension fracture and shear yield:

$$R_{BS} = \left(1.50 - \frac{1}{2} \times \frac{7}{8}\right)0.415(58) + 7.50(0.415)(0.6)36 = 93 \text{ k}$$

The <u>larger</u> value governs.

Step 4—Report the least value as connection strength.
   A307 Bolts $T_u = 42.9$ k, governed by bolt shear.
   A325X Bolts $T_u = 99.0$ k, governed by block shear.

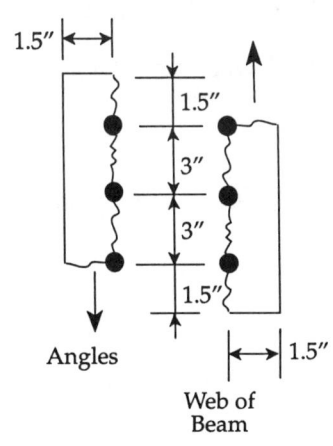

8.4 **LRFD.** Step 1—Use Table 1-A, pg 5-3. For A325 bolt 7/8 in diameter, $T_u = 40.6$ k/bolt. For 4 bolts:

$$T_u = 4(40.6) = 163 \text{ k}$$

ASD. Step 1—Use Table 1-A, pg 4-3. For A325 bolt 7/8 in diameter, $T = 26.5$ k/bolt. For 4 bolts:

$$\text{Allowable } T = 4(26.5) = 103 \text{ k}$$

8.5 **LRFD.** Step 1—Use equations of Table J3-3, pg 6-68. Compute:

$$f_v = \frac{P_u \cos\alpha}{A_b} = \frac{0.8 P_u}{6(0.44)} = 0.303 P_u$$

$$f_t = \frac{P_u \sin\alpha}{A_b} = \frac{0.6 P_u}{6(0.44)} = 0.227 P_u$$

Solve governing equation for P_u:

$$f_t = 85 - 1.4 f_v \leq 68 \text{ ksi}$$

$$0.227 P_u = 85 - 1.4(0.303 P_u)$$

$$P_u = 130 \text{ k}$$

Step 2—Check $f_t = 0.227 P_u = 0.227(130) = 30$ ksi < 68 ksi. ∴ OK.

ASD. Step 1—Use equations of Table J3-3, pg 5-74.

$$f_v = \frac{P \cos\alpha}{A_b} = \frac{0.8P}{6(0.44)} = 0.303 P$$

$$f_t = \frac{P \sin\alpha}{A_b} = \frac{0.6P}{6(0.44)} = 0.227 P$$

Solve governing equation for P:

$$f_t = \sqrt{44^2 - 2.15 f_v^2}$$

$$(0.303P)^2 = 44^2 - 2.15(0.227P)^2$$

$$P = 97.8 \text{ k}$$

8.6 **LRFD.** Step 1—Use Clause J3.5, pg 6-68. Permissible shear = $R_v\left(1 - \dfrac{T}{T_b}\right)$. Obtain $R_v = 17 A_b$ from Table J3.4, pg 6-69. Obtain $T_b = 39$ k/bolt from Table J3.1, pg 6-66. Then,

$$T \cos\alpha = 17 A_b \left(1 - \dfrac{T \sin\alpha}{39 n}\right)$$

where n = number of bolts. Solve for T:

$$T = 17(6 \times 0.44)\left(1 - \dfrac{T \times 0.6}{39 \times 6}\right) / 0.8$$

$$= 49 \text{ k}$$

ASD. Step 1—Use Clause J3-6, pg 5-74. From Table J3.2, $f_v = 17$ ksi, and from Table J3.7, $T_b = 28$ k/bolt.

$$\dfrac{1}{A_b} T \cos\alpha = f_v\left(1 - f_t A_b / T_b\right)$$

$$\dfrac{T(0.8)}{0.44(6)} = 17\left(1 - \dfrac{T(0.6)}{A_b} A_b \times \dfrac{1}{6 \times 28}\right)$$

$$T = 47 \text{ k}$$

8.7 Step 1—Determine properties of bolt group:

$$n = 6 \text{ bolts} = A$$

$$I_x = \sum A y^2$$

$$= 4(4.5 \text{ in})^2 = 81 \text{ bolt} \cdot \text{in}^2$$

$$I_y = \sum A x^2$$

$$= 6(2.75 \text{ in})^2 = 45.4 \text{ bolt} \cdot \text{in}^2$$

Step 2—Determine vertical and horizontal forces on each bolt, using

$$F_{\text{vert}} = \dfrac{P}{n} \pm \dfrac{P e_x}{I_y}$$

$$F_{\text{horiz}} = \pm \dfrac{P e_y}{I_x}$$

Bolts closest to line of force and farthest from centroid of bolts will always resist highest force. Bolts at upper right and lower right corners have maximum force. For bolt on upper right:

$$R_{vert} = \frac{31.5}{6} + \frac{31.5(9)}{45.4} \times 2.75 = 22.4 \text{ k} \downarrow$$

$$R_{horiz} = + \frac{31.5(9) \times 3.0}{81} = 10.5 \text{ k} \rightarrow$$

$$\text{Resultant } R = \sqrt{22.4^2 + 10.5^2} = 24.7 \text{ k}$$

8.8 Step 1—Determine properties of bolt group. (Exactly the same as Step 1 in problem #7.)

$$A = 6 \text{ bolts}$$

$$I_x = 81 \text{ bolt} \cdot \text{in}^2$$

$$I_y = 45.4 \text{ bolt} \cdot \text{in}^2$$

Step 2—Determine vertical and horizontal forces on bolt at upper right corner (closest to line of applied force).

$$R_V = \frac{P_V}{A} + \left(P_V e_x + P_H e_y\right)\frac{x}{I_y}$$

$$= \frac{31.5}{6} + (31.5 \times 9 + 12 \times 6)\frac{2.75}{45.4} = 26.8 \text{ k} \downarrow$$

$$R_H = \frac{P_H}{A} + \left(P_V e_x + P_H e_y\right)\frac{y}{I_x}$$

$$= \frac{12}{6} + (31.5 \times 9 + 12 \times 6)\frac{3}{81.0} = 15.2 \text{ k} \rightarrow$$

$$\text{Resultant } R = \sqrt{26.8^2 + 15.2^2} = 30.8 \text{ k}$$

8.9 Step 1—Determine properties of the welds.

$$A = \text{length of weld} = 2(6 + 10) = 32 \text{ in}$$

$$I_x = \sum Ay^2 = 2(6)(5)^2 + \frac{2(1)(10)^3}{12} = 467 \text{ weld} \cdot \text{in}^3$$

$$I_y = \sum Ax^2 = 2(10)(4)^2 + \frac{2(1)(6)^3}{12} = 356 \text{ weld} \cdot \text{in}^3$$

Step 2—The portion of weld nearest the line of the force and farthest from the weld centroid will be the most highly stressed. At the top right corner,

$$F_V = \frac{P}{A} + \frac{P(e)x}{I_x}$$

$$= \frac{31.5}{32} + \frac{(31.5 \times 9 \times 4)}{467} = 3.40 \text{ k/in} \downarrow$$

$$F_H = \frac{P}{A} + \frac{P(e)y}{I_y}$$

$$= \frac{0}{32} + \frac{(31.5 \times 9 \times 5)}{356} = 3.98 \text{ k/in} \rightarrow$$

$$\text{Resultant } F = \sqrt{3.40^2 + 3.98^2} = 5.24 \text{ k/in}$$

8.10 Step 1—Determine properties of the welds. The same welds as problem #9:

$$A = \text{length of weld} = 32 \text{ in}$$

$$I_x = 467 \text{ weld} \cdot \text{in}^3$$

$$I_y = 356 \text{ weld} \cdot \text{in}^3$$

Step 2—The top of the 10-in vertical weld is nearest the line of the load and farthest from the weld centroid.

$$F_V = \frac{P_V}{A} + (P_V e_x + P_H e_y)\frac{y}{I_x}$$

$$= \frac{31.5}{32} + (31.5 \times 9 + 12 \times 5)\frac{5}{467} = 4.66 \text{ k/in} \downarrow$$

$$F_H = \frac{P_H}{A} + (P_V e_x + P_H e_y)\frac{x}{I_y}$$

$$= \frac{12}{32} + (31.5 \times 9 + 12 \times 5)\frac{4}{356} = 4.23 \text{ k/in} \rightarrow$$

$$\text{Resultant } F = \sqrt{4.66^2 + 4.23^2} = 6.29 \text{ k/in}$$

8.11 **LRFD.** Step 1—Compute ultimate reaction $P_u = 1.2P_D + 1.6P_L = 1.2(28) + 1.6(40) = 98$ k.
Ultimate bearing pressure $= \frac{98}{7 \times 17} = 0.82$ psi. At toe of flange,

$$M_u = 7(0.82)\left(\frac{17}{2} - 1.31\right)^2 \frac{1}{2} = 148 \text{ in} \cdot \text{k}$$

Step 2—Determine plate thickness. Make plate thickness t large enough that $\phi M_n > 148$ in·k:

$$\phi M_n = \phi F_y \frac{bt^2}{6}$$

$$148 \leq 0.9(36)\frac{(7)t^2}{6}$$

$$\therefore t \geq 1.98$$

Make $t = 2$ in. Choose a bearing plate $7 \text{ in} \times 2 \text{ in} \times 1 \text{ ft-5 in}$.

ASD. Step 1—Determine plate thickness such that maximum bending stress is less than $0.75F_y$.
Compute bearing stress $= \frac{P}{A_R} = \frac{28+40}{7 \times 17} = 0.57$ ksi.

$$\text{Max plate } M \text{ at } k \text{ from } \mathcal{C} = 0.57(7)\left(\frac{17}{2} - 1.31\right)^2 \frac{1}{2} = 103 \text{ in} \cdot \text{k}.$$

$$0.75F_y > \frac{6M}{bt^2}$$

$$t > \sqrt{\frac{6 \times 103}{7(0.75)36}} = 1.80 \text{ in}$$

Make $t = 1\frac{7}{8}$. Bearing plate $7 \times 1\frac{7}{8} \times 1$ ft - 5 in.

8.12 **LRFD.** Refer to Fig. 8.20(b) for plate sized to cover the column section. With plate 10 in × 14 in.
Step 1—Compute limit bearing pressure:

$$F_p = 1.7f'_c = 1.7(30) = 5.1 \text{ ksi}$$

Step 2—Compute

$$P_u = 1.2P_D + 1.6P_L$$
$$= 1.2(72) + 1.6(66) = 192 \text{ k}$$

Step 3—Solve for L, Fig. 8.20(b), for 10 in × 14 in plate:

$$A_{PL} = 2BL + 2L(H - 2L)$$

$$\frac{192}{5.1} = 2(10)L + 2L(14 - 2L)$$

$$4L^2 - 48L = 37.6$$

$$L^2 - 12L + 6^2 = -9.4 + 36$$

$$L = 6 - \sqrt{26.6} = 0.84 \text{ in}$$

Step 4—Solve for $t_p = \sqrt{\dfrac{3F_p L^2}{\phi F_y}} = \sqrt{\dfrac{3(5.1)0.84^2}{0.9(36)}} = 0.63$ in. Base plate $10 \times \dfrac{3}{4} \times 1$ ft - 2 in.

ASD. Refer to Fig. 8.20(b) for a plate sized to cover the column section with 10 in × 14 in plate.
Step 1—Compute allowable bearing pressure $F_p = 0.7f'_c = 2.1$ ksi.

$$A_{PL} = \frac{P + L}{F_p} = \frac{(72 + 66)}{2.1} = 66 \text{ in}^2$$

Step 2—Solve for L, Fig. 8.20(b).

$$A_{PL} = 2BL + 2L(H - 2L)$$

$$66 = 2(10)L + 2L(14 - 2L)$$

$$66 = 48L - 4L^2$$

$$L^2 - 12L + 6^2 = -16.5 + 36$$

$$L = 6 - \sqrt{19.5} = 1.58 \text{ in}$$

Step 3—Solve for $t_p = \sqrt{\dfrac{3F_p L^2}{0.66F_y}} = \sqrt{\dfrac{3(3.1)1.58^2}{24}} = 0.98$ in. Base plate $10 \times 1 \times 1$ ft - 2 in.

8.13 **LRFD.** Step 1—Compute required force $P_u = 1.2D + 1.6L = 1.2(227) + 1.6(185) = 568$ k.

Step 2—Determine required plate size, A_{PL}.

$$A_c = \text{Area top of pier} = \frac{\pi}{4}(30)^2 = 707 \text{ in}^2$$

Assume

$F_p = 1.7 f_c'$ and find $A_{PL} = \dfrac{P_u}{\phi F_p} = \dfrac{568}{0.6(1.7 \times 3)} = 186 \text{ in}^2$. Try 13×15, $A_{PL} = 195 \text{ in}^2$. Use Eq. 8.6.9a to check A_{PL}:

$$\phi F_p A_{PL} = \phi A_{PL}(0.85 f_c')\sqrt{\frac{A_c}{A_{PL}}}$$

$$= 0.6(195)(0.85 \times 3)\sqrt{\frac{707}{195}} = 568 \text{ k}$$

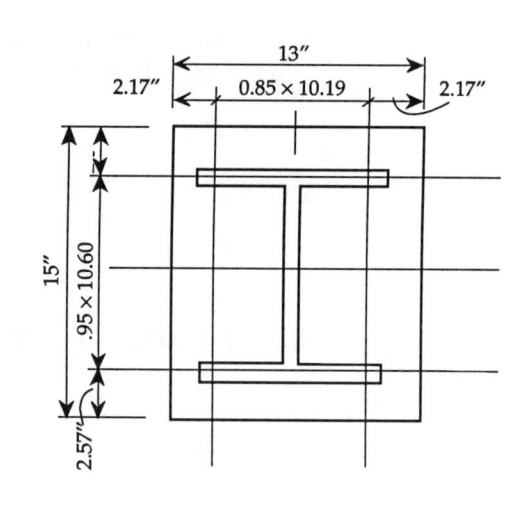

This is exactly enough.

Step 3—Use section dimensions to sketch base plate and to find longer distance = 2.57″:

$$f_p = \frac{P_u}{A_{PL}} = \frac{568}{195} = 2.91 \text{ ksi}$$

Step 4—Determine thickness such that plate will not yield at cantilever:

$$\phi F_y > 3 f_p m^2 / t_p^2$$

$$0.9(36) > 3(2.91)(2.57)^2 / t_p^2$$

$$t_p > 0.77$$

Use plate 13 in $\times \dfrac{7}{8}$ in $\times$ 1 ft - 3 in.

ASD. Step 1—Determine A_{PL}. Compute $A_c = \dfrac{\pi}{4}(30)^2 = 707 \text{ in}^2$. Assume allowable $f_p = 0.7 f_c'$ and solve for A_{PL}:

$$0.7 f_c' A_{PL} = P$$

$$A_{PL} = \frac{(227 + 185)}{0.7(3)} = 197 \text{ in}^2$$

Check Eq. 8.6.9b,

$$A_{PL}\left[0.35 f_c'\sqrt{\frac{A_c}{A_{PL}}}\right] \geq P$$

$$224\left[0.35(3)\sqrt{\frac{707}{224}}\right] = 418 \text{ k} > (227 + 185)$$

So, this is acceptable.

Step 2—Use section dimensions to establish longer cantilever distance = 3.07 in.

Step 3—Determine t_p such that plate will not be overstressed at 3.07 in cantilever:

$$0.66F_y > 3f_p m^2 / t_p^2$$

$$27 > 3\left(\frac{412}{224}\right)3.07^2 / t_p^2$$

$$t_p = 1.39 \text{ in}$$

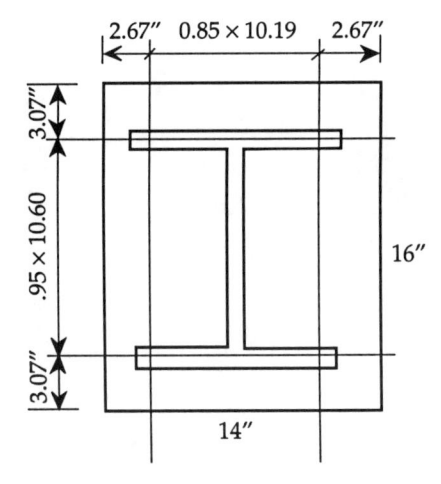

Use plate $14 \text{ in} \times 1\frac{1}{2} \text{ in} \times 1 \text{ ft - 4 in}$.

Solutions to Problems
Multiple Choice (PE Format)

8.14 (b) LRFD

$P_{live} = (20/16)50 = 62.5 \text{ kip}$

$P_u = 1.2 P_{dead} + 1.6 P_{live} = 1.6 \times 62.5 = 100 \text{ kip}$

$\phi F_u A_e \geq P_u;\ A_e = U A_n$

$A_g = A_n \geq P_u / (\phi F_u U) = 100/(0.75 \times 65 \times 0.85) = 2.41 \text{ in}^2$

Try $2L2 \times 2 \times 3/8\,(A_g = 2.72)$; verify that $\phi F_y A_g \geq P_u$

ASD

$P = (20/16)50 = 62.5 \text{ kip}$

$f_t \leq 0.5 F_u;\ f_t = P/(U A_n)$

$A_{g\ req'd} = A_{n\ req'd} = P/(0.5 F_u U)$

$\qquad = 62.5/(0.5 \times 65 \times 0.85) = 2.26 \text{ in}^2$

Try $2L2 \times 2 \times 5/16\,(A_g = 2.30)$; verify that $f_t = P/A_g \leq 0.6 F_y$

8.15 (b) LRFD

$P_{dead} = w_{dead} L/2 = 17.5 \times 16/2 = 140 \text{ kip}$

$P_{live} = (w_{live} L/2) + (12/16) P_{live}$

$\qquad = (5 \times 16/2) + (12/16)\,50 = 77.5 \text{ kip}$

$P_u = 1.2 \times 140 + 1.6 \times 77.5 = 292 \text{ kip}$

$K_y L_y = 1.0 \times 12 = 12 \text{ ft}$

$W8 \times 35$ from column design tables

ASD

$P = wL/2 + (12/16) P_{live}$

$\qquad = (17.5 + 5)16/2 + (12.16)\,50 = 218 \text{ kip}$

$K_y L_y = 1.0 \times 12 = 12 \text{ ft}$

$W10 \times 39$ from column design tables

8.16 (a) LRFD

$P_u = 292 \text{ kip}$

Assume $r_x / r_y = 2$

$(K_y L_y)_{eff} = K_x L_x /(r_x / r_y) = 12/2 = 6 \text{ ft}$

Try $W8 \times 28$ from column design tables. $r_x / r_y = 2.13$

$(K_y L_y)_{eff} = K_x L_x /(r_x / r_y) = 12/2.13 = 5.63 \text{ ft}$

$\phi P_n = 306 \text{ kip} \geq 292.\ \therefore \text{OK.}$

ASD
P = 218 kip
Assume $r_x/r_y = 2$
$(K_yL_y)_{eff} = K_xL_x/(r_x/r_y) = 12/2 = 6$ ft
Try W8×31 from column design tables. $r_x/r_y = 1.72$
$(K_yL_y)_{eff} = K_xL_x/(r_x/r_y) = 12/1.72 = 6.98$ ft
capacity = 234 kip ≥ 180. ∴ OK.

8.17 (c) LRFD
$M_{dead} = wL^2/8 = 17.5 \times 16^2/8 = 560$ kip-ft
$M_{live} = 5 \times 16^2/8 = 160$ kip-ft
$M_u = 1.2 \times 560 + 1.6 \times 160 = 928$ kip-ft
W30×90 from Z_x economy table

ASD
w = 5 + 17.5 = 22.5 kip-ft
$M_{req'd} = mL^2/8 = 22.5 \times 16^2/8 = 720$ kip-ft
W30×99 from S_x economy table

8.18 (d) LRFD
$M_u = 928$ kip-ft
$L_b = 16$ ft; $C_b = 1.14$
$M_u/C_b = 928/1.14 = 814$ kip-ft
W30×90 from beam design charts
$\phi M_{cr} = 827 \geq 814$; $\phi M_p = 1060 \geq 928$

ASD
$M_{req'd} = 720$ kip-ft
unbraced length = 16 ft; $C_b = 1.0$
W33×118 from beam design charts

8.19 (a) LRFD
$h/t_w \leq 418/\sqrt{F_y}$?   43.9 ≤ 59.1.   Yes, so
$\phi V_n = \phi 0.6 F_y A_w = \phi 0.6 F_y d t_w$
$\quad = 0.9 \times 0.6 \times 50 \times 30.31 \times 0.615 = 503$ kip

ASD
$h/t_w \leq 380/\sqrt{F_y}$?   43.9 ≤ 53.7.   Yes, so
$f_v = R/(d t_w) = 180/(30.31 \times 0.615) = 9.7$ ksi

8.20 (e) LRFD
$P_{dead} = w_{dead} L/2 = 17.5 \times 16/2 = 140$ kip
$P_{live} = w_{live} L/2 = 5 \times 16/2 = 40$ kip
$P_u = 1.2 P_{dead} + 1.6 P_{live} = 1.2 \times 140 + 1.6 \times 40 = 232$ kip
$M_{ntx} = P_u e = 232 \times 7/12 = 135$ kip-ft

ASD
$P = wL/2 = (17.5 + 5)16/2 = 180$ kip
$f_{bx} = M/S_x = Pe/S_x = 180 \times 7/35.5 = 35.5$ ksi

8.21 (c) LRFD
$\phi F_v A_b = 43.3$ kip/bolt; $R_u = 232$ kip
$272/43.3 = 5.4$ so use 6 rows (also verify $n\phi 2.4 F_u dt \geq V_u$)
3 lines × 6 rows = 18 bolts

ASD
capacity = 25.3 kip/bolt; required strength = 180 kip
$180/25.3 = 7.1$ so use 8 rows (also verify $V/ndt \leq 1.2 F_u$)
3 lines × 8 rows = 24 bolts

8.22 (d) LRFD
$A_{nv} = 2 \times 3/8 \times [21 - 7 \times (7/8 + 1/16 + 1/16)] = 10.5$ in^2
$\phi 0.6 F_u A_{nv} = 0.75 \times 0.6 \times 65 \times 10.5 = 307$ kip

ASD
$A_{nv} = 2 \times 3/8 \times [21 - 7 \times (7/8 + 1/16 + 1/16)] = 10.5$ in^2
$f_v = R/A_{nv} = 180/10.5 = 17.1$ ksi

8.23 (c) LRFD
$P_u = 100$ kip
design strength per inch of weld
    = lesser of weld shear strength or base metal shear strength
$\phi 0.6 F_{Exx} A_w / l_w = 0.75 \times 0.60 \times 70 \times (.707 \times 3/16) = 4.18$ kip/in
$\phi 0.6 F_u A_{BM} / l_w = 0.75 \times 0.6 \times 65 \times 1/4 = 7.31$ kip/in
$100/4.18 = 23.9$ inches

ASD
$P = 62.5$ kip
capacity per inch of weld
    = lesser of weld shear capacity or base metal shear capacity
$0.3 F_{Exx} A_w / l_w = 0.3 \times 70 \times (.707 \times 3/16) = 2.78$ kip/in
$0.3 F_u A_{BM} / l_w = 0.3 \times 65 \times 1/4 = 4.88$ kip/in
$62.5/2.78 = 22.5$ inches

9. Reinforced Concrete

by Richard W. Furlong

Concepts and recommended practice in this chapter for reinforced concrete structures are consistent with the *Building Code Requirements for Reinforced Concrete* (ACI318-89)(Revised 1992), available from the American Concrete Institute, Detroit. Hereafter it will be called ACI318 for simplicity. It is imperative that designers of concrete structures possess and become familiar with that document.

Concrete is a manufactured form of stone. A powder of portland cement is mixed with water, sand (fine aggregate) and rock (coarse aggregate) to be placed into forms in order to define the shape of structural elements such as slabs, beams, columns, and walls. Concrete quality is measured with a compression test on standard cylinders of 6 in diameter and 12 in height. A capping material is applied to provide a uniform stress at each end of cylinders, and loads are applied until the cylinders fail to resist applied force. As compression load shortens the length of the cylinders, the diameter tends to expand (Poisson's ratio). However, circumferential tension stress near midheight causes the concrete to rupture, and longitudinal cracks along the sides of the cylinder signal that the cylinder is failing. The maximum compression stress from the cylinder is called f'_c. When one orders a batch of 4000-lb concrete, the number means that the nominal strength of standard cylinders should be at least 4000 psi. Concrete weighs approximately 145 lb/cu ft or 4000 lb/cu yd. Transit mix concrete of reliable quality is available in every urban area. The commercial grade of concrete for residential foundations and slabs is 3000 psi, but $f'_c = 4000$ psi is typical for structural frames.

Reinforcement for concrete is in the form of steel bars which have an intentional pattern of deformations on the bar surface. Reinforcement quality is called steel Grade. The actual definition of steel Grade is the steel stress reached at a tension strain of 0.5%. Since the bar surface is rough, "round" bars have an area and diameter that varies along the length. Nominal bar area is established as the weight of a specimen divided by the density of steel (0.283 lb/cu in) and the length of the specimen. Nominal diameter is the actual diameter of a smooth cylinder with the nominal bar area. The most common grade of steel is Grade 60, for which the yield strength is at least 60,000 psi (60 ksi). Grade 40 and Grade 50 steel bars usually can be obtained by special order. Table 9.1 contains data for bars available in the United States. The reader should note that bar numbers coincide with the number of eighths of an inch in the nominal bar diameter of sizes #8 and smaller. For bars #9 and larger, each area

Table 9.1 ASTM Standard Reinforcing Bars

Bar size, No.	Nominal diameter, in.	Nominal area, in.²	Nominal weight, lb/ft
3	0.375	0.11	0.376
4	0.500	0.20	0.668
5	0.625	0.31	1.043
6	0.750	0.44	1.502
7	0.875	0.60	2.044
8	1.000	0.79	2.670
9	1.128	1.00	3.400
10	1.270	1.27	4.303
11	1.410	1.56	5.313
14	1.693	2.25	7.650
18	2.257	4.00	13.600

is the same as the area of a square, i.e., 1-in square for #9, $1\frac{1}{8}$-in square for #10, $1\frac{1}{4}$-in square for #11, $1\frac{1}{2}$-in square for #14, and 2-in square for #18. The stiffness (Young's modulus) for reinforcing bars is $E_s = 29 \times 10^6$ psi.

9.1 Strength Design Concept

9.1.1 Load Effects

Most concrete structures are assembled as monolithic, rigidly connected frames. Many concrete structures involve cast-in-place concrete with reinforcement continuous through all connections. The determination of internal forces (shears, moments, and axial forces) requires an understanding and use of tools for indeterminate structural analysis. ACI318 accepts results from any complete and consistent elastic analysis that is based on logical assumptions of element stiffness. The Commentary to the ACI318 Code suggests that allowance be made for cracking of beam and slab elements due to service load and retrained shrinkage. Column and wall elements are less likely to be similarly cracked.

For concrete, the nominal stiffness E_c in psi units can be computed as

$$E_c = 33w^{1.5}\sqrt{f_c'} \qquad (9.1.1)$$

where

w = density of concrete in lb/ft^3
f_c' = cylinder strength of concrete in psi

The following relationships are acceptable for analysis of concrete frames:

$$EI = 0.30E_c I_g \quad \text{for slabs} \qquad (9.1.2a)$$
$$= 0.35E_c I_g \quad \text{for beams} \qquad (9.1.2b)$$
$$= 0.70E_c I_g \quad \text{for columns and walls} \qquad (9.1.2c)$$
$$EA_g = 1.00E_c A_g \quad \text{for all members} \qquad (9.1.2d)$$

where

A_g = area of gross concrete outline of element cross section
I_g = moment of inertia of gross concrete outline of element cross section

Load factors are applied to estimated service loads anticipated during the life of any structure. The load factors reflect both the relative accuracy of the estimated loads and the relative consequence of erroneous estimates. Thus, even though one should be able to estimate forces due to permanent loads (Dead load) with an accuracy of ±15%, a load factor of 1.4 is applied to estimated force effects from Dead load. Consequences of an erroneous estimate could be severe indeed if the estimate actually were 15% in error permanently. Actual service load member forces from Dead load alone could reach values 1.4(1 − 0.15) = 1.19 or within 20% of the limit strength of the structural element. The following utilizes the load factors as set forth in ACI318 Section 9.2:

$$U = 1.4D + 1.7L \qquad (9.1.3a)$$
$$= 0.9D + 1.3W \qquad (9.1.3b)$$
$$= 0.75(1.4D + 1.7L + 1.7W) \qquad (9.1.3c)$$
$$= 0.75(1.4D + 1.7L + 1.9E) \qquad (9.1.3d)$$
$$= 0.75(0.9D + 1.4E) \qquad (9.1.3e)$$

where

U = required ultimate force
D = service load force from dead load
L = service load force from live load
W = service load lateral force from wind or earth pressure
E = force from design earthquake action

9.1.2 Strength Effects

The strength of cross sections must be estimated according to acceptable and in many cases specified strength criteria. Such strength is based on nominal material properties, dimensions, and largely empirical theories. Strength reduction factors ϕ must be applied to nominal estimates of strength in order to allow for under strength material, errors in dimensions or bar placements, reliability of strength theories, and consequences from the failure of a section. The consequence of problems from the sagging of a ductile beam are less severe than problems associated with, for example, a crushing failure in a concrete column. The ϕ factor for columns is significantly lower than the ϕ factor for beams. A ductile sag from flexure is less severe than local rupture from diagonal shear, and ϕ for shear is lower than ϕ for ductile flexure.

The fundamental relationship that must be satisfied for all elements in concrete structures requires that reduced nominal strength or resistance ϕR be greater than factored service load U:

$$\phi R > U \tag{9.1.4}$$

9.2 Flexural Forces—Beams

Typical stress-strain curves for concrete and for steel are shown in Fig. 9.1. Steel in tension and in compression is assumed to exhibit a linear relationship between stress and strain until the nominal yield strength for the steel is reached. Then steel is assumed to maintain constant stress for all strains beyond the yield strain. Concrete is assumed to have a rather linear increase in compression stress as strain increases to strains near 0.1%. At larger strains, the slope of the stress-strain curve decreases to a value of zero at strains near 0.18%. The zero slope continues to somewhat larger strains before the curve descends to a rupture strain, usually higher than 0.3%. Concrete in tension has a curved stress-strain function that ruptures at strains in the order of 0.1%. The maximum value of concrete strength in compression is called f'_c, and the maximum tensile strength is less than $0.15 f'_c$.

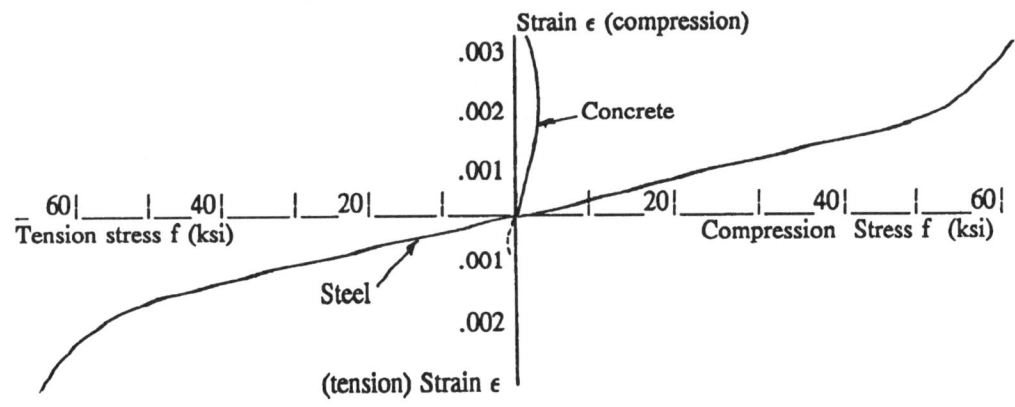

Figure 9.1 Typical stress-strain curves for steel and concrete.

Flexural forces cause beams to bend as suggested by the diagram of Fig. 9.2a. In the absence of an axial force, there will be within the beam a line called a neutral axis along which there is no strain. All fibers above the line are compressed and all fibers below the line are stretched—the amount of compression or stretch increasing linearly with distance from the line of zero strain (neutral axis). The stress-strain graph for concrete in Fig. 9.1 indicates that the curve ends at a strain somewhat larger than 0.3%. The limit flexural strength of reinforced concrete cross sections can be computed as the sum of moments on the cross section when concrete reaches its limit strain. Limit strains are shown in Fig. 9.2b, and limit stresses are shown in Fig. 9.2c. The sum of moments due to forces from the limit stresses equals the ultimate moment on the section.

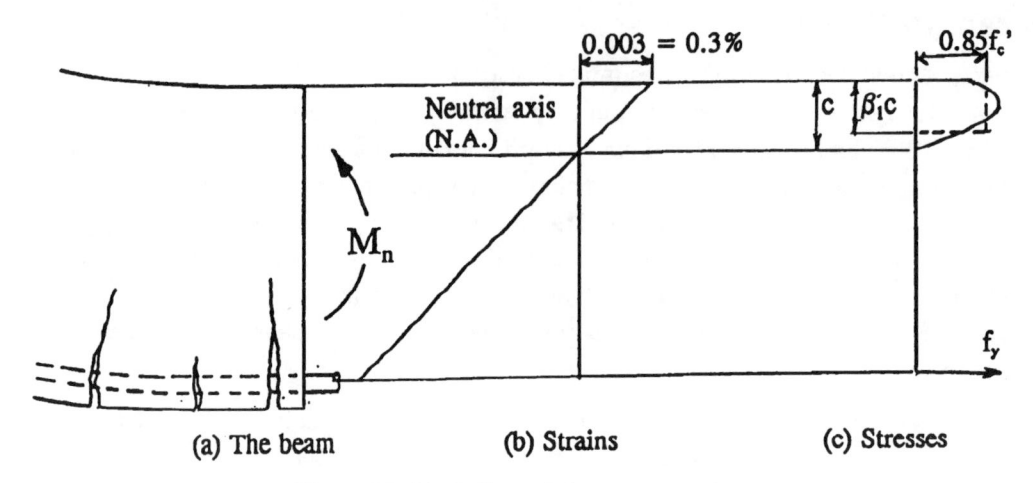

Figure 9.2 Limit flexural force on a section.

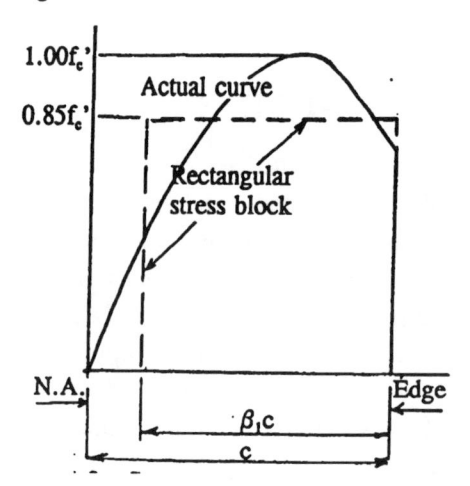

Figure 9.3 Concrete rectangular stress block.

The nominal limit force on tension steel is simply the product of yield strength and bar area, $T_n = A_s f_y$. The nominal limit compression force C_n on concrete equals T_n since tensile stress on concrete is neglected. The complex challenge of locating a centroid for the concrete stress-strain diagram is simplified through the use of a rectangular stress block to represent the actual behavior of concrete. ACI318 Section 10.2.7.3 permits the stress block described in Fig. 9.3. The intensity of stress is taken as a constant value $0.85 f_c'$. The ratio β_1 between depth of stress block and depth c to the neutral axis decreases as the value of f_c' exceeds 4000 psi:

$$0.85 > \beta_1 = 0.85 \frac{(f_c' - 4000)}{20,000} > 0.65 \qquad (9.2.1)$$

The centroid of the rectangular stress block is located a distance $0.5\beta_1 c$ from the edge that is in maximum compression. The total compression force is the product of the stress $0.85 f_c'$ acting on an area $\beta_1 c$ deep and as wide as the compression edge of the section. Calling the section width b, and equating compression force C_n to the tension force $T_n = A_s f_y$, the magnitude of $\beta_1 c$ can be computed as

$$\beta_1 c = \frac{A_s f_y}{0.85 f_c' b} \qquad (9.2.2))$$

The nominal moment M_n at the limit state of flexural strength is the sum of moments on the section. Define the section depth d as the distance from the compression edge to the center of tension steel, and then take moments about the centroid of the compression steel to compute M_n:

$$M_n = A_s f_y (d - 0.5\beta_1 c) = A_s f_y \left(d - \frac{A_s f_y}{1.7 b f_c'} \right) \qquad (9.2.3)$$

A capacity reduction factor ϕ must be applied to the nominal strength, and ACI318 Section 9.3.2.1 gives the ϕ factor for beams as 0.90. The subscript u will be used for the required ultimate force (from factored load). The basic strength relationship then requires that $\phi M_n \geq M_u$. When an area of tension steel is to be determined, Eq. 9.2.3 can be solved for A_s after M_u/ϕ is substituted for M_n.

The strain diagram of Fig. 9.2b shows no specific value for the concrete failure strain at the compression edge. The actual limit strain at failure can vary from as little as 0.0025 to as much as 0.0100, but ACI318 Section 10.2.3 specifies that maximum usable strain in concrete is only 0.0030.

Flexural ductility, the property of a section to crack and bend without breaking, is a desirable characteristic for reinforced concrete beams. The diagram of Fig. 9.2b shows section strains at failure, and suggests that the tension steel has a strain of 0.009, several times greater than the yield strain of the bar, which is $f_y/E_s \approx 0.002$. If the area of tension steel were doubled, the stress block depth $\beta_1 c$ would have to double as well, and the strain at the tension steel would be about 0.003, only 1.5 times larger than the yield strain. The amount of post-yield flexing before failure of the beam would be only about one-third as much as for the section with half as much tension steel. ACI318 Sections 10.3.2 and 10.3.3 forbid the use of more than 75% of the tension steel that would reach its yield strain when concrete reaches its limit strain of 0.003. Actually, even 75% of the "balanced" area of tension steel is too much to use in ordinary circumstances. Desirable ductility is assured if the area of tension steel is kept below 30% to 40% of the "balanced" amount.

It is convenient to use a reinforcement ratio $\rho = A_s/(bd)$ as a parameter for general conditions and design formulae. The limit ratio, ρ_{max}, that is permitted in a rectangular section can be expressed as the value of ρ when A_s is 75% of the "balanced" amount of steel:

$$\rho_{max} = \frac{55,500 \beta_1 f_c'}{87,000 f_y + f_y^2} \qquad (9.2.4a)$$

and for purposes of selecting a section with a ductile proportion of steel:

$$\rho_{good} \approx \frac{27{,}000\beta_1 f'_c}{87{,}000 f_y + f_y^2} \qquad (9.2.4b)$$

Eq. 9.2.3 can be rewritten with ρ and M_u/ϕ as a design expression for rectangular section dimensions b and d to resist a required moment M_u:

$$bd^2 = \frac{M_u}{\rho\phi f_y\left(1 - \rho f_y/1.7 f'_c\right)} \qquad (9.2.5)$$

The denominator of Eq. 9.2.5 can be expressed as a stress ϕk which can be multiplied by the section shape parameters bd^2 to give a value equal to a required moment M_u. After material properties f'_c and f_y have been selected for design, a value for ρ_{good} can be used in Eq. 9.2.5 to determine the value of ϕk. For example, if $f'_c = 4000$ psi (for which $\beta_1 = 0.85$) and $f_y = 60{,}000$ psi, then Eq. 9.2.4b gives $\rho_{good} = 0.0104$, and $\phi k = 567$ psi. A value of $bd^2 \approx M_u/567$ would produce a section with good ductility. M_u must be in units of in-lb if stress values are in psi units. The actual thickness of a cross section will be 2.5 to 3 in deeper than the flexural depth d, as reinforcement must be covered with concrete at least 1.5 in thick outside the bars themselves.

If flexural forces cause a section to crack in tension, it is desirable that there be enough tension reinforcement to pin together the crack without yielding the tension steel. That requirement can be satisfied if the value of ρ is greater than $3\sqrt{f'_c}/f_y = \rho_{min}$.

Table 9.2 contains values of reinforcement ratios ρ and ductile stress values ϕk_{good} for rectangular sections reinforced with Grade 60 steel. The values are useful for proportioning sections and for estimating the relative ductility of rectangular sections. If the reinforcement ratio A_s/bd is near ρ_{good} the section is ductile, but ductility decreases as the value A_s/bd approaches ρ_{max}.

Table 9.2 Proportioning rectangular sections with Grade 60 reinforcement.

f'_c psi	ρ_{good}	ρ_{min}	ρ_{max}	ϕk_{good} psi
3000	.0078	.0027	.0160	383
4000	.0104	.0032	.0214	510
5000	.0122	.0035	.0250	602
6000	.0138	.0039	.0284	685

Once a beam width b and flexural depth d are established, in most cases there will be a variety of different values of required moment at different critical sections along the beam. Critical sections for positive moment occur near midspan, and critical sections for negative moment occur at the face of supports. When values b, d, and M_u are known, Eq. 9.2.3 can be solved for the steel area A_s. The quantity $(d - 0.5\beta_1 c)$ can be approximated as $0.9d$ in order to make the equation for required steel area become $A_s \approx M_u/(0.9d\phi f_y)$. ACI318 Section 9.3.2.2 gives $\phi = 0.90$ for flexure. Grade 60 reinforcement will be used in most cases. Then when M_u is expressed in k-ft units, the expression for approximate steel area is

$$A_s = \frac{12{,}000 M_u}{\phi f_y 0.9d} = \frac{12{,}000 M_u}{0.9(60{,}000)0.9d} \approx \frac{M_u}{4d} \qquad (9.2.6)$$

Eq. 9.2.6 is sufficiently accurate for the selection of longitudinal reinforcement within the limited number of bar sizes available. Examples in Section 9.6 will illustrate this type of beam problem.

9.3 Compression Forces—Columns

Reinforced concrete columns are made with longitudinal bars around the periphery of cross sections and transverse (tie) bars at vertical intervals to hold longitudinal bars in place and to bind the cross section together. Fig. 9.4 displays graphs of compression force and compression deformation distributed uniformly against a 16-in square reinforced concrete column. The graphs of force and deformation can be translated into graphs of stress and strain by dividing force ordinates by the area of the material, and by dividing the deformation values by the initial length of the column.

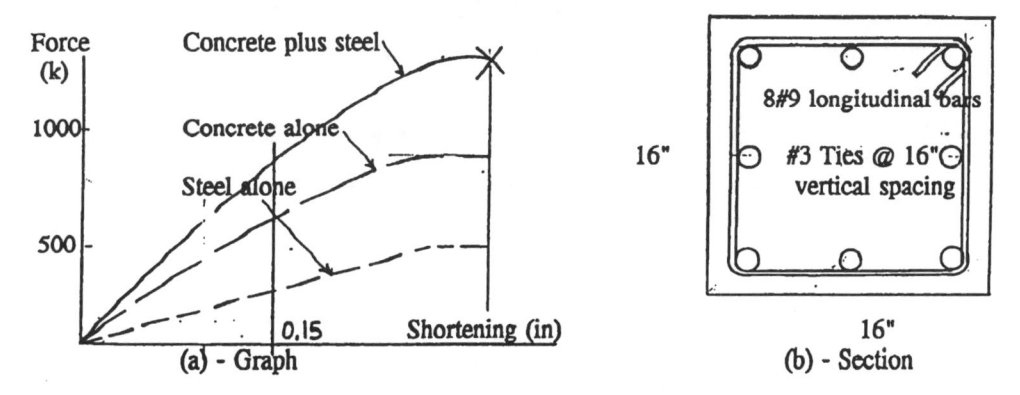

Figure 9.4 Force and compression deformation of a column.

The total force on the column is the sum of the force on concrete and the force on steel. There are 248 in² of concrete that support a total of 843 k before the limit of concrete strength is reached, while only 8 in² of the much stiffer steel supports 480 k before the steel yields. Limit strength is reached at a deformation indicated as 0.3 in. When the deformation is 0.15 in, the steel resists 180 k and the concrete resists 600 k, each material increasing its resistance in an almost linear increase with deformation. Thereafter, concrete begins to lose its capacity to resist compression. As steel yields, its stiffness disappears. It tries to "flow" under further compressive displacement, restrained from buckling only by the transverse ties and the concrete cover over the bars. Of course the concrete is forced to split away, and the longitudinal bars buckle between tie bars. The compression failure of a concrete column can be a rather explosive event.

The ultimate strength or limit axial compression strength of a column is given the symbol P_o. It will be called the *squash* load. It is the sum of strength from concrete and from steel:

$$P_o = 0.85 f'_c A_c + f_y A_{st} = 0.85 f'_c A_g + A_{st}(f_y - 0.85 f'_c) \tag{9.3.1}$$

where A_{st} is the total area of longitudinal steel.

Since reinforced concrete columns are almost always a part of continuous, rigid frame structures, ACI318 requires that all concrete columns be designed as if required axial forces are eccentric about the centroid of the column section. ACI318 Section 10.11.5.4 sets the minimum eccentricity of axial force as at least $(0.6 + 0.03h)$ inches, where h is the section thickness in inches. It is most common for moments from the analysis of load actions to exceed by far the product of required axial force P_u and minimum eccentricity.

Applying the same logic that was specified for flexural strength, concrete in columns cannot be considered useful when compressed to strains greater than 0.3%. Thus, it is assumed that failure occurs when the strain at a "compression" edge reaches 0.003. After a neutral axis is set at a distance c from the compression edge, the compression edge strain and the neutral axis (zero strain) establish a plane of failure strain. Stresses corresponding to the strain at each point then are integrated in order to determine a nominal limit load P_n and nominal limit moment M_n for the plane of failure strain. The procedure is illustrated with Fig. 9.5 and Table 9.3 for a 16-in square section with 8 #9 Grade 60 longitudinal bars and f'_c = 4000 psi. Neutral axis locations c are:

a) at c = 13.5 inches such that zero strain occurs at tension face bars
b) at c = 8 inches such that zero strain occurs at midheight of section
c) at c = 5.51 inches and tension face bars yield when compression edge strain = 0.3%
d) at c = 4 inches such that the tension strain at midheight is 0.3%
e) at c = 2.5 inches such that the tension force on compression face steel is zero.

Table 9.3 Column section strength calculations.

c	(in)	13.50	10.75	8.00	5.25	2.50
$\beta_1 c$	(in)	11.48	9.14	6.80	4.46	2.12
$h/2 - \beta_1 c$	(in)	2.26	3.43	4.60	5.77	6.94
ε @ [1]		0	−0.00061	−0.00296	−0.00471	−0.01320
ε @ [2]		0.00122	0.00061	0	−0.00157	−0.00660
ε @ [3]		0.00244	0.00230	0.00206	0.00157	0
Stress @ [1]	(ksi)	0	−17.7	−60.0	−60.0	−60.0
Stress @ [2]	(ksi)	35.4	17.7	0.0	−45.5	−60.0
Stress @ [3]	(ksi)	60.0	60.0	60.0	45.5	0.0
Conc. P	(k)	624.5	497.2	369.9	242.6	115.3
Force @ [1]	(k)	0.0	−53.1	−180.0	−180.0	−180.0
Force @ [2]	(k)	70.8	34.4	0.0	−91.0	−120.0
Force @ [3]	(k)	180.0	180.0	180.0	136.5	0.0
Sum Forces	(k)	885.3	658.5	369.9	108.1	−184.7
M from conc	(k-ft)	1411.	1705.	1702.	1400.	800.
M from [1]	(k-ft)	0.	292.	990.	990.	990.
M from [2]	(k-ft)	0.	0.	0.	0.	0.
M from [3]	(k-ft)	990.	990.	990.	751.	0.
Sum Moments	(k-ft)	2401.	2987.	3682.	3141.	1790.

Six sets of coordinates were calculated in order to define the strength interaction diagram shown in Fig. 9.5(b). Curvature due to flexure is the ratio 0.003/c. As curvature increases, moment capacity increases and axial load capacity decreases until curvature reaches the balanced curvature, the curvature for which bars in the tension face reach their yield strain when compressed concrete in the opposite face reaches its failure strain of 0.003. At balanced curavture the axial force is the *balanced thrust* P_b and the moment is the *balanced moment* M_b. Both moment and axial load capacity decrease as curvatures are increased beyond the balanced curvature. Failures are classified as compression failures for curvatures less than balanced curvature (axial force larger than P_b), as concrete spalls (surface flakes), and crushes before any steel yields in tension. Failures are called tension failures when curvatures are greater than balanced curvature (axial force less than P_b), and steel yields before concrete reaches its crushing-spalling strain.

Does it seem reasonable that the concrete column under an axial load of 300 k should possess more moment capacity when there is no axial force on the column? In actuality, it is reasonable. The presence of axial force restrains tension fracture of concrete and the section can sustain moment forces greater than those it can sustain without the axial pressure. This principal, with compression forces sustained by very high strength (250 ksi) bars or steel cable tendons, is the basis for prestressed concrete.

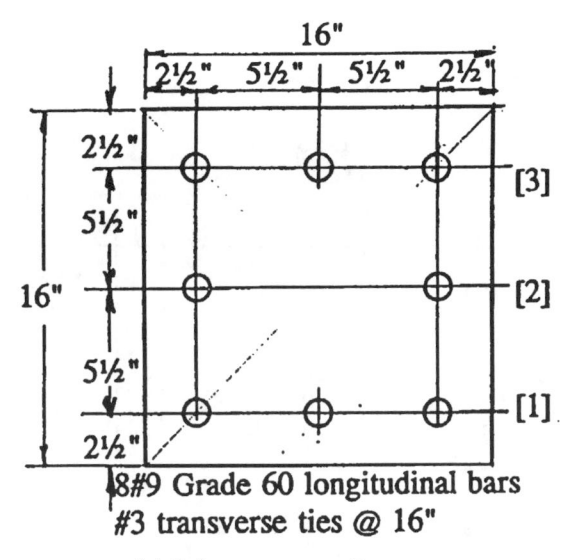

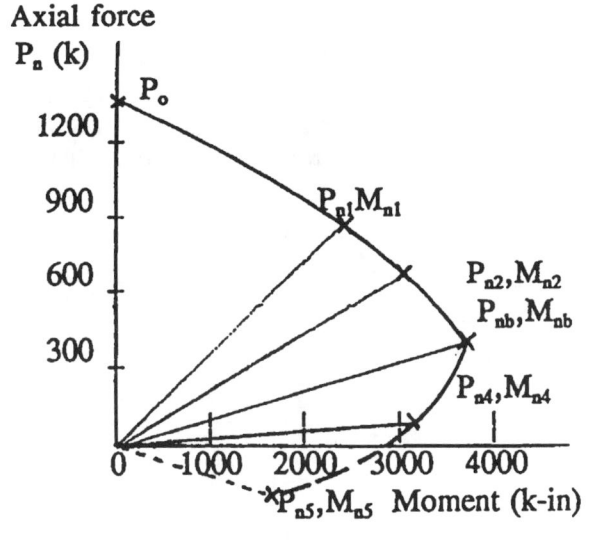

Figure 9.5 Column section for strength analysis and section interaction diagram.

(a) Column cross section — 16" × 16", 8#9 Grade 60 longitudinal bars, #3 transverse ties @ 16"

(b) Interaction diagram

The determination of section strength for concrete columns is tedious and it is highly desirable to use general charts or tables in design practice. Also, a capacity reduction factor $\phi = 0.70$ applied to nominal axial force P_n and nominal moment M_n must provide strengths greater than required axial force P_u and required moment M_u. Furthermore, ACI318 Section 10.3.5 restricts maximum axial force to 0.8 P_o for tied columns. These reductions to nominal strength values are readily incorporated into design charts such as Fig. 9.6. Graphs of values of ϕP_n and ϕM_n are shown for a 16-in square section with $f'_c = 4$ ksi and Grade 60 longitudinal bars. Graphs are shown for longitudinal bar sizes #6 through #11, all the sizes allowed for the 8-bar arrangement on this section. All the graphs have a similar shape, with capacity increasing with bar size. If required strength values are inside a specific graph, the section capacity is adequate with the specific amount of longitudinal steel. For example, if required forces were $P_u = 450$ k and $M_u = 150$ k-ft, 8 bars size #8 or larger would provide adequate strength.

The slope of diagonal lines with respect to the axial force are lines of equal eccentricity, $\phi M_n / \phi P_n = e$, for all points along the same diagonal line. The column strength Table 9.4 displays values ϕP_n for increasing amounts of eccentricity e and various specific sets of longitudinal bars. The table shows corner bars to be in both the end face and the side face.

Thus the section with 3 bars in each end face and 3 bars in each side face actually contains 8 bars in the same arrangement as that shown in Table 9.4. For a given eccentricity of required axial force P_u, the value in Fig. 9.7 must be larger than P_u. If $P_u = 450$ k and $M_u = 150$ k-ft, the eccentricity $e = M_u/P_u = 150(12/450) = 4$ in. With $e = 4$ in., the tabulated value for 3 #8 in the end faces and 3 #8 in the side faces is 460 k, indicating that section to be adequate for the specified required loading.

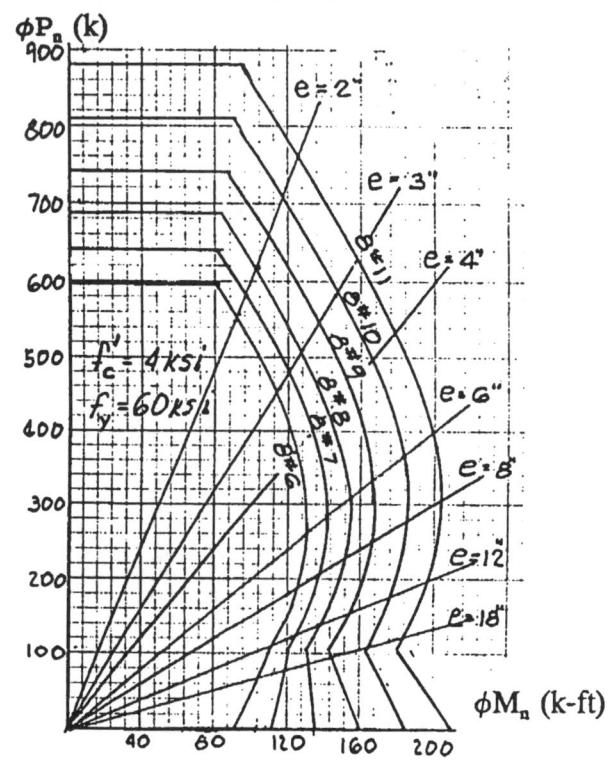

Figure 9.6 Design aid for 16" × 16" section.

Table 9.4 Design aid for 16" × 16" columns.

Strength of cross sections (no length or slenderness considered)
$f_c' = 4000$ psi and steel yield strength = 60,000 psi
Tabulated values have been reduced by a capacity reduction factor $\phi = 0.7$
Cross section width $b = 16$ in; thickness $h = 16$ in; and concrete cover = 1.5 in

Bar Size	Quantity End	Quantity Side	ρ %	Max P_u (k)	Eccentricities M/P (in) 2	3	4	6	8	12	18	e_t (in)	e_b (in)	ϕP_b (k)	ϕM_o (k-ft)
#8	2	2	1.23	588	542	456	377	279	195	105	67	3.02	7.71	206	87
#9	2	2	1.56	614	564	473	393	297	215	121	77	3.07	8.59	196	104
#10	2	2	1.98	648	591	495	412	318	237	140	87	3.12	9.86	184	124
#11	2	2	2.44	685	619	518	432	338	256	159	96	3.16	11.43	169	144
#14	2	2	3.52	773	710	608	524	431	344	228	139	3.24	13.98	192	211
#18	2	2	6.25	995	834	684	567	407	368	253	167	3.28	97.31	29	290
#6	3	2	1.03	571	530	446	369	268	181	98	61	2.97	7.18	215	76*
#7	3	2	1.41	602	556	468	389	292	209	115	73	3.03	8.13	205	97
#8	3	2	1.85	638	585	493	411	316	234	136	85	3.09	9.37	193	119
#9	3	2	2.34	678	617	519	434	340	258	158	96	3.15	10.93	179	142
#10	3	2	2.98	729	658	552	462	368	284	182	110	3.20	13.33	161	170
#11	3	2	3.66	784	700	586	491	395	309	203	127	3.24	16.66	140	198
#14	3	2	5.27	915	836	719	622	479	427	298	194	3.42	20.61	167	301
#6	4	2	1.38	599	554	467	389	292	209	114	72	3.01	8.02	208	95
#7	4	2	1.88	640	589	496	415	321	238	138	86	3.09	9.39	195	121
#8	4	2	2.47	688	628	529	444	350	267	165	99	3.15	11.26	180	149
#9	4	2	3.13	741	671	565	474	379	295	190	115	3.21	13.76	162	178
#10	4	2	3.97	809	725	609	512	414	327	217	138	3.27	17.96	138	214
#6	2	3	1.03	571	524	435	351	237	156	90	57	2.76	6.20	227	74
#7	2	3	1.41	602	548	453	367	255	174	100	66	2.76	6.70	223	91
#8	2	3	1.85	638	575	474	384	273	193	113	75	2.75	7.31	217	108
#9	2	3	2.34	678	604	497	402	291	211	126	83	2.74	8.02	211	125
#10	2	3	2.98	729	642	525	424	312	230	141	91	2.72	8.98	202	145
#11	2	3	3.66	784	680	554	447	333	248	156	99	2.70	10.11	192	163
#14	2	3	5.27	915	804	671	563	436	338	223	142	2.67	11.99	224	241
#6	3	3	1.38	599	548	456	373	265	182	103	68	2.81	6.97	221	92
#7	3	3	1.88	640	581	482	395	288	206	120	78	2.83	7.80	213	113
#8	3	3	2.47	688	618	512	420	312	230	138	89	2.83	8.85	204	135
#9	3	3	3.13	741	658	544	446	337	252	157	98	2.83	10.13	193	158
#10	3	3	3.97	809	709	583	478	366	278	179	111	2.82	12.00	179	184
#11	3	3	4.88	883	762	624	511	396	303	198	125	2.81	14.40	162	210
#14	3	3	7.03	1058	933	786	668	536	425	290	190	2.88	17.31	199	323
#5	4	3	1.21	586	539	450	369	262	178	100	64	2.83	6.87	222	84*
#6	4	3	1.72	627	573	478	394	289	206	119	77	2.86	7.79	214	108
#7	4	3	2.34	678	614	512	422	317	234	140	89	2.89	9.00	204	134
#8	4	3	3.09	738	661	550	454	348	262	164	101	2.90	10.61	191	162
#9	4	3	3.91	804	712	590	488	379	289	187	116	2.91	12.65	176	190
#10	4	3	4.96	890	776	641	530	416	321	212	135	2.91	15.89	156	224
#6	2	4	1.38	599	541	443	360	236	159	95	63	2.61	6.04	234	87
#7	2	4	1.88	640	571	465	378	252	175	104	71	2.58	6.46	231	104
#8	2	4	2.47	688	605	489	399	270	192	117	79	2.53	6.96	228	121
#9	2	4	3.13	741	642	515	422	287	208	128	86	2.48	7.51	225	137
#10	2	4	3.97	809	687	547	449	309	227	142	93	2.41	8.25	220	155
#5	3	4	1.21	586	534	441	364	241	161	94	61	2.69	6.21	231	82
#6	3	4	1.72	627	566	466	386	262	182	107	72	2.67	6.78	227	103
#7	3	4	2.34	678	604	495	398	285	204	123	81	2.65	7.50	222	124
#8	3	4	3.09	738	648	529	425	309	227	140	91	2.61	8.40	215	146
#9	3	4	3.91	804	696	564	453	335	249	156	99	2.57	9.44	208	167
#10	3	4	4.96	890	756	609	487	365	274	176	111	2.52	10.90	197	192
#5	4	4	1.45	605	551	457	372	260	179	103	68	2.73	6.74	227	94
#6	4	4	2.06	655	591	488	397	286	205	121	80	2.72	7.57	220	118
#7	4	4	2.81	716	638	525	427	315	232	141	91	2.71	8.64	212	143
#8	4	4	3.70	788	692	567	461	346	259	163	102	2.69	10.02	202	170
#9	4	4	4.69	868	751	612	497	378	286	185	116	2.66	11.71	190	198
#10	4	4	5.95	970	824	668	542	417	318	209	134	2.62	14.25	174	230

Table 9.4 contains additional information useful in concrete column design. The eccentricity of load for the condition of zero stress in the bars at the tension face $e_t = M_{n1}/P_{n1}$ is listed. The eccentricity $e_b = M_{nb}/P_{nb}$ as well as the value ϕP_{nb} are listed. These eccentricity values are helpful regarding the type of compression splice required for the longitudinal bars. ACI318 Section 12.15.1 requires that Class B splices be used if more than half the bars in tension must be spliced for the development of f_y. If applied to columns designed for axial force plus bending, that requirement can be interpreted to mean that if the eccentricity of required axial force is greater than the average $(e_t + e_b)/2$, a Class B splice is required since the tension bars would require splicing for more than half the yield strength of those bars.

Values for ϕM_o, the moment capacity in the absence of axial force, are listed in the right-hand column of Table 9.4. ACI318 Section 9.3.2.2 allows the value of ϕ to increase linearly from 0.7 to 0.9 as the axial force decreases from $0.10 f'_c A_g$ to zero.

Requirements for transverse ties are given in Section 7.10.5 of ACI318. Ties must be #3 bars unless longitudinal bars are #11 or larger, in which case ties must be at least #4 bars. Transverse ties "shall be arranged such that every corner and alternate longitudinal bar shall have lateral support provided by the corner of a tie with an included angle of not less than 135°, and no bar shall be farther than 6 in. clear on each side along the tie from such a laterally supported bar." This statement requires the use of interior cross ties at alternate bars or at every bar for which the clear space to the edge of the adjacent bar exceeds 6 in. The vertical spacing of sets of ties must be no greater than the least dimension of a section, 16 diameters of longitudinal bars or 48 diameters of tie bars. Typical arrangements for column ties are illustrated below in Fig. 9.7.

Spirally reinforced columns are similar to tied columns except that a continuous, closely spaced transverse tie is used around the longitudinal bars. Tests have demonstrated that excessively loaded spirally reinforced columns can survive the spalling away of concrete cover over ties. Spirally reinforced columns survive the loss of the surface concrete through transverse triaxial confinement of core concrete by the spiral. Requirements of volume of steel in the spiral itself are given by ACI318 Section 10.9.3.

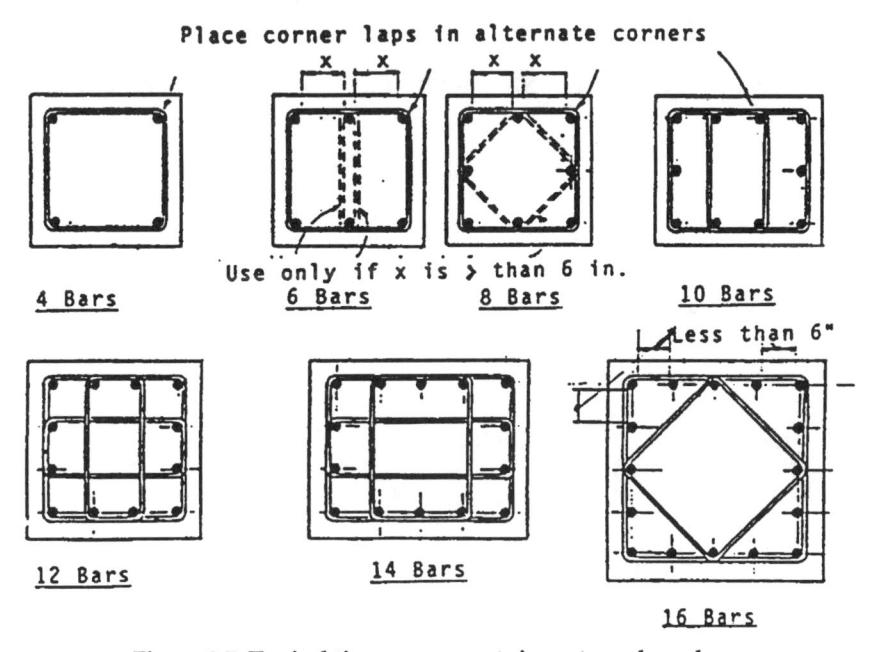

Figure 9.7 Typical tie arrangements in rectangular columns.

Tables such as 9.5 contain values for spiral size and pitch necessary for circular sections. The toughness and the durability of spirally reinforced columns is a characteristic required for survival in earthquakes. Most applications for spiral columns today are related to ductility requirements for structures in seismic zones.

Table 9.5 Spiral size and pitch for round columns.

Column* Diameter (inches)	$f'_c = 4,000$ psi	$f'_c = 5,000$ psi	$f'_c = 6,000$ psi	$f'_c = 8,000$ psi
12	⅜Φ @ 2"	½Φ @ 2¾"	½Φ @ 2¼"	½Φ @ 1¾"
14	⅜Φ @ 2"	½Φ @ 3"	½Φ @ 2½"	½Φ @ 1⅞"
16	⅜Φ @ 2"	½Φ @ 3"	½Φ @ 2½"	½Φ @ 2"
18	⅜Φ @ 2"	½Φ @ 3"	½Φ @ 2½"	⅝Φ @ 3"
20	⅜Φ @ 2"	½Φ @ 3"	½Φ @ 2½"	⅝Φ @ 3"
22	⅜Φ @ 2"	½Φ @ 3"	½Φ @ 2½"	⅝Φ @ 3"
24	⅜Φ @ 2"	⅜Φ @ 1¾"	½Φ @ 2½"	⅝Φ @ 3"
26	⅜Φ @ 2¼"	⅜Φ @ 1¾"	½Φ @ 2½"	⅝Φ @ 3¼"
28	⅜Φ @ 2¼"	⅜Φ @ 1¾"	½Φ @ 2½"	⅝Φ @ 3¼"
30	⅜Φ @ 2¼"	⅜Φ @ 1¾"	½Φ @ 2¾"	⅝Φ @ 3¼"
32	⅜Φ @ 2¼"	⅜Φ @ 1¾"	½Φ @ 2¾"	⅝Φ @ 3¼"
34	⅜Φ @ 2¼"	⅜Φ @ 1¾"	½Φ @ 2¾"	⅝Φ @ 3¼"
36	⅜Φ @ 2¼"	⅜Φ @ 1¾"	½Φ @ 2¾"	

* Spiral size-pitch combinations shown are for standard 1½ in. cover. Outside spiral diameter (core diameter) is 3 in. less than column diameter.

EXAMPLE 9.1

Column Design. Select a size and reinforcement required for a square column at the lowest story of a building where $P_u = 774$ k with $M_u = 103$ ft·k. Use $f'_c = 4000$ psi and Grade 60 steel. (In order to reuse column forms for upper story columns, make the base level column as small as is practical by using ρ_g as high as 5% or so.)

Solution. Step 1—Assume minimum eccentricity governs this heavily loaded column and solve Eq. 9.3.1 for A_g when $\dfrac{A_{st}}{A_g} = 0.05$ and

$$P_u = 0.8\phi P_0 = 0.8\phi A_g \left[0.85 f'_c + \frac{A_{st}}{A_g}\left(f_y - 0.85 f'_c\right) \right]$$

$$774 = 0.8(0.7) A_g \left[0.85(4.0) + 0.05(60 - 0.85 \times 4.0) \right]$$

$$\therefore A_g = 222 \text{ in}^2$$

For a square column, $h = \sqrt{A_g} = \sqrt{222} = 14.9$ in. Select $h = 16$ in, $A_g = 256$ in².

Step 2—Compute actual A_{st} using $P_u = 0.8\phi P_0$ for 16 in × 16 in section:

$$P_u = 0.8\phi P_0 = 0.8\phi \left[0.85 f'_c A_g + A_{st}\left(f_y - 0.85 f'_c\right) \right]$$

$$774 = 0.8(0.7)\left[0.85(4.0)(256) + A_{st}(60 - 0.85 \times 4.0) \right]$$

$$\therefore A_{st} = 9.04 \text{ in}^2$$

Try 8 #10 bars. Tie spacing < least lateral dimension, 16 bar diameter, 48 tie diameter, or 18 inches.

Step 3—Use Table 9.4 Design aid to verify (or improve) choice of reinforcement:

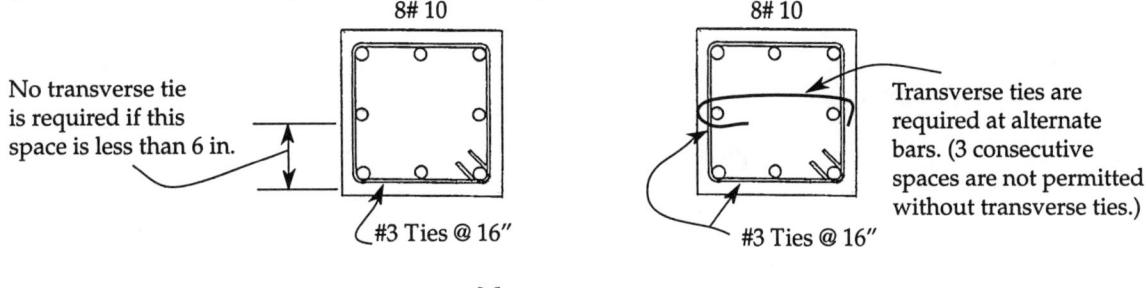

No transverse tie is required if this space is less than 6 in.

8# 10
#3 Ties @ 16"

8# 10
Transverse ties are required at alternate bars. (3 consecutive spaces are not permitted without transverse ties.)
#3 Ties @ 16"

$$e = \frac{M_u}{P_u} = \frac{103 \times 12}{774} = 1.60 \text{ in}$$

Table 9.4 shows $\phi P_n = 709$ k·ft for 8 #10 with 3 bars each face or $\phi P_n = 725$ k·ft for 8 #10 with 4 bars each end face. $\phi P_0 = 809$ k for both bar arrangements.

EXAMPLE 9.2

Column Design. Select reinforcement for upper-floor columns above the column of Example 9.1. Columns are 16 in × 16 in, $f'_c = 4000$ psi, steel is Grade 60, and required loads are tabulated:

	Level	P_u(k)	M_u(ft·k)	$e = \frac{12M_u}{P_u}$
A)	1st floor	686	132	2.30 in
B)	2nd floor	571	122	2.56 in
C)	3rd floor	460	122	3.18 in
D)	Roof	114	149	15.7 in

Solution. Step 1—Use Table 9.4, interpolating ϕP_n for each set of bars.

A) For required $P_u = 686$ k and $e = 2.30$ in, try 8 #10 with 4 bars in each end face,

$$\phi P_n = 725 - \frac{(2.30-2)}{1}(725-609) = 690 > 686 \text{ k}$$

This is acceptable; continue same reinforcement as used for Example 9.1.

B) For required $P_u = 571$ k and $e = 2.56$ in, try 8 #8 with 4 bars in each end face,

$$\phi P_n = 628 - \frac{(2.56-2)}{1}(628-529) = 573 > 571 \text{ k}$$

This is acceptable. Lap splice each #8 above 2nd floor. Since $e = 2.56''$ is less than $e_t = 3.15''$, compression splice length is needed. Eq. 9.5.3 gives us

$$\ell_{db} = 0.02 d_b f_y / \sqrt{f'_c}$$

$$= 0.02(1)60,000/\sqrt{4000} = 19 \text{ in}$$

Extend #8 bars 20 in below top of #10 bars.

C) For required $P_u = 460$ k and $e = 3.18$ in, try 6 #8 with 3 #8 bars in each end face:

$$\phi P_n = 493 - \frac{(3.18-3)}{1}(493-411) = 478 > 460 \text{ k}$$

Continue 3 #8 bars in each end face from floor below.

D) For required $P_u = 114$ k and $e = 15.7$ in, try 6 #9 with 3 #9 bars in each end face:

$$\phi P_n = 158 - \frac{(15.7-12)}{(18.0-12)}(158-96) = 120 > 114 \text{ k}$$

Since $e = 15.7 > 10.93$ inches for "balanced" strains, a full tension splice (Class B) is required. Eq. 9.5.1a with multiplier 1.3 for Class B splice:

$$\ell_{db} = 1.3\left[0.04 A_b f_y / \sqrt{f'_c}\right]$$

$$= 1.3\left[0.04(1.00)60,000/\sqrt{4000}\right] = 49.3 \text{ in}$$

Extend #9 bars 50 inches below top of bars from column below.

9.4 Shear Forces

Shear forces are related to the moments developed in a beam. Two models of the mechanisms that resist shear are shown in Fig. 9.8. ACI318 Section 11.1.1 recognizes two sources of shear strength. Shear resistance from concrete in the compression zone of a section will be designated V_c and shear resistance from stirrups (steel bars) will be designated V_s. The nominal strength V_{cn} of concrete in shear is taken generally as specified in ACI318 Section 11.3.1.1 to be

$$V_{cn} = 2\sqrt{f'_c} b_w d \tag{9.4.1}$$

where

- b_w = width of web of section (stem of T shape or width of rectangular shape)
- d = depth of section from compression edge to tension steel centroid
- f'_c = design strength of standard concrete cylinders in psi

The units require that the coefficient 2 includes dimensionally correct factors to make $2\sqrt{f'_c}$ a stress value in psi units. A more complex formulation for V_c is offered in ACI318 Section 11.3.2.1, generally allowing larger values for V_c in the vicinity of reactions. That Code section can be useful for the designer in special circumstances for which slightly higher values of V_c will show a selected section to be adequate without modification. Otherwise, it is too complex for general use.

(a) Top of section in compression

(b) Bottom of section in compression

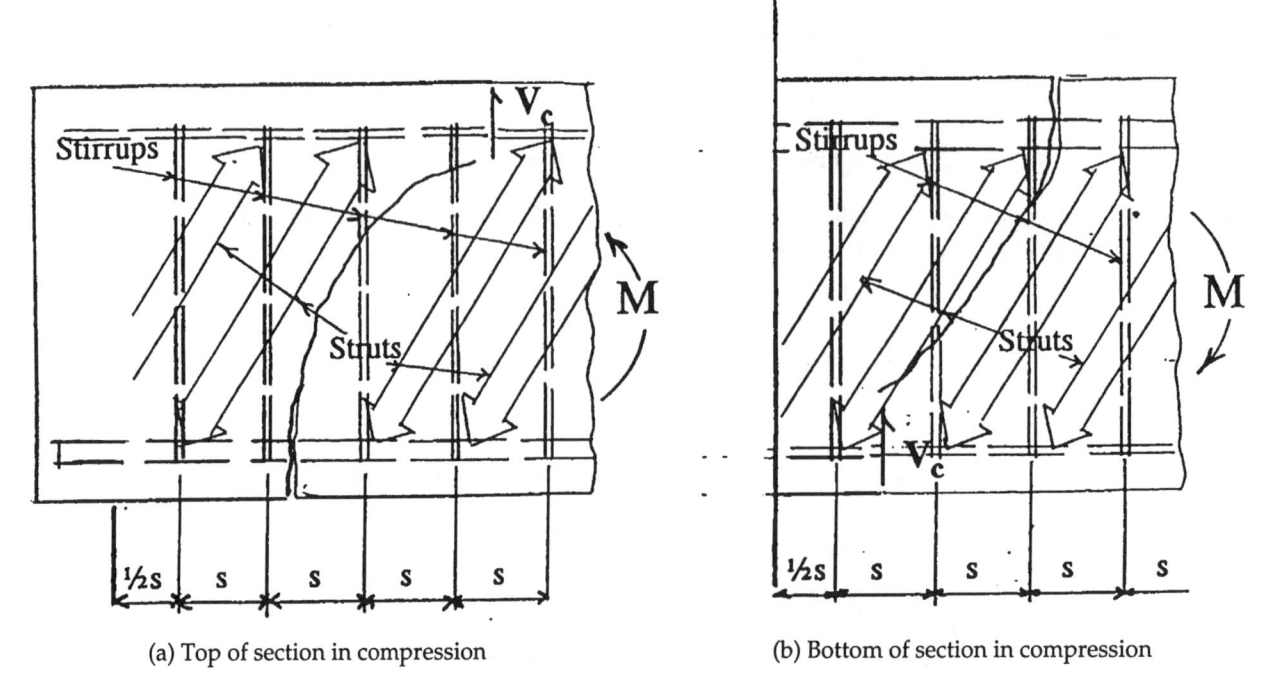

Figure 9.8 Shear forces in beams.

ACI318 Section 11.5.5 requires that shear reinforcement in the form of stirrups must be used for all flexural members in which V_u exceeds $0.5\phi V_{cn}$ except for a) slabs and footings, b) concrete joists, and c) beams for which d is less than 10 in, 2.5 times the flange thickness, or $0.5b_w$. The nominal strength for stirrups acting as tension hangers with concrete that serves as compression bars (struts) is designated V_{sn}. ACI318 Section 11.5.6.2 specifies for stirrups:

$$V_{sn} = A_v f_y d / s \qquad (9.4.2)$$

where

A_v = area of vertical steel bar stirrup; generally A_v will be the area of 2 legs of a U-shaped stirrup

f_y = yield strength of stirrup steel

s = horizontal spacing of vertical stirrups

Since the stirrups are effective in shear only with concrete to provide diagonal compression force components, ACI318 Section 11.5.6.8 limits the amount of V_{sn} to $4V_{cn}$, a stress index level at which the concrete struts fail in compression, and no larger amount of stirrup reinforcement can increase the shear strength. In order to insure that a vertical stirrup intercepts diagonal cracks, the maximum spacing permitted is $s \leq d/2$ according to ACI318 Section 11.5.4.1. Furthermore, ACI318 Section 11.5.4.3 requires that stirrup spacing be not more than $d/4$ if V_{sn} must exceed $4V_{cn}$. If stirrups are required at all, ACI318 Section 11.5.5.3 requires a stirrup area $A_v \geq 50 b_w s$.

The capacity reduction factor for shear specified in ACI318 Section 9.3.2.3 is $\phi = 0.85$.

Design for shear can be made convenient and clear through the superposition of required shear diagrams on a section strength diagram. Each diagram is a graph showing strength on the vertical axis and distance from the face of support on the horizontal axis. Fig. 9.9 illustrates the design method. First, the values for strength ϕV_{cn}, $0.5\phi V_{cn}$, ϕV_{sn} with $s = d/2$, and other possible values of ϕV_{sn} at smaller values of s. Insert horizontal lines at $0.5\phi V_{cn}$, ϕV_{cn}. Then add shear strength values ϕV_{sn} to values of ϕV_{cn} to locate more horizontal lines for shear capacity with stirrups. These lines are shown in Fig. 9.9 as solid lines with labels to indicate each strength line.

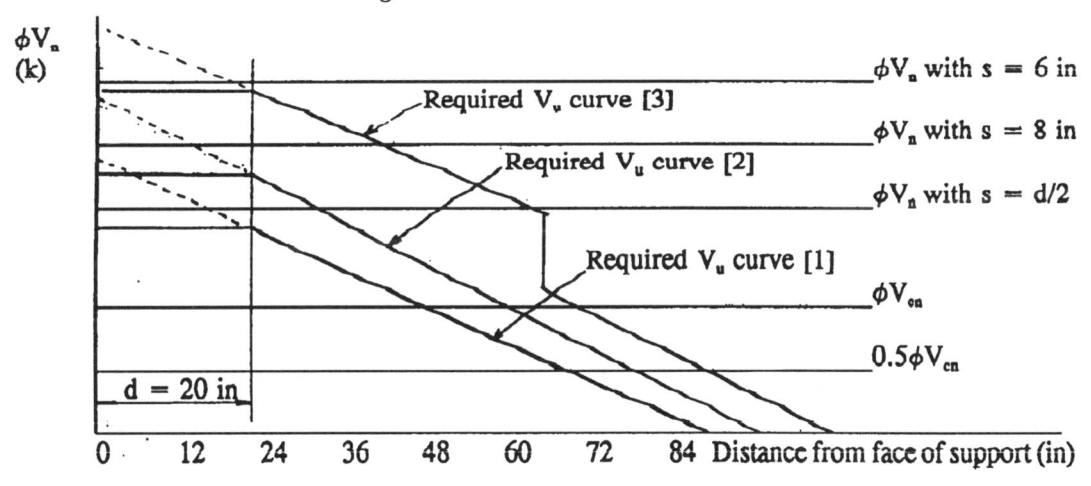

Figure 9.9 Design for shear.

Required shear curves are shown as dashed lines in Fig. 9.9. These curves are typical for the type of required shear required at several different ends of the same continuous beam in a concrete structure. Note that each diagram for required shear is shown with a horizontal line from the face of support to the distance d. ACI318 Section 11.1.3.1, recognizing that diagonal cracks cannot penetrate into a support, allows the designer to consider the shear force at a distance d from the support as the maximum value for which shear strength must be provided. Thus the required shear diagrams are truncated at the distance d from the support.

It is the designer's responsibility to designate required stirrup spacing, generally as a series of dimensions which begin at the face of support. Each stirrup supports shear for a distance $s/2$ each side of the stirrup, and the first space will be $s/2$. For required shear curves of Fig. 9.9:

- Required V_u curve [1]—stirrups at a spacing $d/2$ are adequate for the maximum amount of required shear, and the shear becomes smaller than $0.5\phi V_{cn}$ at a distance of 68" from face of support. For this example, specify spacing as 5", 6 spaces @ 10".

- Required V_u curve [2]—stirrups at 8" spacing are needed for a distance of 31" from the support, and then the $d/2 = 10"$ spacing is adequate until the shear becomes smaller than $0.5\phi V_{cn}$ at a distance of 77" from the support. Specify spacing as 4", 4 spaces @ 8", 4 spaces @ 10".

- Required V_u curve [3]—stirrups at 6" spacing are required for a distance of 42" from the support, then two spaces of 8" would reach the zone for which 10" spacing is adequate until required shear is less than $0.5\phi V_{cn}$ at a distance of 86" from the support. Instead of using 3 different spacings, specify spacing of 3", 9 @ 6", 3 @ 10".

Shear strength of columns can be determined in much the same way as shear strength for beams except that the presence of axial force increases V_{cn} considerably, and required shear has the same value throughout the height of a column. ACI318 Section 11.3.1.2 increases the value of V_{cn} from Eq. 9.4.1 by a factor that includes axial stress:

$$V_{cn} = 2(1 + 0.005 N_u/A_g)\sqrt{f'_c} b_w d \tag{9.4.3}$$

with

A_g = gross area of the column in sq in
N_u = required axial force in the column in lb

Required shear in columns rarely exceeds V_{cn} except for applications in high seismic zones. Column ties can be considered effective as stirrups, although the maximum spacing must not exceed $d/2$ when the ties are needed as stirrups.

───── EXAMPLE 9.3 ─────

T-Beam. Design reinforcement required for the beam shown to support a dead load of 1.8 k/ft and a live load of 1.5 k/ft if $f'_c = 4000$ psi and steel is Grade 50.

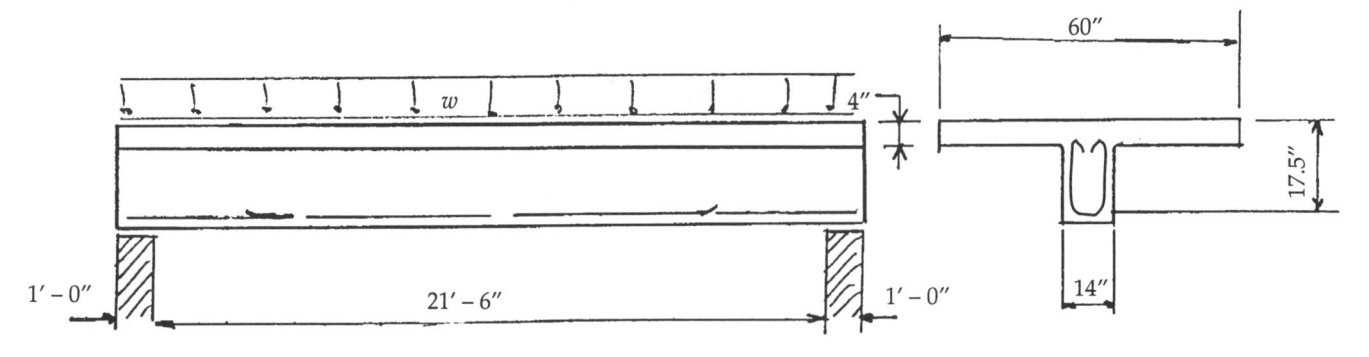

Solution. Step 1—Compute required V_u and M_u (from Eq. 9.1.3a):

$$w_u = 1.4w_D + 1.7w_L$$
$$= 1.4(1.8) + 1.7(1.5) = 5.1 \text{ k/ft}$$

ACI318, Section 11.1.3.1, at d from face of support,

$$V_u = w_u\left(\frac{\ell}{2} - \frac{d}{12}\right) = 5.1\left(\frac{21.5}{2} - \frac{17.5}{12}\right) = 47.4$$

and at midspan

$$M_u = \frac{w_u \ell^2}{8} = \frac{5.1(21.5)^2}{8} = 295$$

Step 2—Select longitudinal bars for flexure (Eq. 9.2.6) modified for Grade 50 steel:

$$A_s = \frac{60}{f_y} \frac{M_u}{4d} = \frac{60}{50} \times \frac{295}{4(21.5)} = 5.06 \text{ in}^2$$

Try 4 #10:
$$A_s = 4(1.27) = 5.08 \text{ in}^2$$

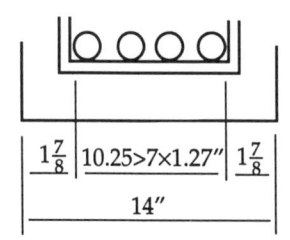

$1\frac{7}{8}$ | 10.25>7×1.27" | $1\frac{7}{8}$
14"

Check, using Eq. 9.2.3:

$$\phi M_n = \frac{\phi A_s f_y}{12}\left[d - \frac{A_s f_y}{1.7 b f_c'}\right]$$

$$= \frac{0.9(5.08)50}{12}\left[17.5 - \frac{5.08(50)}{1.7(60)4.0}\right] = 322 \text{ ft·k} > 295$$

This is acceptable; use 4 #10 18 ft long, starting 2" from each support.

Step 3—Design shear reinforcement using #3 stirrups. Using Eq. 9.4.1,

$$\phi V_c = \phi\left(2\sqrt{f_c'}\right)b_w d$$

$$= 0.85(2)\sqrt{4000}(14)17.5 = 26{,}300 \text{ lb}$$

Using Eq. 9.4.2, compute ϕV_s for $s = \frac{d}{2} \approx 9$ in:

$$\phi V_s = \phi A_v f_y \frac{d}{s} = 0.85(2)0.11(50{,}000)\frac{17.5}{9} = 18{,}200 \text{ lb}$$

Construct shear strength graph (from face of support use s = 3 in, 3 @ 6 in, 8 @ 9 in):

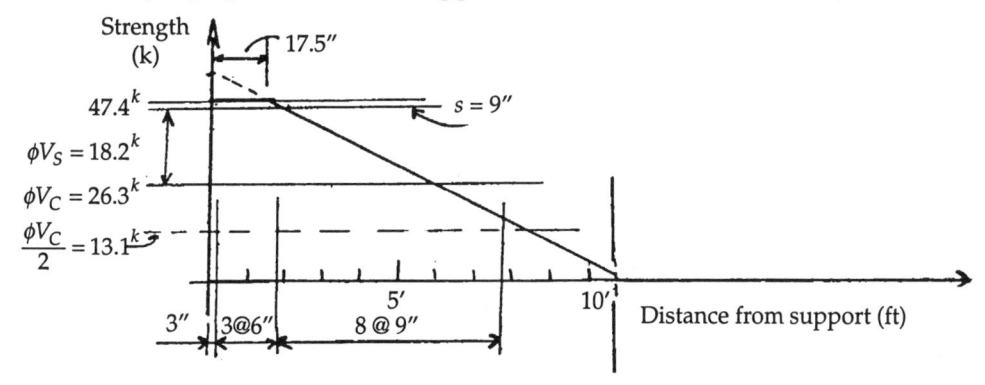

9.5 Bar Development—Concrete/Steel Force Transfer

Every reinforcing bar, from the point at which yield stress is required, must extend into concrete an embedment distance long enough to transfer from concrete to steel the yield force for the bar. As a bar is pulled or pushed in concrete, the deformations on the bar press against the concrete medium, causing shear transfer along the bar while a bursting stress pushes surrounding concrete away from the bar. In order to develop the yield force in a bar, the bar must be anchored into concrete along a distance called the *development length* of the bar. Bars in "favorable" installation conditions for flexural tension require *tension development basic lengths* ℓ_{db} specified in ACI318 Section 12.2.2 as

$$\ell_{db} = 0.04 A_b f_y / \sqrt{f_c'} \qquad \text{for bars \#11 and smaller} \qquad (9.5.1a)$$

$$= 0.085 f_y / \sqrt{f_c'} \qquad \text{for \#14 bars} \qquad (9.5.1b)$$

$$= 0.125 f_y / \sqrt{f_c'} \qquad \text{for \#18 bars} \qquad (9.5.1c)$$

ACI318 Section furthermore specifies a minimum tension bar development length as

$$\ell_{db} \geq 0.03 d_b f_y / \sqrt{f_c'} \quad \text{or } 12'' \qquad (9.5.2)$$

where

A_b = bar area in sq in
d_b = diameter of bar in inches

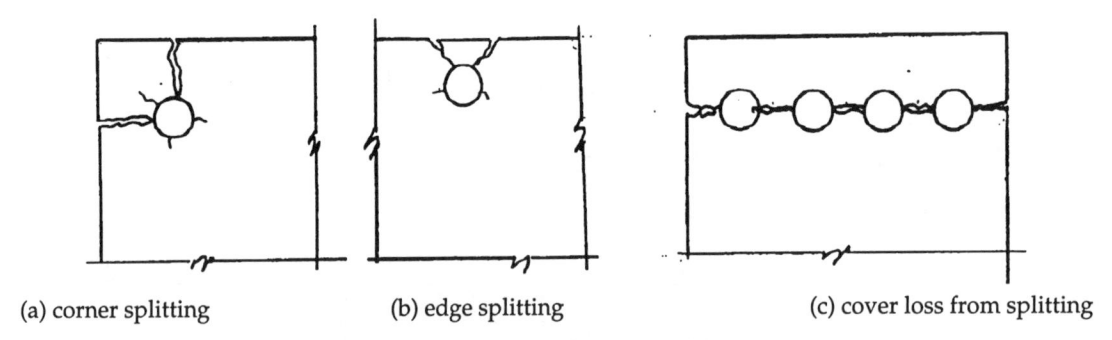

(a) corner splitting   (b) edge splitting   (c) cover loss from splitting

Figure 9.10 Anchorage failure mechanisms.

Some anchorage failure modes caused by the bursting pressure along bars are illustrated in Fig. 9.10. Bars not enclosed by ties or stirrups near the corner of a section as in Fig. 9.10(a) can break out from the concrete more readily than bars buried well within a section. Bars closer than d_b to the edge of a section suggested in Fig. 9.10(b) can split the concrete cover. Closely spaced bars can cause separation of the entire surface cover as shown in Fig. 9.10(c). ACI318 Section 12.2.3 specifies edge and spacing conditions for which development length $\ell_d = 2\ell_{db}$ where ℓ_{db} comes from Eq. 9.5.1:

a) bars have clear spacing less than $2d_b$.
b) bars have clear cover less than d_b.

Values of development length $\ell_d = 1.4\ell_{db}$ (if clear spacing between bars is less than $3d_b$, even though bars are enclosed by ties or stirrups.

As concrete is cast for beams and for thick slabs, the compaction of concrete around bars located near the top of the concrete is less dense than for bars deeper in the concrete. ACI318 Section 12.2.3.3 requires that "top cast bar" development length $\ell_d = 1.3\ell_{db}$, using Eq. 9.5.1 and Eq. 9.5.2, if more than 12 inches of concrete is cast beneath the bars.

Tension lap splices are identified as Class A or Class B reflecting whether or not half the full tension steel yield strength is to be developed through the lap splice. If more than half the yield strength of reinforcement is to be transferred, the splice is a Class B splice. Class A splice lengths are the same as the length ℓ_d. Class B splices with a length $1.3\ell_d$ are permitted for tension bars #11 and smaller. Throughout the length of a lapped bar tension splice, determination of clear spaces between bars must include each of the bars being spliced. Bars #11 and larger may not be lap-spliced for tension.

Compression bar development length can be shorter than for tension, as concrete in longitudinal compression will not be stressed in bi-axial tension. ACI318 Section 12.2.3 specifies compression development lengths as

$$\ell_{db} = 0.02 d_b f_y / \sqrt{f_c'}, \text{ but not less than } 0.0003 d_b f_y, \text{ or } 8". \tag{9.5.3}$$

ACI318 Section 12.14.2.1 permits lap splices of length ℓ_{db} from Eq. 9.5.3 for column bars #10 and smaller. Compression splices for larger bars must be made with mechanical devices that transfer bar forces by direct bearing.

Hooked bars may be used to develop tension forces within distances significantly shorter than those required for straight bars. Force transfer takes place primarily at the bearing of the hooked portion of bar against concrete. Retention of full tension strength in bars after cold bending requires that bend radii be no less than values specified in ACI318 Section 7.2.1, Table 7.2 as $6d_b$ for bars #3 through #8, $8d_b$ for bars #9 through #11, and $10d_b$ for #14 and #18 bars. ACI318 Sections 12.5.2 and 12.2.3 specify development length ℓ_{hb} for bars with standard hooks:

$$\ell_{hb} = 0.02 d_b f_y / \sqrt{f_c'} \tag{9.5.4}$$

If side cover over the region of a hook is greater than $2\frac{1}{2}$ in, ℓ_{hb} can be multiplied by 0.70.

The use of epoxy coated reinforcement requires that development lengths specified in Eq. 9.5.1 through 9.5.3 be increased by a factor of 1.2, and the use of lightweight concrete requires that development lengths specified in Eq. 9.5.1 through 9.5.3 be increased by a factor of 1.3.

EXAMPLE 9.4

Continuous Beam. Select a stem size and all Grade 60 reinforcement for the continuous symmetric beam ($f_c' = 3600$ psi), as sketched on the following page.

Solution. Step 1—Select b_w and h. See Table 9.2 for $\phi k_{good} = 460$ psi when $f_c' = 3600$ psi. Compute $b_w d^2$ at point of max. negative M_u using $\phi k = 500$ psi:

$$\phi k b d^2 = M_u$$

$$b_w d^2 = \frac{264 \times 12,000}{500} = 6336 \text{ in}^3$$

Since columns are 18 in wide, select $b_w = 14"$ to avoid interference of corner bars of beam and column. With $b_w = 14$,

$$d = \sqrt{\frac{6336}{14}} = 21.3 \text{ in}$$

Allow at least 2.5 inches cover to centerline of bars. Make $h = 24$ inches.

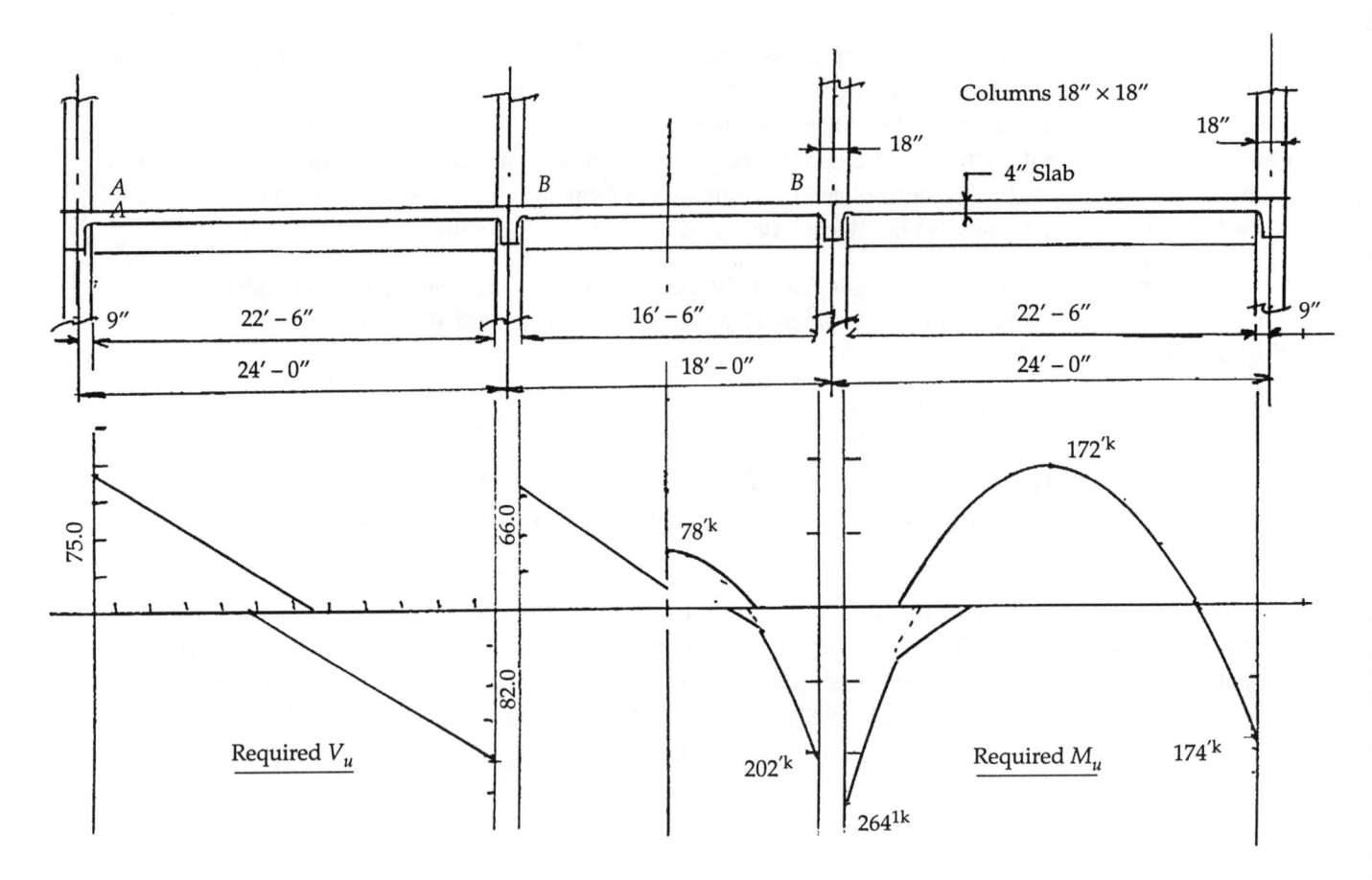

Step 2—Select flexural reinforcement using $d \approx 21.5$ in and Eq. 9.2.6:

$$A_s = \frac{M_u}{4d} = \frac{M_u}{4(21.5)} = \frac{M_u}{86}$$

Moment (ft·k)	A_s (in²)	Use
78	0.91	2 #7, 18 ft extend 9 in into each column
−264	3.07	3 #9, 12 ft long
172	2.00	2 #9, 24 ft extend 9 in into each column
−174	2.02	3 #8 hooked bars, 6 ft long

Extend 1/3 of top bars $\frac{\ell}{16}$ beyond point of inflection.

For #9 top bar using Eq. 9.5.1a with the 1.3 multiplier,

$$\ell_{db} = 1.3\left[0.04 A_b f_y / \sqrt{f'_c}\right]$$

$$= 1.3\left[0.04(1)60,000/\sqrt{4000}\right] = 49.3 \text{ in}$$

For #8 top bar,

$$\ell_{db} = 1.3\left[0.04(0.79)60,000/\sqrt{4000}\right] = 39.0 \text{ in}$$

Using Eq. 9.5.4,

$$\ell_{hb} = \left[0.02 d_b f_y / \sqrt{f'_c}\right] 0.7 \leftarrow \text{confined by column}$$

$$= \left[0.02(1)60,000/\sqrt{4000}\right] 0.7 = 13.3 \text{ in}$$

This is acceptable; use 18 in column.

$$\text{Hooked Bar length} = 15 \text{ in col.} + 39 \text{ in to pt. of infl.} + \frac{24}{16} \text{ ft} = 72 \text{ in}$$

Step 3—Determine spacing for #3 stirrups. Construct shear strength graph.

Using Eq. 9.4.1,

$$\phi V_c = 0.85\left(2\sqrt{f_c'}\right)b_w d$$

$$= 0.85\left(2\sqrt{3600}\right)14(21.5) = 30,700 \text{ lb}$$

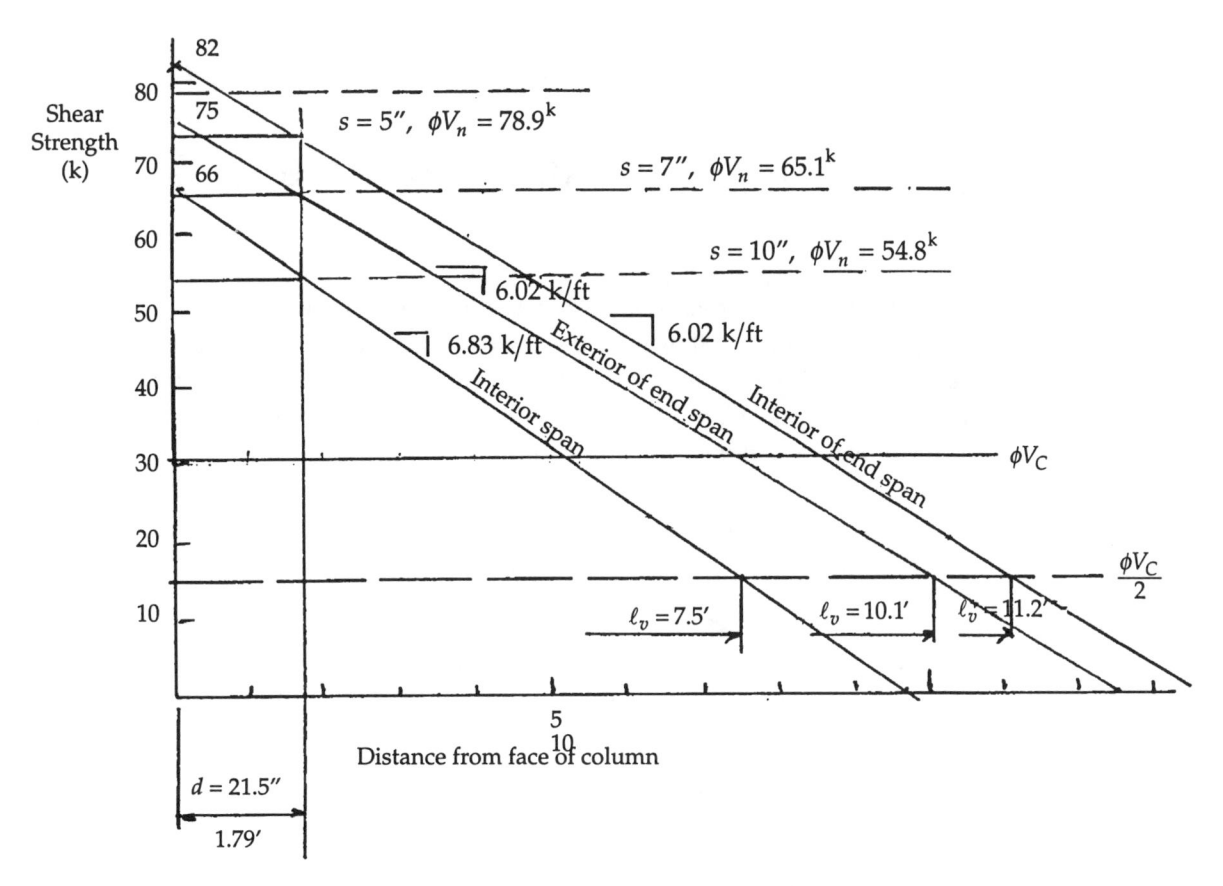

Since $d = 21.5$ in, maximum stirrup spacing $= 10.7$ in. Use max. stirrup spacing of 10 in. Use Eq. 9.4.2 to compute ϕV_s values for $s = 10$ in, 7 in, and 5 in,

$$\phi V_s = \phi A_v f_y d / s$$

$$= 0.85(2)0.11(60,000)21.5/s = 241,000/s$$

For $s = 10$ in, $\phi V_s = 24,100$ and $\phi V_n = 30,700 + 24,100 = 54,800$ lb

$s = 7$ in, $\phi V_s = 34,400$ and $\phi V_n = 30,700 + 34,400 = 65,100$ lb

$s = 5$ in, $\phi V_s = 48,200$ and $\phi V_n = 30,700 + 48,200 = 78,900$ lb

Interior span: graph shows that $s = 10"$ makes $\phi V_n >$ all values V_u. Use $s = 5$ in, 8 spa @ 10 in (extends for $\ell_v = 90$ in).

Exterior end of end span: make $s = 7"$ until $s = 10"$ is adequate. Use $s = 3$ in, 6 @ 7 in (extends for $\ell_v = 120$ in).

Interior end of end span: make $s = 5"$ until $s = 7"$ is adequate, then make $s = 7$ in until $s = 10$ in is adequate. Use s = 2 in, 6 @ 5 in, 4 @ 7 in, 7 @ 10" (extends for ℓ_v = 135 in).

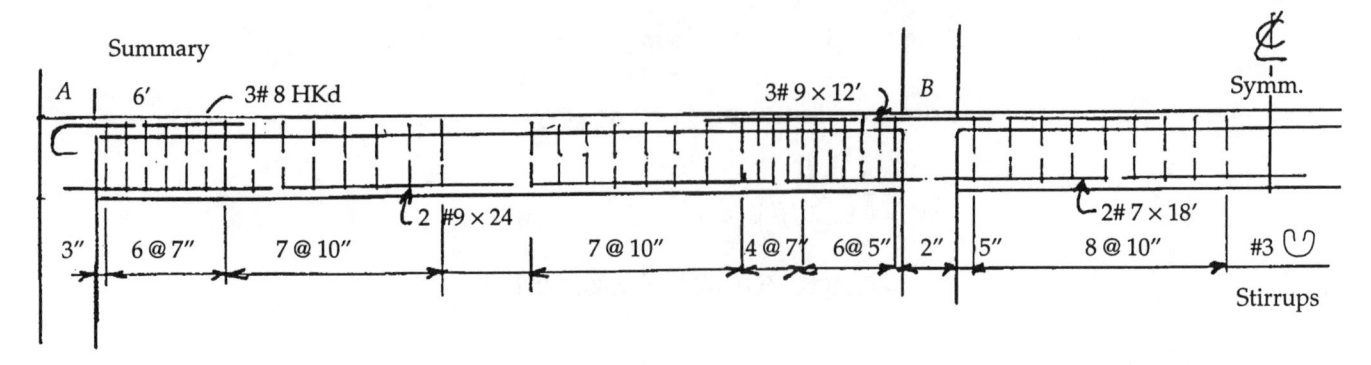

Summary

[Beam reinforcement diagram: From support A to centerline B (symm.): top bars 3#8 HKd at 6', 2#9×24 and 3#9×12'; bottom bars 2#7×18'; stirrups #3 U-stirrups spaced 3", 6@7", 7@10", 7@10", 4@7", 6@5", 2", 5", 8@10"]

9.6 Components of Concrete Structures

9.6.1 Slabs

Slabs are designed as wide beams, but each is analyzed as a beam of unit width—generally 1 ft. For most conditions of loading, slab thickness will be governed by provision of flexural stiffness adequate to limit vertical deflections under service loading. Table 9.5(a) of ACI318 Section 9.5.2.1 is a tabulation of minimum thickness values for which slabs can be considered stiff enough for control of deflections. The thickness decreases as end support conditions help limit vertical deflection. For example, if ℓ_n is the clear span between faces of supports, a simply supported slab should have a thickness greater than $\ell_n/20$, whereas, the same slab continuous across an interior support should have a thickness greater than $\ell_n/24$, and as an interior span of a continuous slab, the thickness should be greater than $\ell_n/28$.

As wide beams, slabs do not require shear reinforcement. Very heavy loading (above 1500 psf) may force slab thickness to be increased above that recommended for deflection control such that slab shear strength V_{cn} is adequate to support required heavy shear loads.

ACI318 Section 7.7.1 requires at least 3/4" clear concrete cover over reinforcement for slabs. ACI318 Section 7.6 also limits the spacing of slab bars to 3 times the slab thickness or 18 in, whichever is smaller.

Reinforcement area A_s (sq in per ft of width) can be provided by placing parallel bars at spacings such that the average amount of steel per ft satisfies longitudinal steel requirements. The average area of steel per ft will be $12A_{bar}/s$, where s is the lateral space between bar centerlines. If Eq 9.2.6 indicated a need for 0.77 in^2/ft, then $12A_{bar}/s \geq 0.77$. The spacing of #6 bars (A_{bar} = 0.44 in^2) must be no greater than $s \leq 12A_{bar}/A_s = 12(0.44)/0.77 = 6.9$ in. In the same application, required spacing of #7 bars (A_{bar} = 0.60 in^2) could be $s \leq 12(0.60)/0.77 = 9.35$ in. Specifying #7 bars at 9 inch spacing is an appropriate choice.

EXAMPLE 9.5

Slab. Determine a thickness and all reinforcement required for the loading dock slab supported along one edge and along a wall as shown. Use $f_c' = 4000$ psi and Grade 60 steel. Live load is 1400 psf.

Solution. Step 1—Estimate $DL = 250$ psf (slab wt). Then,

$$1.4w_D = 1.4(250) = 350 \text{ psf}$$

$$1.7w_L = 1.7(1400) = \underline{2380 \text{ psf}}$$

$$\text{Total } w_u = 2730 \text{ psf}$$

Step 2—Analyze 1 ft wide strip for 3 Load cases.

Load Case I, w_{total} A to C:

$$R_B = 2730(21)(10.5)/14.5$$
$$= 41,500 \text{ lb}$$

$$R_A = 2730(21) - 41,500$$
$$= 15,800 \text{ lb}$$

Load Case II, w_{total} A to B:

$$R_B = \left[2730(14.5)(7.25) + 350(6.5)17.75\right]/14.5$$
$$= 22,600 \text{ lb}$$

$$R_A = 2730(14.5) + 350(6.5) - 22,600$$
$$= 19,300 \text{ lb}$$

$$M_{\max} \text{ at } \frac{R_A}{w_u} = \frac{19,300}{2730} = 7.07 \text{ ft}$$

$$M_{\max} = \frac{(19,300)7.07}{2} = 68,200 \text{ ft·lb}$$

Load Case III, w_{total} B to C:

$$R_B = \left[350(14.5)(7.25) + 2730(6.5)17.75\right]/14.5$$
$$= 24,300 \text{ lb}$$

$$R_A = 350(14.5) + 2730(6.5) - 24,300$$
$$= -1500 \text{ lb}$$

$$M_B = -2730(6)(3) = -49,100 \text{ ft·lb}$$

Step 3—Select slab thickness:

A) ACI318 Table 9.5(a) with one end continuous $h > \dfrac{\ell}{24}$:

$$\frac{\ell}{24} = \frac{14 \times 12}{24} = 7 \text{ in}$$

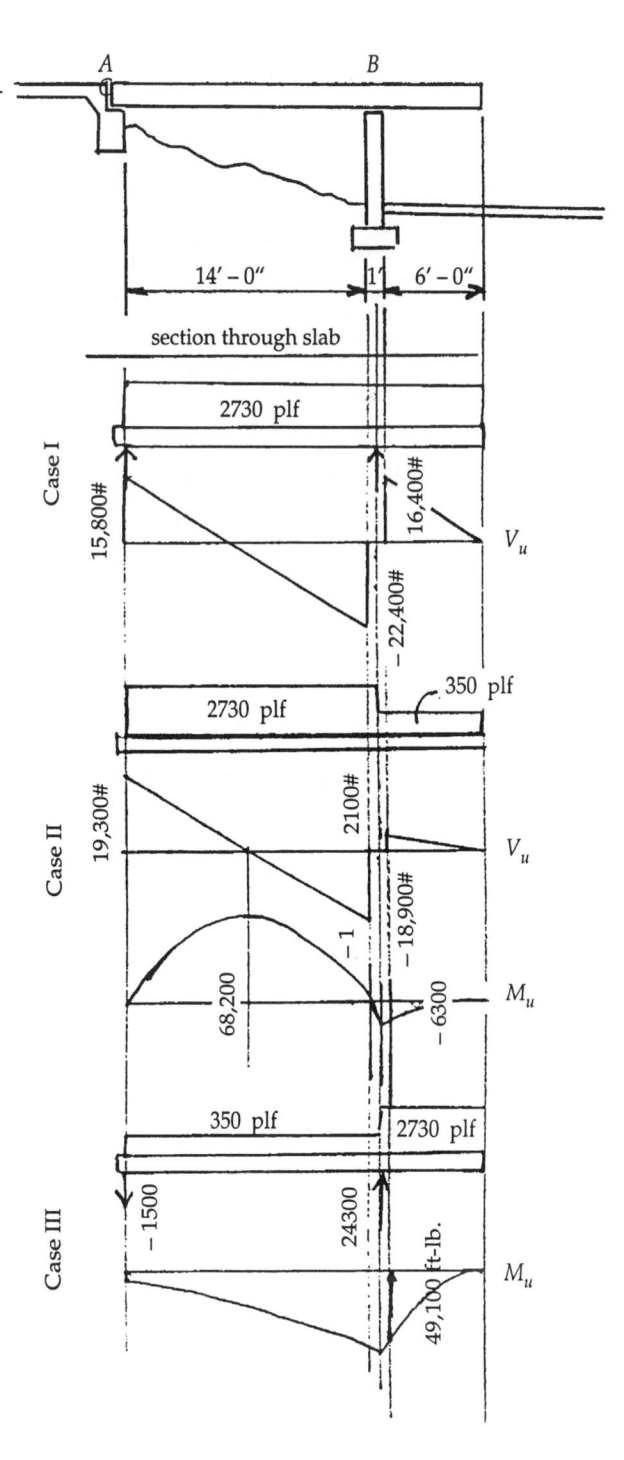

B) For $\phi V_n > V_u$ at about 6 in from B = $22,400 - \frac{2730 \times 6}{12} = 21,000$ lb. Using Eq. 9.4.1

$$\phi V_n = 0.85(2\sqrt{4000})12d > 21,000$$

$$d > 16.3 \text{ in}$$

Try $h = 16$ in, $d \approx 14.8$ in. Since V_u occurs at d from support, $V_u = 22,400 - \frac{2730 \times 14.8}{12} = 19,000$ lb

$$\phi V_n = 0.85(2\sqrt{4000})12(14.8) = 19,100 > 19,000$$

This is large enough to be acceptable. Check estimated self weight, which was 250 psf:

$$\text{actual } w_D = \frac{16 \times 150}{12} = 200 \text{ psf}$$

The estimated value was close enough.

Step 4—Select Reinforcement:

A) In bottom of slab, with $M_u = 68,200$ ft·lb (68.2 ft·k), using Eq. 9.2.6:

$$A_s = \frac{M_u}{4d} = \frac{68.2}{4(14.8)} = 1.15 \text{ in}^2/\text{ft}$$

Use #7 bars @ 6 in c/c, 16 ft long.

B) In top of slab, with $M_u = 49,100$ ft·lb (49.1 ft·k):

$$A_s = \frac{M_u}{4d} = \frac{49.1}{4(14.8)} = 0.83 \text{ in}^2/\text{ft}$$

Use #6 bars @ 6" c/c. Use half of #6 bars 21 ft long and the other half 14 ft long.

C) Shrinkage and temperature steel (see section 9.6.1):

$$A_{min} = 0.0018(12)16 = 0.35 \text{ in}^2/\text{ft}$$

Use #4 at 12" c/c top and bottom.

Step 5—Sketch slab section as designed:

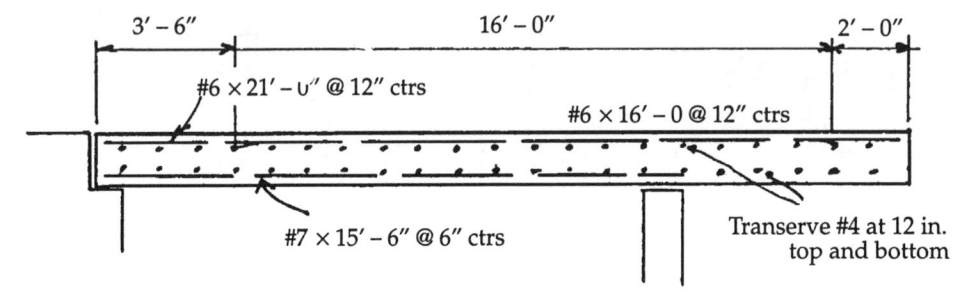

9.6.2 Joists

Joists are ribbed slabs. In current construction practice, standard moulds for joists are placed on a flat plywood deck and concrete is placed over the moulds and deck to form joists. Dimensions of standard forms are shown in Fig. 9.11 with a typical cross section through floor joists. ACI318 Section 8.11 specifies some limit conditions within which floor beams can qualify as joists. Joist ribs must be at least 4 inches wide and have below the slab a depth not more than 3.5 times the width of the ribs. The maximum clear distance between ribs is 30 in, and the slabs between ribs must have a thickness of at least one-twelfth the clear span between ribs but not less than 2 in.

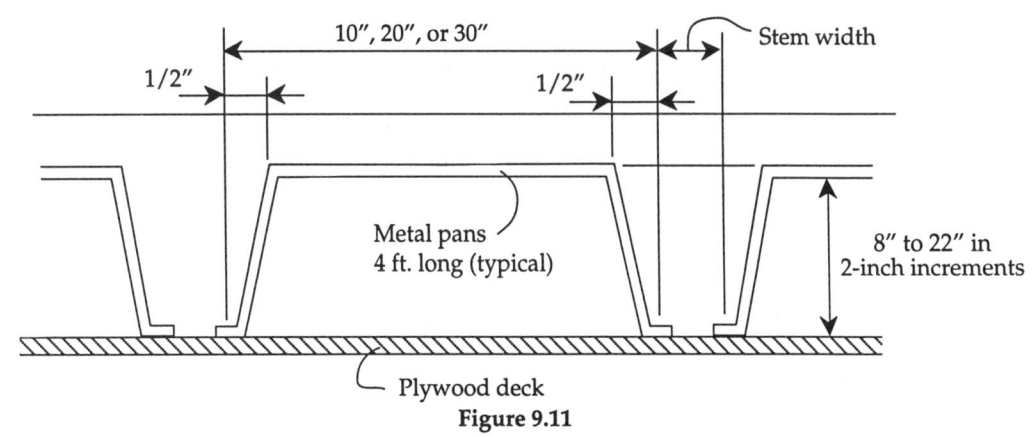

Figure 9.11

ACI318 Section 8.11.8 permits an increase of 10% for joists for the value of V_{cn} according to Eq. 9.4.1. The efficiency of joists would be lost if shear reinforcement were used in joists. Consequently, the joist ribs are made wide enough to support required shears without stirrups, or the rib is made wider near regions of critically high shear. Rib width can be made 4 inches wider at supports by means of tapered forms for the region adjacent to a supporting beam.

The depth of joists is governed largely by service load deflection stiffness requirements. Table 9.5(a) of ACI318 Section 9.5.2.1 contains minimum thickness values for ribbed slabs (joists) if deflections are not to be computed. After joist thickness has been established, the joist stems are made wide enough to sustain required shear without stirrups. Tapered forms for shear are used only in a few regions of a floor system, as wider joists would be less expensive than the frequent use of tapered forms.

Flexural reinforcement for positive moment generally consists of one or two bars, one of which must be continuous along the bottom of the joist and embedded into a support for its tension bar development length ℓ_d (ACI318 Section 12.11.1 and Section 7.13.2). If only one bar is used for positive moment reinforcement, that bar must extend into each end support far enough for its tension bar development length ℓ_d. Flexural reinforcement for negative moment may consist of one to 3 bars directly over the stem of the joist, or the flexural steel can be uniformly distributed in the slab each side of the joist centerline. ACI318 Section 12.2.3 requires that at least 1/3 of the bars for negative moment must be extended into the span for a distance 1/16th of the span beyond the point of inflection. Clear cover of concrete for joists is the same as that for slabs according to ACI318 Section 7.7.1.

EXAMPLE 9.6

Floor Joist. Design a floor to span 19 ft–8 in clear between concrete masonry walls with f'_c = 4000 psi, Grade 60 reinforcement, and 30-inch pan joist forms. Allow 12 psf for floor cover and utilities in addition to joist self weight plus a live load of 120 psf.

Solution. Step 1—Use ACI318 Table 9.5(a) to find $h > \ell/16$ for ribbed one-way slabs and beams:

$$h > \frac{19.67(12)}{16} = 14.75 \text{ in}$$

Use a 3-inch slab on 12"×30" pans such that $h = 12 + 3 = 15$ in.

Step 2—Estimate self weight for 6-in stems $= \dfrac{15 \times 36 - 12 \times 30}{144} \times 150$

$$= 188 \text{ lb/ft}$$

Additional $DL = 12 \times 3 = 36$ lb/ft

Total $DL = 224$ lb/ft

Total $LL = 120 \times 3 = 360$ lb/ft

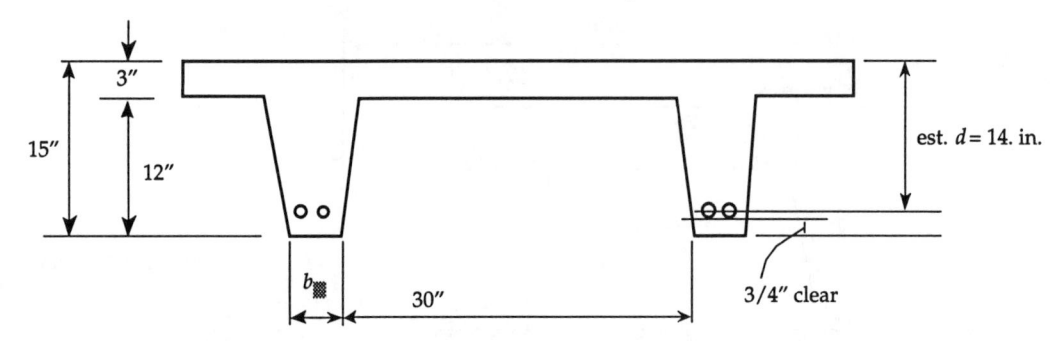

Step 3—Compute required load:
$$w_u = 1.4 w_D + 1.7 w_L$$
$$= 1.4(224) + 1.7(360) = 926 \text{ lb/ft}$$

Compute V_u at a distance 14 in from support:
$$V_u = w_u\left(\dfrac{\ell}{2} - \dfrac{d}{12}\right) = 926\left(\dfrac{19.67}{2} - \dfrac{14}{12}\right) = 8030 \text{ lb}$$

Compute M_u at midspan:
$$M_u = \dfrac{w_u \ell^2}{8} = \dfrac{926(18.67)^2}{8} = 40{,}300 \text{ ft} \cdot \text{lb}$$

Step 4—Select b_w such that $\phi V_c > V_u$ (see Eq. 9.4.2 and Section 9.6.2):

$$(1.10) 2\phi \sqrt{f_c'} b_w d > V_u$$
$$b_w > \dfrac{8030}{(1.10)2(0.85)\sqrt{4000}(14)} = 4.85 \text{ in. Use 5 in}$$

Step 5—Use Eq. 9.2.6 to determine required A_s:

$$A_s = \dfrac{M_u(\text{ft} \cdot \text{k})}{4d} = \dfrac{40{,}300}{1000(4)14} = 0.72 \text{ in}^2$$

Use 2 #6 bars 21 ft long. Check:

$$\rho = \dfrac{A_s}{b_w d} = \dfrac{2 \times 0.44}{5 \times 14} = 0.0126 > \rho_{\min} = \dfrac{3\sqrt{f_c'}}{f_y} = \dfrac{3\sqrt{4000}}{60{,}000} = 0.0031$$

Step 6—Since bar radius makes actual $d = 15 - \dfrac{3}{4} - \dfrac{3}{8} = 13.88$ in, check both V_u and M_u. Required V_n is

$$V_n = V_u/\phi = 8030/0.85 = 9450 \text{ lb}$$

But

$$V_n = b_w d(2\sqrt{f_c'})1.10$$
$$= 5(13.87)(2\sqrt{4000})1.10 = 9650 \text{ lb.} \quad \therefore \text{ OK}$$

Required M_n is

$$M_n = M_u/\phi = 40,300/0.90 = 44,800 \text{ ft·lb}$$

Using Eq. 9.2.3

$$M_n = \frac{1}{12} A_s f_y \left[d - \frac{A_s f_y}{1.7 b f_c'} \right]$$

$$= \frac{0.88(60,000)}{12} \left[13.87 - \frac{0.88(60,000)}{1.7(36)4000} \right] = 60,100 \text{ ft·lb.} \quad \therefore \text{ OK}$$

9.6.3 Beams—Reinforcement for Structural Continuity

Beams require stirrup reinforcement and it is good practice to place at least one bar in each corner of the beam section. ACI318 Section 12.11.1 and 7.13.2 apply for beams as well as joists. In order to insure that at least 1/3 of the positive moment reinforcement is extended into supports, it is convenient to use 3 or more bars, 1/3 of which extend into one support and the remainder into the opposite support. Bar embedment through supports must be for a bar length ℓ_d for tension. At exterior supports, it is necessary to use hooked bars in order to develop most top (negative moment) bars. In order to maintain a bar in each corner of a section, top bars should be spliced near midspan.

Stirrups should be extended and bent around and tied to the corner longitudinal bars, both for development length anchorage of the stirrup and for alignment of stirrups during casting.

ACI318 Section 7.6.1 requires that bars be spaced no closer than 1 inch or one bar diameter for bars larger than #8. If bars are placed in more than one vertical layer, bars in each layer must be placed directly above bars in lower layers to allow concrete to flow between vertical rows of bars during casting. ACI Section 7.7.1 requires at least $1\frac{1}{2}$ in clear concrete cover over beam stirrups.

9.6.4 Columns—Slenderness Effects and Reinforcement Details

ACI318 Section 7.6.1 requires $1\frac{1}{2}$ in clear cover over column ties or spirals, and Section 7.6.3 requires that the lateral clear space between reinforcement be at least 1.5 diameters of longitudinal bars or $1\frac{1}{2}$ in. Longitudinal column bars should be spliced near midheight where required moments are generally smaller than those at the top or bottom ends.

Concrete columns that have a clear height less than 12 times the thickness of the column will experience no detrimental effect from slenderness. A concrete column with a height more than 12 times its thickness can be made to bend away from the line of action of its axial compression force enough to make the moments between column ends larger than the moment at either end. Analysis will show that the largest moments in columns occur when live load is placed on spans that frame into the same side of a column. That loading creates moments of virtually identical magnitude at the top and at the bottom of the column and there is a point of inflection (zero moment) at midheight. ACI318 Section 10.11.5 allows the estimate of a ratio δ_b between maximum moment and end moment as follows:

$$\delta_b = \frac{0.6}{1 - P_u/(\phi P_c)} \geq 1.0 \tag{9.6.1}$$

where

P_u = required (factored) load on column
P_c = effective Euler buckling load = $\pi^2 E_c I_g / [2.5(1+\beta_d)L_n^2]$
I_g = moment of inertia of gross outline of column section
L_n = clear height of column
β_d = creep factor = $1.4 P_D/P_u$

If the value of δ_b is less than 1 from Eq. 9.6.1, the maximum moment remains at the end of the column, and none of the moments between the ends will be larger than the larger of the two end moments.

EXAMPLE 9.7

Column Slenderness. Compute δ_b, given $P_D = 62$ k, $P_L = 80$ k, $\ell_n = 19$ ft – 6 in, and column 14 in × 14 in section with $f'_c = 5000$ psi.

Solution. Step 1—Compute:

$$P_u = 1.4 P_D + 1.7 P_L$$

$$= 1.4(62) + 1.7(80) = 223 \text{ k}$$

$$I_g = \frac{bh^3}{12} = \frac{14(14)^3}{12} = 3201 \text{ in}^4$$

$$E_c = 33 w^{1.5} \sqrt{f'_c}$$

$$= 33(145)^{1.5} \sqrt{5000} = 4,070,000 \text{ psi}$$

$$\beta_d = 1.4 P_D/P_u$$

$$= 1.4(62)/223 = 0.39$$

Step 2—Compute:

$$P_c = \frac{\pi^2 E_c I_g}{2.5(1+\beta_d)\ell_n^2}$$

$$= \frac{\pi^2 (4,070,000) 3201}{2.5(1+0.39)(19.5 \times 12)^2} = 676,000 \text{ lb}$$

Step 3—Compute:

$$\delta_b = \frac{0.6}{1 - P_u/\phi P_c}$$

$$= \frac{0.6}{1 - 223/0.7(676)} = 1.13$$

9.6.5 Walls

Reinforced concrete walls can serve as columns to support vertical loads, and their high in-plane stiffness and strength can be used to resist lateral forces. Many walls are installed as partitions for fire resistant barriers (stairwells, elevator shafts) as their primary purpose. The vertical and in-plane forces on such walls are very small with respect to the limit strength available in the wall.

ACI318 Section 14.3 requires that all walls contain at least enough vertical and horizontal reinforcement to control cracking due to restrained shrinkage. Empirical rules dictate that walls 10 or more inches thick contain reinforcement in two faces, not more than 2 inches from the exterior face of wall nor less than 3/4 inch from the interior face. It is common practice to use $1\frac{1}{2}$ inch clear cover over identical sets of bars in each face. Vertical reinforcement area must be at least 0.0012 times the gross concrete area if Grade 60 bars size #5 or smaller are used. The ratio between area of vertical steel and gross concrete area must be increased to 0.0015 if bars larger than #5 or Grade less than 60 is used. Horizontal reinforcement area must be at least 0.002 times gross concrete area if Grade 60 bars of size #5 or smaller are used, and if bars of size greater than #5 or Grade less than 60 are used, the ratio must be increased to 0.0025. The spacing of bars must be no greater than 3 times the wall thickness nor 18 inches.

Walls that are intended as compression members under factored vertical loads and for which the resultant of all factored vertical loads is within the middle third of the wall thickness can be designed on the basis of the empirical relationship

$$\phi P_{nw} = 0.55 \phi f'_c A_g \left[1 - (k\ell_c / 32h)^2\right] \quad (9.6.2)$$

where

P_{nw} = nominal axial force capacity of wall
h = wall thickness
ℓ_c = clear height of wall
k = effective length factor taken as 1 if top and bottom of wall are unrestrained from rotation, taken as 0.8 if top and bottom of wall are restrained from rotation, and taken as 2 if top of wall is free to move laterally
ϕ = 0.70

The walls must have a thickness at least $\ell_c/25$ or 4 inches, whichever is smallest.

Walls on which the resultant of the applied vertical force acts outside the middle third of the thickness must be designed as if the wall were a column (Section 9.3), and longitudinal bars must be tied laterally in accordance with the rules for tied rectangular columns.

Walls used for resistance of lateral forces on a frame are called shear walls. Vertical reinforcement and horizontal reinforcement both contribute to shear strength. The shear strength of concrete is taken the same as for Eq. 9.4.1 or Eq. 9.4.2, and the strength V_s of horizontal reinforcement A_v at a vertical spacing of s_2 is computed by

$$V_s = A_v f_y d / s_2 \quad (9.6.3)$$

The space s_2 must not exceed $3h$, $\ell_w/5$, or 18 inches. The ratio ρ_n of vertical reinforcement area to gross concrete area of horizontal section shall not be less than

$$\rho_n \geq 0.0025 + 0.5(2.5 - h_w/\ell_w)(\rho_h - 0.0025) \quad (9.6.4)$$

where h_w = height of wall from base to top

ℓ_w = horizontal length of wall

ρ_h = ratio between horizontal bar area and gross concrete area vertical section

Frequently, the forces P_u and M_u from factored loads create a total combined axial and flexural stress that is well within the strength of concrete (compression stress less than $0.5 f'_c$). The creation of an interaction diagram for low axial forces on a shear wall may not be necessary to verify the strength of the wall section. Instead, the longitudinal elastic stresses on the gross cross section under factored loads can be estimated as

$$f = \frac{P_u}{A_g} \pm \frac{M_u}{24h^2} \quad \text{per foot of wall length} \tag{9.6.5}$$

The volume of the tension stress wedge will indicate the force for which longitudinal tension reinforcement can be provided. The centroid of longitudinal bars should coincide with the centroid of the tension stress wedge. This "elastic stress" procedure will be safe and it will provide for somewhat more reinforcement than that determined with the limit strength interaction diagram and compatible strains when maximum compression strain = 0.003 for the wall concrete.

EXAMPLE 9.8

Empirical Design of Wall. Select a thickness and specify the reinforcement required for a wall 18 ft high from slab to joist bearings if the joist load $w_u = 3.2$ k/ft. Use $f'_c = 3000$ psi and Grade 60 reinforcement.

Solution. Step 1—Use thickness limit $\frac{\ell}{25} = h$ to determine h:

$$h = 18 \times 12/25 = 9 \text{ in}$$

Step 2—Check load capacity with $h = 9$ in, using Eq. 9.6.2:

$$\phi P_{nw} = 0.55\phi f'_c A_g \left[1 - \left(k\ell_c/32h\right)^2\right]$$

$$= 0.55(0.70)3000(12)(9)\left[1 - \left(\frac{18 \times 12}{32 \times 9}\right)^2\right]$$

$$= 54{,}600 > 3200 \text{ lb/ft}. \quad \therefore \text{ OK}$$

Step 3—Minimum reinforcement will be adequate:

$$\text{vertical steel} = 0.0012 A_g$$

$$= 0.0012(12)9 = 0.13 \text{ in}^2/\text{ft}$$

#4 @ 18 in provides 0.13 in²/ft.

$$\text{horizontal steel} = 0.0020 A_g$$

$$= 0.0020(12)9 = 0.22 \text{ in}^2/\text{ft}$$

#4 @ 10 in provides 0.24 in²/ft.

Use 9-inch thick wall with #4 @ 18 in vertical bars and #4 @ 10 in horizontal bars in each face.

9.6.6 Footings

The purpose of a footing is to distribute the concentrated forces at the base of a column over an area large enough to keep bearing stresses below a serviceable bearing stress limit for the soil beneath the footing. It is common practice today for a geotechnical consultant to prepare a report on subgrade conditions at a building site, recommending values of allowable bearing stress in the soils. Since vertical displacement from bearing stress is more significant and more predictable than actual bearing "failure", the geotechnical report considers only the probable service (unfactored) load conditions anticipated for a structure. Consequently, the geotechnical report gives values for allowable stress on subgrade material under design conditions of full dead load and maximum design live load as if all load factors were taken to be 1.0. Footing sizes are determined for service load conditions, and then required load factors are applied to column reactions. Bearing stresses are recomputed for the factored forces on the footing in order to compute required shear and moment forces on concrete and reinforcement in the footing. Most footings are called *isolated* footings, as each footing supports one column, but *combined* footings can be used to support two or more columns.

Design soil bearing pressures on footings are in the order of several thousand lbs/ft², and as suggested in Section 9.6.1, for such high pressures the thickness of slabs is governed by shear requirements. After footing thickness has been established for required shear strength, reinforcement for flexure can be computed. Isolated footings deform in the shape of a shallow dish that is deepest beneath the column. Flexural bars at right angles to one another must extend in both directions from the face of columns. Bars must be selected such that there is adequate bar development length between the face of column and the end of each bar.

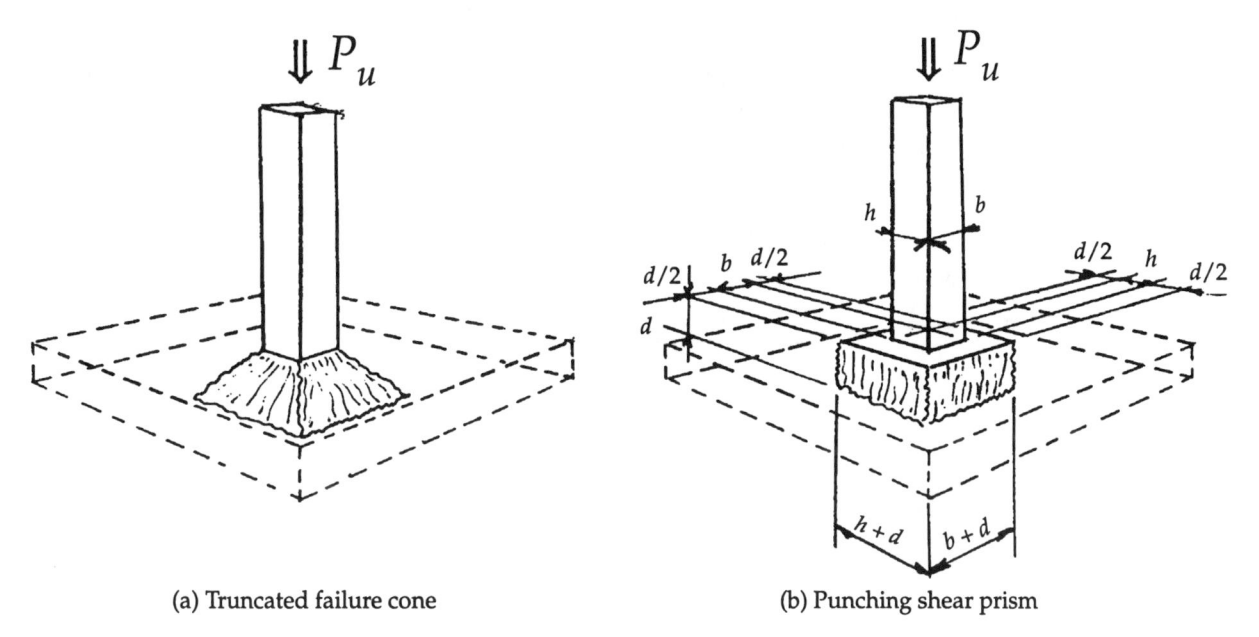

(a) Truncated failure cone   (b) Punching shear prism

Figure 9.12 Punching shear failure of footing.

A shear failure called *punching* or two-way shear can occur as illustrated in Fig. 9.12. The column separates along the tension failure surface which appears to be a truncated pyramid. If d is taken as the average depth of the footing, the failure force can be estimated as a limit shear stress of $4\sqrt{f'_c}$ acting on the prism located at a distance $d/2$ from the face of the column. The punching shear strength relationship becomes

$$V_u \leq (2 + 4/\beta_c)\phi\sqrt{f'_c}[2(b+d) + 2(h+d)]d \leq 4\phi\sqrt{f'_c}b_0 d \qquad (9.6.6)$$

where

b = width of column

b_0 = perimeter of shear prism = $2(b + d) + 2(h + d)$

h = thickness of column

d = distance from top of footing to top of bars in lower layer of bars

β_c = ratio between long side and short side of column, significant only if $\beta_c > 2$

ϕ = 0.85

The "face" of a circular column is considered to be a face for a rectangle of the same area as the circular section.

Beam shear can also occur. Strength is computed with Eq. 9.4.1 for a section at a distance d from the face of the support (column) and a width b_w equal to the width of the footing (Eq. 9.4.1). It can be shown that punching shear always governs required thickness of square or round columns on square footings. Both punching shear and beam shear must be checked if the column is rectangular or if the footing is rectangular. Beam shear is critical only for the shorter width of a footing.

Flexural steel must be provided for moments in the cantilevered slab outside each face of the column. Footings are flexural members, and the rules for minimum reinforcement in Section 9.2 apply also for footings. For reinforcement in the short direction of rectangular footings, an alternate minimum reinforcement rule of ACI318 Section 10.5.2 applies. That requirement states that reinforcement is adequate if it is at least 1/3 more than that required for computed required loads.

EXAMPLE 9.9

Square Footing. Design a square footing to support a 16 in × 16 in column for which $P_D = 312$ k, $P_L = 196$ k, $M_D = 0$, and $M_L = 52$ ft·k. Allowable bearing pressure = 6000 psf. Use Grade 60 reinforcement and $f'_c = 4000$ psi. Assume no overburden on top of footing.

Solution. Step 1—Determine required size of footing. Estimate self weight as 300 psf such that toe pressure less self weight can be 6000 − 300 = 5700 psf. Equate toe pressure to stress caused by P and M from column. Call the footing width B and length L:

$$\frac{5700 \text{ psf}}{1000 \text{ k/ft}} < \frac{P}{BL} + \frac{6M}{BL^2}$$

$$= \frac{(312 + 196)}{BL} + \frac{6(52)}{BL^2}$$

$L = B$, so $5.700B^3 = 508B + 312$. Trial and error:

if $B = 10$, $5700 \neq 5392$

if $B = 11$, $7587 \neq 5900$

if $B = 9.6$, $5043 \neq 5189$

Use footing 9 ft 9 in × 9 ft 9 in.

Step 2—Determine required thickness. Compute net bearing pressures:

$$f_{brg} = \frac{1.4P_D + 1.7P_L}{BL} \pm \frac{(1.4M_D + 1.7M_L)6}{BL^2}$$

$$= \frac{(1.4 \times 312 + 1.7 \times 196)}{9.75^2} \pm \frac{(0 + 1.7 \times 52)6}{9.75^3}$$

$$= 8.10 \pm 0.57 = 7.52 \text{ psf at heel and } 8.67 \text{ psf at toe}$$

If punching (two way) shear governs, the required shear acts on the prism defined at a distance $d/2$ from the face of the column:

$$V_u = 8.10\left[9.75^2 - \left(\frac{16+d}{12}\right)^2\right]$$

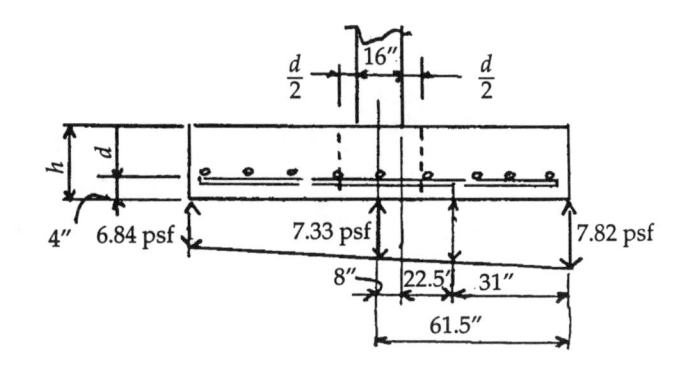

From Eq. 9.6.6,

$$\phi V_n = 4\phi\sqrt{f_c'}b_0 d$$

$$= \frac{4(0.85)\sqrt{4000}(4)(16+d)d}{1000}$$

$$= 0.860(16+d)d$$

Equate

$$8.10\left[9.75^2 - \left(\frac{16+d}{12}\right)^2\right] = 0.860(16+d)d$$

and solve by trial and error:

$$\text{if } d = 20 \text{ in}, \quad 697 > 619$$

$$\text{if } d = 22 \text{ in}, \quad 689 < 719$$

$$\text{if } d = 21.5 \text{ in}, \quad 691 < 693$$

Clear concrete cover must be at least 3 in beneath reinforcement. Make footing thickness at least 4 in more than d, allowing for bar diameter plus 3 in cover. Round off thickness to next value in inches. Use h - 26 in. For bars in the upper layer, assume $d_b = 1$ in to make $d_{short} = 26 - 3 - 1.5 = 21.5$ in. Check one way (beam shear). One way shear is on the plane located at a distance d from the face of the column. The sketch above shows net pressures on the bottom of the footing. Required shear will be the product of net bearing pressure times the area of footing from the edge to the plane located at d from the face of column. At 21.5 in. from the face of column, net pressure = $8.10 + 0.57(8 + 21.5)/58.5 = 8.39$ ksf. The required V_u is

$$V_u = 9.75\frac{(58.5 - 29.5)}{12}\frac{(8.39 + 8.67)}{2} = 201 \text{ k}$$

The one-way shear capacity is

$$\phi V_n = \phi\left(2\sqrt{f_c'}\right)bd$$

$$= 0.85\frac{\left(2\sqrt{4000}\right)}{1000}9.75(12)21.5 = 270 \text{ k} > 201 \text{ k}. \quad \therefore \text{ OK}$$

Step 3—Select flexural steel. It is customary to use the same set of flexural bars in each direction for square footings in order that there be no hazard from confusion with bar orientation. Required moment occurs at the face of the column, and it will be a maximum in the direction of maximum net pressure. At the face of column, net pressure = $8.10 + 0.57(8)/58.5 = 8.18$ ksf. The required M_u is

$$M_u = 9.75\frac{(58.5-8)}{12(2)}\left[3.67\frac{(58.5-8)2}{12(3)} + 8.18\frac{(58.5-8)}{12(3)}\right] = 734 \text{ k-ft}$$

From Eq. 9.2.6,

$$A_s = \frac{M_u}{4d} = \frac{734}{4(21.5)} = 8.53 \text{ in}^2$$

Try 11 #8 bars, checking first to see that available development length of 58.5 − 8 − 3 = 47.5 in is adequate. From Eq. 9.5.1a,

$$\ell_{db} = 0.04 A_b f_y / \sqrt{f'_c}$$

$$= 0.04(0.79)60,000/\sqrt{4000} = 30 \text{ in} < 47.5 \quad \therefore \text{OK}$$

Check ϕM_n with 11 #8 bars. From Eq. 9.2.3,

$$\phi M_n = \phi A_s f_y \left[d - A_s f_y/(1.7 b f'_c)\right]\frac{1}{12}$$

$$= 0.9(11)(0.79)60\left[21.5 - \frac{11(0.79)60}{1.7(117)4}\right]\frac{1}{12}$$

$$= 815 > 734 \text{ ft·k}. \quad \therefore \text{OK}$$

Summary: footing 9 ft 9 in × 9 ft 9 in and 26 m thick with 11 #8 × 9 ft 3 in each way. Allow 3 in clear cover at bottom.

―――― **EXAMPLE 9.10** ――――

Rectangular Footing. Design a footing not more than 7 ft wide to support an 18-in diameter column load $P_D = 218$ k, $P_L = 164$ k, negligible moment, with $f'_c = 3000$ psi, Grade 60 reinforcement and allowable soil bearing pressure of 5000 psf. Allow 200 psf for soil overburden.

Solution. Step 1—Determine footing size. Deduct 200 psf overburden and 300 psf self weight allowance from allowable bearing pressure. Then,

$$\frac{(5000-200-300)}{1000} \leq \frac{(P_D + P_L)}{B \times L}$$

$$4.5 \leq \frac{(218+164)}{7 \times L}$$

$$L \geq 12.1 \text{ ft}$$

Use footing 7 ft by 12 ft.

Step 2—Determine required thickness. Compute ultimate net pressure:

$$p_{net} = \frac{P_u}{B \times L}$$

$$= \frac{1.4(218) + 1.7(164)}{7 \times 12} = 6.95 \text{ k/ft}$$

Since footing is long and narrow, beam shear probably governs instead of punching shear.

"Equivalent" square for circular column,

$$h_c = \sqrt{\frac{\pi D_c^2}{4}}$$

$$= \sqrt{\frac{\pi(18)^2}{4}} = 15.9 \text{ in}$$

From Eq. 9.4.1,

$$\phi V_{cn} > V_u$$

$$\phi(2\sqrt{f'_c})bd > p_{net}(B)\left(\frac{L}{2} - \frac{h_c}{24} - \frac{d}{12}\right)$$

$$0.85\left(\frac{2\sqrt{3000}}{1000}\right)84d > 6.95(7)\left(\frac{12}{2} - \frac{15.9}{24} - \frac{d}{12}\right)$$

$$7.82d \leq 259.7 - 4.05d$$

$$d > 21.9 \text{ in}$$

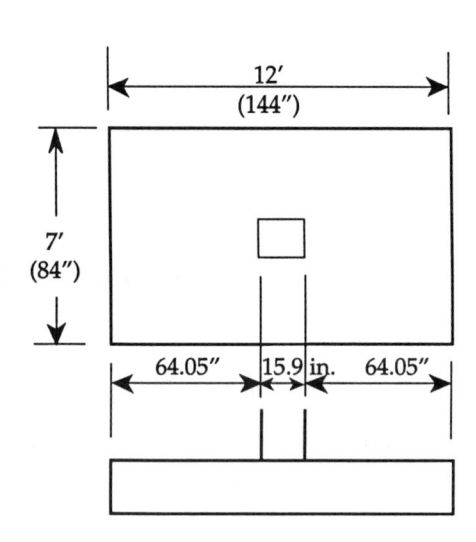

Check punching shear if $h = 26$ in, $d_{avg} = 22$ in. From Eq. 9.6.6,

$$\phi V_{cn} = \phi \frac{(4\sqrt{f'_c})}{1000} \pi \times (D_c + d) \times d$$

$$= 0.85 \frac{(4\sqrt{3000})}{1000} \pi \times (18 + 22) \times 22$$

$$= 515 \text{ k}$$

$$V_u = p_{net}\left[B \times L - \left(\frac{D_c + d}{12}\right)^2 \frac{\pi}{4}\right]$$

$$= 6.95\left[7 \times 12 - \left(\frac{18 + 22}{12}\right)^2 \frac{\pi}{4}\right]$$

$$= 523 > 515 \text{ k} \quad \text{by less than 2\%}$$

Try to use flexural bars less than 1 in diameter in order to make effective $d > 22$ in.

Step 3—Design flexural steel. Bars in long direction:

$$M_u = p_{net}B\left(\frac{L}{2} - \frac{h_c}{24}\right)^2 \frac{1}{2}$$

$$= 6.95(7)\left[\frac{12}{2} - \frac{15.9}{12}\right]^2 \frac{1}{2} = 532 \text{ ft} \cdot \text{k}$$

Using Eq. 9.2.6,

$$A_s = \frac{M_u}{4d}$$

$$= \frac{532}{4(22.5)} = 5.91 \text{ in}^2$$

Use 10 #7 11 ft 6 in long in long direction. Bars in short direction:

$$M_u = p_{net}L\left(\frac{B}{2} - \frac{h_c}{24}\right)^2 \frac{1}{2}$$

$$= 6.95(12)\left[\frac{7}{2} - \frac{15.9}{12}\right]^2 \frac{1}{2} = 197 \text{ ft} \cdot \text{k}$$

$$A_s = \frac{M_u}{4d} = \frac{197}{4(21.5)} = 2.29 \text{ in}^2$$

Compute the minimum A_s:

$$A_s = \frac{4}{3}\frac{M_u}{4d} = \frac{4}{3}(2.29) = 3.06 \text{ in}^2$$

$$\text{or } A_s = \frac{3\sqrt{f_c'}}{f_y} \times (12L)d = \frac{3\sqrt{3000}}{60,000} \times (12 \times 12)21.5 = 8.48 \text{ in}^2$$

3.06 governs 8.48. Therefore use 12 #5 6 ft 6 in long in short direction.

9.6.7 Cantilever Retaining Walls

Cantilever type retaining walls of the type illustrated in Fig. 9.13 are practical for grade changes of 5 to 25 ft. The stem of the wall is designed as a cantilevered slab anchored to the base. The base is a one-way footing that develops a maximum soil pressure at the toe of the base while supporting the weight of soil backfill above the base. The soil behind retaining walls must be drained, as the accumulation and retention of water behind a wall will increase the hydrostatic lateral pressure of soil against the stem.

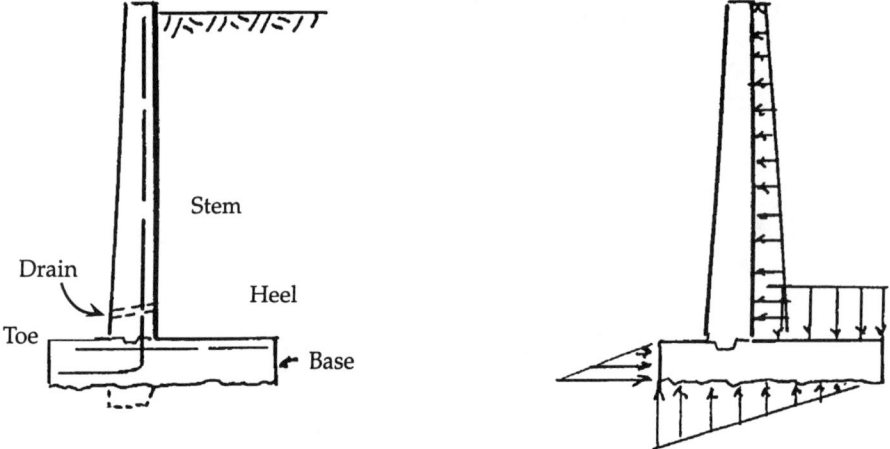

Figure 9.13 Retaining wall and service loads.

Design is based on strength to resist soil pressures against the stem and against the base. Soil pressure against the stem is taken generally as a hydrostatic pressure from a fluid that has a density the same as the soil and a horizontal pressure that increases with a rate that is between 30% and 50% of the soil density. An equivalent fluid pressure of 40% of the vertical weight is considered to be safe for well drained backfill. Vertical pressure on the heel of the stem is simply the weight of backfill over the base. An additional weight of backfill called *surcharge* is used to represent actions caused by loads which act on the backfill. Surcharge loads increase both the horizontal pressure against the stem and the vertical pressure on the heel of the base.

Reaction pressure from the soil beneath the base is thought to vary linearly from a maximum value at the toe of the base to lower pressure or even no pressure at the heel of the base. The assumption of a linearly varying reaction pressure on the base can be used in equilibrium equations to determine a value for maximum bearing pressure at the toe. All pressures, active and reactive are considered as a serviceability condition. After satisfactory service load pressures are established for a given configuration of the retaining wall, it is recommended that service load bearing pressures and horizontal soil pressures against the stem be increased by a load factor of 1.7 when concrete strength requirements are computed. A load factor of 1.4 is appropriate for the service load weight of soil on the heel.

Horizontal reactions to equilibrate soil pressure against the stem consist of passive soil pressure against the base and shear acting along the bottom of the base. Some designers favor the use of a shear key below the base as suggested by the dashed lines of Fig. 9.13.

EXAMPLE 9.11

Cantilever Retaining Wall. Check the toe pressure against an allowable value of 4500 psf for the retaining wall shown. Indicate all reinforcement required if f'_c = 3000 psi and steel is Grade 50.

Soil weight = 110 lb/ft³

Pressure coefficient = 0.40

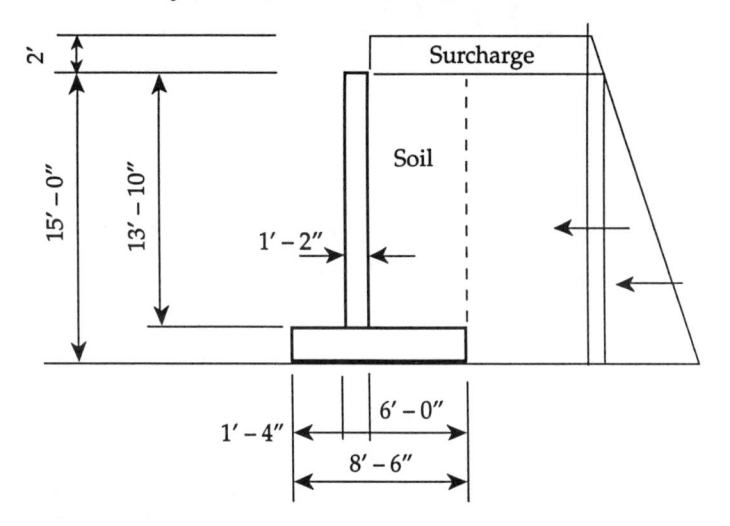

Solution. Step 1— Determine pressures on retaining wall 1 ft strip. Lateral earth pressures:

Surcharge pressure: q = 0.40 (110 lb/ft²) 2 ft = 88 lb/ft

Soil at bottom of stem: q = 0.40 (110) (13.83) = 608.5 lb/ft

Soil at bottom of base: q = 0.40 (110) (15.00) = 660 lb/ft

Vertical forces and moments about heel of base:

Force (lbs)	Moment arm (ft)	Moment (lb-ft)
Surcharge 110 × 2 × 6 = 1320	3	3960
Soil 110 × 13.83 × 6 = 9130	3	27380
Stem 150 × 13.83 × 1.17 = 2430	6.58	15970
Base 150 × 1.17 × 8.50 = 1490	4.25	6340
Total Vertical 14,370 lb		53,650 ft-lb

Moment from lateral pressures:

Surcharge: $88 \times 15 \times 15 \times \frac{1}{2}$ = 9900

Soil: $660 \times 15 \times \frac{1}{2} \times 15 \times \frac{1}{3}$ = 24,750

Total moment about heel = 88,300 ft-lb

Resultant located @ $\frac{88,300}{14,370}$ = 6.14 ft from heel.

Since the resultant is not within the middle third of the base, bearing pressure is a triangular diagram:

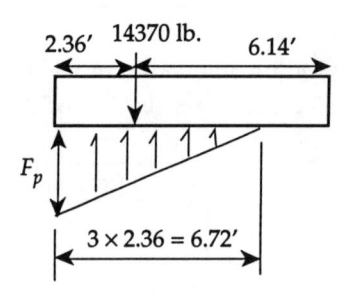

$$\frac{1}{2}F_p(6.72) = 14{,}370$$
$$\therefore F_p = 4280 \text{ psf}$$

Step 2—Select stem reinforcement. ACI318—7.12.2 Minimum $A_s = 0.002(14)(12) = 0.337 \text{ in}^2$. #6 @ 15 in → 0.35 in²/ft. Required M_u at base of stem = 1.7 (service load moment):

$$M_u = 1.7\left[88(13.83)\frac{13.83}{2} + 608.5\frac{13.83}{2}\frac{13.83}{3}\right] = 47{,}300 \text{ ft·lb}$$

Eq. 9.2.6 with $d \approx 11$ and Grade 50 steel:

$$A_s = \frac{M_u}{4d} \times \frac{60}{50} = \frac{47.3}{4(11)} \times \frac{60}{50} = 1.29 \text{ in}^2$$

Try #7 @ 6 in → 1.20 in²/ft. Actual $d = 14 - $ (clear cover) $-$ (bar radius) $= 14 - 2 - 0.44 = 11.56$ in.
Using Eq. 9.2.3,

$$\phi M_n = \phi A_s f_y \left[d - \frac{A_s f_y}{1.7 b f_c'}\right]\frac{1}{12}$$

$$= 0.9(1.20)50\left[11.56 - \frac{0.9 \times 1.20(50)}{1.7(12)5}\right]\frac{1}{12} = 49.6 \text{ k·ft}$$

This is acceptable. Moment capacity with minimum $A_s = 0.35 \text{ in}^2/\text{ft}$:

$$\phi M_n = 0.9(0.35)50\left[11.6 - \frac{0.9 \times 0.35(50)}{1.7(12)5}\right]\frac{1}{12} = 15 \text{ k·ft}$$

At 9 ft from top (4.83 above bottom):

$$M_u = 1.7\left[88\frac{(9)^2}{2} + 44\frac{(9)^3}{6}\right] = 15{,}100 \text{ ft lb}$$

Use #6 @ 15 in for top 9 ft of wall. Extend #7 bars up wall at least ℓ_{db} above end of #6 bars. From Eq. 9.5.1a,

$$\ell_{db} = 0.04 A_b f_y / \sqrt{f_c'}$$

$$= 0.04(0.6)50{,}000/\sqrt{3000} = 22 \text{ in}$$

Then $\left(4.83 + \frac{22}{12}\right) = 6$ ft 8 in above top of base.

Step 3—Select heel reinforcement.

$$A_s = \frac{M_u}{4d} \times \frac{60}{50} = \frac{34.14}{4(11)} \times \frac{60}{50} = 0.93 \text{ in}^2$$

Use #7 @ 8 in → 0.90 in² / ft. Using Eq. 9.4.1, check:

		Arm	Moment
1.7 (1320) Surcharge	= 2240	3'	6720
1.4 (9130) Soil	=12780	3'	38340
-1.4 (5540)	= -7760	4.22'/3	-10920
V_u	= 7260	M_u =	34,140

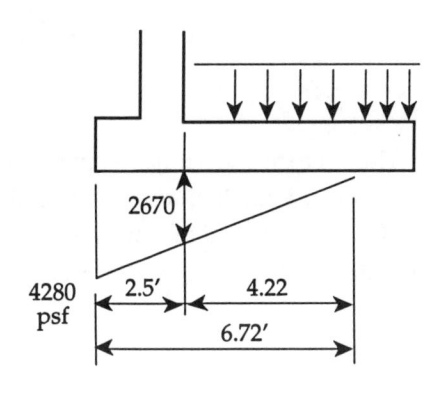

Use Eq. 9.2.6 with Grade 50 steel:

$$A_s = \frac{M_u}{4d} \times \frac{60}{50} = \frac{34.14}{4(11)} \times \frac{60}{50} = 0.93 \text{ in}^2$$

Use #7 @ 8 in → 0.90 in² / ft. Using Eq. 9.4.1, check:

$$\phi V_{cn} = 2\phi\sqrt{f'_c}\,bd$$

$$= 2(0.85)\sqrt{3000}\,(12)11 = 12,300 > 7260 \text{ lb}$$

Summary:

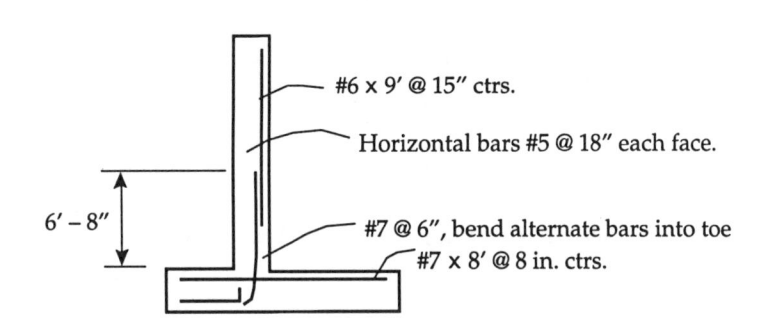

#6 x 9' @ 15" ctrs.

Horizontal bars #5 @ 18" each face.

6' – 8"

#7 @ 6", bend alternate bars into toe
#7 x 8' @ 8 in. ctrs.

Practice Problems (PE-Format)

The T beam shown has a span of 26 ft. (25 ft. clear plus half of each support wall thickness of 12 in.). Service dead load, including the weight of the structure, is taken to be 0.75 kips/ft. Service live loads consist of a uniform load of 1.30 kips/ft and a concentrated load of 25 kips applied at mid-span. The material strengths are $f'_c = 4$ ksi, and $f_y = 60$ ksi.

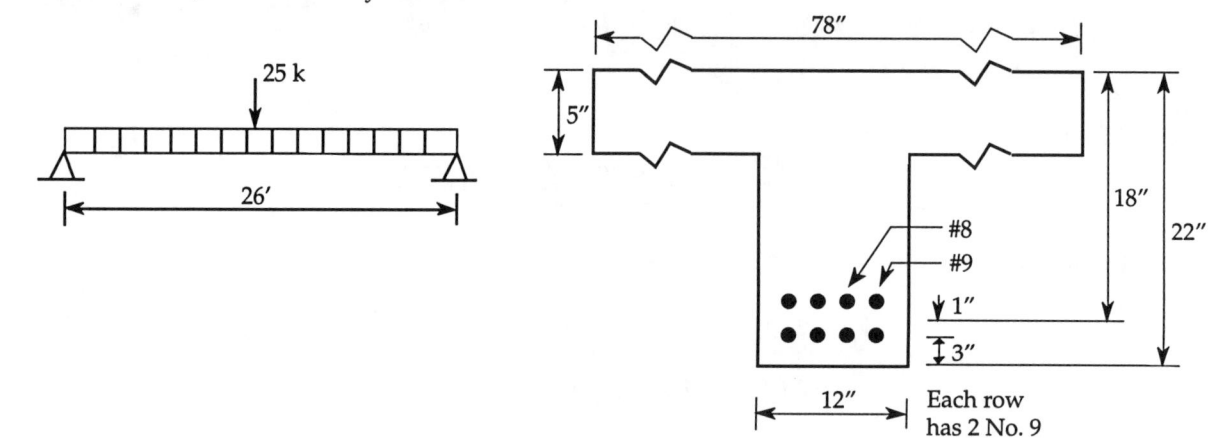

Each row has 2 No. 9 and 2 No. 8 bars.

9.1 Calculate the maximum design moment M_u.
   a) 650 ft-k   b) 550 ft-k   c) 450 ft-k   d) 350 ft-k   e) 250 ft-k

9.2 For the design cross-section shown, compute the moment capacity.
   a) 650 ft-k   b) 550 ft-k   c) 450 ft-k   d) 350 ft-k   e) 250 ft-k

9.3 For the purpose of considering shearing strength, determine the maximum shear force.
   a) 42 k   b) 46 k   c) 51 k   d) 57 k   e) 62 k

9.4 Assuming No. 3 stirrups, calculate the spacing needed at the maximum shear section.
   a) 4.5 in   b) 5 in   c) 6 in   d) 6.5 in   e) 7 in

9.5 Determine the minimum length of the top layer of flexural steel reinforcing bars.
   a) 14 ft   b) 15 ft   c) 16 ft   d) 17 ft   e) 18 ft

The column section shown is symmetric and designed with $f'_c = 3.5$ ksi, and $f_y = 60$ ksi. Consider possible bending about the y-y axis.

$h = 20''$, $2.5''$ from each side to bar, overall $12''$ deep with $2.5''$ cover, #9 bars.

ℓ_u with end moments 120 ft-k (top) and 60 ft-k (bottom).

9.6 Calculate the balanced axial load P_b.
    a) 315 k   b) 340 k   c) 355 k
    d) 370 k   e) 385 k

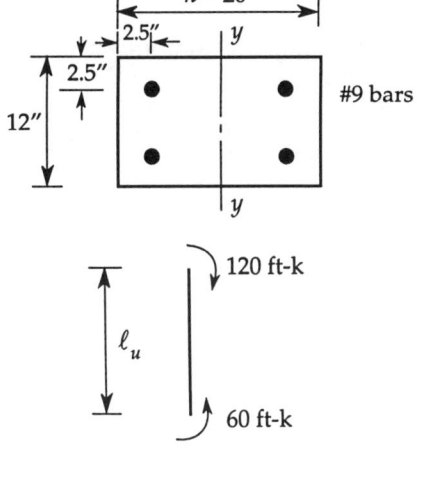

9.7 Calculate the balance moment M_b.
    a) 200 ft-k   b) 250 ft-k   c) 300 ft-k
    d) 350 ft-k   e) 400 ft-k

9.8 Determine the axial load capacity.
    a) 750 k   b) 800 k   c) 850 k   d) 900 k   e) 950 k

9.9 Determine the spacing needed for 3/8 in ties.
    a) 12 in   b) 16 in   c) 18 in   d) 24 in   e) 30 in

9.10 Assuming that the column is part of a braced frame, with end moments as shown, what is the maximum length of the column for which the slenderness effects may be neglected in design calculations?
    a) 10 ft   b) 12 ft   c) 13 ft   d) 14 ft   e) 15 ft

9.11 Suppose the section is to carry an axial load of 600 kips. Based on the results obtained above, and ignoring strength reduction factors ϕ, give the best estimate of the maximum eccentricity with which this load may be applied to the section.
    a) 3.6 in   b) 3.3 in   c) 3.1 in   d) 2.9 in   e) 2.7 in

Solutions to Problems

9.1 b) Step 1—Compute factored load:
$$1.4 \times w_d = 1.4 \times 0.75 = 1.05 \text{ k/ft}$$
$$1.7 \times w_d = 1.7 \times 1.30 = 2.21$$
$$w_u = 3.26 \text{ k/ft}$$
$$P_u = 1.7 \times 25 = 42.5 \text{ k}$$

Step 2—Compute moment at mid-span:
$$M_u = \frac{1}{8} w_u \ell^2 + \frac{P_u \ell}{4}$$
$$= \frac{1}{8} \times 3.26 \times 26^2 + 42.5 \times \frac{26}{4} = 552 \text{ ft-k}$$

9.2 b) Step 1—Compute $A_s = 4 \times (1.00 + 0.79) = 7.16 \text{ in.}^2$

Step 2—Compute $a = \frac{A_s f_y}{0.85 f'_c b} = \frac{7.16 \times 60}{0.85 \times 4 \times 78} = 1.62$

Step 3—Moment capacity = $\phi M_n = 0.90 \times A_s f_y \left(d - \frac{a}{2}\right)$
$$= 0.90 \times 7.16 \times 60 \left(18 - \frac{1.62}{2}\right) \times \frac{1}{12} = 554 \text{ ft-k}$$

9.3 d) Step 1—Determine the critical shear section at a distance d from face of support (wall). Since face of wall is $12/2 = 6''$ from center of support, the critical shear section from the center of support is $d + 6'' = 18 + 6 = 24''$.

Step 2—Reaction $= \frac{42.5 + 26 \times 3.26}{2} = 63.6 \text{ k}$

Step 3—Shear at critical section: $v_u = 63.6 - 3.26 \times \frac{24}{12} = 57.1 \text{ k}$

9.4 c) Step 1—Maximum spacing for #3 stirrups:
$$s_{max} = \text{Min.} \left(\frac{d}{2} = 9'', 24'', A_v f_y / 50 b_w\right)$$
$$\frac{A_v f_y}{50 b_w} = \frac{0.22 \times 60\,000}{50 \times 12} = 22''$$
$$\therefore s_{max} = 9''$$

Step 2—
$$\phi V_c = 0.85 \times 2 \sqrt{f'_c} \, bd$$
$$= 0.85 \times 2\sqrt{4000} \times 12 \times 18 = 23,200 \text{ lb} = 23.2 \text{ k}$$

Step 3—
$$s = \frac{\phi A_v f_y d}{V_u - \phi V_c} = \frac{0.85 \times 0.22 \times 60 \times 18}{57.1 - 23.2} = 6'' \quad (<s_{max} = 9'')$$
∴ Use 6″

9.5 e) Step 1—Calculate strength at section with only the bottom layer of bars:
$$0.9 M_n = 0.9 A_s \times f_y \left(d - \frac{a}{2}\right)$$
$$A_s = 3.57 \text{ m}^2$$
$$a = \frac{3.57 \times 60}{0.85 \times 4 \times 78} = 0.81''$$
$$0.9 M_n = 0.9 \times 3.57 \times 60 [19.0 - (0.81/2)] \div 12 = 299 \text{ ft-k}$$

Step 2—Find x = distance from support, where moment equals $0.9 M_n$:
$$299 = 63.6 x - \frac{3.26}{2} x^2$$
$$\therefore x = 5.47 \text{ ft}$$

Step 3—Development length of #9 (for clear spacing for 2 bar diameters):
$$\ell_d = 2 \times 3.17 \text{ ft} = 6.34 \text{ ft}$$

Step 4—Cut-off must extend over max. (1.5 ft, $12 d_b$ = 1.5 ft

Step 5—Cut-off point from support = 5.47 ft − 1.5 ft ≅ 4 ft

Step 6—Length of bars = 26′ − 2 × 4 = 18 ft

9.6 a) Step 1—Find c_b at concrete strain = $0.003 = \varepsilon_u$
and steel strain = $\varepsilon_y = 60/29{,}000 = 0.00207$.
$d = 20'' - 2.5'' = 17.5''$
$$c_b = 17.5 \times \frac{\varepsilon_u}{\varepsilon_u + \varepsilon_y} = 17.5 \frac{0.003}{0.00507} = 10.36''$$

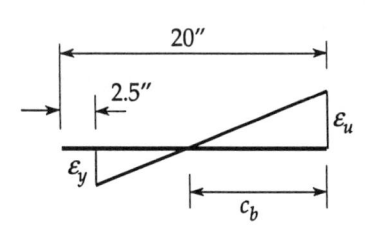

Step 2—Find the strain in the compression steel:
$$= 0.003 \times \frac{10.36 - 2.5}{10.36} = 0.0023 > \varepsilon_y$$
∴ compression steel yielded f'_s = 60 ksi

Step 3—Axial Force Resistance:
$$P_b = C + A'_s f_y - A_s f_y$$
$$= 0.85 \times f'_c \times b \times a + (A'_s - A_s) f_y$$
$$= 0.85 \times 3.5 \times 12 \times (0.85 \times 10.36) - 0 \times f_y$$
$$= 314 \text{ k}$$

9.7 c) The moment resistance is obtained by taking moments of the axial forces about the center line:

$$M_b = 0.85 \times f'_c \times b \times a\left(\frac{h}{2} - \frac{a}{2}\right) + A'_s f'_s\left(\frac{h}{2} - d'\right) + A_s f_s\left(d - \frac{h}{2}\right)$$

$$= 0.85 \times 3.5 \times 8.8 \times 12\left(\frac{20}{2} - \frac{8.8}{2}\right) + 2 \times 60\left(\frac{20}{2} - 2.5\right)$$

$$+ 2 \times 60\left(17.5 - \frac{20}{2}\right) = 296.6 \text{ ft-k}$$

$$\left(\text{eccentricity} = e_b = \frac{M_b}{P_b} = 11.3 \text{ in.}\right)$$

9.8 e) Axial Load Capacity ($e = 0.0$):

$$P_n = 0.85 \times 3.5 \times 12 \times 20 + 4 \times 60 = 954 \text{ k}$$

9.9 a) Maximum spacing = minimum of:

    (a) $48 \times 3/8'' = 18''$

    (b) $16 \times 9/8'' = 18''$

    (c) $12''$ ← governs

9.10 d) Step 1—For braced frame use $k = 1.0$ radius of gyration $= 0.3\,h = 6''$

Step 2 — Max. $\frac{k\ell}{r}$ for "negligible slenderness effect" $= 34 - 12\left(\frac{M_1}{M_2}\right)$

$$\frac{1 \times \ell}{6} = 34 - 12\left(\frac{60}{120}\right) = 28. \quad \ell = 14 \text{ ft}$$

9.11 b) Step 1—Consider the interaction diagram shown. Assume linear variation between points $(0, P_o)$ and (M_b, P_b)

$P_o = 954, \quad P_b = 314, \quad M_b = 297$

Step 2—For $P = 600$ k,

$$M = (954 - 600)\frac{297}{954 - 314} = 164 \text{ ft-k}$$

$$\therefore e = \frac{164}{600} \times 12 = 3.3 \text{ in}$$

10. Indeterminate Structures

by Ronald S. Harichandran

Indeterminate structures are encountered often in civil engineering problems, and a number of methods can be used to analyze them. Before computers became prevalent in engineering practice, the moment distribution method was perhaps the most popular technique for analyzing indeterminate beams and frames. At the present time, however, some academic curricula do not adequately cover moment distribution, and focus instead on the more modern flexibility (force) and stiffness (displacement) methods. This review deals with all three methods for the analysis of indeterminate structures; the reader may use whichever method is most appealing. For particular examples, however, one method may have an advantage over another, and this is pointed out. For indeterminate trusses, the flexibility method is the most suitable one for hand-calculations.

Deformations and forces in indeterminate structures can be caused by loads, thermal effects, support movements and/or prestrain effects (i.e., the stretching or bending of a member to make it fit). Since prestrain effects are not commonly encountered in practice, only loads, thermal effects and support movements are considered in this review.

When using the flexibility method of analysis, it is necessary to compute deflections of determinate structures. A number of methods such as the unit load method, moment-area method, conjugate beam method, etc., are available to compute deflections. Only the unit load method is used in this review since it is the most versatile and can be used for beams, frames and trusses.

10.1 Basic Concepts

10.1.1 Units

In the United States, the English system of units is still prevalent in structural engineering, and is used throughout in this review. However, the reader should be familar with the use of SI units as well.

It is of utmost importance to use consistent units throughout. For example, the maximum deflection of a simply supported beam due to a uniformly distributed load w is given by $\Delta = wL^4/(384EI)$. If the elastic modulus E and moment of inertia I have units of kips/in^2 and in^4, respectively, then the load w and the length L should have units of kips/in and inches, respectively. The computed deflection would then have units of inches. It is common to err by not converting the load w from the customary units of kips/ft to kips/in.

10.1.2 Degree of Static Indeterminacy

For the purpose of identifying whether or not a structure is indeterminate, and in the flexibility method of analysis, it is extremely important to identify the degree of static indeterminacy (DSI). The DSI is the number of unknown "external" forces in excess of the number of equilibrium equations for any collection of free bodies of the structure. Internal forces that are "exposed" when obtaining free bodies are counted as "external" forces. If the DSI = 0, then the structure is statically determinate and all external and internal forces may be found using equilibrium considerations alone. If the DSI is negative, then the structure is not sufficiently constrained by the supports and is unstable.

Beams and Frames

There are three equilibrium equations for each plane (2-D) free body (i.e., $\Sigma F_x = 0$, $\Sigma F_y = 0$ and $\Sigma M_z = 0$), and six equilibrium equations for each 3-D free body (i.e., $\Sigma F_z = 0$, $\Sigma M_x = 0$ and $\Sigma M_y = 0$ in addition to those for 2-D). For structures containing internal releases (e.g., internal hinges), appropriate free bodies should be obtained by cutting through the releases. For multi-story frames with "closed loops," appropriate free bodies not having any closed loops should be obtained.

───── **EXAMPLE 10.1** ─────────────────────────────────────

Find the degree of static indeterminacy of each plane structure.

1.

No. of unknown reactions	= 5 (3 at A and 2 at B)
No. of equilibrium equations	= 3
DSI	= 5 − 3 = 2

Note: The wall at A is capable of resisting a moment and a force, while the support at B is capable of resisting a force only.

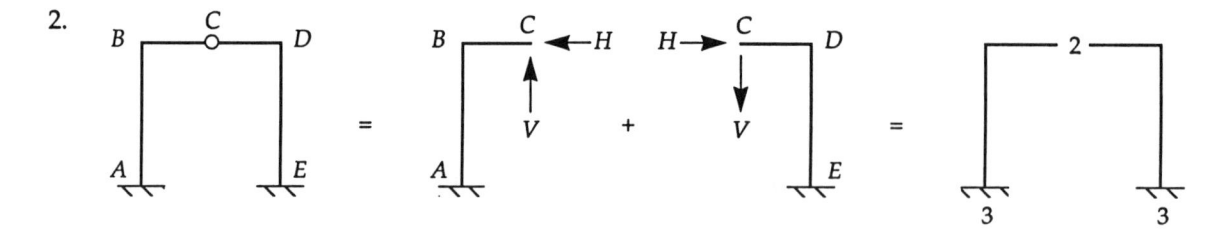

2.

Cut at internal hinge to obtain two free bodies.
No. of unknown "external" forces = 8 (six reactions at A and E and two forces at C)
No. of equilibrium equations = 6 (3 for each free body)
DSI = 8 − 6 = 2

3.

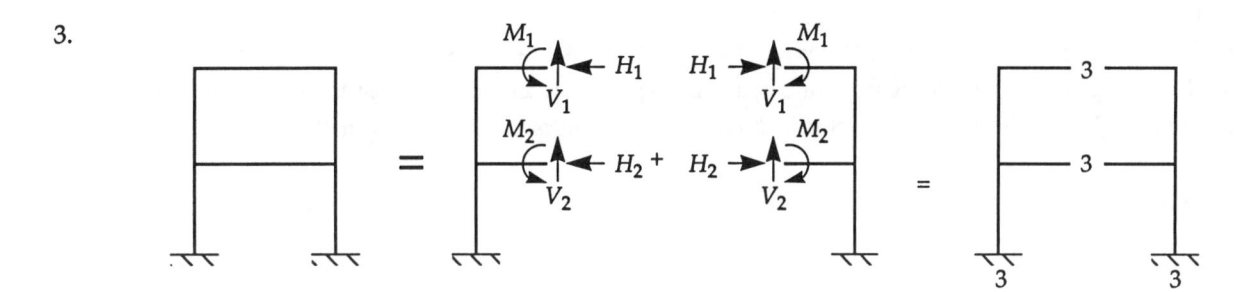

Cut vertically to eliminate closed loops and obtain two free bodies.
No. of unknown "external" forces = 12 (as shown in the figure on the right)
No. of equilibrium equations = 6 (3 for each free body)
DSI = 12 − 6 = 6

Trusses

The following simple equations give the DSI of trusses:

$$\text{DSI for 2-D truss} = m + r - 2j \tag{10.1.1}$$

$$\text{DSI for 3-D truss} = m + r - 3j \tag{10.1.2}$$

in which m = no. of members, r = no. of reactions and j = no. of joints.

―――― **EXAMPLE 10.2** ――――

Find the degree of static indeterminacy of the plane truss.

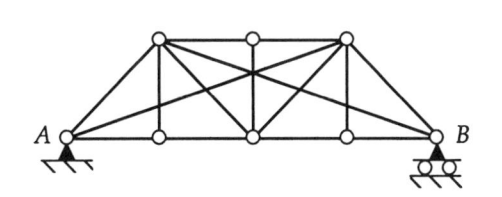

Solution. At A there are two reactions, a horizontal force and a vertical force. At B there is one reaction, a vertical force.

$m = 15, r = 3, j = 8$.

DSI = $15 + 3 - 2(8) = 2$

10.1.3 Degrees-of-Freedom

In the stiffness method of analysis, it is necessary to identify the joint displacements which are considered as the unknowns. The number of free (unknown) joint displacements in a structure are the degrees-of-freedom (DOF) of the structure. Support points, internal hinges and points where straight members intersect, are considered as joints. The treatment of curved members are not covered in this review. For beams and frames, there are in general three displacements at each joint of a 2-D structure (x and y translation and rotation) and six displacements at each joint of a 3-D structure, of which some may be restrained by supports. Often the axial deformation in members is neglected, and in this case the axial displacement at either end of each member must be the same, reducing the DOF by the number of members.

EXAMPLE 10.3

Determine the DOF of the structures in Examples 10.1 when axial deformations are included and when they are neglected. Also determine the DOF of the truss in Example 10.2.

Solution.

1. For the propped cantilever:

   No. of total joint displacements $= 3 \times 2 = 6$

   No. of displacements restrained by supports $= 5$

   DOF $= 6 - 5 = 1$

   Neglecting axial deformations does not change the DOF.

2. For the hinged portal, C must be considered as a joint:

   No. of total joint displacements $= 3 \times 5 = 15$

   No. of displacements restrained by supports $= 6$

   DOF $= 15 - 6 = 9$

   If axial deformations are neglected then the DOF reduces further by 4 (since there are 4 members):

   (a) The vertical displacements at B and D must be zero since AB and DE are axially rigid and the vertical displacements at A and E are zero. This eliminates two DOF.

   (b) The horizontal displacements at B, C and D must be the same since BC and CD are axially rigid. This reduces the 3 horizontal DOF to 1 (i.e., another reduction of 2 DOF).

   DOF = 9 (for general case) − 4 (reductions due to axial rigidity) = 5.

3. For the two story frame:

   No. of total joint displacements $= 3 \times 6 = 18$

   No. of displacements restrained by supports $= 6$

   DOF $= 18 - 6 = 12$

   If axial deformations are neglected then the DOF reduces further by 6:

   (a) Vertical displacements at the second and third story levels are eliminated due to axial rigidity of the vertical members. This reduces the DOF by 4.

   (b) The horizontal displacements at each story level must be the same due to the axial rigidity of the horizontal members. This reduces the 4 horizontal displacements to 2.

   DOF = 12 (for general case) − 6 (reductions due to axial rigidity) = 6.

4. For the truss in Example 10.2:

   No. of total joint displacements $= 2 \times 8 = 16$

   No. of displacements restrained by supports $= 3$

   DOF $= 16 - 3 = 13$

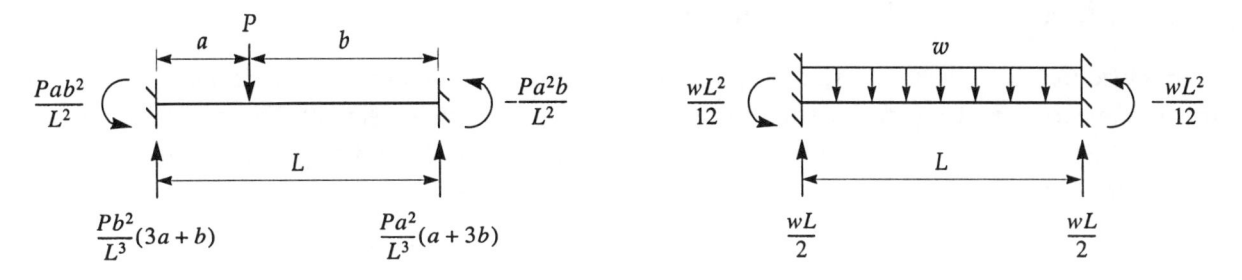

Figure 10.1 Fixed-end forces for common loadings.

10.1.4 Fixed-End Forces

In displacement methods, such as the moment distribution and the stiffness methods, fixed-end forces due to loads applied perpendicular to the members need to be determined. Fixed-end forces are the forces that arise at member ends, when external loads are applied on members and the joints are restrained from rotating (i.e., "locked"). Fig. 10.1 shows the fixed-end forces for two common loadings. Superposition of these fixed-end forces can be used to obtain the fixed-end forces for multiple concentrated loads or a mixture of concentrated and uniform loads. Most textbooks give fixed-end forces for other, more complex, loadings. Note that forces applied directly at the joints are not considered when computing the fixed-end forces.

10.1.5 Equivalent Joint Loads

In displacement methods, such as the moment distribution and the stiffness methods, there are two stages of analysis. First, all joints are restrained and loads applied along the members are placed on the structure. This is the fixed-end case. Second, the structure is analyzed under equivalent joint loads consisting of the sum of the forces applied directly along the DOF (i.e., at the joints) and the *opposite* of the fixed-end forces along the DOF. This two stage analysis is essentially superposition, and is illustrated in Fig. 10.2 for a propped cantilever. The propped cantilever has only one DOF consisting of the rotation at B. In the fixed-end case, the uniform load is applied to the beam and the fixed-end moment $M_B^F = wL^2/12$ is applied at B to prevent it from rotating. In the second stage, an equivalent joint moment is applied at B. The equivalent joint moment must account for the applied moment M_B and cancel the effect of the fixed-end moment M_B^F. Therefore, $M_B^E = M_B - M_B^F$. Since there is no joint rotation for the fixed-end case, the joint rotation at B in the original structure is equal to that due to the equivalent joint moment. It is important to note that reactions and member-end forces in the original structure are the sum of the corresponding reactions and member-end forces from the fixed-end and equivalent joint load cases.

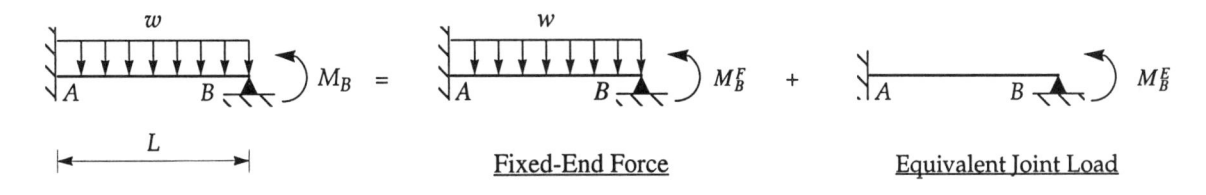

Figure 10.2 Fixed-end and equivalent joint loads.

10.1.6 Displacement Calculations by the Unit Load Method

The unit load (or virtual force) method is a powerful technique for computing the displacements of a determinate structure due to loads, temperature effects, prestrains and support displacements. This review covers the use of this method for calculating flexural displacements due to loads in frames, and displacements due to loads and temperature effects in trusses.

For beams and frames the unit load method consists of the following steps:

1. Determine the distribution of the bending moment M (i.e., the moment diagram) due to the loads.

2. Determine the distribution of the bending moment m due to a unit load corresponding to the displacement that is to be computed. A unit force is applied if a translation is required, or a unit moment is applied if a rotation is required. The unit load is applied at the location where the displacement is required, and the positive direction for the computed displacements will be in the direction of the unit load.

3. Compute the required displacement by performing the following integral over the entire structure:

$$\Delta \text{ or } \theta = \int \frac{Mm}{EI} dx \qquad (10.1.3)$$

The integral in Eq. 10.1.3 is most easily performed by using Table 10.1.

For trusses, the unit load method for computing displacements due to loads and/or temperature effects consists of the following steps:

1. Determine the bar forces P_i due to loads.

2. Determine the bar forces p_i due to a unit load corresponding to the joint displacement that is to be computed.

3. Compute the required joint displacement by performing the following summations for all bars in the truss:

$$\Delta = \sum_i p_i \frac{P_i L_i}{E_i A_i} + \sum_i p_i \alpha_i \Delta T_i L_i \qquad (10.1.4)$$

in which α_i is the coefficient of thermal expansion and ΔT_i is the temperature change in bar i.

EXAMPLE 10.4

Determine the horizontal and vertical translations, and the rotation at D. All members have the same EI.

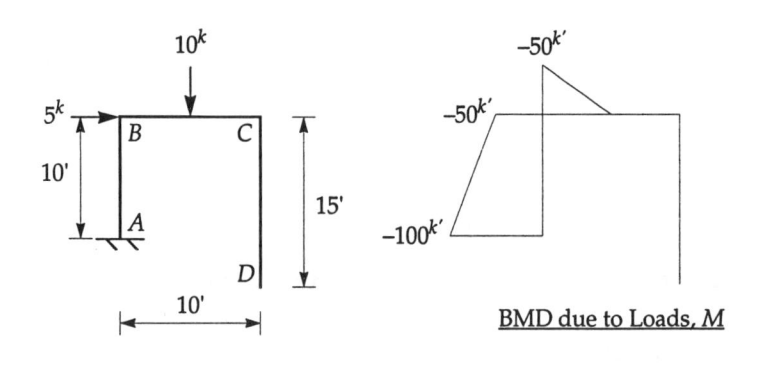

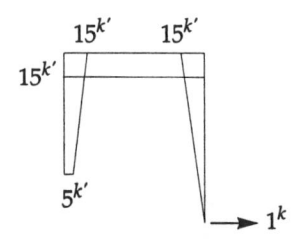

BMD due to Loads, M

BMD due to Unit Horizontal Force, m_1

TABLE 10.1 [a] Product integrals $\int Mm\,dx$

M \ m	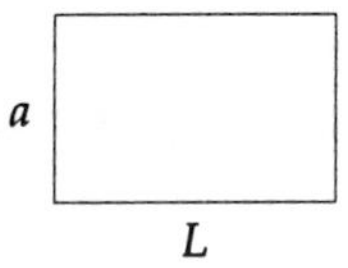 rectangle, a, L	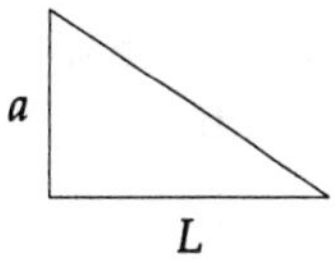 triangle (left), a, L	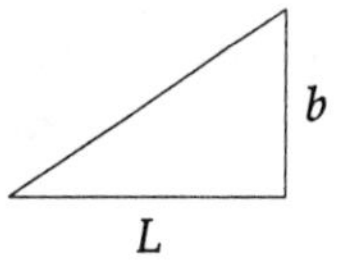 triangle (right), b, L	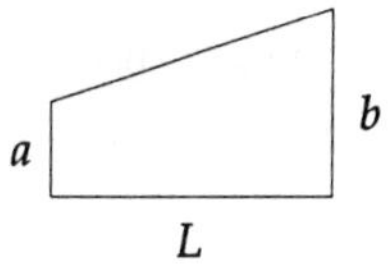 trapezoid, a,b, L
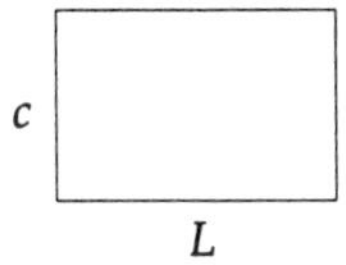 rectangle c, L	Lac	$\frac{L}{2}ac$	$\frac{L}{2}bc$	$\frac{L}{2}c(a+b)$
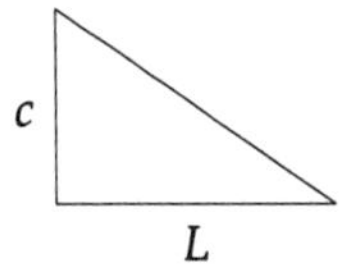 triangle (left) c, L	$\frac{L}{2}ac$	$\frac{L}{3}ac$	$\frac{L}{6}bc$	$\frac{L}{6}c(2a+b)$
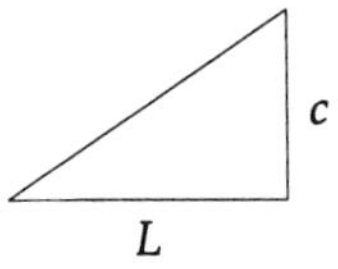 triangle (right) c, L	$\frac{L}{2}ac$	$\frac{L}{6}ac$	$\frac{L}{3}bc$	$\frac{L}{6}c(a+2b)$
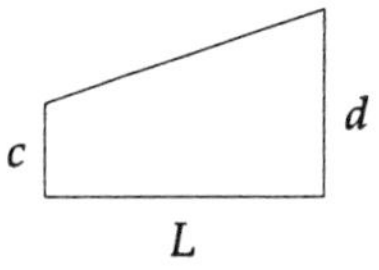 trapezoid c,d, L	$\frac{L}{2}a(c+d)$	$\frac{L}{6}a(2c+d)$	$\frac{L}{6}b(c+2d)$	$\frac{L}{6}[a(2c+d)+b(c+2d)]$
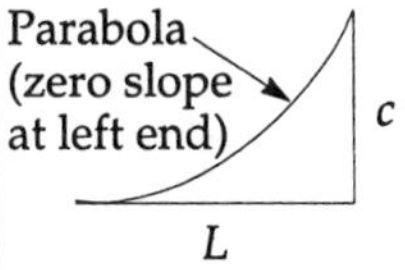 Parabola (zero slope at left end) c, L	$\frac{L}{3}ac$	$\frac{L}{12}ac$	$\frac{L}{4}bc$	$\frac{L}{12}c(a+3b)$
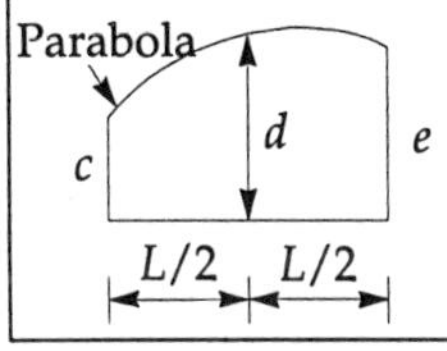 Parabola c,d,e, $L/2$, $L/2$	$\frac{L}{6}a(c+4d+e)$	$\frac{L}{6}a(c+2d)$	$\frac{L}{6}b(2d+e)$ R	$\frac{L}{6}[a(c+2d)+b(2d+e)]$

[a] Notes

1. Insert the factor $1/EI$ and correct signs for a, b, c, etc.
2. Any value a, b, c, d, e can be zero, positive or negative, so that we could have $a > b, c > 0, d < 0$, and so forth.
3. Always go from left to right in *both* diagrams (i.e., the M and m diagrams), or from right to left in *both*. Similarly, go from bottom to top or top to bottom in *both*.
4. For the example below, the linear relation can be used directly, and there is no need to divide into triangles or interpolate:

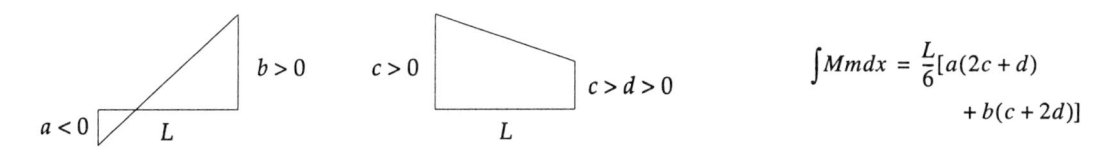

$a < 0$, $b > 0$, L ; $c > 0$, $c > d > 0$, L

$$\int Mm\,dx = \frac{L}{6}[a(2c+d)+b(c+2d)]$$

Solution.

1. Plot the bending moment diagrams (BMDs) due to the applied load and due to the unit loads required for determining the translations and rotation. These are shown below with the convention that moments causing tension on the inside of the frame are positive, and with the moment diagrams drawn on the tension side of the members.

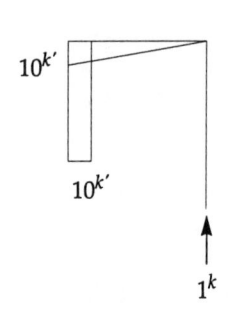

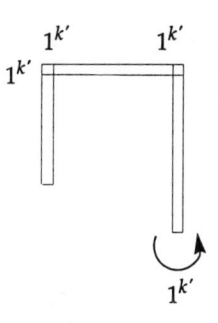

BMD due to Unit Vertical Force, m_2

BMD due to Unit Moment, m_3

2. Use Table 10.1 for computing all the required integrals:

$$\Delta_H = \int \frac{Mm_1}{EI}dx = \frac{1}{EI}\left[\frac{10}{6}[(-100)(2 \times 5 + 15) + (-50)(5 + 2 \times 15)] + \frac{5}{2}(-50)(15)\right]$$

$$= -\frac{8960}{EI}$$

$$\Delta_V = \int \frac{Mm_2}{EI}dx = \frac{1}{EI}\left[\frac{10}{2}(10)(-100 - 50) + \frac{5}{6}(-50)(2 \times 10 + 5)\right] = -\frac{8940}{EI}$$

$$\theta = \int \frac{Mm_3}{EI}dx = \frac{1}{EI}\left[\frac{10}{2}(1)(-100 - 50) + \frac{5}{2}(-50)(1)\right] = -\frac{875}{EI}$$

The negative signs indicate that all displacements are opposite to the assumed senses, i.e., the horizontal displacement is to the left, the vertical displacement is downward and the rotation is clockwise.

EXAMPLE 10.5

Determine the horizontal displacement at C. All bars have a cross sectional area of 1 in² and a modulus of 29,000 k/in². In addition to the load, the temperature of bars AB, BC and CD increases by 50°F. The coefficient of thermal expansion is 6.5×10⁻⁶/°F.

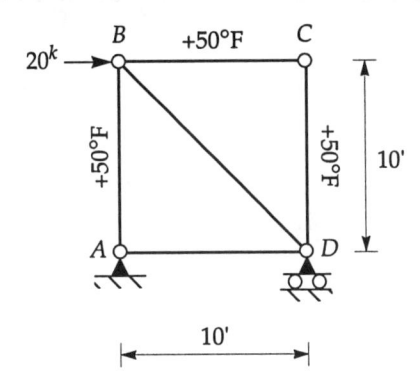

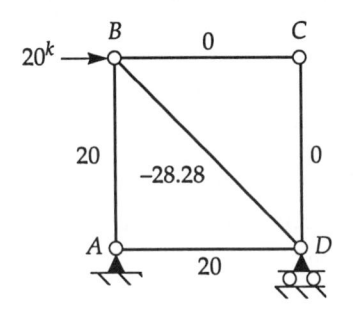

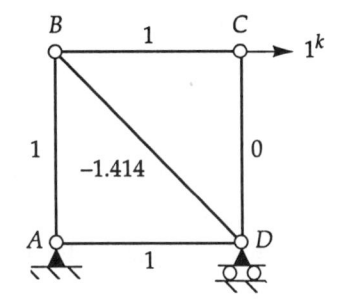

Bar Forces due to Loads, P

Bar Forces due to Unit Load, p

Solution.

1. Determine the bar forces due to the applied load and the unit load. These are shown above.

2. Determine the horizontal displacements at C using Eq. 10.1.4. Use inches for length:

$$\text{Disp. due to load} = \frac{1}{EA}\sum_i p_i P_i L_i$$

$$= \frac{1}{EA}[(1)(20)(120) + (1)(20)(120) + (-1.414)(-28.28)(169.7)]$$

$$= \frac{11585}{(29,000)(1)} = 0.400 \text{ in.}$$

$$\text{Disp. due to temp.} = \alpha\sum_i p_i \Delta T_i L_i = 6.5\times10^{-6}[(1)(50)(120) + (1)(50)(120)]$$

$$= 0.078 \text{ in}$$

Total displacement = 0.400 + 0.078 = 0.478 in.

10.2 The Flexibility Method

The flexibility method is suited for the hand-analysis of structures having a small degree of static indeterminacy. Indeterminate trusses often have a large number of DOF, but may have only a small DSI. The flexibility method also requires the inversion of a matrix of order equal to the DSI of the structure. A calculator capable of inverting matrices is useful for structures with a DSI greater than two.

The flexibility method consists of the following steps:

1. Determine the DSI of the structure.

2. Make releases to the structure such that it becomes determinate and remains stable. The number of releases required will equal the DSI. The released structure is often called the primary structure, and by making different releases different primary structures can be obtained. The forces that are released are called the redundants. For indeterminate beams, by careful choice of releases the displacements required in the flexibility method may sometimes be found in Table 10.2, thereby avoiding their calculation by the unit load method.

3. Compute the displacements of the primary structure corresponding to the redundants and due to all external effects such as loads and temperature effects.

4. Determine the flexibility matrix **F**, with elements F_{ij}, of the primary structure corresponding to the redundants. **F** is always a square symmetric matrix, and its *i*th column contains the displacements corresponding to the redundants due to a unit value of the *i*th redundant.

5. Solve the flexibility (i.e., compatibility) equations for the unknown redundant forces. The flexibility equations are always of the form

$$\mathbf{u} + \mathbf{Fq} = \mathbf{u}_q \tag{10.2.1}$$

in which **u** = vector of the displacements computed in step 3, **q** = vector of unknown redundant forces, and $\mathbf{u}_q$ = vector of displacements corresponding to the redundants in the original structure. In the absence of support movements (which is included later), $\mathbf{u}_q = 0$. This matrix equation is equivalent to a system of linear simultaneous equations and is solved for the unknown redundant forces. In matrix form, the solution is

$$\mathbf{q} = \mathbf{u}_q - \mathbf{F}^{-1}\mathbf{u} \tag{10.2.2}$$

6. Use equilibrium considerations to determine all other unknown reactions and member forces in the original structure.

───── **EXAMPLE 10.6** ───

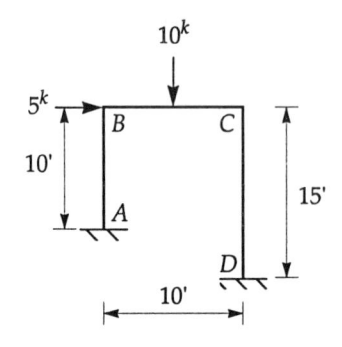

Analyze the frame by the flexibility method.

Solution.

1. The DSI = 3.

2. The support at *D* is removed to obtain the primary structure shown in Example 10.4.

3. The displacements of the primary structure corresponding to the redundants and due to the loads were calculated in Example 10.4. These are the horizontal and vertical displacements and rotation at *D*:

TABLE 10.2 Displacements of prismatic beams

Beam	Downward Translation	Counterclockwise End-Rotations
Simply supported beam A–B with uniform load w, point C at midspan, $L/2 + L/2$	$\Delta_C = \dfrac{5wL^4}{384EI}$	$\theta_B = -\theta_A = \dfrac{wL^3}{24EI}$
Simply supported beam A–B with point load P at midspan C	$\Delta_C = \dfrac{PL^3}{48EI}$	$\theta_B = -\theta_A = \dfrac{PL^2}{16EI}$
Simply supported beam A–B with two point loads P at $L/3$ from each end	$\Delta_C = \dfrac{23PL^3}{648EI}$	$\theta_B = -\theta_A = \dfrac{PL^2}{9EI}$
Simply supported beam A–B with moment M applied at midspan C	$\Delta_C = 0$	$\theta_B = \theta_A = -\dfrac{ML}{24EI}$
Simply supported beam A–B with moment M applied at end B	$\Delta_C = \dfrac{ML^2}{16EI}$	$\theta_B = \dfrac{ML}{3EI}$ $\theta_A = -\dfrac{ML}{6EI}$
Cantilever beam fixed at A, free at B, with uniform load w	$\Delta_B = \dfrac{wL^4}{8EI}$	$\theta_B = -\dfrac{wL^3}{6EI}$
Cantilever beam fixed at A, free at B, with point load P at B	$\Delta_B = \dfrac{PL^3}{3EI}$	$\theta_B = -\dfrac{PL^2}{2EI}$
Cantilever beam fixed at A, free at B, with moment M at B	$\Delta_B = \dfrac{ML^2}{2EI}$	$\theta_B = \dfrac{ML}{EI}$

$$\mathbf{u} = \begin{bmatrix} \Delta_H \\ \Delta_V \\ \theta \end{bmatrix} = -\frac{1}{EI} \begin{bmatrix} 8960 \\ 8940 \\ 875 \end{bmatrix}$$

4. The element F_{ij} is the displacement corresponding to the ith redundant due to a unit value of the jth redundant; referring to the moment diagrams in Example 10.4, this is given by

$$F_{ij} = \int \frac{m_i m_j}{EI} dx$$

Using Table 10.1, these integrals are obtained as follows:

$$F_{11} = \int \frac{m_1 m_1}{EI} dx = \frac{1}{EI} \left[\frac{10}{6}[5(2 \times 5 + 15) + 15(5 + 2 \times 15)] + 10(15^2) + \frac{15}{3}(15^2) \right]$$

$$= \frac{4460}{EI}$$

$$F_{12} = \int \frac{m_1 m_2}{EI} dx = \frac{1}{EI} \left[\frac{10}{2}[10(5+15)] + \frac{10}{2}(10)(15) \right] = \frac{1750}{EI}$$

$$F_{13} = \int \frac{m_1 m_3}{EI} dx = \frac{1}{EI} \left[\frac{10}{2}[1(5+15)] + 10(15)(1) + \frac{15}{2}(15)(1) \right] = \frac{362}{EI}$$

$$F_{22} = \int \frac{m_2 m_2}{EI} dx = \frac{1}{EI} \left[10(10)(10) + \frac{10}{3}(10)(10) \right] = \frac{1333}{EI}$$

$$F_{23} = \int \frac{m_2 m_3}{EI} dx = \frac{1}{EI} \left[10(10)(1) + \frac{10}{2}(10)(1) \right] = \frac{150}{EI}$$

$$F_{33} = \int \frac{m_3 m_3}{EI} dx = \frac{1}{EI}[10(1)(1) + 10(1)(1) + 15(1)(1)] = \frac{35}{EI}$$

$$\mathbf{F} = \frac{1}{EI} \begin{bmatrix} 4460 & 1750 & 362 \\ 1750 & 1333 & 150 \\ 362 & 150 & 35 \end{bmatrix}$$

5. The redundant forces are

$$\mathbf{q} = \begin{bmatrix} X_D \\ Y_D \\ M_D \end{bmatrix} = -\mathbf{F}^{-1}\mathbf{u} = EI \begin{bmatrix} 1.543 \times 10^{-3} & -4.390 \times 10^{-4} & -1.410 \times 10^{-2} \\ -4.390 \times 10^{-4} & 1.573 \times 10^{-2} & -2.195 \times 10^{-3} \\ -1.410 \times 10^{-2} & -2.195 \times 10^{-3} & 0.1840 \end{bmatrix} \frac{1}{EI} \begin{bmatrix} 8960 \\ 8540 \\ 875 \end{bmatrix}$$

$$= \begin{bmatrix} -2.3 \text{ kips} \\ 7.6 \text{ kips} \\ 16.0 \text{ ft-kips} \end{bmatrix}$$

in which X_D and Y_D are the horizontal and vertical reactions at D, and M_D is the moment reaction at D.

6. By taking moments about A in the original structure, the counterclockwise moment at A is found to be

$$M_A = 5(10) + 10(5) + 2.3(5) - 7.6(10) - 16 = 19.5 \text{ ft-kips}$$

Horizontal force equilibrium of the original structure yields

$$R_{A_x} = -2.7 \text{ kips}$$

––––––– EXAMPLE 10.7 –––

Analyze the continuous beam by the flexibility method, using Table 10.2 to obtain all required displacements.

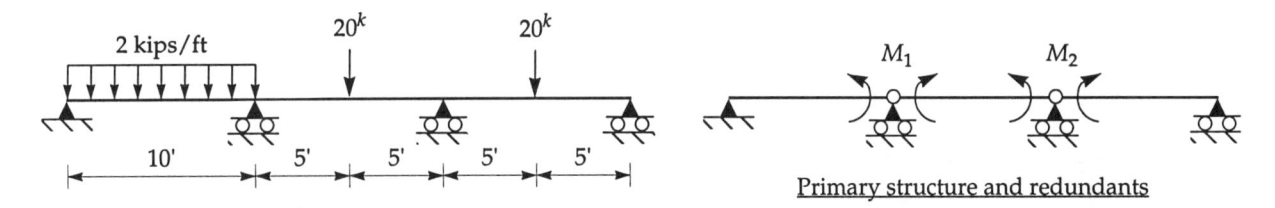

Primary structure and redundants

Solution.

1. The DSI = 2.

2. A convenient primary structure is shown above, with the redundant forces consisting of the internal moments at the two interior supports. The moment releases isolate the bending effects in each span, and we essentially have three simply supported beams.

3. The displacement corresponding to each redundant is the *sum* of the counterclockwise rotation at the left and the clockwise rotation at the right of each interior support. From Table 10.2, the displacements due to the loads are

$$\mathbf{u} = \begin{bmatrix} \dfrac{2 \cdot 10^3}{24EI} + \dfrac{20 \cdot 10^2}{16EI} \\ \dfrac{20 \cdot 10^2}{16EI} + \dfrac{20 \cdot 10^2}{16EI} \end{bmatrix} = \dfrac{1}{EI} \begin{bmatrix} 208.3 \\ 250 \end{bmatrix}$$

4. The flexibilities are the displacements corresponding to the redundants due to *pairs* of moments applied at each interior joint:

$$\mathbf{F} = \begin{bmatrix} \dfrac{L}{3EI} + \dfrac{L}{3EI} & \dfrac{L}{6EI} + 0 \\ 0 + \dfrac{L}{6EI} & \dfrac{L}{3EI} + \dfrac{L}{3EI} \end{bmatrix} = \dfrac{L}{EI} \begin{bmatrix} .667 & .167 \\ .167 & .667 \end{bmatrix}$$

Due to symmetry of the primary structure and the redundants, the flexibility matrix is symmetric about both diagonals. This is a special case and does not usually occur.

5. The internal moments at the interior supports are

$$\mathbf{q} = \begin{bmatrix} M_1 \\ M_2 \end{bmatrix} = -\mathbf{F}^{-1}\mathbf{u} = -\frac{EI}{L}\begin{bmatrix} 1.6 & -0.4 \\ -0.4 & 1.6 \end{bmatrix}\frac{1}{EI}\begin{bmatrix} 208.3 \\ 250 \end{bmatrix} = \begin{bmatrix} -233.3 \text{ ft-kips} \\ -316.7 \text{ ft-kips} \end{bmatrix}$$

The negative signs for the moments indicate that the actual internal moments have opposite senses to that assumed, and therefore cause tension on the upper surface of the beam.

6. All other reactions and internal forces may be obtained using equilibrium considerations.

EXAMPLE 10.8

Analyze the truss by the flexibility method. All bars have a cross sectional area of 1 in² and a modulus of 29,000 k/in². In addition to the load, the temperature of bars AB, BC and CD increases by 50°F. The coefficient of thermal expansion is $6.5 \times 10^{-6}/°F$. BD and AC are not connected.

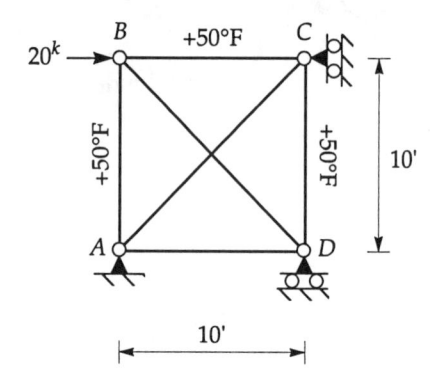

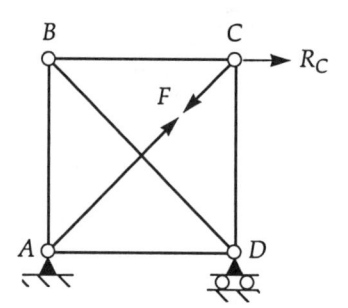

Primary Structure and Redundants

Solution.

1. The DSI = 2.

2. The primary structure shown above is obtained by cutting bar AC and removing the support at C. The redundants are therefore the bar force F and the reaction R_C.

3. The required displacements and flexibilities are conveniently obtained in tabular form as shown in the table below. The bar forces due to the 20 kip load are P_i, those due to a unit value of F (i.e., F = 1 kip) are p_{1i}, and those due to a unit value of R_C are p_{2i}.

Bar	L (in.)	P (kips)	p_1 (kips)	p_2 (kips)	ΔT (°F)	$P + Fp_1 + R_c p_2$ (kips)
AB	120	20	-0.7071	1	50	0.0
BC	120	0	-0.7071	1	50	-20.0
CD	120	0	-0.7071	0	50	12.4
AD	120	20	-0.7071	1	0	0.0
BD	169.7	-28.284	1	-1.4142	0	-0.1
AC	169.7	0	1	0	0	-17.6

Bar	Pp_1L	Pp_2L	$p_1\Delta TL$	$p_2\Delta TL$	p_1^2L	p_1p_2L	p_2^2L
AB	−1697	2400	4243	6000	60	−84.85	120
BC	0	0	4243	6000	60	−84.85	120
CD	0	0	4243	0	60	0	0
AD	−1697	2400	0	0	60	−84.85	120
BD	−4799	6788	0	0	169.7	−240	339.4
AC	0	0	0	0	169.7	0	0
Σ	−8193	11588	12729	12000	579.4	−494.6	699.4

The displacements corresponding to the redundants are the overlap at the cut in bar AC, and the horizontal displacement at C:

$$\mathbf{u} = \begin{bmatrix} \sum_i p_{1i}\frac{P_iL_i}{E_iA_i} + \sum_i p_{1i}\alpha_i\Delta T_iL_i \\ \sum_i p_{2i}\frac{P_iL_i}{E_iA_i} + \sum_i p_{2i}\alpha_i\Delta T_iL_i \end{bmatrix} = \begin{bmatrix} \frac{-8193}{(29000)(1)} + (6.5\times10^{-6})(12729) \\ \frac{11588}{(29000)(1)} + (6.5\times10^{-6})(12000) \end{bmatrix} = \begin{bmatrix} -0.20 \text{ in.} \\ 0.48 \text{ in.} \end{bmatrix}$$

$$\mathbf{F} = \begin{bmatrix} \sum_i \frac{p_{1i}^2L_i}{E_iA_i} & \sum_i \frac{p_{1i}p_{2i}L_i}{E_iA_i} \\ \sum_i \frac{p_{1i}p_{2i}L_i}{E_iA_i} & \sum_i \frac{p_{2i}^2L_i}{E_iA_i} \end{bmatrix} = \frac{1}{(29000)(1)}\begin{bmatrix} 579.4 & -494.6 \\ -494.6 & 699.4 \end{bmatrix}$$

4. The redundants are

$$\mathbf{q} = \begin{bmatrix} F \\ R_C \end{bmatrix} = -\mathbf{F}^{-1}\mathbf{u} = \begin{bmatrix} -17.6 \text{ kips} \\ -32.4 \text{ kips} \end{bmatrix}$$

5. The bar forces in the original truss may be obtained by superposition. Since temperature effects do not induce any bar forces in the determinate primary structure, the force in the ith bar of the original truss is

$$f_i = P_i + Fp_{1i} + R_c p_{2i}$$

These bar forces are shown in the last column of the first table.

10.3 Moment Distribution

Moment distribution is a convenient method for hand-analysis of indeterminate beams and frames. Physically, it is equivalent to "locking" all the joints and then "unlocking" and "re-locking" them one at a time until the structure relaxes to its final deformed configuration. It is an approximate iterative method that converges rather quickly. For large frames, however, the method is rather unwieldy. Moment distribution considers only flexural deformations in members, which is the most dominant in beams and frames.

10.3.1 No Joint Translation

Moment distribution is easier when there are only joint rotations and no load dependent joint translations (i.e., no sidesway in frames and no spring supports), and consists of the following steps:

1. Determine the flexural stiffness of each member. Flexural stiffnesses are:
   - $K = 4EI/L$ for members with both ends framing into other members, or one end fixed.
   - $K = 3EI/L$ for members with one end pinned.
   - $K = 0$ for cantilevers.

2. Determine the distribution factor (DF) for each member end at joints in which multiple members meet. The DF is the ratio of the member stiffness to the sum of the stiffnesses of all members that meet at the joint, i.e., $DF_i = K_i/\Sigma K_j$, in which the summation is taken over all members meeting at the joint.

3. Lock all the joints in place, except for free cantilever ends, apply the loads acting on members, and calculate the fixed-end moments (FEM) at each member end. A global sign convention is used for the moments. In this review, counter-clockwise moments are assumed to be positive.

4. Let pinned joints along the periphery of the structure that support only a single member (including pinned supports with extending cantilevered members) rotate by unlocking these joints. The process of unlocking restores moment equilibrium at the joint and is often called "balancing." This is equivalent to:

   (a) Applying the equivalent joint moments at these joints.

   (b) Carrying over a part of the applied moment to the adjacent locked joints. This occurs due to flexing of the members. For prismatic beams, the carry-over moment is one-half the applied moment.

   The pinned support joints are not relocked, and therefore moments are not carried back to them when adjacent joints are unlocked.

5. Let an interior joint rotate by unlocking it (i.e., balance it). This is equivalent to:

   (a) Applying the equivalent joint moment at this joint. If this joint had been unlocked in a previous iteration, then it is sufficient to consider only the carry-over moments since the last time the joint was unlocked, since all moments prior to the last balancing should add to zero.

   (b) Distributing the applied moment to each member end meeting at the joint. The amount of moment distributed to a particular member end is the applied moment times the DF for that end.

   (c) Carrying over a part of the applied moment to the adjacent locked joints. For prismatic beams, the carry-over moment is one-half the applied moment.

   The joint is relocked and therefore moments will be carried back to the joint when adjacent joints are subsequently unlocked.

6. Proceed to other interior joints one at a time and repeat step 5. When all joints have been unlocked and relocked, proceed back to the first interior joint and begin the process all over again, and keep repeating until the unbalanced moments become sufficiently small.

7. Add up the member end-moments, starting with the initial FEM and including all carry-over and distributed moments. These represent the final member end-moments.

8. Apply equilibrium considerations to each member to determine the shear force and bending moment diagrams.

9. Use joint equilibrium considerations to determine the axial member forces and reactions.

EXAMPLE 10.9

Analyze the frame using moment distribution and determine all the member end-forces. The concentrated loads are applied at the mid-span of the members, the beams have twice the moment of inertia as the columns, and all members have the same modulus of elasticity.

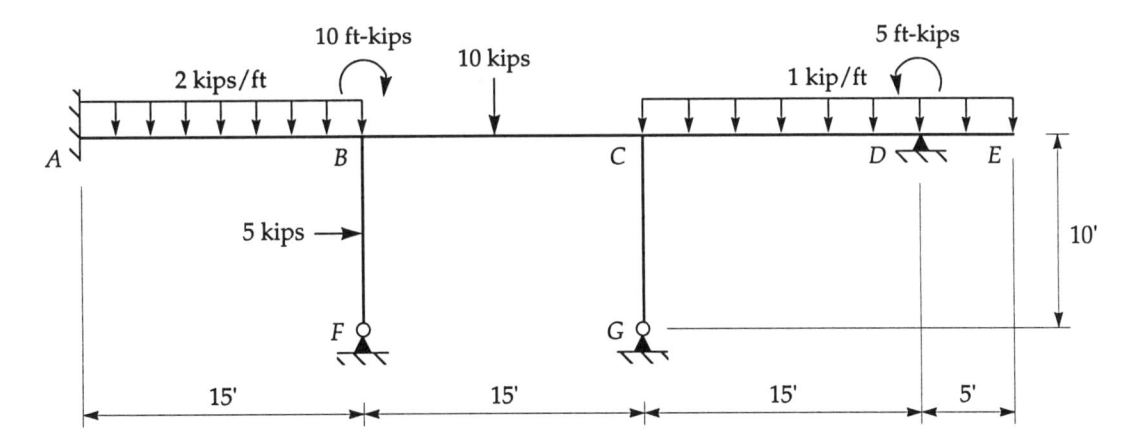

Solution.

1. Member stiffnesses:

$$K_{AB} = K_{BC} = \frac{4E(2I)}{15} = 0.533EI \qquad K_{CD} = \frac{3E(2I)}{15} = 0.4EI$$

$$K_{DE} = 0 \qquad K_{BF} = K_{CG} = \frac{3EI}{10} = 0.3EI$$

2. Distribution factors at joint B:

$$DF_{BA} = \frac{K_{AB}}{K_{AB} + K_{BC} + K_{BF}} = \frac{0.533}{0.533 + 0.533 + 0.3} = 0.390$$

$$DF_{BC} = \frac{K_{BC}}{K_{AB} + K_{BC} + K_{BF}} = \frac{0.533}{0.533 + 0.533 + 0.3} = 0.390$$

$$DF_{BF} = \frac{K_{BF}}{K_{AB} + K_{BC} + K_{BF}} = \frac{0.3}{0.533 + 0.533 + 0.3} = 0.220$$

The DF at the other joints are calculated in a similar way and are shown in the moment distribution table for this example.

3. Fixed-end moments:

$$\text{FEM}_{AB} = -\text{FEM}_{BA} = \frac{wL^2}{12} = \frac{2(15)^2}{12} = 37.5 \text{ ft-kips}$$

$$\text{FEM}_{BC} = -\text{FEM}_{CB} = \frac{PL}{8} = \frac{10(15)}{8} = 18.75 \text{ ft-kips}$$

$$\text{FEM}_{CD} = -\text{FEM}_{DC} = \frac{wL^2}{12} = \frac{1(15)^2}{12} = 18.75 \text{ ft-kips}$$

$$\text{FEM}_{DE} = \frac{wL^2}{2} = \frac{1(5)^2}{2} = 12.5 \text{ ft-kips}$$

$$\text{FEM}_{FB} = -\text{FEM}_{BF} = \frac{PL}{8} = \frac{5(10)}{8} = 6.25 \text{ ft-kips}$$

4. Moment distribution can be performed conveniently in a tabular form as shown in the moment distribution table for this example. The notations "Bal." and "C.O." in the first column indicate whether a balancing (i.e., unlocking of the joint and distribution of moments to the member ends meeting at the joint) or carry-over operation is performed, and the footnotes give more detail regarding the operations. An underline is used each time after a joint is balanced. Arrows are used to indicate where moments are carried over. Since there are no FEM at or carry-over moments to joint G, this joint is omitted from the table. If desired, the iteration may be carried on further to obtain more accurate results.

5. By applying equilibrium considerations to each member the shear forces at the member ends shown below can be determined from the end-moments.

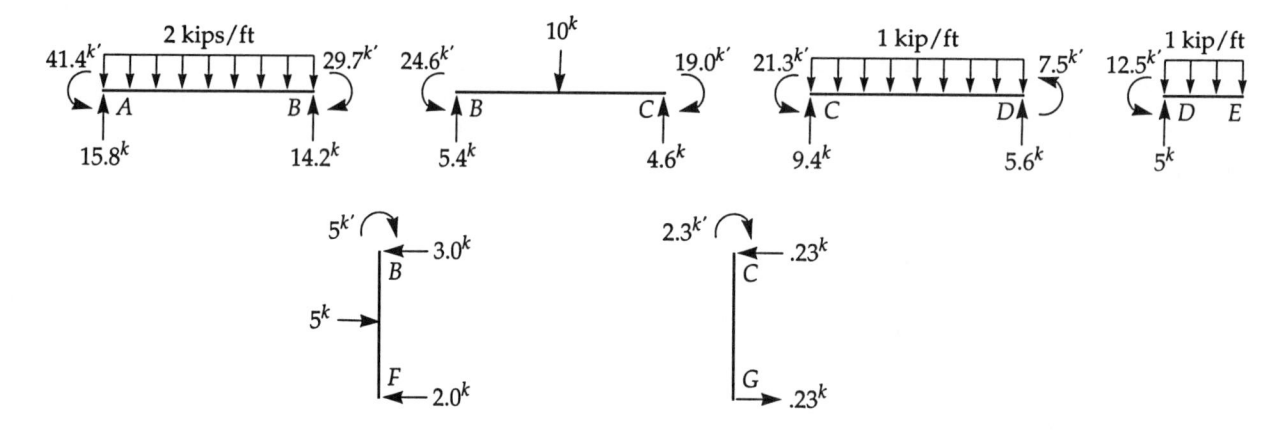

6. By considering force equilibrium at joints D, C and B, the axial member forces can be determined:

   (a) At D: $\Sigma F_x = 0 \Rightarrow$ Axial forces in CD and DE are zero.

   (b) At C: $\Sigma F_x = 0 \Rightarrow$ Axial tension in BC = 0.23 kips
              $\Sigma F_y = 0 \Rightarrow$ Axial compression in CG = 4.6 + 9.4 = 14.0 kips

   (c) At B: $\Sigma F_x = 0 \Rightarrow$ Axial tension in AB = 0.23 + 3.0 = 3.23 kips
              $\Sigma F_y = 0 \Rightarrow$ Axial compression in BF = 14.2 + 5.4 = 19.6 kips

Joint	A	B			C			D		F
End	AB	BA	BF	BC	CB	CG	CD	DC	DE	FB
DF		0.390	0.220	0.390	0.432	0.243	0.324	1	0	
FEM	37.50	−37.50	−6.25	18.75	−18.75		18.75	−18.75	12.50	6.25
Bal. D, F[a]								11.25		−6.25
C.O.			−3.13				5.63			
Bal. B[b]		7.07	3.99	7.07						
C.O.	3.54				3.54					
Bal. C[c]					−3.96	−2.23	−2.97			
C.O.[d]				−1.98						
Bal. B[e]		0.77	0.44	0.77						
C.O.	0.39				0.39					
Bal. C[f]					−0.17	−0.09	−0.13			
Sum[g]	41.43	−29.66	−4.95	24.61	−18.95	−2.32	21.28	−7.50	12.50	0.0
Check[h]		Σ = −10.0			Σ = 0.02			Σ = 5.0		

[a] Balance peripheral pinned joints. Accounting for the applied concentrated moment, the equivalent joint moment at $D = 5 - (-18.75 + 12.50) = 11.25$ ft-kips.
[b] Accounting for the applied concentrated negative (i.e., clockwise) moment, the equivalent joint moment at $B = -10 - (-37.50 - 6.25 + 18.75 - 3.13) = 18.13$. Moment at end $BA = 0.390(18.13) = 7.07$ ft-kips, etc.
[c] Equivalent joint moment at $C = -(-18.75 + 18.75 + 5.63 + 3.54) = -9.17$ ft-kips.
[d] Note that moments are *not* carried-over to the pinned joints.
[e] Equivalent joint moment at $B = 1.98$ ft-kips. Since the previous balancing operation at B establishes moment equilibrium, the sum of the moments at B just after that balancing is zero. The unbalanced moment arises only due to the latest carry-over.
[f] It is best to stop after a balancing operation, since this will ensure moment equilibrium at all joints.
[g] This is the sum of the moments in each column.
[h] Check balancing operations by summing the moments at each joint. The results should add up to the moment applied at the joint except at fixed supports, but allow a small margin for round-off errors.

10.3.2 Joint Translation

Multi-story frames in which joint translation can occur are more tedious to analyze by moment distribution. The problem must be divided into no-sidesway and sidesway cases, and separate moment distributions must be performed for each case. The process is reviewed here only for a single story frame for which two moment distributions must be performed. The following steps outline the procedure:

1. Brace the frame so that it cannot undergo sidesway, apply the loads, and perform a moment distribution. Using equilibrium considerations, determine the magnitude of the horizontal force required to brace the frame.

2. Let the frame sidesway by an amount Δ while all joints are locked. The FEM developed in members that flex are all equal to $6EI\Delta/L^2$. Perform moment distribution starting with these FEM. Using equilibrium considerations, determine the magnitude of the horizontal force required to cause the arbitrary sidesway.

3. For the correct amount of sidesway the horizontal forces determined in steps 1 and 2 must be equal and opposite. Use this criterion to determine the magnitude of Δ.

4. Add the corresponding moments in steps 1 and 2, using the value of Δ computed in step 3.

––––– **EXAMPLE 10.10** –––––

Analyze the frame assuming all members have the same *EI*.

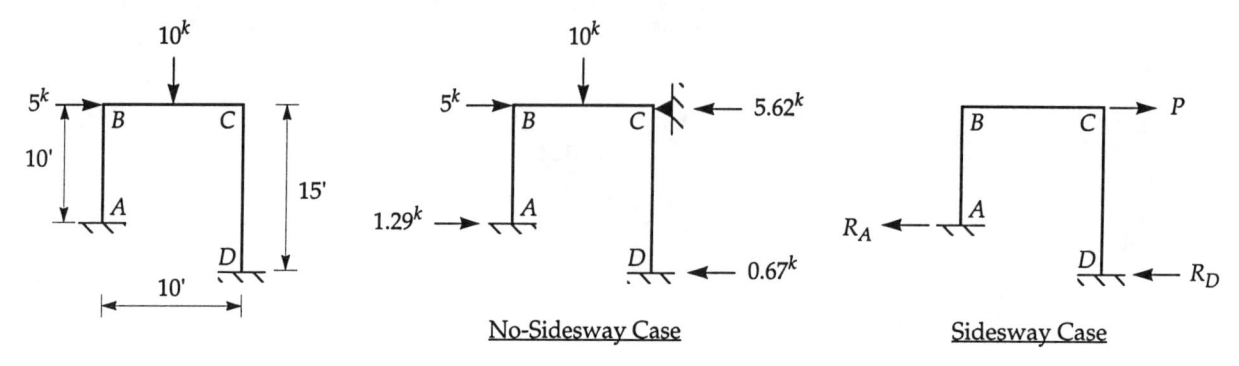

No-Sidesway Case

Sidesway Case

Solution.

1. Moment distribution for the no-sidesway case is shown below. From moment equilibrium of members *AB* and *CD*, the reactions (shears) at *A* and *D* shown in the middle figure above are obtained. From horizontal force equilibrium of the entire structure, the horizontal force required to prevent sidesway is 5.62 kips.

Joint	A	B		C		D
End	AB	BA	BC	CB	CD	DC
DF		0.5	0.5	0.6	0.4	
FEM			12.50	−12.50		
Bal. B		−6.25	−6.25			
C.O.	−3.13			−3.13		
Bal. C				9.38	6.25	
C.O.			4.69			3.13
Bal. B		−2.34	−2.34			
C.O.	−1.17			−1.17		
Bal. C				0.70	0.47	
C.O.						0.23
Sum	−4.30	−8.59	8.60	−6.72	6.72	3.36
Check		Σ = 0.01		Σ = 0		

2. If joints *B* and *C* are both locked and then displaced to the right by an amount Δ, the following FEM will be developed in members *AB* and *CD*:

$$\text{FEM}_{AB} = \text{FEM}_{BA} = \frac{6EI\Delta}{(10)^2} = 0.06 EI\Delta$$

$$\text{FEM}_{CD} = \text{FEM}_{DC} = \frac{6EI\Delta}{(15)^2} = 0.027 EI\Delta$$

Moment distribution for the sidesway case is shown below. For ease of tabulation, all moments that are shown have been divided by $EI\Delta$ and scaled by 100.

Joint	A	B		C		D
End	AB	BA	BC	CB	CD	DC
DF		0.5	0.5	0.6	0.4	
FEM	6.00	6.00			2.70	2.70
Bal. B		−3.00	−3.00			
C.O.	−1.50			−1.50		
Bal. C				−0.72	−0.48	
C.O.			−0.36			−0.24
Bal. B		0.18	0.18			
C.O.	0.09			0.09		
Bal. C				−0.05	−0.04	
C.O.						−0.02
Sum	4.59	3.18	−3.18	−2.18	2.18	2.44
Check		$\Sigma = 0$		$\Sigma = 0$		

From moment equilibrium of members AB and CD, the horizontal reactions at A and D are:

$$R_A = \left(\frac{4.59 + 3.18}{10}\right)\frac{EI\Delta}{100} = 7.77 \times 10^{-3} EI\Delta$$

$$R_D = \left(\frac{2.18 + 2.44}{15}\right)\frac{EI\Delta}{100} = 3.08 \times 10^{-3} EI\Delta$$

The horizontal force required to cause the sidesway Δ is therefore

$$P = R_A + R_D = 10.85 \times 10^{-3} EI\Delta$$

3. Equating P to the horizontal force required to prevent sidesway in the no-sidesway case:

$$10.85 \times 10^{-3} EI\Delta = 5.62 \text{ kips}$$

$$\Rightarrow \Delta = \frac{517.97}{EI}$$

4. Any moment or force may now be obtained using superposition of the no-sidesway and sidesway cases. For example, the moment at support A is

$$M_A = -4.30 + \frac{4.59 EI\Delta}{100} = -4.30 + \frac{4.59(517.97)}{100} = 19.5 \text{ ft-kips}$$

This agrees with the solution obtained by the flexibility method in Example 10.6.

10.4 The Stiffness Method

The stiffness method of analysis is used universally in modern structural analysis computer programs. This method is capable of accounting for flexural, axial and shear deformations in framed members, and formalized techniques are available for programming the method. However, when using this method for hand-calculations it is customary to include only flexural deformations and to use a more intuitive approach as done in this review. The method is exact for the deformations that are considered, but for large structures it requires the inversion of a matrix with dimension equal to the DOF of the structure (or equivalently, the solution of a set of linear simultaneous equations). Modern hand-held calculators are often capable of inverting a matrix of order five or more, and if the reader wishes to use the stiffness method, then it is recommended that such a calculator be used. The stiffness method is not recommended for the hand-analysis of indeterminate trusses, since even the simplest trusses often have many DOF.

The stiffness method consists of the following steps:

1. Lock all joints, apply the external loads and determine all the fixed-end forces.

2. Determine the equivalent joint loads corresponding to the DOF.

3. Determine the stiffness matrix **K** of the structure from individual member stiffnesses. Member stiffnesses are the forces and moments required at the member-ends in order to cause a unit displacement or rotation of a member-end, and are shown in Fig. 10.3. **K** is always a square symmetric matrix, and its ith column contains the forces required along all DOF in order to cause a unit displacement along the ith DOF, while all other DOF are restrained (i.e., locked).

4. Solve the stiffness (i.e., equilibrium) equations for the unknown joint displacements. The structure stiffness equations are of the form

$$\mathbf{Ku} = \mathbf{q} \qquad (10.4.1)$$

in which **u** = vector of the unknown joint displacements, and **q** = vector of equivalent joint loads. This matrix equation is equivalent to a set of linear simultaneous equations, and is solved for the unknown joint displacements. In matrix form, the solution is

$$\mathbf{u} = \mathbf{K}^{-1}\mathbf{q} \qquad (10.4.2)$$

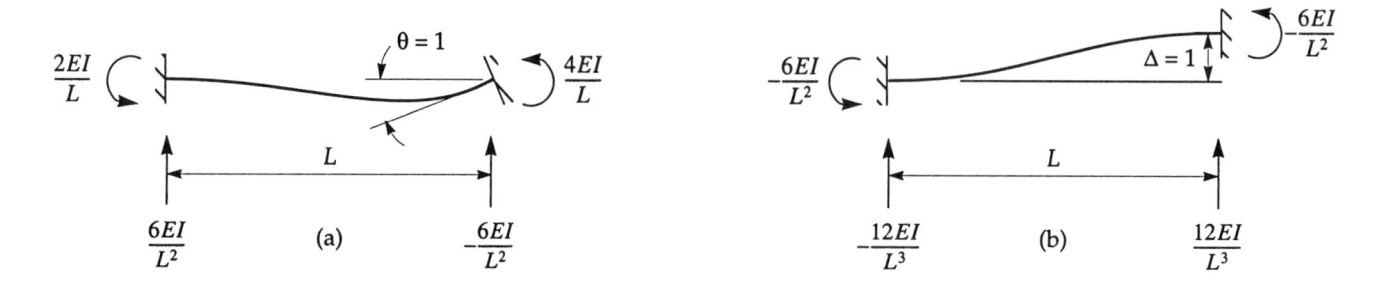

Figure 10.3 Member stiffnesses.

5. Determine the member end-shears and moments by scaling the member stiffnesses shown in Fig. 10.3 by the member end-displacements and adding fixed-end forces to these. This may be written in the following matrix form:

$$\begin{bmatrix} V_L \\ M_L \\ V_R \\ M_R \end{bmatrix} = \begin{bmatrix} V_L^F \\ M_L^F \\ V_R^F \\ M_R^F \end{bmatrix} + \frac{EI}{L^3} \begin{bmatrix} 12 & 6L & -12 & 6L \\ 6L & 4L^2 & -6L & 2L^2 \\ -12 & -6L & 12 & -6L \\ 6L & 2L^2 & -6L & 4L^2 \end{bmatrix} \begin{bmatrix} \Delta_L \\ \theta_L \\ \Delta_R \\ \theta_R \end{bmatrix} \quad (10.4.3)$$

in which V and M denote shear forces and moments, Δ and θ denote translations and rotations, the subscripts L and R denote left and right ends of the member, and the superscript F denotes fixed-end forces. Shear forces and translations are positive when they are upwards, and moments and rotations are positive when they are counter-clockwise. Vertical members should be rotated 90° clockwise so that the bottom end becomes the left end.

6. Determine the axial forces by using joint equilibrium, as was done with moment distribution.

—— EXAMPLE 10.11 ——

Analyze the frame in Example 10.9 by the stiffness method.

Solution.

1. Identify the DOF. The vertical displacement and rotation at E can be taken as the DOF, but this unnecessarily lengthens the problem. The cantilever member DE can be neglected in the analysis since its stiffness is zero, but the load on it must be accounted for. The DOF therefore consist of the rotations at B, C, D, F and G.

2. Determine the equivalent joint loads:

    Equivalent joint moments = Moments applied directly − Fixed-end moments

    The fixed-end moments for each member were computed in Example 10.9. The fixed-end moments at each joint are the sum of the moments at all member ends meeting at that joint:

$$\mathbf{q} = \begin{bmatrix} M_B \\ M_C \\ M_D \\ M_F \\ M_G \end{bmatrix} = \begin{bmatrix} -10 \\ 0 \\ 5 \\ 0 \\ 0 \end{bmatrix} - \begin{bmatrix} -37.5 + 18.75 - 6.25 \\ -18.75 + 18.75 \\ -18.75 + 2.5 \\ 6.25 \\ 0 \end{bmatrix} = \begin{bmatrix} 15 \\ 0 \\ 21.25 \\ -6.25 \\ 0 \end{bmatrix} \text{ft-kips}$$

3. Determine the stiffness matrix, **K**, of the structure. The ith column of the stiffness matrix is determined by imposing a unit rotation along the ith DOF, while all other rotations are zero. The deformed structure when joint B is rotated, along with the moments required along the DOF, are shown below:

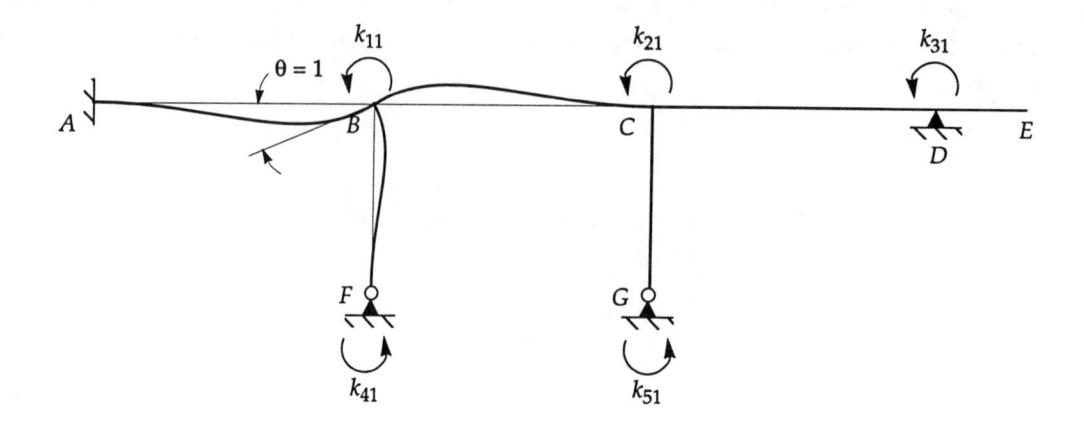

The moments required at the joints in order to cause unit joint rotations can be determined from the members stiffnesses in Fig. 10.3(a). For example, the joint moments required to impose a unit rotation at B constitute the first column of $\mathbf{K}$:

k_{11} = moment required at B = sum of end-moments at BA, BC and BF

$$= \frac{4EI}{L_{AB}} + \frac{4EI}{L_{BC}} + \frac{4EI}{L_{BF}}$$

k_{21} = moment required at $C = \dfrac{2EI}{L_{BC}}$

k_{31} = moment required at $D = 0$

k_{41} = moment required at $F = \dfrac{2EI}{L_{BF}}$

k_{51} = moment required at $G = 0$

The other columns of $\mathbf{K}$ can be determined in a similar manner as shown below:

$$\mathbf{K} = \begin{bmatrix} \frac{4EI}{L_{AB}} + \frac{4EI}{L_{BC}} + \frac{4EI}{L_{BF}} & \frac{2EI}{L_{BC}} & 0 & \frac{2EI}{L_{BF}} & 0 \\ \frac{2EI}{L_{BC}} & \frac{4EI}{L_{BC}} + \frac{4EI}{L_{CD}} + \frac{4EI}{L_{CG}} \frac{2EI}{L_{BC}} & 0 & \frac{2EI}{L_{CG}} \\ 0 & \frac{2EI}{L_{CD}} & \frac{4EI}{L_{CD}} & 0 & 0 \\ \frac{2EI}{L_{BF}} & 0 & 0 & \frac{4EI}{L_{BF}} & 0 \\ 0 & \frac{2EI}{L_{CG}} & 0 & 0 & \frac{4EI}{L_{CG}} \end{bmatrix} = EI \begin{bmatrix} 0.933 & 0.133 & 0 & 0.2 & 0 \\ 0.133 & 0.933 & 0.133 & 0 & 0.2 \\ 0 & 0.133 & 0.267 & 0 & 0 \\ 0.2 & 0 & 0 & 0.4 & 0 \\ 0 & 0.2 & 0 & 0 & 0.4 \end{bmatrix}$$

4. Solve the stiffness equations:

$$\mathbf{u} = \begin{bmatrix} \theta_B \\ \theta_C \\ \theta_D \\ \theta_F \\ \theta_G \end{bmatrix} = \mathbf{K}^{-1}\mathbf{q} = \frac{1}{EI}\begin{bmatrix} 1.234 & -0.215 & 0.107 & -0.617 & 0.107 \\ -0.215 & 1.342 & -0.671 & 0.107 & -0.671 \\ 0.107 & -0.671 & 4.085 & -0.054 & 0.335 \\ -0.617 & 0.107 & -0.054 & 2.809 & -0.054 \\ 0.107 & -0.671 & 0.335 & -0.054 & 2.835 \end{bmatrix}\begin{bmatrix} 15 \\ 0 \\ 21.25 \\ -6.25 \\ 0 \end{bmatrix} = \frac{1}{EI}\begin{bmatrix} 24.65 \\ -18.15 \\ 88.76 \\ -27.95 \\ 9.07 \end{bmatrix}$$

5. Determine the member end-forces and reactions. This is done on a member by member basis. For member AB, the fixed-end shears are $V_A^F = V_B^F = wL/2 = (2)(15)/2 = 15$ kips. The displacement and rotation at A are zero and the displacement at B is also zero (since axial deformations are being neglected). Using Eq. 10.4.3, the member end-forces are:

$$\begin{bmatrix} V_A \\ M_A \\ V_B \\ M_B \end{bmatrix} = \begin{bmatrix} 15 \\ 37.5 \\ 15 \\ -37.5 \end{bmatrix} + \frac{EI}{15^3}\begin{bmatrix} 12 & 6(15) & -12 & 6(15) \\ 6(15) & 4(15)^2 & -6(15) & 2(15)^2 \\ -12 & -6(15) & 12 & -6(15) \\ 6(15) & 2(15)^2 & -6(15) & 4(15)^2 \end{bmatrix}\frac{1}{EI}\begin{bmatrix} 0 \\ 0 \\ 0 \\ 24.65 \end{bmatrix} = \begin{bmatrix} 15.7 \text{ kips} \\ 40.8 \text{ ft-kips} \\ 14.3 \text{ kips} \\ -30.9 \text{ ft-kips} \end{bmatrix}$$

These end-forces are close to those determined by moment distribution, and are in fact more accurate. Other member end-forces can be determined in a similar manner.

EXAMPLE 10.12

Use the stiffness method to analyze the frame in Example 10.10.

Solution.

1. Identify DOF. If axial deformations are neglected, then the rotations at B and C and the horizontal translation of B (or C) are the DOF.

2. Determine the equivalent joint loads. Since the horizontal displacements at B and C are collapsed into a single DOF, care must be taken to lump the horizontal forces at B and C together when determining the equivalent joint loads for the translational DOF:

$$\mathbf{q} = \begin{bmatrix} M_B \\ M_C \\ F_B \end{bmatrix} = \begin{bmatrix} 0 \\ 0 \\ 5 \end{bmatrix} - \begin{bmatrix} 12.5 \\ -12.5 \\ 0 \end{bmatrix} = \begin{bmatrix} -12.5 \text{ ft-kips} \\ 12.5 \text{ ft-kips} \\ 5 \text{ kips} \end{bmatrix}$$

3. Determine the stiffness matrix of the structure. For the translational DOF, care must be taken to lump the horizontal forces at B and C together when determining the corresponding stiffnesses, and both B and C must be translated horizontally (with rotations being restrained) when imposing a unit translation. Using the stiffnesses in Fig. 10.3, and units of feet for the lengths, the structure stiffness matrix is

$$\mathbf{K} = \begin{bmatrix} \dfrac{4EI}{L_{AB}} + \dfrac{4EI}{L_{BC}} & \dfrac{2EI}{L_{BC}} & \dfrac{6EI}{L_{AB}^2} \\ \dfrac{2EI}{L_{BC}} & \dfrac{4EI}{L_{BC}} + \dfrac{4EI}{L_{CD}} & \dfrac{6EI}{L_{CD}^2} \\ \dfrac{6EI}{L_{AB}^2} & \dfrac{6EI}{L_{CD}^2} & \dfrac{12EI}{L_{AB}^3} + \dfrac{12EI}{L_{CD}^3} \end{bmatrix} = EI \begin{bmatrix} 0.8 & 0.2 & 0.06 \\ 0.2 & 0.667 & 0.0267 \\ 0.06 & 0.0267 & 0.0156 \end{bmatrix}$$

4. Solve the stiffness equations:

$$\mathbf{u} = \begin{bmatrix} \theta_B \\ \theta_C \\ \Delta_B \end{bmatrix} = \mathbf{K}^{-1}\mathbf{q} = \frac{1}{EI}\begin{bmatrix} 1.807 & -0.283 & -6.486 \\ -0.283 & 1.655 & -1.746 \\ -6.486 & -1.746 & 92.3 \end{bmatrix}\begin{bmatrix} -12.5 \\ 12.5 \\ 5 \end{bmatrix} = \frac{1}{EI}\begin{bmatrix} -58.55 \\ 15.49 \\ 520.7 \end{bmatrix}$$

5. Determine the member end-forces and reactions. For member AB the fixed-end forces are all zero. When the member is rotated horizontally such that A is on the left and B is on the right, shear forces and translations are considered positive when they are upwards, and hence the rightward translation of B should be considered as being negative for the purpose of evaluating the member end-forces. The member end-forces are therefore

$$\begin{bmatrix} V_A \\ M_A \\ V_B \\ M_B \end{bmatrix} = \frac{EI}{10^3}\begin{bmatrix} 12 & 6(10) & -12 & 6(10) \\ 6(10) & 4(10)^2 & -6(10) & 2(10)^2 \\ -12 & -6(10) & 12 & -6(10) \\ 6(10) & 2(10)^2 & -6(10) & 4(10)^2 \end{bmatrix}\frac{1}{EI}\begin{bmatrix} 0 \\ 0 \\ -520.7 \\ -58.55 \end{bmatrix} = \begin{bmatrix} 2.7 \text{ kips} \\ 19.5 \text{ ft-kips} \\ -2.7 \text{ kips} \\ 7.8 \text{ ft-kips} \end{bmatrix}$$

Note that the value of M_A is identical to the value obtained by the flexibility method in Example 10.6 and by moment distribution in Example 10.10.

10.5 Support Movements

When support movements occur in indeterminate structures, deformations and internal forces are induced. Inclusion of support settlement effects in the three analysis methods discussed in this chapter is briefly reviewed.

10.5.1 Flexibility Method

In the flexibility method, support movements are handled differently depending on whether or not they correspond to the redundants. Support displacements that correspond to the redundants are included in $\mathbf{u}_q$ in Eqs. 10.2.1 and 10.2.2. If there are support movements that do not correspond to the redundants, then the support movements are made to occur in the primary structure and the resulting displacements corresponding to the redundants are calculated and added to the displacements due to the loads to obtain $\mathbf{u}$. The primary structure does not deform under the support movement, and therefore all displacements are due to rigid body motions of the entire structure or of the different parts making up the structure.

Consider the frame in Example 10.6. Let support A settle downward by Δ_A and support D settle downward by Δ_D. The settlement at A does not correspond to a redundant (since the redundants are the three reactions at D), and is accounted for by letting support A of the primary structure settle by Δ_A and determining the displacements corresponding to the redundants. The primary structure will simply move downward as a rigid body, and as a result a vertical displacement will occur at D. The displacement vector $\mathbf{u}$ due to loads and the settlement at A would therefore be

$$\mathbf{u} = \begin{bmatrix} \Delta_H \\ \Delta_V \\ \theta \end{bmatrix} = -\frac{1}{EI} \begin{bmatrix} 8958.33 \\ 8541.67 \\ 875 \end{bmatrix} + \begin{bmatrix} 0 \\ -\Delta_A \\ 0 \end{bmatrix}$$

The settlement at D, however, corresponds to a redundant (the vertical reaction at D) and is included in $\mathbf{u}_q$ ($\mathbf{u}_q = [0, -\Delta_D, 0]^T$).

10.5.2 Moment Distribution

Support movements are included in moment distribution by computing the additional fixed-end moments generated in the structure as a result of the support movements when all other joints are prevented from rotating. The fixed-end moments due to loads and support movements are added together to obtain the combined fixed-end moments, and then moment distribution is carried out in the normal way.

Consider the frame in Example 10.9. If support D was to settle downward by an amount Δ, then from Fig. 10.3(b), the additional fixed-end moments generated at the ends of member CD would be

$$\text{FEM}_{CD} = \text{FEM}_{DC} = 6EI\Delta/L^2.$$

10.5.3 Stiffness Method

In the stiffness method, support movements must be appropriately included in each of the steps:

1. The opposite of the fixed-end forces generated by support movements when all other DOF are fixed, are added to the equivalent joint loads due to external loads, to obtain the combined joint loads. The combined joint loads are used in Eqs. 10.4.1 and 10.4.2 when determining the unknown joint displacements.

2. When individual member end-forces are computed using Eq. 10.4.3, member end-displacements due to support movements must be included in the last vector on the right-hand-side. The fixed-end forces at the member ends (the first vector on the right-hand-side of Eq. 10.4.3) are only due to loads.

Consider Example 10.11. Suppose support F settles downward by an amount Δ. When the support settlement occurs, joint B is also moved down by Δ (since axial deformation is neglected), but all joints are prevented from rotating. From Fig. 10.3(b), fixed-end moments equal to $6EI\Delta/L^2$, and fixed-end forces with magnitude $12EI\Delta/L^3$, are generated at the ends of members AB and BC. The opposite of the fixed-end moments at B and C due to support movement (i.e., $-6EI\Delta/L^2$) must be added to the equivalent joint moments due to loads to obtain $\mathbf{q}$. The joint rotations are then obtained by using this $\mathbf{q}$ in step 4. In step 5, the member end-displacements should also include the support movement, i.e.,

$$\begin{bmatrix} \Delta_A \\ \theta_A \\ \Delta_B \\ \theta_B \end{bmatrix} = \begin{bmatrix} 0 \\ 0 \\ -\Delta \\ \theta_B \end{bmatrix}$$

in which θ_B is obtained from the solution of the stiffness equations (step 4).

10.6 Selection of Analysis Method

In closing, it seems appropriate to provide some guidelines regarding the selection of the most convenient method for the hand-analysis of indeterminate structures.

1. For indeterminate trusses, use the flexibility method.

2. For indeterminate beams and frames which cannot undergo sidesway, use moment distribution. If the DSI ≤ 2, then the flexibility method may also be used.

3. For indeterminate frames that can undergo sidesway:

   (a) If DSI is small (say, ≤ 2) and DOF is large (say, ≥ 5), use the flexibility method.

   (b) If DSI is large, use the stiffness method.

If one does not wish to be fluent in the use of both moment distribution and the stiffness method, then one may focus on only one of these methods without any serious handicap.

10.7 Calculation of Stresses

Once the member forces are computed using any of the methods outlined in the previous sections, stresses may be calculated if desired. Stresses may be needed if an elastic design is to follow the analysis. In ultimate strength design, the member forces are used directly in sizing members. For elastic design, the maximum stresses in beam or frame members subjected to an axial force P and bending moment M is given by

$$f_{max} = \frac{P}{A} + \frac{Mc}{I} \tag{10.7.1}$$

in which A is the cross sectional area, I = moment of inertia, and c = distance from the neutral axis of the cross section to the extreme fiber. Note that the stress due to axial load should always be added to that due to bending moment irrespective of whether the axial load is tensile of compressive. This is because the bending moment causes both tensile and compressive stresses, and if P is a tensile load then the tensile stress will be critical, while if P is a compressive load then compressive stress will be critical. For truss members with loads applied at the joints, only the axial stress P/A exists.

Practice Problems – Essay Type

10.1

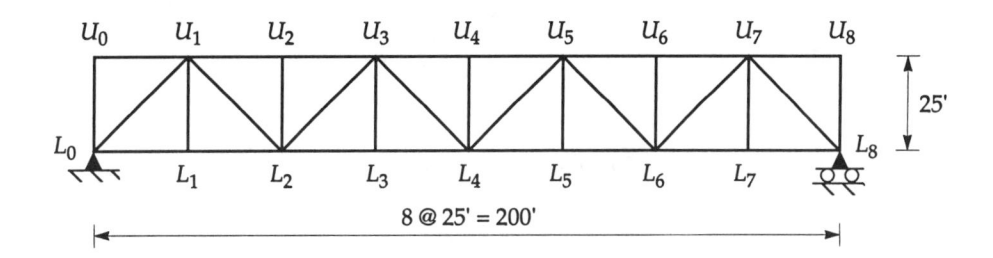

The simply supported steel truss is part of a highway bridge, and supports a concrete deck placed above it. The truss is loaded on the upper chords by uniformly distributed dead and live loads of 3.0 k/ft and 0.8 k/ft, and a concentrated live load of 18 kips. An impact factor of 15% is to be used on all live loads. The truss members have been designed to be symmetric about the centerline U_4L_4, and the following wide flange sections have been chosen:

Upper Chords		Lower Chords		Verticals		Diagonals	
U_0U_1	W14 × 61	L_0L_1	W14 × 90	U_0L_0	W14 × 61	L_0U_1	W14 × 176
U_1U_2	W14 × 176	L_1L_2	W14 × 90	U_1L_1	W14 × 30	U_1L_2	W14 × 90
U_2U_3	W14 × 176	L_2L_3	W14 × 176	U_2L_2	W14 × 61	L_2U_3	W14 × 99
U_3U_4	W14 × 233	L_3L_4	W14 × 176	U_3L_3	W14 × 30	U_3L_4	W14 × 61
				U_4L_4	W14 × 61		

Compute the maximum live load deflection at mid-span.

10.2

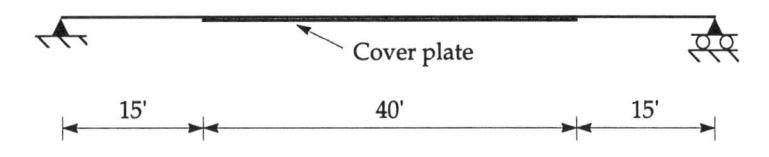

A 70 ft long composite steel/concrete beam is part of seven stringer beams that make up the approach span of a highway bridge. The beam has a 40 ft long 14" × 0.75" cover plate attached to the bottom flange and centered along its length. The moments of inertia of the composite cross sections with and without the cover plates are 27,200 in^4 and 18,300 in^4, respectively. Based on the AASHTO lane loading and the fraction of wheel load carried by the stringer, the live load on the beam consists of a concentrated live load of 11.5 kips and a uniformly distributed live load of 0.4 kips/ft. An impact factor of 25% is to be used with the live loads. Compute the maximum live load deflection of the beam.

10.3

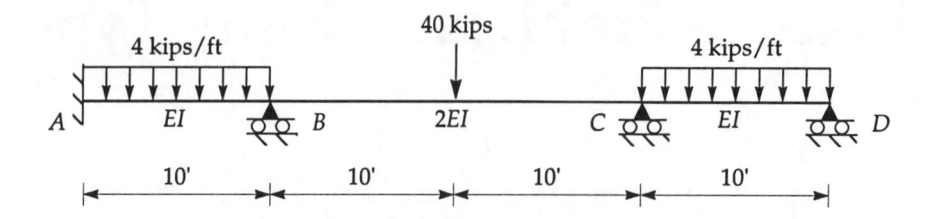

[Figure: Continuous beam A-B-C-D with four 10' spans. Distributed load 4 kips/ft on span A-B (EI), 40 kips point load at midspan of B-C (2EI), distributed load 4 kips/ft on span C-D (EI). Supports at B, C, D.]

Determine the bending moment diagram for the continuous beam.

10.4

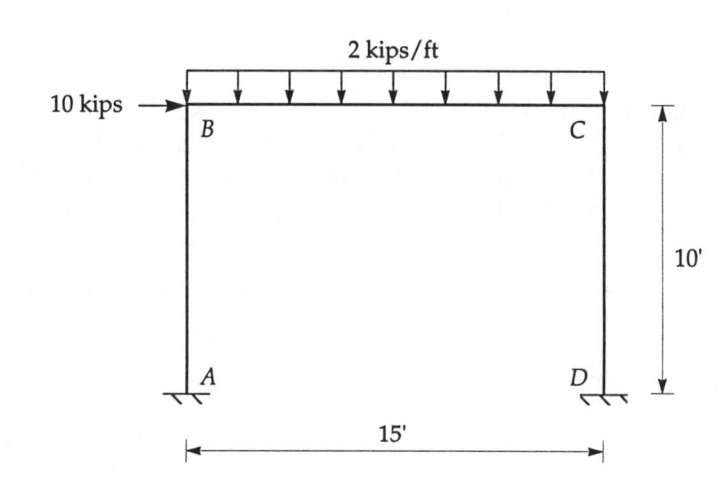

[Figure: Portal frame with columns AB and DC (10' tall) and beam BC (15' long). Horizontal load 10 kips applied at B, distributed load 2 kips/ft on beam BC. Supports at A and D.]

In addition to the loads, support D of the frame settles downward by 0.5 inches. All members have $E = 29{,}000$ ksi, and $I = 500$ in^4. Sketch the shear and bending moment diagrams.

10.5

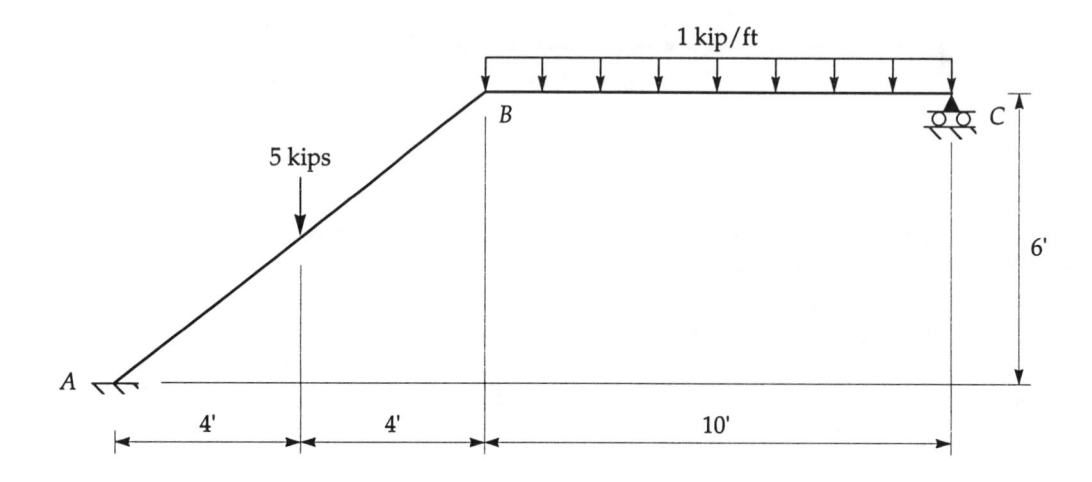

[Figure: Frame with inclined member from A (pin support) rising to B, then horizontal member from B to C (roller support) 10' long. Height at C is 6'. 5 kips point load applied at 4' horizontally from A (4' more to B). 1 kip/ft distributed load on BC.]

Determine all the support reactions for the frame. Both members have the same EI.

10.6

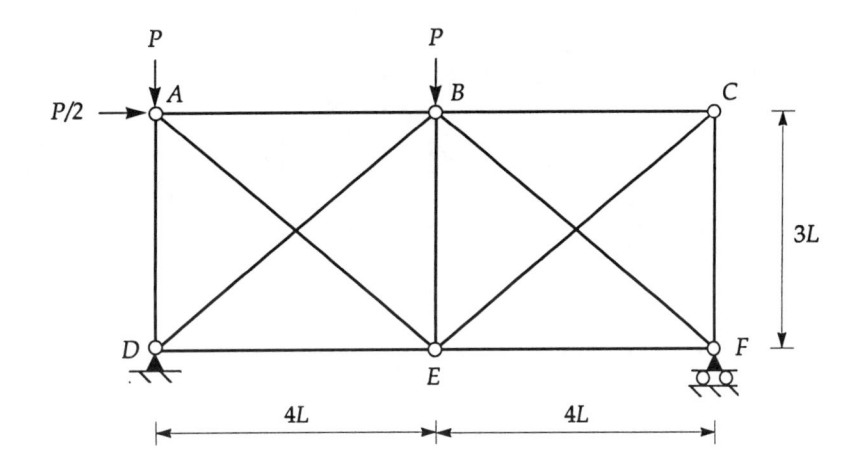

Determine all the truss bar forces. The horizontal and vertical members have cross sectional area A, the diagonals have area $2A$, and all bars have the same elastic modulus.

10.7

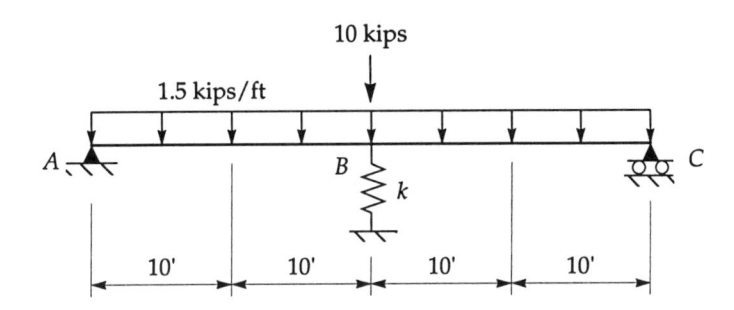

The middle support of the two-span steel beam is flexible and is modeled with a spring having a stiffness $k = 50$ kips/in. The moment of inertia of the beam is 50 in^4. Compute all the reactions and the vertical displacement at B.

10.8

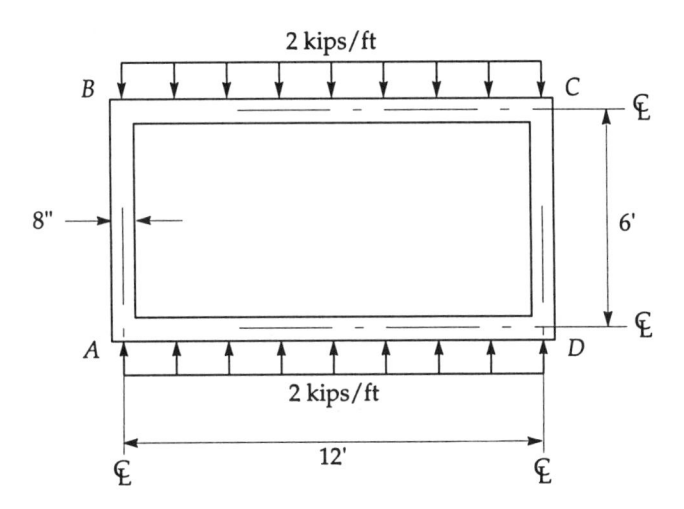

The concrete box culvert has 8 inch thick walls all around and has a modulus of elasticity of 3000 ksi. The loads shown are for a 12" length of the culvert. Determine the axial force, shear force and bending moment variations in members BC and CD.

10.9

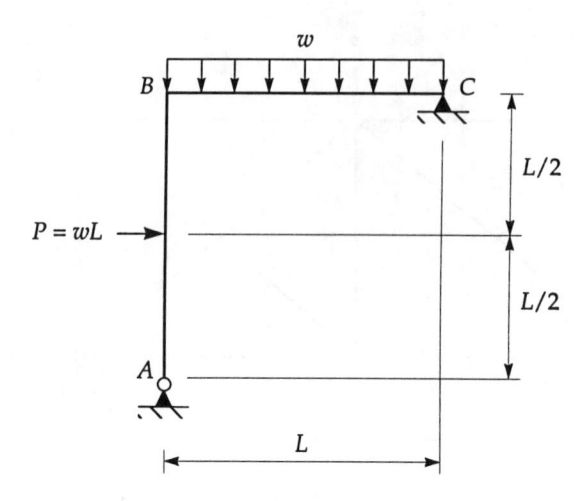

In addition to the loads, support C of the frame settles downward by $\Delta = wL^4/(48EI)$. Analyze the frame to determine the bending moment diagram.

10.10

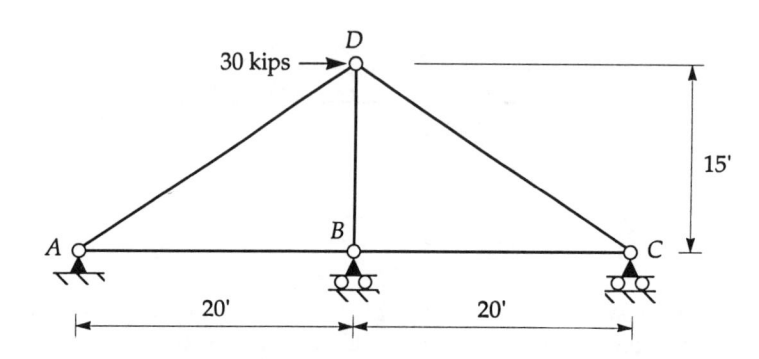

In addition to the load support B settles 0.5 inches and support C settles 0.25 inches. $E = 29,000$ ksi and all bar areas are 0.5 in^2. Compute all the bar forces.

Practice Problems – Multiple Choice

Questions 10.11–10.13

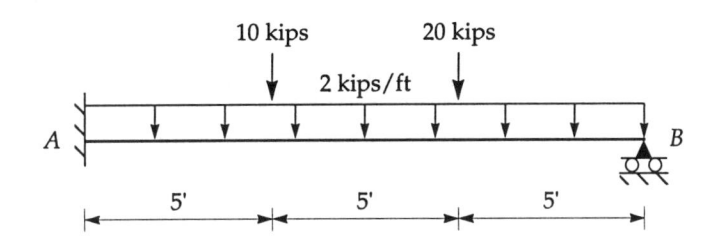

10.11 The counter-clockwise moment at support A, in ft-kips, is most nearly

    (a) 35.3   (b) 81.9   (c) 93.1   (d) 128.5   (e) 175

10.12 The upward reaction at A, in kips, is most nearly

    (a) 24.2   (b) 30   (c) 36.9   (d) 45.5   (e) 60

10.13 The maximum positive moment in span AB, in ft-kips, is most nearly

    (a) 31   (b) 56.3   (c) 67   (d) 90.5   (e) 120

Questions 10.14–10.18

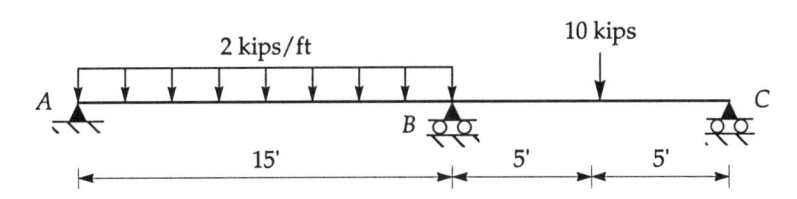

10.14 The value of the bending moment in the beam at B, in ft-kips, is most nearly

    (a) 12.5   (b) 41.25   (c) 50   (d) 56.25   (e) 112.5

10.15 The upward reaction at A, in kips, is most nearly

    (a) 12.25   (b) 13.33   (c) 15   (d) 17.75   (e) 20

10.16 The upward reaction at B, in kips, is most nearly

    (a) 15   (b) 20   (c) 26.9   (d) 32.5   (e) 40

10.17 The maximum positive moment in span AB, in ft-kips, is most nearly

    (a) 37.5   (b) 56.25   (c) 91.9   (d) 125   (e) 225

10.18 The maximum positive moment in span BC, in ft-kips, is most nearly

    (a) 4.4   (b) 8.8   (c) 12.5   (d) 17.75   (e) 25

Questions 10.19–10.24

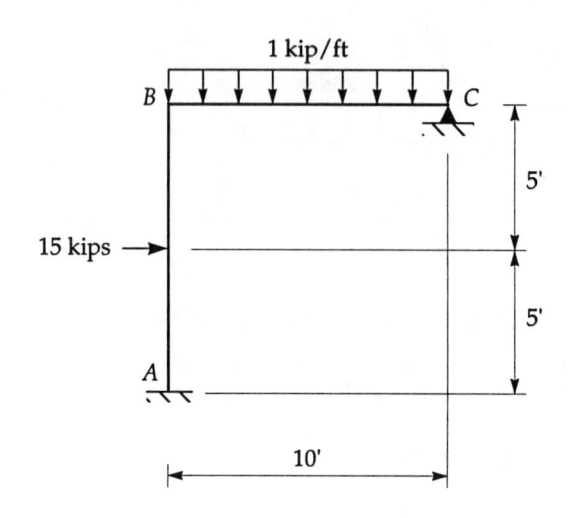

10.19 The moment at A, in ft-kips, is most nearly

    (a) 12.25   (b) 18.75   (c) 20.5   (d) 25   (e) 75

10.20 The moment at B, in ft-kips, is most nearly

    (a) 8.33   (b) 10   (c) 12.5   (d) 15.2   (e) 18.75

10.21 The horizontal (leftward) reaction at A, in kips, is most nearly

    (a) 6   (b) 7.5   (c) 8.5   (d) 10   (e) 15

10.22 The upward reaction at A. in kips, is most nearly

    (a) 2.5   (b) 5   (c) 6.5   (d) 7.5   (e) 10

10.23 The maximum positive moment in span BC, in ft-kips, is most nearly

    (a) 5   (b) 6.1   (c) 7.5   (d) 8.33   (e) 12.5

10.24 The moment at the location of the concentrated load in AB, in ft-kips, is most nearly

    (a) 5   (b) 6.25   (c) 15   (d) 18.75   (e) 30

Solutions to Problems

Question 10.1

1. Determine loads. For maximum deflection, place uniform live load over entire span and concentrated live load at U_4.

   Loads at U_1, U_2, U_3, U_5, U_6 and U_7 due to uniform live load + impact = 0.8(25)(1.15) = 23 kips

   Loads at U_0 and U_8 due to uniform live load + impact = 23/2 = 11.5 kips

   Load at U_4 due to uniform live load + concentrated live load + impact = 23 + 18(1.15)
   = 43.7 kips

2. Tabulate bar forces due to live load P_i and due to unit load p_i applied at U_4, and compute $P_i p_i L_i / A_i$ for bars on the left half of the truss.

Bar	L (in.)	A (in²)	P (kips)	p (kips)	PpL/A
$U_0 U_1$	300	17.9	0	0	0
$U_1 U_2$	300	51.8	−158.7	−1	919.11
$U_2 U_3$	300	51.8	−158.7	−1	919.11
$U_3 U_4$	300	68.5	−225.4	−2	1974.31
$L_0 L_1$	300	26.5	90.85	0.5	514.25
$L_1 L_2$	300	26.5	90.85	0.5	514.25
$L_2 L_3$	300	51.8	135.7	1.5	1178.86
$L_3 L_4$	300	51.8	135.7	1.5	1178.86
$U_0 L_0$	300	17.9	−11.5	0	0
$U_1 L_1$	300	8.85	0	0	0
$U_2 L_2$	300	17.9	−23	0	0
$U_3 L_3$	300	8.85	0	0	0
$U_4 L_4$	300	17.9	−43.7	−1	732.40
$L_0 U_1$	424.26	51.8	−128.48	−0.7071	744.08
$U_1 L_2$	424.26	26.5	95.95	0.7071	1086.21
$L_2 U_3$	424.26	29.1	−63.43	−0.7071	653.90
$U_3 L_4$	424.26	17.9	30.9	0.7071	517.87
					Σ = 10933.21

All members except for $U_4 L_4$ have mirror images on the right side of the truss and must therefore be included twice in the displacement calculation. The vertical deflection at U_4 is

$$\Delta = \frac{1}{E} \sum_i \frac{P_i p_i L_i}{A_i} = \frac{1}{29000}[2(10933) - 732.4] = 0.73 \text{ in.}$$

Question 10.2

1. Determine loads. For maximum deflection apply the uniform load over the entire span and the concentrated load at mid-span.

   Uniform live load + impact = 0.4(1.25)= 0.5 kips/ft

   Concentrated live load + impact = 11.5(1.25)= 14.375 kips

2. Determine the bending moment variations due to the live load and a unit load at mid-span.

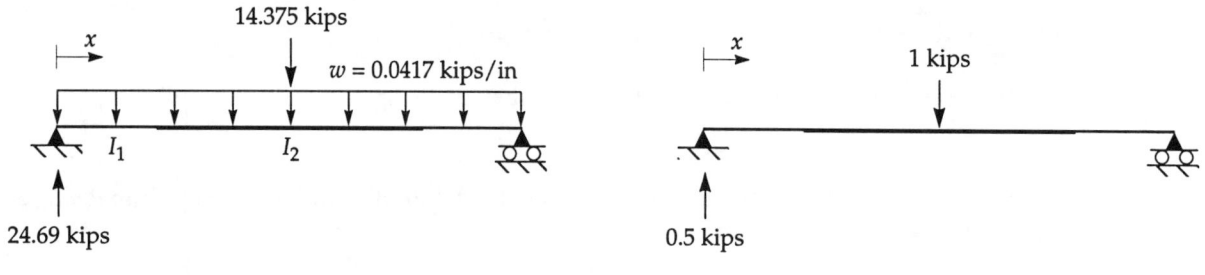

   $M(x) = 24.69x - 0.5x(x/2), \quad 0 \leq x \leq 35'$

   $m(x) = 0.5x, \quad 0 \leq x \leq 35'$

3. Calculate the midspan deflection.

   $E = 29000(144) = 4.176 \times 10^6 \text{ k/ft}^2$

   $I_1 = 18,300/12^4 = 0.8825 \text{ ft}^4 \qquad I_2 = 27,200/12^4 = 1.3117 \text{ ft}^4$

   $$\Delta = \int \frac{mM}{EI} dx = \frac{2}{EI_1} \int_0^{15'} 0.5x(24.69x - 0.25x^2)dx + \frac{2}{EI_2} \int_{15'}^{35'} 0.5x(24.69x - 0.25x^2)dx$$

   $$= \frac{2}{(4.176 \times 10^6)(0.8825)} [4.115x^3 - 0.03125x^4]\Big|_0^{15}$$

   $$+ \frac{2}{(4.176 \times 10^6)(1.3117)} [4.115x^3 - 0.03125x^4]\Big|_{15}^{35}$$

   $= 0.00334 + 0.02140 = 0.0247 \text{ ft} = 0.30 \text{ in.}$

(The integration could also be done using Table 10.1, if preferred.)

Question 10.3

Since there is no sidesway involved, moment distribution is a suitable method for this problem. Since the DOF = 3, and the DSI = 3 (for vertical loading only), the stiffness and flexibility methods may also be used if preferred. This solution uses moment distribution.

1. Compute fixed-end moments.

$$\text{FEM}_{AB} = -\text{FEM}_{BA} = \text{FEM}_{CD} = -\text{FEM}_{DC} = \frac{wL^2}{12} = \frac{4(10)^2}{12} = 33.33 \text{ ft-kips}$$

$$\text{FEM}_{BC} = -\text{FEM}_{CB} = \frac{PL}{8} = \frac{40(20)}{8} = 100 \text{ ft-kips}$$

2. Compute stiffnesses and distribution factors.

$$K_{AB} = \frac{4EI}{10} = 0.4EI \qquad K_{BC} = \frac{4(2EI)}{20} = 0.4EI \qquad K_{CD} = \frac{3EI}{10} = 0.3EI$$

$$\text{DF}_{BA} = \text{DF}_{BC} = \frac{0.4}{0.4 + 0.4} = 0.5 \qquad \text{DF}_{CB} = \frac{0.4}{0.4 + 0.3} = 0.571 \qquad \text{DF}_{CD} = 1 - 0.571 = 0.429$$

3. Perform moment distribution calculations.

Joint	A	B		C		D
End	AB	BA	BC	CB	CD	DC
DF		0.5	0.5	0.571	0.429	
FEM	33.33	−33.33	100	−100	33.33	−33.33
Bal. D						33.33
C.O.					16.67	
Bal. C				28.55	21.45	
C.O.			14.28			
Bal. B		−40.48	−40.48			
C.O.	−20.24			−20.24		
Bal. C				11.56	8.68	
C.O.			5.78			
Bal. B		−2.89	−2.89			
C.O.	−1.44			−1.44		
Bal. C				0.82	0.62	
C.O.			0.41			
Bal. C		−0.21	−0.21			
Sum	11.65	−76.91	76.89	−80.75	80.75	0.0
Check		Σ = −0.02		Σ = 0.0		

4. Sketch moment diagram.

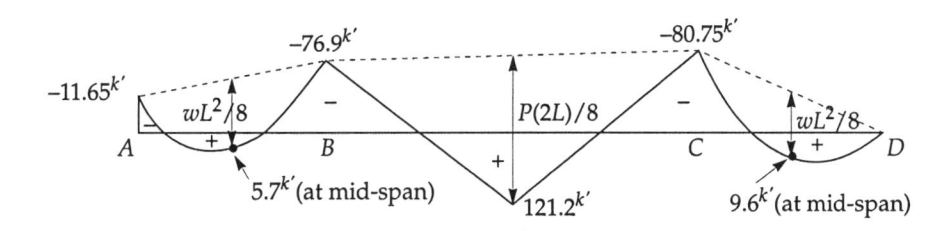

−11.65$^{k'}$ −76.9$^{k'}$ −80.75$^{k'}$

$wL^2/8$ $P(2L)/8$ $wL^2/8$

A   B   C   D

5.7$^{k'}$ (at mid-span)   121.2$^{k'}$ 9.6$^{k'}$ (at mid-span)

Question 10.4

Since the frame can undergo sidesway, moment distribution would be tedious. The DOF = 3 and the DSI = 3. If the flexibility method was used, the required displacements would have to be found by the unit load method, since the use of Table 10.2 is not straightforward. Therefore, the stiffness method is perhaps most suitable. The steps are similar to those in Example 10.12.

1. The DOF are the rotations at B and C, and the horizontal (rightward) displacement at B.

2. Determine the equivalent joint loads. Use units of kips and inches. $w = 2$ k/ft $= 0.167$ k/in.

    (a) The fixed-end moments due to loads are:

    $$\text{FEM}_{BC} = -\text{FEM}_{CB} = \frac{wL^2}{12} = \frac{0.167(180)^2}{12} = 450 \text{ in-kips}$$

    (b) Since axial deformations are neglected, the support settlement at D causes joint C to also move downwards by 0.5". The following additional fixed-end moments are therefore generated in member BC due to support movement:

    $$\text{FEM}_{BC} = \text{FEM}_{CB} = \frac{6EI\Delta}{L^2} = \frac{6(29000)(500)(0.5)}{180^2} = 1342.6 \text{ in-kips}$$

    (c) The total fixed-end moments are therefore, $\text{FEM}_{BC} = 450 + 1342.6 = 1792.6$ in-kips and $\text{FEM}_{CB} = -450 + 1342.6 = 892.6$ in-kips, and the equivalent joint loads are:

    $$\mathbf{q} = \begin{bmatrix} M_B \\ M_C \\ F_B \end{bmatrix} = \begin{bmatrix} 0 \\ 0 \\ 10 \end{bmatrix} - \begin{bmatrix} 1792.6 \\ 892.6 \\ 0 \end{bmatrix} = \begin{bmatrix} -1792.6 \text{ in-kips} \\ -892.6 \text{ in-kips} \\ 10 \text{ kips} \end{bmatrix}$$

3. Determine the stiffness matrix. Using units of kips and inches

    $$\mathbf{K} = \begin{bmatrix} \frac{4EI}{L_{AB}} + \frac{4EI}{L_{BC}} & \frac{2EI}{L_{BC}} & \frac{6EI}{L_{AB}^2} \\ \frac{2EI}{L_{BC}} & \frac{4EI}{L_{BC}} + \frac{4EI}{L_{CD}} & \frac{6EI}{L_{CD}^2} \\ \frac{6EI}{L_{AB}^2} & \frac{6EI}{L_{CD}^2} & \frac{12EI}{L_{AB}^3} + \frac{12EI}{L_{CD}^3} \end{bmatrix} = EI \begin{bmatrix} 805600 & 161100 & 6042 \\ 161100 & 805600 & 6042 \\ 6042 & 6042 & 201 \end{bmatrix}$$

4. Solve for the unknown displacements.

    $$\mathbf{u} = \begin{bmatrix} \theta_B \\ \theta_C \\ \Delta_B \end{bmatrix} = \mathbf{K}^{-1}\mathbf{q} = \begin{bmatrix} -3.417 \times 10^{-3} \text{ rad} \\ -2.020 \times 10^{-3} \text{ rad} \\ 0.2128 \text{ in.} \end{bmatrix}$$

5. Determine the member end-forces using Eq. 10.4.3.

   (a) For member AB the fixed-end forces are zero and therefore:

   $$\begin{bmatrix} V_A \\ M_A \\ V_B \\ M_B \end{bmatrix} = \frac{29000(500)}{120^3} \begin{bmatrix} 12 & 6(120) & -12 & 6(120) \\ 6(120) & 4(120)^2 & -6(120) & 2(120)^2 \\ -12 & -6(120) & 12 & -6(120) \\ 6(120) & 2(120)^2 & -6(120) & 4(120)^2 \end{bmatrix} \begin{bmatrix} 0 \\ 0 \\ -0.2128 \\ -3.417 \times 10^{-3} \end{bmatrix} = \begin{bmatrix} 0.8 \\ 460 \\ -0.8 \\ 366 \end{bmatrix}$$

   (b) For member BC, the fixed-end reactions due to loads are $V_B^F = V_C^F = wL/2 = 15$ kips. The support movement effect at C is included in the member displacements.

   $$\begin{bmatrix} V_B \\ M_B \\ V_C \\ M_C \end{bmatrix} = \begin{bmatrix} 15 \\ 450 \\ 15 \\ -450 \end{bmatrix} + \frac{29000(500)}{120^3} \begin{bmatrix} 12 & 6(180) & -12 & 6(180) \\ 6(180) & 4(180)^2 & -6(180) & 2(180)^2 \\ -12 & -6(180) & 12 & -6(180) \\ 6(180) & 2(180)^2 & -6(180) & 4(180)^2 \end{bmatrix} \begin{bmatrix} 0 \\ -3.417 \times 10^{-3} \\ -0.5 \\ -2.020 \times 10^{-3} \end{bmatrix} = \begin{bmatrix} 15.3 \\ 366 \\ 14.7 \\ -309 \end{bmatrix}$$

   (c) For member AB the fixed-end forces are zero. Rotating the member 90° clockwise:

   $$\begin{bmatrix} V_D \\ M_D \\ V_C \\ M_C \end{bmatrix} = \frac{29000(500)}{120^3} \begin{bmatrix} 12 & 6(120) & -12 & 6(120) \\ 6(120) & 4(120)^2 & -6(120) & 2(120)^2 \\ -12 & -6(120) & 12 & -6(120) \\ 6(120) & 2(120)^2 & -6(120) & 4(120)^2 \end{bmatrix} \begin{bmatrix} 0 \\ 0 \\ -0.2128 \\ -2.020 \times 10^{-3} \end{bmatrix} = \begin{bmatrix} 9.2 \\ 797 \\ -9.2 \\ 309 \end{bmatrix}$$

6. Draw the shear force and moment diagram, being careful to change from the sign convention used for the member end-forces to that used for the diagrams.

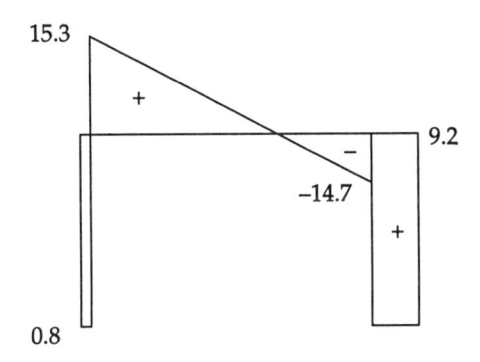

Shear Force Diagram (kips)

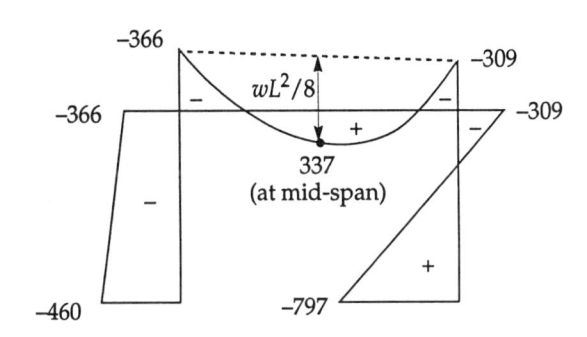

Bending Moment Diagram (in-kips)

Question 10.5

Moment distribution is unsuitable since the frame can undergo sidesway, and the inclined member introduces additional complications. The DOF = 3 (rotations at B and C and horizontal translation at B), so the stiffness method could be used, but the inclined member introduces some complications. The DSI = 1 and hence the flexibility method is perhaps the most suitable.

1. The primary structure is obtained by removing the support at C. The upward reaction at C, R_C, is the redundant.

2. Compute the displacement corresponding to the redundant and due to loads, and the flexibility coefficient. We use the unit load method and Table 10.1.

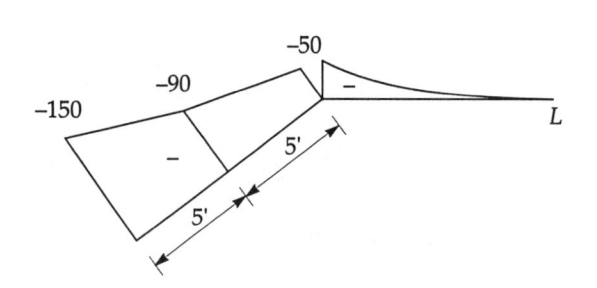

BMD due to load, M (ft-kips) 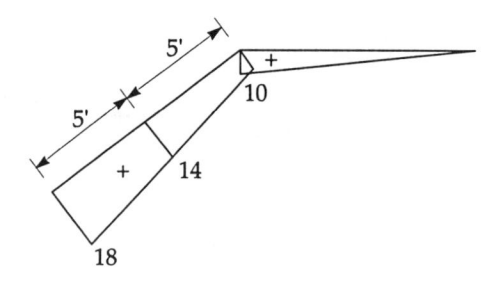   BMD due to unit value of redundant, m

$$u = \frac{1}{EI}\int Mm\,dx = \frac{1}{EI}\bigg[\frac{5}{6}[-150(2\times 18 + 14) - 90(18 + 2\times 14)]$$

$$+ \frac{5}{6}[-90(2\times 14 + 10) - 50(14 + 2\times 10)] + \frac{10}{4}(-50)(10)\bigg]$$

$$= -\frac{15217}{EI}$$

$$F = \frac{1}{EI}\int mm\,dx = \frac{1}{EI}\bigg[\frac{10}{6}[18(2\times 18 + 10) + 10(18 + 2\times 10)] + \frac{10}{3}(10)(10)\bigg] = \frac{2347}{EI}$$

3. Obtain the redundant.

$$q = R_C = -\frac{u}{F} = \frac{15217}{2347} = 6.48 \text{ kips}$$

4. The reactions at A are obtained through equilibrium considerations. The vertical reaction is $R_A = 8.5$ kips, and the counterclockwise moment reaction is $M_A = 33$ ft-kips.

Question 10.6

As with all truss problems, we solve this using the flexibility method.

1. Cut the diagonal bars AE and CE to obtain the primary structure. Since the original structure is symmetric, by choosing redundants that preserve symmetry some work can be saved.

2. Solve the primary truss to determine all the bar forces when loads are applied and when a unit value of each redundant is applied (i.e., bar forces P_i, p_{1i}, and p_{2i}, respectively). These are tabulated below. Due to symmetry, $\Sigma p_2^2 L/A$ is identical to $\Sigma p_1^2 L/A$.

Bar	L	A	P (kips)	p_1 (kips)	p_2 (kips)	$P + F_{AE}p_1 + F_{CE}p_2$ (kips)
AB	4L	A	−0.5P	−0.8P	0	−0.52P
BC	4L	A	0	0	−0.8P	−0.38P
DE	4L	A	0.917P	−0.8P	0	0.90P
EF	4L	A	0.917P	0	−0.8P	0.54P
AD	3L	A	−P	−0.6P	0	−1.02P
BE	3L	A	0	−0.6P	−0.6P	−0.30P
CF	3L	A	0	0	−0.6P	−0.28P
BD	5L	2A	−0.521P	1	0	−0.49P
BF	5L	2A	−1.146P	0	1	−0.68P
AE	5L	2A	0	1	0	0.026P
CE	5L	2A	0	0	1	0.47P

Bar	Pp_1L/A	Pp_2L/A	p_1^2L/A	p_1p_2L/A
AB	1.6PL/A	0	2.56L/A	0
BC	0	0	0	0
DE	−2.93PL/A	0	2.56L/A	0
EF	0	−2.93PL/A	0	0
AD	1.8PL/A	0	1.08L/A	0
BE	0	0	1.08L/A	1.08L/A
CF	0	0	0	0
BD	−1.3PL/A	0	2.5L/A	0
BF	0	−2.86PL/A	0	0
AE	0	0	2.5L/A	0
CE	0	0	0	0
Σ	−0.83PL/A	−5.79PL/A	12.28L/A	1.08L/A

The displacements corresponding to the redundants and due to loads, $\mathbf{u}$, and the flexibility matrix, $\mathbf{F}$, are

$$\mathbf{u} = \frac{1}{E}\begin{bmatrix} \sum_i \frac{P_i p_{1i} L_i}{A_i} \\ \sum_i \frac{P_i p_{2i} L_i}{A_i} \end{bmatrix} = \frac{PL}{AE}\begin{bmatrix} -0.83 \\ -5.79 \end{bmatrix}$$

$$\mathbf{F} = \frac{1}{E}\begin{bmatrix} \sum_i \frac{p_{1i}^2 L_i}{A_i} & \sum_i \frac{p_{1i}p_{2i}L_i}{A_i} \\ \sum_i \frac{p_{1i}p_{2i}L_i}{A_i} & \sum_i \frac{p_{2i}^2 L_i}{A_i} \end{bmatrix} = \frac{L}{EA}\begin{bmatrix} 12.28 & 1.08 \\ 1.08 & 12.28 \end{bmatrix}$$

3. The redundants are

$$\mathbf{q} = \begin{bmatrix} F_{AE} \\ F_{CE} \end{bmatrix} = -\mathbf{F}^{-1}\mathbf{u} = \begin{bmatrix} 0.0263P \\ 0.469P \end{bmatrix}$$

4. The bar forces in the original truss can be obtained as

$$f_i = P_i + F_{AE}p_{1i} + F_{CE}p_{2i}$$

and are given in the last column of the first table.

Question 10.7

This problem is solved by the flexibility method since the DSI = 1.

1. Obtain the primary structure by removing the spring at B. The redundant force is the upward reaction at B.

2. Using Table 10.2, the displacement of the primary structure corresponding to the redundant and due to loads is

$$u = -\frac{5wL^4}{384EI} - \frac{PL^3}{48EI} = -\frac{5(1.5)/12(40\times 12)^4}{384(29000)(5000)} - \frac{10(40\times 12)^3}{48(29000)(5000)} = -0.755 \text{ in}$$

The flexibility coefficient due to a unit upward load at B is

$$F = \frac{1(40\times 12)^3}{48(29000)(5000)} = 1.589\times 10^{-2} \text{ in}$$

3. The compatibility equation must account for the deformation of the spring. If the reaction at B is R_B, then the compression in the spring is R_B/k, and equating the upward displacements we obtain

$$u + FR_B = -\frac{R_B}{k}$$

which yields

$$R_B = -\frac{u}{F + \frac{1}{k}} = \frac{0.755}{1.589\times 10^{-2} + \frac{1}{50}} = 21.0 \text{ kips}$$

4. Using symmetry, the reactions at A and C are

$$R_A = R_C = \frac{1.5(40) + 10 - 21.0}{2} = 24.5 \text{ kips}$$

5. The vertical downward displacement at B is equal to the compression in the spring

$$\Delta_B = \frac{R_B}{k} = \frac{21}{50} = 0.42 \text{ in}$$

Question 10.8

By taking advantage of the symmetry of the structure and the loading, only one-quarter of the culvert needs to be analyzed and the problem may be reduced to that shown on the left below. This structure has a DSI of one and can be analyzed easily by the flexibility method. The structure has 3 DOF, making the stiffness method more involved. Since joints E and F can translate, moment distribution is also unsuitable. The primary structure chosen is shown on the right.

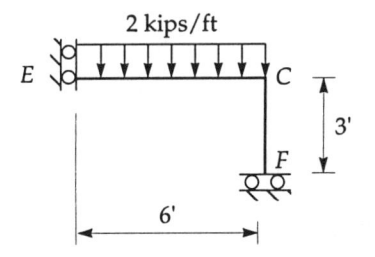

Reduced Problem

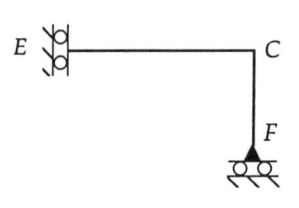

Primary Structure

1. Determine the bending moment diagrams due to external load and a unit counter-clockwise moment at F on the primary structure. These are shown below:

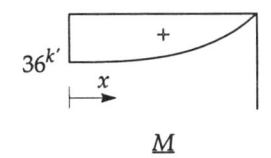

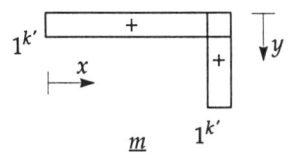

2. Compute the counter-clockwise rotation at F due to external load and the flexibility coefficient. Considering a 12" length of the culvert, the moment of inertia of the cross section is

$$I = \frac{bh^3}{12} = \frac{12 \times 8^3}{12} = 512 \text{ in}^4$$

The flexural rigidity is

$$EI = 3000 \times 512 = 1,536,000 \text{ k-in}^2 = 10,667 \text{ k-ft}^2$$

Indeterminate Structures

The rotation at F due to the load is

$$\theta = \frac{1}{EI}\int Mm\,dx = \frac{1}{EI}\int_0^6 (36 - x^2)(1)\,dx = \frac{1}{10,667}\left[36x - \frac{x^3}{3}\right]_0^6 = 0.0135 \text{ rad}$$

The flexibility coefficient is

$$f = \frac{1}{EI}\int m^2\,dx = \frac{1}{EI}\left(\int_0^6 dx + \int_0^3 dy\right) = \frac{9}{10,667} = 8.438\times10^{-4} \text{ rad}$$

Note: The integral could be obtained using Table 10.1.

3. Use compatibility to compute the value of the redundant force.

$$M_F = -\frac{\theta}{f} = -\frac{0.0135}{8.438\times10^{-4}} = -16 \text{ ft-kips}$$

4. Determine all other reactions using equilibrium.

$$R_E = 0 \text{ (horizontal)}, \quad M_E = 20 \text{ ft-kips (clockwise)}, \quad R_F = 12 \text{ kips (upward)}$$

5. Determine the axial force, shear force and bending moment variations in BC and CD.

Axial force in $BC = 0$. Axial force in $CD = 12$ kip. Shear force in $CD = 0$.

The shear force diagram for BC, and the bending moment diagram for the segment BCD are shown below:

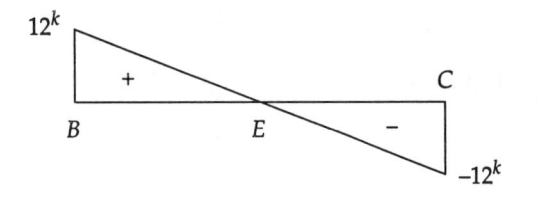

Shear Force Diagram

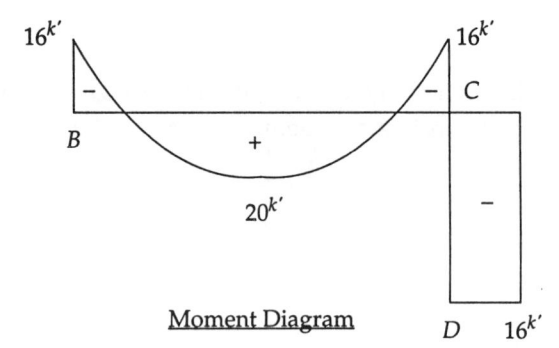

Moment Diagram

Question 10.9

Since the frame cannot undergo sidesway, moment distribution is suitable. Further, since joints A and C are pinned, no iteration is necessary. The DSI=1 and the DOF = 3, so the flexibility method may also be used. In the flexibility method, if the primary structure is obtained by putting a hinge at B, then each member behaves like a simple beam and Table 10.2 can be used to obtain displacements due to loads. This solution outlines the use of moment distribution as well as the flexibility method.

Moment Distribution

1. Compute fixed-end moments due to loads.

$$\text{FEM}_{AB} = -\text{FEM}_{BA} = \frac{PL}{8} = \frac{wL^2}{8} \qquad \text{FEM}_{BC} = -\text{FEM}_{CB} = \frac{wL^2}{12}$$

2. Compute fixed-end moments due to support settlement.

$$\text{FEM}_{BC} = \text{FEM}_{CB} = \frac{6EI\Delta}{L^2} = \frac{6EI}{L^2}\left[\frac{wL^4}{48EI}\right] = \frac{wL^2}{8}$$

3. Unlock A and C, carry over moments to B and add with the FEM.

$$M_{BA} = -\frac{wL^2}{8} + \frac{1}{2}\left[-\frac{wL^2}{8}\right] = -\frac{3wL^2}{16}$$

$$M_{BC} = \left[\frac{wL^2}{12} + \frac{wL^2}{8}\right] + \frac{1}{2}\left[\frac{wL^2}{12} - \frac{wL^2}{8}\right] = \frac{3wL^2}{16}$$

Flexibility Method

1. The primary structure is obtained by putting a hinge at B. The redundant consists of the internal moment at B and is taken as a pair of moments, counter-clockwise at end BA and clockwise at end BC.

2. Using Table 10.2, the displacement corresponding to the redundant and due to loads is

$$u_1 = \theta_{BA} + \theta_{BC} = \frac{PL^2}{16EI} + \frac{wL^3}{24EI} = \frac{5wL^3}{48EI}$$

3. The settlement of C simply causes a rigid body rotation of member BC, and the displacement corresponding to the redundant is

$$u_2 = \theta_{BC} = \frac{wL^3}{48EI}$$

4. Imposing a unit pair of moments at B, the flexibility is

$$f = \frac{L}{3EI} + \frac{L}{3EI} = \frac{2L}{3EI}$$

5. Using compatibility, the value of the redundant is

$$q = M_B = -\frac{u_1 + u_2}{f} = -\frac{3}{16}wL^2$$

6. The bending moment diagram is

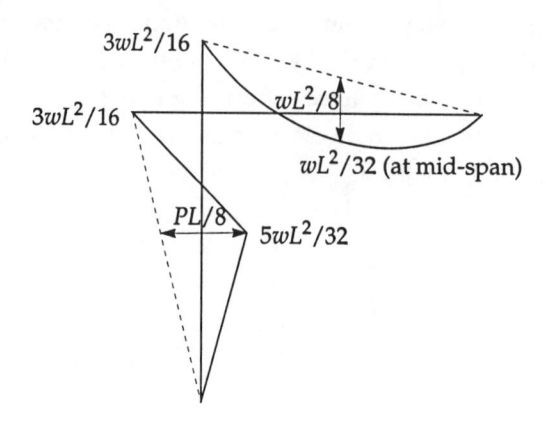

Question 10.10

As with all truss problems, we use the flexibility method.

1. The DSI = 1. The primary structure is obtained by removing the support at B, and the upward reaction at B is taken as the redundant force.

2. The truss bar forces due to the 30-kips load are denoted by P_i, and those due to a unit upward force at B are denoted by p_i. The bar forces and derived quantities are shown in the table below.

Bar	L (in.)	P (kips)	p (kips)	PpL	p^2L	$P + R_Bp$ (kips)
AB	240	15	−0.667	−2400	106.67	15.53
AD	300	18.75	0.833	4687.5	208.33	18.09
BC	240	15	−0.667	−2400	106.67	15.53
CD	300	−18.75	0.833	−4687.5	208.33	−19.41
BD	180	0	−1	0	180	0.79
Σ				−4800	810	

3. The upward displacement at B due to the 30-kip load is

$$u_1 = \frac{1}{EA}\sum_i P_i p_i L_i = -\frac{4800}{29000(0.5)} = -0.331 \text{ in}$$

4. A settlement at C of the primary structure causes it to rotate clockwise about A as a rigid body. The upward displacement at B of the primary structure due to a settlement of 0.25 in at C is

$$u_2 = -\frac{0.25}{2} = -0.125 \text{ in}$$

5. The flexibility coefficient is

$$f = \frac{1}{EA}\sum_i p_i^2 L = \frac{810}{29000(0.5)} = 0.055862 \text{ in}$$

6. Compatibility at B yields

$$u_1 + u_2 + fR_B = -0.5 \text{ in}$$

$$\Rightarrow R_B = \frac{-0.5 - u_1 - u_2}{f} = -0.79 \text{ kips}$$

7. The bar forces in the original truss are $P_i + R_B p_i$, and are tabulated in the last column of the table above.

Questions 10.11–10.13

10.11 d. Using moment distribution and Fig. 10.1, the fixed-end moments are

$$\text{FEM}_{AB} = \frac{2(15)^2}{12} + \frac{10(5)(10)^2}{15^2} + \frac{20(10)(5)^2}{15^2} = 81.94 \text{ ft-kips}$$

$$\text{FEM}_{BA} = -\frac{2(15)^2}{12} - \frac{10(5)^2(10)}{15^2} - \frac{20(10)^2(5)}{15^2} = -93.06 \text{ ft-kips}$$

Unlocking B and carrying half the moment to A yields the moment at A:

$$M_A = 81.94 + \frac{93.06}{2} = 128.5 \text{ ft-kips}$$

10.12 c. Moment equilibrium about B yields:

$$128.47 - R_A(15) + 2(15)(7.5) + 10(10) + 20(5) = 0$$

$$\therefore R_A = 36.90 \text{ kips}$$

10.13 d. The shear force diagram is shown below:

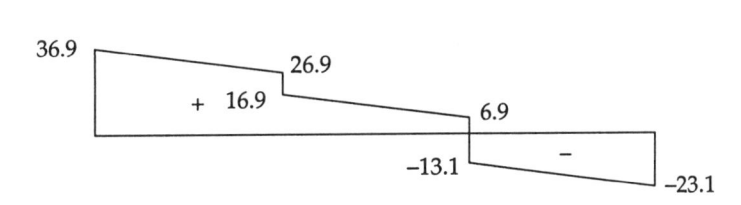

```
36.9 ┐
     │         26.9
     │    + 16.9 ┐
     │           │    6.9
     │           └─────┐
                 -13.1 │  -
                       └─────┐
                             -23.1
```

The maximum positive moment occurs at the point where the shear force diagram crosses the horizontal axis, under the 20 kip load. The moment is equal to the area under the shear force diagram from the right support to the 20 kip load:

$$\text{Maximum positive moment} = \left(\frac{13.1 + 23.1}{2}\right)(5) = 90.5 \text{ ft-kips}$$

Questions 10.14–10.18

10.14 b. Use moment distribution. The fixed-end moments are:

$$\text{FEM}_{AB} = -\text{FEM}_{BA} = \frac{2(15)^2}{12} = 37.5 \text{ ft-kips}$$

$$\text{FEM}_{BC} = -\text{FEM}_{CB} = \frac{10(10)}{8} = 12.5 \text{ ft-kips}$$

The stiffnesses are: $K_{AB} = \frac{3EI}{15} = 0.2$, $K_{BC} = \frac{3EI}{10} = 0.3$

The distribution factors at B are: $\text{DF}_{BA} = \frac{0.2}{0.2 + 0.3} = 0.4$, $\text{DF}_{BC} = 1 - 0.4 = 0.6$

Perform moment distribution calculations.

Joint	A	B		C
End	AB	BA	BC	CB
DF		0.4	0.6	
FEM	37.5	−37.5	12.5	−12.5
Bal. A	−37.5			
C.O.		−18.75		
Bal. C				12.5
C.O.			6.25	
Bal. B		15	22.5	
Σ	0	−41.25	41.25	0

10.15 a. Taking moments about B for the free body AB yields

$$R_A(15) - 2(15)(7.5) + 41.25 = 0$$

$$\therefore R_A = 12.25 \text{ kips}$$

10.16 c. Taking moments about B for the free body BC yields

$$R_C(10) - 10(5) + 41.25 = 0$$

$$\therefore R_C = 0.875 \text{ kips}$$

Vertical equilibrium of the entire structure yields $R_B = 2(15) + 10 - R_A - R_C = 26.9$ kips.

10.17 a. The shear in span AB is zero at

$$x = \frac{R_A}{w} = \frac{12.25}{2} = 6.125 \text{ ft}$$

The maximum moment in AB occurs at this point and is

$$M_{max} = R_A x - w\frac{x^2}{2} = 12.25(6.125) - 2\frac{6.125^2}{2} = 37.52 \text{ ft-kips}$$

10.18 a. The maximum positive moment occurs at the 10 kip load application point.

$$M_{max} = R_C(5) = 0.875(5) = 4.375 \text{ ft-kips}$$

Questions 10.19–10.24

10.19 c. Use moment distribution. The fixed-end moments are:

$$\text{FEM}_{AB} = -\text{FEM}_{BA} = \frac{15(10)}{8} = 18.75 \text{ ft-kips}$$

$$\text{FEM}_{BC} = -\text{FEM}_{CB} = \frac{1(10)^2}{12} = 8.33 \text{ ft-kips}$$

Stiffnesses and DF's are: $K_{AB} = \frac{4EI}{L}$, $K_{BC} = \frac{3EI}{L}$, $\text{DF}_{BA} = \frac{4}{7}$, $\text{DF}_{BC} = \frac{3}{7}$.

Perform moment distribution calculations.

Joint	A	B		C
End	AB	BA	BC	CB
DF		4/7	3/7	
FEM	18.75	−18.75	8.33	−8.33
Bal. C				8.33
C.O.			4.17	
Bal. B		3.57	2.68	
C.O.	1.79			
Σ	20.54	−15.18	15.18	0

10.20 d.

10.21 a. Taking moments about B of member AB yields:

$$R_{A_x}(10) - 15(5) + 15.18 = 0$$

$$\therefore R_{A_x} = 5.98 \text{ kips}$$

10.22 c. The upward reaction at A is equal to the shear at B in member BC. Taking moments about C in member BC yields:

$$V_B(10) - 15.18 - 1(10)(5) = 0$$

$$\therefore V_B = R_{A_Y} = 6.518 \text{ kips}$$

10.23 b. The upward reaction at C: $R_C = 10 - 6.518 = 3.482$ kips.
The shear in BC is zero at a distance of 3.482 ft from C. The moment at this point is

$$M_{max} = R_C(3.482) - 1\frac{(3.482)^2}{2} = 6.06 \text{ ft-kips}$$

10.24 e. $M = R_{A_X}(5) = 5.98(5) = 29.9$ ft-kips

11. Water Treatment

by Mackenzie L. Davis

Selection of water supply sources and conventional water treatment processes for municipalities are reviewed in this chapter. Regulatory requirements of the Safe Drinking Water Act, although an important consideration in the selection of water sources and treatment processes, are not covered here. It is strongly suggested that your preparation for the examination include a review of current federal and state regulatory requirements. The federal requirements may be found in the Code of Federal Regulations (40 CFR parts 141–143).

11.1 Water Source Evaluation

Selection of a suitable water source requires evaluation of the availability of water, the quality of the water and the potential for contamination.

The long term ("safe") yield of surface water is a function of the allowable withdrawal as determined by the regulatory agency and the hydrologic characteristics of the source. Flow duration (or "yield") curves are used to estimate the hydrologic suitability of rivers. For direct withdrawal, drought durations of 30 to 90 days and return periods of approximately 10 years are used to evaluate suitability. When reservoirs are employed, drought durations of 1 to 5 years and return periods of 25 to 50 years are used.

When the source being considered is an unconfined aquifer, the drawdown for sustained pumping (usually taken as 100 days of continuous pumping) as well as the depth of the aquifer and annual precipitation and recharge must be evaluated. Drought durations of 1 to 5 years with return periods of 25 to 50 years are normally used for design estimates. For confined aquifers, the elevation of the existing piezometric surface, the drawdown for sustained pumping and the relationship of these to the top of the aquifer (bottom of the aquiclude or aquitard) must be evaluated. For both the confined and unconfined aquifer, the proximity of other wells and the potential for interference must also be considered.

In addition to the biological, chemical and physical characteristics specified by the regulatory agencies, a "sanitary survey" should be conducted to determine the presence of potential sources of contamination for each water source. Examples of potential surface water sources of contamination include: sewer outfalls, septic tank tile fields, bathing areas, swamps, cultivated areas including orchards, marinas and boat launching areas. Groundwater sources should be examined for the proximity of septic tanks and tile fields, lagoons and seepage pits, well injection of industrial wastes, landfills, and raw material storage piles.

11.2 Demand Estimates

Tank volumes and loading rates in water treatment systems are normally designed on the basis of average demand, while the number of units, chemical feeders and hydraulic performance of the plant must provide for the fluctuations around the mean. The best approaches to estimating demand are based on projections from existing data. A summary of projection methods may be found in <u>Wastewater Engineering</u> by Metcalf and Eddy, Inc. (McGraw-Hill, 1972) and "Population Forecasting by Sanitary Engineers" by F.E. McJunkin (*Journal of the Sanitary Engineering Division*, ASCE, 1964).

In the absence of data, a number of empirical estimating techniques have been developed. For specific facilities (commercial, institutional, industrial, etc.), Metcalf and Eddy provide a series of tables that may be used to estimate demand. For domestic use, either the Federal Housing Administration Standards or data based on national averages may be used. Examples of these data are summarized in Table 11.1. Commonly accepted values for the national average per capita water consumption in urban areas range from 568 to 628 liters per capita per day (Lpcd).

Table 11.1 Estimates of water use for dwelling units.

	Liters per day per dwelling unit		
	Average Day	Maximum Day	Peak Hour
Federal Housing Administration Standards	1515 (1.0)*	3030 (2.0)	7575 (5.0)
National Average:			
Metered Dwellings	1515	3300	8025
Unmetered Dwellings	2610	8900	19 570
Unsewered Dwellings	950	2760	6965

* Ratio to average day

Source: F. P. Linaweaver, J. C. Geyer and J. B. Wolff, "Summary Report on the Residential Water Use Research Project," *Journal of American Water Works Association*, 59, p. 267, 1967.

Fire demand is added to the demand from residential, commercial and industrial use. The American Insurance Association provides standards for various community sizes as well as individual building units.

11.3 Coagulation/Flocculation

The removal of suspended particles from water may be achieved by charge reduction, bridging, or bonding of the particles followed by sedimentation. Coagulation, as referred to here, means charge reduction. Flocculation refers to the contacting of particles to cause them to grow to sufficient size that they will settle by gravity in a reasonable time.

11.3.1 Coagulation Physics

Classical coagulation physics assumes that colloidal particles in suspension have a negative charge. The colloids remain in suspension because the normal mechanical forces that can be applied cannot overcome the electrostatic repulsion forces. The addition of an appropriate cation provides a means for reducing the charge.

The Shultz-Hardy rule states that, in theory, the destabilization of a colloid by an indifferent electrolyte is brought about by ions of opposite charge to that of the colloid and that coagulation effectiveness increases with charge so that monovalent, divalent and trivalent species should be effective approximately in the ratio: 1: 10^2: 10^3. In most practical systems, the Shultz-Hardy rule is "violated" because the electrolytes are not indifferent. Inorganic compounds added to water react with solution electrolytes, form complexes and precipitates. Hence, coagulation power for monovalent, divalent and trivalent species is taken as 1:60:700. The important result of the Shultz-Hardy rule is its implication in the selection of a coagulant.

The characteristics of the coagulant selected must be such that it is non-toxic, has a high charge density and is insoluble in a pH range that is approximately neutral. Conventionally, these conditions lead to the selection of alum ($Al_2SO_4 \cdot 14\ H_2O$), ferric chloride ($FeCl_3$) or ferric sulfate ($Fe(SO_4)_3 \cdot H_2O$) as coagulants.

11.3.2 Coagulation Chemistry

Alkalinity is defined as the sum of all titratable bases down to a pH of about 4.5. It is a measure of buffer capacity. In natural waters, the only significant contributions to alkalinity are the carbonate species. This leads to a working definition of alkalinity as

$$\text{Alkalinity} = [HCO_3] + 2[CO_3] + [OH] - [H] \tag{11.3.1}$$

where the brackets, [ ], refer to concentrations in moles/L. Below a pH of 8.3 it may be assumed that all of the alkalinity is in the bicarbonate form (HCO_3). By convention, alkalinity analyses are reported in units of "mg/L as $CaCO_3$." The conversion from mg/L as the species (for any compound) is

$$\text{mg/L as } CaCO_3 = (\text{mg/L as species}) \frac{\text{EW of } CaCO_3}{\text{EW of Species}} \tag{11.3.2}$$

where EW refers to equivalent weight.

The conventional coagulants form precipitates if the pH of the water is in the correct range. In hard water, aluminum is insoluble in the pH range 5 to 8. In soft water, the effective range of pH is 5 to 7. The iron salts are insoluble in the nominal pH range of 4 to 11. Alum and the iron salts are acid forming when dissolved in water. Thus, a coagulant's effectiveness may be self-inhibiting if insufficient alkalinity is present. The reactions for alum and ferric chloride when alkalinity is present may be summarized as

$$Al_2(SO_4)_3 \cdot 14H_2O + 6HCO_3 \leftrightarrow 2Al(OH)_3 \downarrow + 6CO_2 + 14H_2O + 3SO_4 \tag{11.3.3}$$

$$FeCl_3 + 3HCO_3 \leftrightarrow Fe(OH)_3 \downarrow + 3CO_2 + 3Cl \tag{11.3.4}$$

When no alkalinity is present or the alkalinity is destroyed, the reactions are

$$Al_2(SO_4)_3 \cdot 14H_2O \leftrightarrow 2Al(OH)_3 \downarrow + 3H_2SO_4 + 8H_2O \tag{11.3.5}$$

$$FeCl_3 + 3H_2O \leftrightarrow Fe(OH)_3 \downarrow + 3HCl \tag{11.3.6}$$

——— EXAMPLE 11.1 ———

How much alkalinity (in mg/L as $CaCO_3$) is destroyed from the addition of 1.0 mg/L of alum?

Solution. Using Eq. 11.3.3, note that 6 moles of HCO_3 are consumed for each mole of alum added. Taking the gram molecular weight (GMW) of alum as 594 (note: because the waters of hydration may vary, some take the GMW to be 600), the moles of alum added are

$$\frac{1.0 \text{ mg/L}}{(594 \text{ g/mole})(10^3 \text{ mg/g})} = 1.68 \times 10^{-6} \text{ moles/L}$$

The moles of bicarbonate consumed are

$$6(1.68 \times 10^{-6}) = 1.01 \times 10^{-5} \text{ moles/L}$$

The mass of bicarbonate consumed is

$$(1.01 \times 10^{-5} \text{ moles/L})(61 \text{ g/mole})(10^3 \text{ mg/g}) = 0.62 \text{ mg/L}$$

The equivalent weight of $CaCO_3$ is the gram molecular weight (GMW) divided by the number of exchangeable hydrogen ions, i.e., 2:

$$\text{EW of } CaCO_3 = \frac{100}{2} = 50$$

The equivalent weight of HCO_3 is 61, i.e., the GMW of 61 divided by 1. The alkalinity consumed in mg/L as $CaCO_3$ is

$$(0.62 \text{ mg/L})\left(\frac{50}{61}\right) = 0.51$$

Note: it is often assumed that 1 mg/L of alum destroys 0.5 mg/L of alkalinity.

11.3.3 Selection of Coagulant

The selection of an appropriate coagulant and the determination of the appropriate dose are best determined with jar tests and an economic evaluation of the alternatives. In the absence of testing, the following qualitative observations may be made.

High turbidity, high alkalinity waters are the easiest to coagulate. All of the metal salts are suitable as well as high molecular weight polymers.

High turbidity, low alkalinity waters require special attention to pH control. Addition of base may be required with the addition of metal salts. Polymers may be advantageous because they have minimal impact on the pH.

Low turbidity, high alkalinity waters will require large doses of metal salts. Metal salt precipitates may not settle well. Polymers are not effective alone. Metal salts followed by polymers or preceded by the addition of coagulant aids such as bentonite or kaolinite clays may be required.

Low turbidity, low alkalinity waters are the hardest to coagulate. A combination of metal salts, polymers and pH adjustment is often required. Direct filtration without coagulation/flocculation has proven successful for this case.

Apparent color may be removed by coagulation/flocculation. pH control is very important. The metal salts and cationic polymers are effective at the low end of the operating pH range (4 to 5). The floc that are formed are very fragile. Transport between the flocculation basin, sedimentation basin and rapid sand filter requires special consideration to prevent breakup of the floc.

11.3.4 Rapid Mix

Rapid mixing may be accomplished in a tank with a vertical shaft mixer, in a pipe using an in-line blender, or in a pipe using a static mixer. Only tank mixing is discussed here.

In water and wastewater treatment, the degree of mixing is expressed by the velocity gradient G. The velocity gradient is defined as

$$G = \sqrt{\frac{P}{\mu \forall}} \tag{11.3.7}$$

where G = velocity gradient, s^{-1}
P = power input, W
$\forall$ = volume of water, m^3
μ = dynamic viscosity, $Pa \cdot s$ or $N \cdot s/m^2$

Detention times for mixing as well as other processes is defined as

$$t_o = \frac{\forall}{Q} \tag{11.3.8}$$

where t_o = theoretical detention time, s
$\forall$ = volume of water, m^3
Q = flow into basin, m^3/s

The relationship between detention time and velocity gradient is shown in Table 11.2.

Table 11.2 G values for rapid mixing.

Retention time, s	G, s^{-1}
0.5 (in-line blending)	3500
10-20	1000
20-30	900
30-40	800
Longer	700

Source: M.L. Davis and D.A. Cornwell, *Introduction to Environmental Engineering*, McGraw-Hill, Inc., New York, NY, 1991, p. 183.

Nominal design detention times for rapid mix tanks are in the range of 10 to 30 s with velocity gradients from 600 to 1000 s^{-1}. Tank volumes seldom exceed 8 m^3. Although either turbine or axial flow impellers may be used, turbine impellers are preferred for mixing coagulants because they generate more shear. Axial flow impellers are frequently used to mix lime in softening systems because the pumping action helps to keep the lime slurry in suspension. The mixing tank should be baffled into two or three compartments and vertically baffled to prevent vortexing. Coagulation chemicals should be added directly below the impeller at the point of maximum mixing. Common rules of thumb in setting the mixing tank proportions are shown in Table 11.3. Standard practice is to provide a minimum of 2 mixing units, each sized for the average day and valved so that either may be used. Mixing capacity (or standby units) is designed such that the flow on the maximum day may be treated with one unit out of service.

Table 11.3 Rules of thumb for mixing tank dimensions.

Variable	Proportion
Liquid depth	0.5 to 1.1 times diameter or width
Impeller diameter	0.3 to 0.50 times diameter
Vertical baffles	Extend into tank 0.10 times diameter
Impeller diameter	Generally do not exceed 1 m diameter
Motor power	Power imparted to liquid is ≈ 0.8 times motor power
Exceptions	
Liquid depth	May increase to 1.1 to 1.6 times diameter if dual impellers are employed. Dual impellers are placed two impeller diameters apart on the same shaft. Two straight blade turbine impellers impart 1.9 times as much power to water as one turbine alone for the same motor power.

The relationship between the power imparted to the liquid and the design of the mixer is expressed by Rushton's equation:

$$P = K n^3 D_i^5 \rho \quad (11.3.9)$$

where
- P = power, W
- K = impeller constant
- n = rotational speed, rotations/s
- D_i = impeller diameter, m
- ρ = density of liquid, kg/m³

Values for the impeller constant are given in Table 11.4.

Table 11.4 Values for impeller constant.

Type of Impeller	K
Propeller, pitch of 1, 3 blades	0.32
Propeller, pitch of 2, 3 blades	1.00
Turbine, 6 flat blades, vaned disc	6.30
Turbine, 6 curved blades	4.80
Fan Turbine, 6 blades at 45°	1.65
Shrouded turbine, 6 curved blades	1.08
Shrouded turbine, with stator, no baffles	1.12

Note: The constant assumes baffled tanks having four baffles at the tank wall with a width equal to 10 percent of the tank diameter.

Source: J.H. Rushton, "Mixing of Liquids in Chemical Processing," *Industrial & Engineering Chemistry*, 44, (12), 1952, p. 2931.

EXAMPLE 11.2

A coagulation plant is to be designed to treat a surface water having a temperature of 18 °C to provide water for a community of 4500 dwelling units. Assuming that the community water demand meets the FHA standards and that a detention time of 10 s is adequate, design a mixing basin and select a mixer from one of the models provided in the table below. Identify the type of impeller and determine its diameter.

Rotational speeds, rpm	Power, kW
30, 45	0.37
45, 70	0.56
45, 110	0.75
45, 110	1.12
70, 110	1.50

Solution. Using Table 11.1, estimate the average and maximum day flows. The volume each day is

$$(1.51 \text{ m}^3/\text{d} / \text{dwelling})(4500 \text{ dwellings}) = 6818 \text{ m}^3/\text{d}$$

This provides the average day flow as

$$\frac{6818 \text{ m}^3/\text{d}}{86\,400 \text{ s}/\text{d}} = 0.0789 \text{ m}^3/\text{s}$$

The maximum day flow is then

$$(2)(0.0789 \text{ m}^3/\text{s}) = 0.1578 \text{ m}^3/\text{s}$$

The volume of the rapid mixing basin with a 10 s detention time is

$$\forall = (10s)(0.0789 \text{ m}^3/\text{s}) = 0.789 \text{ m}^3$$

If three of these units are provided then the maximum day demand may be met with one unit out of service. Note that this volume is also well below the nominal maximum of 8 m³.

With a 10 s detention time, a corresponding G of 1000 s^{-1} is selected from Table 11.2. Using the water temperature, the viscosity from Table 4.2 is 1.05×10^{-3} Pa·s (or N·s/m²). The power input may be calculated as

$$P = G^2 \mu \forall$$
$$= (1000)^2 (1.05 \times 10^{-3})(0.789) = 828 \text{ W}$$

For maximum shear and mixing, select a vaned disc turbine with 6 flat blades that has $K = 6.30$. Assuming the power imparted to the water is about 80% of the motor power, then the motor power should be about $828/0.8 = 1040$ W. The nearest motor from the table is one with 1.12 kW and rotational speeds of 45 and 110 rpm. Select the 110 rpm (1.83 rps) rotational speed as a first trial and calculate the diameter of the impeller using Eq. 11.3.9:

$$D_i = \left[\frac{(0.8)(1120)}{(6.30)(1.83)^3 (1000)} \right]^{1/5} = 0.47 \text{ m}$$

This impeller diameter is less than 1 m and is acceptable. Using a ratio of impeller diameter to tank diameter of 0.4, the tank diameter would be

$$\frac{0.47}{0.4} = 1.18 \text{ m}$$

The surface area of the tank would be

$$\frac{\pi(1.18)^2}{4} = 1.08 \text{ m}$$

The depth of the tank would be

$$\frac{0.789 \text{ m}^3}{1.08 \text{ m}^2} = 0.73 \text{ m}$$

Checking the liquid depth to tank diameter ratio gives

$$\frac{0.73 \text{ m}}{1.18 \text{ m}} = 0.62$$

This is within the rule of thumb guidelines of 0.5 to 1.1.

11.3.5 Flocculation

Flocculation is the process of joining two or more particles together to form floc. This may be accomplished by a variety of mechanisms: mechanical stirring to induce collisions, enmeshment in the precipitate formed by the coagulant, bridging between particles by polymers (polyelectrolytes). Modern coagulation/flocculation practice makes use of all of these processes.

Mechanical mixing brings particles together to form floc and provides motion to keep the floc from settling in the flocculation basin. While mixing in the rapid mix tank is violent, mixing for flocculation must be gentle to prevent shearing of the floc. Flocculation mixing may be accomplished with an axial flow impeller, paddle flocculator, or baffle chamber. In newer installations, axial flow impellers are being used because they impart a nearly constant G throughout the tank. Many systems use paddle flocculators.

Common design practice is to provide a flocculation basin divided into three compartments. The velocity gradient is tapered to reduce the G value from the first to last compartment. For axial flow impellers, the rules of thumb used in design are that the impeller diameter is between 0.2 and 0.5 times the width of the chamber and that the maximum impeller diameter is about 3 m. For paddle flocculators, the peripheral velocity of the blades ranges from 0.1 to 1.0 m/s and the velocity of the paddle blade relative to that of the water is taken as approximately 3/4 of the peripheral blade velocity. The total blade area on a horizontal shaft should not exceed 15 to 20 percent of the total basin cross sectional area. Typical values for G and Gt_o are given in Table 11.5.

Table 11.5 G and Gt_o values for flocculation.

Type	G, s^{-1}	Gt_o
Low-turbidity, color removal coagulation	20–70	60 000 to 200 000
High-turbidity, solids removal coagulation	10–150	90 000 to 180 000
Softening, 10% solids	130–200	200 000 to 250 000

Source: M.L. Davis D.A. Cornwell, *Introduction to Environmental Engineering*, McGraw-Hill, Inc., New York, NY, 1991, p. 183.

The relationship between the power imparted to the liquid and the design of impeller mixers is expressed by Rushton's equation given above (Eq. 11.3.9). For paddle flocculators, the power imparted is given by:

$$P = C_D A \rho \left(\frac{v^3}{2} \right) \qquad (11.3.10)$$

where
- P = power, W
- C_D = coefficient of drag (nominally = 1.8)
- A = paddle-blade area perpendicular to flow, m²
- ρ = mass of water, kg/m³
- v = relative velocity of paddle to fluid, m/s

EXAMPLE 11.3

Design a horizontal-shaft, axial-flow, paddle wheel flocculation basin for a high-turbidity water. Assume the paddle wheel will have one arm with six blades as shown in the sketch below. The average flow is 0.0789 m³/s and the water temperature is 18 °C. Determine the following: G and Gt_o, basin dimensions, paddle wheel dimensions, motor power and rotational speed of the paddle.

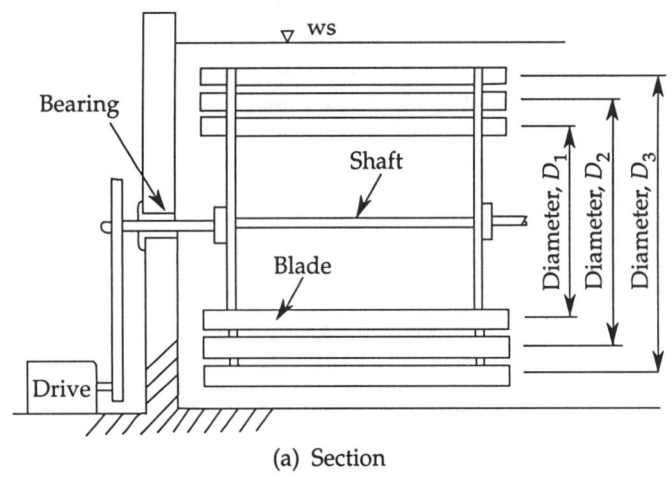

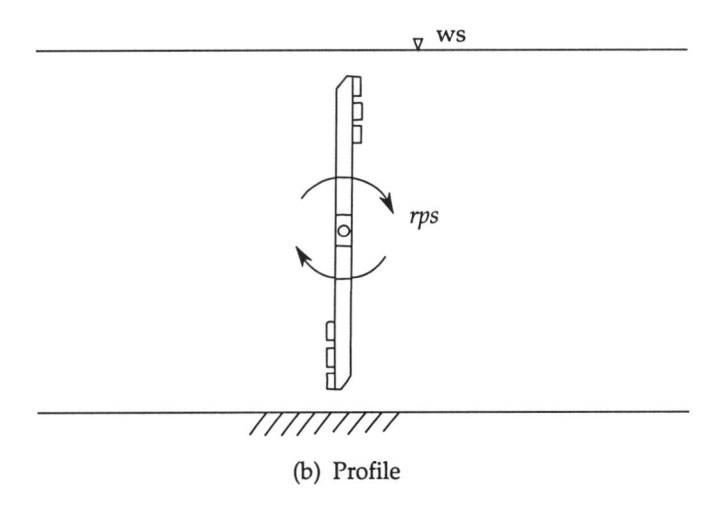

(a) Section

(b) Profile

Solution. From Table 11.5, select a G of 50 s⁻¹ and Gt_o of 90 000 for the high-turbidity water. The detention time is then

$$t_o = \frac{90,000}{50} = 1,800 \text{ s} = 30 \text{ min}$$

The volume of the flocculation basin is then

$$V = (30 \text{ min})(0.0789 \text{ m}^3/\text{s})(60 \text{ s/min}) = 142 \text{ m}^3$$

To provide for redundancy and to handle the maximum day flow, two basins are provided.

For tapered flocculation, the tank is divided into three compartments. Assume the G values for the compartments are 70, 50 and 30 so that the average G is 50 s^{-1}. Assuming a side water depth of 3 m, the surface area is

$$\frac{142 \text{ m}^3}{3 \text{ m}} = 47.3 \text{ m}^2$$

With a square section for each compartment (length = width) and three compartments, the dimension of each compartment is

$$(3x)(x) = 47.3 \text{ m}^2$$

$$\therefore x = 3.97 \text{ or } 4 \text{ m}$$

Using a clearance of 0.5 m on either side of the paddle wheel yields a blade length = 3.0 m. Assume each blade is 10 cm wide. Using the arrangement of blades as shown in the sketch, the blade area is then

$$(0.10 \text{ m})(3.00 \text{ m})(6 \text{ blades}) = 1.80 \text{ m}^2$$

Checking the percent of the cross sectional area:

$$\frac{1.80 \text{ m}^2}{(4.0 \text{ m})(3.0 \text{ m})} = 0.15$$

This is less than the recommended limit of 15 to 20%.

The water temperature is used to find the viscosity (from Table 4.2) as 1.05×10^{-3} Pa·s. The water power required for the first compartment is

$$P = G^2 \mu V$$

$$P = (70)^2 (1.05 \times 10^{-3})(3 \times 4 \times 4) = 247 \text{ W}$$

Assume the blades are spaced as follows: $D_1 = 1.60$ m; $D_2 = 2.00$ m; $D_3 = 2.40$ m.

The blade velocity relative to the water is

$$v = (rps)(\pi)(D)(0.75)$$

where rps is the revolutions per second. The factor of 0.75 is to accent for the nominal assumption of the paddle blade velocity relative to the water. Thus, for the blades at D_3:

$$v_3 = (rps)(\pi)(2.40)(0.75) = 5.65(rps)$$

Similarly, $v_2 = 4.71(rps)$ and $v_1 = 3.77(rps)$. The power for the wheel is then

$$P = C_D A_1 \rho \left(\frac{v_1^3}{2}\right) + C_D A_2 \rho \left(\frac{v_2^3}{2}\right) + C_D A_3 \rho \left(\frac{v_3^3}{2}\right)$$

Since $A_1 = A_2 = A_3 = 0.60$ m², and all the C_D values are equal,

$$P = C_D A \frac{\rho}{2}\left(v_1^3 + v_2^3 + v_3^3\right)$$

With $C_D = 1.8$ and taking the density of water as 1,000 kg/m³:

$$247.67 = (1.8)(0.60)\left(\frac{1\,000}{2}\right)\left[(3.47)^3(rps)^3 + (4.71)^3(rps)^3 + (5.65)^3(rps)^3\right]$$

$$= 540\,[41.78 + 104.49 + 180.36](rps)^3$$

$$\therefore rps = 0.112$$

or 6.72 rpm. The peripheral speed = $(\pi D)(rps)$, so the speed of the outside blade is

$$v_3 = (\pi)(2.40)(0.112) = 0.84 \text{ m/s}$$

This is within the guideline range of 0.1 to 1.0 m/s. These calculations are repeated with g values of 50 s⁻¹ and 30 s⁻¹ for the second and third components to complete the design.

11.4 Softening

11.4.1 Definition of Hardness

Hardness is defined as the sum of all polyvalent cations (in consistent units such as mg/L as $CaCO_3$). For practical purposes, total hardness is taken as the sum of the calcium and magnesium ions in mg/L as $CaCO_3$. Because of the difference in reaction chemistry, total hardness is divided into two subcategories: carbonate hardness and noncarbonate hardness.

Carbonate hardness (also known as temporary hardness) is defined as the amount of hardness equal to the alkalinity or the total hardness, whichever is less. Noncarbonate hardness (also called permanent hardness) is the total hardness in excess of the alkalinity, i.e., the difference between the total hardness and the alkalinity. If the alkalinity is equal to or exceeds the total hardness, there is no noncarbonate hardness.

11.4.2 Softening Chemistry

To remove calcium, the precipitate calcium carbonate ($CaCO_3$) must be formed. The carbonate may come from naturally occurring carbonate in the form of bicarbonate (HCO_3) or may be added as sodium carbonate (Na_2CO_3). Significant removal of $CaCO_3$ is achieved when the pH is above about 9.6 to 10.8. Magnesium hydroxide ($Mg(OH)_2$) must be formed to remove magnesium. The hydroxide is normally provided in the form of slaked lime ($Ca(OH)_2$). Significant removal of $Mg(OH)_2$ is achieved when the pH is above about 10.8 to 11.5.

The primary reactions for softening are summarized below:

$$CO_2 + Ca(OH)_2 \leftrightarrow CaCO_3 \downarrow + H_2O \tag{11.4.1}$$

$$Ca + 2\,HCO_3 + Ca(OH)_2 \leftrightarrow 2CaCO_3 \downarrow + 2H_2O \qquad (11.4.2)$$

$$Mg + 2\,HCO_3 + Ca(OH)_2 \leftrightarrow MgCO_3 + CaCO_3 \downarrow + 2H_2O \qquad (11.4.3)$$

$$MgCO_3 + Ca(OH)_2 \leftrightarrow Mg(OH)_2 \downarrow + CaCO_3 \downarrow \qquad (11.4.4)$$

$$Ca + Na_2CO_3 \leftrightarrow CaCO_3 \downarrow + 2\,Na \qquad (11.4.5)$$

$$Mg + Ca(OH)_2 \leftrightarrow Mg(OH)_2 \downarrow + Ca \qquad (11.4.6)$$

$$Ca\ (\text{from reaction 11.4.6}) + Na_2CO_3 \leftrightarrow CaCO_3 \downarrow + 2\,Na \qquad (11.4.7)$$

Although the reactions proceed simultaneously, it is convenient to visualize the softening reaction chemistry as following a stepwise process. The CO_2 in solution is effectively carbonic acid (H_2CO_3) that consumes the lime (CaO) used to raise the pH. Because the formation of the precipitates requires a high pH, the first step in the reactions is to remove the CO_2 that is present. This may be by lime neutralization (Eq. 11.4.1) or by air stripping. When the concentration of CO_2 exceeds 10 mg/L, the economics of removal by aeration are favored over removal by lime neutralization. The second step is to neutralize the HCO_3 that is present (Eqs. 11.4.2 and 11.4.3). If all of the bicarbonate is not removed or reacted with lime, the high pH values required for the precipitates to form will not be achieved. The remaining softening steps are dependent on whether selective calcium softening, softening to the practical solubility limits or split treatment is to be performed.

Two decision criteria are commonly used in softening calculations. They are a total hardness of the product water of approximately 100 mg/L as $CaCO_3$ (which may range from 80 to 120) and a magnesium concentration of approximately 40 mg/L as $CaCO_3$. The total hardness limit is set because it is economical to achieve and satisfies a majority of the consumers. A magnesium concentration of 40 mg/L as $CaCO_3$ is used as a decision criterion because this is the critical concentration for the formation of scale and the extra cost in removing magnesium below this concentration is not justified for non-commercial use. When the water contains less than 40 mg/L as $CaCO_3$ of magnesium, it may be desirable to remove only a portion of the calcium. In this case, lime is added to react with the CO_2 and HCO_3. If noncarbonate calcium is also to be removed, soda ash (Na_2CO_3) equivalent to the amount of calcium to be removed is added.

The practical limits of calcium and magnesium removal based on solubility considerations may be taken to be 30 mg/L of $CaCO_3$ and 10 mg/L of $Mg(OH)_2$ expressed as $CaCO_3$. Softening to these practical limits is referred to as excess lime softening. In this scheme, lime and soda ash are added according to Eqs. 11.4.1 through 11.4.7.

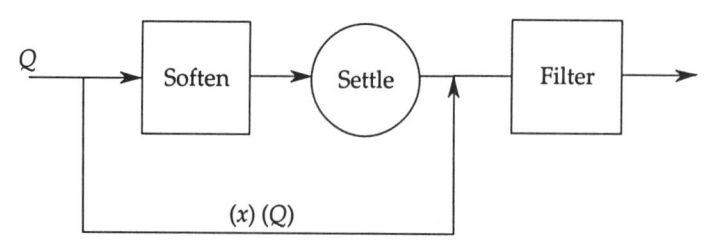

Figure 11.1 Split flow treatment scheme.

Split treatment may be performed when the magnesium concentration exceeds 40 mg/L as $CaCO_3$. As shown in Fig. 11.1, in split treatment a portion of the raw water is bypassed around the softening reaction tank and settling tank. This serves several functions. First, it allows for the tailoring of the product water to 40 mg/L as $CaCO_3$ of magnesium. Second, it allows for a reduction in the capital cost of tankage to treat the entire flow. Third, it minimizes operating costs for chemicals by treating only a fraction of the flow. Fourth, it uses the natural alkalinity of the water to lower the pH of the product water and assist in stabilization. The fractional amount of the split is calculated as follows:

$$x = \frac{Mg_f - Mg_i}{Mg_r - Mg_i} \qquad (11.4.8)$$

where
- x = ratio of bypassed flow to total flow
- Mg_f = final magnesium concentration, mg/L as $CaCO_3$
- Mg_i = magnesium concentration from the first stage, mg/L as $CaCO_3$
- Mg_r = raw water magnesium concentration, mg/L as $CaCO_3$

Thus, the bypassed flow = $(x)(Q)$ and the amount of flow that passes through the first stage is $(1 - x)(Q)$. The first stage is operated to produce the practical limits of softening, i.e., 30 mg/L of $CaCO_3$ and 10 mg/L of $Mg(OH)_2$ expressed as $CaCO_3$. Thus, the values for Mg_f and Mg_i are commonly taken to be 40 and 10 respectively.

Because of the size of the reaction vessels, reaction kinetics and side reactions, lime in excess of the stoichiometric amount is often required. One rule of thumb is to provide no more than 40 mg/L as $CaCO_3$ of excess lime in any instance but to provide the difference between the magnesium concentration in mg/L as $CaCO_3$ and 40 mg/L as $CaCO_3$. If the difference is less than 20 mg/L as $CaCO_3$, no less than 20 mg/L as $CaCO_3$ if the excess lime is provided.

EXAMPLE 11.4

Determine the chemical dosages needed for selective calcium softening of the following water.

Constituent	mg/L	EW	EW $CaCO_3$/EW ion	mg/L as $CaCO_3$	mEq/L
CO_2	6.6	22.0	2.28	15.0	0.30
Ca^{2+}	80.0	20.0	2.50	200	4.00
Mg^{2+}	8.5	12.2	4.12	35.0	0.70
HCO_3	200.0	61.0	0.82	164	3.28
SO_4^{+2}	76.0	48.0	1.04	76.0	1.58

Solution. Since Mg is less than 40 mg/L as $CaCO_3$ removal of magnesium is unnecessary. With Mg less than 40 mg/L as $CaCO_3$, the excess lime = 20 mg/L as $CaCO_3$. The chemical additions are as follows:

Addition = to:	Lime mg/L as $CaCO_3$	Lime mEq/L
CO_2	15.0	0.30
HCO_3	164.0	3.28
	179.0	3.58

With these additions, calcium = 164 mg/L as $CaCO_3$ will be removed. The total hardness remaining will be 200 − 164 = 36 mg/L as $CaCO_3$ of calcium hardness plus the original magnesium hardness of 35 mg/L as $CaCO_3$ or 71 mg/L as $CaCO_3$. This is more than acceptable and no further hardness needs to be removed. Note that lime equal to the bicarbonate must be added to achieve the desired pH for precipitation. The total amount of lime to be added would then be 179.0 mg/L as $CaCO_3$ + 20 mg/L as $CaCO_3$ excess = 199 mg/L as $CaCO_3$. The amount of lime as CaO would be

$$(199)\left(\frac{28}{50}\right) = 111.4 \text{ mg/L as CaO}$$

The problem also may be worked in milliequivalents per liter (mEq/L). The excess lime = 0.40 mEq/L so that the total lime addition is 3.98 mEq/L and the amount of CaO would be

$$(3.98)(28) = 111.4 \text{ mg/L as CaO}$$

EXAMPLE 11.5

Determine the chemical dosages for softening the following water to the practical solubility limits.

Constituent	mg/L	EW	EW CaCO$_3$/EW ion	mg/L as CaCO$_3$	mEq/L
CO$_2$	9.6	22.0	2.28	21.9	0.44
Ca^{2+}	95.2	20.0	2.50	238	4.76
Mg^{2+}	13.5	12.2	4.12	55.6	1.11
Na$^+$	25.8	23.0	2.18	56.2	1.12
Alkalinity				198	3.96
Cl	67.8	35.5	1.41	95.6	1.91
SO$_4^{+2}$	76.0	48.0	1.04	76.0	1.58

Solution. To soften to the theoretical solubility limits, lime and soda must be added as shown below.

Addition = to:	Lime mg/L as CaCO$_3$	Lime mEq/L	Soda mg/L as CaCO$_3$	Soda mEq/L
CO$_2$	21.9	0.44		
HCO$_3$	198	3.96		
Ca − HCO$_3$			40	0.80
Mg	55.6	1.11	55.6	1.11
	275.5	5.51	95.6	1.91

Since the difference Mg − 40 = 15.6 mg/L as CaCO$_3$, the minimum excess lime of 20 mg/L as CaCO$_3$ is selected. The total lime addition is 295.5 mg/L as CaCO$_3$ or 165.5 mg/L as CaO. The soda addition is 95.6 mg/L as CaCO$_3$ or

$$(95.6)\left(\frac{53}{50}\right) = 101.3 \text{ mg/L as Na}_2\text{CO}_3$$

EXAMPLE 11.6

Determine the chemical dosages for split treatment softening of the following water. The finished water criteria is a maximum magnesium hardness of 40 mg/L as CaCO$_3$ and a total hardness in the range 80 to 120 mg/L as CaCO$_3$.

Constituent	mg/L	EW	EW CaCO$_3$/EW ion	mg/L as CaCO$_3$	mEq/L
CO$_2$	11.0	22.0	2.28	25.0	0.50
Ca^{2+}	95.2	20.0	2.50	238	4.76
Mg^{2+}	22.0	12.2	4.12	90.6	1.80
Na$^+$	25.8	23.0	2.18	56.2	1.12
Alkalinity				198	3.96
Cl$^-$	67.8	35.5	1.41	95.6	1.91
SO$_4^{+2}$	76.0	48.0	1.04	76.0	1.58

Solution. In the first stage the water is softened to the theoretical solubility limits, lime and soda must be added as shown below.

Addition = to:	Lime mg/L as CaCO$_3$	Lime mEq/L	Soda mg/L as CaCO$_3$	Soda mEq/L
CO$_2$	25.0	0.50		
HCO$_3$	198	3.96		
Ca − HCO$_3$			40	0.80
Mg	90.6	1.80	90.6	1.80
	313.6	6.26	130.6	2.60

The split is calculated in terms of mg/L as CaCO$_3$:

$$x = \frac{40-10}{90.6-10} = 0.372$$

The fraction of water passing through the first stage is then 1 − 0.372 = 0.628. The total hardness of the water after passing through the first stage is the theoretical solubility limit, i.e., 40 mg/L as CaCO$_3$. Since the total hardness in the raw water is 238 + 90.6 = 328.6 mg/L as CaCO$_3$, the mixture of the treated and bypassed water has a hardness of

$$0.372(328.6) + .628(40) = 147.4 \text{ mg/L as CaCO}_3$$

This is above the acceptable range of 80–120 mg/L as CaCO$_3$, so further treatment is required. Since the split is designed to yield the required 40 mg/L as CaCO$_3$ of magnesium, more calcium must be removed. Removal of the calcium equivalent to the bicarbonate will leave 40 mg/L as CaCO$_3$ of calcium hardness plus the 40 mg/L as CaCO$_3$ of magnesium hardness for a total of 80 mg/L as CaCO$_3$. The additions are as follows.

Constituent	Lime mg/L as CaCO$_3$	Lime mEq/L
CO$_2$	25.0	0.50
HCO$_3$	198.0	3.96
	223.0	4.46

The total chemical additions are in proportion to the flows:

$$\text{Lime} = 0.628(313.6) + 0.372(223) = 280 \text{ mg/L as CaCO}_3$$

$$\text{Soda} = 0.628(130.6) + 0.372(0.0) = 82 \text{ mg/L as CaCO}_3$$

11.4.3 Mixing

As mentioned in the coagulation discussion, mixing of slaked lime differs from mixing of coagulants in that longer rapid mix times are often used and that the impellers selected are frequently of the axial flow type to take advantage of pumping action to maintain the slurry in suspension. Although reaction vessels may be designed along the same configuration as flocculation basins, modern practice is to use a solids contact unit in one of the configurations shown in Fig. 11.2 (next page).

The Great Lakes Upper Mississippi River Board (1978) recommends the following sizing criteria: mixing period of not less than 30 minutes, contact period of 1 to 2 hours, weir loading not to exceed 360 m^3/d · m, and upflow rate not to exceed 100 m^3/d · m^2. The mixing equipment is designed in the same fashion as that for coagulation. Tables 11.2 and 11.5 may be used for selecting G and Gt_o values.

11.5 Sedimentation

11.5.1 Sedimentation Physics

For convenience in describing sedimentation behavior, particle settling is classified into "Types." Type I settling is characterized by particles that settle discretely at a constant settling velocity. In water treatment, Type I settling occurs in pre-sedimentation for sand removal prior to coagulation and in settling of sand and other media during cleaning of rapid sand filters. It also occurs in grit chambers in wastewater treatment plants.

Newton's equation may be used to describe Type I settling for spherical particles:

$$v_s = \left[\frac{4g(\rho_s - \rho)d}{3C_D\rho}\right]^{1/2} \tag{11.5.1}$$

where
- v_s = terminal settling velocity, m/s
- g = acceleration due to gravity, 9.80 m/s²
- ρ_s = density of particle, kg/m³
- ρ = density of water, kg/m³
- d = diameter of particle, m
- C_D = drag coefficient

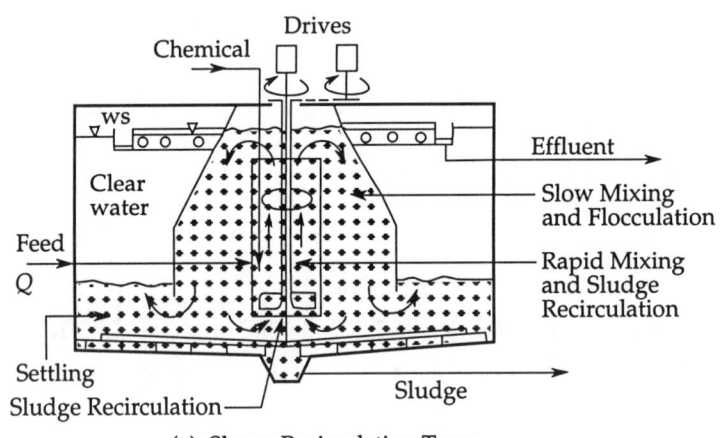

(a) Slurry-Recirculation Type

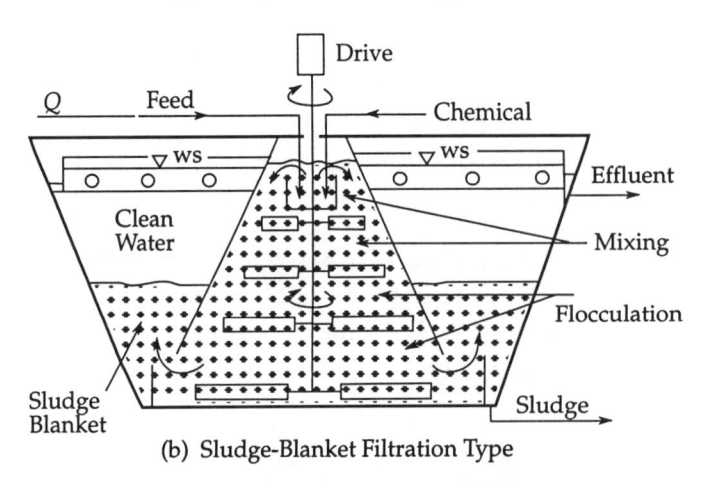

(b) Sludge-Blanket Filtration Type

Figure 11.2 Solids contact units. (Source: T. D. Reynolds, *Unit Operations and Processes in Environmental Engineering*, PWS-Kent Publishing Co., Boston, MA, 1982, p. 55)

The drag coefficient is a function of the flow regime surrounding the particle as it settles. For viscous resistance at low Reynolds numbers ($R < 0.5$) the drag coefficient takes the form

$$C_D = \frac{24}{R} \quad \text{where} \quad R = \frac{\rho v d}{\mu} \qquad (11.5.2)$$

In the laminar flow regime, with the drag coefficient defined by Eq. 11.5.2, the terminal settling velocity of a spherical particle is described by the form of Newton's equation known as Stoke's Law:

$$v_s = \frac{g(\rho_s - \rho)d^2}{18\mu} \qquad (11.5.3)$$

In the transition range between laminar and turbulent flow (R between 0.5 and 10^4) the drag coefficient is expressed as

$$C_D = \frac{24}{R} + \frac{3}{\sqrt{R}} + 0.34 \qquad (11.5.4)$$

For turbulent flow, C_D has a value of about 0.4.

Type II settling is characterized by particles that flocculate as they settle. Because of this flocculation, the particle size is continuously changing and, hence, the settling velocity changes continuously. Floc from either iron or alum coagulation follows a Type II behavior. In wastewater treatment, particles in primary sedimentation tanks and secondary settling tanks that follow trickling filters are described by Type II settling.

Settling column data provide one means of developing a design for Type II settling. The column is filled with the suspension to be analyzed and is allowed to settle. Samples are taken from sample ports at selected time intervals. The concentration of suspended solids is determined and the percent removal is calculated as

$$\%R = 1 - \frac{C_t}{C_o}(100\%) \qquad (11.5.5)$$

where
- $\%R$ = percent removal at one depth at time t
- C_t = concentration at time t and given depth, mg/L
- C_o = initial concentration, mg/L

Percent removal versus depth is then plotted as shown in Fig. 11.3. Isoconcentration lines are plotted at convenient percent removals, i.e., 5 or 10 percent increments.

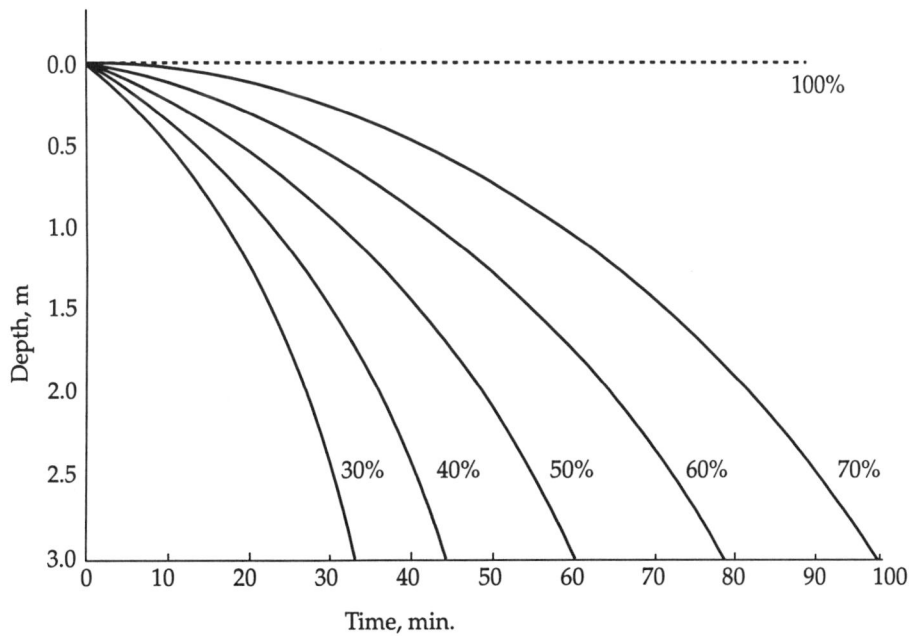

Figure 11.3 Isoconcentration lines for Type II Settling in a 3.0 m deep column.

The intersection of an isoconcentration line and the bottom of the column defines an overflow rate v_o:

$$v_o = \frac{H}{t_i} \tag{11.5.6}$$

where
- H = height of the column, m
- t_i = time defined by the intersection of an isoconcentration line and the bottom of the column. The subscript refers to the first, second, third, etc., intersection points. The units are days.

The computation procedure is to construct a vertical line from each t_i to intersect all the isoconcentration lines above it. The midpoints between the isoconcentration lines are used to define a series of heights (H_1, H_2, H_3) that are used to calculate the fraction of solids removed:

$$R_{Ta} = R_a + \frac{H_1}{H}(R_b - R_a) + \frac{H_2}{H}(R_c - R_b) + \cdots \tag{11.5.7}$$

where
- R_{Ta} = total fraction removed for settling time t_a
- R_a, R_b, R_c = isoconcentration fractions

The overflow rates, removal fractions and detention times are used to prepare design curves such as those shown in Figs. 11.4 and 11.5. Scale-up factors of 0.65 for overflow rate and 1.75 for detention time are used to correct for the more ideal settling that occurs in the test column.

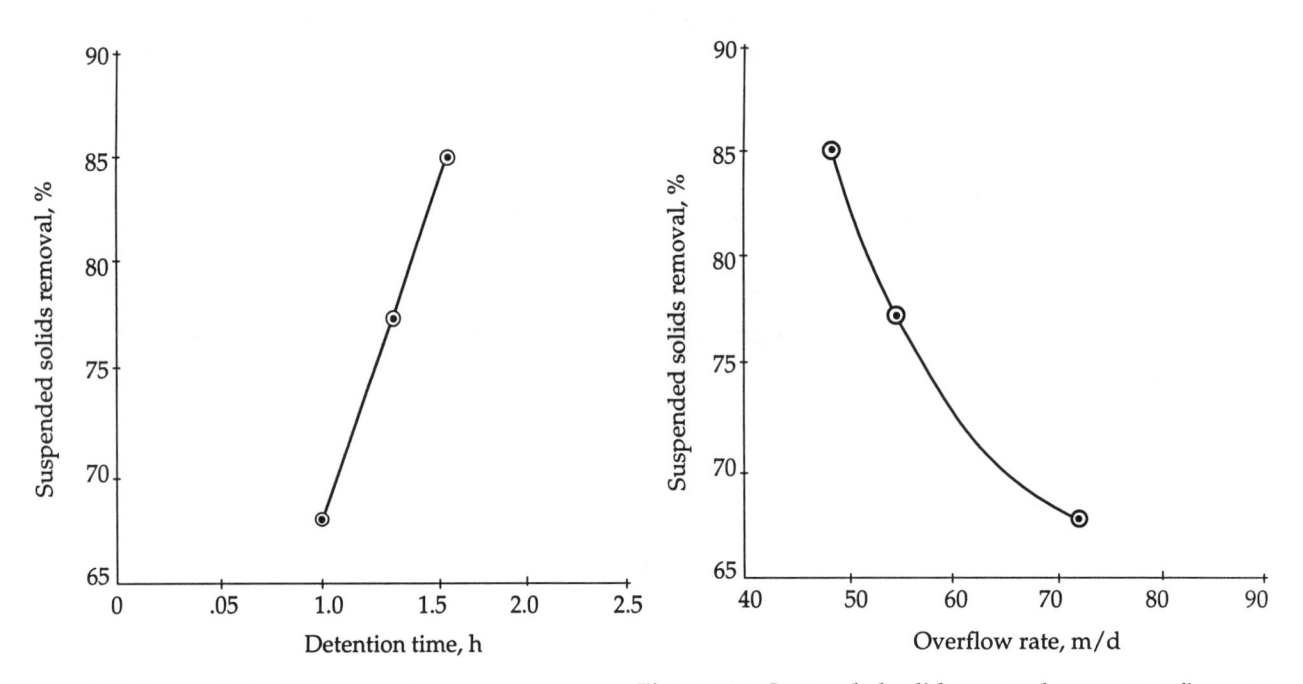

Figure 11.4 Suspended solids removal versus detention time.

Figure 11.5 Suspended solids removal versus overflow rate.

EXAMPLE 11.7

Using the settling test data in Fig. 11.3, determine the design detention time and design overflow rate to achieve an effluent concentration of 20 mg/L from the clarifier. The influent concentration is 100 mg/L. The design flow rate is 0.0789 m³/s.

Solution. Using the test data plotted in Fig. 11.3, calculate the removal fraction and overflow rate for each point where an isoconcentration line intersects the bottom of the column. For example, the 50% line intersects 3.0 m depth at 60 minutes. The time (in this case 60 minutes) is taken as the detention time. The overflow rate is calculated as

$$v_o = \left(\frac{3 \text{ m}}{60 \text{ min}}\right)(1440 \text{ min/d}) = 72.0 \text{ m/d}$$

A vertical line is plotted from the intersection of the 50% line and the bottom of the column through all of the isoconcentration lines. The mid-point between each %-removal curve is determined. The percent removal is calculated as

$$R_T = 50 + \left(\frac{2.375}{3}\right)(60-50) + \left(\frac{1.4}{3}\right)(70-60) + \left(\frac{0.525}{3}\right)(100-70)$$

$$= 50 + 7.92 + 4.67 + 5.25 = 67.8\%$$

Note, for example, that 2.375 is the mid-point depth between the 50 and 60 percent isoconcentration lines. This process is repeated for each intersection point. The detention time and overflow rates are plotted for the corresponding percent removals. These are shown in Figs. 11.4 and 11.5.

If the influent concentration is 100 mg/L and the effluent is to be 20 mg/L, then the percent removal required is:

$$\left(\frac{100-20}{100}\right)(100\%) = 80\%$$

Using Figs. 11.4 and 11.5, enter each graph at 80% suspended solids removal and find that the corresponding overflow rate and detention times are 52 m³/m²·d and 1.45 hours. Multiply these by the appropriate safety factors to find the design overflow rate and detention time:

$$v_o = (0.65)(52 \text{ m}^3/\text{m}^2 \cdot \text{d}) = 33.8 \text{ m/d}$$

$$t_o = (1.75)(1.45 \text{ h}) = 2.5 \text{ h}$$

When particles are at a high concentration (> ~1000 mg/L) they tend to settle as a mass creating a distinct clear zone above the sludge mass. This settling behavior is described as Type III or zone settling. Type III settling occurs with lime softening sludges, activated sludge and in gravity sludge thickeners.

As in Type II settling, experimental data may be used to design a clarifier for zone settling. A batch settling curve is developed using a settling column. The height of the interface is plotted versus time (Fig. 11.6). The straight line portion of the curve is used to calculate the slope (H/t) which defines the overflow rate. Tests should be performed over a range of concentrations that might be expected in the full-scale plant because the suspended solids concentration affects the settling rate. A scale-up factor of 2.0 is used for the design overflow rate.

Frequently the clarifier may also serve as a gravity thickener as well as a solids separation device. A batch flux curve may be constructed from batch settling curve data for various suspended solids concentrations (Fig. 11.7) to determine whether the surface area required for clarification or that for thickening governs.

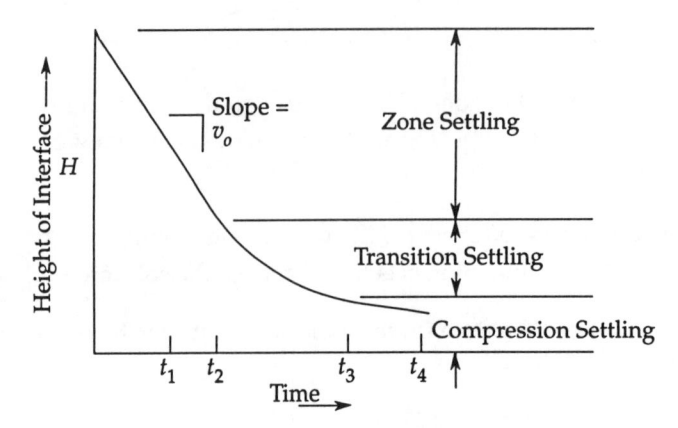

Figure 11.6 Height of interface versus time for Type III settling.

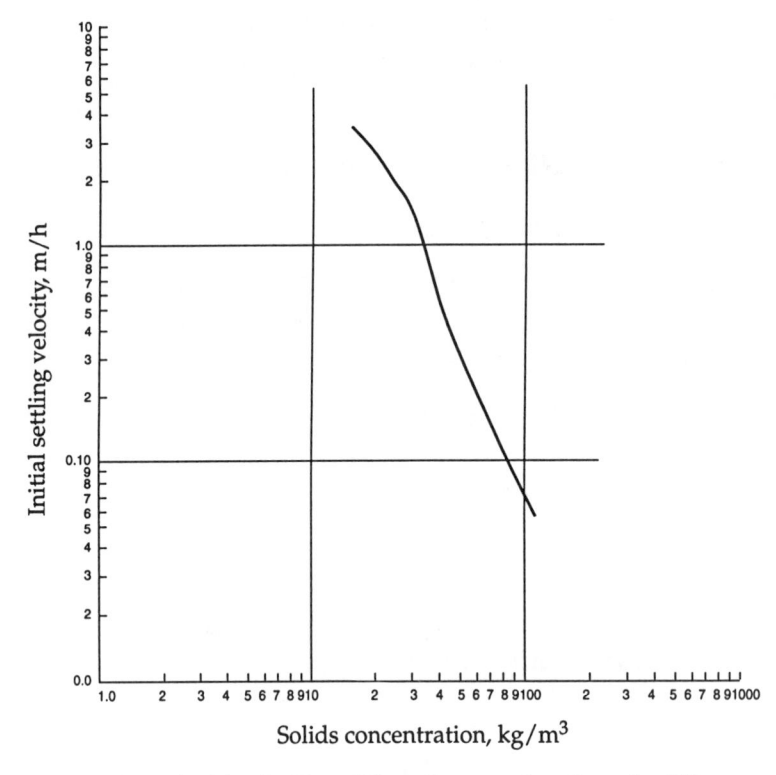

Figure 11.7 Settling velocities for Type III settling as a function of solids concentration.

The batch flux curve is constructed by first selecting convenient suspended solids concentrations (e.g., 100, 80, 60 kg/m³, etc.) and the corresponding settling velocities from a plot such as that in Fig. 11.7. The solids flux is then calculated as

$$F_s = C_u v \tag{11.5.8}$$

where F_s = solids flux, kg/m² · d
  C_u = concentration of solids in underflow, kg/m³
  v = underflow velocity, m/d

The batch flux curve is a plot of solids flux versus suspended solids concentration (Fig. 11.8). A line drawn from the design underflow solids concentration tangent to the solids flux curve and extrapolated to ordinate yields the desired solids flux. A scale-up factor of 0.67 is used for the design flux.

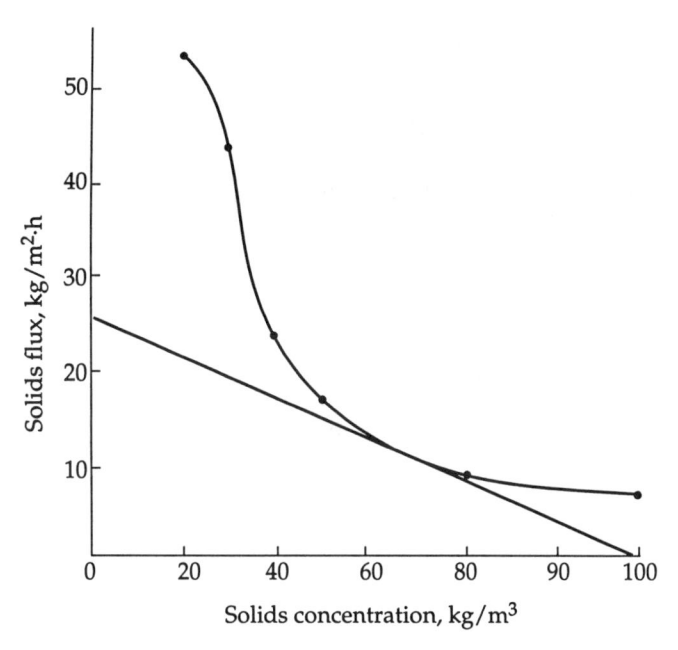

Figure 11.8 Batch flux curve.

EXAMPLE 11.8

A clarifier is to be designed for a lime softening sludge. The initial concentration of the sludge is 2%. The design underflow concentration is 10%. The design flow rate is 0.0789 m³/s. Determine the area required for clarification, the area required for thickening and the design diameter. Assume that Figs. 11.7 and 11.8 apply.

Solution. An initial sludge concentration of 2% is equivalent to 20 g/L. Using Fig. 11.7, find a subsidence velocity of about 64.8 m/d. Thus, the design overflow rate for clarification must not exceed 64.8 m/d or 64.8 m³/m² · d. Using a scale-up factor of 2.0, the surface area required for clarification is then

$$A_c = 2.0 \frac{(0.0789 \text{ m}^3/\text{s})(86\,400 \text{ s})}{64.8 \text{ m}^3/\text{d}\cdot\text{m}^2} = 210.4 \text{ m}^2$$

Figure 11.7 is used to construct a batch flux curve. At convenient values of solids concentration, find the corresponding settling velocities and calculate the flux according to Eq. 11.5.8. These are tabulated below.

Solids Concentration, kg/m³	Initial Settling Velocity, m/h	Solids Flux kg/m² · h
20	2.7	54
30	1.46	43.8
40	0.58	23.2
50	0.33	16.5
80	0.11	8.8
100	0.069	6.9

These data were used to plot Fig. 11.8. Plot a straight line from the desired underflow solids concentration of 100 kg/m³ tangent to the solids flux curve. Extrapolate to the ordinate to find the controlling solids flux of 25.5 kg/m² · h.

The initial solids concentration of 2% is equivalent to 20 kg/m³. The solids loading to the clarifier is

$$(0.0789 \text{ m}^3/\text{s})(3600 \text{ s/h})(20 \text{ kg/m}^3) = 5681 \text{ kg/h}$$

With a scale-up factor of 0.67, the surface area required for thickening is

$$A_t = \frac{5681 \text{ kg/h}}{25.5 \text{ kg/m}^2 \cdot \text{h}} = 222.8 \text{ m}^2$$

The controlling area is then 332.5 m². The design diameter is then

$$\left(\frac{4(222.8)}{\pi}\right)^{1/2} = 16.8 \text{ or } 17 \text{ m}$$

11.5.2 Sedimentation Tank Design

The fundamental design parameter for sedimentation tank design is overflow rate. The overflow rate and the design flow rate determine the surface area of the clarifier:

$$v_o = \frac{Q}{A_s} \qquad (11.5.9)$$

where
- v_o = overflow rate, m/d
- Q = flow rate, m³/d
- A_s = surface area of tank, m²

Theoretically, in an ideal horizontal flow settling tank, the fraction of particles with a given settling velocity that will be removed is determined by the ratio of the settling velocity to the overflow rate:

$$\% \text{ Removal} = \frac{v_s}{v_o}(100\%) \qquad (11.5.10)$$

In an upflow clarifier or, for example, during backwashing of a rapid sand filter, this equation does not apply. Rather, the removal is either 100% for particles with settling velocities greater than the overflow rate or 0% for particles with settling velocities less than the overflow rate.

In the absence of laboratory test data, the values shown in Table 11.6 may be used to estimate overflow rates. Typical detention times range from 2 to 8 hours for coagulation of iron or alum floc and from 4 to 8 hours for softening floc. Standard practice is to provide no less than two sedimentation basins. In small facilities one of the basins may serve as a backup system or be used to handle high flow situations rather than using both for average conditions. Side water depth (SWD) should be 4.5 to 5.5 m for basins more than 30 m long or 20 m in diameter. Smaller basins should have a minimum SWD of 2.5 m. Rectangular basins are preferred. Common length-to-width ratios range from 2:1 to 5:1 with lengths seldom exceeding 100 m. Nominal weir loading rates are shown in Table 11.7. The troughs should cover at least one-third, and preferably, up to one-half, of the clarifier length. They should be evenly distributed.

Table 11.6 Typical sedimentation tank overflow rates.

Treatment	Rectangular & Circular Overflow Rate, m³/d·m²	Upflow Contact Overflow Rate, m³/d·m²
Lime softening		
Low magnesium	70	130
High magnesium	60	100
Alum or iron coagulation		
Turbidity removal	40	50
Color removal	30	35
High algae content	20	

Source: American Society of Civil Engineers and American Water Works Association, *Water Treatment Plant Design Manual*, second edition, McGraw-Hill Publishing Co., New York, NY, 1990.

Table 11.7 Typical weir loading rates.

Type of Floc	Loading rate, m³/d·m²
Light alum floc	
Low turbidity water	140 – 180
Heavier alum floc	
High turbidity water	180 – 270
Lime softening floc	270 – 320

Source: Walker Process Equipment, Inc., Division of Chicago Bridge and Iron Co., *Walker Process Circular Clarifiers*, Bulletin 9-W-65, Aurora, IL., 1973.

EXAMPLE 11.9

Using nominal design values, design a sedimentation tank system for a design flow of 0.0789 m³/s for light alum floc. Determine the tank dimensions and length of weir.

Solution. From Table 11.6 select an overflow rate of 30 m/d for coagulation of color floc (a light floc). The surface area of the tank is then

$$A_s = \frac{(0.0789 \text{ m}^3/\text{s})(86,400 \text{ s/d})}{30 \text{ m/d}} = 227.23 \text{ m}^2$$

Assuming a rectangular tank with a length to width ratio of 2:1, width of the tank is

$$(2x)(x) = 227.2 \text{ m}^2$$

$$x = 10.66 \text{ m}$$

The length of the tank would be 21.32 m. For standard dimensions, choose a width of 10.5 m and a length of 21.5 m to yield a surface area of 225.75 m² and an overflow rate of 30.2 m/d.

Select a side water depth of 3.0 m (the minimum for this size basin would be 2.5 m). The tank volume is then

$$V = (10.5)(21.5)(3.0) = 677.2 \text{ m}^3$$

Check the detention time:

$$t_o = \frac{677.2 \text{ m}^3}{(0.0789 \text{ m}^3/\text{s})(3600 \text{ s/h})} = 2.38 \text{ h}$$

This is within the nominal detention time of 2 to 8 hours.

Assuming a weir loading rate of 140 m/d from Table 11.7, the required weir length is

$$\frac{(0.0789 \text{ m}^3/\text{s})(86\,400 \text{ s/d})}{140 \text{ m/d}} = 48.69 \text{ m}$$

Use 50 m of weir. Since the tank is 21.5 m long, finger weirs will be used. Two finger weirs 12.5 m long will provide 50 m of weir length since both sides of the weir are included. This arrangement provides troughs that cover slightly more than half the length. An alternative would be to provide 3 troughs each 8.33 m in length.

11.5.3 Sludge Volume Calculations

Theoretically, the rate of sludge generation for coagulation may be estimated as

$$M_s = 86.40\, Q\, (0.44A + SS + M) \quad (11.5.11)$$

where
- M_s = mass of dry sludge produced, kg/d
- Q = plant flow, m³/s
- A = alum dose, mg/L
- SS = suspended solids in raw water, mg/L
- M = miscellaneous additions such as clay, polymer, etc., mg/L

For softening, sludge generation may be estimated as

$$M_s = 86.40\, Q\, (Ca_R + 0.58\, Mg_R + L_A) \quad (11.5.12)$$

where
- Ca_R = calcium hardness removed, mg/L as $CaCO_3$
- Mg_R = magnesium hardness removed, mg/L as $CaCO_3$
- L_A = lime added, mg/L as $CaCO_3$

The volume of sludge may be estimated if the specific gravity of the sludge and the percent solids are known. The specific gravity of the sludge may be estimated from the specific gravity of the solids and the percent solids:

$$S_{sl} = \frac{S_s}{P_s + S_s P_w} \quad (11.5.13)$$

where
- S_{sl} = specific gravity of sludge
- S_s = specific gravity of solids
- P_s = percent solids as a decimal fraction
- P_w = percent water as a decimal fraction

The volume of sludge V_{sl} generated may be estimated from the mass as

$$V_{sl} = \frac{M_s}{\rho S_{sl} P_s} \quad (11.5.14)$$

If sludge dewatering is performed, the dewatered sludge volume may be approximated by

$$V_2 = V_1 \frac{P_1}{P_2} \quad (11.5.15)$$

Coagulation sludge is normally generated at solids concentrations on the order of 1 to 2 percent and often may be less than 1 percent. Lime softening sludge underflow concentrations are nominally at 10 percent solids.

11.5.4 Sludge Dewatering

In addition to gravity thickening, several other sludge dewatering techniques are commonly employed. A summary of potential cake concentrations is given in Table 11.8. When space is available, sludge lagoons are economically favored. In sludge lagoon operation, the surface water must be removed to achieve satisfactory solids concentrations. This water is often returned to the head end of the plant for reprocessing.

Water Treatment

Table 11.8 Potential cake solids obtainable by dewatering.

	Lime sludge, %	Coagulation sludge, %
Gravity thickening	15-30	3-4
Centrifuge	55-65	10-20
Belt filter press		10-15
Vacuum filter	45-65	
Pressure filter	55-70	30-45
Sand drying beds	50	20-25
Storage lagoons	50-60	7-15

Source: M.L. Davis and D.A. Cornwell, *Introduction to Environmental Engineering*, McGraw-Hill, Inc., New York, NY, 1991, p. 235.

EXAMPLE 11.10

River water containing 43.0 mg/L of suspended solids is treated with 60.00 mg/L of alum for coagulation. The design flow is 0.0789 m³/s. Determine the mass of sludge generated, the volume of sludge if the specific gravity of the solids is 1.52 and the percent solids is 1.3, and the volume of sludge if it is dewatered to 30 percent solids with a pressure filter.

Solution. Using Eq. 11.5.11, the mass of sludge generated each day is

$$M_s = 86.40(0.0789)[0.44(60.0) + 43.0 + 0] = 473.10 \text{ kg/d}$$

The specific gravity of the sludge is

$$S_{sl} = \frac{1.52}{0.013 + 1.52(0.987)} = 1.004$$

The volume generated each day is

$$V_{sl} = \frac{473.10}{(1000)(1.004)(0.013)} = 36.23 \text{ or } 36 \text{ m}^3/\text{d}$$

After dewatering to 30% solids, the volume would be

$$V_2 = (36)\frac{0.013}{0.30} = 1.56 \text{ m}^3/\text{d}$$

11.6 Filtration

11.6.1 Media Selection

Sand has been the traditional media for granular filtration of water. Slow sand filters having a uniform sand diameter of 0.2 mm were operated at loading rates of 3 to 7.5 m/d. These filters were replaced by rapid sand filters having graded sand with grain size distributions designed to optimize the flow of water while retaining particulate matter. Rapid sand filters are operated at loading rates from 120 to 235 m/d. The sand depth varies between 0.5 and 0.75 m.

Dual-media filters consisting of a layer of coarse coal on top of a layered sand were developed to increase loading rates to 300 m/d. The depth of sand in the filter is about 0.3 m and the coal is about 0.45 m in depth. Because the specific gravity of the coal is less than that of the sand, the coal remains on the surface after backwashing. In contrast to the rapid sand filter, the particles removed in a dual media filter tend to penetrate deeply into the bed.

Deep bed monofilters of anthracite coal have been introduced to further increase the loading rates to 800 m/d. Monofilters typically consist of 1.5 to 2.5 m deep beds of 1.0 to 1.5 mm diameter coal.

11.6.2 Grain Size Distribution

Grain size is determined from a sieve analysis using the U.S. Standard Sieve Series summarized in Table 11.9. The two most common parameters used to define the grain size distribution of the filter media are the effective size and the uniformity coefficient. The effective size (E or P_{10}) is defined as the 10-percentile grain size, i.e., cumulative mass of 10 percent less than or equal to the stated grain size. The uniformity coefficient U is defined as the ratio of the 60-percentile grain size to the 10-percentile grain size (P_{60}/P_{10}).

Table 11.9 U.S. standard sieve sizes.

Sieve Number	Size of Opening, mm	Sieve Number	Size of Opening, mm
200	0.074	20	0.84
140	0.105	(18)	1.00
100	0.149	16	1.19
70	0.210	12	1.68
50	0.297	8	2.38
40	0.42	6	3.36
30	0.59	4	4.76

The effective size for silica sand should be in the range of 0.35 to 0.55 mm with a recommended maximum of 1.0 mm. Smaller effective sizes will yield a product water with a lower turbidity but at the cost of higher pressure losses in the filter and shorter operating cycles between cleaning. The uniformity coefficient should be in the range of 1.3 to 1.7.

When anthracite coal alone is used as the filter, media specifications typically call for an effective size of 0.45 to 0.55 mm with a uniformity coefficient not greater than 1.65. In dual media applications, the effective size of the coal is in the range 0.8 to 1.2 mm with a uniformity coefficient not greater than 1.85.

The equivalent diameter for mixed media may be found from the following relationship:

$$\frac{d_m}{d_s} = \left[\frac{\rho_s - \rho}{\rho_m - \rho}\right]^{2/3} \tag{11.6.1}$$

where
- d_m = diameter of the coal or other media
- d_s = diameter of the sand
- ρ_s = density of the sand
- ρ_m = density of the coal or other media
- ρ = density of the water

Typical specific gravities of filter media are: coal = 1.5, sand = 2.5, garnet = 4.2.

11.6.3 Head Loss During Filtration

Head loss during filtration is a function of the filtration rate, media characteristics and the accumulation of filtered particles. There is no satisfactory theory for prediction of head loss with the accumulation of filtered particles because of the varying nature of the particles and the rate at which they may be loaded on the bed. The initial clean bed head loss should not exceed about 0.6 m. Common practice is to backwash when the head loss reaches about 1.8 to 2.4 m.

Initial estimates of the head loss through a clean bed may be made to establish an acceptable design of the grain size distribution. For beds having a relatively uniform diameter media (not stratified), the Rose equation in the following form may be used:

$$h_L = \frac{1.067 v_a^2 D C_D}{\phi g \varepsilon^4 d} \tag{11.6.2}$$

where:
- h_L = head loss, m
- ϕ = shape factor
- C_D = drag coefficient
- D = depth of bed, m
- v_a = approach velocity, m/s
- ε = porosity of the bed
- d = particle diameter, m
- g = acceleration due to gravity, m/s²

The drag coefficient is a function of Reynolds number and takes one of the forms described for Type I sedimentation (Eq. 11.5.2, 11.5.4 or $C_D = 0.4$).

For stratified beds with uniform porosity, the Rose equation has the following form:

$$h_L = \frac{1.067 v_a^2 D}{\phi g \varepsilon^4} \sum \frac{C_D f}{d} \qquad (11.6.3)$$

where f = mass fraction for particle size d.

11.6.4 Backwash Hydraulics

The filter media is cleaned (backwashed) by pumping water into the bottom of the filter bed at such a rate that the media is fluidized. The flow rate is regulated to prevent the loss of media but is rapid enough to overcome the settling velocity of the entrapped particles. Backwash rates normally vary between 800 and 1200 m/d. However, the choice of backwash rate is governed by the terminal settling velocity of the smallest grain size of the media that is to be retained in the filter.

The head loss during backwash determines the elevation of the backwash troughs by limiting the depth of the expanded filter bed. An approximation of the depth of the expanded bed may be made using the following equation:

$$D_e = D(1-\varepsilon) \sum \frac{f}{1-\varepsilon_e} \qquad (11.6.4)$$

where
- D_e = depth of the expanded bed, m
- ε = porosity of the bed
- ε_e = porosity of the expanded bed
- f = mass fraction of sand with expanded porosity

The porosity of the expanded bed may be estimated as

$$\varepsilon_e = \left(\frac{v_b}{v_s}\right)^{0.22} \qquad (11.6.5)$$

where
- v_b = velocity of backwash, m/s
- v_s = settling velocity of sand particles in fraction, m/s

The calculation of D_e requires an initial estimate of the particle settling velocity so that a Reynolds number may be estimated for calculating C_D in the settling velocity equation (Eq. 11.5.1). Figure 11.9 may be used for the initial estimate.

11.6.5 Filter Box Design

A minimum of two filter boxes are constructed for redundancy. For plants with flows greater than 0.5 m³/s, a minimum of four filter boxes is recommended. The surface area is generally restricted to about 100 m² although larger boxes have been successful. The depth of the box is determined by the design head loss before backwashing (i.e., 1.8 to 2.4 m) plus freeboard.

The elevation of the outlet from the filter box controls the hydraulic grade line through the plant. Writing Bernoulli's equation between the center of the effluent line and the water surface in the filter box (Fig. 11.10) yields an equation for the pressure in the effluent line:

$$\frac{p_2}{\gamma} = z_1 - \frac{v_2^2}{2g} - H_L \tag{11.6.6}$$

Most regulations require that the effluent pressure not fall below zero as negative pressures have a tendency to pull particulate matter out of the bed. Knowing the depth of the underdrains, pipe gallery and media, and with an initial estimate of the clean bed head loss (all summed to yield z_1), the maximum allowable head loss can be calculated by setting $p_2/\gamma = 0$.

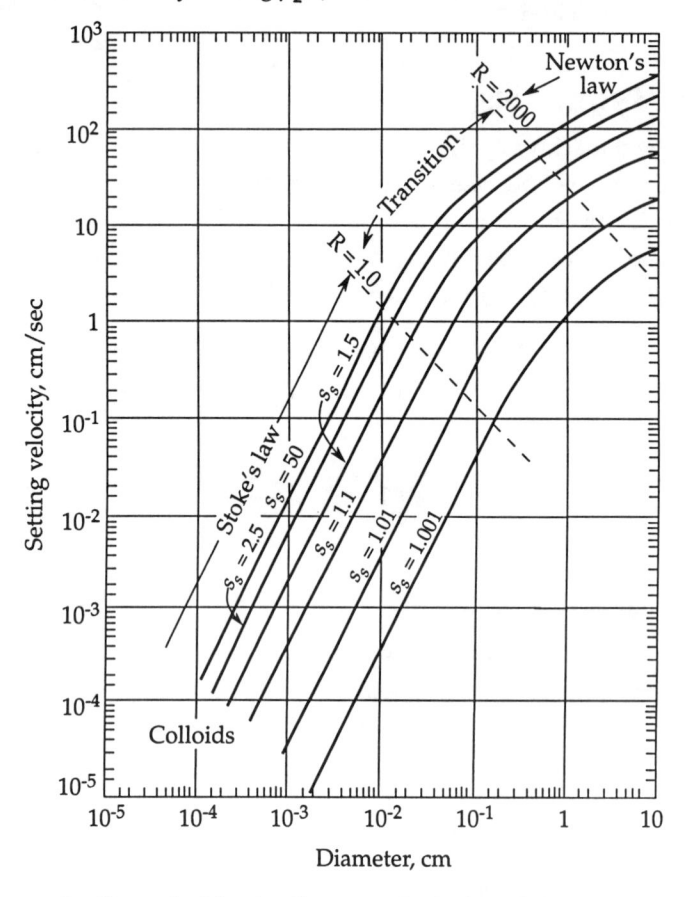

Figure 11.9 Settling velocities for discrete spherical particles in quiescent water.
Source: G. M. Fair, J. C. Geyer and D. A. Okun, *Elements of Water Supply and Wastewater Disposal*, second edition, John Wiley & Sons, Inc., New York, NY, 1971, p. 371.)

The backwash troughs are placed at the same height above the filter bed with the trough lips horizontal. The distance from the bottom of the trough to the top of the filter bed is usually taken to be equal to the depth (D_e) of the expanded bed plus a margin of safety of about 0.15 m. The troughs are usually spaced no more than 1.8 m apart and are arranged to run either the length or width of the filter. One rule-of-thumb is that the troughs should be spaced such that the maximum horizontal distance the suspended particles must travel does not exceed 0.9 m and that they are spaced such that each trough serves approximately the same area of filter.

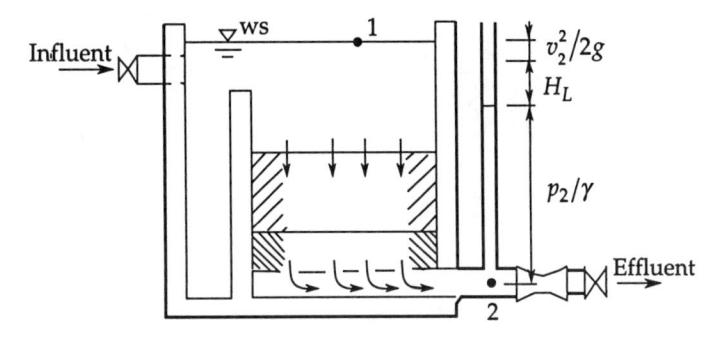

Figure 11.10 Filter schematic with notation for Bernoulli's equation.
Source: T. D. Reynolds, *Unit Operations and processes in Environmental Engineering*, Brooks/Cole Engineering Division, Wadsworth, Inc., Belmont, CA, 1982, p. 134.)

──────── **EXAMPLE 11.11** ────────

Design a rapid sand filter for a design flow of 0.0789 m³/s. Size the filter boxes. Determine the effective size and uniformity coefficient of the sand. Estimate the clean filter head loss and determine the height the backwash troughs must be placed above the filter bed. The following assumptions may be used in the design:

Loading rate: 120 m³/d·m²
Water temperature: 10 °C
Specific gravity of sand: 2.5
Shape factor: 0.88

Bed porosity: 0.40
Sand depth: 0.5 m
Backwash rate: 1200 m/d

The grain size analysis for the sand is as follows:

U.S. Standard Sieve No.	Cumulative Mass % Passing
50	0.06
40	1.30
30	14.00
20	51.00
16	87.50
12	98.60
8	99.94

Solution. The cumulative percent passing is plotted against the diameter of the sieve opening as shown below. At the 10-percentile, the sieve size is 0.55 mm. This is the effective size. The 60-percentile size is 0.90 mm. The uniformity coefficient is then

$$U = \frac{P_{60}}{P_{10}} = \frac{0.90}{0.55} = 1.64$$

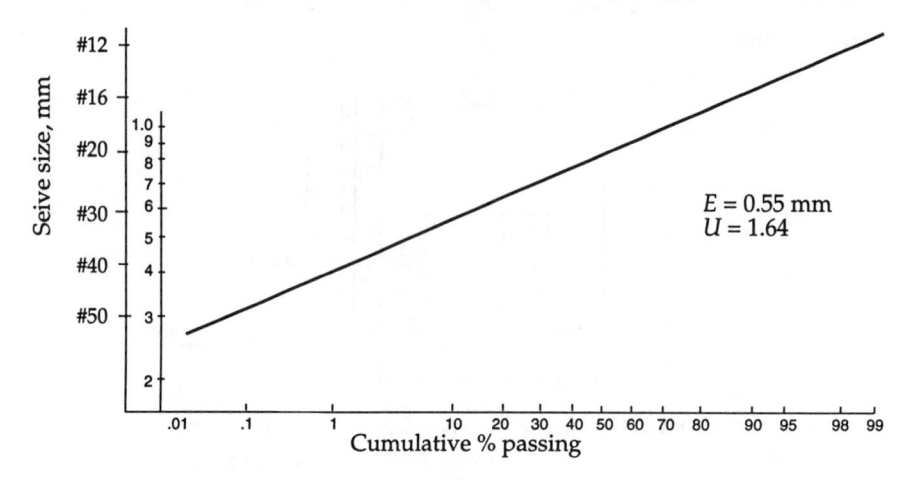

Figure for Ex. 11.11 Grain size analysis.

The computations for determining the clean bed head loss are shown on the next page.

Sieve No.	Fraction Retained	d, m	R	C_D	$\dfrac{C_D f}{d}$
8-12	0.0134	0.002	1.872	15.35	103
12-16	0.111	0.00142	1.329	21.00	1 642
16-20	0.365	0.001	0.936	29.08	10 610
20-30	0.370	0.000714	0.668	39.93	20 690
30-40	0.127	0.000505	0.473	55.48	13 950
40-50	0.0124	0.000357	0.334	77.36	2 687
					$\Sigma = $ 49 700

In the first two columns, the grain size distribution is recalculated to show the fraction retained between sieves. The third column is the geometric mean diameter of the two sieve sizes. The fourth column is the Reynolds number. With a correction for sphericity, it is

$$R = \frac{(0.88)(0.002)(0.00139)}{1.307 \times 10^{-6}} = 1.872$$

The velocity is calculated from the loading rate:

$$v_a = \frac{120 \text{ m}^3/\text{d} \cdot \text{m}^2}{86\ 400 \text{ s/d}} = 0.00139 \text{ m/s}$$

The kinematic viscosity is determined from the water temperature.

The drag coefficient is calculated with either Eq. 11.5.2 or 11.5.4. For the first row:

$$C_D = \frac{24}{1.872} + \frac{3}{\sqrt{1.872}} + 0.34 = 15.35$$

The sixth row is the product of fraction retained and the drag coefficient divided by the diameter:

$$\frac{(15.3549)(0.0134)}{0.002} = 103$$

The head loss is calculated from Eq. 11.6.3:

$$h_L = \frac{1.067(0.00139)^2(0.5)}{(0.88)(9.8)(0.40)^4}(49\,700) = 0.23 \text{ m}$$

This is less than 0.6 m and, therefore, acceptable.

The computations to determine the height of the backwash trough are shown in the table below. The rows are arranged by sieve size as they were for the head loss in a clean bed.

Estimated v_s, m/s	R	C_D	Calculated v_s, m/s	ε_e	$\dfrac{f}{1-\varepsilon_e}$
0.30	404	0.5487	0.267	0.522	0.028
0.20	191	0.6825	0.202	0.555	0.249
0.15	101	0.8762	0.150	0.593	0.897
0.10	48.1	1.272	0.105	0.641	1.031
0.07	23.8	1.963	0.071	0.699	0.421
0.05	12.0	3.202	0.047	0.766	0.053
					$\Sigma = 2.68$

The estimated settling velocities are from Fig. 11.9. The Reynolds number was computed with this estimated velocity. The shape factor, diameter and viscosity are the same as that used for the head loss calculation. The drag coefficient is calculated as shown above. The settling velocity is calculated using Eq. 11.5.1. For the first row:

$$v_s = \left[\frac{(4)(9.8)(2500-1000)(0.002)}{(3)(0.5487)(1000)}\right]^{1/2} = 0.2673 \text{ m/s}$$

The density of the sand grain is calculated from the specific gravity and the density of water:

$$(2.5)(1000 \text{ kg/m}^3) = 2500 \text{ kg/m}^3$$

The expanded bed porosity is calculated with Eq. 11.6.5:

$$\varepsilon_e = \left(\frac{0.0139}{0.2673}\right)^{0.22} = 0.5218$$

The last column is calculated with the fraction (f) from the previous table:

$$\frac{0.0134}{1-0.5218} = 0.0280$$

The depth of the expanded bed is

$$D_e = (1 - 0.40)(0.50)(2.68) = 0.804 \text{ m}$$

Assuming a margin of safety of 0.15 m, the bottom of the backwash trough should be placed at

$$(D_e - D) + 0.15 = (0.80 - 0.5) + 0.15 = 0.45 \text{ m}$$

or about 0.5 m above the top of the sand bed.

Note that the backwash rate (1200 m/d) does not exceed the settling velocity of the finest grain size in the bed. Thus, the filter media will not be washed out.

11.7 Stabilization

Internal corrosion of piping and valves in the water distribution system may be inhibited by deposition of a thin film of calcium carbonate. This deposition process is controlled by altering the carbonate equilibrium. In waters that have been softened by lime-soda processes, there is a tendency for continued precipitation in the distribution system that ultimately plugs the lines. This reaction is also altered by adjusting the carbonate equilibrium.

The Langelier saturation index (SI) is used to estimate the stability of water to deposit or dissolve calcium carbonate in the pH range 6.5 to 9.5. The stability index is defined as:

$$SI = pH - pH_s = pH - \left[(pK_2 - pK_s) + pCa + pAlk\right] \quad (11.7.1)$$

where
- pH = measured pH of the water
- pH_s = pH at $CaCO_3$ saturation (equilibrium)
- $(pK_2 - pK_s)$ = constants based on ionic strength
- pCa = negative logarithm of the calcium ion concentration, moles/L
- $pAlk$ = negative logarithm of the total alkalinity, equivalents/L

The ionic strength may be calculated as

$$\text{Ionic strength} = (0.5)[(C_1)(Z_1) + (C_2)(Z_2) + \cdots] \quad (11.7.2)$$

where
- C = concentration of ionic species, moles/L
- Z = valence of an individual ion

Values of the difference $(pK_2 - pK_s)$ may be obtained from Table 11.10.

Table 11.10 Values of the difference $(pK_2 - pK_s)$.

Ionic Strength	Total Dissolved Solids, mg/L	0°C	10°C	20°C	30°C
0.000		2.45	2.23	2.02	1.86
0.001	40	2.58	2.36	2.15	1.99
0.002	80	2.62	2.40	2.19	2.03
0.003	120	2.66	2.44	2.23	2.07
0.004	160	2.68	2.46	2.25	2.09
0.005	200	2.71	2.49	2.28	2.12
0.006	240	2.74	2.52	2.31	2.15
0.007	280	2.76	2.54	2.33	2.17
0.008	320	2.78	2.56	2.35	2.19
0.009	360	2.79	2.57	2.36	2.20
0.010	400	2.81	2.59	2.38	2.22
0.011	440	2.83	2.61	2.40	2.24
0.012	480	2.84	2.62	2.41	2.25
0.013	520	2.86	2.64	2.43	2.27
0.014	560	2.87	2.65	2.44	2.28
0.015	600	2.88	2.66	2.45	2.29
0.016	640	2.90	2.68	2.47	2.31
0.017	680	2.91	2.69	2.48	2.32
0.018	720	2.92	2.70	2.49	2.33
0.019	760	2.92	2.70	2.49	2.33
0.020	800	2.93	2.71	2.50	2.34

A positive value for the stability index indicates the water is over saturated with calcium carbonate and will continue to deposit $CaCO_3$. A negative SI indicates a corrosive water. The pH of the product water may be adjusted by the addition of lime or soda ash if the pH is too low or with carbon dioxide (recarbonation) or an acid if the pH is too high. A slight oversaturation is desirable to deposit a protective film on the pipes. The recommended range is a carbonate oversaturation of 4 to 10 mg/L as $CaCO_3$.

Polyphosphates are often added to inhibit crystal nucleation. This inhibits the precipitation of excess calcium but allows the water to remain saturated and prevent dissolution of the protective film.

EXAMPLE 11.12

The composition of a treated water is as shown below. The water temperature is 8 °C and the pH is 9.95. Determine whether or not the water is stable.

Constituent	Valence	moles/L
Ca	+2	6.90×10^{-4}
Mg	+2	4.00×10^{-4}
Na	+1	2.11×10^{-3}
CO_3	−2	8.50×10^{-4}
Cl	−1	4.23×10^{-4}
SO_4	−2	1.41×10^{-3}

Solution. Calculate the ionic strength using Eq. 11.7.2:

$$Ca = (0.5)(6.90 \times 10^{-4})(2)^2 = 1.38 \times 10^{-3}$$

$$Mg = (0.5)(4.00 \times 10^{-4})(2)^2 = 8.00 \times 10^{-4}$$

$$Na = (0.5)(2.11 \times 10^{-3})(1)^2 = 1.06 \times 10^{-3}$$

$$CO_3 = (0.5)(8.50 \times 10^{-4})(2)^2 = 1.70 \times 10^{-3}$$

$$Cl = (0.5)(4.23 \times 10^{-4})(1)^2 = 2.12 \times 10^{-4}$$

$$SO_4 = (0.5)(1.41 \times 10^{-3})(2)^2 = \underline{2.82 \times 10^{-3}}$$

$$7.97 \times 10^{-3}$$

From Table 11.10, at 8°C the value of $(pK_2 - pK_s)$ is approximately 2.56. The value for *pCa* is

$$pCa = -\log(6.90 \times 10^{-4}) = 3.16$$

The value for *pAlk* is

$$pAlk = -\log(8.50 \times 10^{-4}) = 3.07$$

The stability index is then

$$SI = 9.95 - (2.56 + 3.16 + 3.07) = +1.16$$

The positive value indicates the water is not stable. It is oversaturated and precipitation of $CaCO_3$ will result.

11.8 Disinfection

11.8.1 Disinfectant Selection

The common choices for disinfecting water are: chlorine (Cl_2), chlorine dioxide (ClO_2), ozone (O_3), advanced oxidation processes (AOPs), and ultraviolet radiation.

Of these alternatives, chlorination is the most common in the U.S. The chlorine is supplied under pressure in liquified form in cylinders, ton containers and railroad cars. The gas (Cl_2) is vaporized into a slip stream of treated water that is then mixed with the primary flow. Chlorine is selected because it is inexpensive and provides a residual in the distribution system that can be monitored to ensure protection against contamination. It has the major disadvantage that it reacts with organic precursors to form trihalomethanes (THMs) that are known or potential carcinogens.

Chlorine dioxide (ClO_2) is formed on-site by combining chlorine and sodium chlorite. Chlorine dioxide is selected because it does not form THMs. However, it's reaction byproducts include chlorite and chlorate. These compounds may have a human health risk. This, combined with potential taste and odor problems, as well as a relatively high cost have limited it's application.

Because of its instability, ozone is manufactured on-site. Dry air or oxygen is used as the source of the oxygen molecules. Either a discharge electrode or ultraviolet light are used to create atomic oxygen which rapidly combines with surrounding oxygen molecules to form ozone. The ozone gas is bubbled into a slip stream of treated water and then mixed with the primary flow. Ozone is widely used in Europe and is gaining favor as the primary disinfectant in the U.S. It is favored because it does not form THMs. It does not provide a residual in the distribution system and so is often followed by chloramine addition in the treatment process.

AOPs are a combination of disinfectants that result in production of hydroxyl radicals (OH•). One of the more common AOP processes is the addition of ozone and hydrogen peroxide (H_2O_2). The advantage of the AOPs is that they destroy many organic compounds and do not form THMs. The disadvantage is that they do not provide a residual in the distribution system.

Ultraviolet radiation may be used for disinfection by passing a shallow depth of water over ultraviolet lamps. The depth of light penetration is about 50 to 80 mm. Although UV performs well under these limited conditions and generates no THMs, it leaves no residual protection and is very expensive.

11.8.2 Free and Available Chlorine

When chlorine is added to water, the gas dissolves to form hypochlorous and hydrochloric acid:

$$Cl_2 \text{ (g)} + H_2O \leftrightarrow HOCl + H^+ + Cl^- \tag{11.8.1}$$

In dilute solutions at pH levels above 1, the reaction proceeds rapidly to the right and little Cl_2 exists. The hypochlorous acid is a weak acid that dissociates as shown below:

$$HOCl \leftrightarrow H^+ + OCl^- \tag{11.8.2}$$

The dissociation is pH dependent. Below pH levels of about 6, the predominant form is HOCl. Above pH 9, the predominant form is hypochlorite ion (OCl^-). Both forms are know as free chlorine and are effective disinfectants. The optimum pH for disinfection is in the range 6.5 to 7.5.

Hypochlorite salts may be used to provide free chlorine because they dissociate in water to yield hypoclorite ions:

$$NaOCl \leftrightarrow Na^+ + OCl^- \quad (11.8.3)$$

$$Ca(OCl)_2 \leftrightarrow Ca^{2+} + 2OCl^- \quad (11.8.4)$$

Chloramines are reaction products of ammonia and hypochlorous acid. Mono-, di- and trichloramines are formed as follows:

$$NH_3 + HOCl \leftrightarrow NH_2Cl + H_2O \quad (11.8.5)$$

$$NH_2Cl + HOCl \leftrightarrow NHCl_2 + H_2O \quad (11.8.6)$$

$$NHCl_2 + HOCl \leftrightarrow NCl_3 + H_2O \quad (11.8.8)$$

Chloramines retain some of the oxidizing power of chlorine and have a longer half-life than chlorine. The chloramines are called combined chlorine rather than free chlorine.

The comparative oxidizing power of disinfectants may be made based on their ability to exchange electrons. The percent available chlorine is calculated as

$$\% \text{ Available Chlorine} = \frac{\text{Equiv. Wt. of Chlorine}}{\text{Equiv. Wt. of Compound}} \quad (11.8.9)$$

The equivalent weight of chlorine is 35.5. The equivalent weights of other oxidizing compounds may be calculated based on their half-reactions (Table 11.11).

Table 11.11 Half-reactions for disinfectants.

Calcium hypochlorite (high test hypochlorite, HTH)
$Ca(OCl)_2 + 2H^+ + 4e \leftrightarrow CaO + H_2O + 2Cl^-$

Chlorine
$Cl_2 + 2e \leftrightarrow 2Cl^-$

Chlorine dioxide
$ClO_2 + e \leftrightarrow ClO_2$

Chlorinated lime
$CaOCl_2 + 2e \leftrightarrow 2Cl^- + CaO$

Dichloramine
$NHCl_2 + 2H^+ + 4e \leftrightarrow 2Cl^- + NH_3$

Hypochlorous acid
$HOCl + H^+ + 2e \leftrightarrow H_2O + Cl^-$

Ozone
$O_3 + 2H^+ + 2e \leftrightarrow H_2O + O_2$

Sodium hypochlorite
$NaOCl + H^+ + 2e \leftrightarrow H_2O + Cl^-$

EXAMPLE 11.13

Calculate the percent available chlorine for dichloramine.

Solution. The equivalent weight of dichloramine ($NHCl_2$) is calculated based on its half-reaction. The equivalent weight is then

$$\frac{86}{4} = 21.5$$

Note that the number of electrons in the half-reaction determines the value of the denominator.

The percent available chlorine is calculated using Eq. 11.8.9:

$$\% \text{ Available Chlorine} = \frac{35.5}{21.5}(100\%) = 165\%$$

11.8.3 Safety

Because chlorine is a very toxic gas, special safety precautions are required. The chlorination facility must be provided with proper isolation and ventilation. In general this means that the room housing the chlorine cylinders and the room housing the feed device do not open directly into the rest of the plant. Because chlorine gas is dense, ventilation duct entrances are placed a the floor level. The exhaust system is not connected to the heating/ventilating system for the rest of the plant.

In the event of an emergency, a glass inspection panel is placed such that the feed device and storage containers may be viewed without entering the rooms. Self contained breathing apparatus are placed on the exterior wall (outside) of the storage and feed rooms. The apparatus must have a complete face mask to protect the eyes as well as provide for respiration.

Practice Problems

11.1 Jar test data indicate that an alum dose of 40 mg/L provides the optimum turbidity removal. If the water initially contains 16.39 mg/L as $CaCO_3$ of bicarbonate alkalinity, is there "enough" alkalinity at the optimum coagulant dose?

11.2 Determine the impeller speed in rpm for a marine impeller (3 blade, pitch of 2) for a flocculation basin having liquid dimensions ($W \times H \times L$) of 30' 9" × 24' 9" × 61' 6". Assume that two impellers on two separate shafts will be used. Also assume that the design $G = 35$ s^{-1}, that the impeller will be 30% of the width, the water temperature is 10 °C, and that baffles will be present.

11.3 Design a split treatment softening process (flow scheme/split, chemical dose in mg/L as $CaCO_3$, and final hardness) for the following water. Compounds are given in mg/L as the ion stated unless otherwise specified.

CO_2	42.7	HCO_3	344.0 mg/L as $CaCO_3$
Ca	102.0	SO_4	65.0
Mg	45.2	Cl	32.0
Na	21.8		

11.4 Given the following water analysis (all in mEq/L), design a softening process to soften the water (flow scheme/split; amount of lime and/or soda required in mg/L as CaO and Na_2CO_3, respectively; and final hardness):

CO_2	0.40	Mg	1.12
Ca	2.16	HCO_3	2.72

11.5 Sedimentation tanks are to be designed for a coagulation water treatment plant with an average design flow of 0.800 m³/s and a maximum day flow that is 1.5 times the average day. Assuming an overflow rate of approximately 15 m/d, a length to width ratio of 5:1, a maximum length of 100 m, a side water depth of 4.5 m, and a weir loading rate of 150 m/d, determine the dimensions of the tanks (length, width, depth), the detention time, the number of tanks, and the length of weir in each tank.

11.6 Using the figure shown below, compute the design overflow rate (in m/d) and percent removal at a detention time of 4.2 hours.

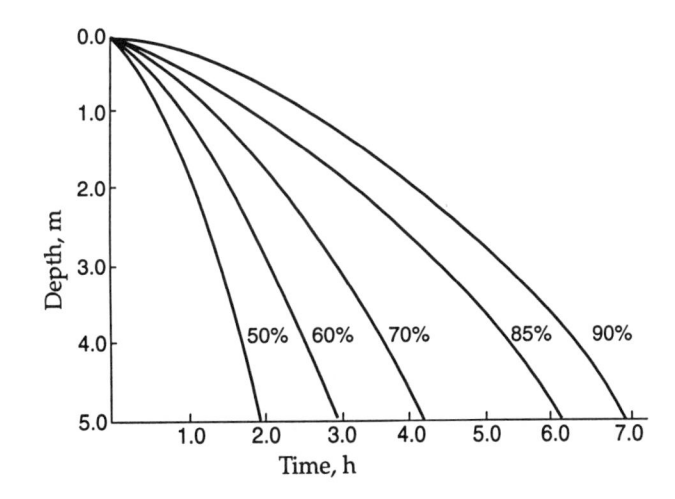

11.7 A lime softening plant removes 135.0 mg/L as $CaCO_3$ of calcium hardness and 29.9 mg/L as $CaCO_3$ of magnesium hardness by the addition of 237.5 mg/L as $CaCO_3$ of lime. The specific gravity of the softening solids is approximately 2.75 and the sludge is generated at about 10% solids. If the design flow of the plant is 0.10 m³/s, and the sludge is dewatered to 60% solids, estimate the annual volume of sludge generated.

11.8 For the following sand size distribution, determine:

A) The head loss through the clean sand bed. (Note: you must finish the table first. You may assume the other values in the table are correct.)

B) The minimum height of the backwash troughs above the sand bed. (Note: you must finish the table first. You may assume the other values in the table are correct)

C) The smallest diameter (in feet) anthracite coal particle (specific gravity = 1.55) that can be used to form a dual media filter.

The depth of the sand bed is 24 inches; the specific gravity of the sand is 2.65; the filtration rate is 2.5 gpm/sq ft; the porosity ratio is 0.40; the water temperature is 10°C; the shape factor is 0.95; the rate of backwash is 20 gpm/sq ft.

FILTERING

Sieve No.	d, ft	% Wt. Retained	C_D	$C_D(f)/d$
14-20	0.003283	1.10	18.5	62
20-28	0.002333	6.60	26.4	761
28-32	0.001779	15.94	34.3	3090
32-35	0.001500	18.60	40.0	4930
35-42	0.001258	19.10	49.0	7480
42-48	0.001058	17.60	57.2	9600
48-60	0.000888	14.30	68.6	11 150
60-65	0.000746	5.10	80.0	5450
65-100	0.000583	1.66	??	??

BACKWASHING

Sieve No.	% Wt. Retained	ε_e	$f/1-\varepsilon_e$
14-20	1.10	0.378	0.0177
20-28	6.60	0.414	0.1125
28-32	15.94	0.443	0.287
32-35	18.60	0.460	0.344
35-42	19.10	0.483	0.370
42-48	17.60	0.502	0.353
48-60	14.30	0.521	0.299
60-65	5.10	0.555	0.1145
65-10	01.66	??	??

11.9 Determine whether or not the following water is stable. (Is it corrosive or will it deposit Ca?) Assume all alkalinity is carbonate. What chemicals should be added?

$$Ca = 25.00 \text{ mg/L as } CaCO_3$$
$$Alkalinity = 40.00 \text{ mg/L as } CaCO_3$$
$$pH = 8.60$$
$$Temperature = 10°C$$
$$TDS = 80.00 \text{ mg/L}$$

11.10 A salesperson has proposed that the local water treatment plant replace the current gaseous chlorine disinfection system with a chlorine dioxide disinfection system because it would be economical based on "percent available chlorine." Assuming the same plumbing can be used for either disinfectant, that chlorine gas costs $300 per ton, and that chlorine dioxide costs $600 per ton, show by calculation whether or not this is true. Discuss other issues that should be considered in switching from chlorine to chlorine dioxide.

11.11 A water plant supplies 2200 MGD of water on an average day. The equivalent residential population is in the range:

a) 10×10^6 to 15×10^6

b) 3×10^6 to 6×10^6

c) 20×10^6 to 25×10^6

d) 6×10^6 to 10×10^6

e) 20×10^6 to 40×10^6

11.12 If 0.1 mole/L of sodium is sufficient to destabilize a colloid, then which of the following doses will be equally effective

a) 1×10^{-3} mole/L of ferric chloride

b) 1×10^{-4} mole/L of alum

c) 1.43×10^{-3} mole/L of ferric chloride

d) 1.43×10^{-4} mole/L of alum

e) 1.43×10^{-5} mole/L of ferric chloride

11.13 What type of instrument(s)/control system(s) would you specify to ensure that a softening plant precipitation process (i.e., the reaction chemistry) was operating correctly?

a) Flow meter with proportional controller

b) pH meter with proportional controller

c) Weighing belt feeder

d) Hardness meter with proportional controller

e) None of the above

11.14 Which of the following waters would be a candidate for air stripping prior to softening?

a) One containing 10^{-4} moles/L of CO_2

b) One containing 20.00 mg/L as $CaCO_3$ of CO_2

c) One containing 0.40 milliequivalents/L of CO_2

d) None of the above

e) All of the above

11.15 In preparation for the design of a new water treatment plant, jar tests with a slightly turbid water (4 TU) show that it is difficult to coagulate/flocculate. Which of the following design alternatives is the most appropriate?

a) Increase the safety factors in design of the sedimentation basin

b) Provide for substantial additions of polymer and coagulant aid (bentonite clay)

c) Provide for direct filtration without coagulation/flocculation

d) Use the minimum G values in design of the flocculation system

e) None of the above

11.16 The overflow rate in an ideal horizontal flow settling tank is 77.5 m/d. Under ideal conditions, approximately what percent of particles having a settling velocity of 0.0579 cm/s would be expected to be removed?

a) 100%

b) 64.6%

c) 0%

d) 1.08%

e) 0.0747%

11.17 The operator of a softening plant has stated that the sludge in the lagoons reaches 75% solids before it is excavated and removed from the lagoons. If the sludge from the clarifiers was initially at 10% solids, what percent reduction in volume occurs during dewatering?

a) 13.33%

b) 86.67%

c) 73.34%

d) 7.50%

e) 2.65%

11.18 The saturation pH of a finished water is 9.80. The actual pH is 10.2.

a) This water will precipitate $CaCO_3$ and more lime should be added.

b) This water is corrosive and lime should be added.

c) This water is corrosive and carbon dioxide should be added.

d) This water will precipitate $CaCO_3$ and carbon dioxide should be added.

e) This water is corrosive; $CaCO_3$ and more lime should be added.

11.19 Which of the following compounds does not provide free chlorine?

a) NaOCl

b) HOCl

c) $NHCl_2$

d) $CaOCl_2$

e) All of the above

11.20 In the design of chlorination facilities for a water treatment plant, the duct inlets for the air handling system should be placed:

a) Near the floor

b) Near the ceiling

c) Midway between the floor and ceiling

d) In the ceiling

e) Anywhere between the floor and ceiling

Solutions to Problems

11.1 Calculate the mmoles of alum ($Al_2(SO_4)_3 \cdot 14H_2O$):

$$\frac{40 \text{ mg/L}}{594 \times 10^3 \text{ mg/mole}} = 6.73 \times 10^{-5} \text{ moles/L}$$

This is equivalent to 0.0673 millimoles/L. From the reaction given in Eq. 11.3.3, calculate the mmoles of alkalinity consumed. Since 6 moles of HCO_3 are consumed for each mole of alum:

$$6(0.0673) = 0.4040 \text{ mmoles/L of alkalinity}$$

Convert to mg/L:

$$(0.4040 \text{ mmoles/L})(61 \text{ mg/mmole}) = 24.6 \text{ mg/L as } HCO_3 \text{ are consumed}$$

Convert the initial alkalinity to mg/L as HCO_3. The equivalent weight of $CaCO_3$ is

$$\frac{GMW}{n} = \frac{100.09}{2} = 50.04$$

The equivalent weight of HCO_3 is

$$\frac{61.02}{1} = 61.02$$

The mg/L of HCO_3 is

$$(16.39 \text{ mg/L as } CaCO_3)\left(\frac{\text{EW of Species}}{\text{EW of } CaCO_3}\right)$$

$$(16.39)\left(\frac{61.02}{50.04}\right) = 19.98 \text{ or } 20 \text{ mg/L}$$

Since 20 mg/L is less than the 24.6 mg/L required, there is not enough alkalinity.

11.2 Compute the volume of the basin:

$$V = (30.75)(24.75)(61.5) = 46{,}805 \text{ ft}^3$$

In SI units:

$$V = (46{,}805 \text{ ft}^3)(0.02832 \text{ m}^3/\text{ft}^3) = 1326 \text{ m}^3$$

The viscosity of water at 10 °C is 1.307×10^{-3} Pa·s. The power input required is

$$P = (1.307 \times 10^{-3})(35)^2(1326) = 2123 \text{ or } 2120 \text{ W}$$

For two propellers on separate shafts, the power for one shaft is:

$$0.5(2120) = 1060 \text{ W}$$

Select a $K = 1.00$ from Table 11.4. Select a D using the assumption that it is 30% of the width:

$$D = (0.30)(30.75 \text{ ft})(0.3048 \text{ m/ft}) = 2.81 \text{ m}$$

Solve for n in Eq. 11.3.9:

$$n = \left[\frac{P}{K\rho D^5}\right]^{1/3} = \left[\frac{1{,}060}{(1.00)(1000)(2.81)^5}\right]^{1/3} = 0.182 \text{ rps or } 10.9 \text{ rpm}$$

11.3 Convert mg/L as the ion to mg/L as $CaCO_3$.

Constituent	mg/L	EW	EW $CaCO_3$/EW	mg/L as $CaCO_3$	mEq/L
CO_2	42.7	22.0	2.28	97.36	1.94
Ca^{2+}	102.0	20.0	2.50	255	5.10
Mg^{2+}	45.2	12.2	4.12	186.22	3.70
Na^+	21.8	23.0	2.18		
HCO_3	420.0	61.0	0.820	344	6.88
Cl	32.0	35.5	1.41		
SO_4	65.0	48.0	1.04		

Note that Na^+, Cl and SO_4 were not converted since they do not enter into the computations.

Calculate the split:

$$x = \frac{40 - 10}{186.22 - 10} = 0.17$$

In the first stage the water is softened to the theoretical solubility limits; lime and soda must be added as shown below.

Addition = to:	Lime mg/L as $CaCO_3$	Lime mEq/L	Soda mg/L as $CaCO_3$	Soda mEq/L
CO_2	97.36	1.94		
HCO_3	344	6.88		
Mg	186.22	3.70		
(Ca + Mg)– HCO_3			97.22	1.92
	627.58	12.52	97.22	1.92

The fraction of water passing through the first stage is 1 − 0.17 = 0.83. The total hardness of the water after passing through the first stage is the theoretical solubility limit, i.e., 40 mg/L as $CaCO_3$. Since the total hardness in the raw water is 255 + 186 = 441 mg/L as $CaCO_3$, the mixture of the treated and bypass water has a hardness of

$$0.17(441) + 0.83(40) = 108 \text{ mg/L as } CaCO_3$$

This is within the acceptable range of 80 to 120 mg/L as $CaCO_3$, so no further treatment is required. The split treatment flow scheme is shown below.

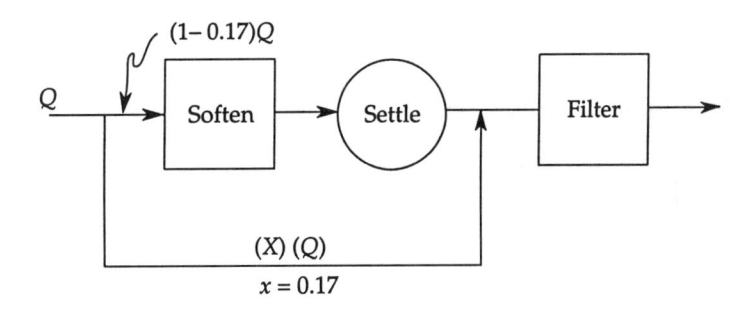

Water Treatment

[Flow diagram: Q → Soften → Settle → Filter, with (1−0.17)Q entering Soften, and bypass (X)(Q), x = 0.17]

11.4 In the first stage the water is softened to the theoretical solubility limits; lime and soda must be added as shown below.

Addition = to:	Lime mEq/L	Soda mEq/L
CO_2	0.40	
HCO_3	2.72	
Mg	1.12	
$(Ca + Mg)- HCO_3$		0.56
	4.24	0.56

Convert mEq/L of Mg to mg/L as $CaCO_3$:

$$(1.12 \text{ mEq/L})(50 \text{ mg } CaCO_3/\text{mEq}) = 56 \text{ mg/L as } CaCO_3$$

The split is

$$x = \frac{40-10}{56-10} = 0.65$$

The fraction of water passing through the first stage is $1 - 0.65 = 0.35$. The total hardness of the water after passing through the first stage is the theoretical solubility limit, 40 mg/L as $CaCO_3$. Multiply the equivalent weight of $CaCO_3$ times the mEq/L to find total hardness in the raw water:

$$(2.16)(50) + (1.12)(50) = 164 \text{ mg/L as } CaCO_3$$

The mixture of the treated and bypass water has a hardness of

$$0.65(164) + .35(40) = 120.6 \text{ mg/L as } CaCO_3$$

This is within the acceptable range of 80 to 120 mg/L as $CaCO_3$, so no further treatment is required.

The chemical dose in mg/L as CaO and Na_2CO_3 is

$$\text{Lime} = (4.24 \text{ mEq/L})(28 \text{ mg CaO/mEq}) = 118.72 \text{ or } 120 \text{ mg/L as CaO}$$

$$\text{Soda} = (0.56 \text{ mEq/L})(53 \text{ mg } Na_2CO_3/\text{mEq}) = 29.7 \text{ or } 30 \text{ mg/L as } Na_2CO_3$$

The split treatment flow scheme is shown below:

```
        (1−0.65)Q
          ↓
Q  →  [Soften]  →  (Settle)  →  [Filter]  →
      ↑_____|
              (X)(Q)
              x = 0.65
```

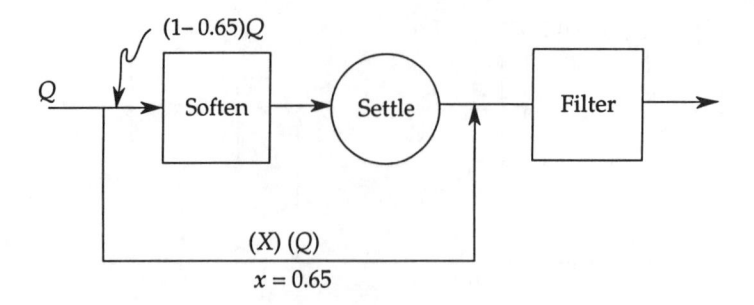

11.5 The total surface area of the tanks is

$$\frac{(0.800 \text{ m}^3/\text{s})(86\,400 \text{ s}/\text{d})}{15 \text{ m}^3/\text{d}\cdot\text{m}^2} = 4608 \text{ m}^2$$

With a L/W of 5:1, a single tank would be

$$(5x)(x) = 4608$$

$$x = (921.60)^{1/2} = 30.36 \text{ m wide}$$

and 5(30.36) = 151.8 m long. This exceeds the maximum length of 100 m. If four tanks, each 15 m × 75 m were used, the total surface area would be

$$4(15)(75) = 4500 \text{ m}^2$$

The overflow rate would be

$$\frac{(0.800 \text{ m}^3/\text{s})(86\,400 \text{ s}/\text{d})}{4500 \text{ m}^2} = 15.36 \text{ m}^3/\text{d}\cdot\text{m}^2$$

This is sufficiently close to the desired 15 m/d.

With a side water depth of 4.5 m, the volume of the four tanks is

$$(4.5)(4500) = 20\,250 \text{ m}^3$$

The detention time is

$$t_o = \frac{20\,250 \text{ m}^3}{(0.800 \text{ m}^3/\text{s})(3600 \text{ s}/\text{h})} = 7 \text{ h}$$

The length of weir in each tank is

$$\frac{(0.800 \text{ m}^3/\text{s})(86\,400 \text{ s}/\text{d})}{(4 \text{ tanks})(150 \text{ m}/\text{d})} = 115 \text{ m}$$

To provide for the maximum day flow, provide 1.5(4) = 6 tanks.

11.6 At 4.2 hours the overflow rate is

$$v_o = \left(\frac{5.0 \text{ m}}{4.2 \text{ h}}\right)(24 \text{ h}/\text{d}) = 28.57 \text{ or } 29 \text{ m}/\text{d}$$

Begin the calculation for the percent removal by plotting a vertical line from the intersection of the 70% line and the bottom of the test column (i.e., at 4.2 h and 5.0 m) and determining the mid-point between each of the isoconcentration lines. The percent removal is then calculated as

$$R_T = 70 + \frac{3.9}{5.0}(85-70) + \frac{2.5}{5.0}(90-85) + \frac{1.1}{5.0}(100-90)$$

$$= 70 + 11.7 + 2.5 + 2.2 = 86.4\%$$

11.7 The mass of sludge generated each day is

$$M_s = (86.40)(0.10)(135.0 + (0.58)(29.9) + 237.5) = 3368 \text{ kg/d}$$

The specific gravity of the sludge is

$$S_{sl} = \frac{2.75}{0.10 + (2.75)(0.90)} = 1.07$$

The volume of the sludge is

$$V_{sl} = \frac{3368}{(1000)(1.07)(0.10)} = 31.48 \text{ m}^3/\text{d}$$

The approximate volume after dewatering is

$$V_2 = 31.48\left(\frac{0.10}{0.60}\right) = 5.25 \text{ or } 5.3 \text{ m}^3/\text{d}$$

On an annual basis the volume is

$$\mathcal{V} = (5.25 \text{ m}^3/\text{d})(365 \text{ d/y}) = 1916 \text{ or } 1900 \text{ m}^3/\text{y}$$

11.8 A) The Head Loss Through a Clean Bed:

To compute the head loss, the last row in the table must be completed. Note that English units are used.

The Reynolds number is required to calculate the drag coefficient. Using the temperature of the water, the kinematic viscosity is found to be $1.408 \times 10^{-5} \text{ ft}^2/\text{s}$. The velocity is found from the filtration rate:

$$v_a = (2.5 \text{ gal/min} \cdot \text{ft}^2)(0.1337 \text{ ft}^3/\text{gal})\left(\frac{1}{60 \text{ s/min}}\right) = 0.00557 \text{ ft/s}$$

For a sand particle with a sphericity of 0.95 and a diameter of 0.000583 ft, the Reynolds number is

$$R = \frac{(0.95)(0.000583)(0.00557)}{1.408 \times 10^{-5}} = 0.219$$

Because the Reynolds number is less than 0.5, calculate the drag coefficient as

$$C_D = \frac{24}{0.219} = 109.6$$

The product $C_D(f)/d$ for the last row is

$$\frac{(109.6)(0.0166)}{0.000583} = 3120$$

The sum of the last column is 45 640. The head loss is

$$h_L = \frac{(1.067)(2.0)(0.00557)^2}{(0.95)(32.2)(0.40)^4}(45\,640) = 3.86 \text{ ft}$$

This head loss exceeds 0.6 m (2 ft) and would be considered excessive.

B) Height of Backwash Troughs:

To compute the depth of the expanded bed, the last row in the table must be completed. Note that English units are used.

Estimate the settling velocity of the sand particle as 1.4 cm/s using Fig. 11.7 with a diameter of 0.178 cm and a specific gravity of 2.5.

The Reynolds number is calculated as

$$R = \frac{(0.95)(0.000583)(1.4 \text{ cm})\left(\frac{1}{30.5 \text{ cm/ft}}\right)}{1.408 \times 10^{-5}} = 1.806$$

Because the Reynolds number is in the transition region, the drag coefficient is estimated as

$$C_D = \frac{24}{1.806} + \frac{3}{\sqrt{1.806}} + 0.34 = 15.86$$

The settling velocity is calculated as

$$v_s = \left[\frac{4(32.2)(2.65-1)(0.000583)}{3(15.86)}\right]^{1/2} = 0.0510 \text{ ft/s}$$

The velocity of the backwash is:

$$v_b = (20 \text{ gpm/ft}^2)(0.1337 \text{ ft}^3/\text{gal})\left(\frac{1}{60 \text{ s/min}}\right) = 0.0446 \text{ ft/s}$$

Since v_b is less than settling velocity of the smallest sand grains, filter media will not be lost and the backwash rate is acceptable.

The porosity of the expanded bed is

$$\varepsilon_e = \left(\frac{0.0446}{0.0510}\right)^{0.22} = 0.970$$

The fraction $(f/(1-\varepsilon_e))$ for the last row is

$$\frac{0.0166}{1-0.970} = 0.5533$$

Summing the last column, the depth of the expanded bed is

$$D_e = (1-0.40)(2.0)(2.45) = 2.94 \text{ ft}$$

The bottom of the backwash trough must be

$$2.94 - 2 + 0.5 = 1.44 \text{ ft}$$

or about 1.5 ft above the surface of the sand bed.

C) Smallest Size Anthracite Coal Particle:

Using Eq. 11.6.1 and the specific gravities of the coal and sand, the diameter of the smallest coal particle must be such that it settles on top of the smallest sand particle (0.000583 ft):

$$d_A = (0.000583)\left[\frac{2.65-1.00}{1.55-1.00}\right]^{2/3} = 0.001213 \text{ ft}$$

11.9   From Table 11.10 at a temperature of 10 °C and TDS of 80 mg/L find

$(pK_2 - pK_s) = 2.40$

Convert Ca to moles/L:

$$(25.00 \text{ mg/L as CaCO}_3)\left(\frac{20}{50}\right) = 10.00 \text{ mg/L as Ca}$$

$$(10.00 \text{ mg/L})\left(\frac{1}{(40 \times 10^3 \text{ mg/mole})}\right) = 2.50 \times 10^{-4} \text{ moles/L}$$

Calculate pCa:

$$pCa = -\log(2.5 \times 10^{-4}) = 3.60$$

Convert alkalinity to moles/L assuming all alkalinity is CO_3 since the pH is > 8.3:

$$(40 \text{ mg/L as CaCO}_3)\left(\frac{30}{50}\right) = 24.0 \text{ mg/L as CaCO}_3$$

$$(24.0 \text{ mg/L})\left(\frac{1}{60 \times 10^3 \text{ mg/mole}}\right) = 4.00 \times 10^{-4} \text{ moles/L}$$

Calculate $pAlk$:

$$pAlk = -\log(4.00 \times 10^{-4}) = 3.40$$

The stability index is

$$SI = 8.60 - 2.40 - 3.60 - 3.40 = -0.80$$

Therefore the water is corrosive. Add lime.

11.10   The available chlorine from chlorine gas is 100%. The cost of available chlorine is then $300 per ton.

For chlorine dioxide, using the half reaction from Table 11.11, the equivalent weight is

$$EW = \frac{67.5}{1} = 67.5$$

The percent available chlorine is

$$\% \text{ Available Chlorine} = \frac{35.5}{67.5}(100\%) = 52.6\%$$

The cost of available chlorine is then

$$\frac{\$600/\text{ton}}{0.526} = \$1140/\text{ton of available chlorine}$$

This is obviously much more expensive.

Other considerations are the fact that chlorine dioxide does not form THMs and that its reaction byproducts include chlorite and chlorate. While the elimination of THMs is desirable to reduce cancer risk, chlorite and chlorate compounds may have a human health risk. Chlorine dioxide also has the potential of generating taste and odor problems.

11.11 a) Using the national per capita water consumption of 628 Lpcd, the estimated population is

$$\frac{(2200 \times 10^6 \text{ gal/d})(3.785 \text{ L/gal})}{628 \text{ L/capita} \cdot \text{d}} = 13\ 259\ 554 \text{ people}$$

11.12 d) The Shultz-Hardy rule states that monovalent, divalent and trivalent species should be effective approximately in the ratio $1:10^{-2}:10^{-3}$. In most practical systems, the Shultz-Hardy rule is "violated" because the electrolytes are not indifferent. Inorganic compounds added to water react with solution electrolytes and form complexes and precipitates. Hence, coagulation power for monovalent, divalent and trivalent species is taken as 1:60:700. The ratio for alum or ferric chloride would be

$$\frac{0.10 \text{ mole/L}}{700} = 1.4 \times 10^{-4} \text{ mole/L}$$

11.13 b) Significant removal of $CaCO_3$ is achieved when the pH is above about 9.6 to 10.8. Significant removal of $Mg(OH)_2$ is achieved when the pH is above about 10.8 to 11.5. Because the water quality (hardness) from the source may vary with time, controlling neither the flow rate nor the feed rate precisely guarantees that the desired pH will be achieved. Only pH control will ensure the softening precipitation process operates correctly. There is no such thing as a hardness meter.

11.14 d) Convert each of the units to mg/L as CO_2:

$$(10^{-4} \text{ mole/L})(44 \times 10^3 \text{ mg/mole}) = 4.40 \text{ mg/L}$$

$$(20 \text{ mg/L as } CaCO_3)(22/50) = 8.80 \text{ mg/L}$$

$$(0.40 \text{ mEq/L})(22 \text{ mg/mEq}) = 8.80 \text{ mg/L}$$

Since 10 mg/L is the guideline for air stripping, none of these waters is a candidate for air stripping.

11.15 c) Increasing the safety factor for the sedimentation tank will not improve performance if the water cannot be coagulated/flocculated. Likewise, minimum G values will have little effect because the particles are too dispersed and are already having trouble coming into contact. Additions of polymer and coagulant aid may improve performance but sludge production will increase and cost will be high. The most appropriate solution is direct filtration. This saves the cost of building coagulation/flocculation tanks and the cost of chemicals.

11.16 b) Calculate the settling velocity in m/d:

$$(0.0579 \text{ cm/s})\left(\frac{1}{100 \text{ cm/m}}\right)(86\ 400 \text{ s/d}) = 50.03 \text{ m/d}$$

Calculate the percent removal using Eq. 11.5.10:

$$\frac{50.03}{77.5}(100\%) = 64.55 \text{ or } 64.6\%$$

11.17 b) Percent reduction may be calculated a

$$\left(\frac{v_1 - v_2}{v_1}\right)(100\%) = \left(1 - \frac{v_2}{v_1}\right)(100\%)$$

The ratio of volumes based on Eq. 11.5.15 is

$$\frac{v_2}{v_1} = \frac{P_1}{P_2} = \frac{0.10}{0.75} = 0.1333$$

The percent reduction is:

$$\% \text{ Reduction} = (1 - 0.1333)(100\%) = 86.67\%$$

11.18 d) The saturation index is calculated as

$$SI = pH - pH_s = 10.2 - 9.80 = +0.40$$

This water will precipitate $CaCO_3$ but lime will only raise the pH further and continue to precipitate $CaCO_3$. The pH of the water must be lowered to stop the precipitation. Recarbonation with carbon dioxide will lower the pH.

11.19 c) Free chlorine is defined as HOCl and OCl. Only dichloramine ($NHCl_2$) will not provide free chlorine.

11.20 a) Because chlorine gas is more dense than air the duct inlets should be placed near the floor.

12. Wastewater Treatment

by Mackenzie L. Davis

Selection of conventional wastewater treatment processes for municipalities are reviewed in this chapter. Regulatory requirements of the Clean Water Act, although an important consideration in the selection of water sources and treatment process, are not covered here. It is strongly suggested that your preparation for the examination include a review of current Federal and State regulatory requirements. The Federal requirements may be found in the Code of Federal Regulations.

12.1 Water Quality Management

The design of a wastewater treatment facility is, fundamentally, a function of the degree of protection required to ensure that the receiving body water quality is not degraded to unacceptable levels. Although a wide range of pollutants may be discharged to a municipal sewer system, the primary pollutants of concern are oxygen demanding materials, suspended solids, nutrients and pathogens.

Although regulatory requirements are established on a case by case basis, for the purpose of this discussion secondary treatment is taken as the minimum desirable treatment. The Environmental Protection Agency defines secondary treatment as providing treatment such that the discharge contains a BOD_5 less than or equal to 30 mg/L and suspended solids less than or equal to 30 mg/L. During periods when water contact may be expected, secondary treatment may also require that disinfection be provided. Because disinfection processes may result in discharge of compounds that destroy pathogen predators and other biota that contribute to the health of the receiving body, this requirement is seasonal.

Nitrogen in the form of ammonia and nitrates may also require treatment. Ammonia creates an oxygen demand by the reaction

$$NH_4 + 2\,O_2 \leftrightarrow NO_3 + H_2O + 2\,H \qquad (12.1.1)$$

From this reaction, theoretically, 4.57 g of oxygen is required per gram of nitrogen oxidized. In addition, nitrogen serves as a source of nutrients that stimulate algal bloom. Phosphorus removal may be required because it also contributes to the growth of algae. When the algae die they become a source of oxygen demand.

12.1.1 Biochemical Oxygen Demand

Biochemical oxygen demand (BOD) is the most common method for expressing the measurement of oxygen demand. Because the measurement technique is a biochemical assay, the kinetics of the reaction are an important consideration in assessing data for design. The exertion of oxygen demand by microorganisms may be described by the equation

$$\text{BOD}_t = L(1 - e^{-kt}) \qquad (12.1.2)$$

where
- BOD_t = BOD at time t, mg/L
- L = oxygen equivalent of organics at time = 0, mg/L
- k = reaction rate constant, d^{-1} (d = day)

L is often referred to as the ultimate BOD.

The rate constant is a function of the nature of the waste, the ability of the organisms in the system to degrade the waste, and the temperature. Typical values for the rate constant are shown in Table 12.1.

Table 12.1 Typical values for the BOD rate constant.

Waste	k at 20 °C, d^{-1}
Raw sewage	0.35 – 0.70
Well-treated sewage	0.12 – 0.23

The rate constant may be expressed as a function of temperature:

$$k_T = k_{20} \theta^{T-20} \qquad (12.1.3)$$

where
- T = temperature of interest, °C
- k_T = BOD rate constant at temperature of interest, d^{-1}
- k_{20} = BOD rate constant determined at 20°C, d^{-1}
- θ = temperature coefficient

The temperature coefficient has a value of 1.135 for temperatures between 4° and 20°C and 1.056 for temperatures between 20° and 30°C.

12.1.2 DO Sag Curve

The classical Streeter-Phelps model for prediction of the dissolved oxygen in a stream is used in many forms as a basis for establishing allowable BOD discharge levels. The fundamental form of the equation is

$$D = \frac{k_d L_a}{k_r - k_d}\left(e^{-k_d t} - e^{-k_r t}\right) + D_a\left(e^{-k_r t}\right) \qquad (12.1.4)$$

where
- D = oxygen deficit after exertion of BOD for time t, mg/L
- L_a = initial ultimate BOD after river and wastewater have mixed, mg/L
- k_d = deoxygenation rate, d^{-1}
- k_r = reaeration rate, d^{-1}
- t = time of travel of wastewater discharge downstream, d
- D_a = initial deficit after river and wastewater have mixed, mg/L

Other terms may be added to this equation to account for nitrogenous oxygen demand, algal respiration, benthic demand, and algal photosynthetic oxygen production. The terms k_d and k_r may be corrected for temperature using Eq. 12.1.3. The temperature coefficient for k_d is the same as noted below Eq. 12.1.3. For k_r the coefficient is 1.024 for all temperatures.

The dissolved oxygen concentration is found by subtracting the oxygen deficit from the saturation concentration:

$$D = \text{DO}_s - \text{DO} \qquad (12.1.5)$$

The saturation value is a function of water temperature. Typical values may be found in Table 12.2.

The initial ultimate BOD after the river and wastewater have mixed may be taken to be the weighted average of the ultimate BOD of the wastewater and the ultimate BOD of the river:

$$L_a = \frac{Q_w L_w + Q_r L_r}{Q_w + Q_r} \quad (12.1.6)$$

where

Q_w = flow rate of wastewater, m³/s
L_w = ultimate BOD of wastewater, mg/L
Q_r = flow rate of river, m³/s
L_r = ultimate BOD of river, mg/L

In a similar fashion, the initial deficit after mixing is:

$$D_a = \text{DO}_s - \frac{Q_w \text{DO}_w + Q_r \text{DO}_r}{Q_w + Q_r} \quad (12.1.7)$$

The lowest point on the DO sag curve, called the critical point, occurs at the critical time defined by

$$t_c = \frac{1}{k_r - k_d} \ln\left[\frac{k_r}{k_d}\left(1 - D_a \frac{k_r - k_d}{k_d L_a}\right)\right]$$

(12.1.8)

The critical deficit D_c is calculated using $t = t_c$ in Eq. 12.1.4.

Table 12.2 Typical values for the BOD rate constant.

Saturation values of dissolved oxygen in fresh water exposed to a saturated atmosphere containing 20.9% oxygen under a pressure of 101.325 kPa[a]

Temperature (°C)	Dissolved oxygen (mg/L)	Saturated vapor pressure (kPa)
0	14.62	0.6108
1	14.23	0.6566
2	13.84	0.7055
3	13.48	0.7575
4	13.13	0.8129
5	12.80	0.8719
6	12.48	0.9347
7	12.17	1.0013
8	11.87	1.0722
9	11.59	1.1474
10	11.33	1.2272
11	11.08	1.3119
12	10.83	1.4017
13	10.60	1.4969
14	10.37	1.5977
15	10.15	1.7044
16	9.95	1.8173
17	9.74	1.9367
18	9.54	2.0630
19	9.35	2.1964
20	9.17	2.3373
21	8.99	2.4861
22	8.83	2.6430
23	8.68	2.8086
24	8.53	2.9831
25	8.38	3.1671
26	8.22	3.3608
27	8.07	3.5649
28	7.92	3.7796
29	7.77	4.0055
30	7.63	4.2430
31	7.51	4.4927
32	7.42	4.7551
33	7.28	5.0307
34	7.17	5.3200
35	7.07	5.6236
36	6.96	5.9422
37	6.86	6.2762
38	6.75	6.6264

[a] For other barometric pressures, the solubilities vary approximately in proportion to the ratios of these pressures to the standard pressures.

Source: Calculated by G. C. Whipple and M. C. Whipple from measurements of C. J. J. Fox, Journal of the American Chemical Society, vol. 33, p. 362, 1911.

---- EXAMPLE 12.1 ----

Determine whether or not the wastewater shown below can be discharged into the river described. The saturation value for dissolved oxygen is 8.38 mg/L at 25.0°C.

Wastewater Discharge Characteristics

Flow	0.0500 m³/s
Ultimate BOD	129.60 kg/d
DO	0.900 mg/L
Temperature	25.0°C

River Characteristics

Flow	0.500 m³/s
Ultimate BOD	19.00 mg/L
DO	5.85 mg/L
Temperature	25.0°C
Speed	0.100 m/s
k_d at 20°C	0.1201 d⁻¹
k_r at 20°C	0.1548 d⁻¹

Solution. Since the ultimate BOD of the discharge is given on a mass basis this must be converted to a concentration using the following relationship:

$$\frac{(kg/d)(1\times 10^6 \text{ mg/kg})}{(Q \text{ m}^3/s)(86\,400 \text{ s/d})(10^3 \text{ L/m}^3)}$$

The conversion is then

$$\frac{(129.60)(10^6)}{(0.0500)(86\,400)(10^3)} = 30.00 \text{ mg/L}$$

This is L_w. The mixed ultimate BOD is

$$L_a = \frac{(0.0500)(30.00)+(0.500)(19.00)}{0.0500+0.500} = 20.00 \text{ mg/L}$$

The DO saturation at 25°C is 8.38 mg/L. The initial deficit is

$$D_a = 8.38 - \frac{(0.0500)(0.900)+(0.500)(5.85)}{0.0500+0.500} = 2.98 \text{ mg/L}$$

Since the deoxygenation and reaeration coefficients are at 20°C, they must be converted to the stream temperature:

$$k_d = (0.1201)(1.056)^{25-20} = 0.1577 \text{ d}^{-1}$$

$$k_r = (0.1548)(1.024)^{25-20} = 0.1743 \text{ d}^{-1}$$

The critical time is

$$t_c = \frac{1}{0.1743-0.1577}\ln\left\{\frac{0.1743}{0.1577}\left[1-2.98\frac{0.1743-0.1577}{(0.1577)(20.00)}\right]\right\}$$

$$= 60.24\ln\{1.0879\} = 5.08 \text{ d}$$

The critical deficit is

$$D_c = \frac{(0.1577)(20.00)}{0.1743 - 0.1577}\left[e^{-(0.1577)(5.08)} - e^{-(0.1743)(5.08)}\right] + (2.98)e^{-(0.1743)(5.08)}$$

$$= (190)[0.4488 - 0.4125] + 1.229 = 8.13 \text{ mg/L}$$

The DO is determined by subtracting the deficit from the saturation value:

$$DO = 8.38 - 8.13 = 0.25$$

or about 0.3 mg/L. Obviously, this is not satisfactory.

12.2 Pretreatment

12.2.1 Bar Racks

Bar racks function to protect pumps, valves and other mechanical equipment in the plant. Current practice is to design the racks for mechanical cleaning. They may be placed either upstream or downstream of grit chambers but, generally, they are placed ahead of grit chambers. A minimum of two are provided for redundancy. The space between the bars may be in the range 15 to 150 mm with a nominal opening of 25 mm being common. Each bar is 25 to 40 mm deep and 5 to 15 mm wide. An approach velocity of at least 0.5 m/s is recommended to keep solids in suspension. The maximum velocity should not exceed 1.0 m/s at peak flow to prevent debris from passing through the rack. The operating cycle of the cleaning mechanism should maintain the head loss below 15 cm.

Bar rack screenings may range from 0.004 to 0.04 m^3/m^3 of sewage. They are quite obnoxious. They should be collected directly from the bar rack to a water tight container. The entire facility should be enclosed. Screenings may be disposed of in a sanitary landfill or incinerated.

12.2.2 Grit Chambers

Of the three types of grit chambers (horizontal flow gravity, aerated, and vortex or cyclone) only the aerated grit chamber is discussed here.

The design rules-of-thumb for aerated grit chambers include detention times of 2 to 5 minutes at peak hourly flow and nominal air flows of 0.15 to 0.45 m^3/min per meter of tank length. A variable air flow system is preferred to allow the operator to adjust the system to local conditions. Chambers may be square or rectangular. Rectangular chambers have length to width ratios ranging from 2.5:1 to 5:1 with depths on the order of 2 to 5 m. A grit hopper about 1 m deep is placed under the diffusers. The diffusers are placed 0.45 to 0.6 m above the plane of the bottom of the tank. As with bar racks, a minimum of two are provided in municipal systems.

Grit removal may be by clam buckets, screw conveyors, jet pumps or air lifts. Because of the abrasive nature of the grit, mechanical equipment will be subject to very high rates of wear and should be selected with this in mind. Grit quantities vary over a wide range (0.004 to 0.2 m^3/m^3 of sewage) depending on whether or not the sewer system is combined or separated. Generally, grit is disposed of in sanitary landfills.

EXAMPLE 12.2

Design a rectangular aerated grit chamber facility for a wastewater treatment plant with an average daily flow of 0.0789 m³/s and a peak hourly flow of 0.2067 m³/s. Assume a 3 minute detention time at peak flow, a depth of 3 m and a length to width ratio of 2.5:1.

Solution. With a 3 minute detention time at peak flow the volume is

$$V = (0.2067 \text{ m}^3/\text{s})(60 \text{ s/min})(3 \text{ min}) = 37.2 \text{ m}^3$$

With a depth of 3 m, the surface area is

$$A_S = \frac{37.2 \text{ m}^3}{3 \text{ m}} = 12.4 \text{ m}^2$$

A rectangular chamber will have a width w of

$$(2.5w)(w) = 12.4 \text{ m}^2$$

$$\therefore w = 2.2 \text{ m}$$

and a length of $(2.5)(2.2) = 5.5$ m.

The air supply should be variable between

$$(0.15 \text{ m}^3/\text{min} \cdot \text{m})(5.5 \text{ m}) = 0.8 \text{ m}^3/\text{min}$$

and

$$(0.45)(5.5) = 2.5 \text{ m}^3/\text{min}$$

Two such chambers should be provided to allow for one out of service.

12.2.3 Equalization

Equalization tanks to dampen flow variations may be either in-line or off-line. In-line systems that provide equalization for diurnal flow variations are based on the fundamental mass balance equation:

$$\frac{dS}{dt} = Q_{in} - Q_{out} \tag{12.2.1}$$

where dS/dt = change in storage per unit of time, m³/s
  Q_{in} = flow rate into the basin, m³/s
  Q_{out} = flow rate out of the basin, m³/s

For design purposes the flow rate out of the basin is set to the average flow and is rewritten as:

$$\Delta S = Q_{in}(\Delta t) - Q_{avg}(\Delta t) = (Q_{in} - Q_{avg})\Delta t \tag{12.2.2}$$

The solution may be found either numerically or graphically. The numerical technique is demonstrated in Example 12.3.

In practice, the volume provided is 10 to 25 percent greater than that calculated to allow for contingency, and to allow for continuous operation of mixing equipment at low flow. Mixing must be provided to keep suspended material from settling. Rule-of-thumb suggestions are for a power input in the range of 0.004 to 0.008 kW/m³ of storage. In addition, air should be supplied at a rate of 0.01 to 0.015 m³/min · m³ of storage. Mechanical aerators are often used to serve both functions. If floating aerators are used, provision must be made to protect the units from "grounding" at low flow and when the basin is drained for maintenance.

EXAMPLE 12.3

Design an equalization basin for an average flow rate of 0.0789 m³/s for flow variation shown in the table below. Determine the volume of the equalization basin with a conservative factor of safety, minimum power of the mixing equipment and minimum air required.

Time, h	% of Avg. Flow	Time, h	% of Avg. Flow
00-01	80	12-13	112
01-02	65	13-14	111
02-03	55	14-15	108
03-04	49	15-16	109
04-05	49	16-17	114
05-06	54	17-18	120
06-07	74	18-19	126
07-08	105	19-20	128
08-09	131	20-21	124
09-10	130	21-22	114
10-11	122	22-23	107
11-12	115	23-00	98

Solution. For computational purposes it is convenient to rearrange the data such that the first entry is the first flow rate greater than the average after the sequence of low flows (night time in this instance) as shown in the table below. With this arrangement, the last row of the computation should result in a storage value of zero. The tabular solution is shown below. Calculations are explained after the table.

Time, h	Fraction of Avg. Flow	Volume in, m³	Volume out, m³	Storage m³	Cum. storage m³
07-08	1.05	298	284	14	14
08-09	1.31	372	284	88	102
09-10	1.30	369	284	85	187
10-11	1.22	347	284	62	250
11-12	1.15	327	284	43	293
12-13	1.12	318	284	34	327
13-14	1.11	315	284	31	358
14-15	1.08	307	284	23	381
15-16	1.09	310	284	26	406
16-17	1.14	324	284	40	446
17-18	1.20	341	284	57	503
18-19	1.26	358	284	74	577
19-20	1.28	364	284	80	656
20-21	1.24	352	284	68	724
21-22	1.14	324	284	40	764
22-23	1.07	304	284	20	784
23-00	0.98	278	284	-6	778
00-01	0.80	227	284	-57	721
01-02	0.65	185	284	-99	622
02-03	0.55	156	284	-128	494
03-04	0.49	139	284	-145	349
04-05	0.49	139	284	-145	204
05-06	0.54	153	284	-131	74
06-07	0.74	210	284	-74	0

The third column is the product of the flow rate and the selected time increment of one hour. For the first row,

$$Q_{in}(\Delta t) = (1.05)(0.0789 \text{ m}^3/\text{s})(3600 \text{ s/h}) = 298 \text{ m}^3$$

The flow out in each time increment is the average flow:

$$Q_{in}(\Delta t) = (0.0789 \text{ m}^3/\text{s})(3600 \text{ s/h}) = 284 \text{ m}^3$$

The fifth column is the difference:

$$Q_{in}(\Delta t) - Q_{avg}(\Delta t) = 14$$

The sixth column is simply the cumulative sum of the values in the fifth column. The highest value in column 6, 784 m³, is selected as the theoretical value.

Using a 25% excess "safety factor," the design volume of the equalization basin is
$$V = (1.25)(784) = 980 \text{ m}^3$$

The minimum power required is
$$P = (0.004 \text{ kW/m}^3)(980 \text{ m}^3) = 4 \text{ kW}$$

The minimum air supplied should be
$$Q_{air} = (0.01 \text{ m}^3/\text{min} \cdot \text{m}^3)(980) = 9.80 \text{ or } 10 \text{ m}^3/\text{min}$$

12.3 Primary Treatment

12.3.1 Sedimentation Physics

As discussed in chapter 11, settling in a primary sedimentation tank may be described as Type II settling. The procedure for design using column test data is the same as that described in chapter 11.

12.3.2 Primary Settling Tank Design

As in all settling tank design, the primary design parameter is overflow rate. The overflow rate and the design flow rate determine the surface area of the clarifier (see Eq. 11.5.9). In the absence of laboratory test data, the values shown in Table 12.3 may be used to estimate overflow rates. Detention times range from 1.5 to 2.5 hours with a typical time of 2 hours. Standard practice is to provide no less than two sedimentation basins. Although either rectangular or circular tanks may be used, rectangular tanks are preferred because of economy of space. Side water depths (SWD) may be 3 to 4.5 m with typical depths of 3.5 m. Widths range from 3 to 24 m with lengths of 15 to 90 m. Typical widths are in the range of 5 to 10 m with lengths of 25 to 40 m. The Great Lakes-Upper Mississippi River Board of State Sanitary Engineers (GLUMRB) recommends that weir loading rates not exceed 120 m³/d·m of weir length for plants with average flows less than 0.04 m³/s. For larger flows, the recommended maximum rate is 190 m³/d·m.

Table 12.3 Typical primary sedimentation tank overflow rates.

Treatment	Range of Overflow Rates, m³/d·m²	Typical Overflow Rate, m³/d·m²
Primary followed by secondary		
Average flow	32.6 – 48.8	
Peak hourly	81.4 – 122	101
Primary with waste activated sludge return		
Average flow	24.4 – 32.6	
Peak flow	48.8 – 69.2	61

Source: Metcalf & Eddy, *Wastewater Engineering: Treatment, Disposal and Reuse*, McGraw-Hill, Inc., New York, NY, 1991, p. 475.

Scum collection is provided in rectangular tanks by arranging the sludge scraping system such that the return flights pass over the top of the tank. In circular tanks a separate skimming arm is provided. The scum is collected by means of a surface boom or baffle. It may be drawn off by a variety of techniques including a slotted pipe or wiper blade. It is generally disposed of with the sludges produced in the plant.

Empirical relationships such as those shown in Figs. 12.1 and 12.2 may be used to estimate treatment efficiency in the absence of test column data.

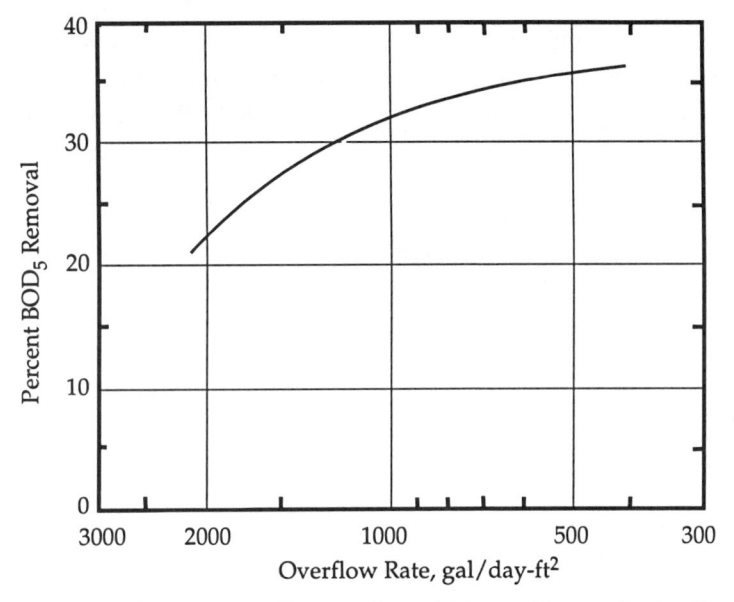

Figure 12.1 Percent BOD_5 removal versus overflow rate for primary settling tanks treating municipal wastewater. Source: Great Lakes-Upper Mississippi River Board of State Sanitary Engineers, *Recommended Standards, for Sewage Works*, 1978.

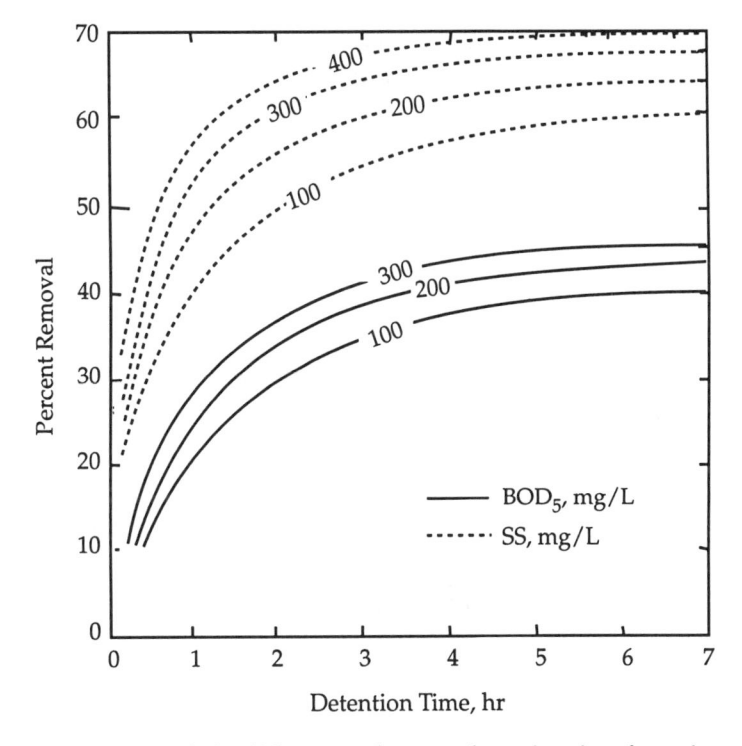

Figure 12.2 Percent BOD_5 and suspended solids removal versus detention time for primary settling tanks, treating municipal wastewater. Source: E. W. Steel, *Water Supply and Sewage*, 4th edition, McGraw-Hill Book Co., Inc., New York, NY, 1960.

EXAMPLE 12.4

Design the sedimentation tank facilities for a wastewater that has a BOD_5 of 135 mg/L, a suspended solids concentration of 200 mg/L, and a flow rate of 0.0789 m³/s. Waste activated sludge is not to be returned to the primary tank. Assume an overflow rate of 40 m³/d · m² and a width to depth ratio of 2:1. Determine the overflow rate, detention time, BOD_5 and suspended solids removal efficiency, dimensions of the tank and length of weir.

Solution. Using Fig. 12.1 and the overflow rate $[(40 \text{ m}^3/\text{d} \cdot \text{m}^2)(24 \cdot 545) = 982 \text{ gal}/\text{d} \cdot \text{ft}^2]$, find the BOD removal efficiency is approximately 33%. Plot a line from 33% removal efficiency to a BOD_5 of 135 mg/L in Fig. 12.2. Read down from this point to find a detention time of approximately 2.2 h. Read up from this point to a suspended solids concentration of 200 mg/L to find a suspended solids removal efficiency of approximately 57%.

The settling tank volume is

$$V = (0.0789 \text{ m}^3/\text{s})(2.2 \text{ h})(3600 \text{ s/h}) = 624.89 \text{ m}^3$$

The surface area of the tank is

$$A_s = \frac{Q}{v_o} = \frac{(0.0789 \text{ m}^3/\text{s})(86\,400 \text{ s/d})}{40 \frac{\text{m}^3}{\text{d} \cdot \text{m}^2}} = 170.42 \text{ m}^2 \, S = (Q_{in} - Q_{avg})\Delta t$$

The depth of the tank is

$$h = \frac{V}{A_s} = \frac{624.89}{170.42} = 3.67 \text{ m}$$

This is within the typical range of SWDs. With a width to depth ratio of 2:1, the width is

$$w = 2(3.67 \text{ m}) = 7.34 \text{ m}$$

The length is

$$L = \frac{A_s}{w} = \frac{170.42 \text{ m}^2}{7.34 \text{ m}} = 23.21 \text{ m}$$

Final dimensions would then be 3.7 m deep by 7.3 m wide by 23 m long. With a weir loading rate of 190 m³/d · m, the weir length is

$$L_W = \frac{(0.0789)(86\,400)}{190} = 35.88 \text{ m}$$

12.4 Secondary Treatment

12.4.1 Activated Sludge

Of the fifteen possible activated sludge processes and process modifications, only the conventional plug flow and completely mixed flow processes will be discussed in detail here. For further discussion of the variations see *Wastewater Engineering*, by Metcalf & Eddy, Inc., McGraw-Hill, 1991.

12.4.1.1 Operation of Activated Sludge Units

For the biochemical processes of waste degradation to operate, the microorganisms must take the waste material through the cell wall. Passage is limited to soluble material. Waste material that is not soluble may sorbed to the surface of the biological floc and be solubilized by enzyme activity of the microorganism, or simple hydrolysis reactions may be sufficient to make it available to the organism. Once the waste is incorporated into the cell, it may be oxidized to carbon dioxide and water or it may be used to synthesize new cell material.

The control of the fraction of material that is either oxidized or synthesized is regulated by adjusting the food to microorganism ratio (F/M):

$$\frac{F}{M} = \frac{QS_o}{VX} = \frac{\text{mg BOD}_5/\text{d}}{\text{mg MLVSS}} = d^{-1} \tag{12.4.1}$$

where
- Q = flow rate into the plant
- S_o = influent soluble BOD$_5$ per day
- V = volume of aeration tank
- X = mixed liquor volatile suspended solids (MLVSS)

Since the influent flow rate, soluble BOD and volume (V) of the tank cannot be regulated, the quantity of biomass X becomes the only operational parameter that may be used to regulate the process. The biomass quantity is controlled by removing or "wasting" a fraction of the biomass from the system. Low wasting results in a low F/M ratio and high oxidation. Thus, power requirements for oxygen supply are increased as a trade-off for a low rate of sludge generation. Conversely, a high wastage will result in high synthesis and a high rate of sludge production with a decreased oxygen requirement.

In the conventional plug-flow reactor, the oxygen demand is greatest at the head end of the tank because of the large amount of available food and "starved" nature of the microorganisms. This operational difficulty can be overcome by altering the design of the process to provide more air at the head end of the tank. This process modification is called tapered aeration.

Although it is more common with industrial wastes than municipal wastes, nutrients may have to be provided. Rule-of-thumb requirements are for nitrogen to be available in proportion to the BOD$_5$ in the ratio 1:32 (N:BOD$_5$). For phosphorous, the ratio is 1:150 (P:BOD$_5$).

Sludge volume index (SVI) is used as a measure of the performance of the secondary settling tanks. It is defined as

$$\text{SVI} = \frac{SV}{\text{MLSS}} \times 1000 \, \frac{\text{mg}}{\text{g}} \tag{12.4.2}$$

The units of SVI are mL of sludge/g of solids. The mixed liquor suspended solids (MLSS) includes both the volatile and non-volatile solids. Values of the SVI less than 100 generally indicate adequate performance. The SVI can be misleading if the MLSS being carried in aeration tanks is high.

12.4.1.2 Design of Plug Flow Reactors

A schematic of a plug flow reactor with recycle is shown in Fig. 12.3. The symbols on the schematic are used in defining variables in the model presented below. If one assumes the concentration of microorganisms in the influent to the reactor is approximately the same as that in the effluent (generally, this assumption applies only if the ratio of mean cell residence time to hydraulic residence time is less than 5), and that the rate of substrate utilization is described by the expression

$$r_{su} = \frac{\left(\frac{\mu_m}{Y}\right)SX_m}{K_s + S} \tag{12.4.3}$$

where
- r_{su} = rate of substrate utilization
- μ_m = maximum specific growth rate, d^{-1}
- Y = maximum yield coefficient, mg cells/mg substrate

S = concentration of growth-limiting substrate in solution, mg/L
X_m = average concentration of microorganisms in reactor, mg/L
K_s = half-velocity constant, mg substrate/L

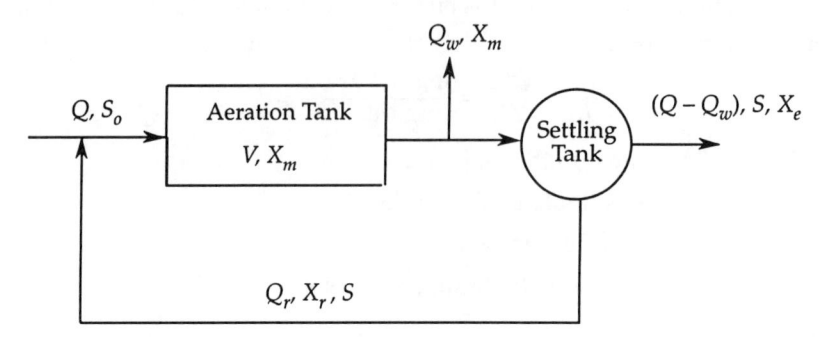

Figure 12.3 Plug-flow reactor with recycle.

Then, a kinetic model of the process yields the following:

$$\frac{1}{\theta_c} = \frac{\mu_m(S_o - S)}{(S_o - S) + (1+\alpha)K_s \ln\left(\frac{S_i}{S}\right)} - k_d \qquad (12.4.4)$$

where
S_o = influent concentration, mg/L
S = effluent concentration, mg/L
k_d = decay rate of microorganisms, d^{-1}
S_i = influent concentration after dilution with recycle flow

and

$$S_i = \frac{S_o + \alpha S}{1 + \alpha} \qquad (12.4.5)$$

where
α = recycle ratio (Q_r/Q)
Q_r = return sludge flow rate, m³/s
Q = flow rate into aeration tank, m³/s

If the suspended solids X_e in the influent and effluent are neglected and wastage is from the aeration tank, the return sludge flow rate is

$$Q_r = \frac{X'_m Q - X'_r Q_w}{X'_r - X'_m} \qquad (12.4.6)$$

where
X'_m = mean MLSS concentration in aeration tank, mg/L
X'_r = mean MLSS concentration of sludge in return line, mg/L

If the suspended solids in the influent and effluent are neglected, but wastage is from the return line instead of the aeration tank, the wastage flow rate is

$$Q_{wr} \approx \frac{V X_m}{\theta_c X_r} \qquad (12.4.7)$$

where
Q_{wr} = waste sludge flow rate from the return sludge line, m³/d
V = aeration tank volume, m³

The average concentration of microorganisms in the reactor is

$$X_m = \frac{\theta_c(Y)(S_o - S)}{\theta(1 + k_d \theta_c)} \quad (12.4.8)$$

where θ = hydraulic retention time, d.

The mass of volatile waste activated sludge that must be disposed of each day is

$$P_x = Y_{obs} Q(S_o - S) \quad (12.4.9)$$

The observed yield coefficient Y_{obs} is defined as

$$Y_{obs} = \frac{Y}{1 + k_d(\theta_c)} \quad (12.4.10)$$

The theoretical oxygen requirement for removal of carbonaceous BOD is computed as

$$\text{kg } O_2/d = \frac{Q(S_o - S)(86\,400 \text{ s/d})(10^{-3} \text{ kg/g})}{f} - 1.42(P_x) \quad (12.4.11)$$

where f = conversion factor for converting BOD_5 to BOD_L.

EXAMPLE 12.5

Design a plug flow activated sludge process using the design assumptions provided below. The discharge standard is 20 mg/L BOD_5 and 30 mg/L for suspended solids (SS). Determine the volume of the aeration tank, the mass of sludge that must be wasted each day and the required air flow rate.

Design Assumptions

Variable	Assumption
Flow rate	0.0789 m³/s
Influent soluble BOD_5	135.0 mg/L
MLVSS	2500 mg/L
MLSS	1.43(MLVSS)
K_s	100 mg BOD_5/L
μ_m	2.5 d^{-1}
k_d	0.050 d^{-1}
Y	0.50 mg VSS/mg BOD_5 removed
X_r	10 000 mg/L
X_e	30 mg/L
BOD_5 of effluent SS	0.63(X_e)
Wastage	From aeration tank
BOD_5/BOD_L	0.68
O_2 transfer efficiency	8%

Solution. The allowable soluble BOD_5 in the effluent is the difference between the total allowable BOD_5 and the BOD_5 contained in the suspended solids that are in the effluent:

$$S = 20.0 - (0.63)(30.0) = 1.1 \text{ mg/L}$$

With the assumption that wastage is from the aeration tank, the return sludge flow rate is estimated as

$$Q_r = \frac{X'Q}{X'_r - X'}$$

$$= \frac{(1.43)(2500)(0.0789)}{10\,000 - (1.43)(2500)} = \frac{282}{6425} = 0.0439 \, \frac{m}{s}$$

The recycle ratio is

$$\alpha = \frac{0.0439 \, \frac{m^3}{s}}{0.0789 \, \frac{m^3}{s}} = 0.5564$$

The influent concentration S_i after dilution with recycle flow is

$$S_i = \frac{135 + (0.5564)(1.1)}{1 + 0.5564} = 87.1 \, \frac{mg}{L}$$

The mean cell residence time is

$$\frac{1}{\theta_c} = \frac{2.5(135 - 1.1)}{(135 - 1.1) + (1 + 0.5564)(100) \ln\left(\frac{87.1}{1.1}\right)} - 0.05 = 0.3611$$

$$\therefore \theta_c = 2.77 \, d$$

The hydraulic detention time is found as follows:

$$2500 = \frac{2.77(.5)(133.9)}{\theta(1 + (0.05)(2.77))} = \frac{185.45}{\theta(1.1385)}$$

$$\therefore \theta = 0.065 \, d = 1.56 \, hrs$$

The volume of the aeration basin is

$$V = (0.0789 \, m^3/s)(1.56 \, h)(3600 \, s/h) = 445.9 \text{ or } 446 \, m^3$$

The observed yield is

$$Y_{obs} = \frac{0.5}{1 + (0.05)(2.77)} = 0.4392$$

Noting that $g/m^3 = mg/L$, the mass of sludge produced each day is

$$P_x = (0.4392 \, g/g)(0.0789 \, m^3/s)(135 \, g/m^3 - 1.1 \, g/m^3)(86\,400 \, s/d)(10^{-3} \, kg/g) = 400 \, kg/d$$

The amount of sludge to be wasted is this mass minus that lost in the clarifier effluent:

$$400 \, kg/d - (30 \, g/m3)(0.0789 \, m^3/s)(86\,400 \, s/d)(10^{-3} \, kg/g) = 195.5 \, kg$$

The theoretical oxygen requirement is

$$kg \, O_2 = \frac{(0.0789)(135 - 1.1)(86\,400 \, s/d)(10^{-3} \, kg/d)}{0.68} = 1342 \, \frac{kg}{d}$$

Assuming that air has a density of 1.20 kg/m³ and contains 23.2% oxygen by mass, the theoretical air requirement is

$$\text{Air} = \frac{1\,342}{(1.20)(0.232)} = 4820 \ \frac{m^3}{d}$$

With an oxygen transfer efficiency of 8%, the air required is

$$\text{Air Transfer} = \frac{4\,820 \ m^3/d}{0.08} = 60\,255 \ \frac{m^3}{d}$$

Notes:

1. If sludge is wasted from the return sludge line, the solution requires either an assumption about the recycle ratio (α) or a trial and error solution using Eq. 12.4.7.

2. The hydraulic detention time is lower than nominal values discussed below because the development of Eq. 12.4.3 assumes the aeration tank operates as an ideal plug flow reactor.

3. The computed volume, sludge production, and air requirement are based on average day flows. Even with equalization in place, some extra capacity should be provided for the sustained peak flow and back-up for equipment taken out of service for repairs.

12.4.1.3 Design of Completely Mixed Flow Reactors

A schematic of a completely mixed flow reactor with recycle is shown in Fig. 12.4. The symbols on the schematic are used in defining variables.

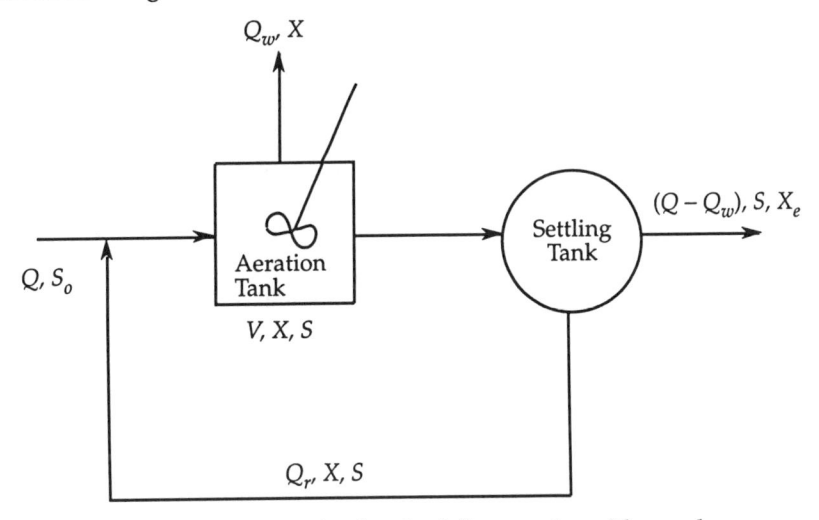

Figure 12.4 Completely mixed flow reactor with recycle.

The effluent substrate concentration from a completely mixed reactor may be estimated as

$$S = \frac{K_s(1+k_d\theta_c)}{\theta_c(\mu_m - k_d) - 1} \tag{12.4.12}$$

where the terms have the same meaning as defined for the plug flow reactor. Equations 12.4.6 through 12.4.11 apply to completely mixed reactors as well as plug flow reactors.

12.4.1.4 Design Criteria and Limits

GLUMRB recommends that if flow rates exceed 0.0044 m³/s, multiple units capable of independent operation be provided. It is desirable to have a parallel system of aeration tanks, secondary settling tanks and return sludge lines such that shut down of one does not necessitate the shut down of all the others.

For conventional plug flow or mixed flow systems, aeration tank widths range from 1.5 times the depth to 2.15 times the depth. The air requirement is greater for wider channels. Lengths range from 8 to 18 times the depth with "end-around" channels to achieve long lengths. GLUMRB recommends liquid depths greater than 3 m and less than 9 m. Oxygen transfer efficiency, soil conditions and economics lead to nominal depths around 4 to 5 m.

Completely mixed aeration tanks are frequently square in plan. Depths are governed by the same considerations as rectangular tanks.

GLUMRB has provided the guidance in Table 12.4 for use when process design calculations are not provided. The values apply for peak to average diurnal loadings from 2:1 to 4:1. Use of equalization basins to reduce the diurnal peak organic load may provide justification for exceeding the values shown.

Table 12.4 Permissible aeration tank capacities and loadings.

Process	Aeration tank Organic Loading, g BOD_5/d · m³	F/M Ratio, kg BOD_5/d per kg MLVSS	MLSS* mg/L
Conventional Step Aeration Complete Mix	640	0.2 – 0.5	1000 – 3000
Contact Stabilization	800**	0.2 – 0.6	1000 – 3000
Extended Aeration Oxidation Ditch	240	0.05 – 0.1	3000 – 5000

* MLSS values are dependent on the surface area provided for sedimentation and the rate of sludge return as well as the aeration process.

** Total aeration capacity including both the contact and reaeration capacities. Normally, the contact zone is about 30 to 35% of the total aeration capacity.

Source: Great Lakes-Upper Mississippi River Board of State Sanitary Engineers, *Recommended Standards, for Sewage Works*, 1978.

In rectangular tanks, diffusers are placed on one side of the channel approximately 30 to 60 cm off the floor. Discharge rates range from 35 to 150 m³/h · m² of effective diffuser area. In square tanks used for completely mixed systems, mechanical aerators often are of choice. (The sizing of mechanical aerators is discussed under lagoons below.) Regardless of the means of supplying oxygen, the system must be capable of maintaining dissolved oxygen concentrations above 2.0 mg/L. It is common to provide excess blower capacity at 150 to 200% of that required to meet the average day demand. Blowers should be provided in multiple units in such capacities that the maximum air demand may be met with the largest unit out of service.

GLUMRB has suggested that in the absence of experimentally determined values, the design oxygen requirements for all activated sludge processes except extended aeration be 1.1 kg of O_2/kg peak BOD_5 or 93.5 m³ of air/kg of BOD_5. For extended aeration the values are 1.8 kg of O_2/kg peak BOD_5 or 125 m³ of air/kg of BOD_5. The nitrogenous oxygen demand is taken as 4.6 times the peak Total Kjeldahl Nitrogen (TKN).

EXAMPLE 12.6

Using the data in Example 12.5 determine whether or not the design meets the GLUMRB guidelines for organic loading, F/M ratio, MLSS and oxygen supply.

Solution. The organic loading is determined by dividing the mass of BOD_5 entering the aeration tank divided by its volume:

$$\text{Organic Load} = \frac{(135 \text{ g/m}^3)(0.0789 \text{ m}^3/\text{s})(86\,400 \text{ s/d})}{446 \text{ m}^3} = 2063 \frac{\text{g}}{\text{d} \cdot \text{m}^3}$$

The F/M ratio as defined by Eq. 12.4.1 is

$$\frac{QS_o}{VX} = \frac{(0.0789)(86\,400)(135)}{(446)(2500)} = 0.825$$

The MLSS is

$$(1.43)(2500) = 3575 \text{ mg/L}$$

The oxygen supplied is

$$\frac{1342}{(135)(0.0789)(86\,400)(10^{-3})} = 1.46 \frac{\text{kg O}_2}{\text{kg BOD applied}}$$

for the average BOD_5. If the peak BOD_5 is twice the average, then oxygen is supplied at

$$\frac{1.46}{2} = 0.73 \frac{\text{kg O}_2}{\text{kg BOD}}$$

From these calculations, it is evident that the design is not conservative by GLUMRB guidelines. As noted in Example 12.5, the assumption of plug flow yielded numbers substantially below the norms for conventional processes because they seldom operate as plug flow units.

12.4.2 Trickling Filter

Only two of the "practical" fixed film models in common use will be discussed here: first order equations and the National Research Council (NRC) equations. The first order equations are generally applicable to synthetic media. The NRC equations were developed from observations on the performance of rock media filters at military installations during World War II.

12.4.2.1 Synthetic Media Filters

The removal of organic matter in a synthetic media filter may be described by an equation of the form

$$\frac{L_e}{L_o} = \exp\left[-\frac{k_{20}D}{Q_L^n}\right] \qquad (12.4.13)$$

where
- L_e = total BOD_5 of settled effluent from filter, mg/L
- L_i = total BOD_5 of wastewater applied to the filter, mg/L
- k_{20} = treatability constant at filter depth, D at 20°C, $(m/d)^n/m$
- D = depth of filter, m
- Q_L = hydraulic loading rate per unit area, $m^3/d \cdot m^2$
- n = empirical constant based on the media, frequently taken as 0.5

The k value is determined empirically from test data. It is very specific to local conditions including temperature and depth. Constants from the literature should be used with great caution.

For a given waste, the effect of temperature may be accounted for by adjusting k_{20} with the following equation:

$$k_T = k_{20}\theta^{T-20} \qquad (12.4.14)$$

where the value of θ is 1.035.

The following equation may be used for depths other than the one used to develop the value for k_{20}:

$$k_2 = k_1\left(\frac{D_1}{D_2}\right)^x \qquad (12.4.15)$$

where
x = 0.5 for vertical and rock media
= 0.3 for cross flow plastic media

Recirculation may be accounted for in a synthetic media filter by the relationship

$$\frac{L_e}{L_i} = \frac{\exp\left[-\frac{kD}{Q^n}\right]}{(1+r)-r\exp\left[\frac{-kD}{Q^n}\right]} \qquad (12.4.16)$$

where r = recirculation ratio.

12.4.2.2 Rock Media Filters

The NRC study of military installations with trickling filters lead to two empirical relationships to describe the efficiency of the filter. For a single-stage or the first of two stages, the efficiency is

$$E_1 = \frac{100}{1+0.0085\sqrt{\frac{W}{VF}}} \qquad (12.4.17)$$

where
E_1 = efficiency of BOD removal at 20°C including recirculation and sedimentation %
W = BOD loading to filter, lb/d
V = volume of filter media, acre-ft
F = recirculation factor

The recirculation factor is defined as

$$F = \frac{1+R}{(1+(0.1)(R))^2} \qquad (12.4.18)$$

where R = recirculation ratio (Q_r/Q). The NRC found that the optimum recirculation ratio is 8.

For the second-stage filter, the efficiency is

$$E_2 = \frac{100}{1+\frac{0.0085}{1-E_1}\sqrt{\frac{W'}{VF}}} \qquad (12.4.19)$$

where W' = BOD loading applied to the second stage, lb/d.

Because the data base used to develop the NRC equations reflects high BOD loads and low wastewater flow rates, the application of these design equations to more dilute municipal wastewaters will result in a conservative design.

12.4.2.3 Design Criteria and Limits

A summary of design variable ranges for trickling filter processes is presented in Table 12.5. Rock media filters may range up to 60 m in diameter. Synthetic media filters may be much deeper than rock media filters because the structural load on the collection works below the media is less and because the media structure may be opened up to allow aeration at greater depths.

Table 12.5 Typical design criteria for trickling filters.

Filter Class	Hyd. Load $m^3/m^2 \cdot d$	Org. Load $kg/m^3 \cdot d$	Depth m	Power $kW/10^3 \, m^3$	Recirc. Ratio
Low rate	1 – 4	0.08 – 0.32	1.5 – 3.0	2 – 4	0
Intermediate	4 – 10	0.24 – 0.48	1.25 – 2.5	2 – 8	0 – 1
High rate	10 – 40	0.32 – 1.0	1.0 – 2.0	6 – 10	1 – 3
Roughing	40 – 200	0.80 – 6.0	4.5 – 12	10 – 20	1 – 4

(Source: *Process Design Manual for Upgrading Existing Wastewater Treatment Plants*, U. S. Environmental Protection Agency, Washington, DC, 1974.)

EXAMPLE 12.7

Determine the diameter of a 10 m deep rock media filter for a municipal wastewater having the characteristics described in the table below. The treatability constant was developed using a 5 m test column operated at a temperature of 20°C. The discharge standard is 30 mg/L.

Variable	Value
Flow rate	0.0789 m³/s
BOD_5 to filter	135 mg/L
k_{20}	1.8 $(m/d)^n/m$
Temperature	15°C

Solution. Begin by correcting the treatability constant for temperature and depth. The temperature correction is

$$k_{15} = (1.8)(1.035)^{15-20} = 1.52$$

The depth correction is

$$k_{10} = 1.52 \left(\frac{5 \, m}{10 \, m}\right)^{0.5} = 1.07$$

Since the hydraulic load is defined as

$$\text{Hydraulic Load} = \frac{Q}{A}$$

Eq. 12.4.13 may be rewritten as

$$\ln \frac{L_e}{L_i} = -\frac{kDA^n}{Q^n}$$

Converting the flow rate to m³/d:

$$Q = (0.0789)(86\,400) = 6817 \, m^3/d$$

Solving the rewritten expression for A, the area of the filter is

$$A = Q\left[-\frac{\ln\frac{L_e}{L_i}}{kD}\right]^{1/n}$$

$$= 6\,817\left[-\frac{\ln\frac{30}{135}}{1.07(10)}\right]^{1/0.5} = 134.7\ \text{m}^2$$

The diameter is

$$\frac{\pi D^2}{4} = 134.7$$

$$\therefore D = 13.1\ \text{m}$$

12.4.3 Lagoons

The term lagoon may apply to a variety of treatment processes. Other names include oxidation ponds and stabilization ponds. For convenience, the following classification is used in this discussion:

1. *Aerobic ponds.* Shallow ponds, less than 1 m in depth to maintain dissolved oxygen (DO) throughout the entire depth.

2. *Facultative ponds.* Ponds 1 to 2.5 m deep that have an anaerobic lower zone, a facultative middle zone and anaerobic upper zone.

3. *Anaerobic ponds.* Deep ponds that receive very high organic loadings and are anaerobic throughout their depth.

4. *Tertiary ponds.* Those ponds that follow other biological treatment processes. DO may be provided either naturally through surface reaeration and photosynthesis or by mechanical means.

5. *Aerated lagoons.* DO is provided through mechanical or diffused aeration.

Of these classes, facultative ponds are the most common for small communities (5,000 people or less). Aerated lagoons are often used in lieu of concrete structures for conventional or completely mixed treatment systems. These two systems will be discussed in more detail.

12.4.3.1 Facultative Ponds

Facultative ponds are designed on the basis of rule-of-thumb guidelines. Commonly, three cells of equal size are used. These may be operated either in series or in parallel. Loading rates generally do not exceed 22 kg BOD_5/ha·d on one cell. In cold climates, the detention time for all three lagoons is nominally six months. As mentioned above, depths are limited to about 1 to 2.5 m.

12.4.3.2 Aerated Lagoons

Aerated lagoons are designed as completely mixed reactors without recycle. The effluent $BOD_5(S)$ may be calculated using Eq. 12.4.12 with θ substituted for θ_c. The concentration of biological solids produced may be estimated as

$$X = \frac{\mu_m(S_o - S)}{k(1 + k_d\theta)} \qquad (12.4.20)$$

This is also the concentration of microorganisms wasted each day. The theoretical oxygen demand maybe calculated using Eq. 12.4.11.

Mechanical aerators are rated in terms of kilograms of oxygen transferred per kilowatt-hour. Because manufacturer's ratings are not at field conditions, reported transfer rates must be corrected to these conditions with the following equation:

$$N = N_o \left(\frac{\beta C_{walt} - C_L}{C_{s_{20}}} \right) 1.024^{T-20} \alpha \qquad (12.4.21)$$

where
- N = kg O_2/kW·h transferred under field conditions
- N_o = kg O_2/kW·h transferred in water at 20°C, and zero dissolved O_2
- β = salinity-surface tension correction factor (usually taken as 1)
- C_{walt} = oxygen saturation for water at given temperature and altitude, mg/L
- $C_{s_{20}}$ = oxygen saturation for water at 20°C
- C_L = operating oxygen concentration, mg/L
- T = temperature, °C
- α = oxygen transfer correction factor for waste (usually taken as 0.8 to 0.85 for wastewater)

The oxygen solubility factor correction for elevation is shown in Fig. 12.5.

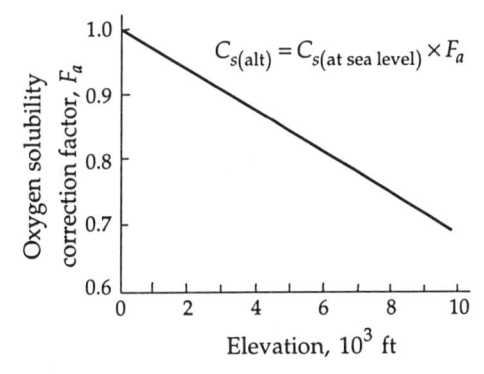

Figure 12.5 Oxygen solubility factor versus elevation. Source: Metcalf & Eddy, *Wastewater Engineering: Treatment, Disposal and Resue*, McGraw-Hill, Inc., New York, NY, 1991, p. 573.

EXAMPLE 12.8

Determine the volume of an aerated lagoon to meet a discharge standard of 30 mg/L BOD_5. Assume that 30 mg/L of suspended solids are discharged. Determine the power required to meet the oxygen requirements if the aerators are rated at 1.8 kg O_2/kW·h.

Design Assumptions

Variable	Assumption
Flow rate	0.0789 m³/s
Influent soluble BOD_5	135.0 mg/L
K_s	100 mg BOD_5/L
μ_m	2.5 d⁻¹
k_d	0.050 d⁻¹
Y	0.50 mg VSS/mg BOD_5 removed
X_e	30 mg/L
BOD_5 of effluent SS	0.63(X_e)
BOD_5/BOD_L	0.68
Temperature of wastewater	15°C
α	0.85
β	1.0
C_L	2.0 mg/L

Solution. The allowable soluble BOD_5 in the effluent is the difference between the total allowable BOD_5 and the BOD_5 contained in the suspended solids that are in the effluent:

$$S = 30.0 - (0.63)(30.0) = 11.1 \text{ mg/L}$$

The hydraulic detention time to meet this soluble BOD limit is

$$11.1 = \frac{(100)(1+\theta(0.05))}{\theta(2.5-0.05)-1} = \frac{100+5.0\theta}{2.45\theta-1}$$

$$\therefore \theta = 4 \text{ d}$$

The volume of the aerated lagoon is

$$V = (0.0789 \text{ m}^3/\text{s})(86{,}400 \text{ s/d})(4 \text{ d}) = 27\,267.8 \text{ m}^3 \quad \text{or} \quad 27\,000 \text{ m}^3$$

The concentration of biological solids produced is

$$X = \frac{0.5(135-11.1)}{1+(0.05)(4)} = 51.6 \text{ mg/L} \quad \text{or} \quad 52 \text{ mg/L}$$

The mass of solids produced is

$$P_x = (52 \text{ g/m}^3)(0.0789 \text{ m}^3/\text{s})(86\,400 \text{ s/d})(10^{-3} \text{ kg/g}) = 354 \text{ kg/d}$$

The theoretical oxygen requirement is

$$\text{kg O}_2 = \frac{(0.0789)(135-11.1)(86{,}400)(10^{-3})}{0.68} - 1.42(354)$$

$$= 740 \, \frac{\text{kg O}_2}{\text{d}}$$

From the temperature and the oxygen saturation values in Table 12.2, note that the oxygen saturation concentration is 10.15 mg/L. At sea level the correction factor is 1.0. The oxygen saturation concentration at 20°C is 9.17 mg/L. The field transfer rate for the mechanical aerator is

$$N = 1.8 \left[\frac{(1.0)(10.15-2.0)}{9.17} (1.024^{15-20})(0.85) \right]$$

$$= (1.8)(0.67) = 1.21 \, \frac{\text{kg O}_2}{\text{kW·h}}$$

$$= 1.21 \, \frac{\text{kg O}_2}{\text{kW·h}} \times 24 \, \frac{\text{h}}{\text{d}} = 29 \, \frac{\text{kg O}_2}{\text{kW·d}}$$

The power required is

$$P = \frac{740}{29} = 25.5 \quad \text{or} \quad 26 \text{ kW}$$

Note that the volume required is substantially greater than that for the plug flow system in Example 12.4 even allowing for the ideal nature of the plug flow.

12.4.4 Secondary Settling

12.4.4.1 Sedimentation Physics

As discussed in chapter 11, settling in a secondary sedimentation tank following an activated sludge unit may be described as Type III settling; Type II settling characterizes sedimentation following trickling filters. The procedures for design using column test data is the same as those described in chapter 11.

12.4.4.2 Secondary Settling Tank Design

As in all settling tank design, the primary design parameter is overflow rate. However, in secondary clarifiers following activated sludge units, the surface area required for thickening may control. The overflow rate and the design flow rate determine the surface area of the clarifier (see Eq. 11.5.9). In the absence of laboratory test data, the values shown in Table 12.6 may be used to estimate overflow rates. Detention times range from 1.5 to 2.5 hours with a typical time of 2 hours. Standard practice is to provide no less than two sedimentation basins. Although either rectangular or circular tanks may be used, circular tanks are preferred following activated sludge units because they are more efficient and can incorporate sludge thickening more readily. Recommended side water depths (SWD) are shown in Table 12.7. Deeper tanks are preferred as they offer advantages in flexibility of operation as well as a margin of safety. For plants with average flows less than $0.04 \text{ m}^3/\text{s}$, the recommended weir loading rate is $125 \text{ m}^3/\text{d} \cdot \text{m}$. Weir loading rates in large tanks should not exceed $250 \text{ m}^3/\text{d} \cdot \text{m}$ of weir length.

Table 12.6 Typical secondary settling tank design criteria*.

Treatment Process	Overflow Rate $\text{m}^3/\text{d} \cdot \text{m}^2$	Loading $\text{kg}/\text{m}^2 \cdot \text{h}$
Following Activated Sludge (excluding extended air)	16 – 32	3 – 6
Following Extended Air	8 – 16	1 – 5
Following Trickling Filter	16 – 24	2 – 5

* For average flow

Source: U.S. Environmental Protection Agency, Process Design Manual for Upgrading Existing Wastewater Treatment Plants, 1974.

Table 12.7 Final settling basin side water depths.

Tank Diameter, m	Recommended Side Water Depth, m
< 12	3.4
12 to 20	3.7
20 to 30	4.0
30 to 42	4.3
> 42	4.6

Source: Joint Committee of the Water Pollution Control Federation and the American Society of Civil Engineers, *Wastewater Treatment Plant Design. Manual of Practice 8, 1977.*

12.4.4.3 Operational Problems

A bulking sludge (noted by pin point floc in the clarifier) is one that will not settle well. This may be the result of either the growth of filamentous organisms or the inclusion of water in the microorganism floc. These conditions may result from one or more of the following: low available ammonia nitrogen when the organic load is high, low pH, or lack of macro-nutrients.

A rising sludge is one that floats after it has settled. Rising sludge results from denitrification and the production of nitrogen gas that lifts globs to the surface. Increasing the return sludge flow rate, increasing the speed of the sludge scraper mechanism, and decreasing the mean cell residence time will alleviate the problem.

12.5 Tertiary Treatment

The tertiary treatment processes (often called advanced wastewater treatment or AWT) are designed to follow secondary treatment for removal of special categories of pollutants. With the exception of carbon adsorption and ammonia stripping, the unit operations are special applications of those already discussed.

12.5.1 Filtration

The geometric and hydraulic considerations used in the design of rapid sand filters for water treatment are also applicable to gravity filtration units used for wastewater. Filters for wastewater are not designed to achieve effluent turbidities as low as those used for water treatment. As a consequence, the effective size (E or P_{10}) tends to be larger. Typical design data for mono-, dual- and multi-media filters are summarized in Tables 12.8 and 12.9.

Because solids loadings to wastewater filters are higher than those for water filters, more frequent backwashing will be required. In addition, several operational problems not encountered in water filters may arise. These include turbidity breakthrough, mudball formation, grease buildup, and cracks and contraction of the bed. Turbidity breakthrough may be reduced by the addition of coagulants ahead of the filter. Mudball formation, grease buildup, cracks and contraction often result from inadequate cleaning of the bed during backwash. Provision of rigorous air scour and surface washing generally alleviates these problems.

12.5.2 Phosphorus Removal

Although conventional wastewater treatment processes such as primary settling and activated sludge may remove from 10 to 50% of the phosphorus, treatment limits are frequently more stringent (for example, some states require 90% removal or effluent limits less than 1 mg/L as P). To achieve the more stringent levels, precipitation followed by filtration may be required.

The common precipitants are aluminum, iron salts or lime. Because lime will react with any hardness present to form excessive amounts of precipitate, it is generally less desirable than the aluminum or iron salts. Exclusive of side reactions, the reactions for precipitation of phosphorus are

$$Al_2(SO_4)_3 \cdot 14H_2O + 2PO_4 \leftrightarrow 2AlPO_4 + 14 H_2O + 3 SO_4 \tag{12.5.1}$$

$$FeCl_3 + PO_4 \leftrightarrow FePO_4 + 3Cl \tag{12.5.2}$$

Other salts of iron such as ferrous sulfate and spent pickle liquor have also been used successfully.

The most appropriate point for addition of the precipitant must be determined experimentally. Addition prior to the primary settling tank may improve settling but it may limit the phosphorus availability for downstream biological processes. Addition at the tail end of a conventional activated sludge process provides the opportunity for mixing as well as enhanced settling in the secondary clarifier.

Table 12.8 Typical design criteria for mono-media filters.

Characteristic	Range	Typical
Shallow-bed (stratified)		
Sand		
Depth, cm	25 – 30	28
Effective size, mm	0.35 – 0.6	0.45
Uniformity Coefficient	1.2 – 1.6	1.5
Filtration rate, $m^3/d \cdot m^2$	120 – 350	175
Anthracite		
Depth, cm	30 – 50	40
Effective size, mm	0.8 – 1.5	1.3
Uniformity Coefficient	1.3 – 1.8	1.6
Filtration rate, $m^3/d \cdot m^2$	120 – 350	175
Conventional (stratified)		
Sand		
Depth, cm	50 – 75	60
Effective size, mm	0.4 – 0.8	0.65
Uniformity Coefficient	1.2 – 1.6	1.5
Filtration rate, $m^3/d \cdot m^2$	120 – 350	175
Anthracite		
Depth, cm	60 – 90	75
Effective size, mm	0.8 – 2.0	1.3
Uniformity Coefficient	1.3 – 1.8	1.6
Filtration rate, $m^3/d \cdot m^2$	120 – 470	235
Deep Bed (unstratified)		
Sand		
Depth, cm	90 – 180	120
Effective size, mm	2 – 3	2.5
Uniformity Coefficient	1.2 – 1.6	1.5
Filtration rate, $m^3/d \cdot m^2$	120 – 590	290
Anthracite		
Depth, cm	90 – 210	150
Effective size, mm	2 – 4	2.75
Uniformity Coefficient	1.3 – 1.8	1.6
Filtration rate, $m^3/d \cdot m^2$	120 – 470	290

Sources: S. L. Bishop and B. W. Behrman, *"Filtration of Wastewater Using Granular Media,"* presented at the 1976 Thomas R. Camp Lecture Series on Wastewater Treatment and Disposal, Boston Society of Civil Engineers, Boston, 1976 and G. Tchobanoglous, *"Filtration of Secondary Effluent for Reuse Applications,"* presented at the 61st Annual Meeting of the Water Pollution Control Federation, Dallas, Tx, Oct. 1988.

12.5.3 Nitrogen Control

Nitrogen in the form of NH_3 and NH_4 may be removed either biologically or by air stripping. Nitrite and nitrate may be removed biologically. The biological processes may be broadly classified as nitrification/denitrification.

Nitrification is used to convert ammonium to nitrate:

$$NH_4 + 2O_2 \leftrightarrow NO_3 + H_2O + 2H \qquad (12.5.3)$$

This reaction is mediated by microorganisms. To denitrify, the microorganisms are placed in an anoxic environment:

$$2NO_3 + \text{organic matter} \leftrightarrow N_2 + CO_2 + H_2O \qquad (12.5.4)$$

Table 12.9 Typical design criteria for dual- and multi-media filters.

Characteristic	Range	Typical
Dual-media		
Anthracite		
Depth, cm	30 – 75	60
Effective size, mm	0.8 – 2.0	1.3
Uniformity Coefficient	1.3 – 1.8	1.6
Sand		
Depth, cm	15 – 30	30
Effective size, mm	0.4 – 0.8	0.65
Uniformity Coefficient	1.2 – 1.6	1.5
Filtration rate, $m^3/d \cdot m^2$	120 – 590	290
Multi-media		
Anthracite (top of quad layer)		
Depth, cm	20 – 50	40
Effective size, mm	1.3 – 2.0	1.6
Uniformity Coefficient	1.5 – 1.8	1.6
Anthracite (2nd of quad layer)		
Depth, cm	10 – 40	20
Effective size, mm	1.0 – 1.6	1.1
Uniformity Coefficient	1.5 – 1.8	1.6
Anthracite (3rd of quad layer)		
Depth, cm	20 – 50	40
Effective size, mm	1.0 – 2.0	1.4
Uniformity Coefficient	1.4 – 1.8	1.6
Sand		
Depth, cm	20 – 40	25
Effective size, mm	0.4 – 0.8	0.5
Uniformity Coefficient	1.3 – 1.8	1.6
Garnet or ilmenite		
Depth, cm	5 – 15	10
Effective size, mm	0.2 – 0.6	0.3
Uniformity Coefficient	1.5 – 1.8	1.6
Filtration rate, $m^3/d \cdot m^2$	120 – 590	290

Sources: S. L. Bishop and B. W. Behrman, *"Filtration of Wastewater Using Granular Media,"* presented at the 1976 Thomas R. Camp Lecture Series on Wastewater Treatment and Disposal, Boston Society of Civil Engineers, Boston, 1976 and G. Tchobanoglous, *"Filtration of Secondary Effluent for Reuse Applications,"* presented at the 61st Annual Meeting of the Water Pollution Control Federation, Dallas, Tx, Oct. 1988.

Many process alternatives, either fixed film or suspended growth, may be operated to achieve either nitrification or denitrification or both.

Nitrogen in the form of ammonia also may be removed by air stripping. Raising the pH converts ammonium ion to ammonia:

$$NH_4 + OH \leftrightarrow NH_3 + H_2O \qquad (12.5.5)$$

The hydroxide is usually supplied by adding lime. Air is then passed through the wastewater to drive off the ammonia.

Air stripping is not commonly practiced because the addition of the lime results in scaling. Low temperatures reduce air stripping effectiveness or make it impractical as the water nears the freezing point.

12.5.4 Carbon Adsorption

Refractory organics (those organic compounds resistant to microbial degradation) may be removed by passing them through a bed of activated carbon or by adding carbon to the activated sludge process. When carbon columns are used, they are placed downstream of a filtration process to prevent the column from being plugged with suspended solids.

Both granular activated carbon (GAC) and powered activated carbon (PAC) appear to adsorb low molecular weight polar organics poorly. Lack of pH control, temperature control and flow rate control will also affect performance.

12.6 Sludge Treatment

Up to fifty percent of the cost of operating a municipal wastewater treatment plant may be for sludge treatment and disposal. For primary treatment, the volume generated ranges from 0.25 to 0.35 percent by volume of the wastewater treated. Secondary treatment by activated sludge processes can raise the volume to 1.5 to 2.0 percent of the volume of wastewater treated. Use of chemicals for phosphorus removal can raise the volume another 1.0 percent. Primary sludges contain 3 to 8 percent solids by mass. Waste activated sludge is typically 0.5 to 2 % solids and trickling filter sludge ranges from 2 to 5 % solids.

Because the sludges are predominantly water, thickening is often the first process used in sludge treatment. In some instances the final clarifier in an activated sludge process may be designed to incorporate thickening. Thickener design was discussed under Type III sedimentation in chapter 11.

The putrescible nature of sludges requires stabilization to more inert forms before they may be applied to the land. If the sludges are to be incinerated, they are not stabilized. Aerobic digestion and anaerobic digestion stabilization processes are discussed below.

Dewatering by either heat or mechanical means is often part of sludge treatment. Filter press design and sand drying bed design are discussed as examples of dewatering processes.

12.6.1 Conventional Aerobic Digestion

12.6.1.1 Operation of Aerobic Digestion Units

A primary function of digestion is the destruction of solids. The process functions in a similar fashion to the activated sludge process with the significant difference that the substrate is supplied by microorganism degradation of their own cell tissue. Approximately 75 to 80 percent of the cell may be oxidized. The remainder is inert. The aerobic oxidation process may be summarized by the following reactions where $C_5H_7NO_2$ represents microorganism cells:

$$C_5H_7NO_2 + 5O_2 \leftrightarrow 5CO_2 + NH_3 + 2H_2O \tag{12.6.1}$$

$$NH_3 + 2O_2 \leftrightarrow NO_3 + H_2O + H^+ \tag{12.6.2}$$

These reactions illustrate two requirements of the aerobic digestion process that differ from conventional activated sludge: the need for oxygen to satisfy the oxygen demand for oxidation of the cell tissue and the need for alkalinity to maintain the pH as H^+ is released from the oxidation of ammonia to nitrate. Theoretically, approximately 2.3 kg of oxygen per kg of cells is required for complete oxidation. If primary sludge is aerobically digested, an additional 1.6 to 1.9 kg of oxygen per kg of BOD_5 destroyed is required. On a stoichiometric basis, 7.1 kg of $CaCO_3$ alkalinity is destroyed per kg of ammonia oxidized.

12.6.1.2 Aerobic Digester Design

Solids destruction is a function of the liquid temperature and the solids retention time (sludge age) as shown in Fig. 12.6. The digester tank volume may be calculated as

$$V = \frac{(Q_i)(X_i + fS_i)}{X\left(K_d P_v + \dfrac{1}{\theta_c}\right)} \tag{12.6.3}$$

where
- Q_i = influent average flow rate to digester, m³/d
- X_i = influent suspended solids, mg/L
- f = fraction of influent BOD$_5$ consisting of raw primary sludge
- S_i = influent BOD$_5$, mg/L
- X = digester suspended solids, mg/L
- K_d = reaction rate constant, d⁻¹
- P_v = volatile fraction of digester suspended solids
- θ_c = solids retention time, d

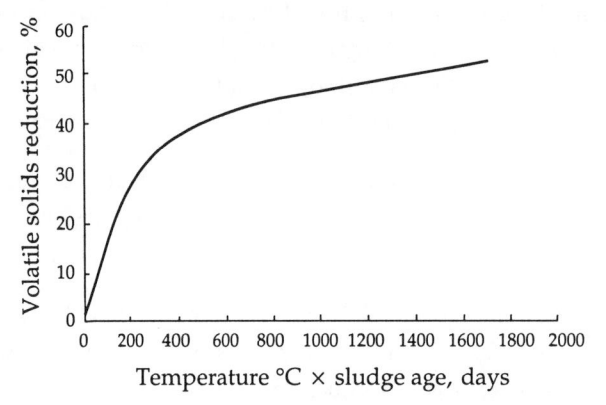

Figure 12.6 Volatile solids reduction in an aerobic digester as a function of digester liquid temperature and sludge age. Source: Water Pollution Control Federaton,*Sludge Stabilizatioin*, Manual of Practice FD-9, 1985.

The reaction rate constant is a function of the type of sludge, temperature and solids concentration. Typical values for waste activated sludge range from 0.05 d⁻¹ to 0.15 d⁻¹. Metcalf & Eddy in *Wastewater Engineering* (McGraw-Hill Book Company, New York, 1991) recommend that Eq. 12.6.3 not be used where significant nitrification will occur.

Typical design criteria are listed in Table 12.10.

Table 12.10 Typical design criteria for aerobic digesters.

Characteristic	Typical Values
Hydraulic detention time (at ~ 20°C), d	
Waste activated sludge (WAS) only	10-15
Primary plus WAS or trickling filter	15-20
Solids loading, kg/m³·d	1.6–4.8
Oxygen, kg O$_2$/kg solids destroyed	
Cell tissue	~2.3
BOD$_5$ in primary sludge	1.6-1.9
Energy requirements for mixing	
Mechanical aerators, kW/1000 m³	20-40
Diffused air mixing, m³/m³·min	0.02-0.04
Dissolved oxygen residual, mg/L	1-2
Reduction in volatile suspended solids, %	40-50

Source: Joint Committee of the Water Pollution Control Federation and the American Society of Civil Engineers, *Wastewater Treatment Plant Design. Manual of Practice 8*, 1977.

EXAMPLE 12.9

Design an aerobic digester to treat 195 kg/d of waste activated sludge. Determine the sludge age, volatile solids reduction, oxygen requirements, and volume of the digester. Use the design assumptions given below.

Design Assumptions

Variable	Assumption
Season	Winter
Activated sludge solids concentration	2%
Specific gravity of waste sludge	1.03
Volatile solids reduction required	40%
Liquid temperature	10°C
Digester suspended solids concentration	15 000 mg/L
Reaction rate constant	0.05 d^{-1}
Volatile fraction of digester suspended solids	0.70

Solution. The sludge age is computed using Fig. 12.6 and the design temperature. From the figure, at 40% solids reduction, note the product of temperature and sludge age to be 500 degree days. The sludge age is then

$$\theta_c = \frac{500°\text{C} \cdot \text{d}}{10°\text{C}} = 50 \text{ d}$$

The volatile solids reduction is based on the total mass of VSS in the influent. This mass is

$$(0.70)(195 \text{ kg/d}) = 136.5 \text{ kg/d}$$

With 40% reduction, the VSS reduced is

$$(136.5)(0.40) = 54.60 \text{ kg/d}$$

Using the rule-of-thumb oxygen requirements in Table 12.10 and assuming no primary sludge, the oxygen required is

$$(2.3 \text{ kg/kg VSS destroyed})(54.60) = 125.6 \text{ kg/d}$$

The volume of sludge entering the digester each day is

$$Q_i = \frac{195 \text{ kg/d}}{\left(1\,000 \text{ kg/m}^3\right)(1.03)(0.02)} = 9.47 \text{ m}^3/\text{d}$$

At 2% solids concentration, the influent suspended solids concentration is 20,000 mg/L. With no primary sludge, the volume of the digester is

$$V = \frac{(9.47)(20\,000 + 0)}{(15\,000)\left[(0.05)(0.70) + \left(\frac{1}{50}\right)\right]} = 230 \text{ m}^3$$

12.6.2 Completely Mixed High Rate Anaerobic Digestion

Of the several operating modes for anaerobic digesters, the completely-mixed, high-rate digester without recycle has been selected for discussion here. Other configurations are discussed in detail in *Wastewater Engineering*, by Metcalf & Eddy, Inc., McGraw-Hill, 1991.

12.6.2.1 Operation of Completely Mixed High Rate Anaerobic Digesters

As in all anaerobic digesters, the sludge is protected from exposure to oxygen. It is intimately mixed by recirculation of the digester gas or by mechanical mixers. To achieve optimum digestion rates, the sludge is heated. Suggested mean cell residence times as a function of temperature are given in Table 12.11. External heaters are provided to achieve design temperatures above ambient. It is common practice to provide two tanks in series where the second tank serves as a settling tank.

Table 12.11 Mean cell residence times for completely mixed anaerobic digesters.

Operating Temperature, °C	θ_c, d
18	28
24	20
30	14
35	10

Source: P. L. McCarty, "Anaerobic Waste Treatment Fundamentals," *Public Works*, 95, nos. 9-12, 1964.

Operating variables for anaerobic digesters include alkalinity, pH, volatile acids and gas composition. Typical alkalinities should be approximately 2500 mg/L as $CaCO_3$. The range of alkalinities may be from 1000 to 5000 mg/L as $CaCO_3$. Alkalinities below 1500 mg/L as $CaCO_3$ may be of concern because of incipient pH drop. The normal operating pH range is from 6.5 to 7.5. Volatile acid concentrations are normally in the range of 50 to 250 mg/L. The gas composition is usually 30 to 40% carbon dioxide (CO_2) and 60 to 70% methane (CH_4). Theoretically, at standard temperature and pressure (0°C and 1 atmosphere), 0.35 m^3 of CH_4 is produced for each kg of COD oxidized. Other gases of consequence are mercaptans and hydrogen sulfide (H_2S). In addition to having very strong odors, these gases are very corrosive and, in the case of H_2S, very toxic.

The term "stuck digester" has been used to describe progressive deterioration in the performance of an anaerobic digester. Increasing volatile acids and low alkalinity lead to a decreasing pH. The methane-forming microorganisms are inhibited and methane production decreases. A significant increase in CO_2 and decrease in CH_4 implies that the methane formers are being adversely affected. If not corrected, methane production will cease and the digester will be "stuck." The operational remedies include cessation of loading, increasing alkalinity by concentrating the sludge and ammonia and, as a last resort, addition of alkalinity by the addition of carbonate. Lime is not recommended since pH control is too difficult and over addition may result.

The term "ammonia toxicity" is used to describe deterioration in digester performance that results from excess NH_3 in solution. This results in a high pH in the digester, high volatile acids and no gas production. Hydrochloric acid has been used to remedy this problem. Because of the lag time in response, the dosing must proceed very cautiously. Sulfuric acid should not be used because the SO_4 released is toxic to the microorganisms.

12.6.2.2 Design of Completely Mixed Anaerobic Digesters

Four methods are used to determine the volume of a digester: population basis, volume reduction, loading factors, and mean cell residence time. Of these methods only the last two will be described here.

The loading factor method is analogous to the F/M ratio method of designing activated sludge units. For high-rate digesters, loading rates vary from 1.6 to 4.8 kg of volatile solids/m³·d. Hydraulic detention times from 10 to 20 days are reasonable. The relationship between hydraulic detention time, sludge concentration and solids loading is summarized in Table 12.12, by Metcalf & Eddy, Inc., *Wastewater Engineering*, by McGraw-Hill, 1979.

Table 12.12 Sludge concentration, hydraulic detention time and volatile solids loading for anaerobic digestion.

Sludge Concentration, %	Volatile Solids Loading Factor, kg of volatile solids/m³·d			
	10 d	12 d	15 d	20 d
4	3.06	2.55	2.04	1.53
5	3.83	3.19	2.55	1.91
6	4.59	3.83	3.06	2.30
7	5.36	4.46	3.57	2.68
8	6.12	5.10	4.08	3.06
9	6.89	5.74	4.59	3.44
10	7.65	6.38	5.10	3.83

Source: Metcalf & Eddy, Inc., *Wastewater Engineering*, McGraw-Hill Book Company, New York, 1979, p. 618.

Mean cell residence time may also be used as a design method. For a completely mixed high rate digester without recycle, the biomass synthesized each day may be estimated as

$$P_x = \frac{Y[(S_o - S)(Q)](10^{-3} \text{ kg/g})(86\,400 \text{ s/d})}{1 + k_d \theta_c} \tag{12.6.4}$$

where
- P_x = mass synthesized, kg/d
- Y = yield coefficient, kg/kg
- S_o = ultimate BOD in influent, mg/L
- S = ultimate BOD in effluent, mg/L
- Q = flow rate, m³/d
- k_d = endogenous coefficient, d⁻¹
- θ_c = mean cell residence time, d

The production of methane, in cubic meters, may estimated as

$$V_{CH_4} = 0.35[(S_o - S)(Q) - 1.42 P_x] \tag{12.6.5}$$

where all the terms are as described in Eq. 12.6.4.

EXAMPLE 12.10

Design a two-stage, completely mixed, high rate anaerobic digester to treat 195 kg/d of waste activated sludge. Determine the sludge age, volume of the digester, volatile solids reduction, solids synthesized, and methane production. Use the design assumptions given below.

Design Assumptions

Variable	Assumption
Season	Winter
Operating temperature	35 °C
Activated sludge solids concentration	2%
Specific gravity of waste sludge	1.03
Volatile fraction of digester suspended solids	0.70
Volatile solids reduction required	40%
Yield coefficient	0.05 kg/kg
Endogenous rate constant	0.05 d⁻¹

Solution. With a design operating temperature of 35°C, select a sludge age (mean cell residence time) of 10 d from Table 12.11.

The sludge flow rate may be estimated as

$$Q_{sl} = \frac{195 \text{ kg/d}}{(1000 \text{ kg/m}^3)(1.03)(0.02)} = 9.47 \text{ m}^3/\text{d}$$

The volume of the digester is then

$$V = Q\theta_c$$
$$= (9.47 \text{ m}^3/\text{d})(10 \text{ d}) = 94.7 \text{ m}^3$$

Assume that the ultimate BOD of the influent (S_o) may be estimated as the equivalent of the volatile fraction of the sludge:

$$S_o = (0.70)(195 \text{ kg/d}) = 136.5 \text{ kg/d}$$

Ignoring synthesis of anaerobic microorganisms, for 40% volatile solids reduction,

$$S = (1 - 0.4)(S_o) = (0.6)(136.5) = 81.9 \text{ kg/d}$$

The solids synthesized are

$$P_x = \frac{(0.05)(136.5 - 81.9)}{1 + (0.05)(10)} = 1.82 \text{ kg/d}$$

The volume of methane produced each day is

$$V_{CH_4} = (0.35)[(136.5 - 81.9) - 1.42\,(1.82)] = 18.2 \text{ m}^3/\text{d}$$

Note: when synthesis is included, the percent stabilization is

$$\% \text{ Stabilization} = \frac{(135.6 - 81.9) - 1.42(1.82)}{136.5}(100\%) = 38.1\%$$

rather than the 40% initially called for in the design criteria.

12.6.3 Sludge Dewatering

The fundamental relationships regarding sludge (specific gravity, volume and volume reduction) are the same as those discussed in chapter 11.

Numerous methods are available for dewatering sludges. Of these, sand drying beds are among the oldest and most commonly used methods in the U.S. Solids concentrations on the order of 50 to 60% can be achieved on drying beds. Because of the comparatively large land area required, the necessity for the sludge to be stabilized before dewatering, and the limitations of climate, mechanical dewatering devices have been employed extensively in recent years. A relative comparison of the performance of mechanical dewatering devices is shown in Table 12.13. Typical performance of a belt press for various types of sludges is shown in Table 12.14. The design of a sand drying bed and the selection of a plate filter press are discussed below to illustrate the techniques.

Table 12.13 Comparative performance of mechanical dewatering devices.

Type	Cake Solids[1], %	Recovery, % TSS	Polymer Cost[3]
Belt press	X	90 – 95[2]	Y
Centrifuge	$X \pm 2$	90 – 95	$0.8Y$
Vacuum filter	$X - 4$	85 – 90	$0.9Y$
Filter press – Lo P	$X + 8$	98 +	$1.1Y$
Filter press – Hi P	$X + 10$	98 +	$1.1Y$
Filter press – Diaph	$X + 12$	98 +	$1.1Y$
Screw press	$X - 2$	90 +	$1.2Y$
Low press drum/belt	$X - 10$	90 +	$0.8Y$

1. Relative to belt press, X denotes base level
2. Controlled by polymer dosage
3. Relative to belt press, Y denotes base level

Source: U.S. environmental Protection Agency, *Design Manual: Dewatering Municipal Wastewater Sludges*, Pub. No. EPA/625/1-87/014, SEP 1987, p. 80.

Table 12.14 Typical data for various sludges dewatered on belt filter press.

Type of Sludge	Feed solids, %	Solids loading, kg/h·m belt width	Polymer dose, g/kg	Cake solids, %
Raw				
Primary	3 – 10	360 – 680	1 – 5	28 – 44
WAS	0.5 – 4	45 – 230	1 – 10	20 – 35
P + WAS	3 – 6	180 – 590	1 – 10	20 – 35
P + TF	3 – 6	180 – 590	2 – 8	20 – 40
Anaerobically digested				
P	3 – 10	360 – 590	1 – 5	25 – 36
WAS	3 – 4	40 – 135	2 – 10	12 – 22
P + WAS	3 – 9	180 – 680	2 – 8	18 – 44
Aerobically digested				
P + WAS	1 – 3	90 – 230	2 – 8	12 – 20
P + TF	4 – 8	135 – 230	2 – 8	12 – 30
Oxygen activated				
WAS	1 – 3	90 – 180	4 – 10	15 – 23
Thermally conditioned				
P + WAS	4 – 8	290 – 910	0	25 – 50

Source: U.S. environmental Protection Agency, *Design Manual: Dewatering Municipal Wastewater Sludges*, Pub. No. EPA/625/1-87/014, SEP 1987, p. 80.

12.6.3.1 Design of a Sand Drying Bed

The common operational procedure for drying beds is as follows:

- While adding chemical conditioners, pump 20 to 30 cm of stabilized sludge onto the bed
- Allow the sludge to dry
- Remove the sludge mechanically
- Repeat the cycle.

Conventional sand drying beds have dimensions of 4.5 to 18 m by 15 to 50 m. Ten to 20 cm of sand is placed over 20 to 50 cm of graded gravel. The effective size of the sand is in the range 0.3 to 1.2 mm and the uniformity coefficient is less than 5.0. The gravel is graded from 0.3 to 2.5 cm effective diameter. Perforated underdrain pipe, no less than 10 cm in diameter, is placed 2.5 to 6 m on centers with a minimum slope of 1%.

Sludge dewaters primarily through drainage. Evaporation accounts for a smaller fraction of water loss. Precipitation, of course, impedes the drying process and is a critical consideration in the design. Although there is no "rational" engineering design procedure, the general approach is to select a common bed loading (see Table 12.15 for example) and then perform a water balance to determine if the area selected will allow all of the sludge to be processed.

Table 12.15 Typical area requirements for open sludge drying beds.

Type of Sludge	Area, $m^2/10^3$ persons	Sludge loading rate kg dry solids/ $m^2 \cdot y$
Primary digested	90 – 140	120 – 200
Primary and activated digested	160 – 275	60 – 100
Primary & chem. precipitated & digested	185 – 230	100 – 160

*Note: For covered beds the area may be 70 to 75% of that required for open beds.

Source: Metcalf & Eddy, Inc., *Wastewater Engineering: Treatment Disposal Reuse*, McGraw-Hill Book Company, New York, 1972, p. 655.

A relatively new concept in drying beds is to have the bottom completely paved. Sludge is pumped to the bed, allowed to settle and the supernatant is decanted. Metcalf & Eddy suggest the following equation to estimate the area of this type of drying bed:

$$A = \frac{(1.04)(S)\left[\left(\frac{1-s_d}{s_d}\right)-\left(\frac{1-s_e}{s_e}\right)\right]+(1\,000)(P)(A)}{(10)(k_e)(E_p)} \tag{12.6.6}$$

where
- A = bottom area of paved bed, m^2
- S = annual sludge production, dry solids, kg
- s_d = percent dry solids in sludge after decant
- s_e = percent dry solids required for final disposal
- P = annual precipitation, m
- k_e = reduction factor for evaporation from sludge versus free water surface
- E_p = free water pan evaporation rate for the area, cm/y

A trial value of 0.6 for k_e is suggested for preliminary estimates, and field testing is recommended for final design.

12.6.3.2 Design of a Filter Press

Two alternative designs are common: the gasketed, recessed-chamber (Fig. 12.7) and the membrane or diaphragm plate press. Both have essentially the same operating cycle. High pressure pumps force the sludge into a feed inlet where it is dispersed into the space between the plates. Sludge deposits and builds up particles on the filter cloth while the filtrate passes through the cake and cloth. The filtrate is channeled out of the press through ports in the plate. When the chambers become filled with solids, the effluent discharge stops. The feed pump is turned off. If the press is a membrane or diaphragm type, a back pressure is then placed on the membrane to squeeze the cake and further dewater it. The press is then opened by either manually or mechanically separating the plates and allowing the cake to drop into hoppers.

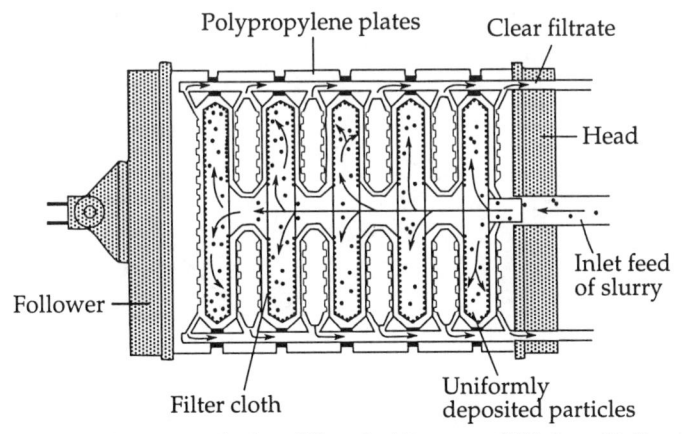

Figure 12.7 Filter press during fill cycle (Courtesy JWI, Inc., Holland, MI)

The fundamental design equation for sizing a filter press is an application of the principle of mass balance:

$$V_p = \frac{(V_f)(P_1)(SG)(\rho_w)}{(\rho_c)(P_2)} \quad (12.6.7)$$

where
- V_f = volume of feed per cycle, m³
- V_p = filter press volume, m³
- P_1 = solids content of feed as decimal fraction
- SG = specific gravity of feed slurry
- ρ_w = density of water, kg/m³
- ρ_c = density of wet cake, kg/m³
- P_2 = solids content of filter cake, as decimal fraction

Practice Problems (Essay)

12.1 Given the following information about a stream, calculate the DO at an observation point 33.16 km downstream from the waste discharge point:

Speed of river = 0.500 m/s $\quad Q_r + Q_w = 6.500$ m³/s
$K_d = 1.370$ d⁻¹ $\quad\quad\quad\quad\quad$ $K_r = 4.830$ d⁻¹
River temp. = 24.0 °C $\quad\quad$ Ultimate BOD after mixing = 2324.0 kg/d
Deficit at the discharge point after mixing = 0.0 mg/L

Assume that the rate constants are at the river temperatures.

12.2 Design an equalization basin for the following flow variation. Determine the volume of the tank, the dimensions, and the minimum power of the mixing equipment.

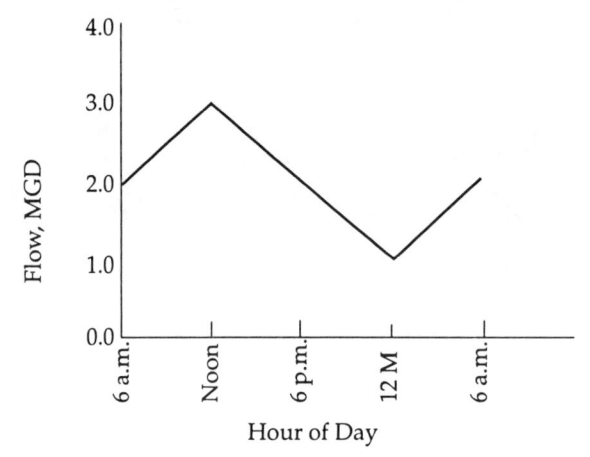

12.3 You have been asked to evaluate the ability of a horizontal-flow gravity grit chamber to remove particles having a diameter of 1.71×10^{-4} m. The depth of the grit chamber is 1.0 m. The detention time of the liquid in the grit chamber is 60 s. The particle density is 1.83 g/cm³. The water temperature is 12°C. Assume the density of the water is 1 000 kg/m³.

12.4 The community using the sedimentation facility designed in Example 12.4 is considering a request for a permit for a new industrial discharge to their wastewater treatment plant. The industry will contribute an average daily flow of 0.0395 m³/s to the plant. The BOD₅ of the industrial waste is 330 mg/L. The wastewater from the industry contains no suspended solids. Provide a discussion of the impacts of this additional waste stream on the BOD and suspended solids removal efficiency of the sedimentation facility. What is the BOD (in mg/L) of the wastewater leaving the primary sedimentation tank?

12.5 Determine the diameter of a high-rate, single-stage, rock media trickling filter for a flow of 1 MGD (before recirculation). Assume the NRC equations apply. The BOD₅ applied to the filter is 125 mg/L and the effluent must be 25 mg/L. Assume the following: hydraulic loading rate is 15 mgad, recirculation ratio is 12.0, filter depth is 6 ft.

12.6 The raw BOD entering a wastewater plant is 135 mg/L. Using the design assumptions shown below, determine the volume of a plug flow activated sludge reactor to treat this waste.

Design Basis:
    a. Effluent soluble BOD concentration = 12.0 mg/L
    b. Primary treatment removes 33% of the BOD
    c. Plant flow = 0.1065 m³/s
    d. Plug flow reactor:

(1) $K_s = 100$ mg BOD_5/L
(2) $\mu_m = 2.5$ d^{-1}
(3) $k_d = 0.05$
(4) $Y = 0.50$ g cell/g BOD removed
(5) $X = 2500$ mg/L
(6) $\alpha = 0.25$

12.7 Design a conventional activated sludge plant for 85% BOD removal efficiency using the assumptions shown below. The plant is to handle a flow of 10 MGD with a BOD_5 of 150 mg/L. Determine the tank volume, MLVSS, air required in ft^3/d, and SVI. Assume 1/3 of the BOD_5 is removed in the primary tank, the design F/M is 0.21, the design aeration tank loading is 40 lb/1000 ft^3, 800 ft^3 of air is required per lb of BOD_5 removed, and a return sludge concentration of 8000 mg/L can be achieved.

12.8 Determine the blower capacity (m^3/s) for an 0.250 m^3/s activated sludge treatment process with the following characteristics:
(1) $K_s = 100$ mg BOD_5/L
(2) $\mu_m = 2.5$ d^{-1}
(3) $k_d = 0.05$
(4) $Y = 0.50$ g cell/g BOD
(5) $S_o = 250$ mg/L
(6) $S = 6$ mg/L
(7) $\theta_c = 8$ d

Assume that BOD_5 is equal to 0.68 BOD_L, that air has a density of 1.185 kg/m^3 and is 23.2% oxygen by mass, that the oxygen transfer efficiency is 8 percent, and that a safety factor of 2 is used in sizing the blowers.

12.9 Design a high rate anaerobic digester for a 1 MGD activated sludge process. The influent suspended solids are 200 mg/L and the BOD is 300 mg/L. Determine the volume of the digester and the volume of methane produced each day using the following design assumptions: primary tanks remove 53% of suspended solids and 33% of the BOD_5, the activated sludge process synthesizes 50% of the BOD, 1 lb of MLVSS = 0.6 lb of BOD, effluent suspended solids from secondary clarifier are 20 mg/L, digester retention time is 15 days, primary sludge is 5% solids, secondary sludges are 2% solids, and the specific gravity of all sludges is 1.10.

12.10 Determine the size of sand drying beds to achieve sludge dried to 50% solids for a 10 MGD activated sludge plant that produces 13,460 lb/d of dry solids of mixed digested sludge at 6% solids. The solids reduction in the digester was 50%. Use the following design assumptions: annual precipitation = 40 inches/y, annual evaporation = 60 inches/y, sludge production is equivalent to 0.2 lb/capita/d (before digestion), design basis 2.0 ft^2 drying bed /capita, sludge is placed in 8 inch depths, sludge drains to 18% solids, sludge absorbs 57% of precipitation, sludge evaporates to 75% of potential evaporation.

Practice Problems (Multiple Choice)

12.11 The reaeration rate (k_r) of a stream that is narrow and deep would be expected to be...
    a) ...greater than...
    b) ...less than...
    c) ...the same as...
    d) ...unrelated to...
    e) ...usually the same but sometimes less than...

...that of a river that is wide and shallow.

12.12 Combined sewers are those that
    a) are designed to carry both domestic and industrial wastewater.
    b) operate under pressure and flow by gravity.
    c) are designed to carry both domestic wastewater and storm water.
    d) are designed to carry domestic and industrial wastewater and storm water.
    e) are designed to carry domestic wastewater and storm water, alternately.

12.13 The primary sedimentation tank for a plant treating 0.100 m³/s of wastewater should have an equivalent weir length of about
    a) 72 m
    b) 190 m
    c) 45 m
    d) 460 m
    e) 1 900 m

12.14 The controlling design parameter in the design of all settling tanks is
    a) detention time
    b) percent removal of BOD_5
    c) percent removal of suspended solids
    d) overflow rate
    e) depth

12.15 The F/M ratio is controlled in the activated sludge process by
    a) adjusting the amount of sludge wasting
    b) adjusting the BOD_5
    c) adjusting the volume of the aeration tank
    d) adjusting the flow rate
    e) adjusting the flow rate and the BOD_5

12.16 A high rate trickling filter will have a hydraulic loading in the range
    a) 0.02 to 0.06 gal/ft²·min
    b) 0.06 to 0.16 gal/ft²·min
    c) 0.16 to 0.64 gal/ft²·min
    d) 0.20 to 1.20 gal/ft²·min
    e) 0.80 to 3.20 gal/ft²·min

12.17 The amount of nitrogen that must be added to a phenolic industrial wastewater having an equivalent BOD_5 of 800 mg/L is
   a) 5.3 mg/L
   b) 25 mg/L
   c) 32 mg/L
   d) 150 mg/L
   e) 250 mg/L

12.18 A secondary effluent that has "pinpoint" floc escaping over the weir indicates
   a) over aeration
   b) too low of a return sludge flow
   c) nitrification
   d) rising sludge
   e) bulking sludge

12.19 Phosphorus can be removed from a wastewater by
   a) ferric chloride addition
   b) chlorine addition
   c) air stripping
   d) carbon adsorption
   e) rapid sand filtration

12.20 Which of the following mechanical sludge dewatering devices will give the highest percent cake solids?
   a) belt press
   b) centrifuge
   c) vacuum filter
   d) high pressure filter press
   e) diaphragm filter press

Solutions to Problems

12.1 Note that capital K implies that the rate constants are for base 10. This problem may be worked in base 10 or the rate constants may be converted to base e. Base 10 is used in this solution.

Because the deficit after mixing is 0.0, $D_a = 0$.

Calculate the travel time downstream from the discharge point:
$$t = \frac{(33.16 \text{ km})(1000 \text{ m/km})}{(0.5 \text{ m/s})(86\,400 \text{ s/d})} = 0.7676 \text{ d}$$

Convert the ultimate BOD from a mass discharge to mg/L:
$$L_a = \frac{(2324 \text{ kg/d})(1 \times 10^6 \text{ mg/kg})}{(6.500 \text{ m}^3/\text{s})(86\,400 \text{ s/d})(10^3 \text{ L/m}^3)} = 4.138 \text{ mg/L}$$

Calculate the deficit:
$$D = \frac{(1.370)(4.138)}{4.830 - 1.370}\left[10^{-(1.370)(0.7676)} - 10^{-(4.830)(0.7676)}\right] = 0.145 \text{ mg/L}$$

Calculate the DO by first determining the saturated DO from Table 12.2 using a temperature of 24° C: $DO_s = 8.53$ mg/L:
$$DO = DO_s - D$$
$$DO = 8.53 - 0.145 = 8.385 \text{ or } 8.39 \text{ mg/L}$$

12.2 From the figure, one may observe that the average flow is 2.0 MGD. The storage required may be recognized as the area under the curve from 6 am to 6 pm. The volume of the equalization basin may be calculated by numerical integration or, because of the geometric symmetry, the volume may be calculated as the area of a triangle:
$$V = \left(\frac{1}{2}\right)(b)(h) = (0.5)(12 \text{ h})(1 \times 10^6 \text{ gal/d})\left(\frac{1 \text{ d}}{24 \text{ d}}\right) = 250{,}000 \text{ gal}$$

Add 25% for safety:
$$V = (1.25)(250{,}000) = 312{,}500 \text{ gal}$$

In cubic feet:
$$V = (312{,}500)(0.1337 \text{ ft}^3/\text{gal}) = 41{,}781 \text{ ft}^3$$

For a depth of 10 ft the area is:
$$A_s = \frac{41{,}781 \text{ ft}^3}{10 \text{ ft}} = 4{,}178 \text{ ft}$$

If the length and width are equal:
$$L = W = (4{,}178)^{1/2} = 64.63 \text{ or } 65 \text{ ft per side}$$

The minimum power required is:
$$(0.02 \text{ hp}/1000 \text{ gal})(312{,}500 \text{ gal}) = 6.25 \text{ hp}$$

12.3 Convert the particle density to units of kg/m^3:

$$\rho_s = \left(\frac{1.83 \text{ g}}{\text{cm}^3}\right)\left(\frac{10^{-3} \text{ kg}}{\text{g}}\right)\left(\frac{10^6 \text{ cm}^3}{\text{m}^3}\right) = 1830 \text{ kg/m}^3$$

Calculate the settling velocity using Stokes Law. The viscosity is determined using the temperature of 12° C in Table 4.2:

$$v_s = \frac{(9.80)((1830-1000)(1.71\times 10^{-4})^2}{(18)(1.235\times 10^{-3})} = 1.07\times 10^{-2} \text{ m/s}$$

The time for a particle to fall through the depth of the grit chamber is:

$$t = \frac{h}{v_s} = \frac{1.0 \text{ m}}{1.07\times 10^{-2} \text{ m/s}} = 93.5 \text{ s}$$

Since the detention time in the chamber is only 60 s, the particle will not be captured in the chamber.

12.4 Begin by calculating the new overflow rate. The surface area of the tank is computed in Example 12.4 as 170.42 m².

The new overflow rate is

$$v_o = \frac{Q}{A_s} = \frac{0.0789 + 0.0395}{170.42 \text{ m}^2}(86,400 \text{ s/d}) = 60.0 \text{ m}^3/\text{d}\cdot\text{m}^2$$

This is equivalent to 1,473 gal/d·ft². From Fig. 12.1, the percent BOD_5 removal is 27%.

Because the industrial discharge BOD is all soluble, the increased flow rate will decrease the BOD removal of the fraction associated with the municipal flow. Furthermore, the concentration of BOD in the municipal flow will be diluted since these graphs assume BOD removal is a result of suspended solids. The new equivalent BOD and suspended solids of the raw waste water is:

$$\text{BOD} = \frac{(0.0395)(0)+(135)(0.0789)}{0.0395+0.0789} = 90 \text{ mg/L}$$

$$\text{SS} = \frac{(0.0395)(0)+(200)(0.0789)}{0.0395+0.0789} = 133 \text{ mg/L}$$

Using the 27% and 90 mg/L BOD_5 from Fig. 12.2:

$$t = 1.67 \text{ h}$$

This yields 48% efficiency of suspended solids removal.

The BOD leaving the sedimentation tank is equivalent to all of the BOD from the industrial process plus the fraction of municipal BOD not removed in the primary tank, i.e.:

$$(1-0.27)(90 \text{ mg/L}) = 65.7 \text{ mg/L}$$

The BOD concentration is the weighted average:

$$\text{BOD} = \frac{(0.0395)(330)+(0.0789)(65.7)}{0.1184} = 153.9 \text{ or } 154 \text{ mg/L}$$

12.5 Calculate the required removal efficiency:

$$E = \frac{125-25}{125}(100\%) = 80\%$$

Calculate F:

$$F = \frac{1+12}{(1+(0.1)(12))^2} = 2.69$$

Solve Eq. 12.4.17 for the volume:

$$V = \frac{w}{(F)\left(\dfrac{\dfrac{100}{E_1}-1}{0.0085}\right)^2}$$

Calculate the volume of the filter:

$$V = \frac{(125 \text{ mg/L})(1 \text{ MGD})(8.34 \text{ lb/gal})}{2.69\left[\dfrac{\dfrac{100}{80}-1}{0.0085}\right]^2} = 0.448 \text{ acre-ft}$$

Calculate the area of the filter:

$$A = \frac{0.448 \text{ acre-ft}}{5.52 \text{ ft}} = 0.0812 \text{ acre or } 3{,}535 \text{ ft}^2$$

The diameter of the filter is then

$$\frac{(\pi)(D)^2}{4} = 3{,}535$$

$$D = 67 \text{ ft}$$

Note: The recycle ratio of 12 far exceeds the NRC optimum of 8.

12.6 The BOD_5 after primary settling is

$$S_o = (1-0.33)(135) = 90.45 \text{ mg/L}$$

S_i is then calculated:

$$S_i = \frac{90.45+(0.25)(12)}{1+0.25} = 74.76$$

Calculate the mean cell residence time:

$$\frac{1}{\theta_c} = \frac{(2.5)(90.45-12)}{(90.45-12)+\left[(1.25)(100)\ln\left(\dfrac{74.76}{12}\right)\right]} - 0.05$$

$$\theta_c = 1.699 \text{ d}$$

With the MLVSS, solve for the hydraulic detention time:

$$2{,}500 = \frac{(1.699)(0.5)(90.45-12)}{\theta(1+(0.05)(1.699))}$$

$$\theta = 0.0246 \text{ d}$$

The volume of the plug flow reactor is then

$$V = \theta Q = (0.0246 \text{ d})(0.1065 \text{ m}^3/\text{s})(86\,400 \text{ s/d}) = 226.4 \text{ or } 226 \text{ m}^3$$

12.7 Assuming 1/3 of BOD is removed in the primary tank, the pounds of BOD entering the aeration tank are

$$(2/3)(150 \text{ mg/L})(10 \text{ MGD})(8.34 \text{ lb/gal}) = 8340 \text{ lb BOD}_5/\text{d}$$

For a loading of 40 lb/1,000 ft³, the tank volume is

$$V = \frac{8{,}340 \text{ lb BOD/d}}{0.04 \text{ lb/ft}^3} = 208{,}500 \text{ ft}^3 \text{ or } 1.56 \times 10^6 \text{ gal}$$

The MLVSS is calculated from the F/M ratio and the aeration tank volume of 1.56 MG:

$$\frac{F}{M} = 0.21 = \frac{8{,}340}{M}$$

$$M = \frac{8{,}340}{0.21} = 39{,}714$$

$$M = (x \text{ mg/L})(1.56 \text{ MG})(8.34 \text{ lb/gal}) = 39{,}714$$

$$x = 3{,}052 \text{ or } 3{,}100 \text{ mg/L}$$

Calculate the air required by first determining the BOD_5 to be removed:

$$\text{lb of } BOD_5 \text{ to be removed} = 0.85(8{,}340) = 7{,}089$$

Assume 800 ft³ air/lb BOD_5 removed to determine the volume of air at 0° C and 760 mm Hg:

$$(800 \text{ ft}^3/\text{lb})(7{,}089) = 5.67 \times 10^6 \text{ ft}^3$$

Calculate the volume of air at 20° C:

$$5.67 \times 10^6 \left(\frac{293 \text{ °C}}{273 \text{ °C}}\right) = 6.09 \times 10^6 \text{ ft}^3$$

Calculate the volume of settled sludge:

$$V_{sl} = \frac{3{,}052 \text{ mg/L}}{8{,}000 \text{ mg/L}}(1{,}000 \text{ mL}) = 381 \text{ mL or } 38.1\%$$

The sludge volume index is calculated as

$$SVI = \frac{38.1\%}{0.3052\%} = 125$$

12.8 Determine the observed yield coefficient:

$$Y_{obs} = \frac{0.5}{1 + (0.05)(8)} = 0.3571$$

Determine the increase in mass of MLVSS:

$$P_x = (0.3571)(0.250 \text{ m}^3/\text{s})(250 \text{ g/m}^3 - 6 \text{ g/m}^3)(86\,400 \text{ s/d})(10^{-3} \text{ kg/g}) = 1882 \text{ kg/d}$$

The theoretical oxygen requirement is

$$\frac{(0.250)(250-6)(86\,400)(10^{-3})}{0.68} - (1.42)(1882) = 5078 \text{ kg/d}$$

Assuming air has a density of 1.185 kg/m³ and contains 23.2% oxygen by mass, the theoretical air required is

$$\frac{5078}{(1.185)(0.232)} = 18\,470 \text{ m}^3/\text{d}$$

With an oxygen transfer efficiency of 8% and a safety factor of 2:

$$\frac{18\,471}{0.08}(2) = 461\,800 \text{ m}^3/\text{d}$$

In units of m³/s:

$$\frac{461\,800}{86\,400} = 5.34 \text{ m}^3/\text{s}$$

12.9 Calculate the solids removed in the primary settling:

$$(1 \text{ MG})(8.34 \text{ lb/gal})(0.53)(200 \text{ mg/L}) = 884 \text{ lb/d dry solids}$$

Assuming the primary sludge water content = 95%, the lbs of water sent to the digester is

$$\frac{884}{0.05} = \frac{x}{0.95}$$

$$x = 16,800 \text{ lb water/d}$$

The volume of sludge pumped to the digester is then

$$\frac{884 \text{ lbs}}{(1.10)(62.4 \text{ lb/ft}^3)} + \frac{16,800 \text{ lbs}}{62.4 \text{ lb/ft}^3} = 282 \text{ ft}^3/\text{d}$$

The solids removed in the secondary settling tank are

$$[(200 \text{ mg/L})(1.00 - 0.53) - 20 \text{ mg/L}](1 \text{ MGD})(8.34 \text{ lb/gal}) = 617 \text{ lb}$$

where $(1.00 - 0.53)$ = the suspended solids removed in the primary settling tank and the 20 mg/L is the effluent suspended solids from the secondary clarifier.

Assuming the secondary sludge water content = 98%, the lbs of water sent to the digester is:

$$\frac{617}{0.02} = \frac{x}{0.98}$$

$$x = 30,230 \text{ lb}$$

The volume pumped to the digester is then

$$\frac{617}{(1.10)(62.4)} + \frac{30,230}{62.4} = 493 \text{ ft}^3/\text{d}$$

The wastage from the activated sludge process is

$$(0.50)(1.00 - 0.333)(300 \text{ mg}/\text{L})(1 \text{ MGD})(8.34 \text{ lb}/\text{gal}) = 834 \text{ lb}/\text{d}$$

where (0.5) is the fraction of BOD synthesized and (1.00 − 0.33) = the fraction of BOD removed in the primary tank.

With a water content of 98%, the lbs of water pumped to the digester is

$$\frac{834}{0.02} = \frac{x}{0.98}$$

$$x = 40,900 \text{ lb/d}$$

The volume of sludge pumped is then

$$\frac{834}{(1.10)(62.4)} + \frac{40,900}{62.4} = 667 \text{ ft}^3$$

The total volume of sludge pumped to the digester is then

$$V_{sl} = 282 + 493 + 667 = 1,442 \text{ ft}^3/\text{d}$$

With a 15-day detention time, the storage volume is

$$(1,442 \text{ ft}^3/\text{d})(15 \text{ d}) = 21,630 \text{ ft}^3$$

If sludge is withdrawn once per week, an additional storage volume is required:

$$(1,442)(7 \text{ d}) = 10,090 \text{ ft}^3$$

The total sludge storage volume for the digester is then

$$21,630 + 10,090 = 31,800 \text{ ft}^3$$

The volume of digester gas is a function of the destruction of COD. Assuming that COD destroyed = BOD and that 6.35 ft^3 of methane is produced for each pound of COD destroyed, we may estimate the gas production as follows:

From the primary tank the digester receives 885 lb/d of dry solids of which

$$(1/3)(300 \text{ mg/L})(1 \text{ MGD})(8.34 \text{ lb/gal}) = 834 \text{ lb/d of BOD}$$

Using the assumption that 1 lb of MLVSS = 0.6 lb of BOD, the waste activated sludge yields:

$$(834 \text{ lb/d})(0.6) = 500 \text{ lb/d of BOD}$$

The volume of methane produced is

$$(834 + 500)(6.35 \text{ ft}^3 \text{ CH}_4/\text{lb COD}) = 8,473 \text{ ft}^3/\text{d at } 37°\text{C}.$$

12.10 Determine equivalent population:

$$\frac{13,460 \text{ lb/d}}{(0.5)(0.2 \text{ lb/capita} \cdot \text{d})} = 134,600 \text{ people}$$

Note: 0.5 = 50% solids reduction.

Required area at 2.0 ft^2 drying bed/capita:

$$(2.0)(134,600) = 269,200 \text{ ft}^2$$

Assume 270,000 ft² for a working estimate. Check capacity based on climate:

At 6% solids and 8 inch depth: $(270{,}000 \text{ ft}^2)(0.67 \text{ ft})(7.48 \text{ gal/ft}^2) = 1{,}353{,}132 \text{ gal}$

Sludge volume accumulation: $(13{,}400 \text{ lb/d})\left(\dfrac{94\% \text{ water}}{6\% \text{ solids}}\right)\left(\dfrac{1}{8.34 \text{ lb/gal}}\right) = 25{,}284 \text{ gal/d of sludge}$

Number of days of capacity: $\dfrac{1{,}353{,}132 \text{ gal}}{25{,}284 \text{ gal/d}} = 53.52 \text{ or } 54 \text{ d capacity}$

Loss due to drainage:

At 6% solids: $(0.94)(8 \text{ inches}) = 7.52$ inches of water

Note: 0.94 = 94% water; 8 inches = placement depth given as an assumption.

Assuming the sludge drains to 18% solids, the depth of sludge plus water is

$$\dfrac{0.06}{0.18}(8 \text{ inches}) = 2.67 \text{ inches}$$

Of the 2.67 inches, $(1.00 - 0.18)(2.67) = 2.19$ inches is water. Therefore, anticipated loss due to drainage is $7.52 - 2.19 = 5.33$ inches of water.

For the sludge to achieve 50% solids, the depth of sludge plus water is

$$\dfrac{0.06}{0.50}(8) = 0.96 \text{ inches}$$

Of the 0.96 inches, 50% is water, so the required evaporation is

$$2.19 - 0.48 = 1.71 \text{ inches}$$

The actual moisture balance is estimated as follows:

Calculate rainfall absorbed: $(0.57)(40 \text{ inches/y}) = 23 \text{ inches/y}$

The estimated evaporation rate: $(0.75)(60 \text{ inches/y}) = 45 \text{ inches/y}$

The net evaporation: $45 - 23 = 22 \text{ inches/y}$

Note: 0.57 = assumed absorption of precipitation; 0.75 = assumed evaporation potential.

Check the bed cycle rate:

Beds can be cycled at: $\dfrac{22 \text{ inches}}{1.71 \text{ inches}} = 13 \text{ times/y}$ or $\dfrac{365 \text{ d/y}}{13 \text{ times/y}} = 28 \text{ d}$

for a drying cycle. Since the bed capacity is 54 d, there is a safety factor of

$$\dfrac{54 \text{ d}}{28 \text{ d}} = 1.9$$

All of this assumes the annual average precipitation and evaporation occurs over the whole year. This is not likely. More refined analysis requires monthly moisture balance and allowance for storage. Covering the sand beds in winter helps tremendously.

12.11 b) Narrow streams have less surface area than wide shallow ones and oxygen transfer from the atmosphere is limited by the reduced surface area in narrow streams. The depth of the stream reduces the diffusion of oxygen to the lower reaches.

12.12 c) Combined sewers carry both storm water and domestic sewage.

12.13 c) The recommended loading rate for primary settling tanks is 190 m³/d·m of weir length. The equivalent weir length for 0.100 m³/s of wastewater would be

$$\frac{(0.100 \text{ m}^3/\text{s})(86\,400 \text{ s/d})}{190 \text{ m}^3/\text{d} \cdot \text{m}} = 45.47 \text{ or } 45 \text{ m}$$

12.14 d) Settling design is controlled by overflow rate.

12.15 a) F/M ratio is defined as

$$\frac{F}{M} = \frac{QS_o}{VX} = \frac{\text{mg BOD}_5/\text{d}}{\text{mg MLVSS}} = \text{d}^{-1}$$

The influent BOD to the plant is considered to be a "constant," that is, it cannot be changed by the operator. The mass of microorganisms (MLVSS) can be regulated by the operator by adjusting the amount of sludge wasting. Greater sludge wasting will lower "M" in the F/M ratio.

12.16 c) A high rate trickling filter will have a hydraulic loading rate of 0.16 to 0.64 gal/ft²·min.

12.17 b) Nitrogen must be supplied in the ratio of 1:32 (N:BOD$_5$). For 800 mg/L BOD equivalent, the amount of nitrogen would be

$$N = \frac{800 \text{ mg/L}}{32} = 25 \text{ mg/L}$$

12.18 e) A "pinpoint" floc is indicative of a bulking sludge. This results from under aeration, low pH and low nitrogen and phosphorus.

12.19 a) Phosphorus is removed by chemical precipitation. Common precipitants are ferric chloride, alum and lime.

12.20 e) The diaphragm filter press yields the highest cake solids.

13. Highway Design

by Thomas L. Maleck

Highway design as presented in this chapter is directed toward the design of the geometric elements of a highway. The principal reference is "A Policy on Geometric Design of Highways and Streets" (1984 and 1990 editions) by the American Association of State Highway and Transportation Officials. This publication is the authoritative guide for the design of highways and is commonly known as the "Green Book". The recommendations and expectations change over time and are a function of the classification of the highway. One method of classification is by route numbering (i.e., Interstate, U.S., State, and County). There is also a hierarchy of movement; such as, Primary, Transition, Distribution, Collection and Terminal Access. In general, roads with the higher classification and hierarchy of movement have better design standards. The Michigan Manual of Traffic Control Devices and the 1960 Highway Survey and Design Manual by the Michigan State Highway Department were also used as references.

The design of a highway is also a function of the physical characteristics and performance of numerous designs vehicles. The following table is an example of the types of design vehicles and their dimensions.

Table 13.1 Types of design vehicles with dimensions in feet (p. 21; from *A Policy on Geometric Design of Highways and Streets*, 1984, by the Am. Assoc. of State Highway and Trans. Officials, Wash., DC, by permission).

Design Vehicle Type	Symbol	Overall			Overhang	
		Height	Width	Length	Front	Rear
Passenger car	P	4.25	7	19	3	5
Single unit truck	SU	13.5	8.5	30	4	6
Single unit bus	BUS	13.5	8.5	40	7	8
Articulated bus	A-BUS	10.5	8.5	60	8.5	9.5
Combination trucks						
Intermediate semi-trailer	WB-40	13.5	8.5	50	4	6
Large semi-trailer	WB-50	13.5	8.5	55	3	2
"Double Bottom" semi-trailer full-trailer	WB-60	13.5	8.5	65	2	3
Recreation vehicles						
Motor home	MH		8	30	4	6
Car and camper trailer	P/T		8	49	3	10
Car and boat trailer	P/B		8	42	3	8

Design vehicles are generally considered to be either in one of two classes; passenger cars or trucks. Considerations for pedestrians, bicyclists and equestrians may also affect the geometric design of a highway.

Driver performance is a primary consideration in the design of a highway. Of the numerous considerations in driver performance, perception and reaction time and the height-of-eye are the most important.

The traffic volumes and the distribution of the traffic demand over time is also a consideration. The operating speed is related to the ratio of the traffic volumes to the capacity of the highway. The capacity of a highway is the maximum traffic volume that can be reasonably expected over a period of time and for the prevailing conditions. Traffic volumes and highway capacity are usually expressed in vehicles per hour. Congestion is a restriction to a driver's freedom to maneuver and is often expressed as the ratio of the traffic volume to the capacity of the highway (v/c ratio). The expected operating speed is a function of traffic volume, as illustrated in Figure 13.1.

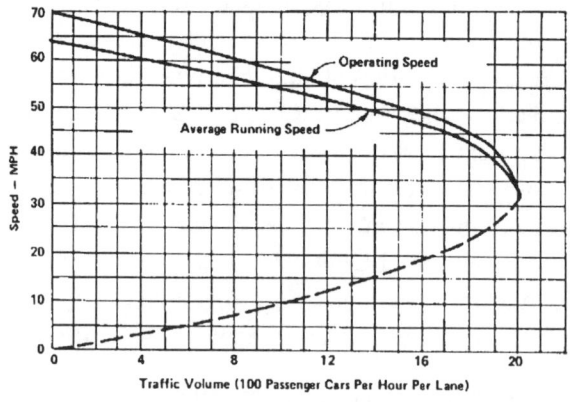

Figure 13.1 Operating speed of a vehicle (p. 79 from *A Policy on Geometric Design of Highways and Streets*, 1984, by the Am. Assoc. of State Highway and Trans. Officials, Wash., DC, by permission).

An extension of this is the concept of Level-of-Service. *Level-of-Service* is a way of quantifying the qualitative aspects of highway travel. The levels go from the highest level A to the lowest level F and represent the expected delays, operating speeds and volume-to-capacity ratios. Levels-of-Service B and C are often used for design purposes.

The traffic volumes that can be accommodated for a given design at each Level-of-Service are expressed as service volumes. The design volume of a highway is the volume of traffic expected to use the facility (normally estimated for a design year of 10 to 20 years in the future).

Another important element in the consideration of the geometric design is the *Design Speed*: the maximum safe speed that can be expected for the geometric features of the highway under favorable conditions. If possible the Design Speed should equal or exceed the operating speed and the "posted speed limit."

It is not unusual in the design of a highway to have known, estimated or assumed a Design Year, Design Traffic Volumes, Design Vehicle, Design Speed, Percent Commercial (trucks), and a Design Year Storm.

13.1 Capacity

The capacity of a highway is its ability to satisfy travel demand. There are many procedures for determining the capacity of a highway. The Highway Capacity Manual (HCM) is the most recognized procedure. The original capacity manual was published by the Bureau of Public Roads (presently the Federal Highway Administration) in 1950. The second edition of the HCM was published by the Highway Research Board (presently the Transportation Research Board) in 1965. The current version of the HCM was published in 1985 by the Transportation Research Board.

Analyses of capacity are usually separated into two categories, interrupted flow and uninterrupted flow. Uninterrupted flow is for traffic under free-flow conditions where drivers are not expected to stop and yield to another flow of traffic. Two common examples of uninterrupted flow are freeways and rural trunklines. Interrupted flow is for conditions where the driver is expected to stop and yield to another flow of traffic. A common example is an urban highway with numerous stop-and-go traffic lights.

Capacity is general thought of as the maximum number of vehicles that can be carried by a facility for a given period of time (usually an hour) under prevailing conditions of geometric features, traffic flow, control (i.e., signal progression) and environment (i.e., urban/rural).

A qualitative measure of flow is often used for the following Level-of-Services:
- A) free-flow conditions, higher speeds and low congestion.
- B) stable flow with slight interference from other traffic.
- C) stable flow but with significant conflicts with other vehicles and a resulting reduction in speed.
- D) the upper limit of stable flow with high density traffic.
- E) at or near capacity.
- F) forced flow and/or stop-and-go conditions.

13.1.1 Uninterrupted Flow

As shown in Figure 13.1, capacity is reached at a speed of 30 mph and a flow rate of 2000 passenger cars per hour per lane. The value of 2000 is a widely accepted standard for capacity under ideal conditions. Thus an eight lane freeway would be expected to accommodate up to 8000 vehicles per hour, in one direction of travel (4 lanes x 2000 veh/lane).

13.1.2 Interrupted Flow

Interrupted flow is much more difficult to analyze than uninterrupted flow. Normally the location being analyzed is an intersection of two arterials. The intersection is controlled by (or assumed to be controlled by) a stop-and-go traffic signal. A traffic signal has a cycle length and various phases. Each phases consists of two or more intervals. A cycle length is the total time required for an approach to complete a green, amber and red. Most signals have two phases, one for each direction of travel (phase 1 is the north and south movements and phase 2 is the east and west movements). An example of a three phase signal is one that has an additional phase for left-turning traffic. An interval is a period of time during which no signal face has a change in the color of a lens.

The 1985 HCM has an excellent automated procedure for analyzing the capacity of a simple intersection. There is also a manual procedure. The manual procedure is labor intensive and not used as much. The HCM procedure is difficult to use for analyzing complex geometric designs with numerous closely-spaced intersections.

A manual process often used is the *Summation of Critical Movements*, which is based upon the assumed discharge of traffic from a standing queue. *Greenshield's headways* are the average time to discharge from a queue as a function of the position in the queue. The assumed values are:

Vehicle in Queue	1	2	3	4	5
Headway	3.8	3.1	2.7	2.4	2.2 sec
Lost Time	1.7	1.0	0.6	0.3	0.1 sec

For the discharge of a queue of 5 or more vehicles the following equation gives the time needed for the discharge:

$$t = \frac{2.1 x M_c}{N_c} + 3.7 \quad (13.1.1)$$

where t is the required time for the interval in seconds, M_c is the *critical movement* (vehicles/hr) and N_c is the cycles/hr. The amber or clearance time is given by

$$Y = t_r + \frac{V}{2a + 64.4G} + \frac{W+L}{V} \quad (13.1.2)$$

where t_r is the reaction time, V is the vehicle speed, a is the vehicle deceleration, G is the percent grade, W is the intersection width, and L is the vehicle length.

The manual process can be used for analyzing complex designs.

EXAMPLE 13.1

Determine the capacity of the intersection of Main Street with Division Road. The Design Year Traffic Volumes are shown in Figure 13.2. The diagonal solid lines are estimates of right turns and the broken diagonal lines are estimates of left turns. The speed limit for both roads is 30 mph. The cycle length is 60 seconds. The grade is level for all roads. Lane widths are assumed to be 12 feet. Use the manual method.

Solution. We will analyze each interval.

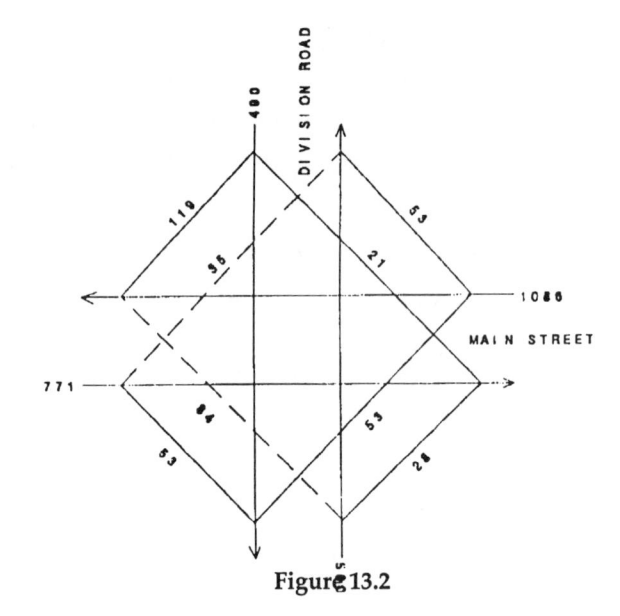

Figure 13.2

Interval 1:

Interval 1 is illustrated in Figure 13.3. During this interval of time the north and southbound movements have the green light and all others are stopped. Traffic assumed to be moving in this interval is assigned to the appropriate lanes. The movement which will require the most time is the northbound movement. Using Eq. 13.1.1 the time required for this movement is

$$t = \frac{2.1 M_c}{N_c} + 3.7$$

$$= \frac{2.1(476)}{60} + 3.7 = 20.4 \text{ sec}$$

where $M_c = 476$ vehicles/hr and $N_c = 60$ cycles/hr (1 cycle/min).

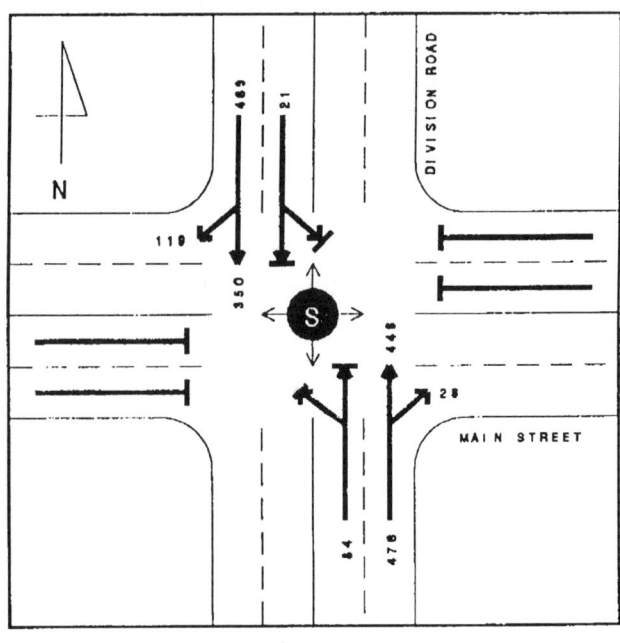

Figure 13.3

Interval 2:

Interval 2 is illustrated in Figure 13.4. This is the amber or clearance time. The approach speed is assumed to be 30 mph. From Eq. 13.1.2 the time for this interval is

$$Y = t_r + \frac{V}{2a + 64.4G} + \frac{W+L}{V}$$

$$= 1 + \frac{44}{2 \times 14} + \frac{48 + 20}{44} = 4.1 \text{ sec}$$

where we have assumed $t_r = 1$ sec, $a = 14$ ft/sec², and $G = 0$. With four lanes, $W = 48$ ft, $V = 30$ mph (44 fps), and $L = 20$ ft.

Please note that the left turning vehicles are assumed to complete this maneuver on amber. It is assumed that 2 vehicles can turn left on each amber. Therefore, 120 left turns can occur during one hour.

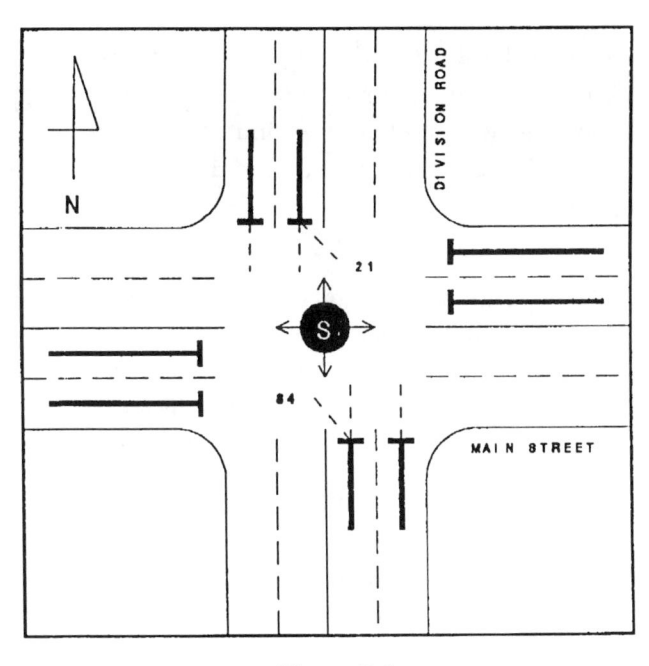

Figure 13.4

Interval 3:

Interval 3 is illustrated in Figure 13.5. This interval is the green time for east and westbound traffic. The time required for this interval is

$$t = \frac{2.1 M_c}{N_c} + 3.7$$

$$= \frac{2.1(910)}{60} + 3.7 = 35.5 \text{ sec}$$

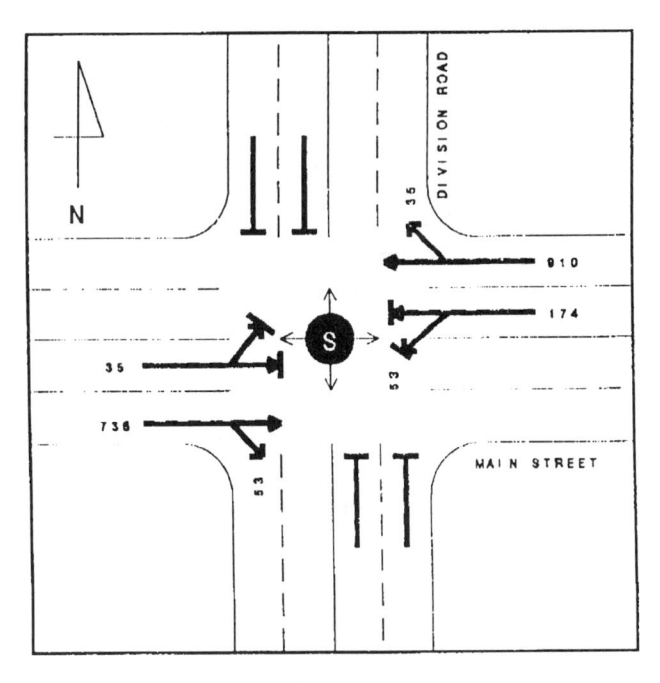

Figure 13.5

Interval 4:

Interval 4 is illustrated in Figure 13.6. This interval is the amber time for the east and west movements. Since the speed limit is the same for both roads the amber time is the same as in Interval 2.

$$Y = 4.1 \text{ sec}$$

The capacity of this geometric configuration is the ratio of the time required to accommodate the demand with respect to the total time available. The time required is the summation of the critical times for each interval:

Interval 1:	20.4
Interval 2:	4.1
Interval 3:	35.5
Interval 4:	4.1
Total	64.1 sec

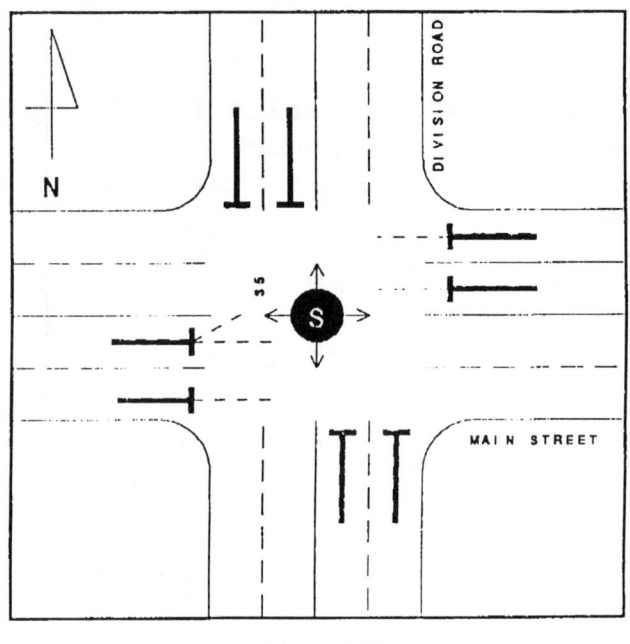

Figure 13.6

Time Available = 60 sec   Capacity = 64.1/60 = 1.07 or 107%

The volume-to-capacity ratio is over 1.0. Therefore the Level-of-Service is "F".

EXAMPLE 13.2

Repeat Example 13.1, but add a northbound left turn lane as shown in Figure 13.7.

Solution. With the exception of Interval 1, everything in this problem is the same as in Example 13.1. Interval 1 is recalculated. The new critical movement is the southbound through traffic. Using $M_c = 364$ we find

$$t = \frac{2.1 \times 364}{60} + 3.7 = 16.4 \text{ sec}$$

The lane assignment for southbound traffic of 364 and 126 is determined as follows:

A) As 238 of northbound traffic discharges per lane, 238 of southbound traffic discharges from the lane not being blocked by left turning traffic.

B) After the northbound traffic clears, southbound left turns can discharge and the remaining traffic will distribute into both southbound lanes.

C) The southbound traffic left after northbound clears = 490 − 338 = 252.

D) Subsequently, 126 vehicles will use the inside lane.

E) The outside lane assignment is 238 + 126 = 364 vehicles.

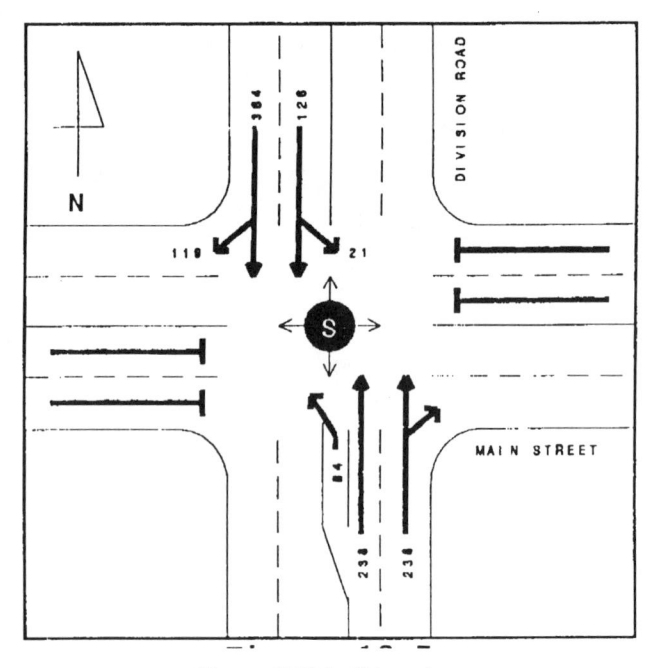

Figure 13.7 Left turn lane.

Time Required:

	Interval 1:	16.4
	Interval 2:	4.1
	Interval 3:	35.6
	Interval 4:	4.1
	Total	60.2 sec

Time Available = 60 sec   Capacity = 60.2/60 = 1.0 or 100%

The volume-to-capacity ratio is 1.0. Therefore the Level-of-Service is approximately "E".

—— EXAMPLE 13.3 ——

Repeat Example 13.1, but add a southbound right turn lane as shown in Figure 13.8.

Solution. All except the first interval remain the same. For Interval 1 the critical movement is the northbound through and right turn movement:

$$t = \frac{2.1 \times 455}{60} + 3.7 = 19.6 \text{ sec}$$

Time Required:

	Interval 1:	19.6
	Interval 2:	4.1
	Interval 3:	35.6
	Interval 4:	4.1
	Total	63.4 sec

Time Available = 60 sec

Capacity = 63.4/60 = 1.06 or 106%

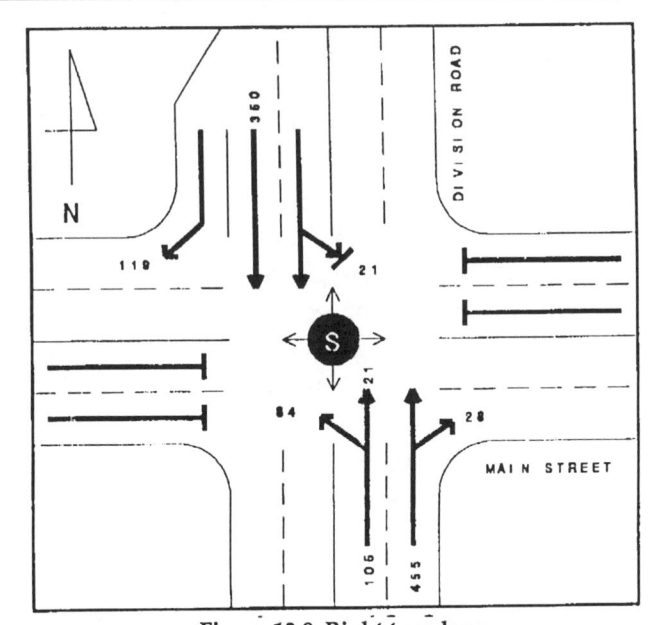

Figure 13.8 Right turn lane.

The volume-to-capacity ratio is over 1.0. Therefore the Level-of-Service is "F".

—— EXAMPLE 13.4 ——

Repeat Example 13.3, but add a northbound left turn lane and a southbound right turn lane as shown in Figure 13.9.

Solution. The critical movement for Interval 1 is now the southbound through movement:

$$t = \frac{2.1 \times 304}{60} + 3.7 = 14.3 \text{ sec}$$

Time Required:

	Interval 1:	14.3
	Interval 2:	4.1
	Interval 3:	35.6
	Interval 4:	4.1
	Total	58.1 sec

Time Available = 60 sec
Capacity = 58.1/60 = 0.97

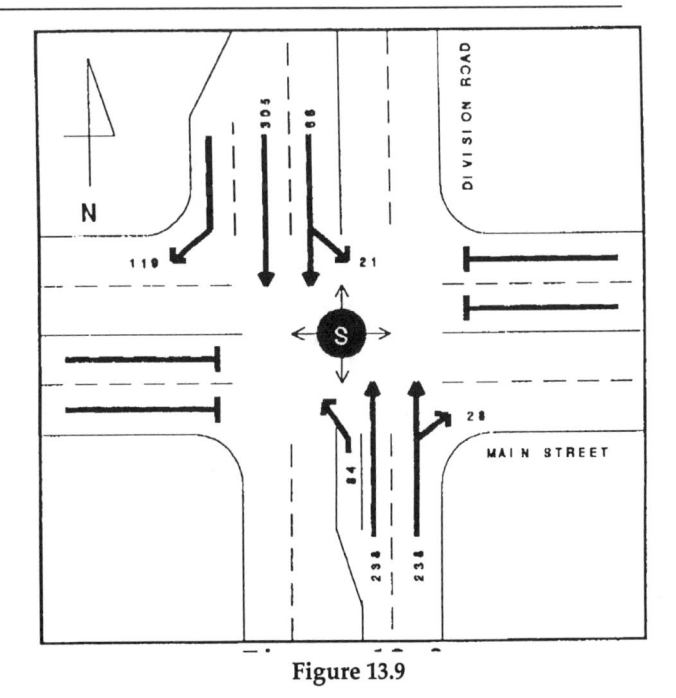

Figure 13.9

The volume-to-capacity ratio is between 0.9 and 1.0. Therefore the Level-of-Service is "E".

13.2 Highway Safety and Accident Analysis

The emphasis on highway safety has increased steadily since the National Highway Safety Act of 1966. The act requires each state to have an approved program to reduce traffic accidents. There have been several additional safety related acts since 1966. In addition most state and local agencies having responsibility for highways have lost their sovereign immunity since 1966. Most road agencies are being sued for tort liabilities.

The accident analysis procedures being used are more of an art than a science. A major component of a highway safety program is the analysis of reported accidents. The difficulty in using this data is that it is highly variable and changes from state to state, year to year, and within the state based upon the rigor of police reporting such as between highly urban areas and rural areas. A common analysis procedure is the development and visual inspection of a collision diagram. An example of a collision diagram is illustrated in Figure 13.10. The engineer inspects this diagram which can be for a one year period or a period of several years. The objective of this inspection is to identify a "correctable" pattern. An example would be a high number of right-angle accidents which may be corrected by a stop sign, a four-way stop, a stop-and-go traffic signal, or improved sight distance. Another example would be a high number or percentage of accidents occurring during wet weather which may be corrected by resurfacing of the pavement.

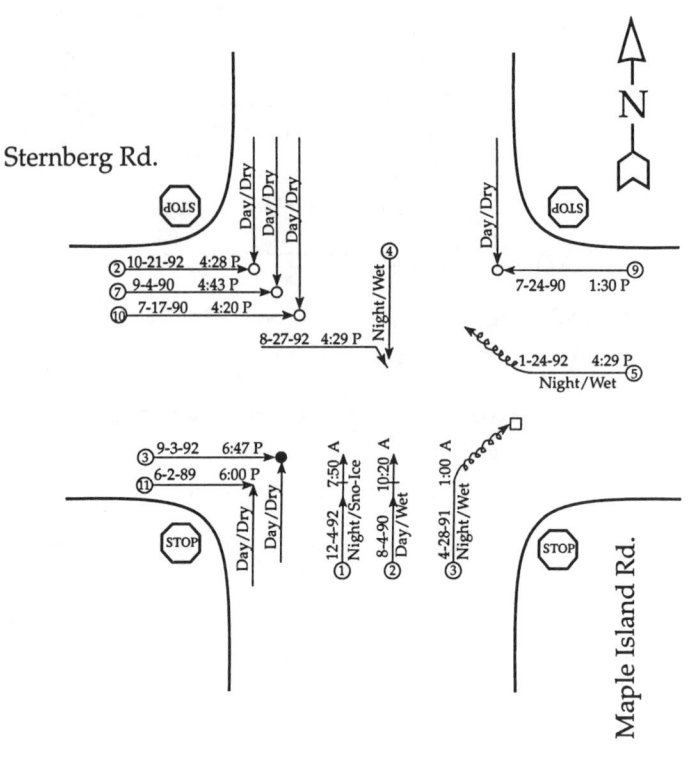

Figure 13.10 A collision diagram.

In addition to inspecting a collision diagram, it is a common practice to calculate the accident rate for a segment of highway or an intersection. This process is more or less a national standard. For a segment of highway this rate is per 100 million vehicle miles of travel (HMVM). The equation for the calculation of the accident rate is

$$R = \frac{A \times 100,000,000}{ADT \times N \times 365 \times L} \tag{13.2.1}$$

where
- R = the accident rate for 100 million vehicle miles
- A = the number of accidents during the period of analysis
- ADT = average daily traffic
- N = time period in years
- L = length of segment in miles

For an intersection the accident rate is per million entering vehicles (MEV). The equation is

$$R = \frac{A \times 1,000,000}{ADT \times N \times 365} \tag{13.2.2}$$

where
- R = the accident rate for one million entering vehicles
- ADT = the average daily traffic entering the intersection from all legs

Accident rates unless specified otherwise are for all accidents. However accident rates may be calculated as a function of severity; such as, Fatal or Injury.

EXAMPLE 13.5

A 10 mile section of a five-lane, two-way highway had the following reported accidents (*PDO* are property damage only accidents):

Year	Fatal	Injury	PDO	ADT
1989	2	48	115	12000
1990	0	53	211	13000
1991	4	61	180	13500
1992	5	73	242	14000
1993	7	98	180	14500

What are accident rates for all accidents, fatal accidents, and injury accidents? What is the severity ratio?

Solution. Use Eq. 13.2.1:

Total number of accidents = Fatal + Injury + *PDO* = 1279
Number of fatal accidents = 18
Number of injury accidents = 351
Injury accidents include fatal accidents
Average *ADT* = 13400

$$R \text{ (all accidents)} = \frac{1279 \times 100,000,000}{13400 \times 5 \times 365 \times 10} = 523 \text{ HMVM}$$

$$R \text{ (fatal accidents)} = \frac{18 \times 100,000,000}{13400 \times 5 \times 365 \times 10} = 7.36 \text{ HMVM}$$

$$R \text{ (injury accidents)} = \frac{351 \times 100,000,000}{13400 \times 5 \times 365 \times 10} = 143.5 \text{ HMVM}$$

The severity ratio is equal to the number of fatal and injury accidents divided by all accidents:

$$\text{Severity Ratio} = \frac{351}{1279} = 0.27$$

13.3 Sight Distance

13.3.1 Stopping Sight Distance

Stopping Sight Distance has two components: Perception and Reaction time and Braking Distance. The *Perception and Reaction Time* is the period of time from when the driver recognizes an object or a hazard on the roadway to the time the driver actually applies the brakes. A time of 2.5 seconds has been determined by the American Association of State Highway and Transportation Officials (AASHTO) to be adequate for the majority of drivers and situations. This time is used to determine the *Perception and Reaction Distance*.

Braking Distance is the distance needed to bring the vehicle to a complete stop after the brakes have been applied. This distance together with the Perception and Reaction Distance equals the necessary *Stopping Sight Distance*.

The height of the driver's eye is 3.5 feet above the road surface. The height of object is considered to be 6 inches above the road surface. The design speed should be at least equal to the posted speed limit. The assumed coefficient of friction for this speed is found in Table 13.2.

Table 13.2 Stopping sight distance (p. 138 from *A Policy on Geometric Design of Highways and Streets*, 1984, by the Am. Assoc. of State Highway and Trans. Officials, Wash., DC, by permission).

Design speed (mph)	Assumed Speed (mph)	Reaction Time (sec)	Reaction Distance (ft)	Design Coefficient of Friction f	Braking Distance on Level (ft)	Stopping Sight Distance Computed for Design (ft)	(ft)
20	20-20	2.5	73.3-73.3	0.40	33.3-33.3	106.7-106.7	125-125
25	24-25	2.5	88.0-91.7	0.38	50.5-54.8	138.5-146.5	150-150
30	28-30	2.5	102.7-110.0	0.35	74.7-85.7	177.3-195.7	200-200
35	32-35	2.5	117.3-128.3	0.34	100.4-120.1	217.7-248.4	225-250
40	36-40	2.5	132.0-146.7	0.32	135.0-166.7	267.0-313.3	275-325
45	40-45	2.5	146.7-165.0	0.31	172.0-217.7	318.7-382.7	325-400
50	44-50	2.5	161.3-183.3	0.30	215.1-277.8	376.4-461.1	400-475
55	48-55	2.5	176.0-201.7	0.30	256.0-336.1	432.0-537.8	450-550
60	52-60	2.5	190.7-220.0	0.29	310.8-413.8	501.5-633.8	525-650
65	55-65	2.5	201.7-238.3	0.29	347.7-485.6	549.4-724.0	550-725
70	58-70	2.5	212.7-256.7	0.28	400.5-583.3	613.1-840.0	625-850

The Perception and Reaction Distance is given by

$$s = 1.47tV \quad (13.3.1)$$

where s = distance (ft)
  t = brake reaction time (seconds)
  V = initial speed (mph)

The Braking Distance is

$$s = \frac{v^2}{2gf} = \frac{V^2}{30f} \quad (13.3.2)$$

where v = initial speed, fps
  V = initial speed, mph
  g = acceleration of gravity, 32.2 fps^2
  f = coefficient of friction between tires and wet pavement

The Braking Distance on a slope is given by

$$s = \frac{V^2}{30(f+G)} \quad (13.3.3)$$

where G = longitudinal slope of the roadway.

EXAMPLE 13.6

You are designing the reconstruction of a rural two lane bituminous highway. The posted speed is 55 miles per hour. Determine the stopping distance. What is the design speed?

Solution. The first assumption is that the pavement surface is wet. The design speed should be at least equal to the posted speed limit. Sixty miles per hour will be used in this solution. The assumed coefficient of friction for this speed is 0.29. This value is found in Table 3.2.

Use Eq. 13.3.1 for the Perception and Reaction Distance:
$$s = 1.47tV = 1.47 \times 2.5 \text{ (sec)} \times 60 \text{ (mph)} = 220.5 \text{ feet}$$

The Braking Distance is determined by using Eq. 13.3.2:
$$s = \frac{V^2}{30f} = \frac{60^2}{30 \times 0.29} = 414 \text{ ft}$$

The total stopping distance would be the sum of the Perception and Reaction Distance plus the Braking distance:
$$220.5 + 413.7 = 634.3 \text{ ft}$$

13.3.2 Passing Sight Distance

Passing sight distance is the length of roadway needed to safely complete normal passing maneuvers. Assumptions about traffic behavior have been established by AASHTO to be used when determining the minimum passing distance. They are as follows:

1. The overtaken vehicle travels at uniform speed.

2. The passing vehicle has reduced speed and trails the overtaken vehicle as it enters a passing section.

3. When the passing section is reached, the driver requires a short period of time to perceive the clear passing section and to react to start the maneuver.

4. Passing is accomplished under what may be termed a delayed start and a hurried return in the face of opposing traffic. The passing vehicle accelerates during the maneuver, and its average speed during the occupancy of the left lane is 5 to 10 mph higher than that of the overtaken vehicle.

5. When the passing vehicle returns to its lane, there is a suitable clearance length between it and an oncoming vehicle in the other lane.

The height of the eye is 3.5 ft and the height of the object is to the top of vehicle or 4.25 ft.

The minimum passing sight distance for two-lane highways is determined by the sum of four distances:

d_1 Distance traversed during perception and reaction time and during the initial acceleration to the point of encroachment on the left lane.

d_2 Distance traveled while the passing vehicle occupies the left lane.

d_3 Distance between the passing vehicle at the end of its maneuver and the opposing vehicle.

d_4 Distance traversed by an opposing vehicle for two-thirds of the time the passing vehicle occupies the left lane, or 2/3 of d_2.

The distance d_1 traveled during the initial maneuver period is calculated using:
$$d_1 = 1.47 t_1 \left(v - m + \frac{a t_1}{2} \right) \tag{13.3.4}$$

where t_1 = time of initial maneuver, sec

a = average acceleration, mph/sec
v = average speed of passing vehicle, mph
m = difference in speeds of the two vehicles, mph

Table 13.3 includes this distance along with d_2, d_3, and d_4.

Table 13.3 Passing distances (p. 151, from *A Policy on Geometric Design of Highways and Streets*, 1984, by the Am. Assoc. of State Highway and Trans. Officials, Wash., DC, by permission).

Speed Group (mph)	30-40	40-50	50-60	60-70
Average passing Speed (mph)	34.9	43.8	52.6	62.0
Initial maneuver:				
a = average acceleration	1.40	1.43	1.47	1.50
t_1 = time (sec)	3.6	4.0	4.3	4.5
d_1 = distance traveled (ft)	145	215	290	370
Occupation of left lane:				
t_2 = time (sec)	9.3	10.0	10.7	11.3
d_2 = distance traveled (ft)	475	640	825	1,030
Clearance length:				
d_3 = distance traveled (ft)	100	180	250	300
Opposing vehicle:				
d_4 = distance traveled (ft)	315	425	550	680
Total distance = $d_1+d_2+d_3+d_4$	1,035	1,460	1,915	2,380

The distance d_2 traveled while the passing vehicle occupies the left lane is computed from

$$d_2 = 1.47 v t_2 \tag{13.3.5}$$

where t_2 = time passing vehicle occupies the left lane, usually found to be from 9.3 to 10.4 sec
v = average speed of passing vehicle, mph

The clearance length d_3 is found in Table 13.3.

The distance d_4 traversed by an opposing vehicle is given as two-thirds of the distance traveled by the passing vehicle. The opposing vehicle is assumed to be traveling at the same speed as the passing vehicle:

$$d_4 = \frac{2 d_2}{3} \tag{13.3.6}$$

───── **EXAMPLE 13.7** ─────

Compute the passing distance if the speed of the passing car is 60 mph.

Solution. Using the above formulas the passing distance is computed to be

$$d_1 = 1.47 t_1 (v - m + a t_1/2)$$

$$= 1.47(4.3)(60 - 5 + 1.47 \times 4.3/2) = 368 \text{ ft}$$

Note: The difference in speeds between the two vehicles was assumed to be 5 mph. We find

$$d_2 = 1.47vt_2$$
$$= 1.47(60)(10.7) = 944 \text{ ft}$$

From Table 13.3 the Clearance Length $d_3 = 250$ feet. Finally,

$$d_4 = 2 \times 944/3 = 629 \text{ ft}$$

Therefore, the total design passing distance is equal to

$$\text{Total distance} = 368 + 944 + 250 + 629 = 2,190 \text{ ft}$$

Note that this is between the values listed in Table 13.3 for the ranges 50–60 and 60–70, as it should be.

13.3.3 Decision Sight Distance

Decision sight distance is the distance required for a driver to detect an unexpected object, information source (traffic signal), or hazard in the roadway and to recognize the hazard, select an appropriate speed and path, and initiate and complete the required safety maneuver. The same assumptions are made here as they were for stopping sight distance. Height of eye is 3.5 feet above the roadway and height of object is 6 inches above the roadway. The total time needed to make the necessary maneuvers is comprised of three distinct elements.

Detection and recognition time is the amount of time required for a driver to detect and recognize that an object or hazard is being approached. *Decision and response initiation time* is the amount of time for the driver to decide on the proper maneuver to be taken and to initiate the required action. The final element is the time required to accomplish a vehicle maneuver, Maneuver Time.

Table 13.4, Decision Sight Distance, lists each of these time periods and the amount of time allocated to each for varying speeds.

Table 13.4 Decision Sight Distance (p. 147; from *A Policy on Geometric Design of Highways and Streets*, 1984, by the Am. Assoc. of State Highway and Trans. Officials, Wash., DC, by permission)

Design Speed (mph)	Time (sec)			Decision Sight Distance (ft)		
	Pre-maneuver					
	Detection & Recognition	Decision & Response Initiation	Maneuver	Summation	Computed	Rounded for Design
30	1.5-3.0	4.2-6.5	4.5	10.2-14.0	449-616	450-625
40	1.5-3.0	4.2-6.5	4.5	10.2-14.0	598-821	600-825
50	1.5-3.0	4.2-6.5	4.5	10.2-14.0	748-1,027	750-1,025
60	2.0-3.0	4.7-7.0	4.5	11.2-14.5	986-1,276	1,000-1275
70	2.0-3.0	4.7-7.0	4.0	10.7-14.0	1,098-1,437	1,100-1,450

13.4 Horizontal Curves

13.4.1 Superelevation and Friction

Horizontal curves are used as a transition between two segments of roadway. Horizontal curves are geometrically designed as circular curves, in contrast to parabolic curves which are used for vertical curves. When a vehicle moves in a circular path it is pulled radially outward by a centrifugal force. This force outward is balanced by a force due to the side friction factor f and the gravity force produced by the rate of superelevation e of the curve. The side friction factor force is the force between the tires and the road surface. The side friction factor and rate of superelevation are directly related to the vehicle speed and radius of the curve as shown by

$$e + f = \frac{V^2}{15R} \qquad (13.4.1)$$

where
- e = rate of roadway superelevation, ft/ft
- f = side friction factor
- V = speed of vehicle, mph
- R = radius of curve, ft

Curved sections of roadway are typically superelevated. This is done by sloping or embanking the outer edge of the curve. The higher the speeds, the greater the amount of banking that is needed. Superelevation is the amount of rise per horizontal foot. Generally, superelevation rates are between 0.02 and 0.16. Areas which include snow and ice have less embankment on the roads due to sliding down the slope and the maximum value is 0.08.

The side friction factor is also known as the lateral ratio, cornering ratio, unbalanced centrifugal ratio, and friction factor. The maximum value for side friction factor can be determined by placing the rate of roadway superelevation at zero in

$$f_{max} = \frac{V^2}{15R} \qquad (13.4.2)$$

The maximum value is representative of when the vehicle is on the verge of a skid. For safe highway design the side friction factor should be designed at its lower limits. The side friction factor can be reduced by either decreasing the velocity or increasing either the superelevation rate or the curve's radius. Side friction factor is dependent on both geometric design and passenger comfort since the passenger as well as the car is pulled radially outward.

There are many combinations of e and f to determine the best overall design for a horizontal curve. There are five methods for counteracting centrifugal forces:

1. Superelevation and side friction are directly proportional to the degree of curve.

2. Side friction is such that a vehicle traveling at design speed has all centrifugal force counteracted in direct proportion by side friction on curves up to those requiring f_{max}. For sharper curves, f remains at f_{max} and e is then used in direct proportion to the continued increase in curvature until e reaches e_{max}.

3. Superelevation is such that a vehicle traveling at the design speed has all centrifugal force counteracted in direct proportion by superelevation on curves up to that requiring e_{max}. For sharper curves, e remains at e_{max} and f is then used in direct proportion to the continued increase in curvature until f reaches f_{max}.

4. Method 4 is the same as method 3, except that it is based on average running speed instead of design speed.

5. Superelevation and side friction are in a curvilinear relation with degree of curve, with values between those of methods 1 and 3.

13.4.2 Design Criteria

Other criteria must be taken into consideration when designing a horizontal curve. The first criteria is *degree of curvature, D*. It is the angle which is subtended at the center of a circular curve by an arc of 100 ft. Its maximum value is

$$D_{max} = \frac{85,950(e+f)}{V^2} \tag{13.4.3}$$

Or, in terms of the radius of curvature R, the more simplified equation is

$$D = \frac{5,729.28}{R} \tag{13.4.4}$$

The minimum safe radius for a design can be determined by using the equation above or the following:

$$R_{min} = \frac{V^2}{15(e+f)} \tag{13.4.5}$$

The sight distance around a horizontal curve is important. It is the distance at which a driver can see an object lying in the roadway ahead. It should equal or exceed the stopping sight distance. Accepted standards place the driver's height of eye at 3.5 ft and the height of object 6 inches above the roadway surface. At night, the driver is unable to see as far because the headlights shine straight ahead. Daytime measurements are normally used, because it is assumed that a driver will drive slower at night.

To provide for adequate sight distance all obstructions must be placed a sufficient distance back from the roadway. Bank setbacks should be a minimum of 2.0 ft.

Other factors which are considered when designing a horizontal curve are length of tangent, external distance, middle ordinate distance, and length of curve. These values are shown in Figure 13.11, and are given by the following equations:

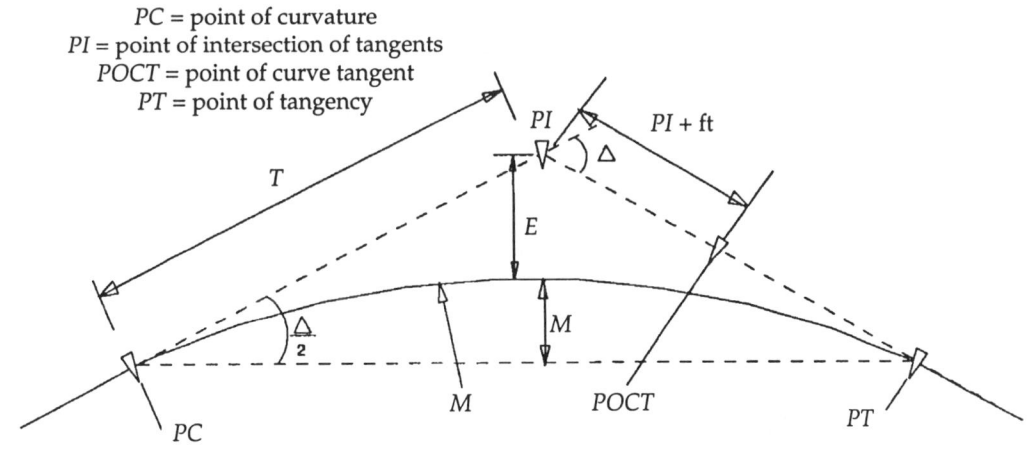

PC = point of curvature
PI = point of intersection of tangents
POCT = point of curve tangent
PT = point of tangency

Figure 13.11 Simple Horizontal Curve.

$$R = \text{Length of Radius of Curve} = 5729.578 / D \tag{13.4.6}$$

$$T = \text{Length of Tangent} = R \tan\left(\tfrac{\Delta}{2}\right) \tag{13.4.7}$$

$$E = \text{External Distance} = T \tan\left(\tfrac{\Delta}{4}\right) \tag{13.4.8}$$

$$M = \text{Middle Ordinate Distance} = E \cos\left(\tfrac{\Delta}{2}\right) \tag{13.4.9}$$

$$L = \text{Length of Curve} = 100(\Delta/D) \tag{13.4.10}$$

EXAMPLE 13.8

Determine the geometric elements of the following horizontal curve:

Given: Δ = 20 degrees 46 minutes
         D = 5 degrees 30 minutes
         PI = 300 + 40 ft (in stations)

Determine: R, T, E, M, L, PC, and PT

Solution. Refer to Fig. 13.12:

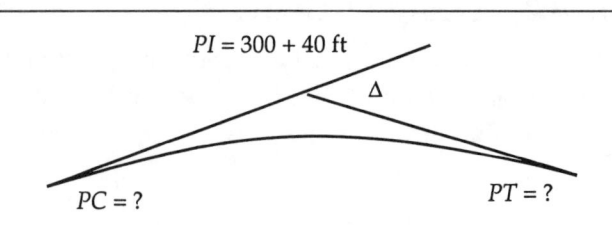

$\Delta = 20\deg\ 46\min$
$\Delta/2 = 10\deg\ 23\min$
$\Delta/4 = 5\deg\ 11\min\ 30\sec$

Figure 13.12

$R = 5729.5780 / D = 1041.74$ ft

$T = R\tan(\Delta/2) = 1041.74 \tan 10d23m = 190.87$ ft

$E = T\tan(\Delta/4) = 190.87 \tan 5d11m30s = 17.54$ ft

$M = E\cos(\Delta/2) = 17.54 \cos 10d23m = 27.60$ ft

$L = 100(\Delta/D) = 100(20d46m)/5d30m = 377.58$ ft

$$
\begin{aligned}
PI &= 300 + 40 \\
-T &= 1 + 90.87 \\
\hline
PC &= 298 + 49.13 \\
+L &= 3 + 77.58 \\
\hline
PT &= 302 + 26.71 \text{ (in stations)}
\end{aligned}
$$

EXAMPLE 13.9

Given:
  PC = 257+31 ft.
  PI = 289+44 ft.
  PT = 301+00 ft.
  D = 2d30m

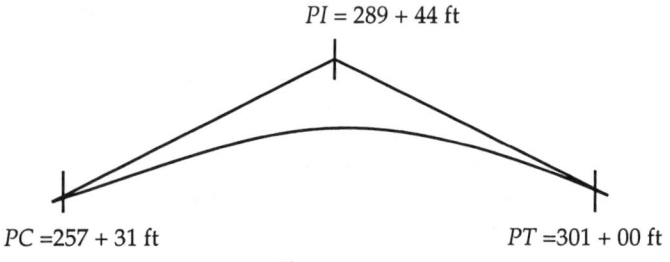

Figure 13.13

Determine: 1) Deflection angles every 100 ft from PC to PI.
            2) Deflection angle between Station 280+00 and PI.

Solution. Refer to Fig. 13.13:

Deflection Angle = ((Station − PC)/100) (D/2)
Angle at 260+00 = ((260+00 − 257+31)/100) (2.5/2) = 3d21m45s
Angle at 270+00 = ((270+00 − 257+31)/100) (2.5/2) = 4d44m15s

Station	Angle
260+00	3d21m45s
270+00	15d51m45s
280+00	28d21m45s
289+00	40d9m45s

Deflection angle between 280+00 and PI: Angle = ((289+44 − 280+00)/100)*(2.5/2) = 11d48m

13.5 Vertical Curves

Vertical curves provide a gradual transition between two different roadway grades. They are designed to enable a vehicle to increase and decrease elevation in a safe and comfortable manner. There are two types of vertical curves, crest and sag.

Considerations for drainage, sight distance, passenger comfort and appearance are essential to the design of every vertical curve. A roadway should have a smooth gradeline. Hidden dips, broken-back gradelines, and roller-coaster crests should be avoided to provide passengers with the greatest possible comfort. Each roadway should be designed to meet the criteria fitting for its terrain and should be similar to surrounding areas.

When continually increasing grades up a hill, higher grades should be located near the bottom of the curve. The change in grades should allow the curve to continually flatten out as it nears the top.

Grades should be reduced, if conditions permit, through intersections which cross through vertical curves. Fig. 13.14 shows a vertical curve.

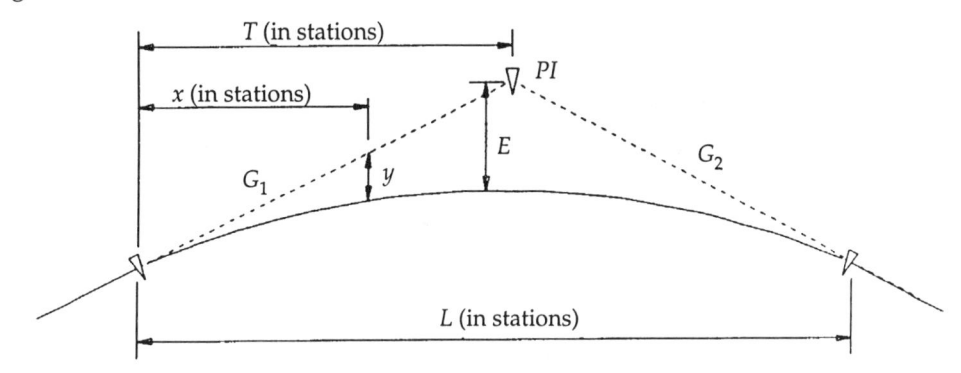

Figure 13.14 A vertical curve.

The following pertain to the parameters in Fig. 13.14:

- G_1 = Grade one (percentage)
- G_2 = Grade two (percentage)
- L = distance from beginning point of vertical curve to point of tangency (in stations)
- T = distance from beginning point of vertical curve to point of highest elevation (in stations)
- PI = Point of Intersection
- E = elevation between the top of curve and Point of Intersection
- y = offset distance between the vertical curve and grade
- x = distance from beginning point on vertical curve to the offset

In terms of the defined parameter, we have

$$E = \frac{(G_1 - G_2)L}{8} \quad (13.5.1)$$

$$\frac{y}{x^2} = \frac{E}{T^2} \quad (13.5.2)$$

13.5.1 Crest Vertical Curve

Vertical curves over crests, see Fig. 13.15, are designed to have the greatest possible length that conditions permit. The minimum length of a crest vertical curve is determined by sight distance, grade difference, the height of eye and the height of object.

Sight distance is the length between a vehicle as it enters a crest curve and the farthest distance visible to a driver to the top of the object. It is typically assumed that the height of eye is at 3.5 ft above roadway surface and the farthest distance visible is at an elevation of 6 inches above the roadway.

There are two basic formulas for stopping sight distance on a crest curve. It is necessary to solve both equations and then determine which solution meets the problem criteria. They are as follows:

$$L = \frac{AS^2}{100\left(\sqrt{2h_1} + \sqrt{2h_2}\right)^2} \qquad S < L \qquad (13.5.3)$$

$$L = 2S - \frac{200\left(\sqrt{h_1} + \sqrt{h_2}\right)}{A} \qquad S > L \qquad (13.5.4)$$

where L = length of vertical curve, ft
 S = sight distance, ft
 A = difference in grades, percent
 h_1 = height of eye, ft
 h_2 = height of object above roadway, ft

G_1 and G_2, Grade
A, Algebraic difference = $G_1 - G_2$
L, length of vertical curve

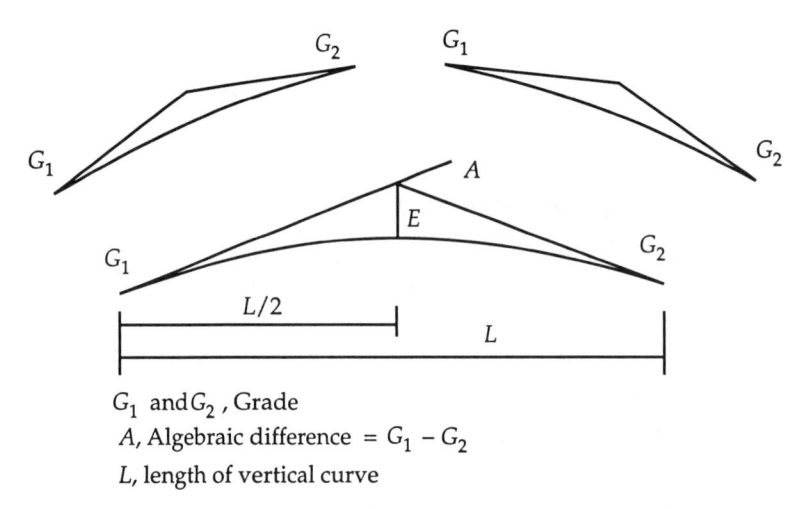

Figure 13.15 Crest vertical curve.

Passing sight distance must also be determined for crest vertical curves. It places height of eye at 3.5 ft, similar to stopping sight distance. The top of the oncoming vehicle is assumed to be 4.25 ft. Two equations for passing sight distance are given. It is necessary to solve both equations and then determine which one satisfies the initial criteria. They are as follows:

$$L = \frac{AS^2}{3,093} \qquad S < L \qquad (13.5.5)$$

$$L = 2S - \frac{3,093}{A} \qquad S > L \qquad (13.5.6)$$

EXAMPLE 13.10

Determine the minimum length of a creast vertical curve and 600 ft of stopping sight distance. The approach grade is 3% and the leaving grade is 0% (flat).

Solution. We know or assume the following:

$$h_1 = 3.5 \text{ ft}$$
$$h_2 = 0.5 \text{ ft}$$
$$S = 600 \text{ ft}$$
$$A = 3 - 0 = 3$$

For $S < L$ use Eq. 13.5.3:

$$L = \frac{3 \times 600^2}{100\left(\sqrt{7.0} + \sqrt{1}\right)^2} = 812.5 \text{ ft}$$

We see that $S < L$ so this is the correct answer.

Check by solving the other possibility $S > L$ using Eq. 13.5.4:

$$L = 2 \times 600 - \frac{200 \times \left(\sqrt{3.5} + \sqrt{0.5}\right)^2}{3} = 756.9 \text{ ft}$$

Here $S < L$; this is not acceptable.

13.5.2 Sag Vertical Curve

The length of a sag vertical curve is determined by four basic criteria. These are headlight sight distance, rider comfort, drainage control, and overall general appearance. General appearance is considered because short curves give a disjointed image in contrast to long curves which provide a more flowing impression.

Headlight sight distance, the first criterion, is the basis for determining the minimum sight distance. The distance that is visible to the driver is dependent on the angle of the headlights to the ground surface and their position on the vehicle. The standard placement for headlights is 2.0 ft above the roadway surface with a 1° upward divergence. Two equations are provided to determine the length of a sag vertical curve; both must be solved to determine which is a true solution. They are:

$$L = \frac{AS^2}{400 + 3.5S} \qquad S < L \qquad (13.5.7)$$

$$L = 2S - \frac{400 + 3.5S}{A} \qquad S > L \qquad (13.5.8)$$

where L = length of sag vertical curve, ft
 S = distance visible from headlights, ft
 A = difference in grades, %

The length of a sag vertical curve should be designed to provide a headlight distance equal to the stopping sight distance.

Comfort is also a concern on sag curves. Comfort on a sag curve is affected by vehicle suspension, tire flexibility, and overall vehicle and passenger weight. It is approximately 50 percent of the distance required by the headlight distance in the above equation. The standard equation for the length of a sag vertical curve due to the comfort criterion is

$$L = \frac{AV^2}{46.5} \tag{13.5.9}$$

EXAMPLE 13.11

Given:

G_1 = –3%
G_2 = 2%
S = 700 ft (Sight Distance)
A = 2 – (–3) = 5

Determine: Minimum Length of vertical sag curve, shown in Fig. 13.16.

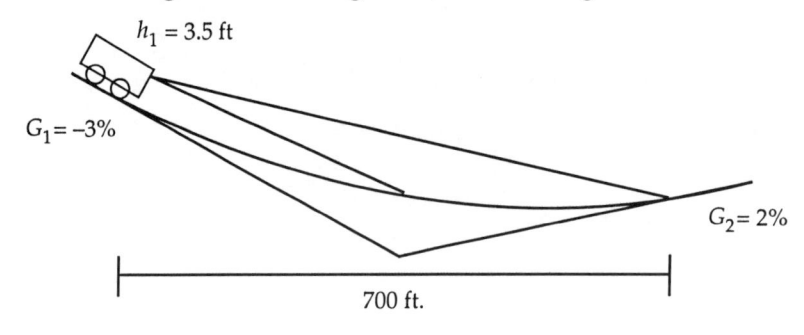

Figure 13.16

Solution. First, use Eq. 13.5.7 assuming $S < L$:

$$L = \frac{AS^2}{400 + 3.5S} = \frac{5 \times 700^2}{400 + 3.5 \times 700} = 860 \text{ ft}$$

Obviously, $S < L$ so this is correct.

Check by solving the other possibility of $S > L$. Using Eq. 13.5.8

$$L = 2 \times 700 - \frac{400 + 3.5 \times 700}{5} = 830.0 \text{ ft}$$

Here $S < L$, so this is not correct.

EXAMPLE 13.12

Given:

G_1 = 2%
G_2 = –3%
L = 400 ft = 4 stations
T = 200 ft = 2 stations

Determine: Offset every 50 ft.

Solution. First, solve for E:

$$E = \frac{(G_1 - G_2)L}{8} = \frac{(2-(-3))4}{8} = 2.5 \text{ ft}$$

$$y = \frac{Ex^2}{T^2} = \frac{2.5x^2}{4^2} = 0.625x^2$$

	Station	Tan. Elevation	x	y	Elevation
PC	10+00	1000.00	0.0	0.000	1000.00
	10+50	1001.00	0.5	0.156	1000.84
	11+00	1002.00	1.0	0.625	1001.37
	11+50	1003.00	1.5	1.406	1001.59
PI	12+00	1004.00	2.0	2.500	1001.50
	12+50	1002.50	1.5	1.406	1001.09
	13+00	1001.00	1.0	0.625	1000.37
	13+50	999.50	0.5	0.156	999.34
PT	14+00	998.00	0.0	0.000	998.00

13.6 Cross Section Design

The design of a highway cross section begins with the determination of the number of lanes and type of facility, such as, a 4-lane, 2-way highway, a 5-lane highway with a 2-way center turn lane, or a 4-lane divided highway with a wide turf median. The configuration is a function of the need for capacity, safety and the classification of the route.

The design of the pavement is a function of the traffic volumes and composition of commercial vehicles, soil characteristics, and life-cycle costs. The cross slope (crown) of the pavement is also part of the cross section design. On high type two lane pavements the cross slope normally varies from 1.5 to 2.0 percent. For an intermediate type pavement the cross slope varies from 1.5 to 3.0 percent. Low type pavement cross slopes vary from 2.0 percent to 6.0 percent.

Lane widths vary from 10 to 13 ft, with 12 foot lanes being desirable for most urban and rural highways. Lesser lane widths are often used for lower-speed routes. Shoulders improve the strength of the pavement, improve recovery for run-off-the -road vehicles and if wide enough, provided adequate room for parked vehicles. Shoulders may be surfaced, gravel, turf or any combination. The minimum shoulder width is 2 ft. Roads with high speeds and heavy truck volumes should have shoulders at least 10 if not 12 ft wide. The cross slope of a shoulder should be 2 to 6 percent for a paved surfaced, 4 to 6 percent for a gravel surface, and approximately 8 percent for a turf surface.

Sideslopes are designed for the stability of the cut or fill section and provide for the recovery of errant vehicles. Foreslopes go from the edge of shoulder to the start of the ditch. Backslopes go from the back of ditch to natural ground. Back slopes usually vary from 2:1 (2 ft horizontal for 1 foot vertical) to 3:1. Foreslopes 4:1 and flatter are desirable. Foreslopes 3:1 and steeper are undesirable and consideration of a roadside barrier is recommended.

Medians are desirable. They may range in width from 4 to 100 ft. A minimum width equal to the width of a lane is often used to allow for a two-way left turn lane. Otherwise widths should vary from 40 to 80 ft. This width is often a function of the minimum turning radius of the design vehicle.

EXAMPLE 13.13

Design the typical cross section for a rural two lane highway with a Design Speed of 60 mph. The Average Daily Traffic is 6000 vehicles with a Design Hour Volume of 700 vehicles and 8 percent commercial traffic. The terrain rolls.
  A) 2 paved lanes at 12 ft each, 10 foot gravel shoulders, the cross slope of the pavement is 2 % and 4% for the shoulders, the foreslope is 6:1.

B) 2 paved lanes at 12 ft each, 10 foot gravel shoulders, the cross slope of the pavement is 2 % and 3% for the shoulders, the foreslope is 5:1.

C) 2 paved lanes at 10 ft each, 8 foot gravel shoulders, the cross slope of the pavement is 2% and 5 % for the shoulders, the foreslope is 4:1.

D) 2 paved lanes at 11 ft each, 8 foot gravel shoulders, the cross slope is 1.5% for the pavement and 5% for the shoulders, the foreslope is 2.5:1.

Solution. The answer is A. The others are wrong because of the following:
- B) Cross slope for a gravel shoulder should be 4 to 6 percent.
- C) Pavement should be 12 ft for a 60 mph Design Speed.
- D) Pavement should be 12 ft for a 60 mph Design Speed and the foreslope should be 4:1 or flatter.

13.7 Geometric At-Grade Intersection Design

An intersection is an area where two or more roads come together. Except for freeways, intersections are at the same grade. The roads approaching the intersection are called the "legs" of the intersection.

Channelized areas help direct traffic and can restrict turns. Flared designs involve widening the entering traffic lanes to permit deceleration out of through lanes, and widening of leaving lanes to provide for acceleration and merging. Multileg intersections (more than four legs), are sometimes used, but should be avoided as much as possible. The legs of the intersection should be at right angles. The intersection should not be more acute than 60 degrees.

There are many factors to consider when designing at-grade intersections. These factors include, human factors, traffic control devices, traffic characteristics, vehicle characteristics, and economic factors.

The three design elements discussed are the design of tapers for auxiliary lanes, design of sight distance, and design of turning radii.

13.7.1 Design Of Auxiliary Lanes

Based on engineering traffic studies, volume and capacity requirements will determine the number of entering lanes required. A capacity analysis is conducted to determine the need for auxiliary lanes. Figure 13.17 shows the difference between an intersection with and without auxiliary lanes. An intersection with auxiliary lanes is often called a flared intersection. Auxiliary lanes move heavy turning movements out of the through traffic lanes. The most common auxiliary lane is for exclusive left turns. Since left turns must yield to oncoming traffic, they often increase delay of through movements. In addition, auxiliary lanes are also used for lane drops, lane shifts in construction zones, and ramps on freeways.

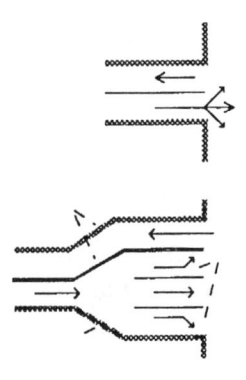

Figure 13.17 An intersection with and without auxiliary lanes.

There are three parts in the design of an auxiliary lane: tapers, deceleration lengths, and capacity. Capacity is the number of cars that need to be stored during peak hours. This is determined from the engineering field study. Deceleration lengths are determined from Table 13.5. Due to cost, many times the deceleration length is eliminated.

Table 13.5 Deceleration length.

On grades of 2 percent or less:

Design Speed (mph)	Deceleration Length (ft)
30	235
40	315
50	435

The design of the taper of Fig. 13.18 is determined from the following equation:

$$L = V \times W \quad \text{for } V \geq 45 \text{ mph} \quad (13.7.1)$$

$$L = \frac{W \times V^2}{60} \quad \text{for } V < 45 \text{ mph} \quad (13.7.2)$$

where
L = length of the taper, ft
V = design speed, mph
W = offset, ft

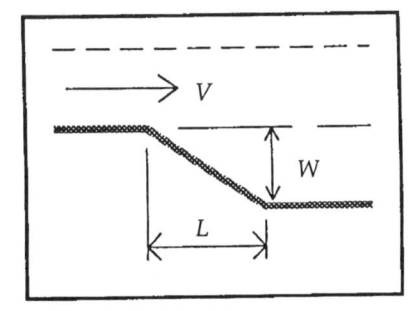

Figure 13.18 Highway taper.

The recommended storage length is based upon the number of vehicles expected to arrive during one cycle length of a traffic signal. The minimum length is recommended to be $1\frac{1}{2}$ times the average arrival rate. The desirable length is recommended to be twice the average arrival rate. The length of a vehicle is considered to average between 20 and 25 ft.

EXAMPLE 13.14

Determine the length of an auxiliary lane with:
A) a capacity of 10 vehicles, a design speed of 50 mph, and an offset of 12 ft;
B) a capacity of 5 vehicles, a design speed of 30 mph, and an offset of 6 ft; and
C) a capacity of 7 vehicles, a design speed of 40 mph, and an offset of 12 ft.

Solution.

A) From Table 13.5, the deceleration length for a design speed of 50 mph equals 435 ft. The length for capacity is 25 ft times 10 vehicles, which equals 250 ft. Using Eq. 13.7.1, with V equal to 50 mph and W equal to 12 ft:

$$L = 50(12) = 600 \text{ ft}$$

The total length of the auxiliary lane is

$$L_{\text{total}} = 435 + 250 + 600 = 1285 \text{ ft}$$

This design is too long for practical reasons and the designer would more than likely eliminate the deceleration length giving

$$L_{\text{total}} = 250 + 600 = 850 \text{ ft}$$

B) Using Table 13.5, the deceleration length for a design speed of 30 mph equals 235 ft. The length for capacity is 25 times 5 vehicles which equals 125 ft. Using Eq. 13.7.2, with V

equal to 30 mph and W equal to 6 ft:

$$L = (6 \times 30^2)/60 = 90 \text{ ft}$$

The total length of the auxiliary lane is

$$L_{total} = 235 + 125 + 90 = 450 \text{ ft}$$

A length of $125 + 90 = 215$ ft is also acceptable.

C) Using Table 13.5, the deceleration length for a design speed of 40 mph equals 315 ft. The length for capacity is 25 times 7 vehicles equals 175 ft. Using Eq. 13.7.2, with V equal to 40 mph and W equal to 12 ft:

$$L = (12 \times 40^2)/60 = 320 \text{ ft}$$

The total length of the auxiliary lane is

$$L_{total} = 315 + 175 + 320 = 810 \text{ ft}$$

A length of 495 ft would also be acceptable.

13.7.2 Design of Intersection Sight Distance

Adequate sight distance at an intersection is important. The driver should have an unobstructed view of the intersection. If the intersection right-of-way is controlled by yield signs, stop signs or a traffic signal, the sight distance may be limited to the area of control.

There should be adequate sight distance along both approaches and across the corner for all four quadrants of the intersection. This sight line is illustrated in Fig. 13.19, from page 741 of the 1990 Green Book.

There are four types of control: no control, yield control, stop control, or signal control.

13.7.2.1 No Control

For no control a driver needs to be able to adjust speed to avoid crossing traffic. The time required for perception, reaction, braking (or acceleration) is assumed to equal 3.0 seconds. Table 13.6 provides the distance required for the various approach speeds.

Table 13.6 Distance required for various approach speeds at an intersection.

Speed (mph)	Distance (ft)
10	45
15	70
20	90
25	110
30	130
35	155
40	180
50	220
60	260
70	310

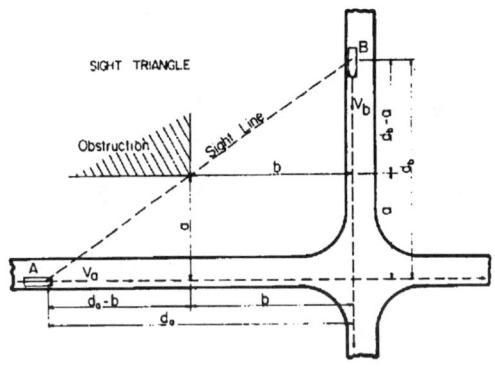

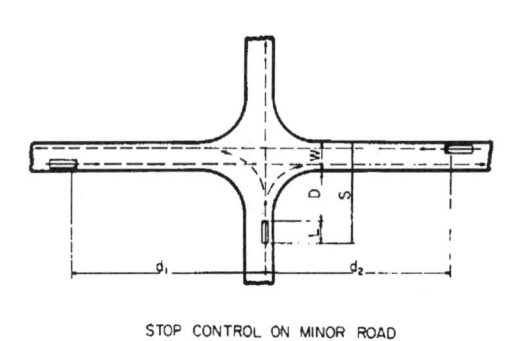

Figure 13.19 Sight distance at an intersection.

─── **EXAMPLE 13.15** ───

Determine the sight triangle for an intersection where A) one road has a design speed of 40 mph and the other road has a design speed of 30 mph, and B) one road has a design speed of 60 mph and the other road has a design speed of 50 mph.

Solution. From Table 13.6:
  A) The distance required in the sight triangle for 30 mph is 130 ft and for 40 mph is 180 ft.
  B) The distance required on the legs of the sight triangle for 50 mph is 220 ft and for 60 mph it is 260 ft.

13.7.2.2 Stop Control On A Minor Road

There are three maneuvers that are assumed to occur at an intersection.

1. The driver stops and crosses the major road.
2. The driver stops and turns left and in front of a vehicle approaching from the right.
3. The driver stops and turns right and in front of a vehicle approaching from the left.

It is assumed that the vehicle on the major road only slows down to 85 percent of its speed in the second and third case.

The sight distance d (ft) along the major road from the intersection is given by the Green Book as

$$d = 1.47V(J + t_a) \qquad (13.7.3)$$

where: V = design speed on the major road, mph
  J = sum of the perception time and the time required to actuate the clutch or an automatic shift, sec

$$t_a = \text{time required to accelerate and traverse the distance } S \text{ to clear the major road, sec}$$

The distance S (ft) that the crossing vehicle must travel to clear the major road is

$$S = D + W + L \tag{13.7.4}$$

where: D = distance from near edge of pavement to the front of a stopped vehicle, ft
          W = pavement width along path of crossing vehicle, ft
          L = overall length of vehicle, ft

Figures 13.20 and 13.21 provide the acceleration time t_a needed in Eq. 13.7.3.

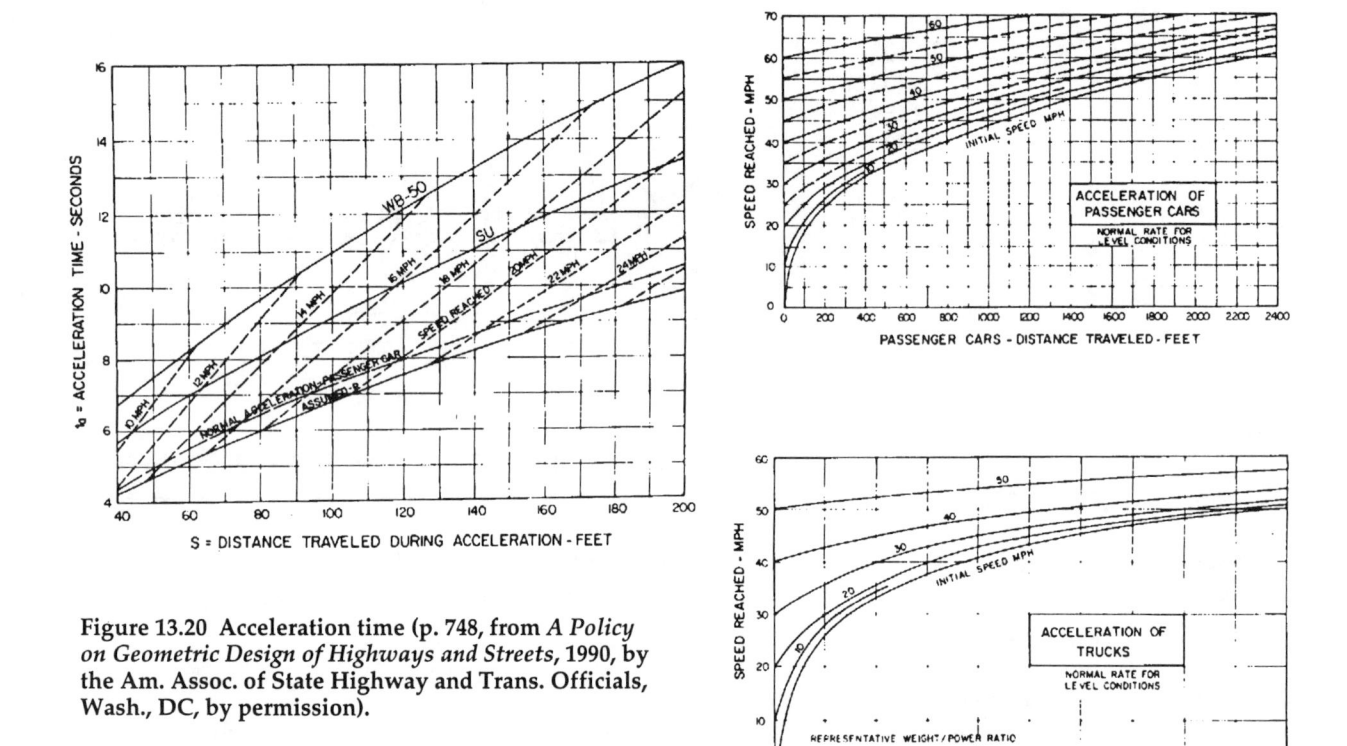

Figure 13.20 Acceleration time (p. 748, from *A Policy on Geometric Design of Highways and Streets*, 1990, by the Am. Assoc. of State Highway and Trans. Officials, Wash., DC, by permission).

Figure 13.21 Acceleration time (p. 749; Green Book).

─────── **EXAMPLE 13.16** ───────

Determine the sight distance required for a crossing maneuver by (a) a passenger car with a design speed of the major road of 50 mph; the major road has five 12-foot lanes and the length of the car is equal to 19 ft, and (b) a WB-50 truck with a design speed on the major road of 40 mph; the major road has four 11-foot lanes and the length of the truck is 55 ft.

Solution.

(a) First, S is determined by using Eq. 13.7.4. For general design purpose, $D = 10$ ft is assumed (measured from the pavement edge to the front of the vehicle). The width of the pavement $W = 5(12) = 60$ ft. The length of the passenger vehicle L is 19 ft. Therefore,

$$S = 10 + 60 + 19 = 89 \text{ ft}$$

This value of S is used to find t_a from Fig. 13.20: with the vehicle being a passenger car (P), $t_a = 6$ sec. For general purpose design, J is assumed to be equal to 2 sec. The design speed V was given as 50 mph. Therefore,

$$d = 1.47(50)(2 + 6) = 588 \text{ ft}$$

(b) First, S is solved by using Eq. 13.7.4. For general design purpose, a D value of 10 ft is assumed (measured from the pavement edge to the front of the vehicle). The width of the pavement is 4 × 11 = 44 ft. The length of the WB-50 truck is 55 ft. Therefore,

$$S = 10 + 44 + 55 = 109 \text{ ft}$$

This value of S is then used to find t_a using Fig. 13.20 with the vehicle being a WB-50 truck, t_a equals about 11.5 sec. The required sight distance is then determined from Eq. 13.7.3 with $J = 2$ sec and $V = 40$ mph:

$$d = 1.47(40)(2 + 11.5) = 794 \text{ ft}$$

─── **EXAMPLE 13.17** ───

A stop control on a minor road controls a car turning left onto a 2-way 2-lane major road. Determine the sight distance required to the right and to the left if the design speed of the major road is 50 mph, each lane is 12 ft wide, and the length of the car equals 19 ft.

Solution. Eq. 13.7.3 is used to determine sight distance to the left where $S = 10 + 19 + (1.5)(12) = 47$ ft. From Figure 13.21, $t_a = 4.5$ sec for a passenger vehicle. With $J = 2$ sec and $V = 50$ mph, the sight distance to the left is

$$d = 1.47(50)(2 + 4.5) = 478 \text{ ft}$$

To calculate the necessary sight distance to the right, determine the distance required for the turning vehicle to reach a speed of 85 percent of the major road design speed. Using Fig. 13.21, with initial speed equal to 0 mph and the final speed equal to 0.85(50) = 42.5 mph, the distance equals 500 ft for a passenger vehicle. Next, the time required to travel this distance t_a is equal to about 14.7 sec. To this time add $J = 2$ sec. Using Eq. 13.7.3 we find

$$Q = 1.47(0.85)(50)(2 + 14.7) = 1043 \text{ ft}$$

From Fig. 13.22 which follows we find

$$h = P - 16 - (1.9)(\text{major road design speed})$$

Therefore,

$$h = 500 - 16 - (1.9)(50) = 389 \text{ ft}$$

The required sight distance to the right is then

$$d = Q - h = 1043 - 389 = 654 \text{ ft}$$

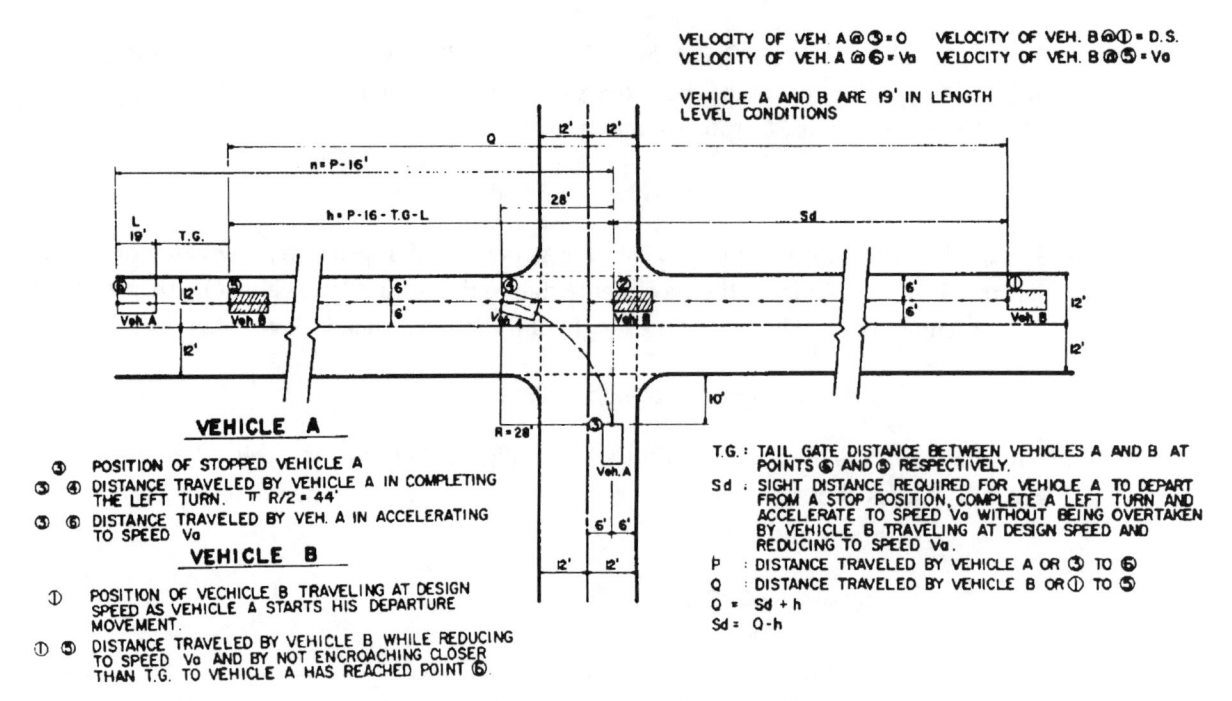

Figure 13.22 p. 756; from *A Policy on Geometric Design of Highways and Streets*, 1990, by the Am. Assoc. of State Highway and Trans. Officials, Wash., DC, by permission.

13.7.3 Design of Turning Radii

The design of the intersection radii is a function of available right-of-way and the minimum turning radius of the design vehicle. Radii larger than the minimum values allow for higher speed turns and a reduced need for deceleration. Table 13.7 and 13.8, on the following two pages, are from page 690 to 693 of the 1990 Green Book.

EXAMPLE 13.18

(a) For a passenger vehicle P, determine the minimum simple curve radius for a 90 degree turn.

(b) For a WB-62 truck and a 75 degree turn, determine the radius, offset, and taper for a simple curve radius with taper.

(c) For a WB-40 truck and a 135 degree turn, determine the curve radius and the asymmetric offset for a 3-centered compound curve.

(d) For a SU truck and a 75 degree turn, determine the curve radius and the symmetric offset.

Solution.

(a) From Table 13.7, the simple curve radius for a passenger vehicle is 30 ft.

(b) From Table 13.7, the simple curve radius with taper has the following characteristics: radius equals 140 ft, the offset equals 4 ft, and the taper is 20:1 (ft:ft).

(c) From Table 13.8, the 3-centered compound curve has the following characteristics: curve radius equals 100-25-180 (ft) and the asymmetric offset equals 3-13 (ft).

(d) From Table 13.8, the 3-centered compound curve has the following description: curve radius equals 120-45-120 (ft) and the symmetric offset equals 2 ft.

Table 13.7 Minimum edge-of-pavement designs for turns at intersections, simple curves.

Angle of Turn (degrees)	Design Vehicle	Simple Curve Radius	Simple Curve Radius with Taper		
			Radius (ft)	Offset (ft)	Taper (ft:ft)
30	P	60	--	--	--
	SU	100	--	--	--
	WB-40	150	--	--	--
	WB-50	200	--	--	--
	WB-62	360	220	3.0	15:1
45	P	50	--	--	--
	SU	75	--	--	--
	WB-40	120	--	--	--
	WB-50	--	120	2.0	15:1
	WB-62	--	140	4.0	15:1
60	P	40	--	--	--
	SU	60	--	--	--
	WB-40	90	--	--	--
	WB-50	--	95	3.0	15:1
	WB-62	--	140	4.0	15:1
75	P	35	25	2.0	10:1
	SU	55	45	2.0	10:1
	WB-40	--	60	2.0	15:1
	WB-50	--	65	3.0	15:1
	WB-62	--	140	4.0	20:1
90	P	30	20	2.5	10:1
	SU	50	40	2.0	10:1
	WB-40	--	45	4.0	10:1
	WB-50	--	60	4.0	15:1
	WB-62	--	120	4.0	30:1
105	P	--	20	2.5	8:1
	SU	--	35	3.0	10:1
	WB-40	--	40	4.0	10:1
	WB-50	--	55	4.0	15:1
	WB-62	--	115	3.0	30:1
120	P	--	20	2.0	10:1
	SU	--	30	3.0	10:1
	WB-40	--	35	5.0	8:1
	WB-50	--	45	4.0	15:1
	WB-62	--	100	5.0	25:1
135	P	--	20	1.5	15:1
	SU	--	30	4.0	8:1
	WB-40	--	30	8.0	6:1
	WB-50	--	40	6.0	10:1
	WB-62	--	80	5.0	20:1
150	P	--	18	2.0	10:1
	SU	--	30	4.0	8:1
	WB-40	--	30	6.0	8:1
	WB-50	--	35	7.0	6:1
	WB-62	--	60	10.0	10:1
180	P	--	15	0.5	20:1
	SU	--	30	1.5	10:1
	WB-40	--	20	9.5	5:1
	WB-50	--	25	9.5	5:1
	WB-62	--	55	10.0	15:1

TABLE 13.8 Minimum edge-of-pavement designs for turns at intersections, compound curves.

Angle of Turn (degrees)	Design Vehicle	3-Centered Compound		3-Centered Compound	
		Curve Radii (ft)	Symmetric Offset (ft)	Curve Radii (ft)	Asymmetric Offset (ft)
30	P	--	--	--	--
	SU	--	--	--	--
	WB-40	--	--	--	--
	WB-50	--	--	--	--
	WB-62	--	--	--	--
45	P	--	--	--	--
	SU	--	--	--	--
	WB-40	--	--	--	--
	WB-50	200-100-200	3.0	--	--
	WB-62	460-240-460	2.0	120-140-500	3.0-8.5
60	P	--	--	--	--
	SU	--	--	--	--
	WB-40	--	--	--	--
	WB-50	200-75-200	5.5	200-75-275	2.0-6.0
	WB-62	400-100-400	15.0	110-100-220	10.0-12.0
75	P	100-75-200	2.0	--	--
	SU	120-45-120	2.0	--	--
	WB-40	120-45-120	5.0	120-45-200	2.0-6.5
	WB-50	150-50-150	6.0	150-50-225	2.0-10.0
	WB-62	440-75-440	15.0	140-100-540	5.0-12.0
90	P	100-20-100	2.5	--	--
	SU	120-40-120	2.0	--	--
	WB-40	120-40-120	5.0	120-40-200	2.0-6.0
	WB-50	180-60-180	6.0	120-40-200	2.0-10.0
	WB-62	400-70-400	10.0	160-70-360	6.0-10.0
105	P	100-20-100	2.5	--	--
	SU	100-35-100	3.0	--	--
	WB-40	100-35-100	5.0	100-55-200	1.0-8.0
	WB-50	180-45-180	8.0	150-40-210	2.0-8.0
	WB-62	520-50-520	15.0	360-75-600	2.0-10.0
120	P	100-20-100	2.0	--	--
	SU	100-30-100	3.0	--	--
	WB-40	120-30-120	6.0	100-30-180	2.0-9.0
	WB-50	180-40-180	8.5	150-35-220	2.0-12.0
	WB-62	520-70-520	10.0	80-55-520	24.0-17.0
135	P	100-20-100	1.5	--	--
	SU	100-30-100	4.0	--	--
	WB-40	120-30-120	6.5	100-25-180	3.0-13.0
	WB-50	160-35-160	9.0	130-30-185	3.0-14.0
	WB-62	600-60-600	12.0	100-60-640	14.0-7.0
150	P	75-85-75	2.0	--	--
	SU	100-30-100	4.0	--	--
	WB-40	100-30-100	6.0	90-25-160	1.0-12.0
	WB-50	160-35-160	7.0	120-30-180	3.0-14.0
	WB-62	480-55-480	15.0	140-60-560	8.0-10.0
180	P	50-15-50	0.5	--	--
	SU	100-30-100	1.5	--	--
	WB-40	100-20-100	9.5	85-20-150	6.0-13.0
	WB-50	130-25-130	9.5	100-25-180	6.0-13.0
	WB-62	800-45-800	20.0	100-55-900	15.0-15.0

13.8 Corridor Design

Traffic signals along an arterial road are usually coordinated in a way that allows for a group of vehicles called "platoons" to travel along the arterial without stopping. This progression of vehicles is important in traffic engineering. Progression reduces delay, travel times, fuel consumption, and emissions. Progression also reduces rear-end type accidents since vehicles do not have to stop as often. The key engineering tool used to design progression is called the space-time diagram. A space-time diagram helps the traffic engineer determine critical variables involved with coordinating the traffic signals. These factors include offsets and band widths.

13.8.1 Data Required

Space-time diagrams require the following data: cycle lengths, effective green time, speed limits, predominant directional flows, and distances between intersections.

In order to keep the traffic signals coordinated, cycle lengths must always be the same, or a multiple of the same duration. For example, a 40 second cycle length is a multiple of an 80 second cycle length. If they are not, as each cycle is completed, the traffic signals will steadily become un-coordinated. A green split is the proportional length of time of the total cycle that a direction of traffic receives a green light. The effective green time is the green split plus the amount of time that traffic may continue into the yellow time. For example, if the green split is 25 seconds and the cars consistently go through 2 of the 4 second yellow time, then the effective green time would be 27 seconds.

The speed limit is needed to determine how long it takes a car to travel from one intersection to the next. However, a better speed would be the operating speed. For example, if the speed limit is 35 mph, but the traffic usually moves at 32 mph, the later speed of 32 mph should be used. If this data can not be collected, the posted speed limit is commonly used.

Predominant directional flows are also an important traffic characteristic needed for space time diagrams. Many arterial roads, during peak hours, will have one direction having more traffic than the opposing direction. For example, a major arterial connecting the suburbs with a large city might have the following characteristics. In the morning most traffic will travel into the city, but the evening rush will have the predominant flow back to the suburbs. If this information is known, a space-time diagram for each peak period of the day can be developed that progresses traffic in the predominant direction.

13.8.2 Space-Time Diagram

In the space-time diagram of Fig. 13.23 the x-axis represents space (distance), and the y-axis represents time (although some people prefer to have space on the x-axis and the time on the y-axis). The solid line extending from the lower left to the upper right of the diagram represents the speed of a car traveling a certain direction. A car traveling 30 mph will take 120 sec to go a mile.

The vertical lines of strips and blanks are where traffic signals are located along a street. The strips of black represent red time. The car must travel through the spaces between these black strips. These spaces are the effective green time. In this diagram the effective green time equals 80

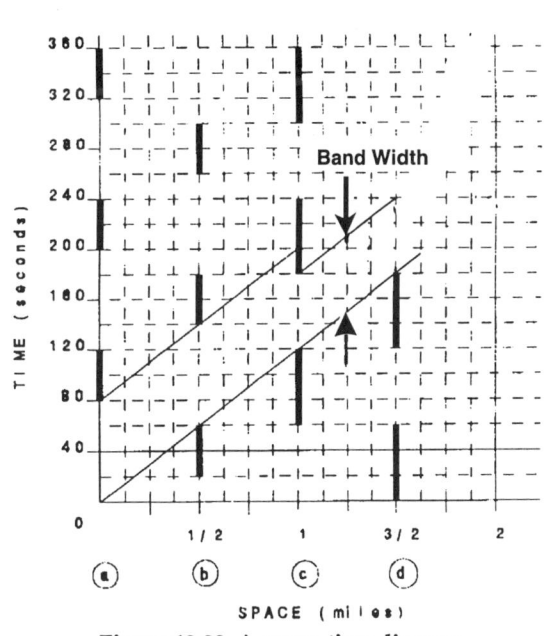

Figure 13.23 A space-time diagram.

seconds, the red time equals 40 seconds, and the cycle length equals 120 seconds for traffic signal "a" and "b". Traffic signals "c" and "d" have effective greens of 60 seconds and red periods of 60 seconds.

An offset is an amount of time from a reference "zero" time point that a traffic signal starts a cycle. The beginning of cycle usually follows the order of the effective green first followed by the red time period. In the above diagram, signal "a" has an offset equal to zero seconds. This will also be used as the reference "zero" time. Signal "b" has an offset of 60 seconds, "c" has an offset of 120 seconds, and "d" has an offset of 180 seconds. However, offsets are not stated longer than the cycle length. For example, the 120 second offset would be zero seconds, and the 180 second offset would be 60 seconds.

Band width is the segment of time that can start at the beginning of the diagram and travel through the diagram without hitting red periods (black strips). The maximum band width is determined by projecting a parallel line, representing speed, from the first traffic signal on the street. If the line hits a red period, the line is moved down to the bottom of the red period and then continues (still parallel to the lower line). This is called "cutting into the band width". This occurs in the above diagram at signal "c". At signal "a" the band width is 80 seconds, but at "c" it is cut down to 60 seconds. It continues at this width the rest of the way. At the end of the diagram, the time between the lines is know as the maximum band width. In this case the maximum band width is 60 seconds.

All of the above also applies if a line (dashed) is drawn from the lower right corner to the upper left corner as in Fig. 13.24. This dashed line represents cars traveling at a certain speed (in this case 30 mph) on the same street but in the opposite direction.

To progress traffic along a street, the following steps should be followed when using the space-time diagram:

1. Obtain the following information: cycle length, effective green times, speed limits, predominant directional flows, and distances between intersections.

2. Locate the position of the signals along the x-axis.

3. Draw the line representing the speed of the car starting in the lower left corner of the diagram.

4. As shown in Fig. 13.23, the beginning of each cycle is on the line for the speed of the car. This means that if the first car travels at the set speed, it will hit the beginning of each green split. This will allow the "platoon" of cars with the first car to pass through on the green light.

5. The offsets are determined by measuring up from the *x*-axis where each signal is located until the line representing speed is met.

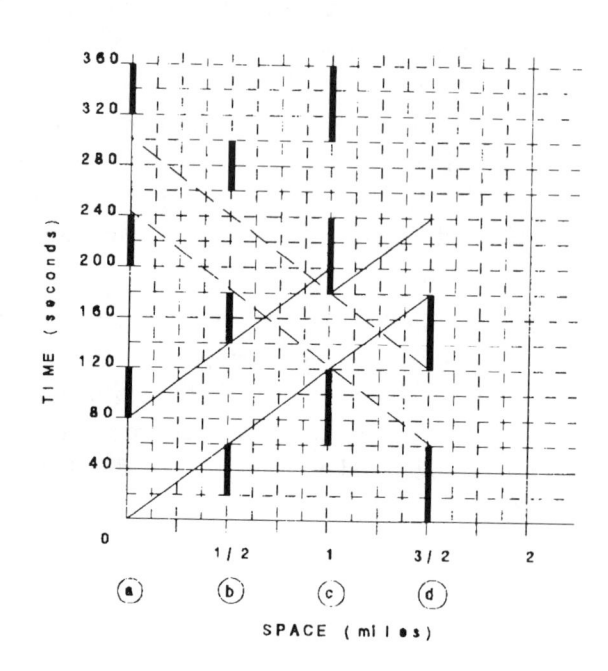

Figure 13.24 A space-time diagram with traffic in both directions.

EXAMPLE 13.19

A system of traffic signals on a two-way corridor are to be timed progressively for an eastbound (left to right on a space-time diagram) speed of 45 mph. There are five intersections spaced at the following mile points: 0, 0.5, 1.5, 3, and 4. The cycle length of all of the signals is 80 seconds. Based on traffic volumes, the effective green times for each signal on the corridor are the following:

$$\text{Signal 1} = 60 \text{ sec}$$
$$\text{Signal 2} = 50$$
$$\text{Signal 3} = 60$$
$$\text{Signal 4} = 40$$
$$\text{Signal 5} = 50$$

a) Draw the space-time diagram representing the timing of signals for eastbound progression.
b) Determine the maximum band width for eastbound traffic.
c) Determine the maximum band width for westbound traffic if the assumed speed is also 45 mph.
d) If a new signal is placed at the 2.5 mile point (effective green equals 40 seconds, 80 second cycle length), what should be the offset to ideally progress the westbound traffic?
e) Does the new signal reduce the band width of the eastbound traffic?

Solution.

Draw a space-time diagram that matches distances and travel times required, as in Fig. 13.25. The next step is to draw a line that has an inverse slope corresponding to 45.

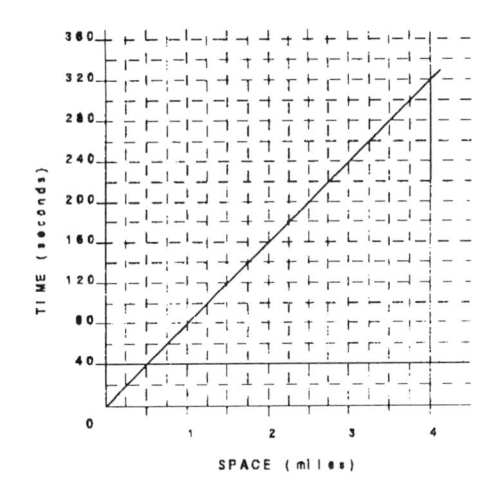

Figure 13.25

a) As shown in Fig. 13.26, locate all of the signals and put in the signal patterns that correspond to eastbound progression. All of the cycles begin on the 45 mph line.

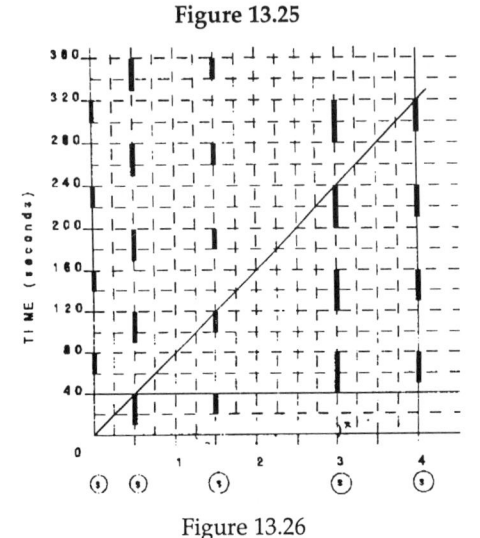

Figure 13.26

b) Fig. 13.27 shows how the maximum band width is determined. Notice how the band starts at 60 sec then is cut to 50 sec at signal 2, then cut to 40 sec at signal 4. The maximum band width = 40 sec. Notice at signal 3 that the band width did not increase. A band width can not increase because cars have already been cut off.

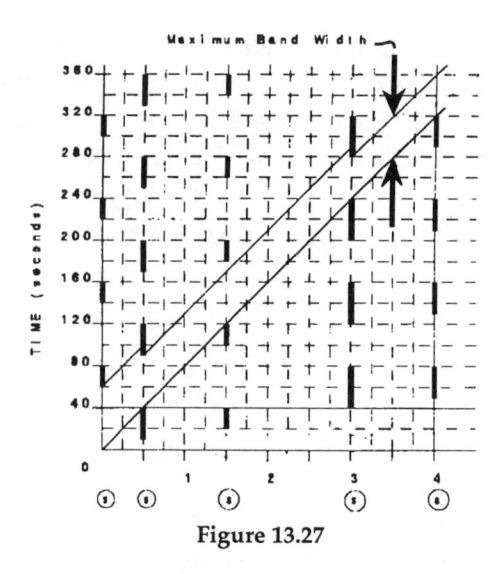

Figure 13.27

c) Fig. 13.28 shows the determinations of the maximum band width for westbound traffic. Again notice that the band width starts out at 50 sec and is cut to 40 sec at signal 2 (in westbound direction). The maximum band width equals 40 sec.

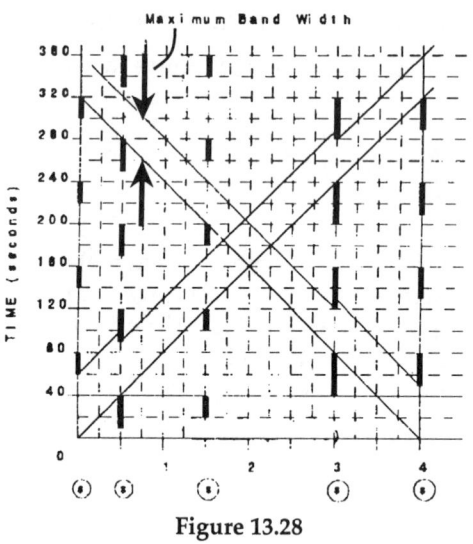

Figure 13.28

d) Fig. 13.29 shows the proper offset with the addition of a new signal at the 2.5 mile point. The offset equals 40 sec. The cycle starts 120 sec from the x-axis, but the offset must always be less than the cycle length. So the cycle length is subtracted from 120 sec until the number is less than 80 sec.

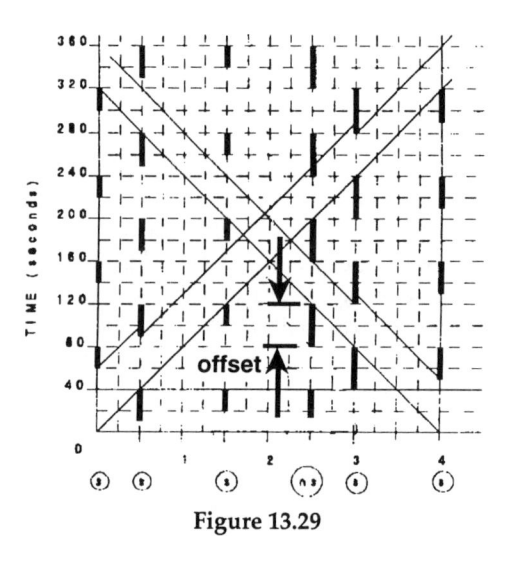

Figure 13.29

e) Yes, the band width in the eastbound directions is reduced (Fig. 13.30). At point "a" the width is 50 sec, but at the new signal the width is cut to 40 sec. Point "b" shows that 10 sec have been cut off the band width. The final band width is 40 sec at point "c". Notice that the band width is the same as in part B. The difference is that the band width was reduced earlier.

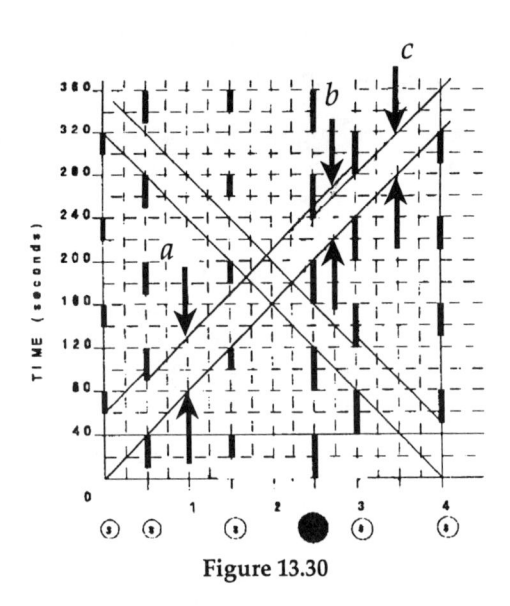

Figure 13.30

13.9 Drainage

The design of drainage facilities usually requires the estimation of the expected runoff for an assumed Design Year Storm. The facility for carrying the water away from the Highway is either a ditch (open channel) or a storm sewer (closed conduit). It should be noted that highway drainage is somewhat different than typical open channel flow.

Design Year Storms are often 10 years for non-freeways and 50 years for freeways. A design for a 50-year storm is expected to have one chance in 50 of being full or over loaded in a given year.

13.9.1 Runoff

The Rational Method is commonly used by highway agencies for estimating the runoff of relatively small drainage areas. The equation used is

$$Q = ciA_d \tag{13.9.1}$$

where Q = runoff, cfs
   c = a runoff coefficient expressing the ratio of rate of runoff to rate of rainfall
   i = Intensity of rainfall, in/hr for a duration equal to the time of concentration
   A_d = Drainage area, acres

Values used for the coefficients of runoff are given in Table 13.9:

Table 13.9 Runoff coefficients.

Drainage Surface	Coefficient (c)
Concrete or Bituminous	0.8-0.9
Gravel	0.4-0.6
Earth	0.2-0.8
Turf	0.1-0.4
Cultivated field	0.2-0.4
Forest	0.1-0.2

If the grade of the surface is relatively flat then the lower range of coefficient is used. As the grade increases a higher value should be used. The Rational Method is based upon two significant assumptions: the basin is in equilibrium, and the rainfall is of uniform intensity.

The design of a highway ditch section is based upon the principles of open channel flow and Manning's Equation:

$$Q = \frac{1.49}{n} A \times R^{2/3} \times S^{1/2} \tag{13.9.2}$$

where
- Q = discharge, cfs
- V = velocity, fps
- n = Manning's roughness coefficient, given in Table 13.10
- R = Hydraulic radius, ft
- S = Slope of the ditch

Table 13.10 Manning's Coefficient as Function of Ditch Lining

Ordinary rock - smooth	0.02
Rough rock	0.04
Rough concrete	0.02
Bituminous	0.02
Grass —flow over 6 inches	0.04
—flow under 6 inches	0.06
Heavy grass	0.10

EXAMPLE 13.20

The section of a north-south highway is 2.5 miles long. The highway has 2, 12-ft-wide paved lanes with 6-ft-wide gravel shoulders. The total width of the right-of-way is 66 ft and except for the traveled portion of the roadway everything has a turf surface. In addition to the runoff within the right-of-way, runoff is expected from farm fields on the west side of the highway (from an average lateral distance of 30 ft). On the east side of the highway there is a paved parking lot of 3.5 acres and 10 acres of forest. What is the expected runoff for a 10-year storm in the Detroit, Michigan area?

Solution. Use the Rational Method for determining runoff. For the west side:

Area 1, paved lane: $c = 0.85$, $A_d = 2.5 \times 5280 \times 12/43{,}560 = 3.64$ acres

Area 2, gravel shoulder: $c = 0.5$, $A_d = 2.5 \times 5280 \times 6/43{,}560 = 1.82$ acres

Area 3, turf right-of-way: $c = 0.4$ (steep side slopes), $A_d = 2.5 \times 5280 \times [33 - (12 + 6)]/43{,}560 = 4.55$ acres

Area 4, farm fields: $c = 0.3$, $A_d = 2.5 \times 5280 \times 30/43{,}560 = 9.09$ acres

Figure 13.31 is a Michigan Department of Transportation graph of rainfall intensity curves for the Detroit area. By using a 15 minute duration of storm (which is a common practice) for a 10-year storm, the intensity of rainfall is 3.8 in/hr. For the estimated runoff for the west side of the highway we have

Area 1: $Q_1 = 0.85 \times 3.8 \times 3.64 = 11.76$ cfs

Area 2: $Q_2 = 0.50 \times 3.8 \times 1.82 = 3.46$ cfs

Area 3: $Q_3 = 0.40 \times 3.8 \times 4.55 = 6.92$ cfs

Area 4: $Q_4 = 0.30 \times 3.8 \times 9.09 = 10.36$ cfs

Total runoff = 11.76 + 3.46 + 6.92 + 10.36 = 32.5 cfs.

For the east side:

Area 5, paved lane: the same as Area 1

Area 6, gravel shoulder: the same as Area 2

Area 7, parking lot: $c = 0.8$, $A_d = 3.5$ acres

Area 8, forest: $c = 0.1$, $A_d = 10$ acres

With an intensity of runoff of 3.8 in/hr an estimate of runoff for the east side is:

$Q_5 = 11.76$ cfs

$Q_6 = 3.46$ cfs

$Q_7 = 0.8 \times 3.8 \times 3.5 = 10.64$ cfs

$Q_8 = 0.1 \times 3.8 \times 10 = 3.80$ cfs

Total runoff = 11.76 + 3.46 + 10.64 + 3.80 = 29.66 cfs

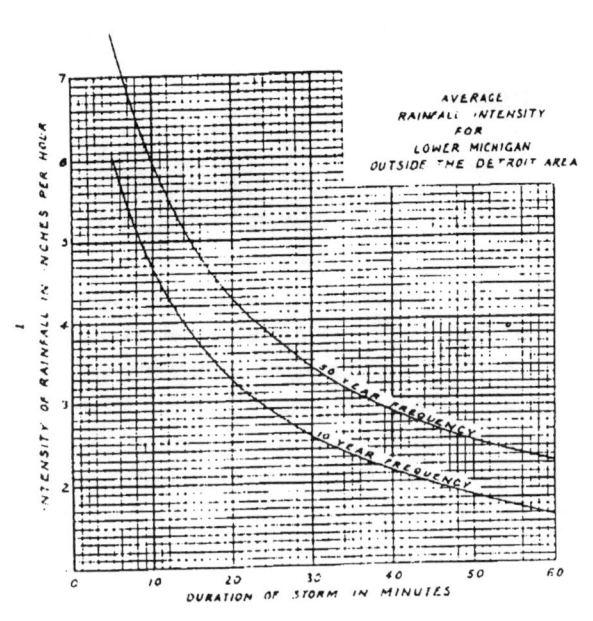

Figure 13.31

EXAMPLE 13.21

For the estimated runoff for the west side of the highway in Example 13.20 design a ditch section. The foreslope is assumed to be 1 on 4 and the back slope is 1 on 2. The bottom of the ditch is assumed to be 3 feet. The slope of the road for the 2.5 mile section has an average of 4 percent. A freeboard of 1/2 foot is desired. A grass lining is expected. The ditch cross-section is shown in Fig. 13.32. Calculate the depth of flow and the average velocity.

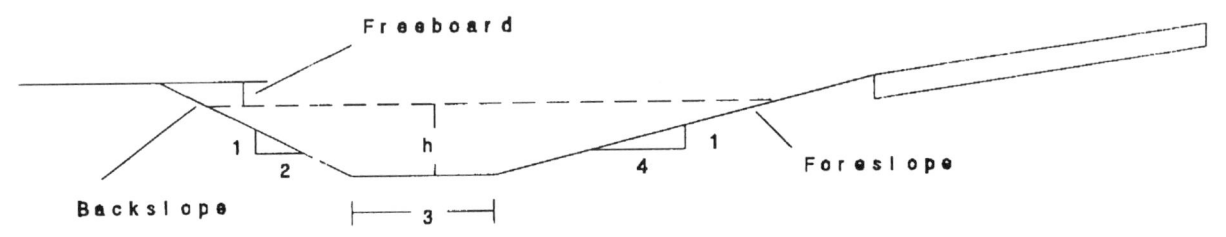

Figure 13.32

Solution. The usual practice is to solve Manning's Equation by trial-and-error. The first step is to reduce the equation to a function of the assumed height of flow. The area is

$$A = 3 \times h + h \times \frac{2h}{2} + h \times \frac{4h}{2} = 3h^2 + 3h$$

The wetted perimeter is

$$P = 3 + \sqrt{h^2 + (2h)^2} + \sqrt{h^2 + (4h)^2}$$

$$= 3 + 6.36h$$

Using $n = 0.04$ and $R = A/P$, Manning's equation is

$$Q = \frac{1.49}{0.04} A \times R^{2/3} \times \sqrt{0.04}$$

$$= 22.35h(1+h)\left[\frac{h(1+h)}{1+2.12h}\right]^{2/3}$$

Assume $h = 1$ ft. Then $Q = 33.2$ cfs. This is higher than the 32.5 cfs of Example 13.24.

Assume $h = 0.9$ ft. This gives $Q = 26.8$ cfs. So we use $h = 1$ ft.

The average velocity is

$$V = \frac{Q}{A} = \frac{32.5}{6} = 5.41 \text{ fps}$$

The maximum velocity for water over grass is 6 fps. So the ditch design is acceptable.

The design of a storm sewer for a highway (especially a major highway, such as an urban freeway) is unique in that the drainage basin is relatively long and narrow. The width of the basin may be less than 100 feet and the length may be several miles. Thus the design of a storm sewer for a highway takes in to consideration the length of time required for the discharge to travel from the beginning of the sewer to its depository. The storage capacity of the sewer is another consideration. The vertical and horizontal alignment of the sewer can and often is independent of the alignment of the highway. Therefore the discharge can be transported much farther before being deposited than with a normal ditch section. With a ditch section the rolling of the vertical alignment and the available right-of-way limits the length of highway that can be serviced by a single ditch section.

The approximate controlling grade of the storm sewer can be estimated by knowing the highest possible elevation at the start of the sewer and the lowest possible grade at the end of the sewer. The highest elevation is usually 4 to 5 feet below the elevation of the center of the highway which allows for the diameter of the pipe and adequate cover. The lowest elevation is usually determined by the bottom of the lake, river, existing drain, or retention basin. By knowing the difference in elevation and the length of the sewer, the controlling grade is defined.

The following steps are common in designing a sewer:

1. Plot the controlling grade on the profile.
2. Plot the alignment of the sewer on the plan view.
3. Locate the manholes and catch basins.
   —Catch basins are located at the bottom of the vertical curves.
   —Catch basins are not needed at the top of the vertical curves.
   —Locate the remaining catch basins such that water does not travel more than 300 ft along the curb and gutter.
   —Manholes are located at approximately 300 feet for small sewers and up to 600 ft for large sewers.
4. Determine the boundaries of the areas contributing to each catch basin.
5. Calculate the area and estimate the runoff coefficient for each area.
6. Using an initial time of 15 minutes determine the runoff for the first area.
7. Select a design of sewer (diameter and slope) appropriate for the first area.
8. Determine the velocity of the flow in the first section of sewer and by knowing the distance to the next catch basin determine the time required (in minutes) to reach the second catch basin.
9. Increase the time of concentration to 15 minutes plus the increment of time determined in step 8.
10. Repeat steps 7 through 9 until the end of the sewer is reached.

The velocity of the flow should increase gradually to prevent sedimentation in the sewer. The velocity needs to be fast enough to keep the sewer clean and slow enough to prevent scouring.

Practice Problems

13.1 For a four-lane freeway with a 60 mph Design Speed and an Average Travel Speed of 45 mph, what is the expected maximum flow rate (passenger cars per hour per lane)?

    a) 1400   b) 1700   c) 2000   d) 2800   e) 3400

13.2 Compute the minimum and maximum total maneuver time for a design speed of 60 mph.

    a) 6.7, 10.0   b) 8.2, 10.5   c) 9.2, 11.5   d) 11.2, 14.5   e) 14.5, 16.7

13.3 Design the typical cross section for a rural divided highway with a Design Speed of 70 mph. The Average Daily Traffic is 40,000 with a Design Hour Volume of 4000 with a peak directional volume of 2500 vehicles. The commercial traffic is 5 percent.

    a) 4 paved lanes at 12 feet each, 8 foot gravel shoulder, the cross slope is 1.5% for the pavement and 6% for the shoulder, the foreslopes are 6:1, there is a turf median of 20 feet.

    b) 4 paved lanes at 12 feet each, 8 foot paved shoulders, the cross slope is 4% for the pavement and 5% for the shoulder, the foreslopes are 4:1, the median is 60 feet of turf.

    c) 6 paved lanes at 12 feet each, 10 foot paved shoulders, the cross slope is 2% for the pavement and 4% for shoulders, the foreslopes are 6:1, and the median is turf and 80 feet wide.

    d) 6 paved lanes at 12 feet each, 10 foot gravel shoulders, the cross slope is 1% for the pavement and 4% for the shoulders, the foreslopes are 6:1, and the median is turf and 60 feet wide.

13.4 Determine the capacity of the intersection of Maple Avenue with the service roads of an urban freeway. The Design Year Traffic Volumes are shown in Figure 13.33. The diagonal solid lines are estimates of right turns and the broken diagonal lines are estimates of left turns. The speed limit for Maple Avenue is 45 and the speed limit for the service roads is 30 mph. The cycle length is 80 seconds. The grade is level for all roads. Lane widths are assumed to be 12 feet. The separation between the centerlines of the two service roads is 260 feet.

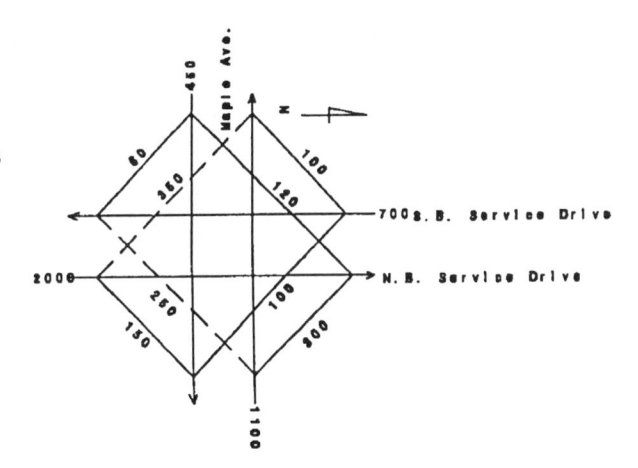

Figure 13.33

13.5 An intersection had the following number of reported accidents:

Year	Fatal	Injury	PDO
1990	2	19	55
1991	0	25	42
1992	3	21	41
1993	1	18	66

The north-south ADT is 6600 vehicles per day and the east-west ADT is 8000 vehicles per day. What are the accident rates for all accidents, fatal accidents, and injury accidents? What is the Severity Ratio?

13.6 Given two reverse horizontal curves with $\Delta_1 = 30°$ and $\Delta_2 = 60°$. The stationing of $PI_1 = 132 + 50.00$. The distance between PI_1 and PI_2 is 1000 ft. Assume a maximum allowable side friction factor of 0.06 and a maximum superelevation rate of 0.06.

a) Determine the maximum velocity (design speed) that can be provided while allowing for a 3 sec tangent distance between curves.

b) Determine D, R, T, E, M and L for both curves.

c) Determine the stationing of the PC and PI for both curves.

13.7 Determine the minimum length of a crest vertical curve for 1200 feet of passing sight distance. The approach grade is 3% and the leaving grade is 0%.

13.8 A stop control on a minor road controls a car turning right onto a 2-way 2-lane major highway. Determine the sight distance required to the left using the following data. The design speed of the major highway is 50 mph, each lane is 12 ft wide, and the length of the car equals 19 ft.

13.9
a) Given a simple curve radius of 180 ft and a turn of 30°, what is the largest vehicle that can make this turn without running off the edge of the pavement?

b) Given a simple curve radius (with taper) of 130 ft on a 90° turn, what is the largest vehicle that can safely make this turn?

c) The design vehicle is a WB-62 truck, there is a 90° turn, and the simple curve radius with a taper is 120 ft. Will a larger radius be needed if the angle of the intersection is increased to 120°? Will the radius need to be increased if the angle of the turn is increased to 180°?

d) Does a WB-50 truck always need a larger radius on a turn with a taper than a SU truck?

13.10 The following data has been collected on an existing eastbound one-way arterial:

Signal	Mile Point	Offset	Effective Green
1	0.00	0 sec	60 sec
2	0.75	40 sec	60 sec
3	2.00	70 sec	60 sec
4	3.25	50 sec	60 sec
5	4.00	20 sec	60 sec

All cycle lengths are 100 sec.

a) Draw the space-time diagram that represents the current signal locations, timings, and offsets.

b) Determine the band width if traffic travels at a speed of 45 mph.

c) Change the offsets to get the largest possible band width. What is the maximum band width?

13.11 Design a storm sewer for the following conditions assuming a 10-year storm in the Detroit area and a critical grade of 2.5 percent:

Catch Basin	Station (ft)	Coefficient of Runoff	New Area (acres)	Slope
A	2+50.00	0.35	1.75	1%
B	5+25.00	0.30	1.90	0.8%
C	8+00.00	0.45	1.50	
D	11+00.00	0.40	1.80	

Practice Problems (PE-Format)

Scenario #1: A horizontal curve is going to be constructed for a paved county road in northern Michigan. The pave-ment is 20 feet wide with 4 ft. shoulders. The legs of the proposed curve are perpendicular. A design speed of 60 mph is desired.

13.12 What is the minimum radius needed to provide a design where there is no reliance on a lateral friction force?
   a) If $e = 0.12$ and $f = 0.10$; $R = 1090.91$ ft.
   b) If $e = 0.12$ and $f = 0.00$; $R = 2000.00$ ft.
   c) If $e = 0.08$ and $f = 0.12$; $R = 1200.00$ ft.
   d) If $e = 0.08$ and $f = 0.00$; $R = 3000.00$ ft.
   e) If $e = 0.06$ and $f = 0.06$; $R = 2000.00$ ft.

13.13 What is the minimum radius that can be provided?
   a) If $e = 0.12$ and $f = 0.10$; $R = 1090.91$ ft.
   b) If $e = 0.12$ and $f = 0.00$; $R = 2000.00$ ft.
   c) If $e = 0.08$ and $f = 0.12$; $R = 1200.00$ ft.
   d) If $e = 0.08$ and $f = 0.00$; $R = 3000.00$ ft.
   e) If $e = 0.06$ and $f = 0.06$; $R = 2000.00$ ft.

13.14 What is the minimum length of the curve?
   a) If $R = 1091.91$ ft., $e = 0.12$ and $f = 0.10$
   b) If $R = 2000.00$ ft., $e = 0.12$ and $f = 0.00$
   c) If $R = 1200.00$ ft., $e = 0.08$ and $f = 0.12$
   d) If $R = 3000.00$ ft., $e = 0.08$ and $f = 0.00$
   e) If $R = 3200.00$ ft., $e = 0.12$ and $f = 0.00$

13.15 For the condition in question # 3, what is the recommended superelevation runoff?
   a) With $e = 0.08$, $L = 215$ ft. with all of the runoff provided in the horizontal curve.
   b) With $e = 0.08$, $L = 215$ ft. with all of the runoff provided in the tangent portion of the approach to the horizontal curve.
   c) With $e = 0.08$, $L = 215$ ft. with 50% of the runoff provided in the horizontal curve and 50% provided in the tangent.
   d) With $e = 0.08$, $L = 180$ ft. with 2/3 of the runoff provided in the tangent section and 1/3 of the runoff provided in the horizontal curve.
   e) With $e = 0.08$, $L = 180$ ft. with 50% of the runoff provided in the horizontal curve and 50% in the tangent.

13.16 For the conditions in questions #3 and #4, what is the recommended length of spiral transition?
   a) With the centripetal acceleration factor equal to 1, length of spiral transition equals 567 ft.
   b) With the centripetal acceleration factor equal to 2, length of spiral transition equals 283.5 ft.
   c) With the centripetal acceleration factor equal to 3, length of spiral transition equals 189 ft.
   d) The spiral distance is equal to the length of the runoff and equals 180 ft.
   e) All of the above.

Scenario #2: A 1 mile section of highway is to be constructed through a national forest. The highway will have a 24 ft. paved surface, with 6 ft. gravel shoulders. There is 66 ft. of right-of-way. There is a decrease in grade of 132 ft.

13.17 What is the design year storm and expected runoff for this highway?
   a) With a 50 year design storm and an intensity of runoff of 6 inches per hour, the runoff is 27.50 cfs.
   b) With a 50 year design storm and an intensity of runoff of 5 inches per hour, the runoff is 22.92 cfs.
   c) With a 10 year design storm and an intensity of runoff of 3.8 inches per hour, the runoff is 17.41 cfs.
   d) With a 10 year design storm and an intensity of runoff of 3.2 inches per hour, the runoff is 14.66 cfs.
   e) With a 15 year design storm and an intensity of runoff of 3.5 inches per hour, the runoff is 16.04

13.18 Assuming that the runoff is 17.41 cfs, What would be the smallest cross-section need to carry this storm water if the sideslopes are 1 ft. vertical for 2 ft. horizontal? The velocity is not to exceed 7 ft/sec in a V bottom ditch.
   a) Bituminous ditch lining with a Manning's coefficient of 0.02 and a resulting depth of flow of 0.85 feet.
   b) Grass ditch lining with a Manning's coefficient of 0.04 and a resulting depth of flow of 2.2 feet.
   c) Grass ditch lining with a Manning's coefficient of 0.06 and a resulting depth of flow of 4.0 feet.
   d) Rough concrete ditch lining with a Manning's coefficient of 0.02 and a resulting depth of flow of 1.5 feet.
   e) Rough concrete ditch lining with a Manning's coefficient of 0.06 and a resulting depth of flow of 2.0 feet.

13.19 Which of the following would provide a design with a minimum rate of flow (velocity)?
   a) A V-bottom ditch with bituminous lining.
   b) A V-bottom ditch with a grass lining.
   c) A 2-ft. wide flat bottom ditch with a bituminous lining.
   d) A 2-ft. wide flat bottom ditch with a grass lining.
   e) A 4-ft. wide flat bottom ditch with a grass lining.

13.20 For the predicted runoff what size storm sewer is needed?
   a) A 20 inch sewer with a velocity of 2.5% 10 ft. per sec. and a discharge of 22 c.f.s.
   b) A 20 inch sewer with a velocity of 9 ft. per sec. and a discharge of 20 c.f.s.
   c) A 20 inch sewer with a velocity of 8 ft. per sec. and a discharge of 18 c.f.s.
   d) A 18 inch sewer with a velocity of 10.2 ft. per sec. and a discharge of 18 c.f.s.
   e) A 22 inch sewer with a velocity of 6.9 ft. per sec. and a discharge of 18 c.f.s.

Solutions to Problems

13.1 d) 2800 pcphpl from Figure 13.1 (2 lanes at 1400 pcphpl).

13.2 d) For a design speed of 60 mph, Detection and Recognition is expected to require from 2.0 to 3.0 seconds; Decision and Response Initiation is expected to require from 4.7 to 7.0 seconds; and the Maneuver is assumed to require 4.5 seconds:

Minimum Decision Sight Distance = 2.0 + 4.7 + 4.5 = 11.2 sec
Maximum Decision Sight Distance = 3.0 + 7.0 + 4.5 = 14.5 sec

Values from Table 13.4 have been used.

13.3 c) The others are wrong because of the following:

a) 6 lanes and a 10 foot shoulder are desirable for the high volumes. The median should be at least 40 feet wide.

b) 6 lanes and a 10 foot shoulder are desirable for the high volumes.

d) The cross slope of the pavement should be at least 1.5%.

13.4 Analyze each interval.

Interval 1:

Interval 1 is illustrated in Figure 13.34. During this interval of time the north and southbound movements have the green light and all others are stopped. Traffic assumed to be moving in this interval are assigned to the appropriate lanes. The movement which will require the most time is the northbound movement. Using Eq. 13.1.1 the time is

$$t = \frac{2.1 M_c}{N_c} + 3.7$$

$$= \frac{2.1(667)}{45} + 3.7 = 34.8 \text{ sec}$$

where M_c = 667 vehicles/hr and N_c = 3600/80 = 45 cycles/hr.

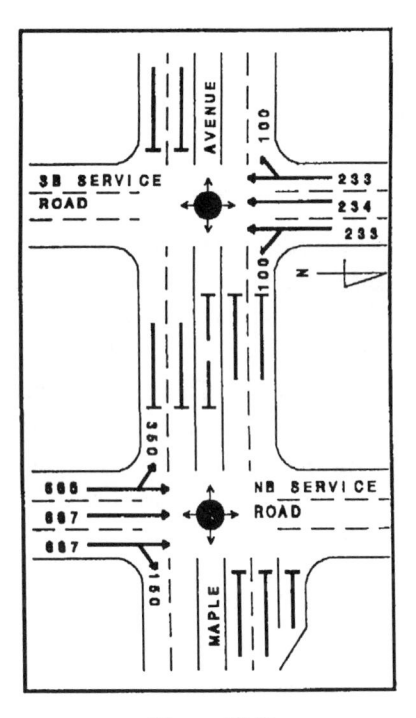

Figure 13.34

Interval 2:

Interval 2 is illustrated in Figure 13.35. This is the amber or clearance time. The amber time is a function of the approach speed which is assumed to be 30 mph. From Eq. 13.1.2 the required time for this interval is

$$Y = t_r + \frac{V}{2a + 64.4g} + \frac{W + L}{V}$$

$$= 1 + \frac{44}{2 \times 14} + \frac{72 + 20}{44} = 4.7 \text{ sec}$$

where we have assumed $t_r = 1$ sec, $a =$ deceleration $= 14$ ft/sec^2, $V = 44$ ft/sec (30 mph speed limit), and $g = 0$. With six lanes at 12 ft, $W =$ intersection width $= 72$ ft and $L =$ length of vehicle $= 20$ ft.

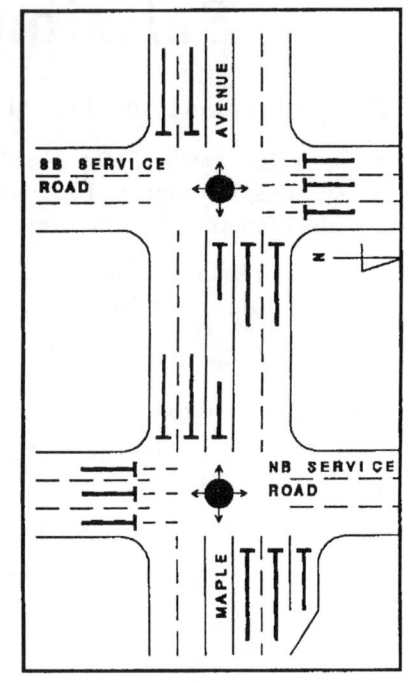

Figure 13.35

Interval 3:

Interval 3 is illustrated in Figure 13.36. This interval is for a lead green to clear the left-turn vehicles (from the service roads) and prevent their interfering with the east and westbound through movements on Maple Avenue. The time required for this interval is equal to the time needed to start the vehicles waiting on Maple Avenue between the two service drives less the travel time for the through traffic arriving on Maple Avenue.

$\frac{175 \text{ vph}}{45 \text{ cycles/hour}} = 3.9$ Vehicles per Cycle. Since this is less than 5 vehicles, Eq. 13.1.1 can not be used. Hence, assume 4 vehicles:

Clearance Time = 3.8 + 3.1 + 2.7 + 2.4 = 12.0 sec.

Travel Time = 260 ft/66 ft/sec = 3.9 sec

Time Required = 12.0 − 3.9 = 8.1 sec

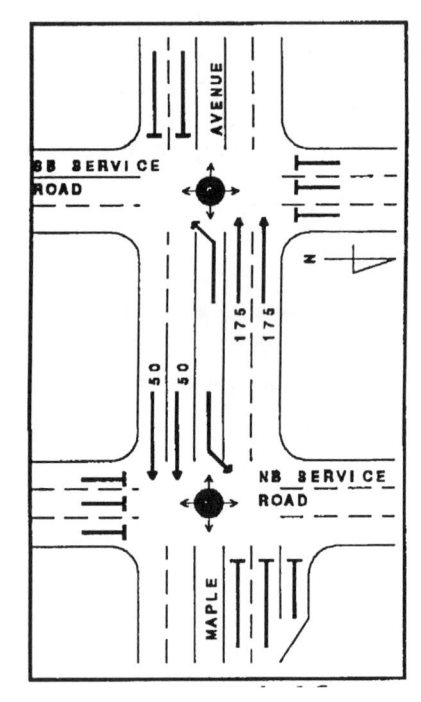

Figure 13.36

Interval 4:

Interval 4 is illustrated in Figure 13.37. During this interval the east and westbound movements occur. The westbound movement is critical because it requires the most time. Using Eq. 13.1.1 the required time for this movement is

$$t = \frac{2.1 \times 400}{45} + 3.7 = 22.4 \text{ sec}$$

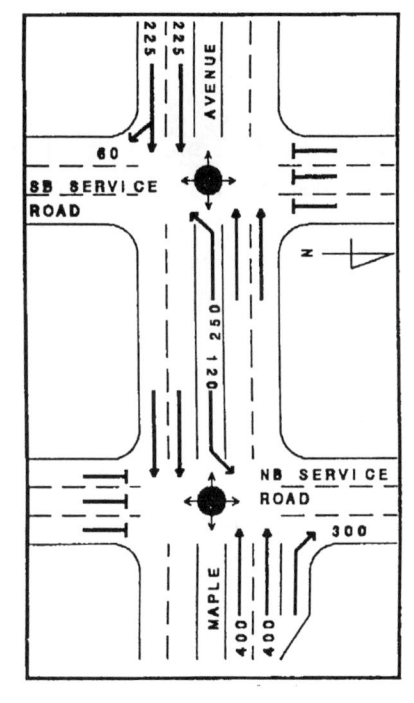

Figure 13.37

Interval 5:

Interval 5 is illustrated in Figure 13.38. This interval is the amber time for the east and westbound movements. From Eq. 13.1.2 this time is

$$Y = t_r + \frac{V}{2a + 64.4g} + \frac{W+L}{V}$$

$$= 1 + \frac{66}{2 \times 14} + \frac{36 + 20}{66} = 4.2 \text{ sec}$$

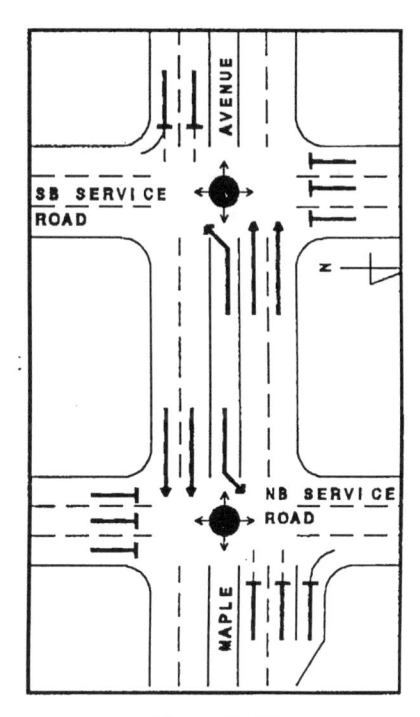

Figure 13.38

Interval 6:

Interval 6 is illustrated in Figure 13.39. This interval is to allow the last vehicle to clear the amber in interval 5 to also clear the amber in interval 7. The time period is a function of the speed limit and the distance to be traveled less the amber time in interval 7:

Clearance Time = 260 ft/ 66 ft/sec = 3.9 sec
Amber Time = 4.2 sec
Time required = 3.9 + 4.2 = –0.3 sec

A negative value means that there will be a time period of 0.3 seconds when both Ambers will be on.

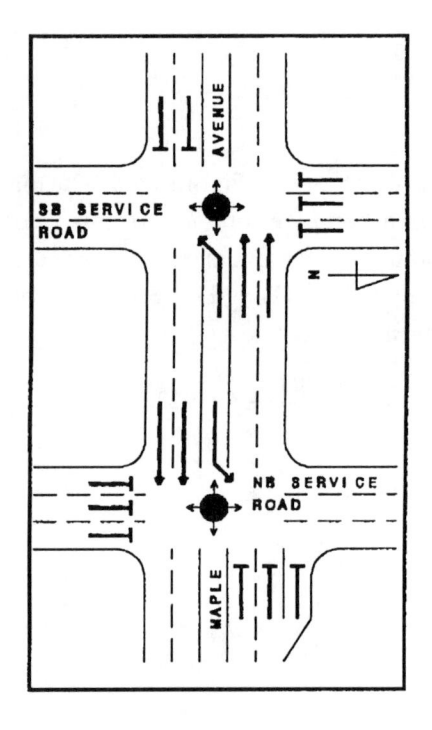

Figure 13.39

Interval 7:

This interval is illustrated in Figure 13.40. It has the same length of time as in interval 5:

$$Y = 4.2 \text{ sec}$$

The capacity of this geometric configuration is ratio of the time required to accommodate the demand with respect to the total time available. The time required is the summation of the critical times for each interval:

Interval 1:	34.8
Interval 2:	4.7
Interval 3:	8.1
Interval 4:	22.4
Interval 5:	4.2
Interval 6:	–0.3
Interval 7:	4.2
Total	78.1 sec

Time Available = 80 sec

Capacity = 78.1/80 = 0.98 or 98%

The volume-to-capacity ratio is over between 0.9 and 1.0. Therefore the Level-of-Service is "E".

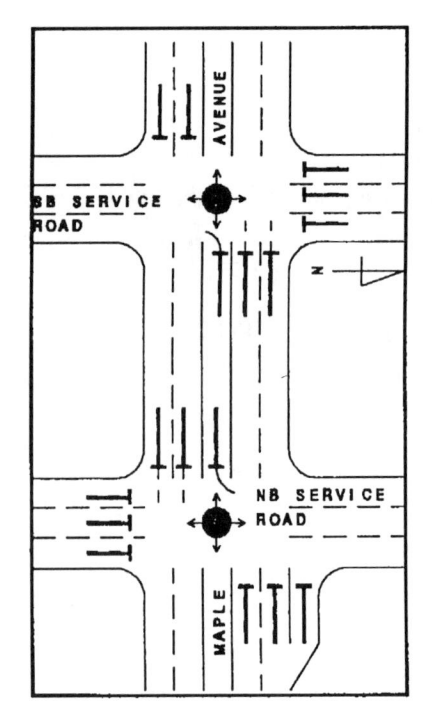

Figure 13.40

13.5 Use Eq. 13.2.2: Total number of accidents = Injury + PDO = 293   ADT = 14600

Number of fatal accidents = 6   Number of injury accidents = 89

$$R \text{ (all accidents)} = \frac{293 \times 1,000,000}{14600 \times 4 \times 365} = 13.7 \text{ MEV}$$

$$R \text{ (fatal accidents)} = \frac{6 \times 1,000,000}{14600 \times 4 \times 365} = 0.28 \text{ MEV}$$

$$R \text{ (injury accidents)} = \frac{89 \times 1,000,000}{14600 \times 4 \times 365} = 4.18 \text{ MEV}$$

$$\text{Severity Ratio} = \frac{89}{293} = 0.30$$

13.6 A) Determine R and T as a function of V using Eq. 13.4.1:

$$R = \frac{V^2}{(e+f)15} = \frac{V^2}{(0.06+0.06)15} = \frac{V^2}{1.8}$$

Tangent leg of Curve #1:

$$T_1 = R \tan \frac{\Delta_1}{2} = \frac{V^2}{1.8}(0.2679) = 0.14886 V^2$$

Tangent leg of Curve #2:

$$T_2 = \frac{V^2}{1.8}(0.57705) = 0.320 V^2$$

Tangent distance t for a 3 sec transition between curves is

$$t = (3)(1.47)V = 4.41V$$

Total available distance is 1000 feet. Therefore,

$$1000 = t + T_2 + T_1$$

$$1000 = 4.41V + 0.320V^2 + 0.149V^2$$

$$V^2 + 9.39V - 2129 = 0$$

$$\therefore V = \frac{-9.39 \pm \sqrt{9.39^2 - 4(-2129)}}{2} = 41.7 \text{ mph}$$

B)

Curve #1
$R = 966.001$

$D = 5729.6 / 966 = 5.93°$

$T_1 = R \tan\left(\frac{\Delta_1}{2}\right) = 258.89$

$E_1 = T_1 \tan\left(\frac{\Delta_1}{4}\right) = 34.08$

$M_1 = E_1 \cos\left(\frac{\Delta_1}{2}\right) = 32.92$

$L_1 = 100(\Delta_1 / D) = 505.90$

Curve #2
$T_2 = R \tan\left(\frac{\Delta_2}{2}\right) = 557.72$

$E_2 = T_2 \tan\left(\frac{\Delta_2}{4}\right) = 149.44$

$M_2 = E_2 \cos\left(\frac{\Delta_2}{2}\right) = 129.42$

$L_2 = 100(\Delta_2 / D) = 1011.80$

C) Stationing

$$
\begin{aligned}
PI_1 &= 132 + 50.00 \\
-T_1 &= \underline{2 + 58.84} \\
PC_1 &= 129 + 91.16 \\
+L_1 &= \underline{5 + 05.90} \\
PT_1 &= 134 + 97.06 \\
+t &= \underline{1 + 83.44} \\
PC_2 &= 136 + 80.50 \\
+L_2 &= \underline{10 + 11.80} \\
PT_2 &= 146 + 92.30
\end{aligned}
$$

13.7 The given data and assumptions are

$h_1 = 3.5$ ft
$h_2 = 4.25$ ft
$S = 1200$ ft
$A = 3 - 0 = 3$

For $S < L$ use Eq. 13.5.3:

$$L = \frac{3 \times 1200^2}{100\left(\sqrt{7.0} + \sqrt{8.50}\right)^2} = 1397 \text{ ft}$$

Since $S < L$, this is the correct answer.

Check by solving the other possibility $S > L$ using Eq. 13.5.4:

$$L = 2 \times 1200 - \frac{200 \times \left(\sqrt{3.5} + \sqrt{4.25}\right)^2}{3} = 1369 \text{ ft}$$

This gives $S < L$, which is not acceptable.

13.8 The sight distance requirement for a right turn maneuver is only one to three feet less than required with the sight distance from the right for a left turn maneuver onto a 2-lane 2-way highway. Therefore the procedure and answer is essentially the same as in Example 13.17.
∴ $d = 654$ ft.

13.9 A) From Table 13.7, the largest vehicle is a WB-40 which only needs a 150-ft radius. The next largest truck which is a WB-50, needs at least a 200-ft radius.

B) From Table 13.7, the largest truck which is a WB-62, only needs a 120-ft radius. Therefore all of the trucks listed can make this turn.

C) No, the radius will not have to be increased. From Table 13.7, the minimum radius actually decreases from 120 to 115 ft. Even when the angle is increased to 180°, the radius still drops to 55 ft.

D) No, as can be seen in Table 13.7 for a 180° turn. The SU truck needs a radius of 30 ft while the WB-50 needs a radius of only 25 ft.

13.10 The first step is to draw a space-time diagram that matches the distances and times required for this problem, as in Fig. 13.41.

  A) Fig. 13.41 shows the space-time diagram with the given information.

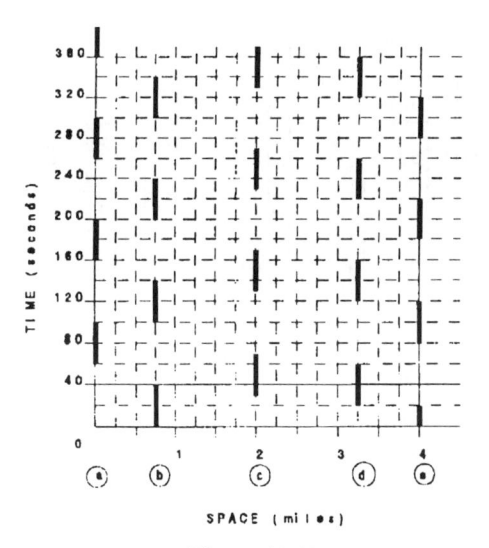

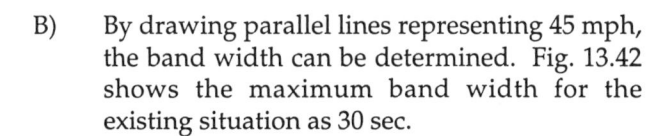

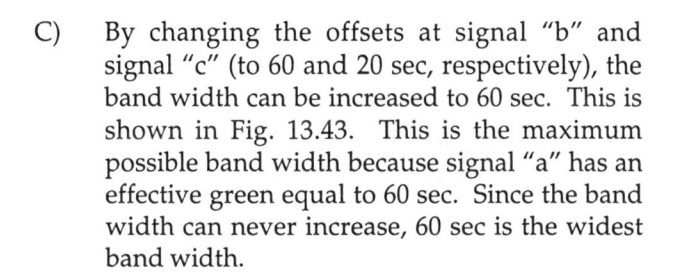

Figure 13.41

  B) By drawing parallel lines representing 45 mph, the band width can be determined. Fig. 13.42 shows the maximum band width for the existing situation as 30 sec.

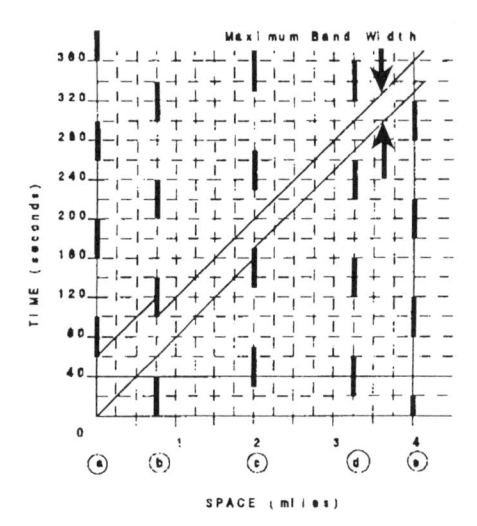

Figure 13.42

  C) By changing the offsets at signal "b" and signal "c" (to 60 and 20 sec, respectively), the band width can be increased to 60 sec. This is shown in Fig. 13.43. This is the maximum possible band width because signal "a" has an effective green equal to 60 sec. Since the band width can never increase, 60 sec is the widest band width.

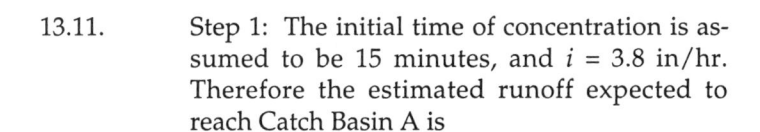

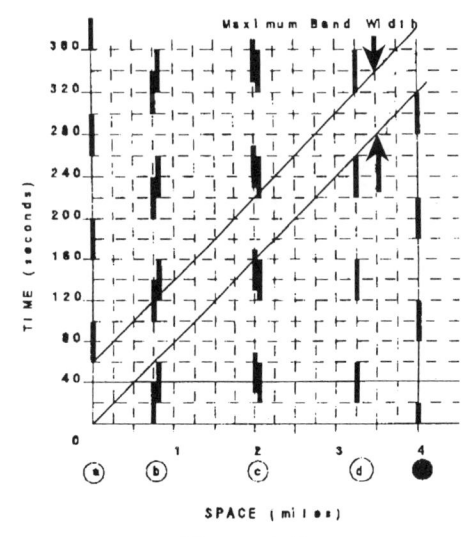

13.11. Step 1: The initial time of concentration is assumed to be 15 minutes, and $i = 3.8$ in/hr. Therefore the estimated runoff expected to reach Catch Basin A is

$$Q_1 = ciA = 0.35 \times 3.8 \times 1.75 = 2.33 \text{ cfs}$$

Figure 13.43

Figure 13.44 Kutters curves.

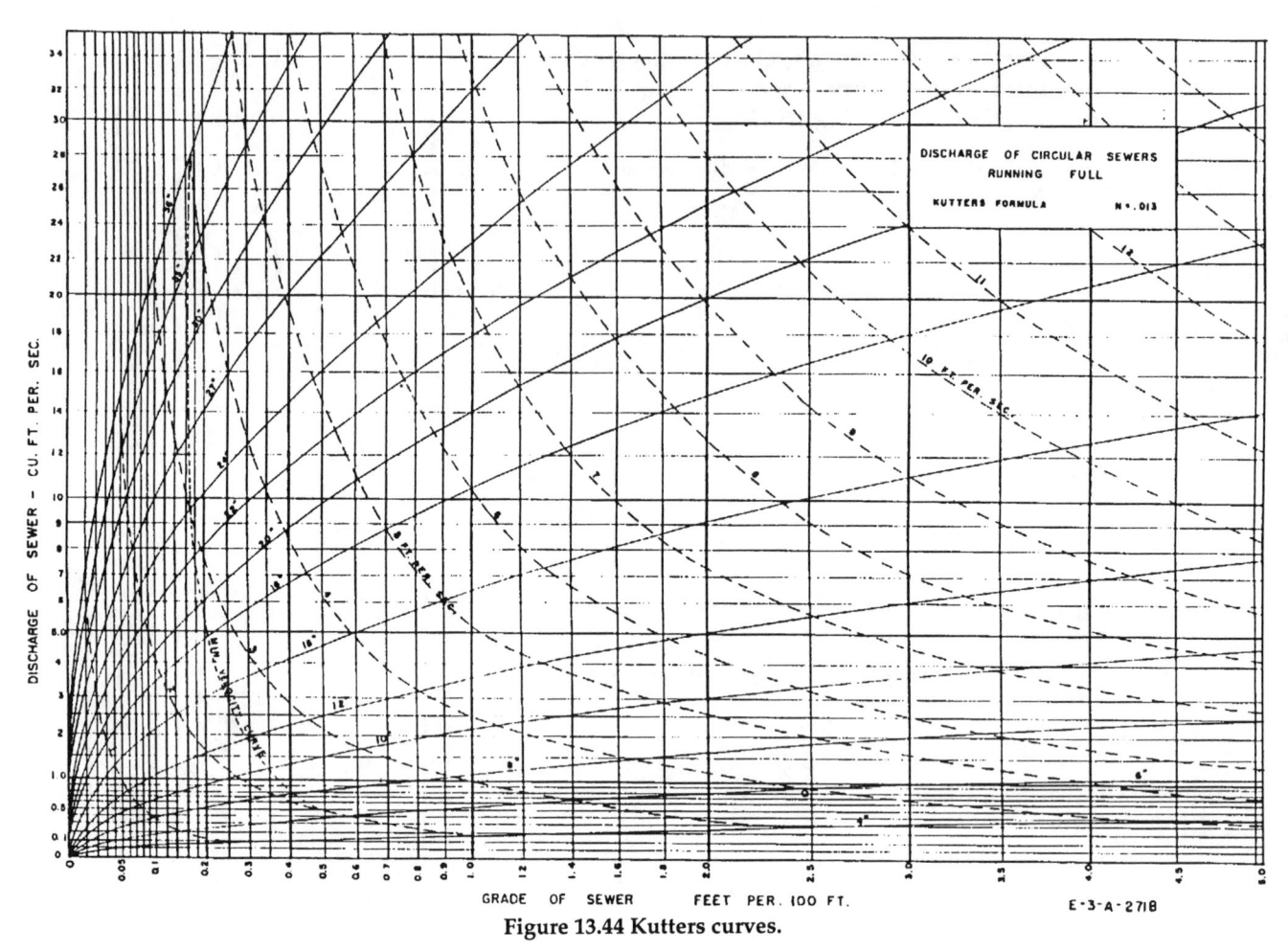

From Fig. 13.44 a 12-inch sewer with a 1% slope has a discharge capacity of 3.6 cfs with a velocity of 4.5 fps. The time for the flow to reach Catch Basin B is

$$275 \text{ ft} / 4.5 \text{ (ft/sec)} = 61 \text{ sec} = 1 \text{ min}$$

Step 2: The time of concentration is now $15 + 1 = 16$ min, and $i = 3.7$ in/hr. Therefore the estimated runoff expected to reach Catch Basin B is

$$Q_2 = ciA = 0.30 \times 3.7 \times 1.90 = 2.11 \text{ cfs}$$

resulting in

$$Q_1 + Q_2 = 2.33 + 2.11 = 4.44 \text{ cfs}$$

From Fig. 13.44 a 15-inch sewer with a 0.8% slope has a discharge capacity of 5.8 cfs and a velocity of 4.7 fps. The time required for the flow to reach Catch Basin C is

$$275 \text{ ft}/(4.7 \text{ ft/sec}) = 59 \text{ sec} = 1 \text{ min}$$

Step 3: The time of concentration is now 17 minutes with $i = 3.6$ in/hr. Therefore, the estimated runoff expected to reach Catch Basin C is

$$Q_3 = ciA = 0.45 \times 3.6 \times 1.50 = 2.43 \text{ cfs}$$

and

$$Q_1 + Q_2 + Q_3 = 4.44 + 2.43 = 6.87 \text{ cfs}$$

From Fig. 13.44, an 18-inch sewer with a 0.7% slope has a discharge capacity of 8.8 cfs with a velocity of 5.0 fps.

Step 4: etc.

It may take several iterations to develop a good design. To this point the design has an acceptable velocity and capacity. The slope and depth of sewer are within the critical values and the velocity has increased slightly; at no point has there been a decrease in the velocity.

Solutions to Problems (PE-Format)

13.12 d) Refer to equation 13.4.1. The maximum superelevation, e, for a cold climate should not exceed 0.08 and with no lateral friction, f is equal to 0.00.

13.13 c) Refer to equation 13.4.1. The maximum value of f is 0.12 (page 154, 1990 Green Book). The minimum radius is provided by using the maximum values of e and f.

13.14 c) Refer to Equations 13.4.4 and 13.4.10. With perpendicular legs the external angle, Δ, equals 90 degrees. The minimum length of curve also requires the minimum radius. Therefore L = 1884.96 ft.

13.15 d) From page 178 of the 1990 Green Book. The length of runoff is 180 ft. for a design speed of 60 mph, e = 0.08 and 10 ft. lanes. The tangent section is to contain 60 to 80% of the runoff.

13.16 e) All of the above are correct. However, d) is the recommended practical control for length of spiral. (Page 174 of the 1990 Green Book).

13.17 c) A 10 year storm is adequate for a county highway. There is 2.91 acres of pavement and a runoff coefficient of 0.85 was assumed. There is 0.73 acres of gravel surface with an assumed runoff coefficient of 0.50. There is 4.36 acres of turf right-of-way with an assumed coefficient of runoff of 0.40. For a 10 year storm and an initial time of concentration of 15 minutes the runoff from Figure 13.31 is estimated as 3.8 inches per hour. It was assumed that little if any significant runoff would come from the forest and the only runoff would be generated from within the limits of the right-of-way.

13.18 a) The most efficient ditch would have the lowest value of Manning's coefficient and have the smallest cross-sectional area. For this case the design flow is 8.7 cfs. (assuming a ditch on each side). The slope is 132/5280 ft/ft. By using Equation 13.9.2, the resulting cross-section area is 1.445 sqft. The wetted perimeter is 3.8 ft. The resulting flow is 8.9 cfs with a velocity of 6.2 fps.

13.19 e) The grass lining has the highest Manning's coefficient and the 4-ft flat bottom ditch produces the lowest Hydraulic radius (R). These two factors increase the cross-section area and decrease the fate of flow for the same runoff.

13.20 a) From Kuthers Curves the required grade of the sewer for answer a) is 2.5% which is appropriate for this location.

14. Soil Mechanics

by Thomas F. Wolff

Soil mechanics deals with the properties, classification, and stress-strain behavior (compressibility and strength) of soil materials and the movement of water through soils. Application of the theory of soil mechanics to problems involving foundations, slopes and retaining structures is covered in this chapter.

14.1 The Phase Diagram

Soil is a complex mixture of two or three materials: solids, water, and perhaps air. Relationships among these components must often be calculated. In an actual volume of soil, the solids, water and air are arranged in a complex fashion as shown in Figure 14.1. To facilitate visualization of the problem and calculations, a phase diagram may be used. In a phase diagram, shown in Figure 14.2, the solid particles are shown as if assembled into a single contiguous mass at the bottom of the diagram. Above the solids is shown the water, and above that, the air. On the left side of the diagram are shown the total volume V, the volume of solids V_s, the volume of water V_w, the volume of air V_a, and the volume of voids V_v, which is the combined volume of water and air. On the right side is shown the total weight W, the weight of solids W_s, and the weight of water W_w. The weight of air is taken as zero.

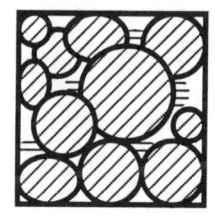

Figure 14.1 The Soil-Water System.

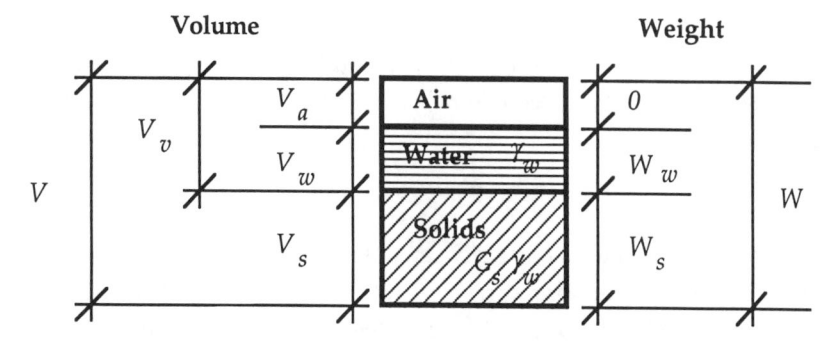

Figure 14.2 A phase diagram.

14.2 Weight-Volume and Mass-Volume Relationships

Relationships regarding soil and water volumes include the *void ratio e*, which is the ratio of the volume of voids to the volume of solids:

$$e = \frac{V_v}{V_s} \tag{14.2.1}$$

the *porosity n*, which is the ratio of the volume of voids to the total volume:

$$n = \frac{V_v}{V} \tag{14.2.2}$$

and the *degree of saturation S*, which is the ratio of the volume of water to the volume of voids, and is usually expressed as a percentage:

$$S = \frac{V_w}{V_v} \times 100\% \tag{14.2.3}$$

The primary relationship involving soil and water weights is the *water content* or *moisture content w*, which is the ratio of the weight of water to the weight of solids,

$$w = \frac{W_w}{W_s} \tag{14.2.4}$$

Relationships involving both weights and volumes are expressed as unit weights. In some references, unit weights are called specific weights or weight densities. The *total unit weight* γ is the total weight of a soil mass (soil, water and air) divided by the total volume occupied, including the pore volumes:

$$\gamma = \frac{W}{V} = \frac{W_s + W_w}{V_s + V_w + V_a} \tag{14.2.5}$$

The *saturated unit weight* γ_{sat} is the total unit weight obtained if the entire void volume were filled with water ($S = 100\%$ and $V_w = V_v$).

The *dry unit weight* γ_d, or *dry density*, is the ratio of the weight of solids to the total volume:

$$\gamma_d = \frac{W_s}{V} = \frac{W_s}{V_s + V_w + V_a} \tag{14.2.6}$$

Although this definition may appear to mismatch the weight of one component with the volume of three components, it is in fact this definition that makes the concept useful. The dry unit weight is used to track the amount of soil solids in the entire occupied volume (e.g., the volume of a truck bed) and is the key principle in many problems involving quantities of earthwork, as will be seen in Example 14.14.

The *buoyant unit weight* or *effective unit weight* γ' of a soil below the water table is the apparent weight of a unit volume of soil if it could be weighed below water. It is equal to the saturated unit weight minus the unit weight of water,

$$\gamma' = \gamma_{sat} - \gamma_w \tag{14.2.7}$$

The buoyant unit weight is sometimes used as a shortcut to calculate effective stresses rather than calculating total stresses and subtracting pore pressures. Above the water table, buoyant unit weight has no physical significance; total unit weights should be used in effective stress calculations.

Problems may be phrased in the context of masses and mass densities rather than weights and unit weights; however, unit weights must ultimately be obtained for stress calculations. Mass density is denoted by ρ. As weight W equals mass m times the acceleration of gravity g ($W = mg$), the unit weight values defined above are in fact the product of mass densities and g:

$$\gamma = \rho g \tag{14.2.8}$$

$$\gamma_{sat} = \rho_{sat} g \tag{14.2.9}$$

$$\gamma_d = \rho_d g \tag{14.2.10}$$

$$\gamma' = \rho' g \tag{14.2.11}$$

In the SI system, mass densities are commonly expressed as Mg/m^3, kg/m^3, or g/ml:
$1\ Mg/m^3 = 1000\ kg/m^3 = 1\ g/ml$

In the English system, mass densities (which would be $slugs/ft^3$) are never used.

───── **EXAMPLE 14.1** ─────

Convert a total mass density of 1900 kg/m^3 to a total unit weight for use in stress calculations.

Solution.
$$\gamma = \rho g$$
$$= (1900\ kg/m^3)(9.807\ m/s^2)$$
$$= 18630\ kg/m^2 s^2 = 18.63\ kN/m^3$$

14.3 Specific Gravity

The *specific gravity* G_s is the ratio of the density of the solid particles to the density of water; hence, the unit weight of solids, γ_{solids}, can be expressed as $G_s \gamma_w$. For most of the minerals comprising soil solids, G_s is in the range 2.65 to 2.75. Recall the mass density and unit weight of water:

Mass density of water:

$$\rho_w = 1\ Mg/m^3 = 1000\ kN/m^3 = 1\ g/ml$$

Unit weight of water:

$$\gamma_w = 9.807\ kN/m^3 = 9807\ N/m^3$$
$$= 62.4\ lb/ft^3$$

───── **EXAMPLE 14.2** ─────

Soil solids have a specific gravity of 2.71. Find the unit weight of solids, γ_{solids}.

Solution.
$$\gamma_{solids} = G_s \gamma_w$$
$$= (2.71)(9.807\ kN/m^3) = 26.58\ kN/m^3$$
$$= (2.71)(62.4\ lb/ft^3) = 169.10\ lb/ft^3$$

EXAMPLE 14.3

A soil mass has a total unit weight of 130 lb/ft³. Find the total unit weight in SI units.

Solution. Although a chain of conversion factors could be used, conversions can be simplified by using the total density of water in both systems as an equality. Thus,

$$\gamma = \left(130 \text{ lb/ft}^3\right)\left(\frac{9.807 \text{ kN/m}^3}{62.4 \text{ lb/ft}^3}\right) = 20.43 \text{ kN/m}^3$$

14.4 The Use of a Phase Diagram

Although weight-volume problems can be worked directly with equations, errors are reduced and checks are provided if such problems are worked by using a phase diagram and performing the following steps.

- Sketch a blank phase diagram.
- Fill in the known weights, volumes, and component densities.
- If no weights or volumes are given, there is no defined quantity of soil and the problem involves only calculating unknown relationships (γ, e, w, S, etc.) from given relationships; assume a value (as further discussed below) for one unknown weight or volume and proceed.
- Multiply known component volumes by the component unit weight to obtain component weights. Divide known component weights by the component unit weight to obtain volumes. Although values in practice are seldom significant to more than three digits, use at least four significant digits in phase diagram calculations to reduce errors and ensure that the same results will be obtained when values are summed in the vertical direction and multiplied or divided in the horizontal direction.
- Where values of relationships (γ, e, w, S, etc.) are given, use the definitions in equations 2.1 through 2.6 to obtain additional weights or volumes from the given values. Continue multiplication and division across the diagram and addition and subtraction up and down the sides until all weights and volumes are known and all quantities check.
- Calculate the required values from the completed and checked diagram.

Weight-volume problems can broadly be grouped into two categories:

- Those where the weights and volumes are determinate; at least one quantity is fixed.
- Those where only relationships are given and only relationships are desired.

For problems in the second category, a fixed weight or volume must be assumed; while any quantity can be assumed, certain assumptions greatly simplify calculations:

- If a water content is given, assume $W_s = 1.00$ or 100.00 (lb or kg); then $W_w = w$ or $100w$.
- If a void ratio is given, assume $V_s = 1.000$ ft³ or m³; then $V_v = e$.
- If a dry unit weight is given, assume $V = 1.000$ ft³ or m³; then $W_s = \gamma_d$.
- If a total unit weight is given, assume $V = 1.000$ ft³ or m³; then $W = \gamma$.

EXAMPLE 14.4

A mold having a volume of 0.10 ft³ was filled with moist soil. The weight of the soil in the mold was found to be 12.00 lb. The soil was oven dried and the weight after drying was 10.50 lb. The specific gravity of solids was known (or assumed) to be 2.70. Find the water content, void ratio, porosity, degree of saturation, total unit weight and dry unit weight density.

Solution: As actual weights and volumes are given, the problem falls into the first category. A phase diagram is drawn showing the known quantities (in bold):

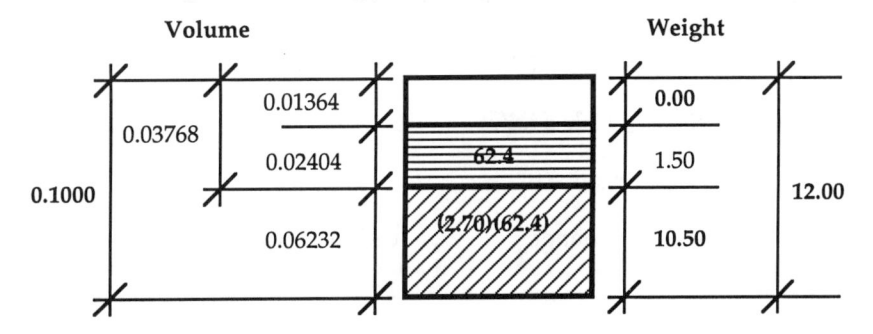

The weight and volume of water are, respectively,

$W_w = W - W_s = 12.00 - 10.50 = 1.50$ lb

$V_w = W_w / \gamma_w = 1.50/62.4 = 0.02404$ ft³

The volume of solids is

$V_s = W_s / G_s \gamma_w = (10.50) / (2.70)(62.40) = 0.06232$ ft³

The volume of air is

$V_a = V - V_w - V_s = 0.10000 - 0.02404 - 0.06232 = 0.01364$ ft³

The volume of voids is found to be

$V_v = V_w + V_a = 0.02404 + 0.01364 = 0.03768$ ft³

The phase diagram is now complete and all desired relationships can be obtained:

$w = W_w/W_s = 1.50 / 10.50 = 0.143 = 14.30\%$

$e = V_v/V_s = 0.03768 / 0.06232 = 0.605$

$n = V_v/V = 0.03768 / 0.1000 = 0.377$

$S = (V_w/V_v) \, 100\% = (0.02404 / 0.03768)(100\%) = 63.8\%$

$\gamma = W / V = 12.00 / 0.100 = 120$ lb/ft³

$\gamma_d = W_s/V = 10.50 / 0.100 = 105.0$ lb/ft³

EXAMPLE 14.5

The soil in example 14.4 becomes saturated by accumulating water. Assuming it retains the same total volume, find the saturated water content, saturated unit weight and dry unit weight.

Solution. The new volume of water is the entire void volume, 0.03768 ft³. Multiplying this value by 62.4 lb/ft³, the new weight of water is 2.35 lb. The phase diagram is modified and calculations follow.

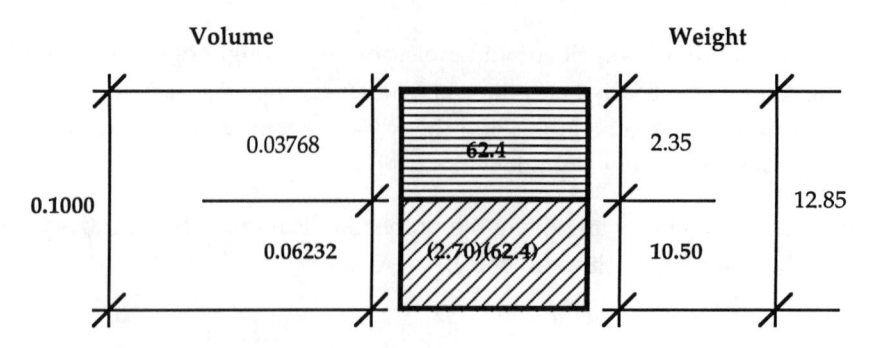

$w = W_w/W_s = 2.35 / 10.50 = 0.2239$ or 22.4%

$\gamma = W/V = (10.50 + 2.35)/(0.100) = 128.5$ lb/ft^3

$\gamma_d = W_s/V = (10.500)/(0.100) = 105.0$ lb/ft^3

Note that the dry unit weight does not change with an addition of water at total constant volume.

EXAMPLE 14.6

A soil has a water content of 20 percent, a void ratio of 0.800, and a specific gravity of 2.65. Find the degree of saturation, porosity, total unit weight, and dry unit weight.

Solution: As no specific quantity of soil is involved, and only relationships are given and desired, the problem is of the second category: any fixed quantity may be assumed in order to draw the phase diagram. Assume that the volume of solids $V_s = 1.000$ ft^3; the phase diagram follows.

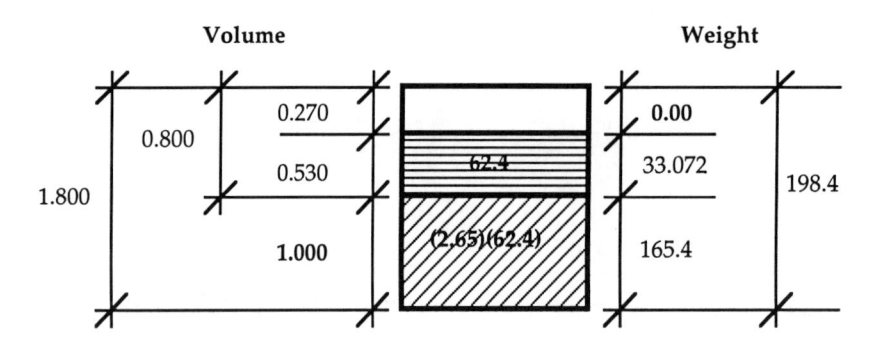

$V_v = eV_s = (0.800)(1.000) = 0.800$ ft^3

$W_s = V_s G_s \gamma_w = (1.000)(2.65)(62.4) = 165.4$ lb

$W_w = wW_s = (0.20)(165.4) = 33.07$

$V_w = W_w/\gamma_w = (33.07)/(62.4) = 0.530$

$V_a = V_v - V_w = 0.800 - 0.530 = 0.270$

At this point, the weights and volumes of all components are known for the assumed 1.000 ft^3 of solids and all desired relationships can be calculated:

$S = (V_w/V_v)100\% = (0.530 / 0.800)(100\%) = 66.3\%$

$n = V_v/V = 0.800 / (1.000 + 0.800) = 0.444$

$\gamma = (165.4 + 33.07)/ 1.800 = 110.2$ lb/ft^3

$\gamma_d = 165.4 / 1.800 = 91.9$ lb/ft^3

EXAMPLE 14.7

A 0.500 ft³ sample of saturated soil weighs 61.0 lb. It is known from experience that the specific gravity can be taken as 2.70 for this material. Determine its water content, void ratio, and dry unit weight.

Solution. The problem is of the first category as the volume of soil is fixed. However, as the relative proportions of soil and water are unknown, two simultaneous equations are required to obtain the solution. It is known the soil is a mixture of two materials, water with a unit weight of 62.4 lb/ft³, and solids with a unit weight of (2.70)(62.4) = 168.48 lb/ft³. The soil has a unit weight of (61.0 lb)/(0.500 ft³) = 122.0 lb/ft³, and only one unique proportioning of solids and water will yield this value.

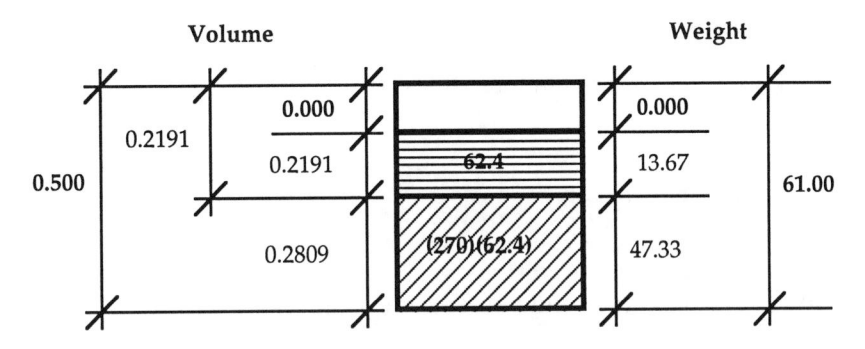

As the soil is saturated, $V_a = 0$. On the volume side, one can write:

$$V_s + V_v = 0.500$$

On the weight side, one has

$$\begin{aligned}
61.0 &= W_s + W_w \\
&= V_s G_s \gamma_w + V_w \gamma_w \\
&= V_s(2.70)(62.4) + V_w(62.4) \\
&= 168.5 V_s + 62.4 V_w
\end{aligned}$$

The volume equation and the weight equation are solved simultaneously:

$$168.5 \, V_s + 62.4(0.500 - V_s) = 61.0$$

$$V_s = 0.2809 \text{ ft}^3 \quad \text{and} \quad V_w = 0.2191 \text{ ft}^3$$

Then,

$$W_s = (0.2809)(2.70)(62.4) = 47.33 \text{ lb}$$

$$W_w = (0.2191)(62.4) = 13.67 \text{ lb}$$

The desired relationships are:

$$w = 13.67 / 47.33 = 0.289 \text{ or } 28.9 \%$$

$$e = 0.2191 / 0.2809 = 0.780$$

$$\gamma_d = 47.33 / 0.5000 = 94.66 \text{ lb/ft}^3$$

14.5 Other Useful Equations for Weight-Volume Problems

It is strongly recommended that weight-volume problems be solved using phase diagrams rather than only formulas, as completing a phase diagram clearly indicates whether sufficient information is known to complete the problem, whether information is insufficient and assumptions must be made, or whether too much information is present and the problem is overconstrained. For example, it may not be immediately apparent from the information given whether a soil is saturated until all quantities are calculated. Nevertheless, following are given additional useful equations that may be used to solve certain classes of weight-volume problems.

A very useful equation relating four different quantities is

$$Se = wG_s \qquad (14.5.1)$$

For saturated soils (S = 100%) there results

$$e = wG_s \qquad (14.5.2)$$

The relationships between the void ratio and porosity are

$$e = \frac{n}{1-n} \qquad (14.5.3)$$

and

$$n = \frac{e}{1+e} \qquad (14.5.4)$$

The total unit weight can be obtained as

$$\gamma = \frac{(G_s + Se)\gamma_w}{1+e} = \frac{(1+w)\gamma_w}{w/S + 1/G_s} \qquad (14.5.5)$$

The dry unit weight can be obtained as

$$\gamma_d = \frac{G_s \gamma_w}{1+e} = \frac{G_s \gamma_w}{1+(wG_s/S)} \qquad (14.5.6)$$

EXAMPLE 14.8

Rework example 14.6 using equations introduced in this section.

Solution.
$$Se = wG_s$$

$$S = wG_s/e = (.20)(2.65)/(0.800) = 0.6625 \text{ or } 66.3\%$$

$$n = \frac{e}{1+e} = \frac{0.800}{1+0.800} = 0.444$$

$$\gamma = \frac{(1+w)\gamma_w}{w/S + 1/G_s} = \frac{(1.20)(62.4)}{0.2/0.6625 + 1/2.65} = 110.2 \text{ lb/ft}^3$$

$$\gamma_d = \frac{G_s \gamma_w}{1+e} = \frac{(2.65)(62.4)}{1+0.800} = 91.9 \text{ lb/ft}^3$$

14.6 Grain-Size Characteristics of Soils

Soils are commonly classified according to the grain size of the solid particles as *gravel, sand, silt,* and *clay*. Gravels and sands are referred to as *coarse-grained soils*; silts and clays are referred to as *fine-grained soils*. For engineering purposes, gravel is usually defined as soil with a particle diameter larger than the opening of a No. 4 sieve (4.75 mm or 0.187 in.) and sand is defined as particles smaller than gravel but larger than the opening of a No. 200 sieve (0.075 mm or about 0.003 in.). A soil passing the number 200 sieve may be silt or clay, but its classification is generally based on plasticity characteristics rather than grain size.

The *grain-size distribution* of coarse-grained soils is determined by a *sieve analysis* (also known as a *mechanical analysis*). In a sieve analysis, a nest of sieves is arranged with the largest sieve on top and the smallest sieve just above a catch pan on the bottom. Opening sizes for commonly used standard sieves are shown in Table 14.1. The soil sample is passed over the sieves and the sieves are shaken until all particles have passed down to a sieve that retains them. For each sieve size, the cumulative weight of all material larger than that size is divided by the total sample weight to obtain the *percentage retained*, and the cumulative weight of all material smaller than that size is divided by the total sample weight to obtain the *percentage passing*. The results are plotted as the percent passing (linear scale) versus the sieve opening or grain-size (log scale). The plot is referred to as a *grain size curve* or a *grain size distribution curve*, a sample of which is presented in Figure 14.3.

TABLE 14.1 Opening Sizes of Standard Sieves

	Sieve Size		Opening	
	1.5	in	37.5	mm
	1		25	
	0.75		19	
	0.5		12.5	
No.	4		4.75	
	10		2.00	
	20		0.850	
	40		0.425	
	70		0.212	
	100		0.150	
	200		0.075	

From the grain size curve, certain sizes of interest may be determined by interpolation. The D_{10} size, sometimes called the *effective size*, is the grain diameter for which 10 percent of the sample (by weight) is finer. The D_{50} size, called the *median grain size*, is the grain diameter for which 50 percent of the sample (by weight) is finer. In general, the notation D_{14} refers to 14 percent of the sample weight being finer than the diameter D, in mm.

Percent passing

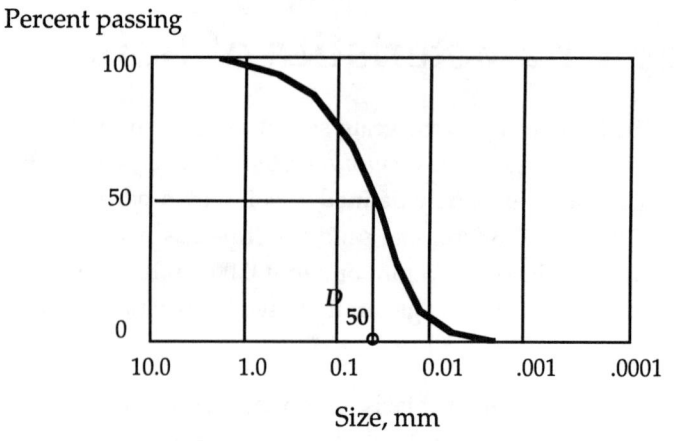

Figure 14.3 Grain size curve.

Several parameters are used to describe the shape of the grain-size curve. The *coefficient of uniformity* C_u is

$$C_u = \frac{D_{60}}{D_{10}} \qquad (14.6.1)$$

The *coefficient of curvature* or *coefficient of gradation* C_z is

$$C_z = \frac{D_{30}^2}{D_{60} \times D_{10}} \qquad (14.6.2)$$

14.7 Atterberg Limits and Plasticity

Atterberg limits are measures of the water contents where soil behavior changes. The upper range of plasticity, above which soil behaves as a liquid, and below which soil behaves as a plastic solid, is the *liquid limit*, denoted LL or w_L. The lower range of plastic behavior, above which the soil behaves as a plastic solid, and below which the soil behaves as a brittle solid, is the plastic limit, denoted PL or w_P. The water content below which the soil no longer reduces in volume with a reduction in water content is the shrinkage limit, SL or w_S. Although Atterberg limits are water contents, and are properly decimals or percentages, they are usually expressed as integers (a percentage).

The *plasticity index*, denoted PI or I_P, is the difference of the liquid limit and the plastic limit:

$$PI = I_P = LL - PL \qquad (14.7.1)$$

The *liquidity index*, a measure of the natural water content relative to the plastic limit and the liquid limit, is

$$LI = I_L = \frac{w - PL}{LL - PL} \qquad (14.7.2)$$

14.8 Soil Classification

The soil classification system commonly used in geotechnical engineering is the *Unified Soil Classification System* (USCS). An alternative system used for highway work is the *AASHTO system*.

In the Unified Soil Classification System, soils are designated by a symbol consisting of two letters, the first identifying the soil, and the second describing its grain size or plasticity characteristics. For example, a well-graded sand is SW and a highly plastic clay is CH. The symbols are presented in Table 14.2.

TABLE 14.2 The USCS soil designation symbols

Soil type	First Letter		Second Letter	
Coarse Grained	G	gravel	W	well graded
	S	sand	P	poorly graded
			C	clayey
			M	silty
Fine Grained	M	silt	H	high plasticity
	C	clay	L	low plasticity
	O	organic		

Soil classification by the Unified System is based on the percents passing the No. 4 and No. 200 sieves, the coefficients of uniformity and the coefficients of curvature, and the liquid limit and plasticity index of the soil fraction passing the No. 40 sieve. The first step in classification is to pass the soil over a No. 200 sieve.

- If more than 50% of the soil is retained on a No. 200 sieve, the soil is *coarse grained*, and the first letter will be *G* or *S*.

- If less than 50% of the soil is retained on a No. 200 sieve (> 50% passing), the soil is *fine grained*, and the first letter will be *M* or *C*.

For coarse grained soils, the *coarse fraction* is the fraction retained on a No. 200 sieve.

- If more than 50 percent of the coarse fraction is retained on a No. 4 sieve, the first letter symbol is *G*.

- If less than 50 percent of the coarse fraction is retained on a No. 4 sieve, the first letter symbol is *S*.

For coarse grained soils, the second letter symbol reflects gradation for clean sands and gravels, and plasticity of the fines for dirty sands and gravels.

- If less than 5 percent of total sample passes the No.200 sieve, the soil is a clean sand or clean gravel (*SW*, *SP*, *GW*, or *GP*).

- If the soil classifies as clean sand, the soil is a well-graded sand (*SW*) if $C_u \geq 6$ and $1 \leq C_z \leq 3$. If both criteria are not met, the soil is a poorly graded sand (*SP*).

- If the soil classifies as clean gravel (*GW* or *GP*), the soil is a well-graded gravel (*GW*) if $C_u \geq 4$ and $1 \leq C_z \leq 3$. If both criteria are not met, the soil is a poorly graded gravel (*GP*).

- If more than 12 percent of the total sample passes the No. 200 sieve, the soil is a clayey sand (*SC*), clayey gravel (*GC*), silty sand (*SM*), or silty gravel (*GM*), and the second letter is assigned based on whether the fines are clay or silt as described for fine grained soils below.

- If the percentage passing the No. 200 sieve is between 5% and 12%, the soil requires a dual classification reflecting both gradation and plasticity, such as *SP-SM*, *SP-SC*, *GW-GC*, etc.

For fine grained soils, classification is based on the results of Atterberg limits determined on that part of the sample passing the No. 40 sieve. The plasticity chart (Figure 14.4) illustrates the liquid limit and plasticity index corresponding to the various classifications. The vertical line at *LL*=50 separates high plasticity soils from low plasticity soils. The "*A*-line," which has the equation

$$PI = 0.73(LL-20) \qquad (14.8.1)$$

separates clay from silt. The *U*-line, which has the equation

$$PI = 0.9(LL-8) \qquad (14.8.2)$$

is not used in classification but is considered to be an upper boundary for natural soils. Data plotting above the *U*-line should be checked for errors.

For fine grained soils:

- Inorganic soils plotting above the *A*-line are clays and are given the first letter *C*; those plotting below the *A*-line are silts and are given the first letter *M*. Soils with an obvious organic content are given the first letter *O*.

- Soils with a liquid limit above 50 are high plasticity (or "fat") and are given the second letter *H*; soils with a liquid limit below 50 are low plasticity (or "lean") and are given the second letter *L*.

- Soils plotting in the shaded zone are borderline silt-clays and are given the dual classification CL-ML.

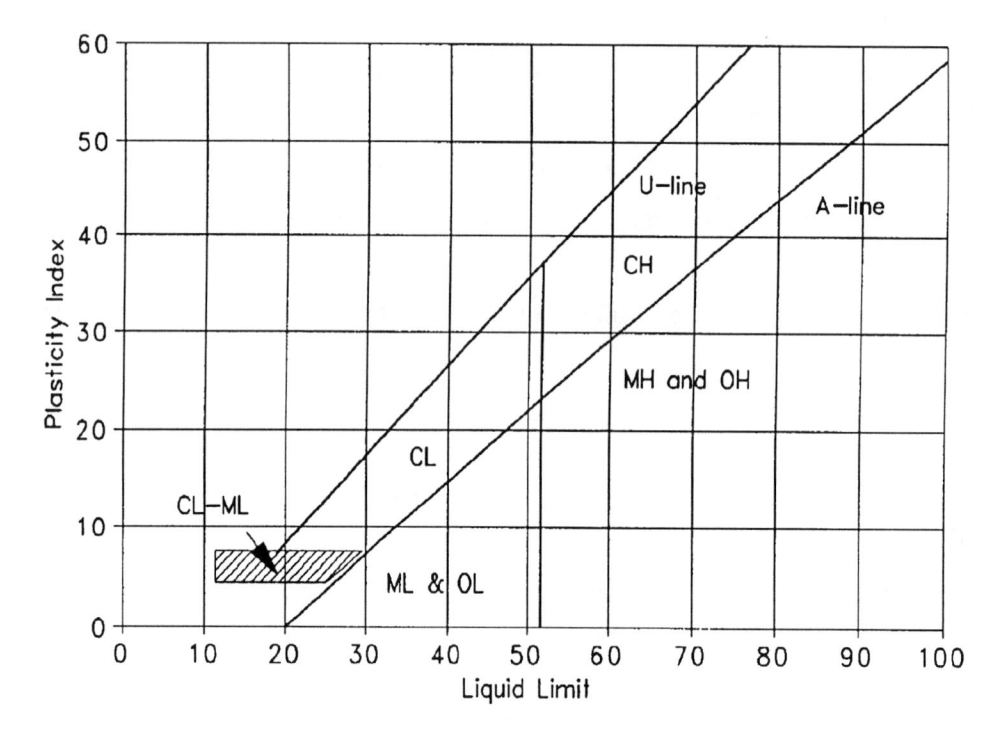

Figure 14.4 Plasticity chart.

EXAMPLE 14.9

Classify the following soil:

Percent passing No. 200 sieve:	3%
Percent passing No. 4 sieve:	100%
D_{10} size:	0.3 mm
D_{30} size:	0.52 mm
D_{60} size:	0.80 mm

Solution: 97 percent of the soil is larger than a No. 200 sieve, so the soil is coarse grained. The entire coarse fraction is smaller than a No. 4 sieve, so the soil is a sand (first letter *S*).

As less than 5 percent passes the No. 200 sieve, the sand is clean (*SP* or *SW*). The coefficient of uniformity is (0.80)/(0.30) = 2.667. As this is less than 6, the sample is poorly graded sand (*SP*) and the coefficient of curvature need not be calculated.

EXAMPLE 14.10

Classify the following soil:

Percent passing No. 200 sieve:	7%
Percent passing No. 4 sieve:	68 %
D_{10} size:	0.15 mm
D_{30} size:	0.90 mm
D_{60} size:	04.00 mm
Liquid Limit	Non-plastic (*NP*)

Solution: 93 percent of the soil is larger than a No. 200 sieve, so the soil is coarse grained. 32 percent (100–68) of the sample is gravel and 61 percent (68 - 7) of the sample is sand. As more of the coarse fraction is sand than gravel, the soil is a sand (first letter *S*).

As the percentage passing the No. 200 sieve is between 5 and 12 percent, a dual classification is required. The coefficient of uniformity is 4.00 / 0.15 = 26.67, which is greater than 6. The coefficient of curvature is 0.90^2 / (4.00 × 0.15) = 1.35, which is between 1 and 3. The sand is well graded, and the first part of the dual classification is *SW*.

Non-plastic indicates that the material would not hold together to perform Atterberg limits tests; hence, the fines are silt and not clay and the second symbol is *SM*. The final classification is *SW-SM*.

EXAMPLE 14.11

Classify the following soil:

Percent passing No. 200 sieve:	60 %
Liquid Limit:	55
Plastic Limit:	25

Solution. As more than 50 percent of the sample passes the No. 200 sieve, the soil is fine-grained and the plasticity chart is used. The plasticity index is 55 - 25 = 30. The coordinates (55,30) plot in the *CH* region, so the soil is a high plasticity clay (*CH*).

14.9 Compaction and Compaction Quality Control

Compaction is the process of densifying soil using mechanical energy to expel air from the soil/water/air mixture. Thus, saturated soils cannot be compacted. (They do change volume by consolidation, the time-dependent expulsion of water, discussed in a later section). Clays, silts, clayey sands and silty sands are often compacted with sheepsfoot rollers. Clean, free-draining sands and gravels are usually better compacted with vibratory rollers and/or the addition of water.

For soils with fines compacted under a given compaction effort, as the water content is increased above zero the dry unit weight first increases, then reaches a peak and decreases. The *maximum dry unit weight* (referred to as the maximum dry density in a number of references) occurs at the *optimum water content*. The test most commonly used to determine the dry unit weight vs water content relationship is the *Standard Proctor Test*, named after the developer of the test and standardized by ASTM D698 and AASHTO T99. Soil is placed in a mold with a volume of 1/30 ft^3 and each of three layers is compacted with 25 tamps of a 5.5-lb hammer dropped 12 in. The dry unit weight after compaction is determined for samples at three to five different water contents and a smooth curve drawn through the results. An example of the resulting compaction curve is shown in Figure 14.5. The coordinates of the peak of the curve are taken as the maximum dry unit weight and the optimum water content. To the extent that the effort applied by the field roller is similar to that applied in the compaction test, the compaction curve indicates the range of water contents at which various dry unit weight values can be achieved.

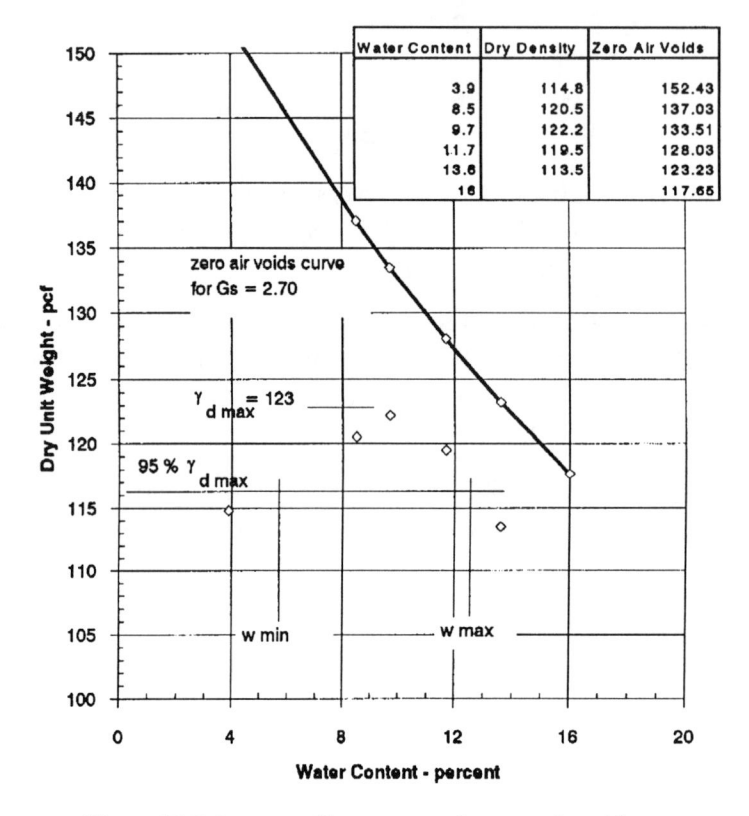

Water Content	Dry Density	Zero Air Voids
3.9	114.8	152.43
8.5	120.5	137.03
9.7	122.2	133.51
11.7	119.5	128.03
13.6	113.5	123.23
16		117.65

Figure 14.5 A compaction curve and a zero-air voids curve.

For any dry unit weight and specific gravity, there is a water content value corresponding to 100% saturation; the water content cannot exceed this value without increasing the volume and thus changing the dry unit weight. The function relating maximum water content and maximum dry unit weight for a given specific gravity is called the *zero-air voids curve*. It is given by

$$\gamma_d = \frac{\gamma_w}{w+(1/G_s)} \qquad (14.9.1)$$

The zero-air voids (ZAV) curve will always be above and to the right of the compaction curve.

―――― **EXAMPLE 14.12** ――――

A soil has the standard compaction results shown in Figure 14.5. Specifications require that the soil be compacted to a minimum of 95 percent of standard Proctor dry unit weight. Assuming that the field compaction effort is similar to the laboratory test effort, what range of water contents is necessary to achieve the specified unit weight?

Solution. The maximum dry unit weight is approximately 123 lb/ft³. The required minimum dry unit weight is (0.95)(123.0) = 116.9 lb/ft³. The compaction curve crosses this unit weight at water contents of approximately 5.5 percent and 12.5 percent. Thus, the field water content must be between these values to achieve the specified unit weight if the field compaction effort corresponds to standard Proctor effort.

For certain compacted fills where a high unit weight is required, such as subgrades for heavily loaded pavements, *Modified Proctor* effort may be specified. In the modified test, the same 1/30 ft³ mold is filled with five layers of soil, each compacted with a 25 tamps of a 10-lb hammer dropped 18 inches.

Specifications commonly require the in-place dry unit weight of a soil to be at least 95 percent of the standard Proctor maximum dry unit weight. In-place unit weight is determined by excavating a sample of soil and determining the volume of the excavation by fluid or sand displacement or direct measurement using a drive sampler of precise volume. The weight of the excavated soil is determined at its natural water content and again after oven-drying. Alternatively, a nuclear density device may be used to directly measure field unit weight.

―――― **EXAMPLE 14.13** ――――

The soil from Figure 14.5 is specified to be compacted to a minimum of 95% of maximum dry unit weight as determined by AASHTO T99. An in-place unit weight test gives the following results:

Volume:	0.0455 ft³
Wet weight:	5.98 lb
Dry Weight:	5.43 lb

Determine whether the fill meets the specifications. Also determine the water content.

Solution. The dry unit weight is

W_s/V = 5.43 lb / 0.0455 ft³ = 119.34 lb/ft³

This is 119.34 / 123.0 = 97 percent of $(\gamma_d)_{max}$; thus, the fill meets the specifications.

The weight of water is

$W_w = W - W_s$ = 5.98 - 5.43 = 0.55 lb

The water content is

W_w/W_s = 0.55 / 5.430 = 10.1 percent

Note that the coordinates (w, γ_d) = (10.1, 119.34) fall below the laboratory compaction curve, but still meet the specifications.

A common problem in practice, as well as on exams, is the determination of quantities of excavation and water required for a compacted fill, given that the unit weight and water content of the borrow area are different from that required for the fill. As shown in the following example, the key to such problems is that weights of solids and water moved are constant, regardless of unit weight and volume. Drawing separate phase diagrams for the borrow area(s), and the fill is recommended.

EXAMPLE 14.14

A compacted fill with a total volume of 40,000 cubic yards is to be constructed using the material in Figure 14.5. Specifications require the soil to be compacted to a minimum of 95 percent of standard Proctor unit weight. Soil in the borrow area has an average unit weight of 108.5 lb/ft^3 and a water content of 4.0 percent. Estimate how many cubic yards of borrow material must be excavated and how much water must be added.

Solution. A phase diagram for the fill is first drawn. The required total volume is 40,000 x 27 = 1,080,000 ft^3. The minimum acceptable dry unit weight is 0.95 x 123 lb/ft^3 = 116.9 lb/ft^3. Thus, the minimum required weight of solids is 1,080,000 ft^3 x 116.9 lb/ft^3 = 126,300,000 lb. To achieve the required unit weight, the water content must be at least 5.5 percent, which corresponds to 6,944,000 lb of water.

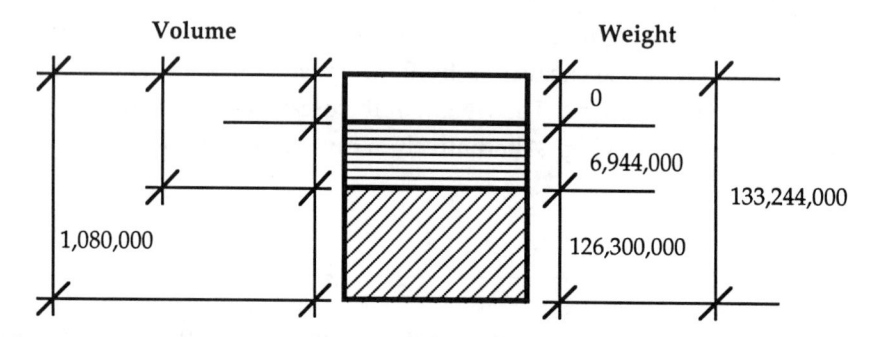

Next a phase diagram for the borrow area is drawn. By conservation of mass, to place 126,300,000 lb of soil solids in the fill, an equal amount of soil solids must be obtained from the borrow area. As the water content of the borrow area is 4.0 percent, (0.04)(126,300,000) = 5,052,000 lb of water will come with the solids, and the total weight of soil and water removed from the borrow area will be 131,352,000 lb. As the total unit weight in the borrow area is 108.5 lb/ft^3, the volume required to be removed is (131,352,000 lb)/(108.5 lb/ft^3) = 1,211,000 ft^3 or 44,900 cubic yards.

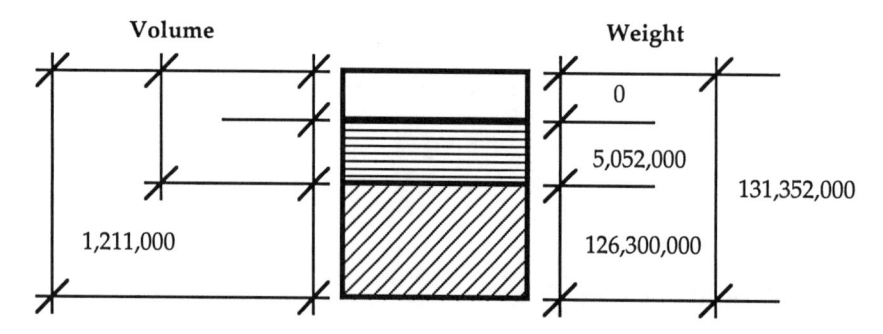

The required amount of additional water needed to be added is 6,944,000 - 5,052,000 = 1,892,000 lb, or 1,893,000 lb / 8.33 lb/gal = 227,000 gallons.

In summary, to provide construct a compacted fill of 40,000 cubic yards meeting the minimum unit weight requirements, a minimum of 44,900 cubic yards must be excavated from the borrow area and a minimum of 227,000 gallons of additional water must be added. To ensure meeting the specifications consistently, a more conservative estimate could be prepared based on the maximum dry unit weight and optimum water content.

The density of clean sands and gravels is commonly specified and measured in terms of the *relative density*, D_R, or *density index*, I_d, which are two names for the same parameter. The term relative density is somewhat of a misnomer as it scales the actual void ratio as a fraction of the range between the maximum and minimum void ratios:

$$D_R = \frac{e_{max} - e}{e_{max} - e_{min}} \qquad (14.9.2)$$

The relative density is usually expressed as a percentage. To determine the relative density using actual, minimum, and maximum dry unit weight values rather than void ratios, the equation is

$$D_R = \frac{\dfrac{1}{\gamma_{d\,min}} - \dfrac{1}{\gamma_d}}{\dfrac{1}{\gamma_{d\,min}} - \dfrac{1}{\gamma_{d\,max}}} \qquad (14.9.3)$$

14.10 The Effective Stress Principle

Normal stresses (generally compressive) in a soil-water mass are carried partly by the soil skeleton and partly by the pore water. Shear stresses must be carried entirely by the soil skeleton since water cannot sustain shear stress. As soil strength and compressibility are largely governed by the stresses in the soil skeleton, it is important to determine the allocation of stresses between the soil skeleton and the pore water. The *effective stress principle*, first proposed by Terzaghi, states that the total stress σ at any point and in any direction in a soil mass is equal to the effective stress σ' plus the pore pressure u:

$$\sigma = \sigma' + u \qquad (14.10.1)$$

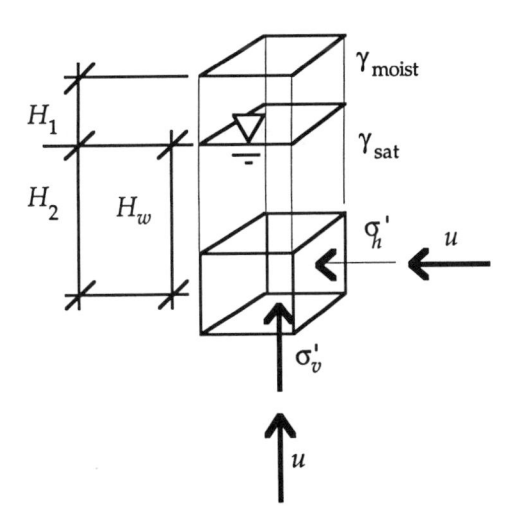

Figure 14.6 An element of soil.

An element of soil at a depth $(H_1 + H_2)$ is shown in Figure 14.6. To determine the stress conditions on the element, the following steps may be followed.

1. The *total vertical stress* σ_v is determined by summing the products of the unit weights times the layer thicknesses:

$$\sigma_v = \Sigma \gamma H \qquad (14.10.1)$$

For the problem shown, there are two layers, one above the water table and one below; hence

$$\sigma_v = \gamma_{moist} H_1 + \gamma_{sat} H_2 \qquad (14.10.2)$$

2. The *pore water pressure u* is determined as the product of the weight density of water and the pressure head. For static water, as in the problem illustrated, the pressure head is simply the height H_w of the water table above the element.

$$u = \gamma_w H_w \qquad (14.10.3)$$

For problems involving flowing water, the pressure head must be determined from a seepage analysis.

3. The *effective vertical stress* σ'_v is obtained by subtracting the porewater pressure from the total vertical stress:

$$\sigma'_v = \sigma_v - u \qquad (14.10.4)$$

4. The *effective horizontal stress* σ'_h is indeterminate except for special cases, such as the active and passive earth pressure conditions. It is related to the vertical effective stress by the coefficient K, where

$$K = \frac{\sigma'_h}{\sigma'_v} \qquad (14.10.5)$$

Thus, if K is known or given, the horizontal effective stress can be determined as

$$\sigma'_h = K \sigma'_v \qquad (14.10.6)$$

5. According to Pascal's Law, the porewater pressure is the same in all directions. Thus, the *total horizontal stress* σ_h is obtained by adding the pore water pressure to the effective horizontal stress:

$$\sigma_h = \sigma'_h + u \qquad (14.10.7)$$

EXAMPLE 14.15

The soil column shown in Figure 14.6 has the following properties:

$H_1 = 4$ ft $\qquad \gamma_{moist} = 115$ lb/ft³
$H_2 = H_w = 6$ ft $\qquad \gamma_{sat} = 125$ lb/ft³
$K = 0.400$

Calculate the vertical and horizontal total and effective stresses and sketch a diagram of stresses versus depth.

Solution:

$$\sigma_v = \Sigma \gamma H$$
$$= (115)(4) + (125)(6)$$
$$= 1210 \text{ lb/ft}^3$$

$$u = \gamma_w H_w$$
$$= (62.4)(6)$$
$$= 374 \text{ lb/ft}^3$$

$$\sigma'_v = \sigma - u$$
$$= 1210 - 374$$
$$= 836 \text{ lb / ft}^3$$

$$\sigma'_h = K\sigma'_v$$
$$= (0.4)(836)$$
$$= 334 \text{ lb / ft}^3$$

$$\sigma_h = \sigma'_h + u$$
$$= 334 + 374$$
$$= 708 \text{ lb / ft}^3$$

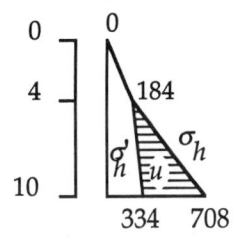

14.11 Water Flow Through Soils

The total energy at a point in a fluid can be expressed as its *total head*, or *potential*. According to Bernoulli's equation, the total head is the sum of the elevation head, the pressure head, and the velocity head. For laminar flow of water through soils the size of fine gravel or smaller, the velocity head is generally negligible: thus, at any and every point:

$$\text{Total head} = \text{elevation head} + \text{pressure head}$$
$$TH = EH + PH \tag{14.11.1}$$

or, in terms of commonly used symbols,

$$\phi = \frac{p}{\gamma_w} + z$$

where ϕ is the total head or potential at the point of interest

(Head is measured as distance above an arbitrary but consistent datum.)

p is the pressure in the water at the point

γ_w is the unit weight of water (62.4 lb/ft³ or 9.81 kN/m³)

z is the elevation of the point above the datum

Water always flows from higher total head to lower total head, and in doing so may flow from a high to low elevation, or vice versa, and may flow from a condition of high pressure to low pressure, or vice versa.

The velocity of flow in a porous medium such as soil is governed by *Darcy's law*, which states that the *discharge velocity v* is the product of the *coefficient of permeability* or *hydraulic conductivity k* times the *hydraulic gradient i* where *i* is the rate of change of total head with respect to length in the direction of flow:

$$v = ki \tag{14.11.2}$$

$$i = \frac{d\phi}{ds}$$

The discharge velocity is the apparent or equivalent velocity over the entire cross-sectional area of soil and voids. In fact, the velocity of water movement in the pores, or *seepage velocity*, must be greater as only the void area is available for flow. The seepage velocity $\bar{v}$, obtained by dividing the discharge velocity by the porosity, is

$$\bar{v} = \frac{ki}{n} \qquad (14.11.3)$$

If k and i can be assumed constant (or taken as an average) over a cross-sectional flow area A the flow rate Q can be determined as

$$Q = vA = kiA \qquad (14.11.4)$$

The coefficient of permeability is a property of the soil, the fluid, and its temperature or viscosity. The value of k may vary from 1 cm/sec for coarse sands and gravels to 1×10^{-9} cm/sec for highly plastic clays. A common equation to estimate permeability of sands from grain size data, first proposed by Hazen, is

$$k = CD_{10}^2 \qquad (14.11.5)$$

where D_{10} is the effective grain size in mm, and C is a constant assuming values from 0.4 to 1.2 cm/sec-mm². A value of 1.0 is often assumed.

────── **EXAMPLE 14.16** ──────

A fine sand with a permeability of 0.01 cm/sec and a porosity of 0.35 is put in a glass tube permeameter from point B to point C and contained with porous screens. (All dimensions are in cm). The tube is 10 cm in diameter. The permeameter is kept full by pouring in water at point A, and water runs out at point D. Determine the total head, pressure head, elevation heads, and pressure at points A, B, C, D and the midpoint of the soil column. Determine the gradient, discharge velocity, seepage velocity and flow rate.

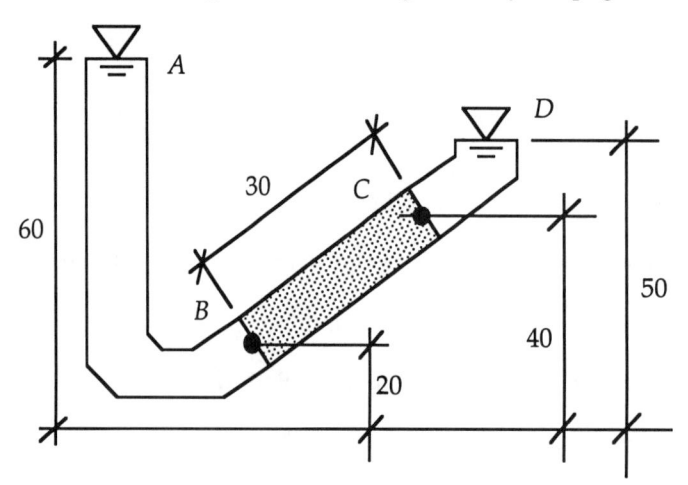

Solution. Point A is at a free surface 60 cm above the datum. Thus $EH = 60$, $PH = 0$, $TH = 60 + 0 = 60$ cm. The pressure is zero.

The resistance to flow through the glass tube from A to B and C to D is negligible in comparison to the resistance to flow through the soil column from B to C. Thus it can be assumed that all energy loss and total head loss occurs through the soil column.

At point B, the total head is 60 cm, the elevation head is 20 cm. Thus the pressure head is 60 - 20 = 40 cm. The pressure p is

$$p = (PH)(\gamma_w) = (0.40 \text{ m})(9.81 \text{ kN/m}^3) = 3.92 \text{ kN/m}^2 = 3.92 \text{ kPa}.$$

At point C, the total head is 50 cm and the elevation head is 40 cm. Hence the pressure head is 10 cm, and the pressure is (0.1)(9.81) = 0.98 kPa.

At point D, the total head and elevation head are 50 cm, the pressure head is zero, and the pressure is zero.

As the cross-section of the soil column is constant, total head is lost uniformly along the length and the total head would be 55 cm at the midpoint. Water in a piezometer tapped into the midpoint of the soil would rise to elevation 55 cm. The elevation head at the midpoint is 30 cm; hence the pressure head is 55 - 30 = 25 cm and the pressure is (0.25)(9.81) = 2.45 kPa.

The gradient is the rate of loss of total head with respect to distance along the flow path. Within the uniform soil column, water loses 10 cm of total head over 30 cm of travel; hence $i = 10/30 = 0.333$.

The discharge velocity is

$$v = ki$$

$$= (0.01 \text{ cm/sec})(0.333) = 0.00333 \text{ cm/sec}$$

The seepage velocity is

$$\bar{v} = \frac{ki}{n} = \frac{(0.01)(0.333)}{0.35} = 0.00951 \text{ cm/sec}$$

The flow rate is

$$Q = kiA = (0.01)(0.333)\frac{\pi(10)^2}{4} = 0.262 \text{ cm}^3/\text{sec} = 0.262 \times 10^{-6} \text{ m}^3/\text{sec}$$

For nonuniform flow regimes, heads and flows can be determined from a *flow net*, a graphical solution of Laplace's differential equation governing total head. A flow net consists of a family of *equipotential lines*, which are contours of equal total head, and *flow lines*, which are curves describing selected flow paths of water particles. If a set of flow lines and equipotential lines are drawn across a homogeneous, isotropic flow regime such that they intersect at right angles at all points and form curvilinear squares, the resulting flow net can be used to determine flows.

The flow through a unit width cross section can be expressed as

$$Q = kH\frac{N_f}{N_e} \qquad (14.11.6)$$

where:

- H is the total head loss across the section
- N_f is the number of flow channels
- N_e is the number of equipotential drops

EXAMPLE 14.17

A sheetpile is driven in a shallow lake and the area to one side is pumped down as shown. The permeability of the sand is 0.1 cm/sec. Below elevation 70 is an impermeable rock layer. A flow net is drawn. Determine the flow per unit length and the heads and pressure at point A.

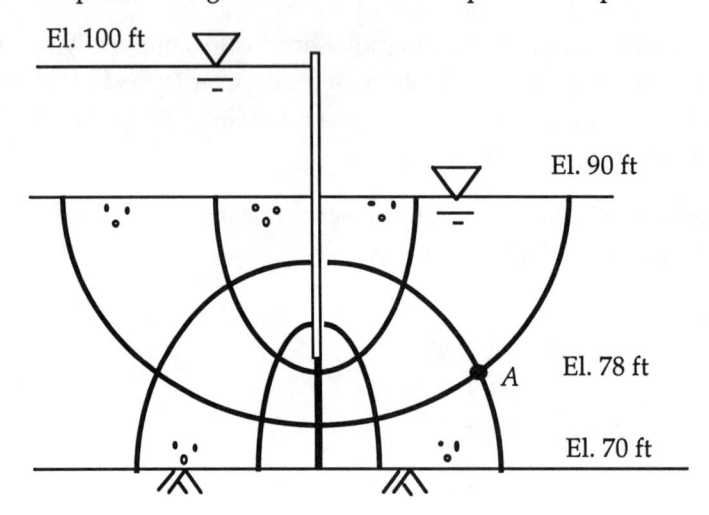

Solution: The number of flow channels N_f is three: one is between the sheet pile and the upper drawn flow line, a second is between the two drawn flow lines, and the third is between the lower drawn flow line and the bedrock. As water flows along a flow line from upstream to downstream, it traverses 6 squares or equipotential drops; $N_f = 6$. The total head loss H is 100 - 90 = 10 ft. Hence

$$Q = kH\frac{N_f}{N_e} = (0.1 \text{cm/sec})\left(\frac{1\text{ft}}{30.48\text{cm}}\right)(10\text{ft})\left(\frac{3}{6}\right) = 0.0164 \text{ ft}^3/\text{sec/ft} \text{ or } 0.984 \text{ ft}^3/\text{min/foot}$$

At point A, 5 of the 6 equipotential drops have been crossed; hence, the total head is

$TH = 100 - (5/6)(100-90) = 100 - 8.33 = 91.7$ ft

Water in a piezometer inserted at point A would rise to elevation 91.7 ft.

The elevation head at A is 78 feet. The pressure head is

$PH = 91.7 - 78 = 13.7$ ft

The pressure at A is

$p_A = (13.67 \text{ ft})(62.4 \text{ lb/ft}^3) = 853 \text{ lb/ft}^2$

14.12 Consolidation Behavior of Saturated Clay

Loading a soil induces a volume change as soil solids are pushed closer together and water is expelled from the pores. For high permeability soils such as sand, volume change is nearly instantaneous. For low permeability soils such as clay, the volume change is time-dependent. This time-dependent volume reduction is termed *consolidation*. The consolidation behavior of a soil can be measured in a *consolidation test* or *oedometer test*. A disk of soil is confined in a rigid ring with a porous material above and below the soil; the apparatus is submerged in a container of water. A vertical load is applied and the reduction in soil height is measured with a dial gauge as a function of time. When compression has essentially stopped another load is applied, typically double the previous load, and the process is repeated. At any time, the void ratio can be calculated knowing the initial soil volume and water content, the height reduction, and the specific gravity.

The test results are analyzed by plotting the final void ratio e under each load on a linear scale versus the applied pressure p (actually an effective vertical stress σ'_v) plotted on a log scale. The result, sometimes termed an e-log p curve, is shown in Figure 14.6.

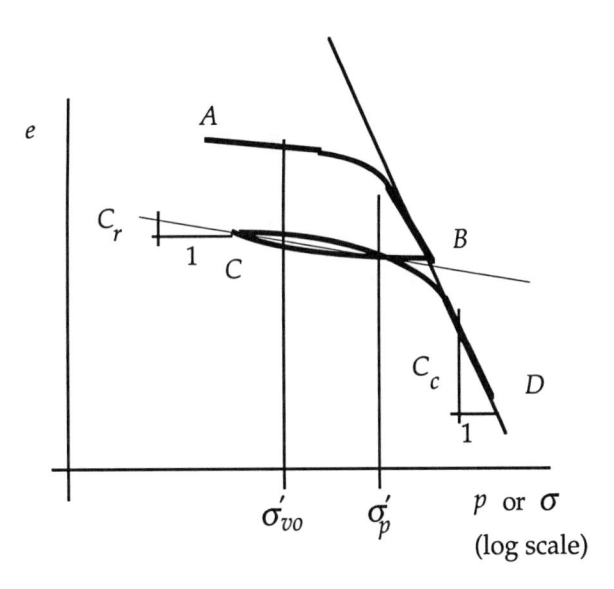

Figure 14.6. e - log p consolidation curve.

The curve describes the void-ratio vs. pressure relationship for a soil where the effective stress was loaded from point A to point B, unloaded from point B to point C, and reloaded from point C to point D. The vertical overburden stress on the sample in the ground was σ'_{vo}. The *preconsolidation pressure* σ'_p or maximum effective stress the soil has been subjected to in the past, is marked by the break in the curve. When the soil is loaded to point B, the effective stress at point B becomes the new preconsolidation pressure. When the soil is unloaded and reloaded below the preconsolidation pressure, the curve has a relatively flat slope, and is termed an *unloading curve* and a *recompression curve*. As the curve passes point B, it steepens slope and becomes a *virgin consolidation curve*.

To simplify settlement calculations, the virgin curve can be approximated as a straight line with a slope of C_c termed the *compression index*. The unloading and recompression curves can be approximated as a straight line with a slope of C_r termed the *recompression index*, or C_s termed the *swell index*. The compression index is found to be

$$C_c = \frac{\Delta e}{\Delta(\log_{10} p)} = \frac{e_2 - e_1}{\log_{10} p_2 - \log_{10} p_1} \qquad (14.12.1)$$

where the subscripts 1 and 2 refer to two points on the virgin curve. Although the compression index is always negative, the sign is not usually shown. The recompression index can be calculated in the same manner using two points on the unloading or recompression curve or an average line between them.

When the preconsolidation pressure is found to be at approximately the same stress as the effective overburden pressure, the soil is *normally consolidated*. The compression index of normally consolidated clays can be estimated as

$$C_c = 0.009(LL - 10) \tag{14.12.2}$$

where LL is the liquid limit expressed as a percentage. The recompression index is commonly on the order of one-fifth to one tenth the compression index.

When the preconsolidation pressure is greater than the vertical effective overburden pressure at the point the soil was sampled, the soil is *overconsolidated*. The *overconsolidation ratio OCR* is the ratio of the preconsolidation pressure to the effective overburden pressure:

$$OCR = \frac{\sigma'_p}{\sigma'_{vo}} \tag{14.12.3}$$

──────── EXAMPLE 14.18 ────────

A normally consolidated clay layer has a liquid limit of 45. The effective stress at the midpoint of the layer is 2000 psf. If placement of a sand fill on the surface over a wide area causes a vertical stress increase of 800 psf, how much will the void ratio reduce at the midpoint due to the stress increase?

Solution. The compression index can be estimated as

$$C_c = 0.009(45-10) = 0.315$$

The void ratio reduction will be:

$$\Delta e = C_c \left[\log_{10}(p + \Delta p) - \log_{10}(p) \right]$$
$$= C_c \log_{10}\left(\frac{p + \Delta p}{p}\right)$$
$$= (0.315) \log_{10}\left(\frac{2000 + 800}{2000}\right)$$
$$= 0.046$$

The void ratio will decrease 0.046 due to the increased stress from the fill.

14.13 Consolidation Settlement Calculations

Total settlement of a clay layer can be calculated by determining the strain as a function of depth and integrating the strain with respect to depth. If a unit volume of soil solids is considered, the void volume is the void ratio e. As shown in Figure 14.7, if consolidation causes a void ratio reduction of Δe, the vertical strain ε_v at a point is

$$\varepsilon_v = \frac{\Delta e}{1+e_0} \tag{14.13.1}$$

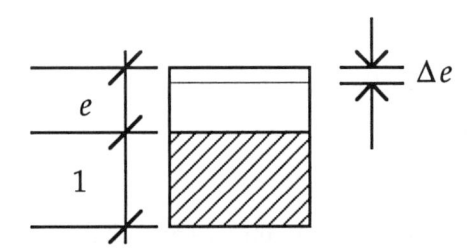

Figure 14.7 Void ratio reduction due to consolidation.

The void ratio change can be determined from points on the e-log p curve as described above.

The total settlement is obtained by integrating the strain with respect to depth:

$$S = \int_{z_1}^{z_2} \varepsilon_z \, dz = \sum_{i=1}^{n} \varepsilon_i \, \Delta z \tag{14.13.2}$$

EXAMPLE 14.19

A clay layer 15 ft thick underlies a 5-ft sand layer. Assume the initial void ratio of the clay is 0.900, the preconsolidation pressure is 1500 psf or the initial overburden stress, whichever is greater, the compression index is 0.20, and the recompression index is 0.04. The water table is at the top of the clay layer. A uniform load of 1000 psf is applied to the surface by a large slab foundation. Calculate the settlement of the layer using the three layers shown. Take the average stresses in the layers as the stresses at the midpoints shown.

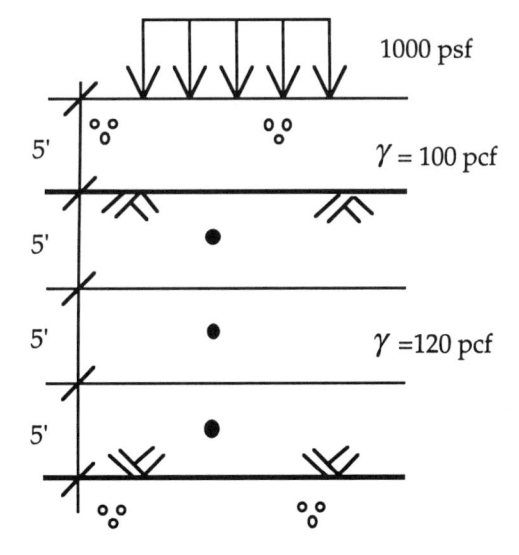

Solution. For the top layer, the existing vertical effective stress is

$$\sigma_v' = \sigma_v - u = (5)(100) + (2.5)(120) - (2.5)(62.4) = 644 \text{ psf}$$

When the foundation is added and consolidation is complete, the vertical effective stress will have increased to

$$\sigma_v' + \Delta\sigma = 644 + 1000 = 1644 \text{ psf}$$

As the preconsolidation pressure is 1500 psf, consolidation will follow the reconsolidation slope on the e- log p curve from 644 to 1500, and will follow the virgin curve from 1500 to 1644. The midpoint strain in the top layer is

$$\varepsilon_v = \frac{\Delta e}{1+e} = \left(\frac{1}{1+e}\right)\left[C_r \log \frac{\sigma_p'}{\sigma_{vo}'} + C_c \log \frac{\sigma_{vo}' + \Delta\sigma}{\sigma_p'}\right]$$

$$= \left(\frac{1}{1.9}\right)\left[0.04 \log \frac{1500}{644} + 0.20 \log \frac{1644}{1500}\right]$$

$$= 0.5263 \, (0.01469 + 0.00796)$$

$$= 0.0119$$

Multiplying the strain by the layer thickness, the settlement of the top layer is

$$S_{top} = (60 \text{ in})(0.0119) = 0.714 \text{ inches}.$$

For the second layer, the midpoint effective stress is

$$\sigma_v' = (5)(100) + (7.5)(120) - (7.5)(62.4) = 932 \text{ psf}$$

The preconsolidation pressure is 1500 psf, and the final effective stress is 1932 psf. The midpoint strain is

$$\varepsilon_v = \frac{1}{1.9}\left[0.04 \log \frac{1500}{932} + 0.20 \log \frac{1932}{1500}\right]$$

$$= 0.5263 \, (0.00827 + 0.02198)$$

$$= 0.01592$$

Multiply the strain by the layer thickness and find the settlement of the middle layer to be

$$S_{middle} = (60 \text{ in})(0.0159) = 0.954 \text{ inches}.$$

For the bottom layer, the midpoint effective stress is

$$\sigma_v' = (5)(100) + (12.5)(120) - (12.5)(62.4) = 1220 \text{ psf}$$

The preconsolidation pressure is 1500 psf, and the final effective stress is 2220 psf. The midpoint strain is

$$\varepsilon_v = \frac{1}{1.9}\left[0.04 \log \frac{1500}{1222} + 0.20 \log \frac{2222}{1500}\right]$$

$$= 0.5263 \, (0.00356 + 0.03413)$$

$$= 0.01984$$

Multiply the strain by the layer thickness and find the settlement of the top layer to be

$$S_{bottom} = (60 \text{ in})(0.0198) = 1.19 \text{ inches}$$

The total settlement is

$$S_{total} = 0.714 + 0.954 + 1.19 = 2.86 \text{ inches}$$

The above examples involved a uniform stress increase due to the addition of a surface load over a very large area. Where the stress increase occurs over a finite area, the stress increase will distribute and thus diminish with depth. For such cases, stress distribution equations (see Chapter 15) must be used.

14.14 Shear Strength of Soils

The shear strength along a plane of failure may depend on the normal stress on the plane. A simple test for determining the shear strength of soil is the *direct shear test*, illustrated in Figure 14.8a. The soil is placed in a split box, and a vertical normal force V is applied. The horizontal force H needed to shear the soil is measured. If V and H are divided by the sample area A the normal stress σ and the shear stress τ are obtained. Plotting τ at failure vs σ for samples tested at several normal stresses, as shown in Figure 8b, provides the strength parameters c and ϕ. The parameter c is commonly termed the *cohesion* and the parameter ϕ is termed the *angle of internal friction*, or simply the *friction angle*.

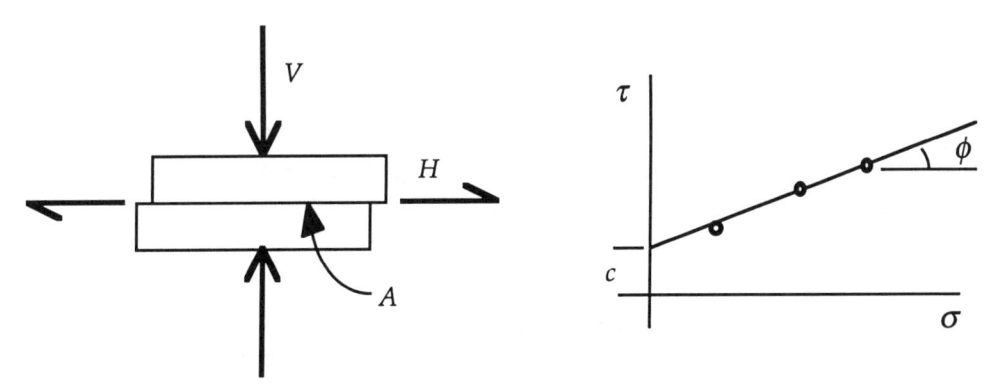

Figure 14.8 Direct shear test.

The strength of soil along the failure plane is represented by the Mohr-Coulomb strength equation:

$$\tau_{ff} = c + \sigma \tan \phi \qquad (14.14.1)$$

where

- τ_{ff} is the shear stress at failure on the failure plane
- σ is the normal stress on the failure plane
- c is the cohesion of the soil
- ϕ is the friction angle of the soil

As written above, the strength parameters are defined in terms of the *total stresses* in the soil-water system. When sheared, soils tend to change volume; for low permeability soils such as clay, shearing faster than the permeability will allow water movement required to accommodate the volume change will cause pore pressure changes. Consequently the soil strength, expressed as c and ϕ, will depend on the rate of loading.

When soils are loaded so rapidly (relative to their permeability) that volume change cannot occur, the pore pressures change and the developed strength is termed the *undrained strength*. When soils are loaded so slowly that no pore pressure change occurs, the developed strength is termed the *drained strength*.

When clay soils are initially loaded, the undrained strength applies. For normally consolidated clays in undrained loading, $\phi = 0$, and the soil is said to be *cohesive*.

Clean sands are so permeable that excess pore pressures cannot be sustained, and thus drained strengths are applicable for all problems except dynamic loads such as earthquakes. Sand strength is entirely frictional; thus sands are termed *cohesionless* soils and $c = 0$.

Expressing soil strength in terms of total stresses reflects the behavior of the soil-water system, and assumes that pore pressure changes during field loading will be the same as in the laboratory tests used to determine the parameters. As only the soil solids (and not the porewater) can sustain shear stresses, it is more correct (but less practical) to express soil strength in terms of *effective stresses*, and the Mohr-Coulomb equation can be re-written as

$$\tau_{ff} = c' + (\sigma - u)\tan\phi' \qquad (14.14.2)$$

where

c' and ϕ' are effective-stress strength parameters

u is the pore pressure at failure

The pore pressure at failure can be measured in a laboratory test. When this is done, it is found that clay behaves essentially as a frictional, cohesionless material ($c' = 0$ for normally consolidated clays and is "small" for overconsolidated clays). Cohesive behavior in undrained or short term loading is largely due to pore pressure changes due to shear. A disadvantage of using effective stress strength parameters in practice is the difficulty of predicting pore pressure changes.

In summary, for practical problems, soil strength can be modeled as follows:

- For clean sands and gravels, use drained effective stress parameters. These soils are cohesionless, so $c' = 0$. The drained friction angle ϕ' is commonly in the range of 28 to 40 degrees and increases with increasing relative density.

- For normally consolidated clays, short-term loading usually governs as the soil gains additional strength with time due to consolidation under an applied load. The undrained strength can be modeled with total stress parameters of $\phi = 0$ and a c value from a laboratory test (commonly the unconfined compression test). Alternatively, the strength may simply be designated as s_u, the undrained strength. The value of c or s_u is a function of the effective stress under which the soil was consolidated, and is often on the order of 25 to 35 percent of this value.

- For overconsolidated clays, the strength envelope is non-linear, but may be approximated by total stress parameters c and ϕ, where both are non-zero.

────── **EXAMPLE 14.20** ──────

A sled is loaded to a weight of 200 lb and pulled across a beach. It takes a horizontal force of 125 lb to pull the sled. What are the strength parameters of the beach sand?

Solution. As the material is sand, $c' = c = 0$. The shear stress on the base of the sled is $\tau = 125/A$, where A is the area of the sled. The normal force on the base of the sled is $\sigma = 200/A$. Applying the Mohr-Coulomb equation:

$$\tau = c + \sigma\tan\phi$$
$$\frac{125}{A} = 0 + \frac{200}{A}\tan\phi$$
$$\tan\phi = \frac{125}{200} = 0.625$$
$$\therefore \phi = 32°$$

14.15 Triaxial Tests and the Unconfined Compression Test

The triaxial test permits more control over applied stress conditions than the direct shear test. A cylinder of soil is subjected to an all-around confining stress σ_3, and a deviator stress $(\sigma_1 - \sigma_3)$ is applied vertically until failure (see Figure 14.9a). The stress conditions on the sample at failure can be determined for all planes by plotting a Mohr's circle (see Figure 14.9b). If the test is repeated for soil samples at different confining stresses, the Mohr-Coulomb strength envelope can be obtained by drawing a line tangent to the failure circles.

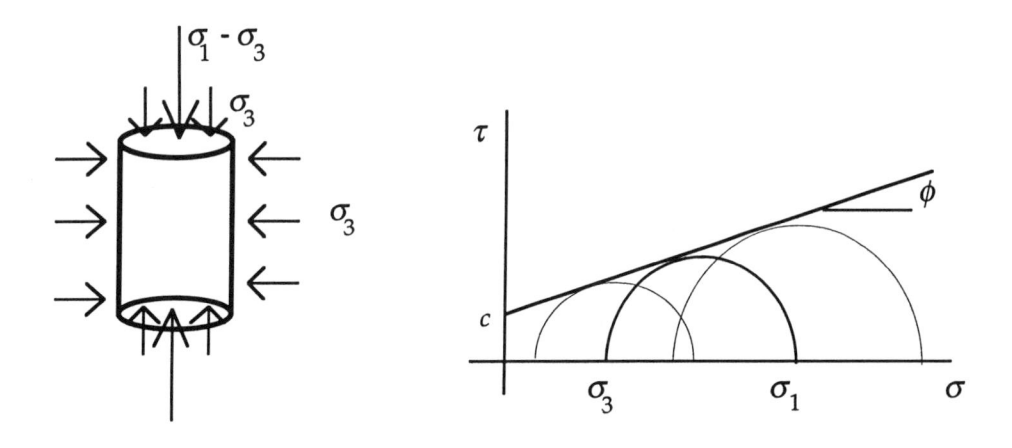

Figure 14.9 Triaxial stress application and Mohr's circle plot.

For saturated clays where $\phi = 0$, it is convenient to test without any confining stress. In the resulting *unconfined compression test* (see Figure 14.10) the cohesion or undrained strength is taken as one-half the applied stress at failure, or unconfined compression strength q_u:

$$c = s_u = \frac{q_u}{2} = \frac{\sigma_1}{2} \tag{14.15.1}$$

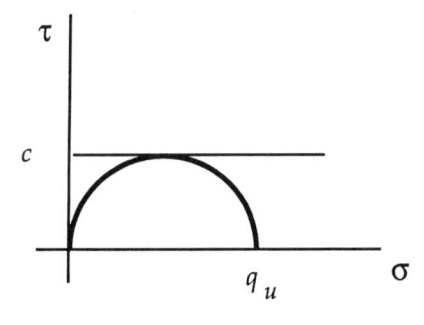

Figure 14.10 Unconfined compression test.

EXAMPLE 14.21

A 2.8-in-dia. cylinder of clay soil is loaded in unconfined compression and fails under a load of 40 lb. Determine the strength parameters.

Solution. For undrained loading of saturated clay, assume $\phi = 0$. The sample area is

$$A = \frac{\pi d^2}{4} = \frac{\pi (2.8)^2}{4(12)^2} = 0.04276 \text{ ft}^2$$

The stress on the sample at failure is, or the unconfined compressive strength is

$$q_u = \sigma_1 = \frac{40 \text{ lb}}{0.04276 \text{ ft}^2} = 935 \text{ lb/ft}^2$$

The cohesion or undrained strength is

$$c = s_u = \frac{q_u}{2} = \frac{935.5}{2} = 468 \text{ lb/ft}^2$$

Practice Problems (Multiple Choice)

14.1 A moist soil sample in a tare can is put on a balance and the mass of the soil plus can is found to be 48.30 g. After oven drying, the mass of the soil plus can is found to be 41.22 g. The mass of the empty tare can is 7.41 grams. Determine the water content of the soil.
  a) 17.3 %
  b) 20.9 %
  c) 14.7 %
  d) 18.0 %
  e) 7.08%

14.2 A saturated soil sample has a total unit weight of 120 lb/ft^3. The specific gravity of solids is 2.70. Determine the dry unit weight.
  a) 91.5 lb/ft^3
  b) 87.6 lb/ft^3
  c) 168.5 lb/ft^3
  d) 57.6 lb/ft^3
  e) 44.4 lb/ft^3

14.3 A saturated soil has a water content of 20 percent, and a specific gravity of 2.75. Determine its void ratio.
  a) 0.20
  b) 0.50
  c) 0.55
  d) 0.27
  e) 1.38

14.4 A soil sample has 70 percent passing the No. 4 sieve and 10 percent passing the No. 200 sieve. The coefficient of uniformity is 4 and the fines are non-plastic. Classify the soil according to the Unified Soil Classification System.
  a) *SP-SM*
  b) *SW-SM*
  c) *SP*
  d) *GW-GM*
  e) *GM*

14.5 A soil sample has liquid limit of 45 and a plastic limit of 30. Classify the soil according to the Unified Soil Classification System
  a) CL-ML
  b) ML
  c) CH
  d) MH
  e) CL

14.6 A dry sand has a minimum unit weight of 105 lb/ft³ and a maximum unit weight of 125 lb/ft³. The unit weight in place is measured at 121 lb/ft³. Calculate the relative density.
   a) 80.0 percent
   b) 82.7 percent
   c) 86.8 percent
   d) 84.0 percent
   e) 121.0 percent

14.7 The soil in a swamp has a total unit weight of 20.0 kN/m³. The water table is at the surface. Calculate the vertical effective stress at a depth of 2 meters.
   a) 58.4 kN/m²
   b) 40.0 kN/m²
   c) 30.2 kN/m²
   d) 20.4 kN/m²
   e) 10.0 kN/m²

14.8 For the conditions in Problem 14.7, calculate the total vertical stress at a depth of 2 meters.
   a) 20.38 kN/m²
   b) 40.00 kN/m²
   c) 30.19 kN/m²
   d) 58.36 kN/m²
   e) 27.00 kN/m²

14.9 For the conditions in Problem 14.7, assume the horizontal earth pressure coefficient $K = 0.5$. Determine the horizontal effective stress at a depth of 2 meters.
   a) 10.2 kN/m²
   b) 20.4 kN/m²
   c) 29.2 kN/m²
   d) 15.1 kN/m²
   e) 10.0 kN/m²

14.10 Refer to the permeameter illustrated in Example 14.16. Change the elevations of points A through D above the datum to:
   A 100 cm
   B 10 cm
   C 60 cm
   D 80 cm

   The length of tube from B to C is 80 cm. What is the hydraulic gradient?
   a) 0.625
   b) 0.250
   c) 0.500
   d) 1.25
   e) 1.00

14. 11   Refer to Problem 14.10. What is the pore water pressure just inside the soil column at point B.
   a) 8.83 kN/m³
   b) 0.98 kN/m³
   c) 1.96 kN/m³
   d) 0.90 kN/m³
   e) 9.81 kN/m³

14. 12   Refer to Example 14.17. The tip of the sheetpile is at elevation 79 ft. What is the total head at the sheetpile tip?
   a) 100 ft
   b) 93 ft
   c) 16 ft
   d) 95 ft
   e) 79 ft

14.13   Refer to Example 14.17. The tip of the sheetpile is at elevation 79 ft. What is the pore water pressure at the sheetpile tip ?
   a) 686 lb/ft²
   b) 312 lb/ft²
   c) 998 lb/ft²
   d) 1310 lb/ft²
   e) 62.4 lb/ft²

14.14   An overconsolidated soil is one where:
   a) The preconsolidation pressure is greater than the effective overburden pressure.
   b) The preconsolidation pressure is greater than the total overburden pressure.
   c) The effective overburden pressure is greater than the preconsolidation pressure.
   d) One that has consolidated under the overburden above it.
   e) The effective vertical stress is greater than the total vertical stress.

14.15   A saturated clay has a cohesion of 400 lb/ft². A cylinder of this soil 6 inches in diameter and 12 inches high is tested in unconfined compression. What force is required to shear it?
   a) 2400 lb.
   b) 78.5 lb
   c) 400 lb.
   d) 200 lb.
   e) 157 lb.

Practice Problems (Essay)

14.16 A saturated clay sample has a mass of 46.95 g. After oven drying, it has a mass of 35.06 g. The specific gravity is 2.70. Calculate the water content, void ratio, porosity, total unit weight and dry unit weight.

14.17 A soil sample has a water content of 20 percent, a degree of saturation of 80 percent, and a specific gravity of 2.68. Determine the total unit weight and dry unit weight in lb/ft^3 and kN/m^3.

14.18 The following data were obtained from a standard compaction test.

Water content	Dry unit weight
6.0	102.0
10.0	107.0
13.0	109.0
16.0	106.0
19.0	102.0

Determine the maximum dry unit weight and optimum water content. If the soil is to be compacted to a minimum of 95 percent of standard maximum dry unit weight, what range of water contents should be used for compaction?

14.19 Five pounds of moist soil has a water content of 20 percent. How many pounds of water must be added to increase the water content to 30 percent?

14.20 A compacted fill is to be constructed with a total volume of 10 000 m^3. The fill is to be compacted to a dry unit weight of at least 20 kN/m^3 and a minimum water content of 10 percent. Soil in the borrow area has a dry density of 19.0 kN/m^3 and an average water content of 8 percent. How many cubic meters of borrow material must be excavated and how many liters of water must be hauled in?

14.21 Plot the vertical and horizontal effective stress vs. depth for the following soil profile:

0-4 ft	Moist sand, $\gamma = 125$ lb/ft^3, $K = 0.40$
4 ft	Groundwater table
4-10 ft	Saturated sand, $\gamma_{sat} = 130$ lb/ft^3, $K = 0.40$

14.22 Given the consolidation test data below, plot an e-log p curve and determine the preconsolidation pressure, compression index, and recompression index.

Pressure, psf	void ratio, e
400	0.930
1000	0.916
2000	0.902
4000	0.885
8000	0.863
16000	0.778
32000	0.680
8000	0.690
2000	0.716

14.23 The average effective stress in a 5-ft-thick layer of saturated clay is reduced from 1800 psf to 1300 psf by excavation above it. The compression index is 0.200, the recompression index is 0.040, and the initial void ratio is 0.900. How much will the clay layer heave due to the unloading?

14.24 Two samples of the same soil are tested in a triaxial shear machine. The stresses at failure are as given below. Determine the strength parameters c and ϕ

Sample	σ_1	σ_3
1	3.0 tsf	1.0 tsf
2	5.0	2.0

14.25 A clean sand with a friction angle of 30 degrees is tested in a direct shear machine. The sample is 3 inches square. The maximum horizontal force the machine can develop is 100 lb. What are the maximum vertical stress and vertical force that can be used if the sample is to be loaded to failure?

Practice Problems (PE-Format)

A cross-section of a sheeted excavation is shown. Two sheetpiles are used to support a 12-ft-deep, 20-ft-wide, very long sewer excavation in a medium sand with void ratio $e = 0.5$ and specific gravity 2.75. The tips of the sheetpiles are driven to elevation 366 ft. The interior of the excavation is dewatered to the bottom elevation of 388 ft. An engineer has drawn the flow net shown to calculate pressures and flow quantities.

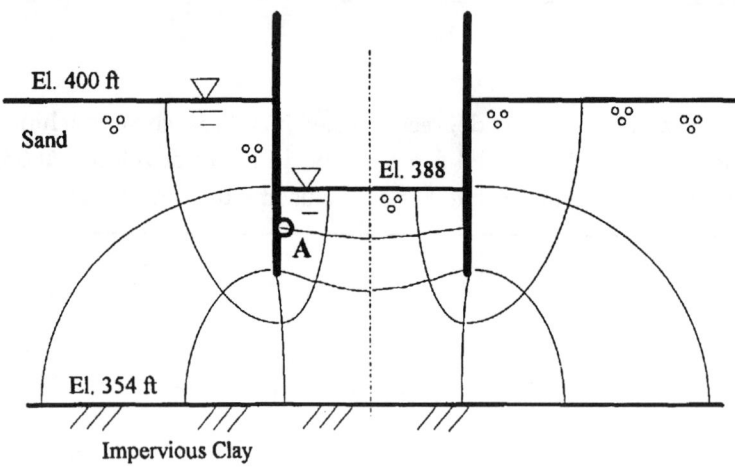

14.26 Calculate the saturated unit weight of the sand.
   a) 124.8 lb/ft³
   b) 135.2 lb/ft³
   c) 133.1 lb/ft³
   d) 72.8 lb/ft³
   e) 62.4 lb/ft³

14.27 If the permeability, k, of the sand is 0.1 ft/min, calculate the total inflow of water, q, into the foundation for a 100 ft length of the excavation.
   a) 0.8 ft³/min
   b) 0.4 ft³/min
   c) 80 ft³/min
   d) 40 ft³/min
   e) 26.67 ft³/min

14.28 Point "A" is located on the inside face of the sheetpile at elevation 380 ft. Determine the total vertical stress at point A.
   a) 582.4 lb/ft²
   b) 2704.0 lb/ft²
   c) 499.2 lb/ft²
   d) 1081.6 lb/ft²
   e) 1580.6 lb/ft²

14.29 Determine the pore water pressure at point A.
- a) 624.0 lb/ft^2
- b) 499.2 lb/ft^2
- c) 873.6 lb/ft^2
- d) 1622.4 lb/ft^2
- e) 98.1 lb/ft^2

14.30 Determine the effective vertical stress at point A.
- a) 728.0 lb/ft^2
- b) 1081.6 lb/ft^2
- c) 582.4 lb/ft^2
- d) 624.0 lb/ft^2
- e) 457.6 lb/ft^2

Solutions to Problems (Multiple Choice)

14.1 b) The weight of moist soil is 48.30 - 7.41 = 40.89 g

The weight of dry soil is 41.22 - 7.41 = 33.81 g

The weight of water is 40.89 - 33.81 = 7.08 g

The water content is 7.08 / 33.81 = 0.209 or 20.9 percent.

14.2 a) As the quantity of soil is not or known, and only relationships are of interest, assume the total volume V is 1.00 ft^3. The total weight W is then 120 lb. As the soil is saturated, the volume of air $V_a = 0$. After sketching a phase diagram, two simultaneous equations can be written, one on the volume side and one on the weight side:

$$V_s + V_w = 1.00$$

$$W_s + W_w = 120.0$$

W_s can be written as $G_s \gamma_w V_s = (2.70)(62.4)V_s$ and W_w can be written as $\gamma_w V_w = 62.4\ V_w$

Hence, there are two simultaneous equations

$$V_s + V_w = 1.00$$

$$168.4\ V_s + 62.4 V_w = 120.0$$

The solution is

$$V_s = 0.543, \quad V_w = 0.456$$

Then

$$W_s = (2.70)(62.4)(0.543) = 91.5 \text{ lb}$$

The required dry unit weight is

$$\gamma_d = W_s / V = 91.5 / 1.00 = 91.5 \text{ lb/ft}^3$$

14.3 c) Use the relationship $Se = wG_s$. Then

$$e = wG_s / S$$

$$e = (0.20)(2.75) / 1 = 0.55$$

14.4 a) Since 10 percent of the sample passes the No. 200 sieve (the *fines*), 90 percent is larger than the No. 200 sieve, and the soil is coarse grained. The first letter of the classification must be *S* for sand or *G* for gravel.

70 percent of the total sample passed the No. 4 sieve and 30 percent was larger than the No. 4 sieve. The sample is 30 percent gravel and 10 percent fines, leaving 60 percent sand. The *coarse fraction* is the 90 percent larger than the No. 200. The coarse fraction is 30 / 90 = 33% gravel and 60 / 90 = 67% sand. Since the coarse fraction contains more than sand than gravel, the first letter of the classification is *S*.

As the percent passing the No. 200 sieve is between 5 and 12, the sample requires a dual classification, with the first symbol being *SP* or *SW*, and the second being *SC* or *SM*.

The coefficient of uniformity must be between 1 and 3 for the soil to be a well-graded *SW*; as it is not, the first symbol is *SP*.

As the soil is stated to be non-plastic, the fines must be silt rather than clay; the second symbol is *SM*.

The soil is classified as *SP-SM*.

14.5 b) If $LL = 45$ and $PL = 30$, then $PI = 45 - 30 = 15$. Referring to the coordinates (45,15) on the plasticity chart, the soil is either *ML* or *OL*. If it is not stated that the soil is significantly organic, it would be classified as *ML*.

14.6 b) Using equation 14.9.3, the relative density is:

$$[(1/105) - (1/121)] / [(1/105) - (1/125)] = 0.827 \text{ or } 82.7\%$$

Calculations involving ratios of differences are numerically sensitive. To obtain three digit accuracy in the solution, four or more digits must be carried through the calculations.

14.7 d) The total vertical stress is

$$\sigma = (2.0)(20.0) = 40.0 \text{ kN/m}^2$$

The pore water pressure is

$$u = (2.0)(9.81) = 19.6 \text{ kN/m}^2$$

The effective vertical stress is

$$\sigma' = 40.0 - 19.6 = 20.4 \text{ kN/m}^2$$

14.8 b) As previously shown, the total vertical stress is 40.00 kN/m².

14.9 a) The horizontal effective stress is K times the vertical effective stress:

$\sigma'_h = (0.5)(20.4) = 10.2$ kN/m²

14.10 b) The total head at B is 100 cm. The total head at C is 80 cm. The head loss from B to C is 100 - 80 = 20 cm. The 20 cm head loss occurs over a length of 80 cm.

For the uniform flow conditions shown, the hydraulic gradient between B and C is the head loss divided by the length of seepage path. Thus

$i = 20/80 = 0.25$

14.11 a) At point B, the total head is 100 cm. The elevation head is 10 cm. Thus, the pressure head is 90 cm or 0.90 m. The water pressure is the pressure head times the unit weight of water, or

$u = (0.90 \text{ m})(9.81 \text{ kN/m}^3) = 8.83$ kN/m²

14.12 d) From the flow net, three of the six equipotential drops have been crossed where the seepage path reaches the tip of the sheetpile. Of the total head loss of 100 - 90 = 10 ft, one half (or 5 ft) has occured by this point. The total head at the tip of the sheetpile is 95 ft.

14.13 c) From problem 14.12, the total head is 95 ft. The elevation head is 79 ft. The pressure head is 95 - 79 = 16 ft. The pore water pressure is the pressure head time the unit weight of water

$(16)(62.4) = 998$ lb/ft²

14.14 a) An overconsolidated soil is one where the preconsolidation pressure is greater than the effective overburden pressure.

14.15 e) For saturated clay tested in undrained shear, the friction angle is zero and = c. For an unconfined compression test, $\sigma_3 = 0$. Since $c = (\sigma_1 + \sigma_3)/2$, the compressive stress required for failure is $\sigma_1 = (2)(400) = 800$ lb/ft². The area of the sample to which the compressive force is applied is $\pi(0.5)^2/4 = 0.1964$ ft². The force required to cause shear failure is

$F = \sigma A = (800)(0.1964) = 157$ lb

Solutions to Problems (Essay)

14.16 The mass of water M_w is

$$46.95 - 35.06 = 11.89 \text{ g or } 0.01189 \text{ kg}$$

The water content is

$$M_w / M_s = 11.89 / 35.06 = 0.339 \text{ or } 33.9 \text{ percent}$$

The volume of solids is

$$V_s = M_s / G_{sw} = 0.03506 \text{ kg} / (2.70)(1000 \text{ kg/m}^3) = 13.0 \times 10^{-6} \text{m}^3$$

The volume of water is

$$V_w = M_w / \gamma_w = 0.01189 / 1000 = 11.9 \times 10^{-6} \text{m}^3$$

As the soil is satruated, $V_A = 0$.

The void ratio is

$$e = V_v / V_s = 11.9 \times 10^{-6} / 13.0 \times 10^{-6} = 0.915$$

The porosity is

$$n = V_v / V = 11.9 \times 10^{-6} / 24.9 \times 10^{-6} = 0.478$$

The total unit weight is

$$\gamma = W/V = Mg/V = (0.04695 \text{ kg})(9.81 \text{ m/s}^2) / 24.9 \times 10^{-6} \text{ m}^3$$

$$= 18\,500 \text{ kg/s}^2\cdot\text{m}^2$$

$$= 18.5 \text{ kN/m}^3$$

The dry unit weight is

$$\gamma_d = W_s/V = M_s g/V = (0.03506 \text{ kg})(9.81 \text{ m/s}^2) / 24.9 \times 10^{-6} \text{ m}^3$$

$$= 13\,800 \text{ kg/s}^2\cdot\text{m}^2$$

$$= 13.8 \text{ kN/m}^3$$

14.17 As the quanitity of soil is not fixed, assume $W_s = 100$ lb. Then

$$W_w = 20 \text{ lb,}$$

$$V_s = 100 / (2.68)(62.4) = 0.598 \text{ ft}^3$$

$$V_w = 20 / 62.4 = 0.321 \text{ ft}^3$$

$$V = 0.598 + 0.321 + 0 = 0.919 \text{ ft}^3$$

The required quantities are

$$\gamma = (100 + 20) / (0.598 + 0.321) = 130.6 \text{ lb/ft}^3, \text{ or}$$

$$130.6\,(9.81 / 62.4) = 20.5 \text{ kN/m}^3$$

$\gamma_d = 100 / (0.598 + 0.321) = 108.8 \text{ lb/ft}^3$, or

$108.8 (9.81 / 62.4) = 17.1 \text{ kN/m}^3$

14.18 Plot the dry unit weight vs water content:

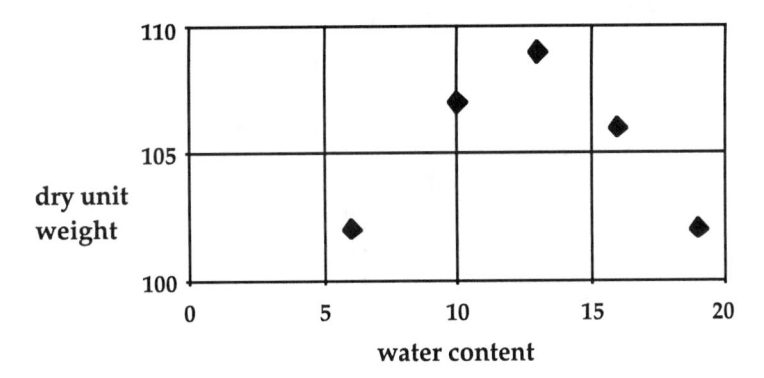

The maximum dry unit weight is approximately 109.0 lb/ft³. The optimum water content, the x value, corresponding to the peak of the curve, is 13 percent. The soil is to be compacted to at least 95 percent of maximum dry unit weight, or 103.6 lb/ft³. Extending a line at 103.6, it is seen that the minimum and maximum water contents for which the minimum specified density can be achieved are about 7 percent and 18 percent.

14.19 From 5 lbs of moist soil with a water content of 20 percent, one can write

$$\left. \begin{array}{l} W_s + W_w = 5.0 \\ W_s + 0.20 W_s = 5.0 \end{array} \right\} \therefore W_s = 4.17 \text{ lb}, \quad W_w = 0.83 \text{ lb}$$

To raise the water content from 20 percent to 30 percent, an additional 10 percent by weight of water must be added, or $(0.10)(4.17) = 0.42$ lb.

14.20 The required minimum weight of soil solids in the fill is $(20)(10\ 000) = 2.0 \times 10^5$ kN. As the soil in the borrow area has a dry unit weight of only 19.0 kN/m³, the required volume of excavation is

$$2.0 \times 10^5 \text{ kN} / 19.0 \text{ kN/m}^3 = 10530 \text{ m}^3$$

The required weight of water in the fill is $(0.10)(2.0 \times 10^5) = 20\ 000$ kN. The weight of water that will come with the borrow soil when hauled is $(0.08)(2.0 \times 10^5) = 16\ 000$ kN. The required water to be added is $20\ 000 - 16\ 000 = 4000$ kN, or $(4000 \text{ kN})/(9.81 \text{kN/m}^3) = 408$ m³ or 4.08×10^5 liters.

14.21 At the ground surface, all quantities are zero.

At 4 ft:

$\sigma_v = (4)(125) = 500 \text{ lb/ft}^2$
$u = 0$ (above water table)
$\sigma'_v = 500 - 0 = 500 \text{ lb/ft}^2$
$\sigma'_h = (500)(0.40) = 200 \text{ lb/ft}^2$
$\sigma_h = 200 + 0 = 200 \text{ lb/ft}^2$

At 10 ft:

$\sigma_v = 500 + (6)(130) = 1280 \text{ lb/ft}^2$
$u = (6)(62.4) = 374 \text{ lb.ft}^2$
$\sigma'_v = 1280 - 374 = 906 \text{ lb/ft}^2$
$\sigma'_h = (0.40)(906) = 362 \text{ lb/ft}^2$
$\sigma_h = 362 + 374 = 736 \text{ lb/ft}^2$

14.22 The *e* - log *p* curve is plotted below.

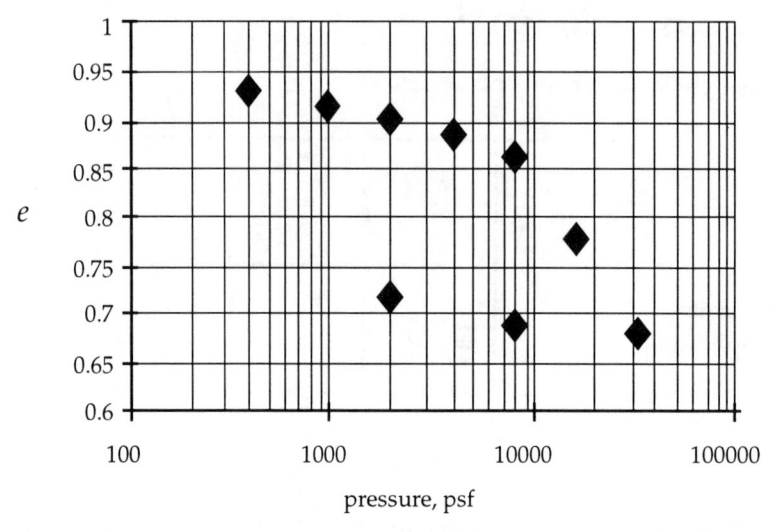

The preconsolidation pressure is approximately 9000 psf.

The compression index is slope of the virgin curve (to the right of 9000 psf). Extending the curve through the points (10 000, 0.85) and (60 000, 0.6) and changing the sign to positive the compression index is

C_c = - (0.85-0.60) / (log10 000 - log60 000) = 0.321

The recompression index is obtained by extending a line through the unloading curve through the points (0.073, 1000) and (0.069, 10 000). The recompression (or swell) index is:

C_s = - (0.073 - 0.069) / (log1000 - log10 000) = 0.04

14.23 As the clay later is being unloaded, the recompression index (or swell index) is used.

The change in void ratio due to reducing the effective vertical stress from 1800 to 1300 psf is

Δe = C_r log [(p + Δp)/p]
= 0.04 log (1800 / 1300) = 0.0057

As the clay is swelling due to unloading, this is an increase in void ratio.

The avearge vertical strain is estimated using the midpoint void ratio change as

ε_v = Δe / (1+e_0)
= 0.0057 / 1.900 = 0.003

The heave is the average strain times the thickness of the layer:

ΔH = (0.003)(5.0)

= 0.015 ft

= 0.18 inches

14.24 Draw two Mohr's circles for the given data and read off the strength as $c = 0.40$ tsf, $\phi = 19° \pm$

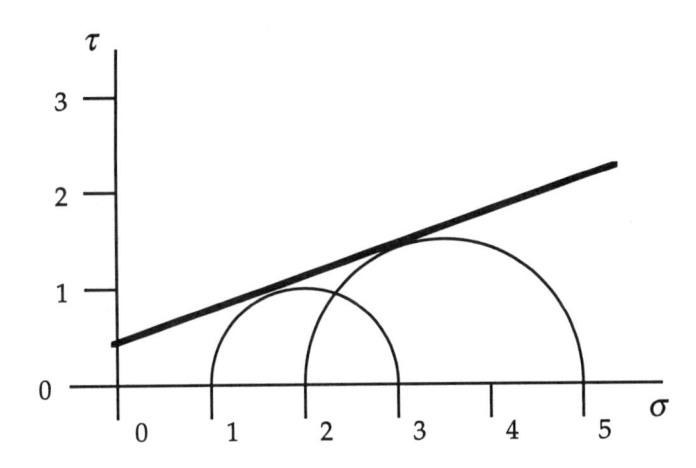

14.25 With 100 lb shear force applied, the shear stress on the sample is

$\tau = 100 / 3^2 = 11.11$ psi

Use the Mohr-Coulomb failure equation:

$\tau = c + \sigma \tan \phi$

$11.11 = 0 + \sigma \tan 30.$ $\therefore \sigma = 19.25$ psi

The maximum vertical force that can be used is (19.25 psi) (3^2 in^2) = 173 lb

Solutions to Problems (PE-Format)

14.26 b) From Equation 14.5.5,

$$\gamma_{sat} = \frac{(G_s + Se)\gamma_w}{1+e} = \frac{[2.75 + 1.0(0.5)]62.4}{1.5} = 135.2 \text{ lb/ft}^3$$

14.27 c) The excavation is symmetrical about the centerline, which is a flow line equivalent to an impervious boundary. Each of the two sides can be considered a separate flow net with $N_f/N_e = 2/6$. From Equation 14.11.6:

$$Q = kH\frac{N_f}{N_e} \times 2 = (0.1 \text{ ft/min})(12 \text{ ft})\left(\frac{2}{6}\right)(2 \text{ sides}) = 0.8 \text{ ft}^3/\text{min/ft}$$

This is the flow per lineal foot of excavation. For a 100 ft long excavation,

$Q_{total} = (0.8 \text{ ft}^3/\text{min/ft})(100 \text{ ft}) = 80 \text{ ft}^3/\text{min}$

14.28 d) Point A is 8 ft below the surface of a soil with a total (saturated) unit weight of 135.2 lb/ft³. From Equation 14.10.1,

$$\sigma_v = \gamma_{sat} H = (135.2 \text{ lb/ft}^3)(8.0 \text{ ft}) = 1081.6 \text{ lb/ft}^3$$

14.29 a) As water is flowing, the pore pressure must be determined from a seepage analysis. As 12 ft of total head is lost over six equipotential drops, each equipotential drop (square) represents 2 ft of total head loss. Point A is one drop above the seepage exit, hence the total head at Point A is:

$$TH = 388+(1/6)(12) = 390 \text{ ft}$$

or

$$TH = 400 - (5/6)(12) = 390 \text{ ft}$$

The elevation head at A is 380. The pressure head is the total head minus the elevation head, or

$$PH = 390 - 380 = 10 \text{ ft}$$

The pore pressure at A is then

$$u = (PH)\,\gamma_w = (10 \text{ ft})(62.4 \text{ lb/ft}^3) = 624 \text{ lb/ft}^2$$

14.30 e) From Equation 14.10.4

$$\sigma'_v = \sigma_v - u = 1081.6 - 624.0 = 457.6 \text{ lb/ft}^2$$

15. Foundations and Retaining Structures

by Thomas F. Wolff

This chapter deals with the application of the principles of soil mechanics (introduced in Chapter 14) to the analysis and design of foundations on soil and retaining structures that hold back soil.

15.1 Estimating Soil Strength from Standard Penetration Test Data

One of the most common methods for sampling soil is to drive a split-spoon or split-barrel sampler with a drop hammer. The *standard penetration test* or SPT (ASTM 1586) is a standard procedure for split-spoon sampling which provides a quantitative measure of driving resistance, the standard penetration resistance N. The value N has been widely correlated with soil properties and is commonly used to estimate density and strength of sands and consistency of clays. For the standard penetration test, the split-spoon sampler has an inside diameter of 1-3/8 inches (35mm) and an outside diameter of 2 inches (51 mm). It is driven 18 inches by repeatedly dropping a 140 lb weight from a height of 30 inches. The number of blows are recorded for each six inches of driving. The blows for the second and third increment are added to obtain the standard penetration resistance N.

––––––– **EXAMPLE 15.1** –––––––

The following data is recorded from a standard penetration test:

Depth, (feet)	Resistance, (blows)
12.0 - 12.5	8
12.5 - 13.0	11
13.0 - 13.5	13

Determine the standard penetration resistance.

Solution. The resistance for the first increment (8) is discarded. The combined resistance for the last one foot is 11 + 13 = 24. Thus

$N = 24$ blows/ft

For sands, N increases with increasing effective stress and increasing relative density (or strength). For a sand deposit with constant relative density (or strength), N will increase with depth due to the stress increase with depth. In order to use standard penetration test data to estimate strength, it must be corrected to a constant vertical effective stress. Several correction equations have been proposed; the Liao and Whitman (1986) equation is simple and fits most research results. The actual or uncorrected N value may be adjusted to the corrected value N_1 expected under a reference stress of 1 ton per square foot (2000 psf) as:

$$N_1 = CN \tag{15.1.1}$$

where

$$C = \sqrt{\frac{1}{\sigma'_v}}$$

The above equation is unit dependent, and σ'_v is the vertical effective stress in tons per square foot.

EXAMPLE 15.2

The data in Example 15.1 was obtained at a depth of 15 feet in a thick sand deposit. The water table is at 10 feet. The unit weight of moist sand above the water table is estimated to be 115 lb/ft³, the total unit weight of saturated sand below the water table is estimated to be 125 lb/ft³. Determine the corrected N value.

Solution. The vertical effective stress is

$$\sigma'_v = (10)(115) + (5)(125) - (5)(62.4)$$

$$= 1460 \text{ lb/ft}^2 \text{ or } 0.73 \text{ tsf}$$

$$C = \sqrt{\frac{1}{0.73}} = 1.17$$

$$N_1 = CN = (1.17)(24) = 28.1 \cong 28$$

The corrected N value is 28. If the sand sample with $N = 24$ under a stress of 0.73 tsf were at a greater depth where the vertical effective stress was 1 tsf, a standard penetration value of 28 would be expected. The 28 value should be used to estimate relative density and strength values.

The friction angle ϕ for sand depends on relative density. It is nearly impossible to sample sands without changing the relative density, so it is common to estimate ϕ from penetration test data rather than laboratory testing. One such correlation adapted from a curve by Peck, et. al., (1974) is

$$\phi = 27.1 + 0.3N_1 - 0.00054N_1^2 \tag{15.1.2}$$

EXAMPLE 15.3

Estimate the friction angle for the sand sample in Examples 15.1 and 15.2.

Solution. We use the corrected value $N_1 = 28$. Thus,

$$\phi = 27.1 + 0.3N_1 - 0.00054N_1^2$$

$$= 27.1 + 0.3(28) - 0.00054(28)^2 = 35.1° \cong 35°$$

The standard penetration test gives a crude indication of the cohesion c (or undrained strength s_u) of clays; the accuracy is not great. The cohesion is estimated as the product of N times a constant, where the constant ranges from 0.075 to 0.25 tsf, and is typically about 0.13. In various units, it is

$$c = s_u = 0.075N \text{ to } 0.25N \text{ tsf} \qquad (15.2.1)$$

$$= 0.15N \text{ to } 0.50N \text{ ksf}$$

$$= 7N \text{ to } 24N \text{ kN/m}^2$$

15.2 Estimating Clay Strength from Cone Penetration Test Data

The standard penetration test is termed an *in-situ* test as it is performed in the ground at the site of the sample. A more sophisticated in-situ test for determining the undrained strength of clay, coming into more common use, is the *cone penetration test* (ASTM D 3441) or *Dutch cone test*. A rod with a conical tip is pushed into the ground. The conical tip has a cross-sectional area of 10 cm² and an apex angle of 60 degrees. The force required to advance the tip is measured mechanically or electrically and divided by the tip area to get the *cone resistance* q_c which has units of pressure. As advancing the cone causes a bearing capacity failure, bearing capacity theory can be used to backcalculate the undrained strength. The undrained strength s_u is calculated as:

$$s_u = \frac{q_c - \sigma_v}{N_k} \qquad (15.2.1)$$

where q_c is the cone resistance

σ_v is the total vertical overburden pressure

N_k is the cone factor, which is site-dependent and empirical

N_k is commonly in the range of 10 to 30, often about 15 or 20.

15.3 Types of Foundations

The function of a foundation is to transmit the loads of a building or other facility to the underlying soil in such a manner that the soil does not excessively compress or yield due to shear. Foundation types are broadly divided into *shallow foundations* and *deep foundations*. Some references classify foundation types based on the ratio of their width B to their depth D. Shallow foundations commonly have a B/D ratio greater than 1.0, and deep foundations commonly have a B/D ratio less than 1/10. Foundations with intermediate ratios are not common. A more logical classification of foundations is according to the manner in which loads are transmitted to the soil. Shallow foundations are those where loads are transmitted to the soil at a depth adjacent to the lowest part of the functional structure, and function by providing sufficient horizontal area to reduce applied nomal stresses to tolerable values. Deep foundations are those where loads are transmitted to the soil at a considerable distance below the lowest part of the functional structure, and function by providing vertical surfaces that carry shear forces and/or horizontal bearing areas at depth that carry normal stresses.

Types of shallow foundations include *footings*, which are horizontal concrete slabs under individual walls or columns, and *rafts or mats*, which are large footings under groups of columns and/or walls or under an entire structure.

Types of deep foundations include *piles*, which are long slender columns of wood, steel or concrete driven into the soil, and *drilled shafts, piers, or caissons*, which are columns (usually concrete) constructed in place in excavated holes.

15.4 Bearing Capacity of Shallow Foundations

Figure 15.1 shows a section of unit length through a continuous footing of width B founded at depth D with a load of Q_{net} per unit length. Continuous infers that the length of the footing L into the page is "long" (i.e., $L/B > 10$). The purpose of bearing capacity analysis is to ensure that the pressures applied to the soil are sufficiently small to prevent shear failure and uncontrolled downward movement of the foundation.

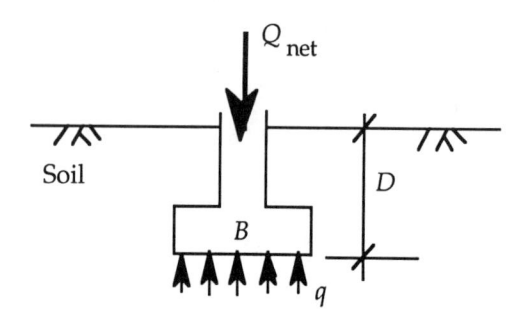

Figure 15.1 Section through continuous footing.

Let Q be the total vertical force on a unit length of the base of the footing, where Q is the sum of the wall or column load Q_{net}, the weight Q_{conc} of the concrete footing and the weight Q_{soil} of the soil above the footing base:

$$Q = Q_{net} + Q_{conc} + Q_{soil} \qquad (15.4.1)$$

Assuming the loads are concentric, the applied pressure at the base of the footing is

$$q = \frac{Q}{B \times L} \qquad (15.4.2)$$

The *ultimate bearing pressure* or *ultimate bearing capacity* q_{ult} is the pressure that would result in shear failure, and is calculated from bearing capacity theory. To avoid shear failure, applied pressures are limited to an *allowable bearing capacity* or *allowable bearing pressure* obtained by dividing the ultimate bearing capacity by a factor of safety, FS:

$$q_a = \frac{q_{ult}}{FS} \qquad (15.4.3)$$

The factor of safety used to determine allowable bearing capacity is commonly in the range 2.0 to 4.0, often 3.0. If the actual applied pressure q is known, the actual factor of safety can be determined as

$$FS = \frac{q_{ult}}{q} \tag{15.4.4}$$

The basis of bearing capacity calculations is shown in Figure 15.2. When the applied pressure reaches q_{ult}, shear failure occurs along a curved surface below and to the sides of the footing. In the Terzaghi bearing capacity theory, the soil above the footing is replaced by a surcharge pressure γD. Soil below the footing has the properties c, ϕ, and γ.

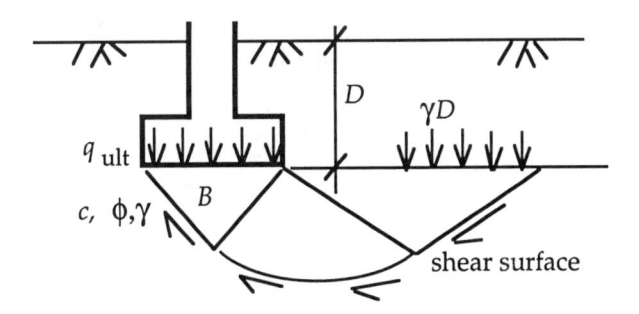

Figure 15.2 Shear surface for bearing failure.

According to Terzaghi's theory, the ultimate bearing capacity q_{ult} for a continuous footing of width B can be expressed as:

$$q_{ult} = cN_c + \gamma D N_q + \tfrac{1}{2}\gamma B N_\gamma \tag{15.4.5}$$

where

- c is the cohesion of the soil below the base of the footing.
- γD is the vertical effective stress at the elevation of the footing base (sometimes also denoted q).
- B is the footing width.
- γ is the unit weight of the soil below the footing base.

$N_c, N_q,$ and N_γ are bearing capacity factors which are functions of the friction angle ϕ of the soil below the footing.

The bearing capacity equation can be thought of as having the following form:

Ultimate capacity = cohesion term + surcharge term + friction term

Researchers have developed various expressions for the bearing capacity factors $N_c, N_q,$ and N_γ. The Vesic (1973) factors are commonly used and easy to calculate:

$$\begin{aligned} N_q &= e^{\pi \tan\phi} \tan^2\left(45 + \tfrac{\phi}{2}\right) \\ N_c &= (N_q - 1)\cot\phi \quad\quad (N_q = 5.14 \text{ for } \phi = 0) \\ N_\gamma &= 2(N_q + 1)\tan\phi \end{aligned} \tag{15.4.6}$$

Table 15.1 summarizes the Vesic bearing capacity factors for a common range of ϕ values.

TABLE 15.1 Vesic's Bearing Capacity Factors

ϕ	N_c	N_q	N_γ
0	5.14	1.00	0.00
5	6.49	1.57	0.45
10	8.34	2.47	1.22
15	11.0	3.94	2.65
20	14.8	6.40	5.39
25	20.7	10.7	10.9
26	22.3	11.9	12.5
27	23.9	13.2	14.5
28	25.8	14.7	16.7
29	27.9	16.4	19.3
30	30.1	18.4	22.4
31	32.7	20.6	26.0
32	35.5	23.2	30.2
33	38.6	26.1	35.2
34	42.2	29.4	41.1
35	46.1	33.3	48.0
36	50.6	37.8	56.3
37	55.6	42.9	66.2
38	61.4	48.9	78.0
39	67.9	56.0	92.3
40	75.3	64.2	109
41	83.9	73.9	130
42	93.7	85.4	156
43	105	99.0	187
44	118	115	225
45	134	135	272

EXAMPLE 15.4

A continuous footing is founded 3 feet below the ground surface in a homogeneous clay. The soil parameters are γ = 125 pcf, c = 1200 psf, and ϕ = 0. Determine the ultimate bearing capacity and recommend an allowable bearing capacity.

Solution.

$$q_{ult} = cN_c + \gamma D N_q + \frac{1}{2}\gamma B N_\gamma$$

$$= (1200 \text{ psf})(5.14) + (125 \text{ pcf})(3 \text{ ft})(1) + (0) = 6540 \text{ psf}$$

Taking the factor of safety as 3.0,

$$q_a = 6540/3 = 2180 \text{ psf} \cong 2200 \text{ psf}$$

15.4.1 Net bearing capacity in clay

Note that, for clay soils where $\phi = 0$, the friction term is zero and the surcharge term is generally small. If the weight of the concrete footing is considered to be similar to the weight of the soil it displaces, the pressure due to the weight of the footing and soil above the footing approximately balances out the surcharge term D. Take the *net ultimate bearing capacity* $q_{\text{ult net}}$ as the additional pressure that can be applied in excess of that due to the weight of the footing and overlying soil and obtain

$$q_{\text{ult net}} + \gamma D = cN_c + \gamma D(1)$$

$$\therefore q_{\text{ult net}} = cN_c \tag{15.4.7}$$

──── **EXAMPLE 15.5** ────

Determine the net ultimate and allowable bearing capacity for the problem in Example 15.4.

$$q_{\text{ult net}} = cN_c$$
$$= (1200 \text{ psf})(5.14)$$
$$= 6170 \text{ psf}$$

Taking the factor of safety as 3.0,

$$q_{a \text{ net}} = 6170/3 = 2060 \text{ psf} \cong 2000 \text{ psf}$$

──── **EXAMPLE 15.6** ────

A continuous footing two feet wide is founded 3 feet below the ground surface in a uniform sand. The soil parameters are $\gamma = 125$ pcf, $c = 0$, and $\phi = 34°$. Determine the ultimate bearing capacity and recommend an allowable bearing capacity.

Solution. Note that the footing width B must be known or assumed when ϕ is greater than zero. For the given $B = 2$ ft,

$$q_{\text{ult}} = cN_c + \gamma D N_q + \frac{1}{2} B \gamma N_\gamma$$
$$= 0 + (3)(125)(29.44) + (.5)(2)(125)(41.06) = 16200 \text{ psf}$$

Taking the factor of safety as 3.0, the allowable bearing capacity is:

$$q_a = 16,200/3 = 5400 \text{ psf}$$

For the two foot wide footing, a working load of (5400 psf)(2 ft) = 10,800 lb per lineal foot could be allowed.

15.4.2 Square and rectangular footings

The equation for continuous footings can be modified to predict the ultimate bearing capacity of rectangular footings by introducing shape factors F_{cs}, F_{qs}, and F_s:

$$q_{\text{ult}} = cN_c F_{cs} + \gamma D N_q F_{qs} + \frac{1}{2} B \gamma N_\gamma F_{\gamma s} \tag{15.4.8}$$

While several versions of the shape factors are available, the following are commonly used:

$$F_{cs} = 1 + \frac{B}{L} \cdot \frac{N_q}{N_c}$$

$$F_{qs} = 1 + \frac{B}{L} \tan \phi \qquad (15.4.9)$$

$$F_{\gamma s} = 1 - 0.4 \frac{B}{L}$$

In the above equations, B is always the smallest dimension of the footing, and L is always the largest dimension.

--- **EXAMPLE 15.7** ---

A two foot by three foot rectangular footing is founded 3 feet below the ground surface in a uniform sand. The soil parameters are $\gamma = 125$ pcf, $c = 0$, and $\phi = 34°$. Determine the ultimate bearing capacity and recommend an allowable bearing capacity.

Solution. $B = 2$ ft, $L = 2$ ft, and $B/L = 2/3$. The shape factors are

$$F_{cs} = 1 + \frac{B}{L} \cdot \frac{N_q}{N_c} = 1 + \frac{2}{3} \cdot \frac{29.44}{41.06} = 1.478$$

$$F_{qs} = 1 + \frac{B}{L} \tan \phi = 1$$

$$F_{\gamma s} = 1 - 0.4 \frac{B}{L} = 1 - 0.4 \cdot \frac{2}{3} = 0.733$$

The ultimate bearing capacity is

$$q_{ult} = cN_c F_{cs} + \gamma D N_q + \frac{1}{2} \gamma B N_\gamma F_{\gamma s}$$

$$= 0 + (3)(125)(29.44)(1) + (.5)(2)(125)(41.06)(0.733) = 14800 \text{ psf}$$

Taking the factor of safety as 3.0, the allowable bearing capacity is

$$q_a = 14{,}800/3 = 4930 \text{ psf} \cong 4900 \text{ psf}$$

For the two foot by three foot rectangular footing, a working load of (4930 psf)(2 ft)(3 ft) = 29,600 (say 30,000) lb per lineal foot could be allowed.

15.4.3 Effect of water table

If the water table is deeper than a distance B (the footing width) below the footing base, the water table is considered to have no effect on the bearing capacity. When it is shallower than this depth, the value of the unit weight γ in the surcharge and friction terms must be modified to account for buoyancy.

If the water table is at the surface, the effective unit weight $\gamma' = \gamma_{sat} - \gamma_w$ should be used in both terms.

If the water table is between the surface and the base of the footing, γD in the surcharge term should be replaced by the vertical effective stress at the level of the footing base and γ' should be used in the friction term.

If the water table is at the level of the base of the footing, use total unit weights to calculate γD and use γ' to calculate the surcharge term.

If the water table is at a depth d below the level of the base of the footing, but within a depth B, use total unit weights to calculate γD and an interpolated γ value $\bar{\gamma}$ in the friction term:

$$\bar{\gamma} = \gamma' + \left(\frac{d}{B}\right)\gamma_w \tag{15.4.10}$$

15.4.4 General bearing capacity equation

The effects of load inclination can be accounted for by introducing load inclination factors F_{ci}, F_{qi}, and $F_{\gamma i}$. The effect of the soil shear resistance above the footing base, neglected in the original Terzaghi formulation, can be accounted for by introducing depth factors F_{cd}, F_{qd}, and $F_{\gamma d}$ however, many designers conservatively take depth factors to be one. Ground slope factors and base tilt factors can also be introduced to account for these respective effects; however, they are beyond the scope of this review. The general bearing capacity equation including shape factors, load inclination factors, and depth factors is

$$q_{ult} = cN_c F_{cs} F_{ci} F_{cd} + \gamma D N_q F_{qs} F_{qi} F_{qd} + \frac{1}{2}\gamma B N_\gamma F_{\gamma s} F_{\gamma i} F_{\gamma d} \tag{15.4.11}$$

Inclination factors for a load acting at an angle i, measured in degrees from the vertical, can be taken as

$$F_{ci} = F_{qi} = \left(1 - \frac{i}{90}\right)^2$$
$$F_{\gamma i} = \left(1 - \frac{i}{\phi}\right)^2 \tag{15.4.12}$$

Several versions of the depth factors have been recommended. It is always conservative to take them as 1.0.

15.4.5 Effect of eccentricity

The above equations are for concentrically loaded footings, i.e., those loaded along their centerline. When footings are loaded eccentrically, the line of action of the load does not coincide with the center of the footing. The effect of eccentricity can be modeled by performing the bearing capacity analysis using a reduced effective footing width B' which is taken as:

$$B' = B - 2e \tag{15.4.13}$$

where e is the eccentricity, or distance between the line of action of the load and the centerline of the footing. Where the load is eccentric in two directions, an effective length is also used:

$$B' = B - 2e_B \tag{15.4.14}$$
$$L' = L - 2e_L$$

The footing dimensions are reduced to their effective dimensions prior to doing any other calculations, including shape factors, bearing capacity, and ultimate load capacity.

The following example summarizes the above. It includes a rectangular footing with inclined eccentric load and water table effect.

EXAMPLE 15.8

Given the retaining wall footing shown, determine the factor of safety in bearing capacity.

The load due to the wall has the following properties:

$Q_{vertical} = 3000$ lb/ft, $Q_{horizontal} = 1000$ lb/ft, $e = 0.75$ ft

The wall is founded on a silty sand with $c = 0$, $\phi = 28°$. The moist density above the water table is 120 lb/ft³, the saturated density below the water table is 125 lb/ft³.

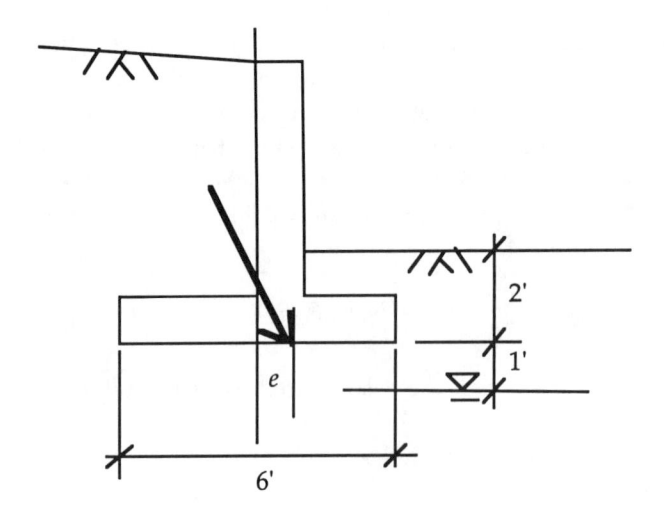

Solution. First the effective footing width is calculated and used for all subsequent calculations:

$B' = B - 2e$

$= 6 - (2)(0.75) = 4.5$ ft

The load inclination angle is determined from the vertical and horizontal components:

$i = \tan^{-1}(1000/3000) = 18.43°$

The load inclination factors are

$$F_{ci} = F_{qi} = \left(1 - \frac{i}{90}\right)^2 = \left(1 - \frac{18.43}{90}\right)^2 = 0.632$$

$$F_{\gamma i} = \left(1 - \frac{i}{\phi}\right)^2 = \left(1 - \frac{18.43}{28}\right)^2 = 0.117$$

The effective soil density below the footing is

$$\bar{\gamma} = \gamma' + \left(\frac{D}{B}\right)\gamma_w = (125 - 62.4) + \left(\frac{1}{4.5}\right)62.4 = 76.5 \text{ lb/ft}^3$$

The ultimate bearing capacity is

$q_{ult} = cN_c F_{cs} F_{ci} F_{cd} + \gamma DN_q F_{qs} F_{qi} F_{qd} + \frac{1}{2}\gamma BN_\gamma F_{\gamma s} F_{\gamma i} F_{\gamma d}$

$= 0 + (2)(120)(14.72)(1)(0.632)(1) + (0.5)(76.5)(16.72)(0.117)(1)$

$= 2300$ psf

The vertical component of the ultimate bearing force is

$$Q_{ult} = q_{ult}B'$$
$$= (2300 \text{ psf})(4.5 \text{ ft})$$
$$= 10{,}350 \text{ lb/ft}$$

The factor of safety in bearing is the ultimate bearing force divided by the vertical component of the applied foundation force:

$$FS = Q_{ult}/Q_v$$
$$= 10350/3000$$
$$= 3.45$$

15.5 Stress Increase under Shallow Foundations

Applying a load on or near the ground surface induces normal and shear stress changes throughout the ground. Solutions for these stress changes in three dimensions for a number of loading conditions are available in many references, and are based on the theory of elasticity and the assumption of a semi-infinite half-space.

For shallow foundations, the load applied by the footing is usually vertical, and interest centers on determining the vertical stress increase as a function of depth. The stress increase may be sought under a number of points, typically the center, edge, and corner of a foundation, or one may wish to estimate an average stress increase. The stress increase values will, in turn, be used to estimate induced strains in the soil required for predicting settlement.

15.5.1 Vertical stress increase due to vertical point load

Around 1885, Boussinesq published a solution for the vertical stress increase at any point in a semi-infinite elastic half-space due to a vertical point load on the surface, as illustrated in Figure 15.3.

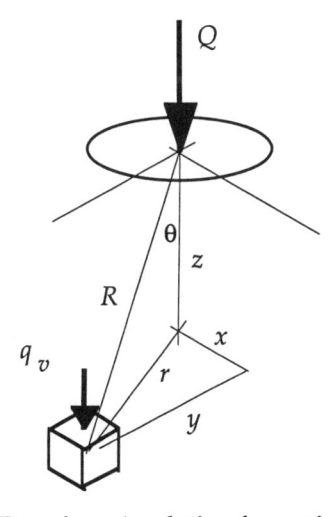

Figure 15.3 Boussinesq's solution for vertical stress increase.

For Boussinesq's solution, the point or element at which the stress is sought is defined as being a distance x, y, and z from the point load on the surface. The radial distance in plan from the point load to the element is r, the radial distance in space is R, and the vertical angle between the element and the z-axis through the load is θ. The vertical stress increase is then

$$q_v = \frac{3Q}{2\pi z^2}\cos^5\theta = \frac{3Qz^3}{2\pi R^5} \qquad (15.5.1)$$

For the special case of stress increase directly below the point load, $R = z$, and the equation becomes

$$q_v = \frac{3Q}{2\pi z^2} \qquad (15.5.2)$$

EXAMPLE 15.9

A point load of 10 kN acts on the surface of the ground. Determine the stress at depths of 2 m and 4 m directly below the load, and at the same depths radially one meter away.

Solution. Below the point load at 2 m,

$$q_v = \frac{3Q}{2\pi z^2} = \frac{(3)(10)}{2\pi(2)^2} = 1.19 \text{ kN}$$

Below the point load at 4 m,

$$q_v = \frac{3Q}{2\pi z^2} = \frac{(3)(10)}{2\pi(4)^2} = 0.298 \text{ kN}$$

One meter away at 2 m depth,

$$R = \sqrt{2^2 + 1^2} = 2.236 \text{ m}$$

$$q_v = \frac{3Qz^3}{2\pi R^5} = \frac{(3)(10)(2)^3}{2\pi(2.236)^5} = 0.683 \text{ kN/m}^2$$

One meter away at 4 m depth,

$$R = \sqrt{4^2 + 1^2} = 4.123 \text{ m}$$

$$q_v = \frac{3Qz^3}{2\pi R^5} = \frac{(3)(10)(4)^3}{2\pi(4.123)^5} = 0.257 \text{ kN/m}^2$$

15.5.2 Vertical stress increase due to uniform surface pressure

A solution for the vertical stress increase under a uniformly loaded area may be derived by integrating the Boussinesq solution over the area. The vertical stress increase q_v under the corner of a uniformly loaded rectangular area of dimensions $B \times L$ is

$$q_v = q_o I \qquad (15.5.3)$$

where $q_o=$ the pressure intensity at the surface and I is an influence factor that is a function of B, L, and z. The solution for I has been derived by Newmark (1935) and plotted by Fadum (1948). To determine I, first the nondimensional parameters m and n are calculated:

$$m = \frac{B}{z}, \quad n = \frac{L}{z} \qquad (15.5.4)$$

where B and L are the width and length of the loaded area and z is the depth below the corner of the area. To simplify calculations, α and α_1 are defined as

$$\alpha = m^2 + n^2 + 1 \qquad (15.5.5)$$

$$\alpha_1 = (mn)^2$$

Then the influence value can be calculated as

$$I = \frac{1}{4\pi}\left[\frac{2mn}{\alpha + \alpha_1} \cdot \frac{\alpha+1}{\sqrt{\alpha}} + \tan^{-1}\left(\frac{2mn\sqrt{\alpha}}{\alpha - \alpha_1}\right)\right] \qquad (15.5.6)$$

If $\alpha_1 > \alpha$, the arc tangent term is negative and π must be added to the term in brackets.

The solution to the above is tabulated in Table 15.2 and plotted in Figure 15.4. If the m and n values plot off the figure, m and n may be interchanged.

TABLE 15.2 Stress Distribution under the corner of a uniformly loaded rectangular area.

						n					
m	0.1	0.2	0.3	0.4	0.5	0.6	0.7	0.8	0.9	1.0	1.2
0.1	0.00470	0.00917	0.01323	0.01678	0.01978	0.02223	0.02420	0.02576	0.02698	0.02794	0.02926
0.2	0.00917	0.01790	0.02585	0.03280	0.03866	0.04348	0.04735	0.05042	0.05283	0.05471	0.05733
0.3	0.01323	0.02585	0.03735	0.04742	0.05593	0.06294	0.06858	0.07308	0.07661	0.07938	0.08323
0.4	0.01678	0.03280	0.04742	0.06024	0.07111	0.08009	0.08734	0.09314	0.09770	0.10129	0.10631
0.5	0.01978	0.03866	0.05593	0.07111	0.08403	0.09473	0.10340	0.11035	0.11584	0.12018	0.12626
0.6	0.02223	0.04348	0.06294	0.08009	0.09473	0.10688	0.11679	0.12474	0.13105	0.13605	0.14309
0.7	0.02420	0.04735	0.06858	0.08734	0.10340	0.11679	0.12772	0.13653	0.14356	0.14914	0.15703
0.8	0.02576	0.05042	0.07308	0.09314	0.11035	0.12474	0.13653	0.14607	0.15371	0.15978	0.16843
0.9	0.02698	0.05283	0.07661	0.09770	0.11584	0.13105	0.14356	0.15371	0.16185	0.16835	0.17766
1.0	0.02794	0.05471	0.07938	0.10129	0.12018	0.13605	0.14914	0.15978	0.16835	0.17522	0.18508
1.2	0.02926	0.05733	0.08323	0.10631	0.12626	0.14309	0.15703	0.16843	0.17766	0.18508	0.19584
1.4	0.03007	0.05894	0.08561	0.10941	0.13003	0.14749	0.16199	0.17389	0.18357	0.19139	0.20278
1.6	0.03058	0.05994	0.08709	0.11135	0.13241	0.15028	0.16515	0.17739	0.18737	0.19546	0.20731
1.8	0.03090	0.06058	0.08804	0.11260	0.13395	0.15207	0.16720	0.17967	0.18986	0.19814	0.21032
2.0	0.03111	0.06100	0.08867	0.11342	0.13496	0.15326	0.16856	0.18119	0.19152	0.19994	0.21235
2.5	0.03138	0.06155	0.08948	0.11450	0.13628	0.15483	0.17036	0.18321	0.19375	0.20236	0.21512
3.0	0.03150	0.06178	0.08982	0.11495	0.13684	0.15550	0.17113	0.18407	0.19470	0.20341	0.21633
4.0	0.03158	0.06194	0.09007	0.11527	0.13724	0.15598	0.17168	0.18469	0.19540	0.20417	0.21722
5.0	0.03160	0.06199	0.09014	0.11537	0.13737	0.15612	0.17185	0.18488	0.19561	0.20440	0.21749
6.0	0.03161	0.06201	0.09017	0.11541	0.13741	0.15617	0.17191	0.18496	0.19569	0.20449	0.21760
8.0	0.03162	0.06202	0.09018	0.11543	0.13744	0.15621	0.17195	0.18500	0.19574	0.20455	0.21767
10.0	0.03162	0.06202	0.09019	0.11544	0.13745	0.15622	0.17196	0.18502	0.19576	0.20457	0.21769
∞	0.03162	0.06202	0.09019	0.11544	0.13745	0.15623	0.17197	0.18502	0.19577	0.20458	0.21770

						n					
m	1.6	1.8	2.0	2.5	3.0	4.0	5.0	6.0	8.0	10.0	∞
0.1	0.03058	0.03090	0.03111	0.03138	0.03150	0.03158	0.03160	0.03161	0.03162	0.03162	0.03162
0.2	0.05994	0.06058	0.06100	0.06155	0.06178	0.06194	0.06199	0.06201	0.06202	0.06202	0.06202
0.3	0.08709	0.08804	0.08867	0.08948	0.08982	0.09007	0.09014	0.09017	0.09018	0.09019	0.09019
0.4	0.11135	0.11260	0.11342	0.11450	0.11495	0.11527	0.11537	0.11541	0.11543	0.11544	0.11544
0.5	0.13241	0.13395	0.13496	0.13628	0.13684	0.13724	0.13737	0.13741	0.13744	0.13745	0.13745
0.6	0.15028	0.15207	0.15326	0.15483	0.15550	0.15598	0.15612	0.15617	0.15621	0.15622	0.15623
0.7	0.16515	0.16720	0.16856	0.17036	0.17113	0.17168	0.17185	0.17191	0.17195	0.17196	0.17197
0.8	0.17739	0.17967	0.18119	0.18321	0.18407	0.18469	0.18488	0.18496	0.18500	0.18502	0.18502
0.9	0.18737	0.18986	0.19152	0.19375	0.19470	0.19540	0.19561	0.19569	0.19574	0.19576	0.19577
1.0	0.19546	0.19814	0.19994	0.20236	0.20341	0.20417	0.20440	0.20449	0.20455	0.20457	0.20458
1.2	0.20731	0.21032	0.21235	0.21512	0.21633	0.21722	0.21749	0.21760	0.21767	0.21769	0.21770
1.4	0.21510	0.21836	0.22058	0.22364	0.22499	0.22600	0.22632	0.22644	0.22652	0.22654	0.22656
1.6	0.22025	0.22372	0.22610	0.22940	0.23088	0.23200	0.23236	0.23249	0.23258	0.23261	0.23263
1.8	0.22372	0.22736	0.22986	0.23334	0.23495	0.23617	0.23656	0.23671	0.23681	0.23684	0.23686
2.0	0.22610	0.22986	0.23247	0.23614	0.23782	0.23912	0.23954	0.23970	0.23981	0.23985	0.23987
2.5	0.22940	0.23334	0.23614	0.24010	0.24196	0.24344	0.24392	0.24412	0.24425	0.24429	0.24432
3.0	0.23088	0.23495	0.23782	0.24196	0.24394	0.24554	0.24608	0.24630	0.24646	0.24650	0.24654
4.0	0.23200	0.23617	0.23912	0.24344	0.24554	0.24729	0.24791	0.24817	0.24836	0.24842	0.24846
5.0	0.23236	0.23656	0.23954	0.24392	0.24608	0.24791	0.24857	0.24885	0.24907	0.24914	0.24919
6.0	0.23249	0.23671	0.23970	0.24412	0.24630	0.24817	0.24885	0.24916	0.24939	0.24946	0.24952
8.0	0.23258	0.23681	0.23981	0.24425	0.24646	0.24836	0.24907	0.24939	0.24964	0.24973	0.24980
10.0	0.23261	0.23684	0.23985	0.24429	0.24650	0.24842	0.24914	0.24946	0.24973	0.24981	0.24989
∞	0.23263	0.23686	0.23987	0.24432	0.24654	0.24846	0.24919	0.24952	0.24980	0.24989	0.25000

Figure 15.4 Stress Distribution under the corner of a uniformly loaded rectangular area.

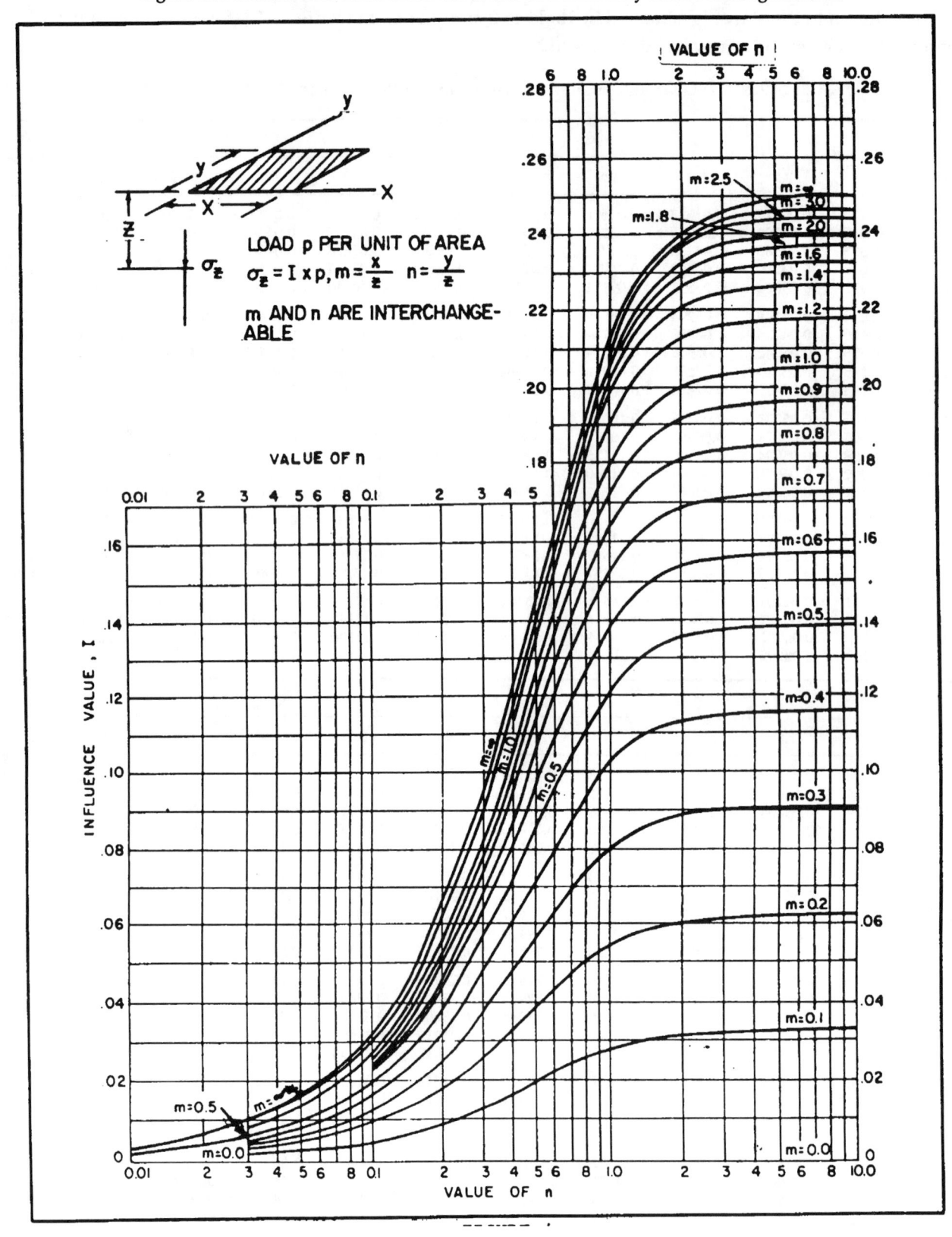

The principal of superposition may be used to solve for points other than the corner and also for irregular areas. The point of interest is taken to be the corner of several smaller areas. The stress increase is calculated for each sub-area and the results are added.

EXAMPLE 15.10

The 10 × 10 ft area shown is loaded with a uniform pressure of 1000 pounds per square foot. Determine the stress increase at point A, which is ten feet below the surface at the location shown.

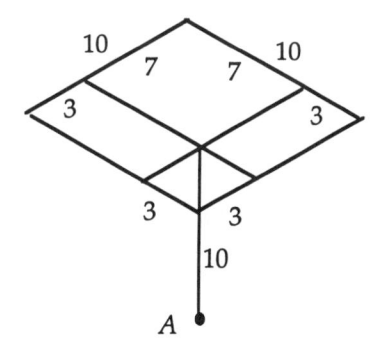

Solution. The point can be taken as being under the corner of four sub-areas, one 7' by 7', one 3' by 3', and two 7' by 3' areas. The influence values for each area can be found in Table 15.1:

$$7 \times 7: \quad m = n = \frac{7}{10} = 0.7, \quad I_1 = 0.12772$$

$$3 \times 3: \quad m = n = \frac{3}{10} = 0.3, \quad I_2 = 0.03735$$

$$7 \times 3: \quad m = 0.3, \quad n = 0.7, \quad I_3 = 0.06858$$

The stress increase is

$$q_v = q_o \Sigma I = (1000)(0.12772 + 0.03735 + 0.06858 + 0.06858) = 302 \text{ psf}$$

15.5.3 Vertical stress increase outside limits of foundation

The principal of superposition can also be used to find the stress increase under points outside the limits of the foundation. Loaded areas are assumed to extend out to the point of interest, and compensating negatively loaded areas are added back to the loaded foundation, leaving a net load of zero outside the foundation.

15.5.4 Vertical stress increase due to non-uniform surface pressure

The principal of superposition can be used to analyze loadings of non-uniform pressure by modeling the loading as the sum (or difference) of uniformly loaded areas.

15.5.5 Average vertical stress increase under a uniformly loaded rectangular area

For settlement calculations of relatively rigid foundations, it is often desired to make a single reasonable estimate of the *average stress increase*, rather than the increase at a number of specific points under the foundation. An empirical solution known as the 2:1 method gives approximate solutions generally consistent with those obtained from more rigorous elasticity methods. In the 2:1 method, the load transferred from the footing to the soil is assumed to be supported by increasingly larger effective soil areas as the depth below the footing increases. The dimensions of the effective support area increase from the foundation dimensions at a ratio of 1 unit horizontal to 2 units vertical. The average pressure increase at a depth z below a uniformly loaded area of dimensions B by L loaded with a pressure intensity of q_o is estimated as

$$q_v = q_o \frac{BL}{(B+z)(L+z)} \qquad (15.5.7)$$

―――― **EXAMPLE 15.11** ――――――――――――――――――――――――――――――

For the same foundation and loading as in Example 15.10, calculate the average stress increase at a depth of 10 ft below the footing.

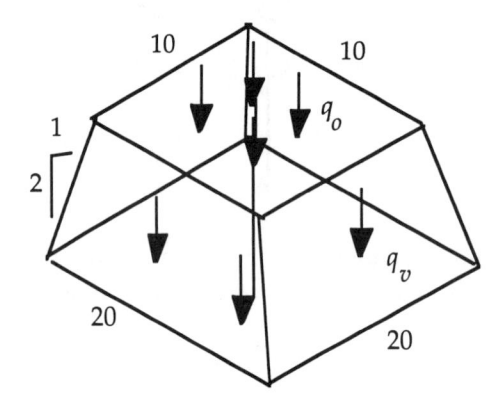

Solution.

$$q_v = q_o \frac{BL}{(B+z)(L+z)} = 1000 \frac{(10)(10)}{(10+10)(10+10)} = 250 \text{ psf}$$

15.6 Settlement of Shallow Foundations

Stress increases in the soil induced by an applied foundation pressure cause vertical strain in the soil, resulting in settlement. All settlement equations, regardless of their various forms, derive from the integral of the vertical strain with respect to depth, taken from zero to infinity, or over the layer for which the settlement is desired to be calculated:

$$S = \int \varepsilon_v dz \qquad (15.6.1)$$

Total settlement may be taken as the sum of up to the following three components:

- **Immediate settlement** is due to the immediate compression and distortion of the soil mass due to the foundation pressure. It is generally calculated using elastic theory.

- **Consolidation settlement** is time-dependent settlement due to a reduction in void ratio and water content. Its rate is dependent on the thickness and permeability of the soil. It is complete when excess pore pressures generated during loading have decreased to their initial values.

- **Secondary compression** is time-dependent settlement due to creep or breakdown of the soil fabric; it continues even after consolidation is complete and no excess pore pressures are present.

For foundations on sand, the immediate settlement usually can be taken as the total settlement. For clay foundations, consolidation settlement usually dominates, although immediate settlement and secondary compression effects may also be significant. For organic soils such as peat, secondary compression may be the dominant component.

15.6.1 Immediate Settlement Calculations for Shallow Foundations

The *immediate settlement of the center of a uniformly loaded flexible rectangular area* on an infinitely thick homogeneous elastic soil deposit can be calculated using elastic theory and expressed as

$$S_e = \frac{Bq_o}{E_s}\left(1-\mu^2\right)\alpha \qquad (15.6.2)$$

where B is the width (least dimension) of the loaded area

q_o is the pressure intensity on the loaded area

E_s is the soil modulus

μ is Poisson's ratio, and

α is an influence factor.

The influence factor α is a function of the width-to-length ratio B/L for the loaded area and is given by

$$\alpha = \frac{1}{\pi}\left[\ln\left(\frac{\sqrt{1+m^2}+m}{\sqrt{1+m^2}-m}\right)+m\ln\left(\frac{\sqrt{1+m^2}+1}{\sqrt{1+m^2}-1}\right)\right] \qquad (15.6.3)$$

where $m = B/L$.

The *immediate settlement of the corner of a loaded flexible area* can be taken as one-half the settlement under the center. Average settlements of flexible footings and settlements of rigid footings are between values for the center and the corner.

The appropriate value for the *soil modulus* E_s is difficult to estimate accurately as the stress-strain behavior is actually non-linear and stress-dependent. For sand E_s increases with depth and is often estimated from *SPT* results as

$$E_s = 8N \text{ tons/ft}^2 = 16N \text{ kips/ft}^2 = 766N \text{ kN/m}^2 \text{ or kPa} \qquad (15.6.4)$$

However, a range from one-half to double that given above might be encountered, with smaller values for silty and fine sands, and greater values for coarse and gravelly sands.

If determined from cone penetration *(CPT)* tests in sand, the soil modulus is often estimated to be

$$E_s = 2 \text{ to } 4\, q_c \qquad (15.6.5)$$

where q_c is the cone penetration resistance and E_s and q_c have the same units.

If either q_c or N is given, it is often desirable to estimate the other. The relationship between q_c and N for sands can be approximated as follows:

q_c (kPa or kN/m²) = $100N$ (silty sands) to $400N$ (sands) to $800N$ (coarse sands)

q_c (ksf) = $2N$ to $4N$ to $8N$ \qquad (15.6.6)

q_c (tsf) = N to $2N$ to $4N$

Soil modulus values for immediate settlement calculations should be based on soil properties in a zone within a depth of $2B$ below the foundation, with more weight given to those within a depth of B below the foundation.

For immediate settlement calculations in clay, the soil modulus E_s is commonly estimated as $500c$ or $500s_u$, with values as low as 100 to $250c$ for soft, normally consolidated clays and values as high as $1000c$ or even $1500c$ for stiff overconsolidated clays.

Poisson's ratio for soils typically ranges from 0.3 to 0.45. The value for an saturated clay which is incompressible immediately after loading is 0.50. As settlement equations are not especially sensitive to the value for μ, a value of 0.40 or 0.45 usually gives reasonable results.

EXAMPLE 15.12

The 10×10 ft foundation in Examples 15.10 and 15.11 is constructed 3 ft below the ground in a dry medium sand deposit for which the density is 110 lb/ft³ and the standard penetration resistance is typically 8. Estimate the settlement of the foundation.

Solution. The footing pressure given is 1000 psf. However, soil strain and resulting settlement can be assumed to be the result of the net pressure increase, $1000 - (3)(110) = 670$ lb/ft².

The soil modulus E_s can be estimated as $16N$ ksf $= (16)(8) = 128$ ksf $= 128{,}000$ psf. Poisson's ratio is assumed as 0.40.

For a square footing, $B/L = 1$. Hence, the influence factor α is

$$\alpha = \frac{1}{\pi}\left[\ln\left(\frac{\sqrt{1+1^2}+1}{\sqrt{1+1^2}-1}\right) + (1)\ln\left(\frac{\sqrt{1+1^2}+1}{\sqrt{1+1^2}-1}\right)\right] = \frac{1}{\pi}\left[\ln\left(\frac{\sqrt{2}+1}{\sqrt{2}-1}\right) + (1)\ln\left(\frac{\sqrt{2}+1}{\sqrt{2}-1}\right)\right] = 1.122$$

The estimated settlement at the center of a flexible foundation would then be

$$S_e = \frac{Bq_0}{E_s}(1-\mu^2)\alpha = \frac{(10)(670)}{128{,}000}(1-0.40^2)(1.122) = 0.0493 \text{ ft} = 0.59 \text{ inches}$$

The estimated settlement at the corner of a flexible foundation would be

0.59 / 2 = 0.3 inches

As the actual foundation is not flexible, but has some rigidity, a good estimate might be 0.4 to 0.5 inches overall.

15.6.2 Consolidation Settlement of Shallow Foundations

Consolidation settlement calculations were discussed in Section 14.12 and an example was given for a surface load over an infinite area. To predict settlements for shallow foundations on clay, the same concept is followed with two modifications:

- The footing transmits a load of finite width so stress-distribution theory must be used to predict induced stress changes.

- Soil may be removed in constructing the foundation, and the net pressure must be used rather than the total footing pressure.

This will be illustrated by an example using the same profile as Example 14.19.

EXAMPLE 15.13

A clay layer 15 ft thick underlies a 5-ft sand layer. Assume the initial void ratio of the clay is 0.900, the compression index is 0.20, the recompression index is 0.04, and the preconsolidation pressure at any depth is 1500 psf or the initial overburden stress, whichever is greater. The water table is at the top of the clay layer. A 10' by 10' footing with a pressure of 1000 lb/ft² is founded 3 ft below the ground. Calculate the settlement of the layer using the three layers shown. Take the average stresses in the layers as the stresses at the midpoints shown.

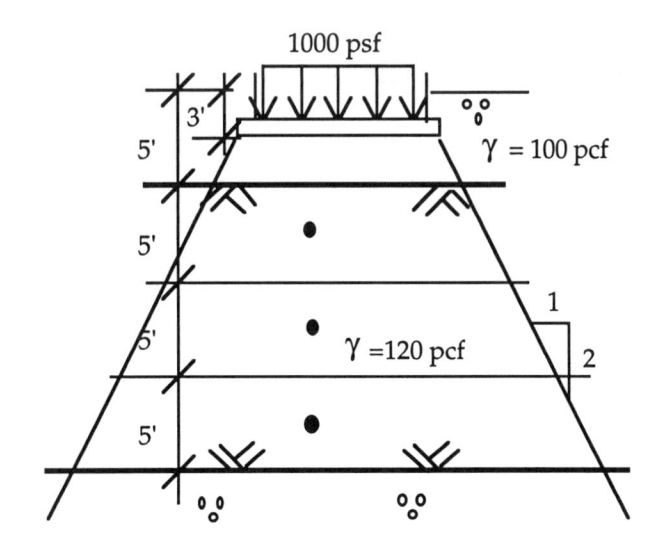

Solution. First calculate the existing vertical effective stress at the middle of the top layer:

$$\sigma'_v = \sigma_v - u = (5)(100) + (2.5)(120) - (2.5)(62.4) = 644 \text{ psf}$$

Construction of the foundation involves a reduction of (3)(100) = 300 psf due to excavation and an increase of 1000 psf due to the footing. Hence, the net increase at the base of the foundation is 1000 - 300 = 700 psf. This increase reduces with increasing depth. The midpoint of the top layer is 4.5 ft below the footing. Using the 2:1 method, the average stress increase at the midpoint of the layer is

$$\Delta\sigma = q_0 \frac{BL}{(B+z)(L+z)} = 700 \frac{(10)(10)}{(10+4.5)(10+4.5)} = 333 \text{ psf}$$

When the foundation is added and consolidation is complete, the vertical effective stress will have increased to

$$\sigma'_v + \Delta\sigma = 644 + 333 = 977$$

As the preconsolidation pressure is 1500 psf, consolidation will follow the reconsolidation slope on the e-log p curve. The midpoint strain in the top layer is

$$\varepsilon_v = \frac{\Delta e}{1+e} = \frac{1}{1+e}\left[C_r \log_{10}\left(\frac{\sigma'_{vo} + \Delta\sigma}{\sigma'_{vo}}\right)\right]$$

$$= \frac{1}{1.9}\left[0.04 \log_{10}\left(\frac{976.9}{644}\right)\right] = 0.0072$$

Multiplying the strain by the layer thickness, the settlement of the top layer is

$$S_{top} = (60 \text{ in})(0.0038) = 0.228 \text{ inches}$$

For the second layer, the initial midpoint effective stress is

$$\sigma'_{vo} = (5)(100) + (7.5)(120) - (7.5)(62.4) = 932 \text{ psf}$$

The stress increase is

$$\Delta\sigma = 700 \frac{(10)(10)}{(10+9.5)(10+9.5)} = 184 \text{ psf}$$

and the final effective stress will be 932 + 184 = 1170 psf. As the preconsolidation pressure is 1500 psf, all consolidation is along the recompression curve. The midpoint strain is

$$\varepsilon_v = \frac{1}{1.9}\left[(0.04)\log_{10}\left(\frac{1166}{932}\right)\right] = 0.0020$$

Multiplying the strain by the layer thickness, the settlement of the middle layer is

$$S_{middle} = (60 \text{ in})(0.0020) = 0.120 \text{ inches}$$

For the bottom layer, the midpoint effective stress is

$$\sigma'_v = (5)(100) + (12.5)(120) - (12.5)(62.4) = 1220 \text{ psf}$$

The stress increase is

$$\Delta\sigma = 700 \frac{(10)(10)}{(10+14.5)(10+14.5)} = 117 \text{ psf}$$

The final effective stress will be 1220 + 117 = 1337 psf, which is below the preconsolidation pressure of 1500 psf. The midpoint strain is

$$\varepsilon_v = \frac{1}{1.9}\left[0.04 \log_{10}\left(\frac{1337}{1222}\right)\right] = 0.0008$$

Multiplying the strain by the layer thickness, the settlement of the bottom layer is

$$S_{bottom} = (60 \text{ in})(0.0008) = 0.048 \text{ inches.}$$

The total settlement is

$$S_{total} = 0.228 + 0.120 + 0.048 = 0.396 \text{ in} \cong 0.4 \text{ in}$$

15.6.3 Settlement vs. Time Calculations for Shallow Foundations

Applying a stress increase to a clay soil induces excess pore pressures, which dissipate with time in accordance with the theory of consolidation, developed by Terzaghi in the 1920's. The time rate of settlement is assumed to coincide with the time rate of pore pressure dissipation averaged throughout the clay. For one-dimensional consolidation of a uniform clay layer, the average percent consolidation at time $U_{avg}(t)$ which has occurred at time t can be expressed as

$$U_{avg}(t) = f(T) \qquad (15.6.7)$$

where

$$T = \frac{c_v t}{H_{dr}^2}$$

In the above equation,

T is a non-dimensional time factor from which $U_{avg}(T)$ can be determined.

c_v is the coefficient of consolidation, with units of length²/time.

H_{dr} is the maximum distance of water drainage in the clay layer.

For a *doubly drained* layer (e.g., sand above and below the clay), it is one-half the clay layer thickness.

For a *singly drained* layer (e.g., impervious shale under the clay), it is the same as the clay layer thickness.

The relationship $U_{avg} = f(T)$ is tabulated in Table 15.3.

TABLE 15.3 Average Percent Consolidation at Time Factor T

U_{avg}, %	T	U_{avg}, %	T
0 %	0.000	70	0.403
10	0.008	75	0.477
20	0.031	80	0.567
30	0.071	85	0.684
40	0.126	90	0.848
50	0.197	95	1.129
55	0.238	99	1.781
60	0.287	100	∞
65	0.340		

EXAMPLE 15.14

The clay from Example 15.13 has a coefficient of consolidation of 0.3 ft²/day. Determine the time when 50% and 90% of the total settlement will occur.

Solution. As the 15-ft-thick clay layer is bounded by a sand deposit on top and bottom, it is doubly drained and $H = 15/2 = 7.5$. For 50% consolidation, $T = 0.197$. Hence,

$$t_{50} = \frac{T_{50} H_{dr}^2}{c_v} = \frac{(0.197)(7.5)^2}{0.3} = 36.9 \approx 37 \text{ days}$$

At 37 days, consolidation will be 50% complete. The estimated settlement will be

$$S_{50} = U_{avg} S_{total} = (0.50)(0.60 \text{ in}) = 0.30 \text{ in}$$

For 90% consolidation

$$t_{90} = \frac{T_{90} H_{dr}^2}{c_v} = \frac{(0.848)(7.5)^2}{0.3} = 159 \text{ days} \approx 5.3 \text{ months}$$

At 5.3 months, the estimated settlement will be

$$S_{90} = U_{avg} S_{total} = (0.90)(0.60 \text{ in}) = 0.54 \text{ in}$$

15.7 Deep Foundations

Deep foundations consist of long, slender, vertical structural elements which develop capacity to support loads either in shear by side resistance and/or in compression by tip resistance. Where driven into the soil, deep foundations are termed *piles;* when drilled and constructed in place, they are termed *drilled shafts* (a term preferred over the older terms of piers or caissons). Predicting the capacity of deep foundations is difficult as the interaction between the pile or shaft is complex and dependent on the installation procedure. For most structures of importance, load tests are performed. Preliminary estimates of capacity may be made using the *static methods* outlined in the next two sections.

The total ultimate capacity of a pile or shaft can be taken as the sum of the *tip resistance* (or point capacity) and the *side resistance* (or skin resistance or shaft capacity):

$$Q_{ult} = Q_{tip} + Q_{side}$$

The tip resistance, calculated from bearing capacity theory, is

$$Q_{tip} = A_{tip} [cN_c^* + qN_q^*] \tag{15.7.1}$$

where

- A_{tip} is the area of the pile tip
- c is the cohesion or undrained strength
- N_c^* is a bearing capacity factor dependent on the friction angle
- q is the overburden pressure at the pile tip
- N_q^* is a bearing capacity factor dependent on the friction angle

The bearing capacity factors are similar in concept to those for shallow foundations but different in value as they include depth effects.

The side resistance is obtained by integrating the unit side resistance (or skin friction) function $f_s(z)$ over the surface area of the pile:

$$Q_{side} = \int p(z) f_s(z) \, dz = \sum p_i f_{si} \Delta z_i \tag{15.7.2}$$

where
- $f_s(z)$ is the unit side resistance at depth z, or
- f_{si} is the unit side resistance in the ith layer
- $p(z)$ is the pile perimeter at depth z, or
- p_i is the pile perimeter in the ith layer

Often the perimeter is constant and p may be moved out of the integral or summation.

The allowable capacity Q_a is determined by dividing the ultimate capacity by a factor of safety *FS*:

$$Q_a = \frac{Q_{ult}}{FS} \tag{15.7.3}$$

For static calculations, *FS* is often taken as 3 or more. With load tests, a value of 2 to 3 is often permitted.

15.7.1 Pile Foundations in Clay

For piles in clay, the undrained ($\phi = 0$) case generally governs, $N_c^* = 9$, and $N_q^* = 1$. The term qN_q^* approximately balances the pile weight and is neglected. The tip capacity is thus taken as

$$Q_{tip} = A_{tip}[cN_c^* + qN_q^*] = 9A_{tip}c \qquad (15.7.4)$$

A number of methods have been proposed to estimate side resistance, including the α method, the λ method, and the β method. Of these, the α method is the simplest and will be described here. In the α method, the unit side resistance f_s is taken as the cohesion or undrained strength c reduced by a factor that accounts for the reduced adhesion between the soil and pile.

$$f_s = \alpha c \qquad (15.7.5)$$

α is considered a function of c and decreases with increasing c. Several functions have been proposed. Table 15.4 below provides typical α values.

TABLE 15.4 Values of α used in the α-method

c, kips/ft²	α
1	0.75 - 1.00
2	0.40 - 0.70
3	0.30 - 0.45
4	0.25 - 0.35
≥ 5	0.20 - 0.30

──────── **EXAMPLE 15.15** ────────

A 12 in-dia concrete pile is driven 30 ft into a soft clay deposit for which $c = 1000$ psf. Estimate the ultimate capacity and recommend an allowable capacity.

Solution. The area of the pile tip is

$$A_{tip} = \frac{\pi(1)^2}{4} = 0.785 \text{ ft}^2$$

The ultimate tip capacity is

$$Q_{tip} = 9A_{tip}c = (9)(0.785)(1000) = 7070 \text{ lb}$$

The pile perimeter is

$$p = (1)(\pi) = 3.14 \text{ ft}$$

Taking α as 0.85, the unit side resistance is

$$f_s = c = (0.85)(1000) = 850 \text{ psf}$$

The ultimate side capacity is

$$Q_{side} = pf_{si}z_i = (3.14)(850)(30) = 80{,}000 \text{ lb}$$

The total ultimate pile capacity is

$$Q_{ult} = Q_{tip} + Q_{side} = 7070 + 80{,}000 = 87{,}070 \text{ lb} \cong 87 \text{ kips}$$

Using a factor of safety of 3, the allowable load could be recommended as

$$Q_a = 87/3 = 29 \text{ kips}$$

15.7.2 Deep Foundations in Sand

A detailed treatment of the prediction of tip and side capacities for deep foundations in sand is beyond the scope of this review. Attempts to correlate the results of load tests with static equations have led to a number of approaches. Although various methods share a similar form of the equations, recommendations for estimating parameter values for the methods are in some cases complex and vary widely. The engineer using any method (e.g., Meyerhof's, Janbu's, Vesic's, or Coyle and Castello's) must be thoroughly familiar with the assumptions.

In general, the tip capacity is expressed in the form

$$Q_{tip} = A_{tip}\, q\, N_q \tag{15.7.6}$$

where q is the effective vertical stress at the pile tip and

N_q^* is a bearing capacity factor.

Because stress transfer at the tip of a pile is complex, estimates of N_q^* vary widely. It has been taken to be a function of the length to width ratio of the pile as well as a function of ϕ. Some methods use the actual effective stress at the pile tip for q and adjust N_q^*, while others limit q to the vertical effective stress at some depth such as 15 pile diameters, even though the pile tip is deeper.

In general, the unit side resistance function is taken to be

$$f_s(z) = K\, q'(z)\, \tan \delta(z) \tag{15.7.7}$$

where K is a coefficient related to the method of installation

$q'(z)$ is the effective vertical stress at depth z

$\delta(z)$ is the angle of friction between the pile and the soil at depth z.

The parameter K is typically in the range 0.5 to 1.25 for non-displacement piles, and 1.0 to 2.0 for displacement piles; however, some methods take K to be a function of pile length to width ratio.

Some methods limit q'_z to a maximum of its computed value at some *critical depth*, often about 15 pile diameters.

The friction angle δ is often taken as $2/3\, \phi$ to $0.9\, \phi$.

15.8 Lateral Earth Pressure

Estimates of the lateral pressure exerted by soil are required for design of retaining walls and braced excavations. At a point in a soil mass with a known vertical stress, the horizontal stress or pressure can assume a wide range of values and hence cannot be calculated from equations of statics alone. The horizontal pressure depends on the horizontal strain conditions in the soil. It is a minimum value when a soil mass is allowed to expand laterally and a maximum value when the soil is compressed laterally (such as when it provides a reaction for external forces).

The relationship between vertical and horizontal stress can be defined by the *lateral earth pressure coefficient K*, where

$$K = \frac{\sigma_h}{\sigma_v} \quad \text{or} \quad K = \frac{\sigma_h'}{\sigma_v'} \tag{15.8.1}$$

The first definition is in terms of total stresses; the second definition is in terms of effective stresses. For cohesionless sands, the effective stress definition should always be used. For undrained loading of clays where $\phi = 0$, strength is expressed in terms of total stresses and the total stress K-value will be 1.0 for all cases. Problems where both c *and* ϕ are non-zero are not common in practice if strength is represented correctly; however, they persist in textbooks and exams.

Under conditions of zero horizontal strain, the soil is said to be in an *at-rest* earth pressure state. The *at-rest earth pressure* is denoted K_o. Although it has an indeterminate value, a common empirical estimation is

$$K_o = \frac{\sigma_h'}{\sigma_v'} = 1 - \sin\phi' \tag{15.8.2}$$

At-rest earth pressure conditions may prevail against very rigid walls and other structures restrained against any movement, such as basement walls restrained at the top, and the sides of box culverts.

A small rotation of a retaining wall away from its backfill will mobilize the shear resistance in the backfill and reduce the earth pressure to its minimum value, the *active earth pressure* condition. Retaining walls are usually designed for active earth pressure conditions. For the special but common case of a frictionless vertical wall backfilled with a cohesionless material with a horizontal backfill surface, the *active earth pressure coefficient* K_a is

$$K_a = \frac{\sigma_h'}{\sigma_v'} = \frac{1-\sin\phi'}{1+\sin\phi'} = \tan^2\left(45° - \frac{\phi'}{2}\right) \tag{15.8.3}$$

Knowing the active earth pressure at any point along a face of a wall, the total earth force on a wall can be obtained by integrating the pressure with respect to depth, or calculating the area of an earth pressure vs. depth diagram.

──────── **EXAMPLE 15.16** ────────

The retaining wall shown is 8 feet high and retains a clean sand for which $\gamma = 120 \text{ lb/ft}^3$ and $\phi = 32°$. Calculate the earth pressure at the base of the wall, the total earth force on the wall, and the overturning moment due to the active earth pressure.

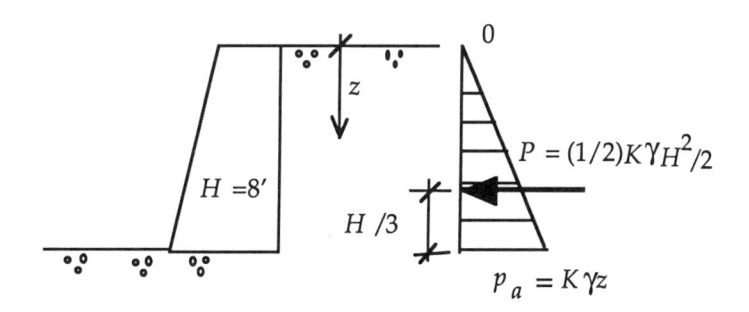

Solution. The earth pressure diagram is shown to the right of the wall. The active earth pressure coefficient is

$$K_a = \frac{1-\sin 32°}{1+\sin 32°} = 0.307$$

The active earth pressure at the base of the 8-ft wall is

$$p_a = K_a z = (0.307)(120)(8) = 295 \text{ psf}$$

The total active earth force is the integral of the earth pressure from $z = 0$ to $8'$:

$$P_a = \frac{1}{2}\gamma K_a H^2 = (0.5)(0.307)(120)(8)^2 = 1180 \text{ lb/ft of wall}$$

The active force acts at the centroid of the earth pressure diagram, in this case, $H/3$. Thus the overturning moment is

$$M_o = P_a \bar{y} = (1179)(8/3) = 3140 \text{ ft-lb/ft of wall}$$

──────── **EXAMPLE 15.17** ────────

Recalculate the pressure, force, and overturning moment on the wall in Example 15.16 assuming that poor drainage behind the wall causes the water table to rise to a height of 4 ft behind the wall.

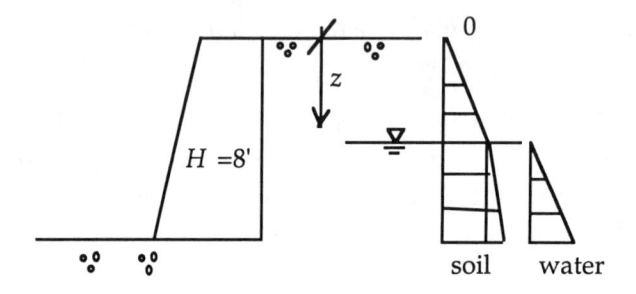

Solution. The active earth pressure at a depth of 4 ft is

$$p_a = K_a p'_v = 0.307[4(120)] = 147 \text{ psf}$$

The active earth pressure at the base of the 8 ft wall is

$$p_a = K_a p'_v = 0.307[4(120) + 4(120 - 62.4)] = 218 \text{ psf}$$

The active earth force is the area of the earth pressure diagram:

$$\begin{aligned} P_a &= (0.5)(4)(147) + (4)(147) + (0.5)(4)(218-147) \\ &= 1024 \text{ lb/ft} \end{aligned}$$

However, there is also a water force of

$$\begin{aligned} P_w &= (0.5)(62.4)(4)^2 \\ &= 499 \text{ lb/ft} \end{aligned}$$

The total lateral force on the wall due to soil and water is

$$P_H = 1024 + 499 = 1523 \text{ lb/ft}$$

The overturning moment due to soil and water is

$$\begin{aligned} M_o &= \Sigma P\bar{y} = 294(4 + 1.33) + 588(2) + 142(1.33) + 499(1.33) \\ &= 1568 + 1176 + 189 + 665 \\ &= 3600 \text{ ft-lb / ft of wall} \end{aligned}$$

A more general solution is provided by *Coulomb's active earth pressure coefficient*. Coulomb solved for the total earth force on a wall by analyzing the statics of an earth wedge behind the wall. As shown in Figure 15.5, the back face of the wall may be non-vertical, the backfill may be inclined, and wall friction may be present.

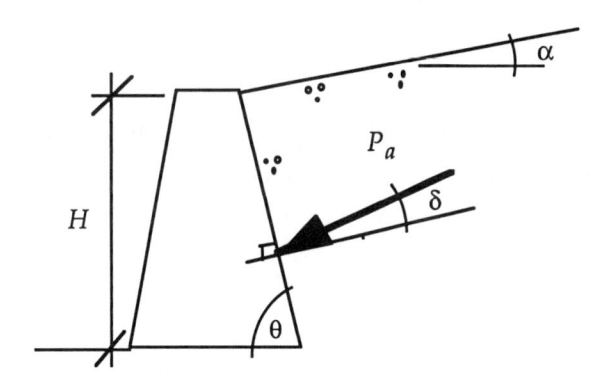

Figure 15.5 Terms used in Coulomb's equation.

For the general case shown, the active earth force per unit length of wall is

$$P_a = \frac{1}{2} K_a \gamma H^2 \tag{15.8.4}$$

where

$$K_a = \frac{\sin^2(\theta + \phi)}{\sin^2\theta \cdot \sin(\theta - \delta)\left[1 + \sqrt{\frac{\sin(\phi + \delta) \cdot \sin(\phi - \alpha)}{\sin(\theta - \delta) \cdot \sin(\alpha + \theta)}}\right]^2}$$

In the above expression,

- θ is the inclination of the back of the wall (90° for vertical)
- ϕ is the friction angle of the backfill (use ϕ' for effective stress analysis or drained conditions)
- δ is the friction angle between the wall surface and backfill
- α is the inclination of the backfill

This equation will reduce to the previous simpler equation when $\theta = 90°$, $\delta = 0$, and $\alpha = 0$.

Where cohesion is present in the backfill, the active earth pressure at a depth z can be expressed as

$$p_h(z) = K_a \gamma z - 2c\sqrt{K_a} \tag{15.8.5}$$

───── **EXAMPLE 15.18** ─────

A cut is made in a saturated clay with an undrained cohesion of c = 800 psf. The clay has a total density of 120 lb/ft³ and $\phi = 0$ for the undrained condition. Find the lateral earth pressure at a depth of 5 ft.

Solution. As $\phi = 0$ for undrained (short term loading) conditions,

$$K_a = \frac{1 - \sin 0°}{1 + \sin 0°}$$
$$= 1$$

Then

$$p_a(z) = K_a \gamma z - 2c\sqrt{K_a}$$
$$= (1)(120)(5) - (2)(800)(1)$$
$$= 600 - 1600$$
$$= -1000 \text{ psf}$$

Note that the earth pressure is negative at this depth. Due to the cohesion of the clay, lateral earth pressures are negative and tend to hold the soil mass together. The soil at this depth will support a vertical cut without a wall so long as the loading remains undrained (so long as negative pore pressures do not dissipate).

───── **EXAMPLE 15.19** ─────

Given the soil conditions in Example 15.18 above, at what depth do the earth pressures become positive?

Solution. Setting $p_h(z) = 0$ in Equation 15.8.5 and solving for z, there results

$$z = \frac{2c}{\gamma\sqrt{K}} = \frac{(2)(800)}{(120)(1)} = 13.3 \text{ ft}$$

If an external force is applied to a soil mass that causes lateral compression, the earth pressure will increase until it reaches the *passive earth pressure condition*. Passive earth pressure is the maximum lateral earth pressure that can occur, and further attempts to increase the pressure result in shear failure. Passive earth pressure is sometimes assumed on the resisting side of retaining walls when calculating stability for design. For a horizontal backfill with zero wall friction and a vertical wall, the passive earth pressure coefficient K_p is

$$K_p = \frac{1+\sin\phi}{1-\sin\phi} = \tan^2\left(45 + \frac{\phi}{2}\right) \tag{15.8.6}$$

Other solutions are available for inclined wall faces, an inclined ground surface, and wall friction. However, the equation above is generally conservative and often used in design.

Where cohesion is present, the passive earth pressure at a depth z is:

$$p_p(z) = K_p \gamma z + 2c\sqrt{K_p} \tag{15.8.7}$$

Where cohesion has the effect of reducing active earth pressure, it has the effect of increasing passive earth pressure.

External forces may be present on the surface of the backfill behind a wall, forming a *surcharge*. Where there is a *uniform surcharge q* of infinite extent, it has the same effect as additional soil providing an equivalent vertical stress increase. The active earth pressure at a depth z below the surface of the ground when loaded with a surcharge q is:

$$p_a(z) = K_a(\gamma z + q) - 2c\sqrt{K_a} \tag{15.8.8}$$

When the surcharge is in the form of a line load parallel to the wall as shown in Figure 15.6 (e.g., a loaded rail), the additional earth pressure Δq, due to line load q is given by

$$\Delta q = \frac{4q}{\pi H} \times \frac{a^2 b}{\left(a^2 + b^2\right)^2} \quad \text{for } a > 0.4$$

$$\Delta q = \frac{q}{H} \times \frac{0.203\, b}{\left(0.16 + b^2\right)^2} \quad \text{for } a \leq 0.4$$

(15.8.9)

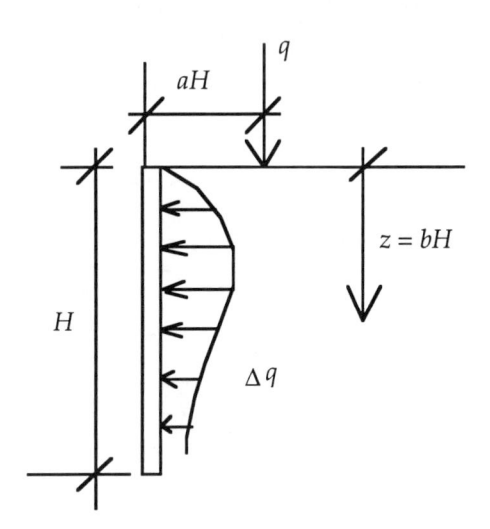

Figure 15.6 A retaining wall with a line load surcharge.

EXAMPLE 15.20

A line load of 2000 lb/ft acts along a line 5 ft behind a 12 ft high wall. Calculate the earth pressure increase at the mid-height of the wall.

Solution. Referring to Figure 15.6, we have

$$a = 5/12 = 0.4167, \qquad b = 6/12 = 0.5$$

The pressure increase at $z = 6$ ft is then

$$\Delta q = \frac{4q}{\pi H} \times \frac{a^2 b}{\left(a^2 + b^2\right)^2} = \frac{(4)(2000)}{(\pi)(12)} \times \frac{(0.4167)^2 (0.5)}{\left(0.4167^2 + 0.5^2\right)^2} = 103 \text{ psf}$$

15.9 Retaining Walls

Using lateral earth pressure theory, retaining walls are designed by trial and error adjustment of the wall dimensions until a design is found that is sufficiently stable against sliding and overturning and is also economical. The base must also be sufficiently large such that the wall is safe in bearing and does not settle excessively. The design checks for retaining walls are best illustrated by an example.

---- **EXAMPLE 15.21** ----

Check the stability of the trial wall design shown below against sliding and overturning. Also determine its base pressure distribution. Assume that $\gamma = 125$ lb/ft³ and $\phi = 30°$. Assume that the base is rough and the friction angle between concrete and sand can be taken as 30°.

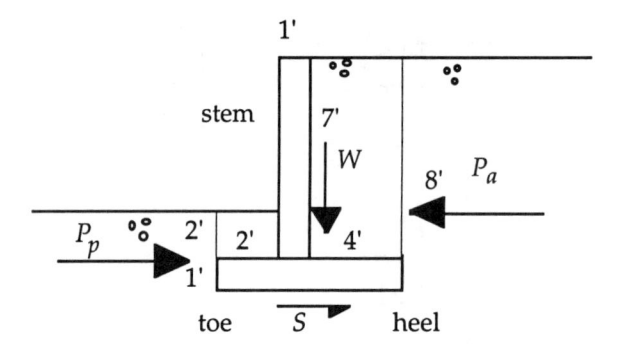

Solution. For $\phi = 30°$, we find $K_a = 0.333$ and $K_p = 3.00$ from Equations 15.8.3 and 15.8.6.

A free body consisting of the wall base, wall stem, and the soil over the toe and heel is analyzed. The active earth force on the driving (right) side of the free body is

$$P_a = \frac{1}{2}K_a H^2 = (0.5)(0.333)(125)(8)^2 = 1333 \text{ lb/ft}$$

The factor of safety against sliding is calculated by comparing P_a to the maximum possible values of S and P_p that can be devploped by the soil, not the values that achieve equilibrium.

The passive earth force is

$$P_p = \frac{1}{2}K_p H^2 = (0.5)(3.0)(125)(3)^2 = 1688 \text{ lb/ft}$$

The maximum base shear force S is the weight of the free body times the tangent of the friction angle:

$$S = W \tan \phi$$

The density of concrete is commonly taken to be 150 lb/ft³. The weight of the free body is

$$\begin{aligned} W &= W_{\text{soil toe}} + W_{\text{stem}} + W_{\text{base}} + W_{\text{soil heel}} \\ &= (2)(2)(125) + (7)(1)(150) + (7)(1)(150) + (7)(4)(125) \\ &= 6100 \text{ lb/ft} \end{aligned}$$

The factor of safety against sliding is defined as the ratio of resisting forces to driving forces:

$$FS = \frac{W \tan \phi + P_p}{P_a} = \frac{(6100)\tan 30 + 1688}{1333} = 3.91$$

Sometimes the passive earth force is coservatively neglected due to the large strains required to develop it and the possibility of soil removal on the passive side. Neglecting the passive earth force, the factor of safety against sliding is

$$FS = \frac{W \tan \phi}{P_a} = \frac{(6100) \tan 30}{1333} = 2.64$$

As a minimum factor of safety of 1.5 to 2.0 is usually sought, the trial design is safe in sliding.

The factor of safety in overturning about the toe is defined as

$$FS = \frac{M_r}{M_o} = \frac{\Sigma W \bar{x} + P_p \bar{y}}{P_a \bar{y}}$$

where M_r is the total resisting moment due to weight and passive earth force, and M_o is the overturning moment due to the active earth force.

The resisting moment is

$$M_r = (500)(1) + (1050)(2.5) + (1050)(3.5) + (3500)(5) + (1688)(1)$$

$$= 26{,}000 \text{ ft-lb/ft}$$

The overturning moment is

$$M_o = (1333)(8/3) = 3555 \text{ ft-lb/ft}$$

The factor of safety is

$$FS = \frac{M_r}{M_o} = \frac{26{,}000}{3555} = 7.31$$

Neglecting the moment due to the passive earth force, the factor of safety is

$$FS = 24{,}300 / 3555 = 6.84$$

As the factor of safety should be above about 2.0, the design is safe.

Additional economy could be effected by reducing the base length and re-analyzing to see if adequate safety factors can still be obtained.

15.10 Braced Excavations

For temporary construction of trenches, for facilities such as sewers, support may be provided by sheeting supported by *struts* braced across the trench. Bracing of such an excavation restrains movement of the top and a triangular pressure diagram does not develop. Empirical earth pressure diagrams used for braced excavations are based on measurements on a number of excavations by Peck, Tschebatarioff, and others. Peck's diagrams are shown in Figure 15.7.

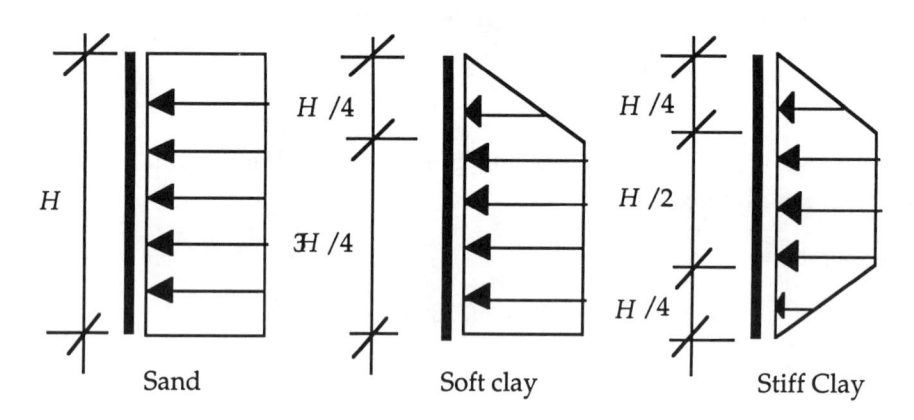

Figure 15.7 Pressure diagram for Braced Excavations.

For sand, the design earth pressure is taken to be uniform with depth, with a value of

$$p = 0.65 \gamma H K_a \tag{15.10.1}$$

Soft clay is defined in a relative sense, in terms of how high the excavation is relative to strength. Soft clay meets the condition

$$\frac{\gamma H}{c} > 4 \tag{15.10.2}$$

For soft clay, the pressure is taken constant in the lower three-fourths of the height as the larger of

$$p = \gamma H \left[1 - \frac{4c}{\gamma H}\right] \quad \text{or} \quad p = 0.3 \gamma H \tag{15.10.3}$$

Stiff clay is that for which

$$\frac{\gamma H}{c} \leq 4 \tag{15.10.4}$$

For stiff clay, the pressure is taken uniform in the middle one-half of the wall height at a value of

$$p = 0.2 \gamma H \quad \text{to} \quad 0.4 \gamma H \tag{15.10.5}$$

---- **EXAMPLE 15.22** ----

A braced excavation in clay is 16 ft deep. The clay has a strength of 400 psf and has a unit weight of 120 lb/ft³. Determine the design pressure diagram and calculate the total earth force.

Solution. First calculate

$$\frac{\gamma H}{c} = \frac{(120)(16)}{400} = 4.8$$

As this term is greater than 4.0, the clay is a soft clay. The maximum earth pressure is the larger of

$$p = \gamma H\left[1 - \frac{4c}{\gamma H}\right] = (120)(16)\left[1 - \frac{4(400)}{120(16)}\right] = 320 \text{ psf}$$

or

$$p = 0.3\gamma H = 0.3(120)(8) = 288 \text{ psf}$$

The design earth pressure is as shown:

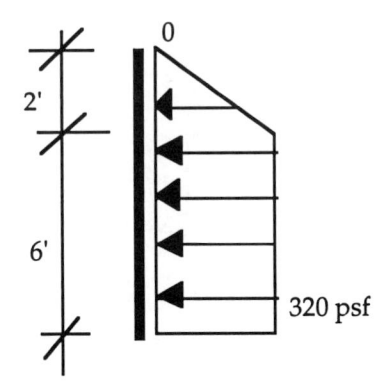

The total earth force is the area of the pressure diagram:

$$\begin{aligned} P &= (1/2)(2)(320) + (6)(320) \\ &= 2240 \text{ lb/ft} \end{aligned}$$

Practice Problems (Multiple Choice)

15.1 A standard penetration test is conducted 10 feet below the surface of a sand for which $\gamma = 120$ lb/ft^3. The water table is at a depth of 15 ft. The number of hammer blows required to drive the sample for each of three six-inch increments is 4, 6, and 8, respectively. The uncorrected standard penetration resistance, N, is:
   a) 18
   b) 14
   c) 10
   d) 12
   e) 9

15.2 For Problem 15.1, the overburden stress correction factor for the N value is:
   a) 0.029
   b) 0.913
   c) 1.29
   d) 1.66
   e) 2.58

15.3 A clay has an unconfined compressive strength of 2000 lb/ft^2. A good estimate of the net ultimate bearing capacity is:
   a) 5140 lb/ft^2
   b) 10,280 lb/ft^2
   c) 2000 lb/ft^2
   d) 1000 lb/ft^2
   e) 6000 lb/ft^2

15.4 A 3 × 3 ft square footing sets on the surface of a sand deposit for which $\phi = 32°$ and $\gamma = 125$ lb/ft^3. Predict the ultimate force that will cause this footing to fail in bearing. The water table is very deep.
   a) 3400 lb
   b) 5660 lb
   c) 30,600 lb
   d) 51,000 lb
   e) 15,300 lb

15.5 A 1000-lb point load acts on the surface of the ground. Determine the stress increase 5 ft directly below the load.
   a) 19 lb/ft^2
   b) 1000 lb/ft^2
   c) 200 lb/ft^2
   d) 400 lb/ft^2
   e) 48 lb/ft^2

15.6 A 10,000 lb load acts uniformly over a 4′ × 4′ footing. Estimate the average pressure distribution on a plane 5 ft below the footing base.
   a) 625 lb/ft²
   b) 123 lb/ft²
   c) 7.7 lb/ft²
   d) 1980 lb/ft²
   e) 16,000 lb/ft²

15.7 The footing in problem 15.6 is founded on a sand with a soil modulus of 200 kips/ft² and a Poisson's ratio of 0.40. Estimate the settlement of the footing.
   a) 0.012 inches
   b) 0.14 inches
   c) 2.26 inches
   d) 1.40 inches
   e) 0.60 inches

15.8 A clay layer 20 ft thick has sand above it and an impervious shale below it. The coefficient of consolidation is 0.4 ft²/day. A uniform stress increase is applied to the soil. Calculate the time when 50% of the settlement will have occurred.
   a) 49 days
   b) 125 days
   c) 1000 days
   d) 197 days
   e) 394 days

15.9 The tip of a 14-inch square pile is founded in a clay with an unconfined compressive strength of 2000 psf. Estimate the ultimate tip capacity.
   a) 18.0 kips
   b) 9.00 kips
   c) 12.3 kips
   d) 24.5 kips
   e) 28.0 kips

15.10 A clay deposit has a unit weight of 100 lb/ft³ and an unconfined compressive strength of 2000 lb/ft². Determine the depth at which the horizontal soil pressure is zero.
   a) 20 ft
   b) 40 ft
   c) 0 ft
   d) 80 ft
   e) 10 ft

15.11 Calculate the total active earth force behind a ten-foot-high retaining wall backfilled with sand with a unit weight of 125 lb/ft³ and a friction angle of 30°.
   a) 208 lb/ft
   b) 2080 lb/ft
   c) 417 lb/ft
   d) 4170 lb/ft
   e) 1250 lb/ft

Practice Problems (Essay)

15.12. A 4 × 4 ft footing is founded 3 ft below the surface in a clean sand deposit for which $\phi = 32°$ and $\gamma = 125$ lb/ft^3. The water table is located at the base of the footing. Determine the ultimate bearing pressure and recommend an allowable design value.

15.13 A 3-ft-wide strip footing is founded 3 ft below the surface in a clean sand deposit for which $\phi = 32°$ and $\gamma = 125$ lb/ft^3. The water table is at the ground surface. Determine the ultimate bearing pressure and recommend an allowable design value.

15.14 A 3-ft-wide strip footing is founded 3 ft below the surface in a clay deposit for which $c = 1500$ lb/ft^2 and $\gamma = 125$ lb/ft^3. The water table is very deep. Determine the ultimate bearing pressure and recommend an allowable design value.

15.15 A square footing of dimension $B \times B$ is founded on the surface of a sand deposit. A second footing of dimension $2B \times 2B$ is founded at the surface on the same sand. How many times more force can be applied to the second footing before a bearing failure occurs?

15.16 A 4 × 4 footing is founded in a sand for which $N = 15$. Estimate the load that can be applied to this footing if settlement is to be kept to 1 inch. Show your work.

15.17 Rework Example 15.13 for a 4 × 4 ft footing loaded with a net pressure of 2 kips per square foot.

15.18 Plot a settlement vs. time curve for Problem 15.17. Take $c_v = 0.2$ ft^2/day.

15.19 A 16 inch diameter pile is driven 40 ft into a clay layer for which $c = 3000$ lb/ft^2. The density of the clay is 120 lb/ft^3. Estimate the ultimate capacity and recommend an allowable working load.

15.20 For the retaining wall shown, recommend a base width B to provide a sliding safety factor of 2.0 (or slightly above).

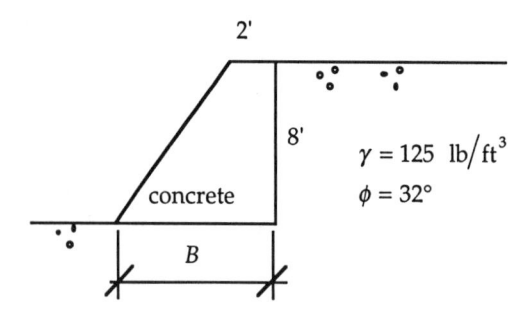

15.21 A braced excavation is cut 12 feet deep in a hard clay with a density of 125 lb/ft^3 and a cohesion of 3000 lb/ft^2. Draw a design earth pressure diagram and determine the total earth force to be supported per foot of wall.

Practice Problems (PE-Format)

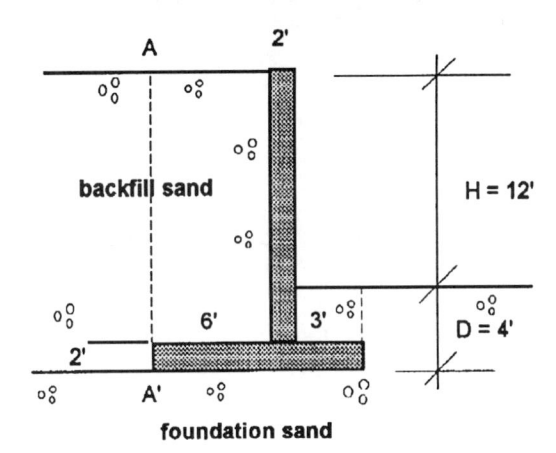

A cross-section of a concrete cantilever retaining wall is shown. The foundation sand, located below the base of the wall, in front of the toe, and over the toe, has a unit weight of 120 lb/ft^3 and a friction angle of $f = 30$ degrees. The backfill sand, placed behind the wall and over the heel, has a unit weight of 125 lb/ft^3 and a friction angle of $f = 34$ degrees. The water table is a considerable distance below the base of the wall and the backfill is well drained. Both sands are cohesionless. The base of the wall is 2 ft thick and 11 ft long, consisting of a 6 ft heel, a 3 ft toe, and 2 ft under the stem of the wall.

15.22 Calculate the horizontal active earth pressure acting at point A' on the vertical plane (AA¢) through the heel of the wall. Assume that the vertical shear on this plane is zero.
   a) 600 lb/ft^2
   b) 542.8 lb/ft^2
   c) 640 lb/ft^2
   d) 2000 lb/ft^2
   e) 565.4 lb/ft^2

15.23 Calculate the horizontal active (driving) earth force acting on plane AA¢ per foot of wall.
   a) 565.4 lb/ft
   b) 4523 lb/ft
   c) 4342 lb/ft
   d) 5119 lb/ft
   e) 2000 lb/ft

15.24 Assuming that full passive pressure is developed in the backfill, calculate the passive earth pressure at point B¢. Assume that the vertical shear on plane BB¢ is zero.
   a) 1698 lb/ft^2
   b) 480 lb/ft^2
   c) 1440 lb/ft^2
   d) 2000 lb/ft^2
   e) 2880 lb/ft^2

15.25 Calculate the horizontal passive (resisting) earth force acting on plane BB¢ per foot of wall.
   a) 3396 lb/ft
   b) 849 lb/ft
   c) 720 lb/ft
   d) 2880 lb/ft
   e) 2000 lb/ft

15.26 Determine the maximum resisting shear force (per foot of wall) that can be developed along the base.
   a) 12,627 lb/ft
   b) 2,880 lb/ft
   c) 4,523 lb/ft
   d) 1643 lb/ft
   e) 10,808 lb/ft

15.27 Determine the factor of safety against sliding. Assume that the full passive earth force can be mobilized.
   a) 2.71
   b) 2.39
   c) 1.00
   d) 3.03
   e) 2.00

15.28 Determine the factor of safety against overturning. Consider full passive earth force.
   a) 3.27
   b) 2.00
   c) 3.10
   d) 1.00
   e) 2.46

15.29. Assume that the full passive earth force is developed and that only a portion of the maximum base shear force is developed, just sufficient to prevent sliding. By summing moments about the toe, calculate the horizontal distance (xbar) from the toe to the line of action of the normal force acting upward on the base.
   a) 5.50 ft
   b) 4.01 ft
   c) 2.92 ft
   d) 3.67 ft
   e) 8.08 ft

Solutions to Problems (Multiple Choice)

15.1 b) The 4 blows for the first six-inch increment are discarded and the blows for the second and third increment are added.

$$N = 6 + 8 = 14$$

15.2 c) At a depth of 10 feet, the total vertical stress is

$$\sigma_v = 10(120) = 1200 \text{ psf}$$

As the water table is at a depth of 15 ft, the pore pressure at a depth of 10 ft is zero, and the effective stress σ_v' is also equal to 1200 psf. To calculate the correction using Equation 15.1.1, the effective overburden stress must be converted to tons per square foot (1200 psf = 0.6 tsf). Then

$$C = \sqrt{\frac{1}{\sigma_v'}} = \sqrt{\frac{1}{0.6}} = 1.29$$

15.3 a) From Equation 14.15.1, the cohesion is one-half the undrained strength; hence $c = 1000$ psf. From Equation 15.4.7, the net ultimate bearing capacity is

$$q_{ult} = 1000(5.14) = 5140 \text{ psf}$$

15.4 c) As sand is cohesionless, the first term of the bearing capacity equation will be zero. From Table 15.1, $N_q = 23.2$ and $N_\gamma = 30.2$. From Equation 15.4.9, the shape factors are

$$F_{qs} = 1 + (3/3)\tan 30° = 1.577$$

$$F_{\gamma s} = 1 - 0.4(3/3) = 0.6$$

From Equation 15.4.8, the ultimate bearing capacity is

$$q_{ult} = 0 + 125(0)(1.577) + 0.5(3)(125)(30.2)(0.6) = 3400 \text{ psf}$$

Multiplying the ultimate bearing capacity by the area of the footing, the ultimate force to cause failure in bearing is

$$Q_{ult} = 3400\ (3^2) = 30,600 \text{ lb}$$

15.5 a) Vertical stress increase due to a point load is calculated using Equation 15.5.1 or 15.5.2. For a point 5 ft directly below the load, $R = z = 5$ ft. Then

$$q_v = 3\ (1000) / [\ 2\pi\ (5^2)\ ] = 19 \text{ psf}$$

15.6 b) The applied pressure at the base of the footing is

$$q_o = p / BL = 10,000 / 4(4) = 625 \text{ psf}$$

The average pressure on a plane 5 ft deep is calculated using Equation 15.5.7

$$q_v = 625(4)(4) / [(4 + 5)(4 + 5)] = 123 \text{ psf}$$

15.7 b) The settlement is calculated using Equation 15.6.2. First the influence factor α is calculated from Equation 15.6.3; from Example 15.12, $\alpha = 1.122$ for the center of square foundations. Then, from Equation 15.6.2,

$$S_e = [4(625) / 200{,}000] (1 - 0.4^2)(1.122) = 0.0118 \text{ ft} = 0.14 \text{ in}$$

15.8 d) Refer to equations 15.8.1. As the layer is singly drained, $H = 20$. Fifty percent of the settlement will have occurred when consolidation is 50 percent complete. From Table 15.2, this occurs when $T = 0.197$. Then

$$t_{50} = T_{50} H^2 / c_v = 0.197(20^2) / 0.4 = 197 \text{ days}$$

15.9 c) The undrained strength or cohesion c is one half the unconfined compressive strength, or 1000 psf. The cross sectional area of the pile tip is

$$A_{tip} = (14/12)^2 = 1.36 \text{ ft}^2$$

From Equation 15.7.4, the tip capacity is

$$Q_{tip} = 9 A_{tip} c = 9(1.36)(1000) = 12{,}240 \text{ lb} = 12.2 \text{ kips}$$

15.10 a) The cohesion is one-half the unconfined compressive strength, or 1000 psf. For short-term loading of clays, undrained conditions apply and $\phi = 0$; hence $K = 1$. From Example 15.19, the depth to zero horizontal pressure is

$$z = 2c / \gamma K^{0.5} = 2(1000) / (100)(1) = 20 \text{ ft}$$

15.11 b) Using equation 15.8.3, for $\phi = 30°$, $K_a = 0.333$. From Example 15.16, the total active earth force is

$$P_a = \tfrac{1}{2} K_a \gamma H^2 = 0.5(0.333)(125)(10^2) = 2080 \text{ lb / ft}$$

Solutions to Problems (Essay)

15.12 As the foundation is clean sand, $c = 0$ and the first term of the bearing capacity equations will drop out. The bearing capacity factors are obtained From Table 15.1; for $\phi = 32°$, $N_q = 23.2$, $N_\gamma = 30.2$. The shape factors are obtained from Equations 15.4.9:

$$F_{qs} = 1 + (4/4) \tan 32° = 1.62$$

$$F_{\gamma s} = 1 - 0.4 (4/4) = 0.6$$

As the water table is at the base of the footing, the total unit weight is used in the second term and the bouyant unit weight is used in the third term of Equation 15.4.8:

$$q_{ult} = 0 + (125)(3)(23.2)(1.62) + (0.5)(4)(125 - 62.4)(30.2)(0.6)$$

$$= 16,400 \text{ psf}$$

Using a factor of safety of 3.0, the allowable bearing pressure would be

$$16,400 / 3 = 5466 \text{ psf} \approx 5500 \text{ psf}$$

15.13 The bearing factors are the same as in the previous problem. For a continuous or strip footing, the shape factors are 1.0. From Equation 15.4.8, the ultimate bearing pressure is

$$q_{ult} = 0 + (125 - 62.4)(3)(23.2) + (0.5)(3)(125 - 62.4)(30.2)$$

$$= 7200 \text{ psf}$$

Using a factor of safety of 3.0, the allowable bearing pressure would be

$$7200 / 3 = 2400 \text{ psf}$$

15.14 For foundation design in clay, the undrained ($\phi = 0$) conditons will govern. For $\phi = 0$, $N_c = 5.14$ and $N_q = 1$. For a continuous or strip footing, the shape factors are 1.0. From Equation 15.4.8, the ultimate bearing pressure is

$$q_{ult} = (1500)(5.14)(1.) + (125)(3)(1) = 8100 \text{ psf}$$

Taking the factor of safety as 3.0, the allowable bearing pressure is

$$q_a = 8100 / 3 = 2700 \text{ psf}$$

15.15 From Equation 15.4.8, the ultimate bearing capacity of a footing on the surface of sand is

$$q_{ult} = 0 + 0 + (.5) B \gamma N_\gamma (0.6)$$

For a second footing twice as wide, the B above would be replace by $2B$; thus the ultimate bearing pressure doubles when the footing width doubles. However, the second footing has four times the base area of the first footing. As it can carry twice as much force per unit area (pressure) over four times the area, the required force to fail the second footing is eight times the force required to fail the first footing.

15.16 Equation 15.6.2 predicts settlement of flexible footings in sand. From Section 15.6, the modulus can be estimated as $16N$ kips/ft², or

$(16)(15) = 240$ kips/ft²

Poisson's ratio μ can be taken as 0.4. From Example 15.12, $\alpha = 1.122$ for the center of a square footing. Setting S_e to 1 inch (0.0833 ft), and rearranging Equation 15.6.2, the bearing pressure for 0.0833 ft settlement is

$$q_o = (0.0833 \text{ ft})(240 \text{ kips/ft}^2) / [(4 \text{ ft})(1 - 0.4^2)(1.122)]$$

$$= 5.30 \text{ kips/ft}^2 = 5300 \text{ psf}$$

For settlement under the corner of a flexible footing, α would be half the above value, and the footing could be loaded to $(2)(5300) = 10{,}600$ psf. As the settlement of a rigid footing is between that at the center and corner of a flexible footing, a bearing pressure between 5300 and 10,600 psf would cause about 1 inch of settlement. Conservatively using 6000 psf, the load that could be applied to a 4 ft × 4 ft footing is

$$Q = 6000(4)(4) = 96{,}000 \text{ lb} = 96 \text{ kips}$$

The above answer could vary considerably depending on the selection of E_s and the rigidity of the footing.

15.17 (Soil from Example 15.13 with a 4 × 4 ft footing with a net load of 2 kips / ft²). The solution uses the same equations as the example.

Top layer (midpoint 4.5 ft below the footing):

$$\Delta\sigma = (2000)(4)(4) / [(8.5)(8.5)] = 443 \text{ psf}$$

$$\sigma' + \Delta\sigma = 644 + 443 = 1087 \text{ psf}$$

Since 1087 is below the preconsolidation pressure of 1500 psf, the recompression index C_r is used.

$$\varepsilon_v = (1 / 1.9) [\, 0.04 \log (1087 / 644) \,] = 0.0048$$

$$S_{top} = 60 \text{ in} (0.0048) = 0.287 \text{ in}$$

Middle layer:

$$\Delta\sigma = (2000)(4)(4) / [(13.5)(13.5)] = 176 \text{ psf}$$

$$\sigma' + \Delta\sigma = 932 + 176 = 1108 \text{ psf}$$

$$\varepsilon_v = (1 / 1.9) [\, 0.04 \log (1108 / 932) \,] = 0.0016$$

$$S_{mid} = 60 \text{ in} (0.0016) = 0.096 \text{ in}$$

Bottom layer:

$$\Delta\sigma = (2000)(4)(4) / [(18.5)(18.5)] = 94 \text{ psf}$$

$$\sigma' + \Delta\sigma = 1220 + 94 = 1314 \text{ psf}$$

$$\varepsilon_v = (1 / 1.9) [\, 0.04 \log (1314 / 1220) \,] = 0.00068$$

$$S_{bot} = 60 \text{ in} (0.00068) = 0.041 \text{ in}$$

The total settlement is the sum of the layer settlements:

$$S = 0.287 + 0.096 + 0.041 = 0.42 \text{ inches}$$

15.18 Settlement versus time is evaluated using Equation 15.6.7. First, detrmine the time to fifty percent of the total settlement. From Table 15.3, The average degree of consolidation U_{avg} is 50% when the time factor T is 0.197. Since the 15 ft clay layer has sand above and below it, it is doubly drained and $H = 15/2 = 7.5$ ft. Rearranging Equation 15.6.7, the time for 50% consolidation is

$$t_{50} = T_{50} H^2 / c_v$$
$$= 0.197(7.5^2) / 0.2 = 55 \text{ days}$$

Similarly

$$t_{20} = T_{20} H^2 / c_v$$
$$= 0.031(7.5^2) / 0.2 = 9 \text{ days}$$
$$t_{90} = T_{90} H^2 / c_v$$
$$= 0.848(7.5^2) / 0.2 = 239 \text{ days}$$

Times for other degrees of consolidations can likewise be determined. The settlements at theses times are obtained by multiplying U_{avg} by the total predicted settlement:

At 9 days,

$$S = 0.20(0.42) = 0.08 \text{ in}$$

At 55 days,

$$S = 0.50(0.42) = 0.21 \text{ in}$$

At 239 days,

$$S = 0.90(0.38) = 0.34 \text{ in}$$

15.19 From Table 15.4, α is taken as 0.40 for $c = 3000$ psf. From Equation 15.7.5, the unit side resistance is

$$f_s = \alpha c = 0.4(3000) = 1200 \text{ psf}$$

The perimeter of the pile is $16\pi / 12 = 4.19$ ft. The surface area of the pile is

$$(40 \text{ ft})(4.19 \text{ ft}) = 168 \text{ ft}^2$$

Multiplying the unit side resistance by the surface area, the total side capacity Q_{side} is

$$1200(168) = 201{,}600 \text{ lb}$$

The tip area of the pile is

$$A_{tip} = (1/4)\pi (16/12)^2 = 1.40 \text{ ft}^2$$

From Equation 15.11.1, the total tip capacity is

$$Q_{tip} = 9(1.40)(3000) = 37{,}800 \text{ lb}$$

The total ultimate pile capacity is the sum of the side and tip capacity:

$$Q_{ult} = 201{,}600 \text{ lb} + 31{,}800 \text{ lb} \approx 240 \text{ kips}$$

The recommended allowable working load is obtainined by applying a factor of safety of 3 or more. Using $FS = 3.0$,

$$Q_a = 80 \text{ kips}$$

15.20 From Equation For $\phi = 32°$, $K_a = \tan^2(45 - 32/2) = 0.307$. The driving force is the active earth force. From Example 15.16, the active earth force for a sand backfill with no water table present is

$$P_a = 0.5 K_a \gamma H^2$$

$$= 0.5 (0.307)(125)(8^2) = 1228 \text{ lb/ft}$$

From Example 15.21, the resisting force is $W \tan \phi = 0.625 W$, where W is the weight of the wall. For a factor of safety of 2.0 against sliding, the resisting force must be twice the driving force, or 2(1228) = 2456 lb/ft. Setting the resisting force to this value,

$$0.625 W = 2456$$

$$W = 3930 \text{ lb/ft}$$

Taking the density of concrete as 150 lb/ft³, the weight of the wall can be expressed as

$$3930 = (0.5)(2+B)(8)(150)$$

Then

$$B = 4.55 \text{ ft}$$

For practicality, recommend a base width of 4.5 or 5.0 ft. For design, the wall should also be checked for overturning and should be founded below the frost depth.

15.21 First check the value of $\gamma H/c$:

$$125(12) / 3000 = 0.5$$

As this value is below 4.0, the clay is a stiff clay. From Equation 15.10.5, the design pressure over the middle half of the height is

$$p = 0.3 (125)(12) = 450 \text{ psf}$$

The design pressure diagram increases linearly from zero to the above value in the top quarter of the wall, and decreases linearly to zero in the bottom quarter. Hence the design pressure diagram is as shown below.

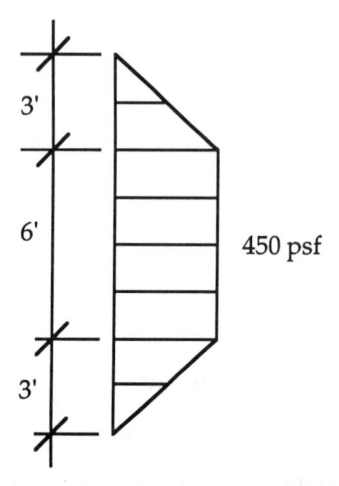

The total design earth force per foot of wall is

$$(0.5)(3)(450) + (6)(450) + (.5)(3)(450) = 4050 \text{ lb/ft}$$

Solutions to Problems (PE-Format)

15.22 e) From Equation 14.10.1, the total vertical stress at point A′ is

$$\sigma_v = \gamma H = (125 \text{ lb/ft}^3)(16 \text{ ft}) = 2000 \text{ lb/ft}^2$$

As there is no groundwater present, the pore pressure is zero, and the effective vertical stress is equal to the total vertical stress.

From Equation 15.8.3, the active earth pressure coefficient is

$$K_a = \tan^2(45 - 34/2) = 0.2827$$

From equation 15.8.1, the horizontal effective stress σ'_h, which is the same as the horizontal active earth pressure p_a, is

$$\sigma'_h = K_a \sigma'_v = (0.2827)(2000 \text{ lb/ft}^2) = 565.4 \text{ lb/ft}^2$$

15.23 b) From Example 15.16, the total active earth force is

$$P_a = 0.5\, \gamma K_a H^2 = (0.5)(125 \text{ lb/ft}^3)(0.2827)(16 \text{ ft})^2 = 4523.2 \text{ lb / ft of wall.}$$

15.24 c) From Equation 15.8.6, the passive pressure coefficient for the material at the toe is

$$K_p = \tan^2(45 + 30/2) = 3.000$$

Similar to problem 15.22, the effective vertical stress at the toe is

$$\sigma'_v = \gamma z - u = (120)(4) - 0 = 480 \text{ lb/ft}^2$$

The effective horizontal stress σ'_h, which is equal to the passive earth pressure is then

$$\sigma'_h = K_p\, \sigma'_v = (3.00)(480) = 1440 \text{ lb / ft}^2$$

15.25 b) Similar to problems 15.23, the total passive earth force is

$$P_p = 0.5\, \gamma K_p H^2 = (0.5)(120 \text{ lb/ft}^3)(3.000)(4 \text{ ft})^2 = 2880 \text{ lb / ft of wall.}$$

15.26 e) Following Example 15.21 and taking the unit weight of concrete to be 150 lb/ft^3, the weight of the free body above the base is the weight of the concrete plus the weight of the soil above the heel and the toe:

$$W = W_{\text{soil heel}} + W_{\text{soil toe}} + W_{\text{base}} + W_{\text{stem}}$$

$$= (6)(14)(125) + (3)(2)(120) + (11)(2)(150) + (2)(14)(150)$$

$$= 10{,}500 + 720 + 3{,}300 + 4{,}200$$

$$= 18{,}720 \text{ lb /ft}$$

The resisting normal force acting upward on the base is equal to this value.

The maximum base shear force is then

$$S = W \tan \phi = (18720 \text{ lb/ft})(\tan 30°) = 10{,}808 \text{ lb / ft}$$

15.27 d) Following Example 15.21, the factor of safety against sliding is

$$FS = \frac{W \tan\phi + P_p}{P_a} = \frac{10808 + 2880}{4523} = 3.03$$

15.28 a) Following Example 15.21 and using the forces from problem 15.26 and 15.25, the resisting moment is

$$M = M\text{soil heel} + M\text{soil toe} + M\text{conc base} + M\text{conc stem} + M \text{ passive}$$
$$= (10520)(5.5) + (720)(1.5) + (3300)(4) + (2880)(4/3)$$
$$= 57{,}860 + 3{,}960 + 13{,}200 + 3{,}840$$
$$= 78{,}860 \text{ ft-lb / ft of wall}$$

The overturning moment is

$$M_o = M_{active}$$
$$= (4523.2 \text{ lb/ft})(16/3 \text{ ft})$$
$$= 24{,}123 \text{ ft-lb / ft of wall}$$

The factor of safety is then

$$FS = M_r / M_o = 78{,}860 / 24{,}123 = 3.27$$

15.29 c) The difference between the resisting moment and the overturning moment must be provided by the moment due to the resisting normal force on the base. The moment due to this force is then

$$M = 78{,}860 - 24{,}123 = 54{,}737 \text{ ft-lb / ft}$$

This moment is due to the normal force of 18,720 lb/ft acting at a moment arm of $x$bar from the toe. The moment arm is then

$$x\text{bar} = 54{,}737 / 18{,}720 = 2.92 \text{ ft}$$

16. Engineering Economy

by Frank Hatfield

Engineering designs are intended to produce good results. In general, the good results are accompanied by undesirable effects including the costs of manufacturing or construction. Selecting the best design requires the engineer to anticipate and compare the good and bad outcomes. If outcomes are evaluated in dollars, and if "good" is defined as positive monetary value, then design decisions may be guided by the techniques known as *engineering economy*. Decisions based solely on engineering economy may be guaranteed to result in maximum goodness only if all outcomes are anticipated and can be monetized (measured in dollars).

16.1 Value and Interest

"Value" is not synonymous with "amount." The value of an amount of money depends on when the amount is received or spent. For example, the promise that you will be given a dollar one year from now is of less value to you than a dollar received today. The difference between the anticipated amount and its current value is called *interest* and is frequently expressed as a time rate. If an interest rate of 10% per year is used, the expectation of receiving $1.00 one year hence has a value now of about $0.91. In engineering economy, interest usually is stated in percent per year. If no time unit is given, "per year" is assumed.

EXAMPLE 16.1

What amount must be paid in two years to settle a current debt of $1,000 if the interest rate is 6%?

Solution.

$$\text{Value after one year} = 1000 + 1000 \times 0.06$$
$$= 1000(1 + 0.06)$$
$$= \$1060$$

$$\text{Value after two years} = 1060 + 1060 \times 0.06$$
$$= 1000(1 + 0.06)^2$$
$$= \$1124$$

Hence, $1,124 must be paid in two years to settle the debt.

16.2 Cash Flow Diagrams

As an aid to analysis and communication, an engineering economy problem may be represented graphically by a horizontal time axis and vertical vectors representing dollar amounts. The cash flow diagram for Example 16.1 is sketched in Fig. 16.1. Income is up and expenditures are down. It is important to pick a point of view and stick with it. For example, the vectors in Fig. 16.1 would have been reversed if the point of view of the lender had been adopted. It is a good idea to draw a cash flow diagram for every engineering economy problem that involves amounts occurring at different times.

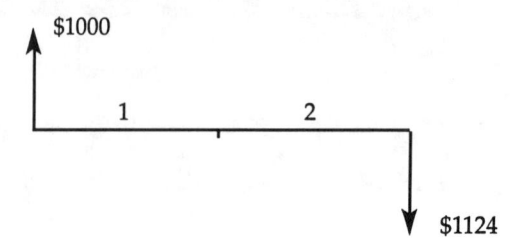

Figure 16.1 Cash flow diagram for Example 16.1.

In engineering economy, amounts are almost always assumed to occur at the ends of years. Consider, for example, the value today of the future operating expenses of a truck. The costs probably will be paid in varied amounts scattered throughout each year of operation, but for computational ease the expenses in each year are represented by their sum (computed without consideration of interest) occurring at the end of the year. The error introduced by neglecting interest for partial years is usually insignificant compared to uncertainties in the estimates of future amounts.

16.3 Cash Flow Patterns

Engineering economy problems involve the following four patterns of cash flow, both separately and in combination.

- *P*-pattern: A single amount P occurring at the beginning of n years. P frequently represents "present" amounts.
- *F*-pattern: A single amount F occurring at the end of n years. F frequently represents "future" amounts.
- *A*-pattern: Equal amounts A occurring at the ends of each of n years. The A-pattern frequently is used to represent "annual" amounts.
- *G*-pattern: End-of-year amounts increasing by an equal annual gradient G. Note that the first amount occurs at the end of the second year. G is the abbreviation of "gradient."

The four cash flow patterns are illustrated in Fig. 16.2.

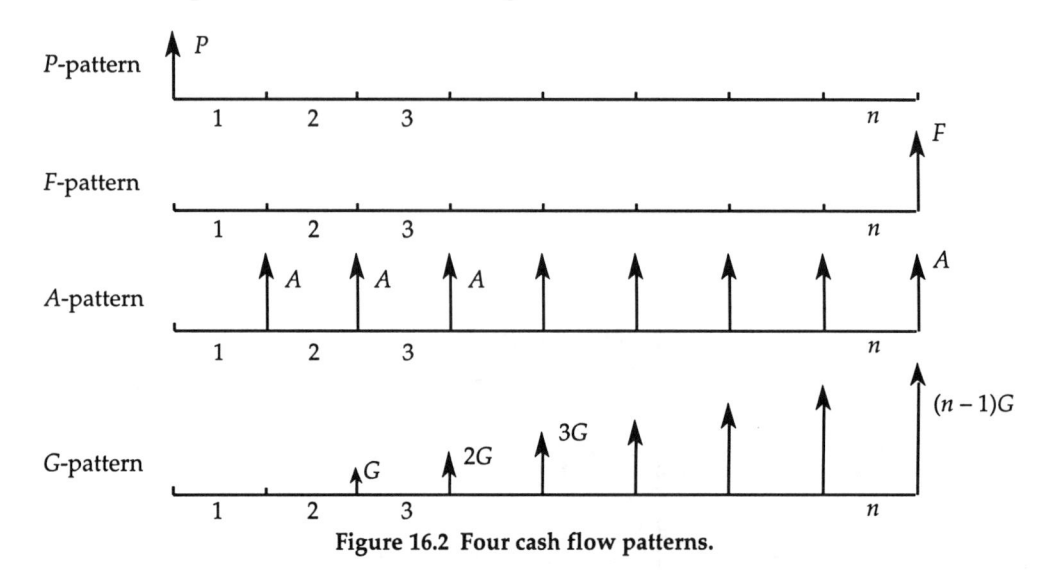

Figure 16.2 Four cash flow patterns.

16.4 Equivalence of Cash Flow Patterns

Two cash flow patterns are said to be equivalent if they have the same value. Most of the computational effort in engineering economy problems is directed at finding a cash flow pattern that is equivalent to a combination of other patterns. Example 16.1 can be thought of as finding the amount in an *F*-pattern that is equivalent to $1,000 in a *P*-pattern. The two amounts are proportional, and the factor of proportionality is a function of interest rate i and number of periods n. There is a different factor of proportionality for each possible pair of the cash flow patterns defined in Section 16.3. To minimize the possibility of selecting the wrong factor, mnemonic symbols are assigned to the factors. For Example 16.1, the proportionality factor is written $(F/P)_n^i$ and solution is achieved by evaluating

$$F = (F/P)_n^i P.$$

To analysts familiar with the canceling operation of algebra, it is apparent that the correct factor has been chosen. However, the letters in the parentheses together with the sub- and super-scripts constitute a single symbol; therefore, the canceling operation is not actually performed. Table 16.1 lists symbols and formulas for commonly used factors. Table 16.2, located at the end of this chapter, presents a convenient way to find numerical values of interest factors. Those values are tabulated for selected interest rates i and number of interest periods n; linear interpolation for intermediate values of i and n is acceptable for most situations.

TABLE 16.1 Formulas for Interest Factors

Symbol	To Find	Given	Formula
$(F/P)_n^i$	F	P	$(1+i)^n$
$(P/F)_n^i$	P	F	$\dfrac{1}{(1+i)^n}$
$(A/P)_n^i$	A	P	$\dfrac{i(1+i)^n}{(1+i)^n - 1}$
$(P/A)_n^i$	P	A	$\dfrac{(1+i)^n - 1}{i(1+i)^n}$
$(A/F)_n^i$	A	F	$\dfrac{i}{(1+i)^n - 1}$
$(F/A)_n^i$	F	A	$\dfrac{(1+i)^n - 1}{i}$
$(A/G)_n^i$	A	G	$\dfrac{1}{i} - \dfrac{n}{(1+i)^n - 1}$
$(F/G)_n^i$	F	G	$\dfrac{1}{i}\left[\dfrac{(1+i)^n - 1}{i} - n\right]$
$(P/G)_n^i$	P	G	$\dfrac{1}{i}\left[\dfrac{(1+i)^n - 1}{i(1+i)^n} - \dfrac{n}{(1+i)^n}\right]$

EXAMPLE 16.2

Derive the formula for $(F/P)_n^i$.

Solution. For $n = 1$,
$$F = (1+i)P$$
that is,
$$(F/P)_1^i = (1+i)^1.$$
For any n,
$$F = (1+i)(F/P)_{n-1}^i P$$
that is,
$$(F/P)_n^i = (1+i)(F/P)_{n-1}^i.$$
By induction,
$$(F/P)_n^i = (1+i)^n.$$

EXAMPLE 16.3

A new widget twister, with a life of six years, would save $2,000 in production costs each year. Using a 12% interest rate, determine the highest price that could be justified for the machine. Although the savings occur continuously throughout each year, follow the usual practice of lumping all amounts at the ends of years.

Solution. First, sketch the cash flow diagram.

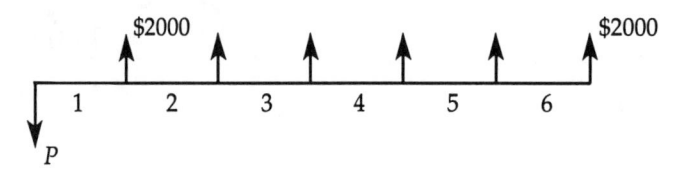

The cash flow diagram indicates that an amount in a P-pattern must be found that is equivalent to $2,000 in an A-pattern. The corresponding equation is
$$P = (P/A)_n^i A$$
$$= (P/A)_6^{12\%} 2000$$

Table 16.2 is used to evaluate the interest factor for $i = 12\%$ and $n = 6$:
$$P = 4.1114 \times 2000$$
$$= \$8223$$

EXAMPLE 16.4

How soon does money double if it is invested at 8% interest?

Solution. Obviously, this is stated as
$$F = 2P.$$

Therefore,
$$(F/P)_n^{8\%} = 2.$$

In the 8% interest table, the tabulated value for (F/P) that is closest to 2 corresponds to $n = 9$ years.

——— EXAMPLE 16.5 ———

Find the value in 1987 of a bond described as "Acme 8% of 2000" if the rate of return set by the market for similar bonds is 10%.

Solution. The bond description means that the Acme company has an outstanding debt that it will repay in the year 2000. Until then, the company will pay out interest on that debt at the 8% rate. Unless otherwise stated, the principal amount of a single bond is $1,000. If it is assumed that the debt is due December 31, 2000, interest is paid every December 31, and the bond is purchased January 1, 1987, then the cash flow diagram, with unknown purchase price P, is:

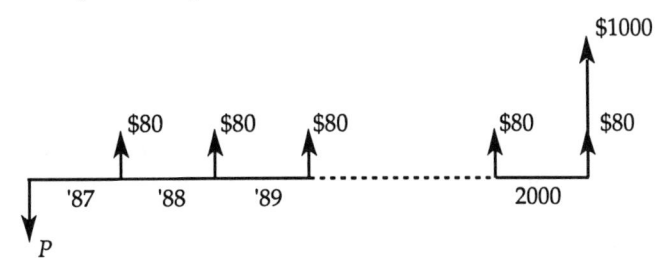

The corresponding equation is

$$P = (P/A)_{14}^{10\%} 80 + (P/F)_{14}^{10\%} 1000$$

$$= 7.3667 \times 80 + 0.2633 \times 1000$$

$$= \$853$$

That is, to earn 10% the investor must buy the 8% bond for $853, a "discount" of $147. Conversely, if the market interest rate is less than the nominal rate of the bond, the buyer will pay a "premium" over $1,000.

The solution is approximate because bonds usually pay interest semiannually, and $80 at the end of the year is not equivalent to $40 at the end of each half year. But the error is small and is neglected.

——— EXAMPLE 16.6 ———

You are buying a new appliance. From past experience you estimate future repair costs as:

First Year	$ 5
Second Year	15
Third Year	25
Fourth Year	35

The dealer offers to sell you a four-year warranty for $60. You require at least a 6% interest rate on your investments. Should you invest in the warranty?

Solution. Sketch the cash flow diagram.

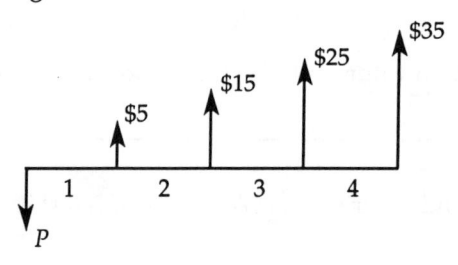

```
              $5    $15   $25   $35
              ↑     ↑     ↑     ↑
        ──────┼─────┼─────┼─────┼──
              1     2     3     4
         ↓
         P
```

The known cash flows can be represented by superposition of a $5 A-pattern and a $10 G-pattern. Verify that statement by drawing the two patterns. Now it is clear why the standard G-pattern is defined to have the first cash flow at the end of the second year. Next, the equivalent amount P is computed:

$$P = (P/A)_4^{6\%} A + (P/G)_4^{6\%} G$$

$$= 3.4651 \times 5 + 4.9455 \times 10$$

$$= \$67$$

Since the warranty can be purchased for less then $67, the investment will earn a rate of return greater than the required 6%. Therefore, you should purchase the warranty.

If the required interest rate had been 12%, the decision would be reversed. This demonstrates the effect of a required interest rate on decision making. Increasing the required rate reduces the number of acceptable investments.

EXAMPLE 16.7

Compute the annual equivalent maintenance costs over a 5-year life of a laser printer that is warranted for two years and has estimated maintenance costs of $100 annually. Use $i = 10\%$.

Solution. The cash flow diagram appears as:

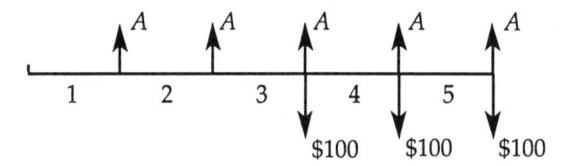

```
        A    A    A    A    A
        ↑    ↑    ↑    ↑    ↑
    ────┼────┼────┼────┼────┼──
        1    2    3    4    5
                  ↓    ↓    ↓
                $100 $100 $100
```

There are several ways to find the 5-year A-pattern equivalent to the given cash flow. One of the more efficient methods is to convert the given 3-year A-pattern to an F-pattern, and then find the 5-year A-pattern that is equivalent to that F-pattern. That is,

$$A = (A/F)_5^{10\%} (F/A)_3^{10\%} 100$$

$$= \$54.$$

16.5 Unusual Cash Flows and Interest Periods

Occasionally an engineering economy problem will deviate from the year-end cash flow and annual compounding norm. The examples in this section demonstrate how to handle these situations.

EXAMPLE 16.8

PAYMENTS AT BEGINNINGS OF YEARS

Using a 10% interest rate, find the future equivalent of:

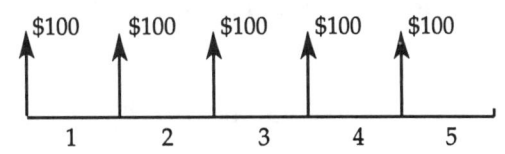

[cash flow diagram: five $100 payments at times 1, 2, 3, 4, 5 (at beginnings)]

Solution. Shift each payment forward one year. Therefore,

$$A = (F/P)_1^{10\%} 100 = \$110.$$

This converts the series to the equivalent A-pattern:

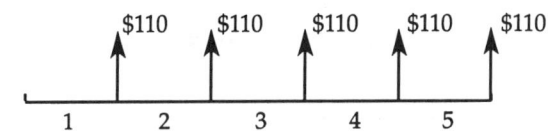

[cash flow diagram: five $110 payments at times 1, 2, 3, 4, 5]

and the future equivalent is found to be

$$F = (F/A)_5^{10\%} 110 = \$672.$$

Alternative Solution. Convert to a six-year series:

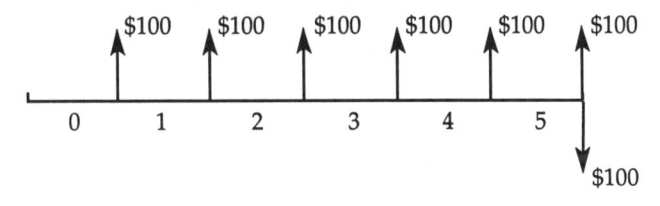

[cash flow diagram: six $100 payments at times 0, 1, 2, 3, 4, 5 with a -$100 at time 5]

The future equivalent is

$$F = (F/A)_6^{10\%} 100 - 100 = \$672.$$

EXAMPLE 16.9

SEVERAL INTEREST AND PAYMENT PERIODS PER YEAR

Compute the present value of eighteen monthly payments of $100 each, where interest is 1/2% per month.

Solution. The present value is computed as

$$P = (P/A)_{18}^{1/2\%} 100 = \$1717$$

EXAMPLE 16.10

ANNUAL PAYMENTS WITH INTEREST COMPOUNDED m TIMES PER YEAR

Compute the effective annual interest rate equivalent to 5% nominal annual interest compounded daily. (There are 365 days in a year.)

Solution. The legal definition of nominal annual interest is

$$i_n = mi$$

where i is the interest rate per compounding period. For the example,

$$i = i_n/m$$
$$= 0.05/365 = 0.000137 \text{ or } 0.0137\% \text{ per day.}$$

Because of compounding, the effective annual rate is greater than the nominal rate. By equating (F/P)-factors for one year and m periods, the effective annual rate i_e may be computed as follows:

$$(1+i_e)^1 = (1+i)^m$$
$$i_e = (1+i)^m - 1$$
$$= (1.000137)^{365} - 1 = 0.05127 \text{ or } 5.127\%$$

EXAMPLE 16.11

CONTINUOUS COMPOUNDING

Compute the effective annual interest rate i_e equivalent to 5% nominal annual interest compounded continuously.

Solution. As m approaches infinity, the value for i_e is found as follows:

$$i_e = e^{mi} - 1$$
$$= e^{0.05} - 1$$
$$= 0.051271 \text{ or } 5.1271\%$$

EXAMPLE 16.12

ANNUAL COMPOUNDING WITH m PAYMENTS PER YEAR

Compute the year-end amount equivalent to twelve end-of-month payments of $10 each. Annual interest rate is 6%.

Solution. The usual simplification in engineering economy is to assume that all payments occur at the end of the year, giving an answer of $120. This approximation may not be acceptable for a precise analysis of a financial agreement. In such cases, the agreement's policy on interest for partial periods must be investigated.

EXAMPLE 16.13

ANNUAL COMPOUNDING WITH PAYMENT EVERY m YEARS

With interest at 10% compute the present equivalent of

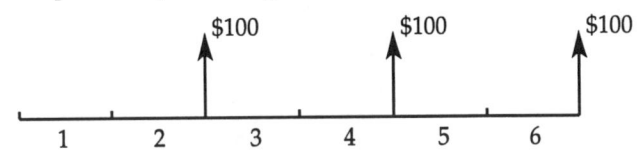

```
              $100        $100        $100
               ↑           ↑           ↑
    |____|____|____|____|____|____|
    1    2    3    4    5    6
```

Solution. First convert each payment to an A-pattern for the m preceding years. That is,

$$A = (A/F)_2^{10\%} \, 100$$
$$= \$47.62$$

Then, convert the A-pattern to a P-pattern:

$$P = (P/A)_6^{10\%} \, 47.62$$
$$= \$207$$

16.6 Evaluating Alternatives

The techniques of engineering economy assume the objective of maximizing net value. For a business, "value" means after-tax cash flow. For a not-for-profit organization, such as a government agency, value may include non-cash benefits, such as clean air, improved public health, or recreation to which dollar amounts have been assigned.

This section concerns strategies for selecting alternatives such that net value is maximized. The logic of these methods will be clear if the following distinctions are made between two different types of interest rates, and between two different types of relationships among alternatives.

TYPES OF INTEREST RATES

Rate of Return (ROR): The estimated interest rate produced by an investment. It may be computed by finding the interest rate such that the estimated income and non-cash benefits (positive value), and the estimated expenditures and non-cash costs (negative value), sum to a net equivalent value of zero.

Minimum Attractive Rate of Return (MARR): The lowest rate of return that the organization will accept. In engineering economy problems, it is usually a given quantity and may be called, somewhat imprecisely, "interest," "interest rate," "cost of money," or "interest on capital."

TYPES OF ALTERNATIVE SETS

Mutually Exclusive Alternatives: Exactly one alternative must be selected.
  Examples: "Shall Main Street be paved with concrete or asphalt?" "In which room will we put the piano?" If a set of alternatives is mutually exclusive, it is important to determine whether the set includes the null (do nothing) alternative. Serious consequences can arise from failure to recognize the null alternative.

Independent Alternatives: It is possible (but not necessarily economical) to select any number of the available alternatives.

Examples: "Which streets should be paved this year?" "Which rooms shall we carpet?"

16.6.1 Annual Equivalent Cost Comparisons

The estimated income and benefits (positive) and expenditures and costs (negative) associated with an alternative are converted to the equivalent *A*-pattern using an interest rate equal to *MARR*. The *A*-value is the *annual net equivalent value (ANEV)* of the alternative. If the alternatives are mutually exclusive, the one with the largest *ANEV* is selected. If the alternatives are independent, all that have positive *ANEV* are selected.

---— **EXAMPLE 16.14**———

A new cap press is needed. Select the better of the two available models described below. *MARR* is 10%.

Model	Price	Annual Maintenance	Salvage Value	Life
Reliable	11,000	1,000	1,000	10 years
Quicky	4,000	1,500	0	5 years

Solution. The *ANEV* is calculated for each model:

Reliable: $ANEV = -(A/P)_{10}^{10\%} 11000 - 1000 + (A/F)_{10}^{10\%} 1000$

$= -\$2730$

Quicky: $ANEV = -(A/P)_{5}^{10\%} 4000 - 1500$

$= -\$2560$

Negative *ANEV* indicates a rate of return less than *MARR*. However, these alternatives are mutually exclusive and the null is not available. The problem is one of finding the less costly way to perform a necessary function. Therefore, *Quicky* is selected. If *MARR* had been much lower, *Reliable* would have been selected. By setting the *MARR* relatively high, the organization is indicating that funds are not available to invest now in order to achieve savings in the future.

16.6.2 Present Equivalent Cost Comparisons

The estimated income and benefits (positive), and expenditures and costs (negative), associated with an alternative are converted to the equivalent *P*-pattern using an interest rate equal to *MARR*. The *P*-value is the *present net equivalent value (PNEV)* of the alternative. If the alternatives are mutually exclusive, the one with the largest *PNEV* is selected. *PNEV* is also called "life cycle cost," "present worth," "capital cost," and "venture worth." If the alternatives are independent, all that have positive *PNEV* are selected.

The present equivalent cost method requires that alternatives be evaluated over the same span of time. If their lives are not equal, the lowest common multiple of the lives is used for the time span, with each alternative repeated to fill the span. A variation, called the *capitalized cost method*, computes the

PNEV for repeated replacement of the alternatives for an infinite time span. The capitalized cost P of an infinite series of equal amounts A is given by

$$P = A(P/A)_\infty^i = A/i.$$

---- **EXAMPLE 16.15** ----

Repeat Example 16.14 using the present equivalent cost method.

Solution. The *PNEV* is calculated for each model:

$$\text{Reliable:} \quad PNEV = -11000 - (P/A)_{10}^{10\%} 1000 + (P/F)_{10}^{10\%} 1000$$

$$= -\$16,800$$

$$\text{Quicky:} \quad PNEV = -4000 - (P/F)_5^{10\%} 4000 - (P/A)_{10}^{10\%} 1500$$

$$= -\$15,700$$

Note that *Quicky* was replaced in order to fill the ten-year time span. As in Example 16.14, *Quicky* is selected. The two methods will always give the same decision if used correctly. Observe that for both alternatives

$$PNEV = (P/A)_{10}^{10\%} ANEV.$$

16.6.3 Incremental Approach

For a set of mutually exclusive alternatives, only the differences in amounts need to be considered. Compute either the *ANEV* or the *PNEV* and base the decision on the sign of that value.

---- **EXAMPLE 16.16** ----

Repeat Example 16.14 using an incremental present net equivalent value approach.

Solution. *Reliable* costs $7,000 more than *Quicky* but saves $500 each year in maintenance expenses and eliminates the need for a $4,000 replacement after five years. In addition, *Reliable* has a $1,000 salvage value whereas *Quicky* has none.

Reliable – Quicky:

$$PNEV = -7000 + (P/A)_{10}^{10\%} 500 + (P/F)_5^{10\%} 4000 + (P/F)_{10}^{10\%} 1000$$

$$= -\$1060$$

The negative result dictates selection of *Quicky*. That is, the additional initial cost required to purchase *Reliable* is not justified.

16.6.4 Rate of Return Comparisons

The expression for *ANEV* or *PNEV* is formulated and then solved for the interest rate that will give a zero *ANEV* or *PNEV*. This interest rate is the rate of return (*ROR*) of the alternative. To apply the rate of return method to mutually exclusive alternatives requires incremental comparison of each possible pair of alternatives; increments of investment are accepted if their rates of return exceed *MARR*. For independent alternatives, all those with ROR exceeding MARR are accepted. The rate of return method permits conclusions to be stated as functions of MARR, which is useful if MARR has not been determined precisely.

EXAMPLE 16.17

A magazine subscription costs $50 for one year or $80 for two years. If you want to receive the magazine for at least two years, which alternative is better?

Solution. The two-year subscription requires an additional initial investment of $30 and eliminates the payment of $50 one year later. The rate of return formulation is:

$$PNEV = 0$$
$$-30 + 50(P/F)_1^i = 0$$

The solution for i is as follows:

$$-30 + 50 \frac{1}{(1+i)} = 0$$
$$i = 0.67 \text{ or } 67\%$$

Therefore, if your MARR is less than 67%, subscribe for two years.

EXAMPLE 16.18

Repeat Example 16.14 using the rate of return method.

Solution. Use the incremental expression derived in Example 16.16, but set PNEV equal to zero and use the interest rate as the unknown:

$$-7000 + (P/A)_{10}^i 500 + (P/F)_5^i 4000 + (P/F)_{10}^i 1000 = 0$$

By trial and error, the interest rate is found to be 6.6%. Therefore, *Reliable* is preferred if, and only if, MARR is less than 6.6%.

16.6.5 Benefit/Cost Comparisons

The benefit/cost ratio is determined from the formula:

$$\frac{B}{C} = \frac{\text{Uniform net annual benefits}}{\text{Annual equivalent of initial cost}}$$

where MARR is used in computing the A-value in the denominator. As with the rate of return method, mutually exclusive alternatives must be compared incrementally, the incremental investment being accepted if the benefit/cost ratio exceeds unity. For independent alternatives, all those with benefit/cost ratios exceeding unity are accepted.

Note that the only pertinent fact about a benefit/cost ratio is whether it exceeds unity. This is illustrated by the observation that a project with a ratio of 1.1 may provide greater net benefit than a project with a ratio of 10 if the investment in the former project is much larger than the investment in the latter. It is incorrect to rank mutually exclusive alternatives by their benefit/cost ratios as determined by comparing each alternative to the null (do nothing) alternative.

The benefit/cost ratio method will give the same decision as the rate of return method, present equivalent cost method, and annual equivalent cost method if the following conditions are met:

1. Each alternative is comprised of an initial cost and uniform annual benefit.
2. The form of the benefit/cost ratio given above is used without deviation.

EXAMPLE 16.19

A road resurfacing project costs $200,000, lasts five years, and saves $100,000 annually in patching costs. *MARR* is 10%. Should the road be resurfaced?

Solution. The benefit/cost ratio is

$$\frac{B}{C} = \frac{100,000}{(A/P)_5^{10\%} \, 200,000} = 1.9.$$

Since the ratio exceeds unity, the resurfacing is justified.

16.6.6 A Note on *MARR*

In engineering economy examination problems, *MARR* is a given quantity. However, the following discussion of the determination of *MARR* will help clarify the logic underlying the various comparison methods.

In general, an organization will be able to identify numerous opportunities to spend money now that will result in future returns. For each of these independent investment opportunities, an expected rate of return can be estimated. Similarly, the organization will be able to find numerous sources of funds for investment. Associated with each source of funds is an interest rate. If the source is a loan, the associated interest rate is simply that charged by the lender. Funds generated by operations of the organization, or provided by its owners (if the organization is a business), or extracted from taxpayers (if the organization is a government agency) can be thought of as being borrowed from the owners or taxpayers. Therefore, such funds can be assigned a fictitious interest rate, which should not be less than the maximum rate of return provided by other opportunities in which the owners or taxpayers might invest.

Value will be maximized if the rates of return of all the selected investments exceed the highest interest rate charged for the money borrowed, and if every opportunity has been taken to invest at a rate of return exceeding that for which money can be borrowed. That rate is the Minimum Attractive Rate of Return. No investments should be made that pay rates of return less than *MARR*, and no loans should be taken that charge interest rates exceeding *MARR*. Furthermore, the organization should exploit all opportunities to borrow money at interest rates less than *MARR* and invest it at rates of return exceeding *MARR*.

To estimate *MARR* precisely would require the ability to foresee the future, or at least to predict all future investment and borrowing opportunities and their associated rates. A symptom of *MARR* being set too low is insufficient funds for all the investments that appear to be acceptable. Conversely, if *MARR* has been set too high, some investments will be rejected that would have been profitable.

16.6.7 Replacement Problems

How frequently should a particular machine be replaced? This type of problem can be approached by varying the life n. For each value of n, the annual costs and salvage value are estimated, and then the *ANEV* is computed. The value of n resulting in the smallest annual equivalent cost is the optimum, or economic, life of the machine. This approach is complicated by technological improvements in replacement machinery, which may make it advantageous to replace a machine before the end of its economic life. In practice, technological advances are difficult to anticipate.

Another form of the replacement problem asks if an existing asset should be replaced by a new (and possibly different) one. Again, the annual equivalent cost method is recommended. The *ANEV* of the replacement is computed, using its economic life for n. However, the annual cost of the existing asset is simply the estimated expense for one more year of operation. This strategy is based on the assumption that the annual costs of the existing asset increase monotonically as it ages.

16.6.8 Always Ignore the Past

Engineering economy, and decision making in general, deals with alternatives. But there is only one past and it affects all future alternatives equally. Therefore, past costs and income associated with an existing asset should not be included in computations that address the question of replacing the asset. Only the estimated cash flows of the future are relevant.

The mistake of counting past costs is common in everyday affairs. For example, a student may say, "I paid $90 for this textbook so I will not sell it for $20." A more rational approach would be to compare the highest offered price to the value of retaining the text.

―――― **EXAMPLE 16.20** ――――

Yesterday a machine was bought for $10,000. Estimated life is ten years, with no salvage value at the end of its useful life. Current book value is $10,000. Today a vastly improved model was announced. It costs $15,000, has a ten-year life and no salvage value, but reduces operating costs by $4,000 annually. The current resale value of the older machine has dropped to $1,000 due to this stunning technological advance. Should the old model be replaced with a new model at this time?

Solution. The purchase price of the old machine, its book value, and the loss on the sale of the old machine are irrelevant to the analysis. The incremental cost of the new machine is $14,000 and the incremental income is $4,000 annually. A rate of return comparison is formulated as follows:

$$-14{,}000 + (P/A)^i_{10}\, 4000 = 0$$

Solving for rate of return gives $i = 26\%$, indicating that the older machine should be replaced immediately if *MARR* is less than 26%.

―――――――――

16.6.9 Break-Even Analysis

A break-even point is the value of an independent variable such that two alternatives are equally attractive. For values of the independent variable above the break-even point, one of the alternatives is preferred; for values of the independent variable below the break-even point, the other alternative is preferred. Break-even analysis is particularly useful for dealing with an independent variable that is subject to change or uncertainty since the conclusion of the analysis can be stated as a function of the variable.

The rate of return method, as applied to mutually exclusive alternatives, is an example of break-even analysis. The independent variable is *MARR*.

EXAMPLE 16.21

An item can be manufactured by hand for $5. Alternatively, the item can be produced by a machine at a fixed annual equivalent cost of $4,000 plus a variable cost of $1 per item. Assume that the cost of laying off and hiring workers is zero. For each of the two manufacturing processes, answer the following questions:

a) For what production rate is one method more economical than the other?
b) If the item is sold for $6, how many must be sold to make a profit?
c) How low must the price fall, in the short term, before production is discontinued?

Solution.

a) Let P be production rate in units per year. Production costs for the two processes are equated:

$$\text{Cost by machine} = \text{Cost manually}$$

$$4000 + 1P = 5P$$

$$\therefore P = 1000$$

If annual production is expected to be less than 1,000 units, the manual process is more economical. For production rates exceeding 1,000 units per year, the machine process is preferred.

b) Setting profit equal to zero is expressed as:

$$\text{gross income} - \text{cost} = 0$$

$$\text{Manual production: } 6P - 5P = 0$$

$$\therefore P = 0$$

$$\text{Machine production: } 6P - (4000 + 1P) = 0$$

$$\therefore P = 800$$

With price maintained at $6, the mechanized operation will be unprofitable if production rate is less than 800 units per year, but the manual operation is profitable at all production rates.

c) Manual production becomes unprofitable if the price drops below $5, and production will cease at that level. For the machine, the $4,000 cost continues whether or not the machine is running. Incremental income is generated so long as the price stays above the variable (per item) cost. Therefore, production will continue at any price over $1, even though a net loss may be sustained. Of course, if it appears that the price and production rate will not soon increase sufficiently to provide a profit, then the operation will be terminated.

16.7 Income Tax and Depreciation

Business pays to the federal government a tax that is a proportion of taxable income. Taxable income is gross revenue less operating costs (wages, cost of materials, etc.), interest payments on debts, and depreciation. Depreciation is different from the other deductions in that it is not a cash flow.

Depreciation is an accounting technique for charging the initial cost of an asset against two or more years of production. For example, if you buy a $50,000 truck for use in your construction business, deducting its total cost from income during the year of purchase gives an unrealistically low picture of income for that year, and an unrealistically high estimate of income for the succeeding years during which you use the truck. A more level income history would result if you deducted $10,000 per year for five years. In fact, the Internal Revenue Service (IRS) requires that most capital assets used in business be depreciated over a number of years rather than being deducted as expenses during the year of purchase.

An asset is depreciable if it is used to produce income, has a determinable life greater than one year, and decays, wears out, becomes obsolete, or gets used up. Examples include tools, production machinery, computers, office equipment, buildings, patents, contracts, franchises, and livestock raised for wool, eggs, milk or breeding. Non-depreciable assets include personal residence, land, natural resources, annual crops, livestock raised for sale or slaughter, and items intended primarily for resale such as stored grain and the merchandise in a department store.

Since depreciation is not a cash flow, it will not directly enter an engineering economy analysis. However, depreciation must be considered when estimating future income taxes, which are cash flows.

The IRS requires that the Modified Accelerated Cost Recovery System (MACRS) be applied to most tangible property placed in service after 1986. In general, MACRS is based on computing depreciation using a declining balance method or the straight line method treating the property as being placed in service and retired from service at midpoints of tax years, and setting salvage value equal to zero. Older methods, such as the Accelerated Cost Recovery System (ACRS) or the straight line method with a non-zero salvage value, may still show up in engineering economy problems, and therefore are included in this discussion. The following notation will be used in defining methods for computing depreciation:

B — The installed first cost, or basis.

n — Recovery period in years.

D_x — Depreciation in year x.

V_x — Undepreciated balance at the end of year x, also called book value. $V_0 = B$.

V_n — Estimated salvage at age n.

In computing depreciation there is no attempt to equate book value with resale value, productive worth, or any other real figure. A business is not obliged to keep an asset for exactly n years, nor to sell it for exactly its book value or estimated salvage value. These, then, are the depreciation methods:

1. *Declining Balance*: Depreciation is taken as a proportion of book value:

$$D_x = V_{x-1} C/n$$

where n is the recovery period. For values of C equaling 1.25, 1.5 and 2 the method is called, respectively: 125% declining balance, 150% declining balance, and double declining balance.

2. *Straight Line Depreciation*: Depreciation is the same for every full year and is calculated as

$$D_x = (B - V_n)/n.$$

3. *Accelerated Cost Recovery System (ACRS)*: An asset is classed as having a recovery period n of 3, 5, 10, or 15 years using IRS guidelines. For each class, a set of annual rates

R_x is specified by the IRS. With 3-year property, for example, $R_1 = 0.25$, $R_2 = 0.38$, $R_3 = 0.37$. Depreciation is calculated as follows:

$$D_x = R_x B.$$

By definition, the salvage value using ACRS is zero.

EXAMPLE 16.22

The purchase price of an over-the-road tractor unit is $100,000, its recovery period is three years, and it can be sold for an estimated $20,000 at that time. Compute the depreciation schedules using each of the methods described.

Solution.

Double Declining Balance (MACRS)

Year	Depreciation	Book Value
		$100,000
1	0.5 × ($100,000 × 2/3) = $33,333	$66,667
2	$66,667 × 2/3 = $44,444	$22,222
3	$22,222 × 2/3 = $14,815	$7,407
4	$7407	$0

*The book value must be zero after three years of service, so the formula was not used for the last year.

Straight Line (MACRS)

Year	Depreciation	Book Value
		$100,000
1	$100,000/3 × 0.5 = $16,667	$83,333
2	$100,000/3 = $33,333	$50,000
3	$100,000/3 = $33,333	$16,667
4	$100,000/3 × 0.5 = $16,667	$0

Straight Line (general)

Year	Depreciation	Book Value
		$100,000
1	(100,000 − 20,000)/3 = $26,667	$73,333
2	$26,667	$46,667
3	$26,667	$20,000

Accelerated Cost Recovery System (ACRS)

Year	Depreciation	Book Value
		$100,000
1	0.25 × $100,000 = $25,000	$75,000
2	0.38 × $100,000 = $38,000	$37,000
3	0.37 × $100,000 = $37,000	$0

Note: The factor 0.5 in the MACRS results from the requirement to use the midpoint of a tax year.

16.8 Inflation

The "buying power" of money changes with time. A decline in "buying power" is experienced due to a general increase in prices, called "inflation."

Inflation, if it is anticipated, can be exploited by fixing costs and allowing income to increase. A manufacturing business can fix its costs by entering long-term contracts for materials and wages, by purchasing materials long before they are needed, or by stockpiling its product for later sale. Income is allowed to respond to inflation by avoiding long-term contracts for the product. Borrowing becomes more attractive if inflation is expected since the debt will be paid with the less valuable cash of the future.

MARR may be adjusted for anticipated uniform inflation using the formula

$$d = i + f + if$$

where d is inflation-adjusted MARR, i is unadjusted MARR, and f is the rate of inflation. This formula facilitates solution of some types of engineering economy problems.

---------- EXAMPLE 16.23 ----------

A machine having a five-year life can replace a worker who is compensated $20,000 per year with 5% annual "cost of living" increases. Operating and maintenance costs for the machine are negligible. MARR is 10%. Find the maximum price that can be justified for the machine if:

    a) general price inflation is 5%

    b) general price inflation is zero

Solution.

    a) Although the worker gets a larger amount of money each year, her raises are exactly matched by increased prices, including those of her employer's product. "Buying power" of her annual compensation remains equal to the current value of $20,000. Hence, the maximum justifiable price for the machine is

$$P = (P/A)_5^{10\%} \, 20{,}000 = \$75{,}816$$

    b) The maximum justifiable price of the machine is equal to the present equivalent value of the annual amounts of compensation:

$$(P/F)_1^{10\%}(1.05)\,20{,}000 = \$19{,}090$$
$$(P/F)_2^{10\%}(1.05)^2\,20{,}000 = \$18{,}224$$
$$(P/F)_3^{10\%}(1.05)^3\,20{,}000 = \$17{,}394$$
$$(P/F)_4^{10\%}(1.05)^4\,20{,}000 = \$16{,}604$$
$$(P/F)_5^{10\%}(1.05)^5\,20{,}000 = \underline{\$15{,}850}$$

$$\text{therefore,} \quad P = \$86{,}162$$

EXAMPLE 16.24

Recompute the value, in terms of 1987 "buying power," of the "Acme 8% of 2000" bond discussed in Example 16.5, but assume 6% annual inflation.

Solution. The cash flow for each year must be divided by an inflation factor, as well as multiplied by an interest factor, and then the factored cash flows are added:

$$(P/F)_1^{10\%} \quad 80/(1.06) = \$\ 69$$
$$(P/F)_2^{10\%} \quad 80/(1.06)^2 = \$\ 59$$
$$(P/F)_3^{10\%} \quad 80/(1.06)^3 = \$\ 50$$
$$\vdots$$
$$(P/F)_{13}^{10\%} \quad 80/(1.06)^{13} = \$\ 11$$
$$(P/F)_{14}^{10\%} \quad 80/(1.06)^{14} = \$\ 9$$
$$(P/F)_{14}^{10\%} \quad 1000/(1.06)^{14} = \underline{\$\ 116}$$
$$\text{therefore,} \qquad P = \$\ 541$$

Note that investors can account for anticipated inflation simply by using increased values of *MARR*. A *MARR* of 16.6% gives the same conclusions as a *MARR* of 10% with 6% inflation.

Alternative Solution. Using inflation-adjusted *MARR* we have

$$d = i + f + if$$
$$= .10 + .06 + .10 \times .06$$
$$= .166 \quad \text{or} \quad 16.6\%$$

The value of the bond is

$$P = (P/A)_{14}^{16.6\%} 80 + (P/F)_{14}^{16.6\%} 1000$$
$$= 5.3225 \times 80 + .1165 \times 1000$$
$$= \$542$$

Formulas from Part 5.4 (a table in the NCEES Handbook) were used to evaluate the interest factors.

Practice Problems

16.1 Which of the following would be most difficult to monetize?
    a) maintenance cost   b) selling price   c) fuel cost
    d) prestige   e) interest on debt

VALUE AND INTEREST

16.2 If $1,000 is deposited in a savings account that pays 6% annual interest and all the interest is left in the account, what is the account balance after three years?
    a) $840   b) $1,000   c) $1,180   d) $1,191   e) $3,000

16.3 Your perfectly reliable friend, Merle, asks for a loan and promises to pay back $150 two years from now. If the minimum interest rate you will accept is 8%, what is the maximum amount you will loan him?
    a) $119   b) $126   c) $129   d) $139   e) $150

EQUIVALENCE OF CASH FLOW PATTERNS

16.4 The annual amount of a series of payments to be made at the end of each of the next twelve years is $500. What is the present worth of the payments at 8% interest compounded annually?
    a) $500   b) $3,768   c) $6,000   d) $6,480   e) $6,872

16.5 Consider a prospective investment in a project having a first cost of $300,000, operating and maintenance costs of $35,000 per year, and an estimated net disposal value of $50,000 at the end of thirty years. Assume an interest rate of 8%.

What is the present equivalent cost of the investment if the planning horizon is thirty years?
    a) $670,000   b) $689,000   c) $720,000   d) $791,000   e) $950,000

If the project replacement will have the same first cost, life, salvage value, and operating and maintenance costs as the original, what is the capitalized cost of perpetual service?
    a) $670,000   b) $689,000   c) $720,000   d) $765,000   e) infinite

16.6 Maintenance expenditures for a structure with a twenty-year life will come as periodic outlays of $1,000 at the end of the fifth year, $2,000 at the end of the tenth year, and $3,500 at the end of the fifteenth year. With interest at 10%, what is the equivalent uniform annual cost of maintenance for the twenty-year period?
    a) $200   b) $262   c) $300   d) $325   e) $342

16.7 After a factory has been built near a stream, it is learned that the stream occasionally overflows its banks. A hydrologic study indicates that the probability of flooding is about 1 in 8 in any one year. A flood would cause about $20,000 in damage to the factory. A levee can be constructed to prevent flood damage. Its cost will be $54,000 and its useful life is thirty years. Money can be borrowed at 8% interest. If the annual equivalent cost of the levee is less than the annual expectation of flood damage, the levee should be built. The annual expectation of flood damage is $(1/8) \times 20{,}000 = \$2{,}500$. Compute the annual equivalent cost of the levee.
    a) $1,261   b) $1,800   c) $4,320   d) $4,800   e) $6,750

16.8 If $10,000 is borrowed now at 6% interest, how much will remain to be paid after a $3,000 payment is made four years from now?
   a) $7,000
   b) $9,400
   c) $9,625
   d) $9,725
   e) $10,700

16.9 A piece of machinery costs $20,000 and has an estimated life of eight years and a scrap value of $2,000. What uniform annual amount must be set aside at the end of each of the eight years for replacement if the interest rate is 4%?
   a) $1,953
   b) $2,174
   c) $2,250
   d) $2,492
   e) $2,898

16.10 The maintenance costs associated with a machine are $2,000 per year for the first ten years, and $1,000 per year thereafter. The machine has an infinite life. If interest is 10%, what is the present worth of the annual disbursements?
   a) $16,145
   b) $19,678
   c) $21,300
   d) $92,136
   e) $156,600

16.11 A manufacturing firm entered into a ten-year contract for raw materials which required a payment of $100,000 initially and $20,000 per year beginning at the end of the fifth year. The company made unexpected profits and asked that it be allowed to make a lump sum payment at the end of the third year to pay off the remainder of the contract. What lump sum is necessary if the interest rate is 8%?
   a) $85,600
   b) $92,700
   c) $122,300
   d) $196,700
   e) $226,000

UNUSUAL CASH FLOWS AND INTEREST PAYMENTS

16.12 A bank currently charges 10% interest compounded annually on business loans. If the bank were to change to continuous compounding, what would be the effective annual interest rate?
   a) 10%
   b) 10.517%
   c) 12.5%
   d) 12.649%
   e) 12.92%

16.13 Terry bought a CD-ROM drive for $50 down and $30 per month for 24 months. The same drive could have been purchased for $675 cash. What nominal annual interest rate is Terry paying?
   a) 7.6%
   b) 13.9%
   c) 14.8%
   d) 15.2%
   e) 53.3%

16.14 How large a contribution is required to endow perpetually a research laboratory which requires $500,000 for original construction, $200,000 per year for operating expenses, and $100,000 every three years for new and replacement equipment? Interest is 4%.
   a) $700,000
   b) $6,400,000
   c) $7,900,000
   d) $10,000,000
   e) $12,490,000

ANNUAL EQUIVALENT COST COMPARISONS

16.15 One of the two production units described below must be purchased. The minimum attractive rate of return is 12%. Compare the two units on the basis of equivalent annual cost.

	Unit A	Unit B
Initial Cost	$16,000	$30,000
Life	8 years	15 years
Salvage value	$2,000	$5,000
Annual operating cost	$2,000	$1,000

   a) A—$5,058; B—$5,270
   b) A—$4,916; B—$4,872
   c) A—$3,750; B—$2,667
   d) A—$1,010; B—$1,010
   e) A—$2,676; B—$4,250

16.16 Tanks to hold a corrosive chemical are now being made of material A, and have a life of eight years and a first cost of $30,000. When these tanks are four years old, they must be relined at a cost of $10,000. If the tanks could be made of material B, their life would be twenty years and no relining would be necessary. If the minimum rate of return is 10%, what must be the first cost of a tank made of material B to make it economically equivalent to the present tanks?
   a) $38,764   b) $42,631   c) $51,879   d) $58,760   e) $92,361

PRESENT EQUIVALENT COST COMPARISONS

16.17 Compute the life cycle cost of a reciprocating compressor with first cost of $120,000, annual maintenance cost of $9,000, salvage value of $25,000 and life of six years. The minimum attractive rate of return is 10%.
   a) $120,000   b) $145,000   c) $149,000   d) $153,280   e) $167,900

16.18 A punch press costs $100,000 initially, requires $10,000 per year in maintenance expenses, and has no salvage value after its useful life of ten years. With interest of 10%, the capitalized cost of the press is:
   a) $100,000   b) $161,400   c) $197,300   d) $200,000   e) $262,700

16.19 A utility is considering two alternatives for serving a new customer. Both plans provide twenty years of service, but plan A requires one large initial investment, while plan B requires additional investment at the end of ten years. Neglect salvage value, assume interest at 8%, and determine the present cost of both plans.

	Plan A	Plan B
Initial Investment	$50,000	$30,000
Investment at end of 10 years	none	$30,000
Annual property tax and maintenance, years 1–10	$ 800	$ 500
Annual property tax and maintenance, years 11–20	$ 800	$ 900

   a) A—$48,780; B—$49,250
   b) A—$50,000; B—$30,000
   c) A—$50,000; B—$60,000
   d) A—$57,900; B—$50,000
   e) A—$66,000; B—$74,000

16.20 The heat loss of a bare stream pipe costs $206 per year. Insulation A will reduce heat loss by 93% and can be installed for $116; insulation B will reduce heat loss by 89% and can be installed for $60. The insulations require no additional expenses and will have no salvage value at the end of the pipe's estimated life of eight years. Determine the present net equivalent value of the two insulations if the interest rate is 10%.
   a) A—$116; B—$90
   b) A—$906; B—$918
   c) A—$1,022; B—$978
   d) A—$1,417; B—$1,406
   e) A—$1,533; B—$1,467

INCREMENTAL APPROACH

16.21 A desalinator is needed for six years. Cost estimates for two are:

	The Life of Brine	The Salty Tower
Price	$95,000	$120,000
Annual maintenance	3,000	9,000
Salvage value	12,000	25,000
Life in years	3	6

With interest at 10%, what is the annual cost advantage of the Salty Tower?

a) 0   b) $4,260   c) $5,670   d) $5,834   e) $56,000

16.22 A motor costs $20,000 and has an estimated life of six years. By the addition of certain auxiliary equipment, an annual savings of $300 in operating costs can be obtained, and the estimated life of the motor extended to nine years. Salvage value in either case is $5,000. Interest on capital is 8%. Compute the maximum expenditure justifiable for the auxiliary equipment.

a) $1,149   b) $1,800   c) $2,700   d) $7,140   e) $13,300

16.23 An existing electrical power line needs to have its capacity increased, and this can be done in either of two ways. The first method is to add a second conductor to each phase wire, using the same poles, insulators and fittings, at a construction cost of $15,000. The second method for increasing capacity is to build a second line parallel to the existing line, using new poles, insulators and fittings, at a construction cost of $23,000. At some time in the future, the line will require another increase in capacity, with the first alternative now requiring a second line at a cost of $32,500, and the second alternative requiring added conductors at a cost of $23,000. If interest rate is 6%, how many years between the initial expenditure and the future expenditure will make the two methods economically equal?

a) 1   b) 3   c) 5   d) 10   e) 25

REPLACEMENT PROBLEMS

16.24 One year ago machine A was purchased at a cost of $2,000, to be useful for five years. However, the machine failed to perform properly and costs $200 per month for repairs, adjustments and shut-downs. A new machine B designed to perform the same functions is quoted at $3,500, with the cost of repairs and adjustments estimated to be only $50 per month. The expected life of machine B is five years. Except for repairs and adjustments, the operating costs of the two machines are substantially equal. Salvage values are insignificant. Using 8% interest rate, compute the incremental annual net equivalent value of machine B.

a) – $877   b) $923   c) $1,267   d) $1,800   e) $2,677

BREAK-EVEN ANALYSIS

16.25 Bear Air, an airline serving the Arctic, serves in-flight snacks on some routes. Preparing these snacks costs Bear Air $5000 per month plus $1.50 per snack. Alternatively, prepared snacks may be purchased from a supplier for $4.00 per snack. What is the maximum number of snacks per month for which purchasing from the supplier is justified economically?

a) 769   b) 1250   c) 2000   d) 3333   e) 4000

*16.26 Bear Air has been contracting its overhaul work to Aleutian Aeromotive for $40,000 per plane per year. Bear estimates that by building a $500,000 maintenance facility with a life of 15 years and a salvage value of $100,000, they could handle their own overhauls at a variable cost of only $30,000 per plane per year. The maintenance facility could be financed with a secured loan at 8% interest. What is the minimum number of planes Bear must operate in order to make the maintenance facility economically feasible?

 a) 5 b) 6 c) 10 d) 40 e) 50

16.27 It costs Bear Air $1,200 to run a scheduled flight, empty or full, from Coldfoot to Frostbite. Moreover, each passenger generates a cost of $40. The regular ticket costs $90. The plane holds 65 people, but it is running only about 20 per flight. The sales director has suggested selling introductory tickets for $50 to people who have never flown Bear Air.

What is the minimum number of introductory tickets that must be sold in order for a flight to produce a profit?

 a) 5 b) 10 c) 15 d) 20 e) 45

What would be the total profit on the flight from Coldfoot to Frostbite if all 65 passengers claimed introductory tickets?

 a) – $800 b) – $550 c) 0 d) $400 e) $500

16.28 Two electric motors are being considered for an application in which there is uncertainty concern-ing the hours of usage. Motor A costs $4,500 and has an efficiency of 90%. Motor B costs $3,000 and has an efficiency of 89%. Each motor has a ten-year life and no salvage value. Electric service costs $18.70 per year per kW of demand and $0.10 per kWh of energy. The output of the motors is to be 75 kW, and interest rate is 8%. At how many hours usage per year would the two motors be equally economical? If the usage is less than this amount, which motor is preferable?

 a) 1800, A b) 1800, B c) 2200, A d) 2200, B e) 2500, A

INCOME TAX AND DEPRECIATION

16.29 A drill press is purchased for $10,000 and has an estimated life of twelve years. The salvage value at the end of twelve years is estimated to be $1,300. Using general straight-line depreciation, compute the book value of the drill press at the end of eight years.

 a) $1,300 b) $3,333 c) $3,475 d) $4,200 e) $4,925

16.30 A grading contractor owns earth-moving equipment that costs $300,000 and is classed as 7-year property. After seven years of use, its salvage value will be $50,000. Using the general straight line method, compute the first two depreciation deductions and the book value at the end of four years.

 a) $35,714; $35,714; $157,143   d) $42,857; $73,469; $93,711
 b) $85,714; $85,714; $0   e) $85,714; $61,224; $78,092
 c) $21,429; $42,857; $150,000

16.31 Rework Prob. 16.30 using the MACRS straight line method to compute the first two depreciation deductions and the book value at the end of the fourth tax year.

 a) $35,714; $35,714; $157,143   d) $42,857; $73,469; $93,711
 b) $85,714; $85,714; $0   e) $85,714; $61,224; $78,092
 c) $21,429; $42,857; $150,000

16.32 Rework Prob. 16.30 using the general double-declining balance method to compute the first two depreciation deductions and the book value at the end of the four years.
   a) $35,714; $35,714; $157,143
   b) $85,714; $85,714; $0
   c) $21,429; $42,857; $150,000
   d) $42,857; $73,469; $93,711
   e) $85,714; $61,224; $78,092

16.33 Rework Prob. 16.30 using the MACRS double-declining balance method to compute the first two depreciation deductions and the book value at the end of the fourth tax year.
   a) $35,714; $35,714; $157,143
   b) $85,714; $85,714; $0
   c) $21,429; $42,857; $150,000
   d) $42,857; $73,469; $93,711
   e) $85,714; $61,224; $78,092

Solutions to Problems

16.1   d) Prestige.

16.2   d) $1000 \times 1.06^3 = \$1191$.

16.3   c) $150 / 1.08^2 = \$129$.

16.4   b) $500(P/A)_{12}^8 = 500 \times 7.536 = \$3768$.

16.5   b), d) $A: \; 300,000 + 35,000(P/A)_{30}^8 - 50,000(P/F)_{30}^8 = \$689,000$
                $B: \; 689,000(A/P)_{30}^8 (P/A)_{\infty}^8 = \$765,000$.

16.6   b) $\left[1000(P/F)_5^{10} + 2000(P/F)_{10}^{10} + 3500(P/F)_{15}^{10}\right](A/P)_{20}^{10} = \$262$.

16.7   d) $54,000(A/P)_{30}^8 = \$4800$.

16.8   c) $10,000(F/P)_4^6 - 3000 = \$9625$.

16.9   a) $18,000(A/F)_8^4 = \$1953$.

16.10   a) $1000(P/A)_{\infty}^{10} + 1000(P/A)_{10}^{10} = \$16,145$. Note: for the first 10 years, this accounts for \$2000/yr.

16.11   a) $20,000(P/A)_6^8 (P/F)_1^8 = \$85,600$.

16.12   b) $e^{0.1} - 1 = 0.10517$ or 10.517%.

16.13   b) $675 = 50 + 30(P/A)_{24}^i$. $\therefore (P/A)_{24}^i = 20.833$.
           $\therefore$ by trial and error $i = 0.0116$. $\therefore 12i = 0.139$ or 13.9%.

16.14   b) $500,000 + \left[200,000 + 100,000(A/P)_3^4\right](P/A)_{\infty}^4 = \$6,400,000$.

16.15   a) $A: \; -16,000(A/P)_8^{12} - 2000 + 2000(A/F)_8^{12} = -\$5058$.
            $B: \; -30,000(A/P)_{15}^{12} - 1000 + 5000(A/F)_{15}^{12} = -\$5270$

16.16   d) $P(A/P)_{20}^{10} = \left[30,000 + 10,000(P/F)_4^{10}\right](A/P)_8^{10}$. $\therefore P = \$58,760$.

16.17   b) $120,000 + 9000(P/A)_6^{10} - 25,000(P/F)_6^{10} = \$145,000$.

16.18   e) $\left[100,000(A/P)_{10}^{10} + 10,000\right](P/A)_{\infty}^{10} = \$262,700$.

16.19   d) $A: \; 50,000 + 800(P/A)_{20}^8 = \$57,900$
            $B: \; 30,000 + 500(A/P)_{20}^8 + \left[30,000 + 400(P/A)_{10}^8\right](P/F)_{10}^8 = \$50,000$.

16.20 b) $A: -116 + 0.93 \times 206 (P/A)_8^{10} = \$906$
$B: -60 + 0.89 \times 206 (P/A)_8^{10} = \$918$

16.21 b) $A: \left[-25{,}000 + 83{,}000(P/F)_3^{10}\right](A/P)_6^{10} - 6000 + 13{,}000(A/F)_6^{10} = \$4260.$

16.22 d) $(20{,}000 + P)(A/P)_9^8 - 300 - 5000(A/F)_9^8 = 20{,}000(A/P)_6^8 - 5000(A/F)_6^8.$ $\therefore P = \$7140.$

16.23 b) $(23{,}000 - 15{,}000) + (23{,}000 - 32{,}500)(P/F)_N^6 = 0.$ $(1.06)^{-N} = 0.84.$ $\therefore N = 3$ yrs.

16.24 b) $-3500(A/P)_5^8 + 12(200 - 50) = \$923.$

16.25 c) $5000 + 1.50n = 4.00n.$ $\therefore n = 2000.$

16.26 b) $40{,}000x = 500{,}000(A/P)_{15}^8 - 100{,}000(A/F)_{15}^8 + 30{,}000x.$ $\therefore x = 5.47.$ Use $x = 6.$

16.27 d) A) $1200 + 40(20 + x) = 90(20) + 50x.$ $\therefore x = 20$
     b) B) $65(50) - 65(40) - 1200 = -550$

16.28 d) $4500(A/P)_{10}^8 + (18.7 + .1x)75/.9 = 3000(A/P)_{10}^8 + (18.7 + .1x)75/.89.$ $\therefore x = 2200$ hr., B

16.29 d) $10{,}000 - (10{,}000 - 1300)8/12 = 4200$

16.30 a) $(300{,}000 - 50{,}000)/7 = 35{,}714$
$300{,}000 - 4(35{,}714) = 157{,}143$

16.31 c) $(300{,}000/7)0.5 = 21{,}419$
$300{,}000/7 = 42{,}857$
$300{,}000 - 3.5(42{,}857) = 150{,}000$

16.32 e) $300{,}000 \times 2/7 = 85{,}714$
$(300{,}000 - 85{,}714)2/7 = 61{,}224$
$300{,}000 - 85{,}714 - 61{,}224 - 43{,}732 - 31{,}237 = 78{,}092$

16.33 d) $(300{,}000 \times 2/7)0.5 = 42{,}857$
$(300{,}000 - 42{,}857)2/7 = 73{,}469$
$300{,}000 - 42{,}857 - 73{,}469 - 52{,}478 - 37{,}484 = 93{,}711$

TABLE 16.2 Compound Interest Factors

$i = \frac{1}{2}\%$

n	(P/F)	(P/A)	(P/G)	(F/P)	(F/A)	(A/P)	(A/F)	(A/G)	n
1	.9950	0.995	0.000	1.005	1.000	1.0050	1.0000	0.000	1
2	.9901	1.895	0.990	1.010	2.005	0.5038	0.4988	0.499	2
3	.9851	2.970	2.960	1.015	3.015	0.3367	0.3317	0.997	3
4	.9802	3.950	5.901	1.020	4.030	0.2531	0.2481	1.494	4
5	.9754	4.926	9.803	1.025	5.050	0.2030	0.1980	1.990	5
6	.9705	5.896	14.655	1.030	6.076	0.1696	0.1646	2.485	6
7	.9657	6.862	20.449	1.036	7.106	0.1457	0.1407	2.980	7
8	.9609	7.823	27.176	1.041	8.141	0.1278	0.1228	3.474	8
9	.9561	8.779	34.824	1.046	9.182	0.1139	0.1089	3.967	9
10	.9513	9.730	43.386	1.051	10.228	0.1028	0.0978	4.459	10
11	.9466	10.677	52.853	1.056	11.279	0.0937	0.0887	4.950	11
12	.9419	11.619	63.214	1.062	12.336	0.0861	0.0811	5.441	12
13	.9372	12.556	74.460	1.067	13.397	0.0796	0.0746	5.930	13
14	.9326	13.489	86.583	1.072	14.464	0.0741	0.0691	6.419	14
15	.9279	14.417	99.574	1.078	15.537	0.0694	0.0644	6.907	15
16	.9233	15.340	113.424	1.083	16.614	0.0652	0.0602	7.394	16
17	.9187	16.259	128.123	1.088	17.697	0.0615	0.0565	7.880	17
18	.9141	17.173	143.663	1.094	18.786	0.0582	0.0532	8.366	18
19	.9096	18.082	160.036	1.099	19.880	0.0553	0.0503	8.850	19
20	.9051	18.987	177.232	1.105	20.979	0.0527	0.0477	9.334	20
21	.9006	19.888	195.243	1.110	22.084	0.0503	0.0453	9.817	21
22	.8961	20.784	214.061	1.116	23.194	0.0481	0.0431	10.299	22
23	.8916	21.676	233.677	1.122	24.310	0.0461	0.0411	10.781	23
24	.8872	22.563	254.082	1.127	25.432	0.0443	0.0393	11.261	24
25	.8828	23.446	275.269	1.133	26.559	0.0427	0.0377	11.N1	25
26	.8784	24.342	297.228	1.138	27.692	0.0411	0.0361	12.220	26
28	.8697	26.068	343.433	1.150	29.975	0.0384	0.0334	13.175	28
30	.8610	27.794	392.632	1.161	32.280	0.0360	0.0310	14.126	30
∞	0	200.000	40000.0	∞	∞	.0050	0	200.00	∞

$i = 2.00\%$

n	(P/F)	(P/A)	(P/G)	(F/P)	(F/A)	(A/P)	(A/F)	(A/G)	n
1	.9804	0.9804	0.0000	1.0200	1.0000	1.0200	1.0000	0.0000	1
2	.9612	1.9416	0.9612	1.0404	2.0200	0.5150	0.4950	0.4950	2
3	.9423	2.8839	2.8458	1.0612	3.0604	0.3468	0.3268	0.9868	3
4	.9238	3.8077	5.6173	1.0824	4.1216	0.2626,	0.2426	1.4752	4
5	.9057	4.7135	9.2403	1.1041	5.2040	0.2122	0.1922	1.9604	5
6	.8880	5.6014	13.6801	1.1262	6.3081	0.1785	0.1585	2.4423	6
7	.8706	6.4720	18.905	1.1487	7.4343	0.1545	0.1345	2.9208	7
8	.8535	7.3255	24.8779	1.1717	8.5830	0.1365	0.1165	3.3961	8
9	.8368	8.1622	31.5720	1.1951	9.7546	0.1225	0.1025	3.8681	9
10	.8203	8.9826	38.9551	1.2190	10.9497	0.1113	0.0913	4.3367	10
11	.8043	9.7868	46.9977	1.2434	12.1687	0.1022	0.0822	4.8021	11
12	.7885	10.5753	55.6712	1.2682	13.4121	0.0946	0.0746	5.2642	12
13	.7730	11.3484	64.9475	1.2936	14.6803	0.0881	0.0681	5.7231	13
14	.7579	12.1062	74.7999	1.3195	15.9739	0.0826	0.0626	6.1786	14
15	.7430	12.8493	85.2021	1.3459	17.2934	0.0778	0.0578	6.6309	15
16	.7284	13.5777	96.1288	1.3728	18.6393	0.0737	0.0537	7.0799	16
17	.7142	14.2919	107.5554	1.4002	20.0121	0.0700	0.0500	7.5256	17
18	.7002	14.9920	119.4581	1.4282	21.4123	0.0667	0.0467	7.9681	18
19	.6864	15.6785	131.8139	1.4568	22.8406	0.0638	0.0438	8.4073	19
20	.6730	16.3514	144.6003	1.4859	24.2974	0.0612	0.0412	8.8433	20
21	.6598	17.0112	157.7959	1.5157	25.7833	0.0588	0.0388	9.2760	21
22	.6468	17.6580	171.3795	1.5460	27.2990	0.0566	0.0366	9.7055	22
23	.6342	18.2922	185.3309	1.5769	28.8450	0.0547	0.0347	10.1317	23
24	.6217	18.9139	199.6305	1.6084	30.4219	0.0529	0.0329	10.5547	24
25	.6095	19.5235	214.2592	1.6406	32.0303	0.0512	0.0312	10.9745	25
26	.5976	20.1210	229.1987	1.6734	33.6709	0.0497	0.0297	11.3910	26
28	.5744	21.2813	259.9392	1.7410	37.0512	0.0470	0.0270	12.2145	28
30	.5521	22.3965	291.7164	1.8114	40.5681	0.0446	0.0246	13.0251	30
∞	.0000	50.0000	2500.0000	∞	∞	0.0200	0.0000	50.0000	∞

TABLE 16.2 Compound Interest Factors (continued)

$i = 4.00\%$

n	(P/F)	(P/A)	(P/G)	(F/P)	(F/A)	(A/P)	(A/F)	(A/G)	n
1	.9615	0.9615	−0.0000	1.0400	1.0000	1.0400	1.0000	−0.0000	1
2	.9246	1.8861	0.9246	1.0816	2.0400	0.5302	0.4902	0.4902	2
3	.8890	2.7751	2.7025	1.1249	3.1216	0.3603	0.3203	0.9739	3
4	.8548	3.6299	5.2670	1.1699	4.2465	0.2755	0.2355	1.4510	4
5	.8219	4.4518	8.5547	1.2167	5.4163	0.2246	0.1846	1.9216	5
6	.7903	5.2421	12.5062	1.2653	6.6330	0.1908	0.1508	2.3857	6
7	.7599	6.0021	17.0657	1.3159	7.8983	0.1666	0.1266	2.8433	7
8	.7307	6.7327	22.1806	1.3686	9.2142	0.1485	0.1085	3.2944	8
9	.7026	7.4353	27.8013	1.4233	10.5828	0.1345	0.0945	3.7391	9
10	.6756	8.1109	33.8814	1.4802	12.0061	0.1233	0.08333	4.1773	10
11	.6496	8.7605	40.3772	1.5395	13.4864	0.1141	0.0741	4.6090	11
12	.6246	9.3851	47.2477	1.6010	15.0258	0.1066	0.0666	5.0343	12
13	.6006	9.9856	54.4546	1.6651	16.6268	0.1001	0.0601	5.4533	13
14	.5775	10.5631	61.9618	1.7317	18.2919	0.0947	0.0547	5.8659	14
15	.5553	11.1184	69.7355	1.8009	20.0236	0.0899	0.0499	6.2721	15
16	.5339	11.6523	77.7441	1.8730	21.8245	0.0858	0.0458	6.6720	16
17	.5134	12.1657	85.9581	1.9479	23.6975	0.0822	0.0422	7.0656	17
18	.4936	12.6593	94.3498	2.0258	25.6454	0.0790	0.0390	7.4530	18
19	.4746	13.1339	102.8933	2.1068	27.6712	0.0761	0.0361	7.8342	19
20	.4564	13.5903	111.5647	2.1911	29.7781	0.0736	0.0336	8.2091	20
21	.4388	14.0292	120.3414	2.2788	31.9692	0.0713	0.0313	8.5779	21
22	.4220	14.4511	129.2024	2.3699	34.2480	0.0692	0.0292	8.9407	22
23	.4057	14.8568	138.1284	2.4647	36.6179	0.0673	0.0273	9.2973	23
24	.3901	15.2470	147.1012	2.5633	39.0826	0.0656	0.0256	9.6479	24
25	.3751	15.6221	156.1040	2.6658	41.6459	0.0640	0.0240	9.9925	25
26	.3607	15.9828	165.1212	2.7725	44.3117	0.0626	0.0226	10.3312	26
28	.3335	16.6631	183.1424	2.9987	49.9676	0.0600	0.0200	10.9909	28
30	.3083	17.2920	201.0618	3.2434	56.0849	0.0578	0.0178	11.6274	30
∞	.0000	25.000	625.0000	∞	∞	0.0400	0.0000	25.0000	∞

$i = 6.00\%$

n	(P/F)	(P/A)	(P/G)	(F/P)	(F/A)	(A/P)	(A/F)	(A/G)	n
1	.9434	0.9434	−0.0000	1.0600	1.0000	1.0600	1.0000	−0.0000	1
2	.8900	1.8334	0.8900	1.1236	2.0600	0.5454	0.4854	0.4854	2
3	.8396	2.6730	2.5692	1.1910	3.1836	0.3741	0.3141	0.9612	3
4	.7921	3.4651	4.9455	1.2625	4.3746	0.2886	0.2286	1.4272	4
5	.7473	4.2124	7.9345	1.3382	5.6371	0.2374	0.1774	1.8836	5
6	.7050	4.9173	11.4594	1.4185	6.9753	0.2034	0.1434	2.3304	6
7	.6651	5.5824	15.4497	1.5036	8.3938	0.1791	0.1191	2.7676	7
8	.6274	6.2098	19.8416	1.5938	9.8975	0.1610	0.1010	3.1952	8
9	.5919	6.8017	24.5768	1.6895	11.4913	0.1470	0.0870	3.6133	9
10	.5584	7.3601	29.6023	1.7908	13.1808	0.1359	0.0759	4.0220	10
11	.5268	7.8869	34.8702	1.8983	14.9716	0.1268	0.0668	4.4213	11
12	.4970	8.3838	40.3369	2.0122	16.8699	0.1193	0.0593	4.8113	12
13	.4688	8.8527	45.9629	2.1329	18.8821	0.1130	0.0530	5.1920	13
14	.4423	9.2950	51.7128	2.2609	21.0151	0.1076	0.0476	5.5635	14
15	.4173	9.7122	57.5546	2.3966	23.2760	0.1030	0.0430	5.9260	15
16	.3936	10.1059	63.4592	2.5404	25.6725	0.0990	0.0390	6.2794	16
17	.3714	10.4773	69.4011	2.6928	28.2129	0.0954	0.0354	6.6240	17
18	.3503	10.8276	75.3569	2.8543	30.9057	0.0924	0.0324	6.9597	18
19	.3305	11.1581	81.3062	3.0256	33.7600	0.0896	0.0296	7.2867	19
20	.3118	11.4699	87.2304	3.2071	36.7856	0.0872	0.0272	7.6051	20
21	.2942	11.7641	93.1136	3.3996	39.9927	0.0850	0.0250	7.9151	21
22	.2775	12.0416	98.9412	3.6035	43.3923	0.0830	0.0230	8.2166	22
23	.2618	12.3034	104.7007	3.8197	46.9958	0.0813	0.0213	8.5099	23
24	.2470	12.5504	110.3812	4.0489	50.8156	0.0797	0.0197	1.87951	24
2s	.2330	12.7834	115.9732	4.2919	54.8645	0.0782	0.0182	9.0722	25
26	.2198	13.0032	121.4684	4.5494	59.1564	0.0769	0.0169	9.3414	26
28	.1956	13.4062	132.1420	5.1117	68.5281	0.0746	0.0146	9.8568	28
30	.1741	13.7648	142.3588	5.7435	79.0582	0.0726	0.0126	10.3422	30
∞	.0000	16.6667	277.7778	∞	∞	0.0600	0.0000	16.667	∞

TABLE 16.2 Compound Interest Factors (continued)

$i = 8.00\%$

n	(P/F)	(P/A)	(P/G)	(F/P)	(F/A)	(A/P)	(A/F)	(A/G)	n
1	.9259	0.9259	−0.0000	1.0800	1.0000	1.0800	1.0000	−0.0000	1
2	.8573	1.7833	0.8573	1.1664	2.0800	0.5608	0.4808	0.4808	2
3	.7938	2.5771	2.4450	1.2597	3.2464	0.3880	0.3080	0.9487	3
4	.7350	3.3121	4.6501	1.3605	4.5061	0.3019	0.2219	1.4040	4
5	.6806	3.9927	7.3724	1.4693	5.8666	0.2505	0.1705	1.8465	5
6	.6302	4.6229	10.5233	1.5869	7.3359	0.2163	0.1363	2.2763	6
7	.5835	5.2064	14.0242	1.7138	8.9228	0.1921	0.1121	2.6937	7
8	.5403	5.7466	17.8061	1.8509	10.6366	0.1740	0.0940	3.0985	8
9	.5002	6.2469	21.8081	1.9990	12.4876	0.1601	0.0801	3.4910	9
10	.4632	6.7101	25.9768	2.1589	14.4866	0.1490	0.0690	3.8713	10
11	.4289	7.1390	30.2657	2.3316	16.6455	0.1401	0.0601	4.2395	11
12	.3971	7.5361	34.6339	2.5182	18.9771	0.1327	0.0527	4.5957	12
13	.3677	7.9038	39.0463	2.7196	21.4953	0.1265	0.0465	4.9402	13
14	.3405	8.2442	43.4723	2.9372	24.2149	0.1213	0.0413	5.2731	14
15	.3152	8.5595	47.8857	3.1722	27.1521	0.1168	0.0368	5.5945	15
16	.2919	8.8514	52.2640	3.4259	30.3243	0.1130	0.0330	5.9046	16
17	.2703	9.1216	56.5883	3.7000	33.7502	0.1096	0.0296	6.2037	17
18	.2502	9.3719	60.8426	3.9960	37.4502	0.1067	0.0267	6.4920	18
19	.2317	9.6036	65.0134	4.3157	41.4463	0.1041	0.0241	6.7697	19
20	.2145	9.8181	69.0898	4.6610	45.7620	0.1019	0.0219	7.0369	20
21	.1987	10.0168	73.0629	5.0338	50.4229	0.0998	0.0198	7.2940	21
22	.1839	10.2007	76.9257	5.4365	55.4568	0.0980	0.0180	7.5412	22
23	.1703	10.3711	80.6726	5.8715	60.8933	0.0964	0.0164	7.7786	23
24	.1577	10.5288	84.2997	6.3412	66.7648	0.0950	0.0150	8.0066	24
25	.1460	10.6748	87.8041	6.8485	73.1059	0.0937	0.0137	8.2254	25
26	.1352	10.8100	91.1842	7.3964	79.9544	0.0925	0.0125	8.4352	26
28	.1159	11.0511	97.5687	8.6271	95.3388	0.0905	0.0105	8.8289	28
30	.0994	11.2578	103.4558	10.0627	113.2832	0.0888	0.0088	9.1897	30
∞	.0000	12.500	156.2500	∞	∞	0.0800	0.0000	12.5000	∞

$i = 10.00\%$

n	(P/F)	(P/A)	(P/G)	(F/P)	(F/A)	(A/P)	(A/F)	(A/G)	n
1	.9091	0.9091	−0.0000	1.1000	1.0000	1.1000	1.0000	−0.0000	1
2	.8264	1.7355	0.8264	1.2100	2.1000	0.5762	0.4762	0.4762	2
3	.7513	2.4869	2.3291	1.3310	3.3100	0.4021	0.3021	0.9366	3
4	.6830	3.1699	4.3781	1.4641	4.6410	0.3155	0.2155	1.3812	4
5	.6209	3.7908	6.8618	1.6105	6.1051	0.2638	0.1638	1.8101	5
6	.5645	4.3553	9.6842	1.7716	7.7156	0.2296	0.1296	2.2236	6
7	.5132	4.8684	12.7631	1.9487	9.4872	0.2054	0.1054	2.6216	7
8	.4665	5.3349	16.0287	2.1436	11.4359	0.1874	0.0874	3.0045	8
9	.4241	5.7590	19.4215	2.3579	13.5795	0.1736	0.0736	3.3724	9
10	.3855	6.1446	22.8913	2.5937	15.9374	0.1627	0.0627	3.7255	10
11	.3505	6.4951	26.3963	2.8531	18.5312	0.1540	0.0540	4.0641	11
12	.3186	6.8137	29.9012	3.1384	21.3843	0.1468	0.0468	4.3884	12
13	.2897	7.1034	33.3772	3.4523	24.5227	0.1408	0.0408	4.6988	13
14	.2633	7.3667	36.8005	3.7975	27.9750	0.1357	0.0357	4.9955	14
15	.2394	7.6061	40.1520	4.1772	31.7725	0.1315	0.0315	5.2789	15
16	.2176	7.8237	43.4164	4.5950	35.9497	0.1278	0.0278	5.5493	16
17	.1978	8.0216	46.5819	5.0545	40.5447	0.1247	0.0247	5.8071	17
18	.1799	8.2014	49.6395	5.5599	45.5992	0.1219	0.0219	6.0526	18
19	.1635	8.3649	52.5827	6.1159	51.1591	0.1195	0.0195	6.2861	19
20	.1486	8.5136	55.4069	6.7275	57.2750	0.1175	0.0175	6.5081	20
21	.1351	8.6487	58.1095	7.4002	64.0025	0.1156	0.0156	6.7189	21
22	.1228	8.7715	60.6893	8.1403	71.4027	0.1140	0.0140	6.9189	22
23	.1117	8.8832	63.1462	8.9543	79.5430	0.1126	0.0126	7.1085	23
24	.1015	8.9847	65.4813	9.8497	88.4973	0.1113	0.0113	7.2881	24
25	.0923	9.0770	67.6964	10.8347	98.3471	0.1102	0.0102	7.4580	25
26	.0839	9.1609	69.7940	11.9182	109.1818	0.1092	0.0092	7.6186	26
28	.0693	9.3066	73.6495	14.4210	134.2099	0.1075	0.0075	7.9137	28
30	.0573	9.4269	77.0766	17.4494	164.4940	0.1061	0.0061	8.1762	30
∞	.0000	10.0000	100.0000	∞	∞	0.1000	0.0000	10.0000	∞

TABLE 16.2 Compound Interest Factors (continued)

$i = 12.00\%$

n	(P/F)	(P/A)	(P/G)	(F/P)	(F/A)	(A/P)	(A/F)	(A/G)	n
1	.8929	0.8929	-0.0000	1.1200	1.0000	1.1200	1.0000	-0.0000	1
2	.7972	1.6901	0.7972	1.2544	2.1200	0.5917	0.4717	0.4717	2
3	.7118	2.4018	2.2208	1.4049	3.3744	0.4163	0.2963	0.9246	3
4	.6355	3.073	4.1273	1.5735	4.7793	0.3292	0.2092	1.3589	4
5	.5674	3.6048	6.3970	1.7623	6.3528	0.2774	0.1574	1.7746	5
6	.5066	4.1114	8.9302	1.9738	8.1152	0.2432	0.1232	2.1720	6
7	.4523	4.5638	11.6443	2.2107	10.0890	0.2191	0.0991	2.5515	7
8	.4039	4.9676	14.4714	2.4760	12.2997	0.2013	0.0813	2.9131	8
9	.3606	5.3282	17.3563	2.7731	14.7757	0.1877	0.0677	3.2574	9
10	.3220	5.6502	20.2541	3.1058	17.5487	0.1770	0.0570	3.5847	10
11	.2875	5.9377	23.1288	3.4785	20.6546	0.1684	0.0484	3.8953	11
12	.2567	6.1944	25.9523	3.8960	24.1331	0.1614	0.0414	4.1897	12
13	.2292	6.4235	28.7024	4.3635	28.0291	0.1557	0.0357	4.4683	13
14	.2046	6.6282	31.3624	4.8871	32.3926	0.1509	0.0309	4.7317	14
15	.1827	6.8109	33.9202	5.4736	37.2797	0.1468	0.0268	4.9803	15
16	.1631	6.9740	36.3670	6.1304	42.7533	0.1434	0.0234	5.2147	16
17	.1456	7.1196	38.6973	6.8660	48.8837	0.1405	0.0205	5.4353	17
18	.1300	7.2497	40.9080	7.6900	55.7497	0.1379	0.0179	5.6427	18
19	.1161	7.3658	42.9979	8.6128	63.4397	0.1358	0.0158	5.8375	19
20	.1037	7.4694	44.9676	9.6463	72.0524	0.1339	0.0139	6.0202	20
21	.0926	7.5620	46.8188	10.8038	81.6987	0.1322	0.0122	6.1913	21
22	.0826	7.6446	48.5543	12.1003	92.5026	0.1308	0.0108	6.3514	22
23	.0738	7.7184	50.1776	13.5523	104.6029	0.1296	0.0096	6.5010	23
24	.0659	7.7843	51.6929	15.1786	118.1552	0.1285	0.0085	6.6406	24
25	.0588	7.8431	53.1046	17.0001	133.3339	0.1275	0.0075	6.7708	25
26	.0525	7.8957	54.4177	19.0401	150.3339	0.1267	0.0067	6.8921	26
28	.0419	7.9844	56.7674	23.8839	190.6989	0.1252	0.0052	7.1098	28
30	.0334	8.0552	58.7821	29.9599	241.3327	0.1241	0.0041	7.2974	30
∞	.0000	8.333	69.4444	∞	∞	0.1200	0.0000	8.3333	∞

$i = 20.00\%$

n	(P/F)	(P/A)	(P/G)	(F/P)	(F/A)	(A/P)	(A/F)	(A/G)	n
1	.8333	0.8333	-0.0000	1.2000	1.0000	1.2000	1.0000	-0.0000	1
2	.6944	1.5278	0.6944	1.4400	2.2000	0.6545	0.4545	0.4545	2
3	.5787	2.1065	1.8519	1.7280	3.6400	0.4747	0.2747	0.8791	3
4	.4823	2.5887	3.2986	2.0736	5.3680	0.3863	0.1863	1.2742	4
5	.4019	2.9906	4.9061	2.4883	7.4416	0.3344	0.1344	1.6405	5
6	.3349	3.3255	6.5806	2.9860	9.9299	0.3007	0.1007	1.9788	6
7	.2791	3.6046	8.2551	3.5832	12.9159	0.2774	0.0774	2.2902	7
8	.2326	3.8372	9.8831	4.2998	16.4991	0.2606	0.0606	2.5756	8
9	.1938	4.0310	11.4335	5.1598	20.7989	0.2481	0.0481	2.8364	9
10	.1615	4.1925	12.8871	6.1917	25.9587	0.2385	0.0385	3.0739	10
11	.1346	4.3271	14.2330	7.4301	32.1504	0.2311	0.0311	3.2893	11
12	.1122	4.4392	15.4667	8.9161	39.5805	0.2253	0.0253	3.4841	12
13	.0935	4.5327	16.5883	10.6993	48.4966	0.2206	0.0206	3.6597	13
14	.0779	4.6106	17.6008	12.8392	59.1959	0.2169	0.0169	3.8175	14
15	.0649	4.6755	18.5095	15.4070	72.0351	0.2139	0.0139	3.9588	15
16	.0541	4.7296	19.3208	18.4884	87.4421	0.2114	0.0114	4.0851	16
17	.0451	4.7746	20.0419	22.1861	105.9306	0.2094	0.0094	4.1976	17
18	.0376	4.8122	20.6805	26.6233	128.1167	0.2078	0.0078	4.Z975	18
19	.0313	4.8435	21.2439	31.9480	154.7400	0.2065	0.0065	4.3861	19
20	.0261	4.8696	21.7395	38.3376	186.6880	0.2054	0.0054	4.4643	20
21	.0217	4.8913	22.1742	46.0051	225.0256	0.2044	0.0044	4.5334	21
22	.0181	4.9094	22.5546	55.2061	271.0307	0.2037	0.0037	4.5941	22
23	.0151	4.9245	22.8867	66.2474	326.2369	0.2031	0.0031	4.6475	23
24	.0126	4.9371	23.1760	79.4968	392.4842	0.2025	0.0025	4.6943	24
25	.0105	4.9476	23.4276	95.3962	471.9811	0.2021	0.0021	4.7352	25
26	.0087	4.9563	23.6460	114.4755	567.3773	0.2018	0.0018	4.7709	26
28	.0061	4.9697	23.9991	164.8447	819.2233	0.2012	0.0012	4.8291	28
30	.0042	4.9789	24.2628	237.3763	1181.8816	0.2008	0.0008	4.8731	30
∞	.0000	5.0000	25.0000	∞	∞	0.2000	0.0000	5.5000	∞

Index

A
accelerated cost recovery system, 16–16
accident rate, 13–8
activated sludge, 12–19
active earth pressure, 15–25
adjoint matrix, 1–13
adjustment factors, –22
advanced wastewater treatment, 12–24
aerated lagoons, 12–20
aerobic digester design, 12–27
aerobic digesters, 12–28
aerobic digestion units, 12–27
aerobic pond, 12–20
alkalinity, 11–3
allowable bearing pressure, 15–4
allowable compression stress, 8–10
allowable stress design, 8–1
alternate depths, 5–4
anaerobic digesters, 12–30
anaerobic ponds, 12–20
anchorage failure modes, 9–18
annual compounding, 16–8
annual net equivalent value, 16–9
antecedent moisture condition, 7–6
aquifers, 7–26
area reduction coefficient, 8–3
at-rest earth pressure, 15–25
Atterberg limits, 14–10
auxiliary lanes, 13–23
available chlorine, 11–35

B
backwash, 11–27
band width, 13–33
bar numbers, 9–1
bar racks, 12–5
batch flux curve, 11–19
batch settling curve, 11–19
beam bearing plates, 8–38
beams, 8–14, 9–4, 10–6
beam shear, 9–32
benefit/cost ratio, 16–12
Bernoulli equation, 4–10
binomial theorem, 1–2
biochemical oxygen demand, 12–1
block shear, 8–36
body forces, 2–1
bolt strength, 8–34
bolted connections, 8–34
bolts in tension and shear, 8–37
Boussinesq's solution, 15–11
branching pipe systems, 6–12
braking distance, 13–10
break-even analysis, 16–14
broad-crested weir, 5–10
buckling, 3–15
buoyant unit weight, 14–2

C
caissons, 15–4
cantilever retaining walls, 9–36
capacity reduction factor, 9–9
capital cost, 16–10
carbon adsorption, 12–26
carbonate hardness, 11–11
cash flow diagrams, 16–1
cash flow pattern, 16–2
cavitation, 4–2, 6–15
center of gravity, 2–17
centrifugal pumps, 6–14
centroid, 2–12
Chezy-Manning equation, 4–16, 5–3
chloramines, 11–35
circumferential stress, 3–13
clarifier, 11–19
clay, 15–32
coagulation, 11–2
coarse fraction, 14–11
coarse-grained soils, 14–9
coefficient of curvature, 14–10
coefficient of friction, 2–10
coefficient of gradation, 14–10
coefficient of permeability, 14–19
coefficient of thermal expansion, 3–2
coefficient of uniformity, 14–10
cofactor, 1–13
cohesion, 15–3, 14–27
cold working, 8–2
collision diagram, 13–8
column, 8–26, 9–7
column base plates, 8–40
combined axial force plus flexure, 8–25
combined footings, 9–31
combined stress, 3–10
compaction, 14–14
compaction curve, 14–14
complex number, 1–11
composite areas, 2–15
composite bar, 3–14
compound interest factors, 16–26
compression buckling, 8–7
compression buckling of web, 8–39

compression flanges, 8–16
compression index, 14–23
compression stress, 9–1
concentrated force, 2–1
concrete column, 9–7
concrete frames, 9–2
concurrent, 2–1
cone penetration test, 15–3
cone resistance, 15–3
conic sections, 1–8
conjugate depths, 5–5
connections, 8–31
consolidation settlement, 15–16
consolidation, 14–23
continuous compounding, 16–8
control, 5–11
control volumes, 4–9
corridor design, 13–32
Coulomb's active earth pressure coefficient, 15–27
Cramer's rule, 1–14
crest vertical curve, 13–18
critical depth, 5–4
critical limit stress, 8–7
critical load, 3–16
critical movement, 13–4
cross product, 1–23
cross slope, 13–22
cross-sectional shapes, 5–1
curve number, 7–6
cylinder, 1–7
cylindrical coordinate system, 1–9

D

Darcy's Law, 7–27, 14–19
Darcy-Weisbach equation, 4–10, 6–1
dead load, 9–3
deceleration lengths, 13–23
decision and response initiation time, 13–14
decision sight distance, 13–14
declining balance, 16–16
deep bed monofilters, 11–25
definite integral, 1–19
degree of curvature, 13–16
degree of saturation, 14–2
degree of static indeterminacy, 10–1
degrees of freedom, 10–3
density, 4–2, 4–3
density index, 14–17
deoxygenation rate, 12–2
depreciation, 16–15
derivative, 1–17
design rainfall, 7–3
design speed, 13–2
design storm, 7–3
design vehicles, 13–1
detection and recognition time, 13–14

detention times, 11–18
determinant, 1–13
dimensionless parameters, 4–7
direct shear test, 14–27
disinfection, 11–34
dissolved oxygen, 12–2
dissolved oxygen concentration, 12–2
distribution factor, 10–16
dot product, 1–23
double declining balance, 16–16
drag coefficient, 11–17
drainage, 13–36
drained strength, 14–27
dry density, 14–2
dry unit weight, 14–2
dual-media filters, 11–25
ductility, 8–2, 9–5
Dupuit approximation, 7–29
Dutch cone test, 15–3

E

eccentricity, 9–9, 15–9
eccentricity of axial force, 9–7
effective horizontal stress, 14–18
effective net area, 8–3
effective porosity, 7–27
effective stress principle, 14–17
effective stresses, 14–28
effective unit weight, 14–2
effective vertical stress, 14–18
efficiency, 6–4
elbows, 4–10
ellipse, 1–8
embedment distance, 9–18
energy equation, 5–3, 6–4
energy grade line, 4–11
enlargements, 4–10
equalization tanks, 12–6
equilibrium, 2–4
equivalent cash flow patterns, 16–3
equivalent joint loads, 10–5
equivalent length, 4–11
equivalent weight, 11–3
estimated salvage, 16–17
Euler buckling stress, 8–7
Euler's equation, 1–11
evapotranspiration, 7–2
exponents, 1–1

F

facultative pond, 12–20
filter boxes, 11–28
filter press, 12–34
filtration, 11–25
fine-grained soils, 14–9

fixed ends, 10–5
flexibility method, 10–9, 10–26
flexural ductility, 9–5
flocculation, 11–2
flood frequency, 7–13
flood probability, 7–13
flood routing, 7–24
flow net, 14–21
food to microorganism ratio, 12–11
footings, 9–31, 15–3
foundations, 15–3
frames, 2–6
free body diagram, 2–4
free chlorine, 11–34
friction angle, 2–10, 14–27, 15–2
friction force, 2–10
Froude number, 5–4

G

geometric similarity, 4–7
gradually varied flow, 5–12
grain size, 11–26
grain-size distribution, 14–9
gravel, 14–9
gravity sludge, 11–19
Greenshield's headway, 13–3
grit chambers, 12–5
groundwater, 7–26
groundwater outflow, 7–2
groundwater storage, 7–2

H

hardness, 11–11
Hazen-Williams, 6–1
head loss, 5–3
head loss during filtration, 11–26
headlight sight distance, 13–20
high rate anaerobic digestion, 12–30
highway capacity, 13–2
highway cross section, 13–22
highway drainage, 13–36
highway safety, 13–8
hooked bars, 9–19
Hooke's law, 3–2
hydraulic conductivity, 14–19
hydraulic grade line, 4–11, 6–4
hydraulic gradient, 14–19
hydraulic jump, 5–7
hydraulic radius, 5–3
hydrographs, 7–10
hydrologic, 7–1
hydrologic soil groups, 7–6
hydrology, 7–1
hyperbola, 1–8
hyperbolic trig functions, 1–6

I

immediate settlement, 15–16
income tax, 16–15
incompressible flow, 4–9
indefinite integral, 1–19
infiltration, 7–5
inflection point, 1–18
initial abstraction, 7–5
independent alternatives, 16–10
inflation, 16–18
installed first cost, 16–17
integration by parts, 1–19
interest, 16–1
interest factors, 16–3
interest tables, 16–29
interrupted flow, 13–3
intersection radii, 13–29
inverse matrix, 1–13
ionic strength, 11–32
isoconcentration line, 11–18
isolated footings, 9–31
isotropy, 3–1

J

joint translation, 10–19
joists, 9–25

K

kinematic viscosity, 4–2, 4–3

L

L'Hospital's rule, 1–18
lagoon, 12–20
lane widths, 13–22
Langelier saturation index, 11–32
lateral buckling, 8–16
lateral earth pressure coefficient, 18–25
law of cosines, 1–6
law of sines, 1–5
level-of-service, 13–2
life cycle cost, 16–10
lime softening sludges, 11–19
limit axial compression strength of a column, 9–7
limit ratio, 9–5
limit stress versus slenderness curve, 8–7
liquid limit, 14–10
liquidity index, 14–10
live load, 9–3
load and resistance factor design, 8–1
load factors, 9–2
logarithms, 1–1

M

Manning roughness coefficient, 5–3
Manning's equation, 7–17, 13–37
mass moments of inertia, 2–18
matrices, 1–13
mats, 15–4
maximum dry unit weight, 14–14
maximum specific growth rate, 12–11
maximum tensile strength, 9–3
maximum yield coefficient, 12–11
median grain size, 14–9
minimum attractive rate of return, 16–9
minor loss, 4–10
mixing, 11–15
modulus of elasticity, 3–2, 8–2
Mohr-Coulomb strength, 14–27
moisture content, 14–2
moment, 2–1
momentum function, 5–5
Moody diagram, 4–10, 6–1
mutually exclusive alternatives, 16–9

N

nappe, 15–10
neutral axis, 3–6, 9–4,
Newton's equation, 11–16
Newtonian fluids, 4–1
nitrogen control, 12–25
nominal compression strength, 8–9
nominal diameter, 9–1
nominal stiffness, 9–2
non-concurrent, 2–1

O

oedometer test, 14–23
open channel, 4–16
optimum water content, 14–14
overconsolidation ratio, 14–24
overflow rates, 11–18
oxidation ponds, 12–20
oxygen deficit, 12–2
oxygen transfer correction factor, 12–21

P

parabola, 1–8
parallel-axis theorem, 2–13
partial fractions, 1–2
passing sight distance, 13–12
passive earth pressure condition, 15–28
Peck's diagrams, 15–31
perception and reaction distance, 13–9
perception and reaction time, 13–9
phase diagram, 14–1, 14–4
phosphorus removal, 12–24
piers, 15–4
piezometric head, 7–27
piles, 15–4, 15–22
plane moments of inertia, 2–13
plane surface, 1–8
plastic moment, 8–15
plasticity index, 14–10
plug flow reactor, 12–11
polar moment of inertia, 2–13
polygon, 1–7
pore water pressure, 14–18
porosity, 14–2
potential cake concentrations, 11–24
precipitation, 7–2,
preconsolidation pressure, 14–23
present net equivalent value, 16–10
present worth, 16–10
prestressed concrete, 9–8
product of inertia, 2–13
pullout strength of bolts, 8–36
pump head, 4–11
pump power, 4–11
punching, 9–31
Pythagorean Theorem, 1–6

Q

quadratic formula, 1–2

R

radii of gyration, 2–13, 8–10
rafts, 15–4
rainfall intensity curves, 13–37
rapid mixing, 11–5
rapidly-varied flow, 5–7
rate of return, 16–9, 16–11
rate of substrate utilization, 12–11
rational formula, 7–14
rational method, 13–36
reaeration rate, 12–2
recirculation factor, 12–18
recirculation ratio, 12–18
recompression index, 14–23
recovery period, 16–16
rectangular footings, 15–7
recurrence interval, 7–13
reduced nominal strength, 9–3
reinforced concrete, 9–1
reinforced concrete columns, 9–7
reinforced concrete walls, 9–29
reinforcement ratio, 9–6
relative density, 14–17
removal fractions, 11–18
Reynolds number, 4–10
ribbed slabs, 9–25
rivets, 8–38

rock media filters, 12–18
Rose equation, 11–27
runoff coefficient, 13–36
runoff curve numbers, 7–7
runoff hydrograph, 7–10
rupture strength limit, 8–3
Rushton's equation, 11–6

S

safety, 13–8
safety factor, 8–2
sag vertical curve, 13–20
salvage, 16–16
sand drying bed, 12–33
sand filters, 11–25
saturated unit weight, 14–2
scalar product, 1–23
secondary compression, 15–16
secondary settling tank design, 12–23
section strength, 9–9
sedimentation, 11–16
sedimentation tank design, 15–22
seepage velocity, 7–27, 14–20
settlement, 15–16
settling tank design, 11–8, 12–9
settling velocity, 11–16, 11–27
sharp-crested weir, 5–10
shear strength of columns, 9–16
shear walls, 9–29
Shultz-Hardy rule, 11–2
side friction factor, 13–15
sidesway web buckling, 8–39
sieve analysis, 14–9
sight distance, 13–19
slabs, 9–22
slenderness, 8–7
slenderness index force, 8–26
slenderness ratio, 3–15, 8–7
slip-critical applications, 8–37
sludge dewatering, 11–24, 12–32
sludge generation, 11–24
sludge treatment, 12–27
sludge volume index, 12–11
softening, 11–11, 11–24
soil modulus, 15–17
space-time diagram, 13–32
specific discharge, 5–4
specific energy, 5–3
specific gravity, 14–3
specific yield, 7–27
sphere, 1–7
spherical (r, q, f) coordinate system, 1–9
spiral columns, 9–12
splices, 9–11, 9–19
squash load, 9–7
stability index, 11–32

stability of water, 11–32
stabilization ponds, 12–20
standard penetration test, 15–1
standard proctor test, 14–14
standard sieve sizes, 11–26
stirrups, 9–14
Stoke's Law, 11–17
stopping sight distance, 13–9,
storage, 7–12
storage length, 13–24
storativity coefficient, 7–27
storm sewer, 13–39
straight line, 1–7
straight line depreciation, 16–16
strain hardening, 8–2
Streeter-Phelps model, 12–2
stress-strain curves for concrete and for steel, 9–3
stress-strain diagram, 8–2
structural steel, 8–1
struts, 15–31
subcritical, 5–4
substrate utilization, 12–11
summation of critical movement, 13–3
supercritical, 5–4
superelevation, 13–15
surcharge, 15–28
surface forces, 2–1
surface runoff, 7–5
swell index, 14–23
synthetic media filters, 12–17

T

taper, 13–24
tapered aeration, 12–11
Taylor's series, 1–18
tension lap splices, 9–19
terminal velocity, 5–2
tertiary treatment, 12–24
Terzaghi bearing capacity, 15–5
Theis method, 7–29
three-force member, 2–4
time of concentration, 7–10
time of recession, 7–18
total head, 14–19
total unit weight, 14–2
total vertical stress, 14–18
traffic signals, 13–32
traffic volumes, 13–2
transition, 5–7
transmissivity, 7–27
transpose, 1–15
transverse stiffeners, 8–19
transverse ties, 9–11
trapezoid, 5–1
trickling filter, 12–17
trigonometry, 1–4

trusses, 10–6
two-force member, 2–4

U

ultimate bearing capacity, 15–4
ultimate bearing pressure, 15–4
ultimate force, 9–3
ultimate moment, 9–4
unconfined compression test, 14–29
undrained strength, 14–27
unified soil classification system, 14–10
uniform flow, 5–2
uniform velocity profiles, 4–9
unit hydrograph method, 7–18
unit hydrograph, 7–11

V

value, 16–1
valves, 4–10
vector product, 1–23
velocity gradient, 11–5
venture worth, 16–10
vertical curves, 13–18
Vesic bearing capacity factors, 15–5
void ratio, 14–2

W

walls, 9–29
wastage flow rate, 12–12
water content, 14–2
water source evaluation, 11–1
water surface profiles, 5–11
web crippling, 8–39
weir, 5–10
weir loading rates, 11–22
welded connections, 8–32
wetted perimeter, 4–11
work, 1–23

Y

yield strain, 8–2
yield strength, 8–2
yield strength index, 8–9
yield strength limit, 8–3

Z

zero-air voids curve, 14–15

The Professional Engineer's Prep Source

✓ Save up to $36.50 per book!

✓ Step #1: Get the Best PE Prep!

○ **Principles & Practice of Civil Engineering**
Edited by Merle C. Potter, PhD, PE (GLP). New 2nd edition of the highly-effective PE review from the publisher of the best FE title. The <u>only</u> PE review written by <u>ten professors</u> from major universities. Concisely covers all critical aspects of the CE/PE. <u>Only</u> PE title with reviews, problems and solutions all in one volume. 693 pages.

✓ *Put me on the waiting list!* ...For: ○ ME/PE (due 10/96) ○ EE/PE (due 5/97)

All-in-one volume!
Only $69.95!

✓ Step #2: Get a Test!

○ **Official NCEES Sample PE Exams—$25 each.** All subjects available. Simply order a sample exam and a discount reference and you have an effective method of preparing for any PE Exam. Just take the exam (open-book) and use the review or handbook to assist you—then you have easy access to all necessary information.

Subjects desired:_____

✓ Step #3: Order a Handbook!

Standard Handbook for Civil Engineers.
By F. Merritt (4th ed., McGraw-Hill). New edition of the most thorough compilation of facts and figures in all areas of civil engineering. Best guide, best buy, and best-seller since 1968! 1,878 pp. ○

Only $99! Reg. $128.50

Reg. $115.50 **Only $95!**

○ **Mark's Standard Handbook for Mechanical Engineers.**
By E. Avallone (9th ed., McGraw-Hill). The book to turn to for practical advice and quick answers on all ME principles, standards and practices. 1,936 pages.

The Electrical Engineering Handbook. Edited by Richard C. Dorf, PhD (CRC Press). The premier EE reference. All-new handbook with up-to-date info in all areas of EE. A must for practicing engineers! 2,661 pages. ○

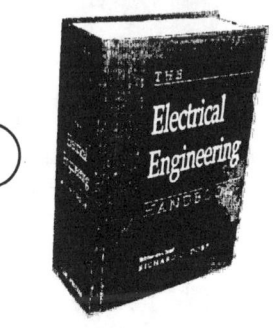

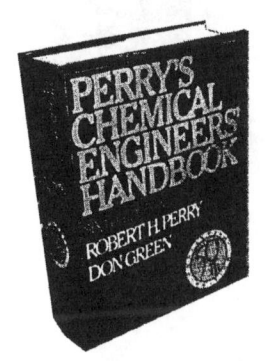

○ **Perry's Chemical Engineer's Handbook.** By R. Perry (6th ed., McGraw-Hill). The best info on all aspects of ChemE, all in one place! The standard for over 40 years! 2,336 pages.

Only $75! Reg. $100

Reg. $129.50 **Only $99!**

Name (print):_____

Street Address:_____

_____

Day phone:_____

PAYMENT

Credit Card: VISA MC (circle one)
Number:
_____

Exp.:_____

Mail this order form, credit info., or check or money order to:

**Great Lakes Press POB 172
Grover, MO 63040-0172**

For info or ordering call 1-800-837-0201/fax 1-314-273-6086

Name of Title	Price	Quan.	Multi-Discount!	Subtotal
PE Sample Exam	$25 x		N.A.	=
PE Exam Review/Civil	$69⁹⁵ x		Take $10 off each book on orders of 3 or more books!	=
CE Handbook	$99 x			=
ME Handbook	$95 x			=
EE Handbook	$75 x			=
ChemE Handbook	$99 x			=
Shipping	$3 x			

(Mich. residents add 6% sales tax.) TOTAL $

CATALOG of titles by GREAT LAKES PRESS

Editor and Principal Author: Dr. Merle C. Potter, Ph.D., P.E.

FUNDAMENTALS OF ENGINEERING, 5th edition. Retail: $49.95

Amazing new edition of the most effective review for the FE/EIT exam. Written by *seven Ph.D. teaching professors*. Now with all new info-rich, easy-to-use text layout and organization. Contains everything you need to pass the exam—no more, no less—for one low price. (Other books contain massive quantities of information *that are not on the exam*.) •Includes hardcover reference book which reviews all 11 engineering subjects tested; with tables and charts; 1100 problems and full solutions; two 8-hour practice exams; NCEES equation summaries and free NCEES Handbook. 600 pages; 8.5"x11".

NEW 5th!

PROFESSIONAL CIVIL ENGINEERING EXAM REVIEW
2nd edition. Retail: $69.95

New second edition of this highly-effective review for the Principles & Practice of Engineering (PE) Civil Exam. Also an excellent desk reference for civil engineers. •The only review written by *Ph.D. professors*—ten experts in their fields. •Includes reviews of all 16 major subject areas tested; hundreds of excellent exam-simulating practice problems with full solutions; equation summaries. The only review with all you need in one volume. 693 pages; hardcover.

NEW 2nd!

GRE TIME•SAVER™, 2nd edition. Retail: $17.95

The most authoritative study guide for the GRE. All authors are *Ph.D. teaching professors* from major universities. Praised by the national Psychology honor society, Psi Chi. •Study same subjects as in other reviews...in *half* the time! •Includes diagnostic test and evaluation to identify your strengths and weaknesses, two practice tests and score-analysis charts, hundreds of problems, full solutions, complete reviews of all 3 subjects, and 3000-word vocabulary list with quick-phrases. Paperback, 510 pages.

GMAT TIME•SAVER™, 2nd edition. Retail: $17.95

The most authoritative study guide for the GMAT Test—authors are *Ph.D. teaching professors* from major universities. Now with fax-in writing evaluation offer lets you maximize the new writing portion of the GMAT. •Study the same material as in other reviews...in *half* the time! •Includes a diagnostic test and evaluation to identify your skills, two practice tests and score-analysis charts, hundreds of problems and full solutions, complete reviews of all subjects. Paperback, 397 pages.

GRE ENGINEERING REVIEW, 1st edition. Retail: $19.95

The only full review for the difficult GRE Engineering Subject Test. •The only prep that reviews all the subjects covered in the exam—and *fully solves all problems*. •The only prep written by nine expert *Ph.D. professors*. •Includes a diagnostic test, two practice tests and score-analysis charts, complete reviews of all subjects, and a self-assessment scale. 297 pages; hardcover.

ORDER FORM

Name (print): _____

Street Address: _____

_____

_____

Day ph.: _____

Credit Card: VISA MC (circle one)

Number: _____ Exp.: _____

Name of Title	Price	Quan.	Multi-Discount!	Subtotal
Civil PE Exam Review	$69.95 x		Take 20% off each book on orders of 3 or more!	
Fundamentals of Engr	$49.95 x			
GRE Time•Saver	$17.95 x			
GMAT Time•Saver	$17.95 x			
GRE/Engr Review	$19.95 x			
Shipping	—		x $3 per item	

(Mich. residents add 6% sales tax.) TOTAL $ _____

Mail this order form, credit info., check or money order to:

Great Lakes Press POB 172 Grover, MO 63040

Ph. 1-800-837-0201/fax 1-314-273-6086

User Remarks

Yes, you! Your suggestions and comments will help us continuously improve this PE review.

Also, we survive and thrive via word-of-mouth. Check the box below and let us tell others your opinion!

Comments: _____
_____
_____
_____

Name: _____

Address: _____ *"You can tell them I said so!"* (check here) ☐

_____ Position: _____

Errata... *I noticed the following errors in your PE...*

error #	proposed correction	pg.

Name & Address _____

New Titles & Deep-Discount Handbooks!

Get on our mailing list for notice of new titles and deep-discount offerings of classic engineering references.

Of particular interest to CE's: we now offer McGraw-Hill's *CE Handbook* in its new 4th edition at 25% off list price. (Special sale price $99; reg. retail $130.)

☐ Put me on your mailing list! My engineering interests are: _____

Name: _____

Address: _____

_____

☐ Yes, I'd like a copy of the *CE Handbook*!   Credit card No.: _____ Exp. date: _____

Mail to: Great Lakes Press, POB 172, Grover, MO 63040-0172

Cut out and mail in!